IMPACT AND EXPLOSION

Analysis and Design

M.Y.H. Bangash

OXFORD

BLACKWELL SCIENTIFIC PUBLICATIONS

LONDON EDINBURGH BOSTON

MELBOURNE PARIS BERLIN VIENNA

© M.Y.H. Bangash 1993

Blackwell Scientific Publications
Editorial Offices:
Osney Mead, Oxford OX2 0EL
25 John Street, London WC1N 2BL
23 Ainslie Place, Edinburgh EH3 6AJ
238 Main Street, Cambridge,
  Massachusetts 02142, USA
54 University Street, Carlton
  Victoria 3053, Australia

Other Editorial Offices:
Librairie Arnette SA
2, rue Casimir-Delavigne
75006 Paris
France

Blackwell Wissenschafts-Verlag
Meinekestrasse 4
D-1000 Berlin 15
Germany

Blackwell MZV
Feldgasse 13
A-1238 Wien
Austria

DISTRIBUTORS

Marston Book Services Ltd
PO Box 87
Oxford OX2 0DT
(*Orders*: Tel: 0865 791155
    Fax: 0865 791927
    Telex: 837515)

Australia
  Blackwell Scientific Publications Pty Ltd
  54 University Street
  Carlton, Victoria 3053
  (*Orders*: Tel: 03 347−5552)

British Library
Cataloguing in Publication Data

A catalogue record for this book is
available from the British Library

ISBN 0−632−02501−8

This book represents information obtained from
authentic and highly regarded sources. Reprinted
material is quoted with permission, and sources
are indicated. A wide variety of references are
listed. Every reasonable effort has been made to
give reliable data and information, but the author
and the publisher cannot assume responsibility
for the validity of all materials or for the
consequences of their use.

First published 1993

Set by Setrite Typesetters, Hong Kong
Printed and bound in Great Britain by
the University Press, Cambridge

CONTENTS

Preface   vii
Acknowledgements   ix
Definition of Terms   xii
Notation   xvii
Conversion Tables   xxiii

1   Accident Survey   1
   1.1   Introduction   1
   1.2   Wind, hurricane and tornado generated missiles   1
   1.3   Impact and explosion at sea   3
   1.4   Car collisions and explosions   7
   1.5   Train collisions and impacts   14
   1.6   Aircraft and missile impacts, crashes and explosions   17
   1.7   Explosions with and without impact   45
   1.8   Nuclear explosions and loss-of-coolant accidents   55
   1.9   The Gulf war   59
   1.10 Recent air crashes: aircraft impact at ground level   59

2   Data on Missiles, Impactors, Aircraft and Explosions   60
   2.1   Introduction   60
   2.2   Types of conventional missiles and impactors   60
   2.3   Military, air force and navy missiles and impactors   103
   2.4   Data on civilian and military aircraft, tanks and marine vessels   125
   2.5   Types of explosion   191
   2.6   Dust explosions   225
   2.7   Underwater explosions   231

3   Basic Structural Dynamics   235
   3.1   General introduction   235
   3.2   Single-degree-of-freedom system   235
   3.3   Two-degrees-of-freedom system   314
   3.4   Multi-degrees-of-freedom systems   321

4   Impact Dynamics   331
   4.1   Introduction   331
   4.2   The impactor as a projectile   331
   4.3   Aircraft impact on structures — peak displacement and frequency   344
   4.4   Aircraft impact: load−time functions   346
   4.5   Impact due to dropped weights   353

4.6	Impact on concrete and steel	367
4.7	Impact on soils/rocks	391
4.8	Impact on water surfaces and waves	398
4.9	Snow/ice impact	416

5 Explosion Dynamics — 426

5.1	Introduction	426
5.2	Fundamental analyses related to an explosion	426
5.3	Explosions in air	432
5.4	Shock reflection	440
5.5	Gas explosions	442
5.6	Dust explosions	448
5.7	Explosions in soils	458
5.8	Rock blasting — construction and demolition	473
5.9	Explosions in water	484

6 Dynamic Finite-element Analysis of Impact and Explosion — 496

6.1	Introduction	496
6.2	Finite-element equations	496
6.3	Steps for dynamic non-linear analysis	507
6.4	Ice/snow impact	523
6.5	Impact due to missiles, impactors and explosions: contact problem solutions	527
6.6	High explosions	529
6.7	Spectrum analysis	529
6.8	Solution procedures	533
6.9	Force or load−time function	535

7 Case Studies — 542

Introduction		542
A:	Steel and composites	542
A.1	Steel structures	542
A.2	Composite structures	552
A.3	Impact analysis of pipe rupture	562
A.4	Explosions in hollow steel spherical cavities and domes	570
A.5	Dropped/impact analysis of a shipping container for radioactive material	574
A.6	Car impact and explosion analysis	578
A.7	Train crash phenomenon	586
B:	Concrete structures	590
B.1	Introduction	590
B.2	Concrete beams	591
B.3	Reinforced concrete slabs and walls	606
B.4	Impact/explosion at roadways and runways	637
B.5	Buildings and structures subject to blast loads	641
B.6	Aircraft crashes on containment vessels (buildings)	646
C:	Brickwork	651
C.1	General introduction	651
C.2	Finite-element analysis of explosion	651
C.3	Bomb explosion at a wall	655
D:	Ice/snow impact	661
D.1	Introduction	661
D.2	Finite-element analysis	661

Contents

E: Nuclear reactors   667
    E.1   PWR: loss-of-coolant accident (LOCA)   667
    E.2   Nuclear containment under hydrogen detonation   684
    E.3   Impact/explosion at a nuclear power station: turbine hall   687
    E.4   Jet impingement forces on PWR steel vessel components   689
F: Concrete nuclear shelters   691
    F.1   Introduction   691
    F.2   Design of a concrete nuclear shelter against explosion and other loads based on Home Office manual   697
    F.3   Design of a nuclear shelter based on the US codes   702
    F.4   Lacing bars   707
    F.5   Finite-element analysis   714
G: Sea environment   719
    G.1   Multiple wave impact on a beach front   719
    G.2   Explosions around dams   724
    G.3   Ship-to-ship and ship-to-platform: impact analysis   729
    G.4   Jacket platform: impact and explosion   734
    G.5   Impact of dropped objects on platforms   742
H: Soil/rock surface and buried structures   748
    H.1   General introduction   748
    H.2   Soil strata subject to missile impact and penetration   748

Bibliography   771
Appendix   827
Index   850

PREFACE

The dynamics of impact and explosion are an important consideration in the design of conventional structures in general and sensitive and unconventional structures in particular. Accidents causing damage and explosion are a matter of growing concern in many areas such as nuclear, chemical, civil, mechanical, electrical, offshore, gas, aeronautical and naval engineering. This book provides, in Chapter 1, a comprehensive, illustrated survey of accidents and explosions, including those caused by aircraft, missiles, bombs, detonators, sea-going vessels, cars, lorries, trains, etc. Gas and nuclear explosions are also covered.

Engineering modelling of impact and explosion requires a great deal of data as an input. This input is needed so that a comprehensive analysis can be carried out on the design of structures. Chapter 2 gives a comprehensive treatment of various types of impact and explosion. Tables and examples are given for tornado-generated, plant-generated and military missiles and civilian and military aircraft. The impactors included may be categorized as follows:

(1) Environmental: jet fluids, snow/ice, falling stones/boulders, trees, poles, pylons and various types of dropped weights.
(2) Military/naval: tanks, tankers, ships, carriers, hydrofoils and hovercrafts.
(3) Civilian: cars, lorries, trains, earth movers and bulldozers.

Data on explosion cover bombs, shells, grenades, explosives, gas leaks, chemical dusts and nuclear and underwater detonations. Tables and graphs are provided to act as inputs to various engineering problems. Both Chapters 1 and 2 will familiarize the reader with the range of types of missiles/impactors and explosions and their disastrous effects in terms of human lives and structural damage.

The modelling of impact and explosion remains one of the most difficult tasks. It involves structural dynamics, load−time relationships, impactor−target interaction, material properties including strain-rate effects and solution/convergence procedures. Before the reader is introduced to numerical models and design/protection techniques, it is essential to emphasize the importance of knowing basic structural dynamics, which is the theme of Chapter 3. This chapter covers all areas of basic structural dynamics, such as elastic and elasto-plastic systems, degrees of freedom, fundamental vibrations, forced vibrations and impact/impulsive loads versus vibrations, and includes tables and graphs covering numerical data with typical examples.

The reader is then in a position to study the dynamics of impact and explosion. Chapter 4 provides an extensive treatment of impact dynamics. It includes vehicle collision mechanics and impact due to dropped weights, water jets, snow/ice, ocean waves, missiles and aircraft. Empirical models are introduced for non-deformable and deformable missiles. Materials considered are steel, concrete, bovine, soil/rock and composites. A special section covers impact on water surfaces. A simplified analysis for load−time is presented. Tables, graphs and line diagrams are used to evaluate parameters and coefficients in the numerical analysis.

The dynamics of explosion, as introduced in Chapter 2, is considered in depth in Chapter 5. Apart from discussing various numerical parameters and major assumptions, this chapter includes detailed analysis and numerical modelling for explosions occurring in air, underground and underwater. Prominent among them are explosions due to nuclear detonations, gas leaks, dust bombs and explosives. Blast loads and their overpressures are fully discussed.

Chapter 6 gives formulations for the dynamic finite-element analysis of impact and explosion. Various boundary conditions are discussed. A step-by-step analysis suggests only the use of higher-order elements representing various materials alone or in combination. Material strain rates, dynamic material modelling simulation and solution techniques are fully discussed. On the other hand, the analysis is flexible enough to include linear, non-linear, plasticity and cracking criteria under both impact and explosion conditions.

Chapter 7 covers impact and blast load design. Case studies are chosen from various engineering disciplines. Each case study is supported by a brief introduction to the background of the relevant areas. Care is taken to include those case studies which are supported by experimental test results and/or site monitoring. This then gives a degree of validation to the analytical results. The major case studies chosen are from the following disciplines: building, civil, mechanical, naval, aerospace, offshore, defence, nuclear, transportation and underwater facilities. Final design recommendations are made for each case study. The text gives a comprehensive bibliography for those who wish to carry out further research in depth.

Impact and Explosion: Analysis and Design will be of use to research and practising engineers, designers, technologists, mathematicians and specialists in computer-aided techniques of structures under transient loads in various engineering disciplines as identified earlier in this text. It will also be of use to non-engineering specialists involved in the manufacture and application of explosives.

M.Y.H. Bangash

ACKNOWLEDGEMENTS

The author is indebted to many individuals, institutions, organizations and research establishments, mentioned in the text, for helpful discussions and for providing useful practical data and research materials.

The author owes a special debt of gratitude to his family, who, for the third time, provided unwavering support.

I also wish to acknowledge private communications from the following:

Aerospace Daily (Aviation and Aerospace Research) 1156 15th Street NW, Washington DC 20005, USA.

Aérospatiale, 37 Boulevard de Montmorency, 75781, Paris Cédex 16, France.

Afghan Agency Press, 33 Oxford Street, London, UK.

Agusta SpA, 21017 Cascina Costa di Samarate (VA), Italy.

Ailsa Perth Shipbuilders Limited, Harbour Road, Troon, Ayrshire KA10 6DN, Scotland.

Airbus Industrie, 1 Rond Point Maurice Bellonte, 31707 Blagnac Cédex, France.

Allison Gas Turbine, General Motors Corporation, Indianapolis, Indiana 46 206−0420, USA.

AMX International, Aldwych House, Aldwych, London, WC2B 4JP, UK.

Bell Helicopter Textron Incorporation, PO Box 482, Fort Worth, Texas 76101, USA.

Boeing, PO Box 3703, Seattle, Washington 98124, USA.

Bofors Noble Industries, S-691 80 Bofors, Sweden.

Bremer Vulkan AG, Lindenstrasse 110, PO Box 750261, D-2820 Bremen 70, Germany.

British Aerospace plc, Richmond Road, Kingston upon Thames, Surrey KT2 5QS, UK.

Catic, 5 Liang Guo Chang Road, East City District (PO Box 1671), Beijing, China.

Chantiers de l'Atlantique, Alsthom-30, Avenue Kléber, 75116 Paris, France.

Chemical and Engineering News, 1155 16th Street NW, Washington DC 20036, USA.

Chinese State Arsenals, 7A Yeutan Nanjie, Beijing, China.

CITEFA, Zufratequ y Varela, 1603 Villa Martelli, Provincia de Buenos Aires, Argentina.

Conorzio Smin, 52, Villa Panama − 00198, Rome, Italy.

Daily Muslim, Abpara, Islamabad, Pakistan.

Daily Telegraph, Peterborough Court, South Quay Plaza, Marshwall, London E14, UK.

Dassault-Breguet, 33 Rue de Professeur Victor Pauchet, 92420, Vaucresson, France.

Défense Nationale, 1 Place Joffre, 75700, Paris, France.

Der Spiegel, 2000 Hamburg 11, Germany.

Etablissement d'Etudes et de Fabrications d'Armement de Bourges, 10 Place G Clémenceau, 92211 Saint-Cloud, France.

Euro fighter Jagdflugzeug GmbH, Arabellastrasse 16, PO Box 860366, 8000 Munich 86, Germany.

Evening Standard, Evening Standard Limited, 118 Fleet Street, London EC4P 4DD, UK.

Financial Times, Bracken House, 10 Cannon Street, London EC4P 4BY.

Fokker, Corporate Centre, PO Box 12222, 1100 AE Amsterdam Zuidoost, The Netherlands.

General Dynamic Corporation, Pierre Laclede Centre, St Louis, Missouri 63105, USA.

General Electric, Neumann Way, Evendale, Ohio 45215, USA.

Grumman Corporation, 1111 Stewart Avenue, Bethpage, New York, 11714−3580, USA.
Guardian, Guardian Newspapers Limited, 119 Farringdon Road, London EC1R 3ER, UK.
Hawker Siddeley Canada Limited, PO Box 6001, Toronto AMF, Ontario L5P 1B3, Canada.
Hindustan Aeronautics Limited, Indian Express Building, PO Box 5150, Bangalore 560 017, India. (Dr Ambedkar Veedhi.)
Howaldtswerke Deutsche Werft, PO Box 146309, D-2300 Kiel 14, Germany.
Independent, 40 City Road, London EC1Y 2DB, UK.
Information Aéronautiques et Spatiales, 6 Rue Galilee, 75116, Paris, France.
Information Resources Annual, 4 Boulevard de l'Empereur, B-1000, Brussels, Belgium.
International Aero-engines AG, 287 Main Street, East Hartford, Connecticut 06108, USA.
The Institution of Civil Engineers, Great George Street, Westminster, London SW1, UK.
The Institution of Mechanical Engineers, 1 Bird Cage Walk, Westminster, London SW1, UK.
The Institution of Structural Engineers, 11 Upper Belgrave Street, London SW1, UK.
Israel Aircraft Industries Limited, Ben-Gurion International Airport, Israel 70100.
Israel Ministry of Defence, 8 David Elazar Street, Hakiryah 61909, Tel Aviv, Israel.
Jane's Armour and Artillery 1988−89, 9th edn, Janes Information Group Ltd, 163 Brighton Road, Coulsdon, Surrey CR3 2NX, UK.
Jane's Fighting Ships 1989−90, Janes Information Group Ltd, 163 Brighton Road, Coulsdon, Surrey CR3 2NX, UK.
KAL (Korean Air), Aerospace Division, Marine Centre Building 18FL, 118-2Ga Namdaemun-Ro, Chung-ku, Seoul, South Korea.
Kangwon Industrial Company Limited, 6-2-KA Shinmoon-Ro, Chongro-ku, Seoul, South Korea.
Kawasaki Heavy Industries Limited, 1−18 Nakamachi-Dori, 2-Chome, Chuo-ku, Kobe, Japan.
Korea Tacoma Marine Industries Limited, PO Box 339, Masan, Korea.
Krauss-Maffei AG, Wehrtocknik GmbH Krauss-Maffei Strasse 2, 8000 Munich 50, Germany.
Krupp Mak Maschinenbau GmbH, PO Box 9009, 2300 Kiel 17, Germany.
Lockheed Corporation, 4500 Park Granada Boulevard, Calabasas, California 91399−0610, USA.
Matra Defense, 37 Avenue Louis-Bréguet BP1, 78146 Velizy-Villa Coublay Cédex, France.
McDonnell Douglas Corporation, PO Box 516, St Louis, Missouri 63166, USA.
Mitsubishi Heavy Industries Limited, 5−1 Marunouchi, 2-Chome, Chiyoda-ku, Tokyo 100, Japan.
NASA, 600 Independence Avenue SW, Washington DC 20546, USA.
Nederlandse Verenigde Scheepsbouw Bureaus, PO Box 16350, 2500 BJ, The Hague, The Netherlands.
Netherland Naval Industries Group, PO Box 16350, 2500BJ The Hague, The Netherlands.
New York Times, 229 West 43rd Street, New York, NY 10036, USA.
Nuclear Engineering International, c/o Reed Business Publishing House Limited, Quadrant House, The Quadrant, Sutton, Surrey SM2 5AS, UK.
Observer, Observer Limited, Chelsea Bridge House, Queenstown Road, London SW8 4NN, UK.
Offshore Engineer, Thomas Telford Limited, Thomas Telford House, 1 Heron Quay, London E14 9XF, UK.
OTO Melara, Via Valdilocchi 15, 19100 La Spezia, Italy.
Pakistan Aeronautical Complex, Kamra, District Attock, Punjab Province, Pakistan.
Patent Office Library, Chancery Lane, London, UK.
Plessey Marine Limited, Wilkinthroop House, Templecombe, Somerset BA8 ODH, UK.
Porsche, Porsche AG, PO Box 1140, 7251 Weissach, Germany. (Dr Ing hcf.)
Pratt and Whitney, East Hartford, Connecticut 06108, USA.
Promavia SA, Chaussée de Fleurs 181, 13−6200 Gosselies Aéroport, Belgium.

RDM Technology, PO Box 913, 3000 Ax, Rotterdam, The Netherlands.
Rockwell International, 100 North Sepulveda Boulevard, Elsegundo, California 90245, USA.
Rolls-Royce plc, 65 Buckingham Gate, London SW1E 6AT, UK.
Rolls-Royce Turbomeca, 4/5 Grosvenor Place, London SW1X 7HH, UK.
Saab-Scania, S-58188 Linköping, Sweden.
SAC (Shenyang Aircraft Company), PO Box 328, Shenyang, Liaoning, China.
Short Brothers plc, PO Box 241, Airport Road, Belfast BT3 9DZ, Northern Ireland.
Sikorsky, 6900 North Main Street, Stratford, Connecticut 06601−1381, USA.
Soloy Conversions Limited, 450 Pat Kennedy Way SW, Olympia, Washington 98502, USA.
Soltam Limited, PO Box 1371, Haifa, Israel.
SRC Group of Companies, 63 Rue de Stalle, Brussels 1180, Belgium.
Tamse, Avda, Rolon 1441/43, 2609 Boulgone sur Mer, Provincia de Buenos Aires, Argentina.
Tanker Market Quarterly, 11 Heron Quay, Docklands, London E4 9YP, UK.
Textron Lycoming, 550 Main Street, Stratford, Connecticut 06497, USA.
Thomson-CSF, 122 Avenue du Général Leclerc, 92105 Boulogne Billancourt, France.
Times Index, Research Publications, PO Box 45, Reading RG1 8HF, UK.
Turbomeca, Bordes, 64320 Bizanos, France.
USSR Public Relations Offices, Ministry of Defence, Moscow, USSR.
Vickers Defence System Limited, Manston Lane, Leeds LS15 8TN, UK.
Vickers FMC, 881 Martin Avenue, PO Box 58123, Santa Clara, California 0502, USA.

DEFINITION OF TERMS

Acceleration: The change of velocity as a function of time.

Ambient pressure: The atmospheric pressure acting at a given altitude.

Angle of incidence: The angle formed by the line which defines the shock propagation and between the centre of explosion and the point on the structure.

Attenuation: The decrease in intensity of a blast wave as a result of absorption of energy.

Barrier: A protective structure.

Blast door: A protective closure for personnel and/or vehicles.

Blast effect (or blast output): The blast pressures and impulse produced by an explosion.

Blast loads: The forces, associated with the pressure output, or pressure acting on structures.

Blast shield: A protective closure device for ventilation openings.

Blast valve: A mechanical and electrical protective closure device for ventilation openings.

Blast wave: A pulse of air in which the pressure increases sharply at the front and is accompanied by blast winds, propagated from an explosion. See **Shock wave**.

Burst: An explosion located on or close to the structure or at a distance away from the structure.

Casing (or container): A metal or any other covering which partially or fully encloses an explosive.

Charge weight (or charge):

    Actual: The total weight of an explosive.

    Effective: The net weight of an explosive involved in the detonation process.

    Equivalent: The weight of a spherical TNT charge which will produce the same blast effects as the explosive under consideration.

Cubicle: An area which is partially or fully enclosed by blast-resistant walls and roof.

Deflagration: Rapid and violent burning.

Degree of freedom: The number of independent displacement variables needed to specify the configuration of a dynamic system.

Degree of protection: A scale which is used to measure the required protection.

Detonation:

    Simultaneous: A detonation of separated quantities of explosives or ammunition occurring very nearly at the same time.

    Mass: Similar to a simultaneous detonation except that the explosives involved are of large quantity.

Direct spalling: The dynamic disengagement of the concrete surface of a structural component.

Donor explosive: The explosive contained in the explosion protective system, called a donor system.

Drag coefficient: The drag pressure divided by the dynamic pressure; it is dependent upon the shape of the structure.

Drag loading (or drag pressure): The force on a structure due to the transient winds accompanying the passage of a blast wave.

Ductile mode: The response of a structural element to attain relatively large inelastic deflections without complete collapse.

Ductility factor: The ratio of the maximum deflection to the equivalent maximum elastic deflection of the structure.

Duration: The interval between the time of arrival of the blast wave and the time for the magnitude of the blast pressures to return to ambient pressure.

Dynamic increase factor: The ratio of the dynamic to static stresses of a material.

Effective mass: The mass of an equivalent single-degree-of-freedom system.

Elastic range: The response range of an element in which its deflections will essentially return to zero after removal of the load.

Elastic rebound: The elastic displacement of an element occurring in an opposite direction to that of the blast load.

Elasto-plastic range: The response range of an element which occurs after the formation of the first yield line.

Element (member): A component of a structure.

Equivalent dynamic system: A system which consists of a number of concentrated masses joined together by weightless springs and subjected to time-dependent concentrated loads.

Explosions:

    Unconfined: An explosion (air or surface burst) occurring exterior to a structure.

    Confined: An explosion occurring within or immediately next to a structure which is subdivided into several vents.

    Multiple: Two or more explosions to be cumulative on the response of the structure.

Explosive: A chemical compound or mechanical mixture subjected to heat, friction and detonation (undergoes a very rapid chemical change) which exerts pressure in the surrounding medium.

Explosive protection system: A storage or a manufacturing facility containing potentially explosive material.

Fragment shield: A device which is composed of steel plates or other mass material which can be attached to or placed a short distance from a protective barrier.

Ground shock:
>   **Air-induced**: Ground motions induced by a blast wave travelling over the ground surface and generating stress waves into the ground.
>   **Direct-transmitted**: Ground motions induced in the ground in the immediate vicinity of the explosion, where it is then dispersed to areas of lower pressure level directly through the ground.

Haunch: A concrete fillet between two intersecting structural elements.

Impulse/impact: Sudden time-dependent load.

Impulse capacity:
>   **Flexural**: The flexural impulse capacity responding in the ductile mode is the area under the resistance—time curve.
>   **Brittle**: The impulse capacity (momentum) of the post-failure fragments must be added to the flexural impulse capacity.

Interior pressure: The pressures from an explosion within a barrier type structure which are amplified due to their multiple reflections.

Kinetic energy: The energy the structural element has by virtue of its motion (translational and/or rotational).

Lacing reinforcement: Continuous, bent diagonal bars which tie together the straight flexural reinforcing bars on each face of an element.

Load factor: The design factor (numerically equal to the resistance factor) by which the total load applied to a structure is multiplied to obtain the equivalent concentrated load for the equivalent single-degree-of-freedom system.

Load-mass factor: Mass factor divided by the load factor.

Mach front or mach stem: The shock front formed by the interaction of the incident and the reflected shock fronts accompanying an air burst.

Magazine: Any structure specifically designated for the storage of explosives, ammunition and loaded ammunition components.

Mass: The weight of a structural component divided by the acceleration due to gravity.

Mass factor: The design factor by which the total distributed mass of a structural element is multiplied to obtain the equivalent lumped mass of the equivalent single-degree-of-freedom system.

Mode of vibration: The dynamic deflection shape of a structure or its components produced by a specific loading.

Modular ratio: The ratio of the modulus of elasticity of structural steel to that of concrete.

Modulus of elasticity: The ratio of stress to strain in the elastic range of response of a structure.

Moment of inertia: The summation of the second moment of the area of a structure about its neutral axis.

Momentum: The product of the mass and velocity of an impactor or structural component.

Multiple reflections: The amplification of the blast wave pressures, produced by a partially confined explosion on various interior surfaces of a structure.

Natural frequency: The number of variations in one second a structure completes in its fundamental mode of vibration.

Natural period of vibration: The time interval during which a structure completes one vibration in its fundamental mode.

Normal distance: The distance measured between the centre of an explosion and a given location of the surface of a structure.

Partial failure: The failure of one or more supports of a structural element resulting in a loss of strength and a reduction in the resistance.

Penetration: A partial perforation but not extending to the other side.

Perforation: A complete penetration or hole from one side to another.

Phase:

   Positive: A period beginning with the arrival of the shock wave and ending with the two positive shock pressures at the level of the ambient pressure.

   Negative: A period from the ambient pressure, described in the 'positive case', to the ending of the negative shock pressures by returning back to the ambient value.

Post-failure fragments: Fragments formed upon failure of a structural element.

Post-ultimate range: The response occurring between partial and incipient failure of the structure.

Potential energy: The strain energy which results from the straining of an element produced by the blast loads.

Primary fragments: Fragments formed from casings, containers and other objects.

Propagation of explosion: A time-dependent movement of the explosion caused by a source.

Protective structure: A structure which is designed to provide protection.

Reflection factor: The ratio of the peak reflected pressure to its corresponding incident pressure.

Resistance: The sum of the internal forces in a structural element whose function is to resist movement of the mass produced by the blast loads. These forces occur at elastic, elasto-plastic, post-ultimate, rebound and ultimate conditions.

Resistance factor: Design factor multiplied by the structural resistance.

Safety distance: A distance at which the explosive can be placed and after detonation causes no destruction or risk of any kind to living beings and their facilities.

Safety factor: A factor used in design to account for unknown factors.

Scabbing: The dynamic disengagement of the concrete surface of an element resulting from a tension failure in the concrete normal to its free surface.

Separation distance: A minimum distance between possible explosions and a facility.

Shelter: A structure which encloses or nearly encloses the receiver system and is used to provide full protection.

Shock front: The sharp boundary between the pressure disturbance created by an explosion and the ambient atmosphere.

Shock wave: A continuously propagated pressure pulse in the surrounding medium produced by an explosion.

Slant distance: The distance between the centre of an explosion and a location on the ground.

Spalling: See **Direct spalling**.

Structural motion: The motion or dynamic displacement of a structure (with velocity and acceleration) caused by an impact or blast or both.

Time:

    Arrival: The finite time interval required for the blast wave to travel from the centre of an explosion to any particular location of the structure.

    Clearing: The finite time interval required for the blast pressures acting on a structure to reduce to some specified intensity.

    Response: The finite time interval required for the maximum deflection of a structure.

    Rise: The finite time interval required for the blast pressures acting on a structure to increase from zero to a maximum value.

TNT equivalent: A measure of the energy released in the explosion of a given quantity of material which would release the same amount of energy when exploded.

Total collapse: The condition of a structural element in which it completely loses its ability to resist the applied blast loads.

Triple point: The intersection of the incident and mach shock fronts accompanying an air burst.

Velocity: A distance travelled in a specific time; area under a distance−time curve.

Work: Force multiplied by distance.

Yield: Weapon detonation capacity.

NOTATION

A	constant
A	projected area, hardening parameter
A_0	initial surface area
A_{ST}	surface area of the enclosure
A_V	vent area
$\bar{A}$	normalized vent area
a	radius of the gas sphere
a_0	loaded length, initial radius of gas sphere
B	burden
$[B]$	geometric compliance matrix
BG	blasting gelatine
b	spaces between charges
b_1	distance between two rows of charges
C_D, C_d	drag coefficient or other coefficients
C_d'	discharge coefficient
C_f	charge size factor, correction factor
C_1	a coefficient which prevents moving rocks from achieving an instant velocity
$[C_{in}]$	damping coefficient matrix
C_p	specific heat capacity at constant pressure
C_r	reflection coefficient
C_v	specific heat capacity at constant volume
$c_a, c_1, c_\psi, c_\theta$	coefficients for modes
D	depth of floater, diameter
$[D]$	material compliance matrix
D_a	maximum aggregate size
D_i	diameter of ice
D_p	penetration depth of an infinitely thick slab
DIF	dynamic increase factor
d	depth, diameter
d_0	depth of bomb from ground surface
E	Young's modulus
E_b	Young's modulus of the base material

ΔE_{cr}	maximum energy input occurring at resonance
E_{ic}	Young's modulus of ice
E_K	energy loss
E_{na}	energy at ambient conditions
E_{ne}	specific energy of explosives
E_R	energy release
E_t	tangent modulus
e	base of natural logarithm
e	coefficient of restitution, efficiency factor
F	resisting force, reinforcement coefficient
F_{ad}	the added mass force
$F_I(t)$	impact
$F(t)$	impulse/impact
F_s	average fragment size, shape factor
f	function
f	frequency (natural or fundamental), correction factor
f_a	static design stress of reinforcement
f_c	characteristic compressive stress
f'_c	static ultimate compressive strength of concrete at 28 days
f^*_{ci}	coupling factor
f_d	transmission factor
f'_{dc}	dynamic ultimate compressive strength of concrete
f_{ds}	dynamic design stress of reinforcement
f_{du}	dynamic ultimate stress of reinforcement
f_{dyn}	dynamic yield stress of reinforcement
f^*_{TR}	transitional factor
f_u	static ultimate stress of reinforcement
f_y	static yield stress of reinforcement
G	elastic shear modulus
G_a	deceleration
G_f	energy release rate
G_m, G_s	moduli of elasticity in shear and mass half space
g	acceleration due to gravity
H	height
H_s	significant wave height
HE	high explosion
HP	horsepower
h	height, depth, thickness
I	second moment of area, identification factor
$[I]$	identity matrix
I_1	the first invariant of the stress tensor
i_p	injection/extraction of the fissure

J_1, J_2, J_3	first, second and third invariants of the stress deviator tensor
J_F, J	Jacobian
K	vent coefficient, explosion rate constant, elastic bulk modulus
$[K_c]$	element stiffness matrix
K_p	probability coefficients
K_s	stiffness coefficient at impact
K_{TOT}	composite stiffness matrix
K_W	reduction coefficient of the charge
K_σ	correction factor
KE	kinetic energy
k_{cr}	size reduction factor
k_r	heat capacity ratio
k_t	torsional spring constant
L	length
L'	wave number
L_i	length of the weapon in contact
$\ln, \log_e$	natural logarithm
l_x	projected distance in x direction
M	Mach number
$[M]$	mass matrix
M'	coefficient for the first part of the equation for a forced vibration
$*M_A$	fragment distribution parameter
M_p	ultimate or plastic moment or mass of particle
m	mass
N	nose shaped factor
N_c	nitrocelluloid
N_f	number of fragments
N'	coefficient for the second part of the equation for a forced vibration
NG	nitroglycerine
n	attenuation coefficient
P_i	interior pressure increment
P_m	peak pressure
P_u	ultimate capacity
PE	potential energy
PETN	pentaerythrite tetra-nitrate
p	explosion pressure
p_a	atmospheric pressure
p_d	drag load
p_{df}	peak diffraction pressure
p_{gh}	gaugehole pressure in rocks

p_{pa}	pressure due to gas explosion on the interface of the gases and the medium
p_r	reflected pressure
p_{ro}	reflected overpressure
p'_s	standard overpressure for reference explosion
p_{so}	overpressure
p_{stag}	stagnation pressure
Q_{sp}	explosive specific heat (TNT)
q_{do}	dynamic pressure
R	distance of the charge weight gas constant, Reynolds number, thickness ratio
R'	radius of the shock front
$\{R(t)\}$	residual load vector
R_T	soil resistance
R_{vd}	cavity radius for a spherical charge
R_w, r_s	radius of the cavity of the charge
r	radius
r_o, r_ψ, r_ϕ	factors for translation, rocking and torsion
S_i	slip at node i
S_{ij}	deviatoric stress
S_L	loss factor
$S_{\eta\eta}(f)$	spectral density of surface elevation
s	$\pm i\omega t$ or distance or wave steepness, width, slope of the semi-log
T	temperature, period, restoring torque
T_a	ambient temperature
T_d	delayed time
T'_i	ice sheet thickness
T_{ps}	post shock temperature
$[T'']$	transformation matrix
TR	transmissibility
t	time
t_A	arrival time
t_{av}	average time
t_c	thickness of the metal
t_d	duration time
t_{exp}	expansion time
t_i	ice thickness
t_p	thickness to prevent penetration, perforation
t_{sc}	scabbing thickness, scaling time
t_{sp}	spalling thickness
U	shock front velocity
u	particle velocity

V	volume, velocity
V'	velocity factor
V_{Rn}	velocity at the end of the nth layer
v_b	fragment velocity or normalized burning velocity
v_{bT}	velocity affected by temperature
v_c	ultimate shear stress permitted on an unreinforced web
v_{con}	initial velocity of concentrated charges
v_f	maximum post-failure fragment velocity
v_{in}, v_0	initial velocity
v_l	limiting velocity
v_m	maximum mass velocity for explosion
v_p	perforation velocity
v_{pz}	propagation velocities of longitudinal waves
v_{RZ}	propagation velocities of Rayleigh waves
v_r	residual velocity of primary fragment after perforation or $\sqrt{E/\rho}$
v_s	velocity of sound in air or striking velocity of primary fragment or missile
v_{so}	blast-generated velocity at initial conditions
v_{su}	velocity of the upper layer
v_{sz}	propagation velocities of transverse waves
v_{szs}	propagation velocity of the explosion
v'_{xs}	initial velocity of shock waves in water
v_z	phase velocity
v_{zp}	velocity of the charge
W	charge weight
$W^{1/3}, Y$	weapon yield
W_t	weight of the target material
w_a	maximum weight
w_f	forcing frequency
X	amplitude of displacement
$\dot{X}$	amplitude of velocity
$\ddot{X}$	amplitude of acceleration
X_f	fetch in metres
$X(x)$	amplitude of the wave at a distance x
X_0	amplitude of the wave at a source of explosion
x	distance, displacement, dissipation factor
x	relative distance
$\dot{x}$	velocity in dynamic analysis
$\ddot{x}$	acceleration in dynamic analysis
$\{x\}^*$	displacement vector
x_{cr}	crushed length
x_i	translation
x_n	amplitude after n cycles

x_p	penetration depth
x_r	total length
Z	depth of the point on the structure
α	cone angle of ice, constant for the charge
$\alpha_a, \alpha_l, \alpha_\psi$	spring constants
α_B	factor for mode shapes
α'	angle of projection of a missile, constant
$\bar{\alpha}$	constant
β	constant for the charge, angle of reflected shock
$\bar{\beta}$	constant
γ	damping factor, viscosity parameter
γ_f	ω_f/ω
δ	particle displacements
$\bar{\delta}$	pile top displacement
δ_{ij}	kronecker delta
$\delta_m, \delta'_m, \delta''_m$	element displacement
δ_{ST}	static deflection
δt	time increment
ε	strain
$\dot{\varepsilon}$	strain rate
ε_d	delayed elastic strain
η	surface profile
θ	deflection angle
θ_g	average crack propagation angle
λ	a constant of proportionality
μ_f	jet fluid velocity
ν	Poisson's ratio
ρ_a	mass density of stone
ρ_w	mass density of water
σ	stress
σ_c	crushing strength
σ_{cu}	ultimate compressive stress
σ_f	ice flexural strength
$(\sigma_{nn})^c$	interface normal stress
$(\sigma_{nt})^c$	interface shear stress
σ_{pi}	peak stress
σ_t	uni-axial tensile strength
τ_o	crack shear strength
ϕ	phase difference
ψ	circumference of projectile
ω	circular frequency
l,m,n,p,q,r,s,t	direction cosines
$X,Y,Z;x,y,z$	Cartesian co-ordinates
(ξ,η,ζ)	local co-ordinates

CONVERSION TABLES

Weight

1 g	= 0.0353 oz	1 oz	= 28.35 g
1 kg	= 2.205 lbs	1 lb	= 0.4536 kg
1 kg	= 0.197 cwt	1 cwt	= 50.8 kg
1 tonne	= 0.9842 long ton	1 long ton	= 1.016 tonne
1 tonne	= 1.1023 short ton	1 short ton	= 0.907 tonne
1 tonne	= 1000 kg	1 stone	= 6.35 kg

Length

1 cm	= 0.394 in	1 in	= 2.54 cm = 25.4 mm
1 m	= 3.281 ft	1 ft	= 0.3048 m
1 m	= 1.094 yd	1 yd	= 0.9144 m
1 km	= 0.621 mile	1 mile	= 1.609 km
1 km	= 0.54 nautical mile	1 nautical mile	= 1.852 km

Area

$1\,cm^2$	$= 0.155\,in^2$	$1\,in^2$	$= 6.4516\,cm^2$
$1\,dm^2$	$= 0.1076\,ft^2$	$1\,ft^2$	$= 9.29\,dm^2$
$1\,m^2$	$= 1.196\,yd^2$	$1\,yd^2$	$= 0.8361\,m^2$
$1\,km^2$	$= 0.386\,sq\ mile$	$1\,sq\ mile$	$= 2.59\,km^2$
1 ha	= 2.47 acres	1 acre	= 0.405 ha

Volume

$1\,cm^3$	$= 0.061\,in^3$	$1\,in^3$	$= 16.387\,cm^3$
$1\,dm^3$	$= 0.0353\,ft^3$	$1\,ft^3$	$= 28.317\,dm^3$
$1\,m^3$	$= 1.309\,yd^3$	$1\,yd^3$	$= 0.764\,m^3$
$1\,m^3$	$= 35.4\,ft^3$	$1\,ft^3$	$= 0.0283\,m^3$
1 litre	= 0.220 Imp gallon	1 Imp gallon	= 4.546 litres
$1000\,cm^3$	= 0.220 Imp gallon	1 US gallon	= 3.782 litres
1 litre	= 0.264 US gallon		

Density

$1\,kg/m^3 = 0.6242\,lb/ft^3$	$1\,lb/ft^3 = 16.02\,kg/m^3$

Force and pressure

1 ton   = 9964 N
1 lbf/ft   = 14.59 N/m
1 lbf/ft^2 = 47.88 N/m^2
1 lbf in   = 0.113 Nm
1 psi   = 1 lbf/in^2 = 6895 N/m^2 = 6.895 kN/m^2
1 kgf/cm^2 = 98070 N/m^2
1 bar   = 14.5 psi = 10^5 N/m^2
1 mbar = 0.0001 N/mm^2
1 kip   = 1000 lb

Temperature, energy, power

$1°C = 5/9 (°F - 32)$ 1 J   = 1 milli-Newton
$0 K = -273.16°C$ 1 HP = 745.7 watts
$0° R = -459.69°F$ 1 W   = 1 J/s
                                        1 BTU = 1055 J

Notation

lb   = pound weight   °R = Rankine
lbf = pound force   °F = Fahrenheit
in   = inch   oz = ounce
cm = centimetre   cwt = one hundred weight
m   = metre   g   = gram
km = kilometre   kg = kilogram
d   = deci   yd = yard
ft   = foot   HP = horsepower
ha = hectare   W   = watt
s   = second   N   = Newton
°C = centigrade = Celsius   J   = Joule
K   = Kelvin

1
ACCIDENT SURVEY

1.1 Introduction

This chapter surveys impact and explosion in various fields ranging from the
domestic environment to military warfare. It covers hurricane and tornado disasters,
aircraft accidents and explosions in cars, houses and military establishments.
Cases of impact on the ground, in water and in the air are given. The work is
supported by numerous tables and photographs.

1.2 Wind, hurricane and tornado generated missiles

Wind, gales, hurricanes and tornadoes have caused disasters in a number of
countries. There is always one country somewhere affected by them at any given
time in any year. The great hurricane of 1987 (120 miles/h) in England and severe
gales in 1990 (100 miles/h), hurricane 'Hugo' (138 miles/h) and many others have
caused death and damage. The extent of the damage to structures ran into
hundreds of thousands. Cars and boats were lifted into the air and dropped on
other vehicles, others were damaged by structural missiles from nearby buildings
or by falling trees. The storms caused widespread flooding, uprooted trees,
damaged buildings, bridges and pylons and stressed air-sea rescue services to the
limit as ships foundered in huge seas.

In Charleston, South Carolina, USA, where sea water was swept 10 miles
inland, a 16 m (50 ft) yacht hit the side of a car parked in a downtown area. The
same hurricane 'Hugo', as shown in Fig. 1.1, continued to ravage Puerto Rico,
forcing three planes at the airport to be twisted by multiple and repeated impacts.
The costs in all three incidents ran into billions and many thousands were killed or
injured. Figure 1.2 shows missiles ejected from a school building after the passage
of hurricane 'Hugo'. In October 1989, hurricane-force winds in the UK caused an
18.3 m (60 ft) steel chimney (Fig. 1.3) to collapse on a car. The same hurricane
caused two cars to collide in France, and in other Western countries 170 vehicles
collided and 27 500 houses, buildings, bridges and other structures were damaged
and many building components were ejected as missiles with speeds ranging
from 5 miles/h to 1500 miles/h. Table 1.1 gives useful data on hurricanes, tornadoes,
typhoons, blizzards and storms.

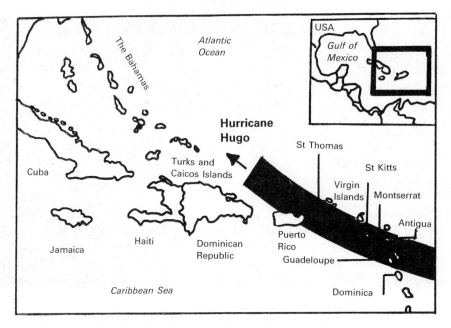

Fig. 1.1 The direction taken by hurricane 'Hugo'.

Fig. 1.2 Missiles ejecting from a school building.

Fig. 1.3 A steel chimney collapsing on a car.

1.3 Impact and explosion at sea

Much can be said on the subject of impact and explosion at sea. As for other incidents or accidents, it is extremely difficult to keep records of their daily or monthly occurrence. In December 1989, the 13 141 tonne North Sea ferry *Hamburg* collided with the 10-year-old roll-on roll-off cargo vessel *Nordec Stream* (8026 tonnes) at the mouth of the River Elbe, about 25 miles off the coast of West Germany. The number of casualties was lower. In the same period two helicopters were lost, one of them was on traffic and weather surveillance duties and the other was the North Sea Chinook helicopter carrying platform workers. In November 1988, the 3500 tonne Swedish vessel *SAMO* smashed into the 120-year-old 250 m swing bridge near Goole. Heavy iron girders on one of the approach spans were buckled. Two of the bridge's five approach spans were pushed up to 10 m out of line, leaving just 50 mm of pier support to prevent one 30 m section falling into the river. The costs exceeded £2.2 million. On 21 October 1988, the 6300 tonne Greek cruiser *Jupiter* collided with an Italian oil tanker less than one mile out of Piraeus harbour at 14.30 GMT. A 3 m hole was created. People were injured. The impact/collision occurred at the side of the ship. After all possible lives had been saved, the ship was allowed to sink.

The collision between the cruise boat *Marchioness*, a 26.6 m long luxury vessel, and the Thames sand dredger *Bowbelle*, 1880 tonnes and 76 m long, occurred on 20 August 1989. The accident happened at the point where the Thames passes through the heart of the city of London. Both were travelling

Table 1.1 Hurricanes, tornadoes, typhoons, blizzards and storms in the USA and other countries.

Notable Tornadoes in the USA

Date	Place	Deaths	Date	Place	Deaths
18/3/1925	Montana, Illinois, Indiana	689	4/6/1958	Northwestern Wisconsin	30
12/4/1927	Rock Springs, Texas	74	10/2/1959	St Louis, Missouri	21
9/5/1927	Arkansas, Poplar Bluff, Missouri	92	5 to 6/5/1960	SE Oklahoma, Arkansas	30
29/9/1927	St Louis, Missouri	90	11/4/1965	Indiana, Illinois, Ohio,	271
6/5/1930	Hill & Ellis Co, Texas	41		Michigan, Wisconsin	
21/3/1932	Alabama (series of tornadoes)	268	3/3/1966	Jackson, Mississippi	57
5/4/1936	Mississippi, Georgia	455	3/3/1966	Mississippi, Alabama	61
6/4/1936	Gainesville, Georgia	203	21/4/1967	Illinois, Michigan	33
29/3/1938	Charleston, S Carolina	32	15/5/1968	Midwest	71
16/3/1942	Central to NE Mississippi	75	23/1/1969	Mississippi	32
27/4/1942	Rogers & Mayes Co, Oklahoma	52	21/2/1971	Mississippi delta	110
23/6/1944	Ohio, Pennsylvania, W Virginia	150	26 to 27/5/1973	South midwest (series)	47
	Maryland		3 to 4/4/1974	Alabama, Georgia, Tennessee,	350
12/4/1945	Oklahoma, Arkansas	102		Kentucky, Ohio	
9/4/1947	Texas, Oklahoma, Kansas	169	4/4/1977	Alabama, Mississippi, Georgia	22
19/3/1948	Bunker Hill & Gillespie, Illinois	33	10/4/1979	Texas, Oklahoma	60
3/1/1949	Louisiana, Arkansas	58	3/6/1980	Grand Island, Nebraska (series)	4
21/3/1952	Arkansas, Missouri, Tennessee	208	2 to 4/3/1982	South midwest (series)	17
	(series)		29/5/1982	S Illinois	10
11/5/1953	Waco, Texas	114	18 to 22/5/1983	Texas	12
8/6/1953	Michigan, Ohio	142	28/3/1984	N Carolina, S Carolina	67
9/6/1953	Worcester and vicinity,	90	21 to 22/4/1984	Mississippi	15
	Massachusetts		26/4/1984	Series Oklahoma to Minnesota	17
5/12/1953	Vicksburg, Mississippi	38	31/5/1985	New York, Pennsylvania, Ohio,	90
25/5/1955	Kansas, Missouri, Oklahoma,	115		Ontario (series)	
	Texas		22/5/1987	Saragosa, Texas	29
20/5/1957	Kansas, Missouri	48			

Table 1.1 *Continued.*

Date	Place	Deaths	Date	Place	Deaths
Hurricanes (H), typhoons (T), blizzards and other storms					
11 to 14/3/1888	Blizzard, eastern USA	400	4 to 12/9/1960	H ('Donna'), Caribbean, eastern USA	148
Aug to Sept 1900	H, Galveston, Texas	6000	11 to 14/9/1961	H ('Carla'), Texas	46
21/9/1906	H, Louisiana, Mississippi	350	31/10/1961	H ('Hattie'), British Honduras	400
18/9/1906	T, Hong Kong	10000	28 to 29/5/1963	Windstorm, Bangladesh	22000
11 to 22/9/1926	H, Florida, Alabama	243	4 to 8/10/1963	H ('Flora'), Caribbean	6000
20/10/1926	H, Cuba	600	4 to 7/10/1964	H ('Hilda'), Louisiana, Mississippi, Georgia	38
6 to 20/9/1928	H, S Florida	1836			
3/9/1930	H, Dominican Republic	2000	30/6/1964	T ('Winnie'), N Philippines	107
21/9/1938	H, Long Island, New York, New England	600	5/9/1964	T ('Ruby'), Hong Kong and China	735
11 to 12/11/1940	Blizzard, USA, northeast, midwest	144	11 to 12/5/1965	Windstorm, Bangladesh	17000
			1 to 2/6/1965	Windstorm, Bangladesh	30000
15 to 16/10/1942	H, Bengal, India	40000	7 to 12/9/1965	H ('Betsy'), Florida, Mississippi, Louisiana	74
9 to 16/9/1944	H, N Carolina to New England	46			
22/10/1952	T, Philippines	300	15/12/1965	Windstorm, Bangladesh	10000
30/8/1954	H ('Carol'), northeast USA	68	4 to 10/6/1966	H ('Alma'), Honduras, southeast USA	51
5 to 18/10/1954	H ('Hazel'), eastern USA, Haiti	347			
12 to 13/10/1955	H ('Connie'), Carolinas, Virginia, Maryland	43	24 to 30/9/1966	H ('Inez'), Caribbean, Florida, Mexico	293
7 to 21/8/1955	H ('Diane'), eastern USA	400	9/7/1967	T ('Billie'), southwest Japan	347
19/9/1955	H ('Hilda'), Mexico	200	5 to 23/9/1976	H ('Beulah'), Caribbean, Mexico, Texas	54
22 to 28/9/1955	H ('Janet'), Caribbean	500			
1 to 29/2/1956	Blizzard, Western Europe	1000	12 to 20/12/1967	Blizzard, southwest USA	51
25 to 30/6/1957	H ('Audrey'), Texas to Alabama	390	18 to 28/11/1968	T ('Nina'), Philippines	63
15 to 16/2/1958	Blizzard, Western Europe	171	17 to 18/8/1969	H ('Camille'), Mississippi, Louisiana	256
17 to 19/9/1959	T ('Sarah'), Japan, S Korea	2000			
26 to 27/9/1959	T ('Vera'), Honshu, Japan	4466			

continued

Table 1.1 *Continued.*

Date	Place	Deaths	Date	Place	Deaths
30/7 to 5/8/1970	H ('Celia'), Cuba, Florida, Texas	31	13 to 27/9/1975	H ('Eloise'), Caribbean, northeast USA	71
20 to 21/8/1970	H ('Dorothy'), Martinique	42	20/5/1976	T ('Olga'), floods, Philippines	215
15/9/1970	T ('Georgia'), Philippines	300	25 and 31/7/1977	T ('Thelma'), T ('Vera'), Taiwan	39
14/10/1970	T ('Sening'), Philippines	583	27/10/1978	T ('Rita'), Philippines	c.400
15/10/1970	T ('Titang'), Philippines	526	30/8 to 7/9/1979	H ('David'), Caribbean, eastern USA	1100
13/11/1970	Cyclone, Bangladesh	3 000 000			
1/8/1971	T ('Rose'), Hong Kong	130	4 to 11/8/1980	H ('Allen'), Caribbean, Texas	272
19 to 29/6/1972	H ('Agnes'), Florida to New York	118	25/11/1981	T ('Irma'), Luzon Island, Philippines	176
3/12/1972	T ('Theresa'), Philippines	169			
June to Aug 1973	Monsoon rains in India	1217	June 1983	Monsoon rains in India	900
11/6/1974	Storm Dinah, Luzon Island, Philippines	71	18/8/1983	H ('Alicia'), southern Texas	17
			2/9/1984	T ('Ike'), southern Philippines	1363
11/7/1974	T ('Gilda'), Japan, South Korea	108	25/5/1985	Cyclone, Bangladesh	10000
19 to 20/9/1974	H ('Fifi'), Honduras	2 000	26/10 to 6/11/1985	H ('Juan'), southeast USA	97
25/12/1974	Cyclone levelled Darwin, Australia	50	25/11/1987	T ('Nina'), Philippines	650

Fig. 1.4 The pleasure boat *Marchioness* after collision with the Thames dredger *Bowbelle*.

down river, eastwards, when the collision took place near Cannon Street rail bridge. The cruise boat *Marchioness*, carrying between 110 and 150 passengers, sank within minutes. A number of casualties were reported within 24 hours of the collision. Figure 1.4 shows the *Marchioness* brought to the surface by two cranes mounted on platform barges.

In the last five decades, the total number of explosions which have occurred in sea-going vessels, including pleasure boats, merchant ships, warships, submarines and others, is around 100 000. In the two World Wars alone the number is 51% of this total value and does not include war damage. The human loss runs into hundred thousands. Table 1.2 gives a list of important sea-going vessels destroyed either by collision or by explosions, excluding war-damaged ones.

1.4 Car collisions and explosions

Car accident statistics vary from country to country, taking into consideration both cars hitting other cars and other objects. Table 1.3 provides useful information on car accidents in chosen countries over a number of years. It is interesting to note that, owing to an increase in terrorist activity in 1980 to 1989, the number of cars blown up by bombs placed underneath or nearby reached 170 000 in 150 countries of the world. The greater part of this total was contributed by Beirut, Afghanistan, Indian Punjab, Northern Ireland and some South American countries. Typical examples of recent collisions and explosions of vehicles are shown in Figs 1.5 and 1.6. Figure 1.5 shows a 5000 gallon petrol tanker that exploded in flames after colliding with an articulated lorry outside Oldbury oil terminal (UK) on

Chapter 1

Table 1.2 Impacts, collisions and explosions of sea-going vessels.

Date	Vessel/location	Deaths	Date	Vessel/location	Deaths
March 1854	*City of Glasgow*; British steamer missing in North Atlantic	480	22/1/1873	*Northfleet*; British steamer foundered off Dungeness, England	300
27/9/1854	*Arctic*; US (Collins Line) steamer sunk in collision with French steamer *Vesta* near Cape Race	285 to 351	1/4/1873	*Atlantic*; British (White Star) steamer wrecked off Nova Scotia	585
23/1/1856	*Pacific*; US (Collins Line) steamer missing in North Atlantic	188 to 286	23/11/1873	*Ville du Havre*; French steamer sunk after collision with British sailing ship *Loch Earn*	226
23/9/1858	*Austria*; German steamer destroyed by fire in North Atlantic	471	7/5/1875	*Schiller*; German steamer wrecked off Scilly Isles	312
27/4/1863	*Anglo-Saxon*; British steamer wrecked at Cape Race	238	4/11/1875	*Pacific*; US steamer sunk after collision off Cape Flattery	236
27/4/1865	*Sultana*; a Mississippi River steamer blew up near Memphis, Tennessee, USA	1450	3/9/1878	*Princess Alice*; British steamer sank after collision on Thames River, Canada	700
27/10/1869	*Stonewall*; steamer burned on Mississippi River below Cairo, Illinois, USA	200	18/12/1878	*Byzantin*; French steamer sank after Dardanelles collision	210
25/1/1870	*City of Boston*; British (Inman Line) steamer vanished between New York and Liverpool	177	24/5/1881	*Victoria*; steamer capsized in Thames River, Canada	200
19/10/1870	*Cambria*; British steamer wrecked off Northern Ireland	196	19/1/1883	*Cimbria*; German steamer sunk in collision with British steamer *Sultan* in North Sea	389
7/11/1872	*Mary Celeste*; US half-brig sailed from New York for Genoa; found abandoned in Atlantic 4 weeks later in mystery of sea; crew never heard from; loss of life unknown		15/11/1887	*Wah Yeung*; British steamer burned at sea	400
			17/2/1890	*Duburg*; British steamer wrecked, in the China Sea	400
			19/9/1890	*Ertogrul*; Turkish frigate foundered off Japan	540
			17/3/1891	*Utopia*; British steamer sank in collision with British ironclad *Anson*, off Gibraltar	562

Table 1.2 *Continued.*

Date	Vessel/location	Deaths	Date	Vessel/location	Deaths
30/1/1895	*Elbe*; German steamer sank in collision with British steamer, *Craithie*, in North Sea	332	5/3/1912	*Principe de Asturias*; Spanish steamer wrecked off Spain	500
11/3/1895	*Reina Regenta*; Spanish cruiser foundered near Gibraltar	400	14 to 15/4/1912	*Titanic*; British (White Star) steamer hit iceberg in North Atlantic	1503
15/2/1898	*Maine*; US battleship blown up in Havana Harbor, Cuba	260	28/9/1912	*Kichemaru*; Japanese steamer sank off Japanese coast	1000
4/7/1898	*La Bourgogne*; French steamer sunk in collision with British sailing ship *Cromartyshire* off Nova Scotia	549	29/5/1914	*Empress of Ireland*; British (Canadian Pacific) steamer sunk in collision with Norwegian collier in St Lawrence River, Canada	1014
26/11/1898	*Portland*; US steamer wrecked off Cape Cod	157	7/5/1915	*Lusitania*; British (Cunard Line) steamer torpedoed and sunk by German submarine off Ireland	1198
15/6/1904	*General Slocum*; excursion steamer burned in East River, New York City, USA	1030	24/7/1915	*Eastland*; excursion steamer capsized in Chicago River, USA	812
28/6/1904	*Norge*; Danish steamer wrecked on Rockall Island, Scotland	620	26/2/1916	*Provence*; French cruiser sank in the Mediterranean	3100
4/8/1906	*Sirio*; Italian steamer wrecked off Cape Palos, Spain	350	3/3/1916	*Principe de Asturias*; Spanish steamer wrecked near Santos, Brazil	558
23/3/1908	*Matsu Maru*; Japanese steamer sunk in collision near Hakodate, Japan	300	29/8/1916	*Hsin Yu*; Chinese steamer sank off Chinese coast	1000
1/8/1909	*Waratah*; British steamer, Sydney to London, vanished	300	6/12/1917	*Mont Blanc, Imo*; French ammunition ship and Belgian steamer collided in Halifax Harbor, Canada	1600
9/2/1910	*General Chanzy*; French steamer wrecked off Minorca, Spain	500			
25/9/1911	*Liberté*; French battleship exploded at Toulon	285			

continued

Table 1.2 *Continued.*

Date	Vessel/location	Deaths	Date	Vessel/location	Deaths
25/4/1918	Kiang-Kwan; Chinese steamer sank in collision off Hankow	500	18/2/1942	Truxtun and Pollux; US destroyer and cargo ship ran aground and sank off Newfoundland	204
12/7/1918	Kawachi; Japanese battleship blew up in Tokayama Bay	500	2/10/1943	Curacao; British cruiser sank after collision with liner Queen Mary	338
17/1/1919	Princess Sophia; Canadian steamer sank off Alaskan coast	398	17 to 18/12/1944	Three US Third Fleet destroyers sank during typhoon in Philippine Sea	790
17/1/1919	Cheonia; French steamer lost in Straits of Messina, Italy	460	19/1/1947	Himera; Greek steamer hit a mine off Athens	392
9/9/1919	Valbanera; Spanish steamer lost off Florida coast, USA	500	16/4/1947	Grandcamp; French freighter exploded in Texas City Harbor, starting fires	510
18/3/1921	Hong Kong; steamer wrecked in South China Sea	1000	Nov 1948	Chinese army evacuation ship exploded and sank off South Manchuria	6000
26/8/1922	Nitaka; Japanese cruiser sank in storm off Kamchatka, USSR	300	3/12/1948	Kiangya; Chinese refugee ship wrecked in explosion south of Shanghai	1100+
25/10/1927	Principessa Mafalda; Italian steamer blew up off Parto Seguro, Brazil	314	17/9/1949	Noronic; Canadian Great Lakes cruiser burned at Toronto dock	130
12/11/1928	Vestris; British steamer sank in gale off Virginia, USA	113	26/4/1952	Hobson and Wasp; US destroyer and aircraft carrier collided in Atlantic	176
8/9/1934	Morro Castle; US steamer, Havana to New York, burned off Asbury Park, New Jersey, USA	134			
23/5/1939	Squalus; US submarine sank off Portsmouth, New Hampshire, USA	26			
1/6/1939	Thetis; British submarine sank in Liverpool Bay, England	99			

Table 1.2 *Continued.*

Date	Vessel/location	Deaths	Date	Vessel/location	Deaths
31/1/1953	*Princess Victoria*; British ferry sunk in storm off Northern Irish coast	134	13/11/1965	*Yarmouth Castle*; Panamanian registered cruise ship burned and sank off Nassau	90
26/9/1954	*Toya Maru*; Japanese ferry sank in Tsugaru Strait, Japan	1172	29/7/1967	*Forrestal*; US aircraft carrier caught fire off North Vietnam	134
26/7/1956	*Andrea Doria* and *Stockholm*; Italian liner and Swedish liner collided off Nantucket	51	25/1/1968	*Dakar*; Israeli submarine vanished in Mediterranean Sea	69
14/7/1957	*Eshghabad*; Soviet ship ran aground in Caspian Sea	270	27/11/1968	*Minerve*; French submarine vanished in Mediterranean Sea	52
8/7/1961	Portuguese ship ran aground off Mozambique	259	May 1968	*Scorpion*; US nuclear submarine sank in Atlantic near Azores	99
8/4/1962	*Dara*; British liner exploded and sank in Persian Gulf	236	2/6/1969	*Evans*; US destroyer cut in half by Australian carrier *Melbourne*, South China Sea	74
10/4/1963	*Thesher*; US Navy atomic submarine sank in North Atlantic	129	4/3/1970	*Eurydice*; French submarine sank in Mediterranean near Toulon	57
10/2/1964	*Voyager*, *Melbourne*; Australian destroyer sank after collision with Australian aircraft carrier *Melbourne* off New South Wales	82	15/12/1970	*Namyong-Ho*; South Korean ferry sank in Korea Strait	308
			1/5/1974	Motor launch capsized off Bangladesh	250

Table 1.3 Data on car accidents (in thousands) in a number of countries.

Year	UK	USA	France	Gulf	Indian subcontinent	Year	UK	USA	France	Gulf	Indian subcontinent
1926	124	110	110			1958	237	161	75	35	13
1927	134	135	135			1959	261	139	90	10	15
1928	148	131	155			1960	272	151	110	11	16
1929	152	131	131			1961	270	310	115	50	15
1930	157	–	–			1962	264	333	110	30	11
1931	181	161	133			1963	272	353	100	40	13
1932	184	131	115			1964	292	164	95	60	15
1933	192	133	115			1965	299	325	99	50	16
1934	205	100	100			1966	292	210	75	30	15
1935	196	75	125			1967	277	255	80	31	17
1936	199	35	110			1968	264	235	90	25	30
1937	196	40	115			1969	262	211	79	26	31
1938	196	–	–			1970	267	219	81	–	–
1939	–	–	–			1971	259	300	130	17	40
1940	–	–	–			1972	265	295	133	16	43
1941	–	–	131			1973	262	330	135	15	45
1942	–	–	–			1974	244	250	170	13	45
1943	–	–	–			1975	246	275	145	20	39
1944	–	115	171			1976	259	310	110	15	41
1945	–	–	–			1977	266	350	109	13	45
1946	–	–	–			1978	265	360	107	12	47
1947	–	131	–			1979	255	159	95	10	50
1948	–	–	–			1980	252	290	97	9	51
1949	147	100	116			1981	248	–	–	–	–
1950	167	150	–			1982	256	245	99	11	47
1951	178	100	89	15	30	1983	243	235	101	20	39
1952	172	75	75	20	15	1984	253	215	105	21	40
1953	186	85	61	15	15	1985	246	159	95	30	35
1954	196	91	65	31	10	1986	248	110	96	32	25
1955	217	110	39	15	9	1987	239	100	93	40	30
1956	216	130	45	71	11	1988	247	90	92	30	25
1957	219	135	79	75	12						

Fig. 1.5 A petrol tanker after a collision resulting in an explosion.

Fig. 1.6 Wrecked vehicles after the explosion of a lorry carrying explosives.

22 November 1989. The A47 near Peterborough (UK) was the scene of an explosion when a 7 tonne lorry carrying explosives exploded on an industrial estate. Figure 1.6 shows wrecked vehicles in a car park of the estate.

1.5 Train collisions and impacts

Trains are subject to accidents. They collide with both one another and other objects. The causes are numerous, mostly signal failure, derailment and human error. Trains can also be subject to missile and rocket attack and can be destroyed by bombs and other detonators. Even trees and other heavy objects such as boulders, pylons and short-span bridge deck components can crash into them. Trains have collided with public vehicles at level crossings on a number of occasions. Hundreds of such cases have occurred. Recent ones are included for the reader's perusal. In the month of October 1989, 10 people were killed and 65 injured when the first ten bogies of the Howrah bound Indian Toofan Express were derailed. The bogies were badly damaged and came in contact with live wires after severe impact with electricity poles. Eight bogies were overturned and one fell into a ditch. In August 1989 a train in Mexico was derailed and fell into the river, causing 100 deaths. It is a classical example of an object impacting a water surface.

In May 1989, 75 deaths were reported in the Karnataka Express accident in India caused, presumably, by mechanical defects of one of the coaches. The train was on its way from Bangalore to Delhi. The 30 tonne locomotives were pulling 20 coaches (a 10 000 tonne load) along a steel track at a speed exceeding 100 km/h. After derailment and impact, the wreckage of the express train is shown in Fig. 1.7. The past few years have been grim ones for the Indian railways. This accident is one of three major tragedies in three years. The other two occurred in Mancherial (Andrha Pradesh) and Perumon in Kerala in July 1988, causing 55 and 105 deaths, respectively. The size and scope of the operations of the Indian Railways are awesome. Accidents are due to a lack of administration and to an overwhelming increase in traffic — around 13 000 trains carry 10.3 million people daily through all manner of terrain and in all kinds of weather.

The Clapham (London) disaster is well remembered. This accident happened on 12 December 1988. The cause of the crash was established as faulty wiring on signalling equipment. The Basingstoke, Bournemouth and Haslemere trains were involved. The Bournemouth train ran into the back of the Basingstoke train. As illustrated in Fig. 1.8, the on-coming Haslemere train prevented worse carnage by absorbing the impact of the Bournemouth carriages and preventing some of them from overturning. In one year alone, more than 800 000 accidents were reported in 120 countries.

On 5 March 1989, two London-bound trains collided outside Purley station. Figure 1.9 illustrates how one of the bogies acted as an impactor on a nearby house, causing serious damage to the house lying under the embankment. On the

Fig. 1.7 Karnataka express train derailment and impacting objects. (Courtesy of the *Front Line*, India.)

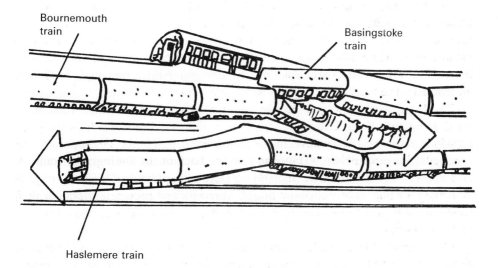

Fig. 1.8 The Clapham train disaster.

Fig. 1.9 The Purley train disaster. (Courtesy of British Rail.)

same day a rail crash in Glasgow (Scotland) at the junction between a branch line and a main line resulted in injuries to a number of people. There has been a spate of accidents since 1984.

In England, 15 people were injured when two express trains collided outside Newcastle upon Tyne Central station on 30 November 1988. In the same month a driver was killed and 18 passengers hurt when a commuter train crashed at St Helens, Merseyside.

In October 1987, four people died when a train fell into the swollen River Towey, in Wales, after a bridge collapsed. Fourteen were injured in the same month when two trains collided at Forest Gate on London's Liverpool Street Line.

On 26 July 1986, nine people died when a passenger train hit a van on a level crossing at Lockington, Yorkshire, UK. In September 1986, 60 people were hurt and one killed in a collision between two express trains at Colwich, Staffordshire, UK. In December 1984, two people were killed when a passenger train hit a tanker train in Salford, UK.

Other major accidents have included 49 killed at Hither Green, southeast London, in November 1967, 90 killed at Lewisham, southeast London, in December 1965 and 112 killed at Harrow, northwest London, in October 1952.

Britain's worst rail crash was on 22 May 1915 when a troop train and a passenger train collided at Gretna Green, killing 227 people.

Another example of impact and direct collision occurred on the London–Bristol railway line when a passenger train overturned after hitting a derailed stone-quarry train near Maidenhead, Berks, England. The damage is shown in Fig. 1.10. Later in the morning a mail van crashed into the parapet, sending

Fig. 1.10 A passenger train after impact with a derailed stone-quarry train. (Courtesy of British Rail.)

masonry tumbling onto the tracks. The Paddington–Penzance train was too close to stop and ran into the same rubble.

The worst train disaster in Europe occurred on 12 December 1917 at Modane, France, killing 543 passengers.

In 1972, 60 people died in a crash near the Punjab town of Liaquatpur. In the nation's worst rail disaster, 225 people were killed and about 400 injured in Southern Pakistan, when a crowded passenger express train ploughed into a stationary freight train, destroying seven packed carriages. The collision occurred on 3 January 1990. The 16-carriage Zakaria–Bahauddin express with 1500 passengers, was travelling at 635 miles/h when it smashed into the freight train. Trains in Pakistan, as in India, are always overcrowded and rail traffic has increased rapidly without a corresponding increase in investment.

Many rail accidents have occurred in the past, the ones described here are exceptional and recent. Table 1.4 gives a historical view of US train accidents.

1.6 Aircraft and missile impacts, crashes and explosions

Aircraft crashes are not uncommon. They happen for various reasons which will be elucidated later on in this section. To begin with, a few recent crashes, with and without explosions, will be discussed.

Table 1.4 Notable US train disasters.

Date	Location	Deaths	Date	Location	Deaths
29/12/1876	Ashtabula, Ohio	92	27/10/1925	Victoria, Missouri	21
11/8/1880	Mays Landing, New Jersey	40	5/9/1926	Waco, Colorado	30
10/8/1887	Chatsworth, Illinois	81	24/8/1928	IRT subway, Times Square, New York	18
10/10/1888	Mud Run, Pennsylvania	55	19/6/1938	Saugus, Montana	47
30/7/1896	Atlantic City, New Jersey	60	12/8/1939	Harney, Nevada	24
23/12/1903	Laurel Run, Pennsylvania	53	19/4/1940	Little Falls, New York	31
7/8/1904	Eden, Colorado	96	31/7/1940	Cuyahoga Falls, Ohio	43
24/9/1904	New Market, Tennessee	56	29/8/1943	Wayland, New York	27
16/3/1906	Florence, Colorado	35	6/9/1943	Frankford Junction, Philadelphia, Pennsylvania	79
28/10/1906	Atlantic City, New Jersey	40			
30/12/1906	Washington DC	53	16/12/1943	Between Rennert and Buie, N Carolina	72
2/1/1907	Volland, Kansas	33	6/7/1944	High Bluff, Tennessee	35
19/1/1907	Fowler, Indiana	29	4/8/1944	Near Stockton, Georgia	47
16/2/1907	New York	22	14/9/1944	Dewey, Indiana	29
23/2/1907	Colton, California	26	31/12/1944	Bagley, Utah	50
20/7/1907	Salem, Michigan	33	9/8/1945	Michigan, North Dakota	34
1/3/1910	Wellington, Washington DC	96	25/4/1946	Naperville, Illinois	45

Date	Location	No.
21/3/1910	Green Mountain	55
25/8/1911	Manchester, New York	29
4/7/1912	East Corning, New York	39
5/7/1912	Ligonier, Pennsylvania	23
5/8/1914	Tipton Ford, Missouri	43
15/9/1914	Lebanon, Missouri	28
29/3/1916	Amherst, Ohio	27
28/9/1917	Kellyville, Oklahoma	23
20/12/1917	Shepherdsville, Kentucky	46
22/6/1918	Ivanhoe, Indiana	68
9/7/1918	Nashville, Tennessee	101
1/11/1918	Brooklyn, New York	97
12/1/1919	South Byron, New York	22
1/7/1919	Dunkirk, New York	12
20/12/1919	Onawa, Maine	23
27/2/1921	Porter, Indiana	37
5/12/1921	Woodmont, Pennsylvania	27
5/8/1922	Sulpher Spring, Missouri	34
13/12/1922	Humble, Texas	22
27/9/1923	Lockett, Wyoming	31
16/6/1925	Hackettstown, New Jersey	50
18/2/1947	Gallitzin, Pennsylvania	24
17/2/1950	Rockville Centre, New York	31
11/9/1950	Coshocton, Ohio	33
22/11/1950	Richmond Hill, New York	79
6/2/1951	Woodbridge, New York	84
12/11/1951	Wyuta, Wyoming	17
25/11/1951	Woodstock, Alabama	17
27/3/1953	Conneaut, Ohio	21
22/1/1956	Los Angeles, California	30
28/2/1956	Swampscott, Massachusetts	13
5/9/1956	Springer, New Mexico	20
11/6/1957	Vroman, Colorado	12
15/9/1958	Elizabethport, New Jersey	48
14/3/1960	Bakersfield, California	14
28/7/1962	Steelton, Pennsylvania	19
28/12/1966	Everett, Massachusetts	13
10/6/1971	Salem, Illinois	11
30/10/1972	Chicago, Illinois	45
4/2/1977	Chicago, Illinois (elevated train)	11
4/1/1987	Essex, Maryland	16

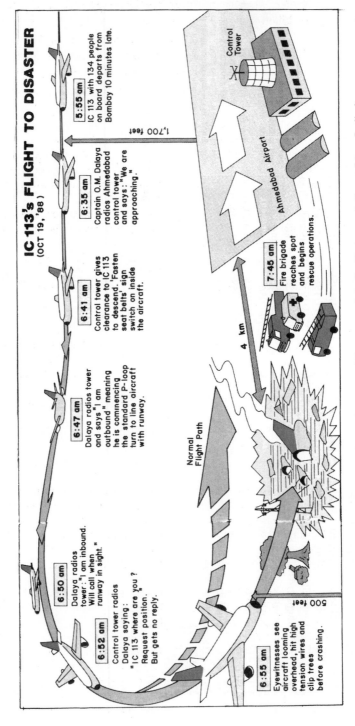

Fig. 1.11 Indian Airlines IC 113's flight route. (Courtesy of *India Today* and *Front Line*, India.)

On 19 October 1988 the first Indian Airlines Boeing 737 aircraft, on a routine flight from Bombay to Ahmedabad, inexplicably crashed into a stream 4 km short of the airport. It killed all but five of the 134 passengers. A chronological breakdown of the flight is given in Fig. 1.11. One hour later and 3000 km away, a Vayudoot Fokker slammed into a hill near Guwahati, killing all 34 passengers. In the same year, two high-tech 2000S two trisonic MIG-25s and two Soviet-built transport AN-12 and AN-32 aircraft crashed while on various missions. Table 1.5

Table 1.5 Notable aircraft disasters.

Date	Aircraft	Site of accident
6/5/1937	German zeppelin Hindenburg	Burned at mooring, Lakehurst, New Jersey, USA
23/8/1944	US Air Force B-24	Hit school, Freckelton, England
28/7/1945	US Army B-25	Hit Empire State building, New York City, USA
30/5/1947	Eastern Airlines DC-4	Crashed near Port Deposit, Michigan, USA
20/12/1952	US Air Force C-124	Fell burned, Moses, Lake Washington, USA
3/3/1953	Canadian Pacific Comet Jet	Karachi, Pakistan
18/6/1953	US Air Force C-124	Crashed, burned near Tokyo
1/11/1955	United Airlines DC-6B	Exploded, crashed near Longmont, Colorado, USA
20/6/1956	Venezuelan Super-Constellation	Crashed in Atlantic off Asbury Park, New Jersey, USA
30/6/1956	TWA Super-Constellation, United DC-7	Collided over Grand Canyon, Arizona, USA
16/12/1960	United DC-8 jet, TWA Super-Constellation	Collided over New York City, USA
16/3/1962	Flying tiger Super-Constellation	Vanished in Western Pacific
3/6/1962	Air France Boeing 707 jet	Crashed on take off from Paris, France
22/6/1962	Air France Boeing 707 jet	Crashed in storm, Guadeloupe, West Indies
3/6/1963	Chartered Northwest Airlines DC-7	Crashed in Pacific off British Columbia, Canada
29/11/1963	Trans-Canada Airlines DC-8F	Crashed after take off from Montreal, Canada
20/5/1965	Pakistani Boeing 720-B	Crashed at Cairo airport, Egypt
24/1/1966	Air India Boeing 707 jetliner	Crashed on Mont Blanc, France/Italy border
4/2/1966	All-Nippon Boeing 727	Plunged into Tokyo Bay
5/3/1966	BOAC Boeing 707 jetliner	Crashed on Mount Fuji, Japan
24/12/1966	US military chartered CL-44	Crashed into village in South Vietnam
20/4/1967	Swiss Britannia turboprop	Crashed at Nicosia, Cyprus
19/7/1967	Piedmont Boeing 727, Cessna 310	Collided in air, Hendersonville, North Carolina, USA

continued

Table 1.5 *Continued.*

Date	Aircraft	Site of accident
20/4/1968	South African Airways Boeing 707	Crashed on take off, Windhoek, Southwest Africa
3/5/1968	Braniff International Electra	Crashed in storm near Dawson, Texas, USA
16/3/1969	Venezuelan DC-9	Crashed after take off from Maracaibo, Venezuela
8/12/1969	Olympia Airways DC-6B	Crashed near Athens, Greece, in storm
15/2/1970	Dominican DC-9	Crashed into sea on take off from Santo Domingo, Dominican Republic
3/7/1970	British chartered jetliner	Crashed near Barcelona, Spain
5/7/1970	Air Canada DC-8	Crashed near Toronto International Airport, Canada
9/8/1970	Peruvian turbojet	Crashed after take off from Cuzco, Peru
14/11/1970	Southern Airways DC-9	Crashed in mountains near Huntington, West Virginia, USA
30/7/1971	All-Nippon Boeing 727 and Japanese Air Force F-86	Collided over Morioka, Japan
4/9/1971	Alaska Airlines Boeing 727	Crashed into mountain near Juneau, Alaska
14/8/1972	East German Illyushin-62	Crashed on take off, East Berlin, Germany
13/10/1972	Aeroflot Illyushin-62	East German airline crashed near Moscow
3/12/1972	Chartered Spanish airliner	Crashed on take off, Canary Islands
29/12/1972	Eastern Airlines Lockhead Tristar	Crashed on approach to Miami International Airport
22/1/1973	Chartered Boeing 707	Burst into flames during landing, Kano Airport, Nigeria
21/2/1973	Libyan jetliner	Shot down by Israeli fighter planes over Sinai, Egypt
10/4/1973	British Vanguard turboprop	Crashed during snowstorm at Basel, Switzerland
3/6/1973	Soviet supersonic TU-144	Exploded in air near Goussainville, France
11/7/1973	Brazilian Boeing 707	Crashed on approach to Orly airport, Paris
31/7/1973	Delta Airlines jetliner	Crashed, landing in fog at Logan Airport, Boston, USA
23/12/1973	French Caravelle jet	Crashed in Morocco
3/3/1974	Turkish DC-10 jet	Crashed at Ermenonville near Paris, France
23/4/1974	Pan American 707 jet	Crashed in Bali, Indonesia
1/12/1974	TWA-727	Crashed in storm, Upperville, Virginia, USA
4/12/1974	Dutch chartered DC-8	Crashed in storm near Colombo, Sri Lanka
4/4/1975	Air Force galaxy C-5B	Crashed near Saigon, South Vietnam, after take off
24/6/1975	Eastern Airlines 727 jet	Crashed in storm, JFK Airport, New York City, USA
3/8/1975	Chartered 707	Hit mountainside, Agadir, Morocco
10/9/1976	British Airways Trident, Yugoslav DC-9	Collided near Zagreb, Yugoslavia

Table 1.5 *Continued.*

Date	Aircraft	Site of accident
19/9/1976	Turkish 727	Hit mountain, southern Turkey
13/10/1976	Bolivian 707 cargo jet	Crashed in Santa Cruz, Bolivia
13/1/1977	Aeroflot TU-104	Exploded and crashed at Alma-Ata, Central Asia
27/3/1977	KLM 747, Pan American 747	Collided on runway, Tenerife, Canary Islands
19/11/1977	TAP Boeing 727	Crashed on Madeira
4/12/1977	Malaysian Boeing 737	Hijacked, then exploded in mid-air over Straits of Johore
13/12/1977	US DC-3	Crashed after take off at Evansville, Indiana, USA
1/1/1978	Air India 747	Exploded, crashed into sea off Bombay, India
25/9/1978	Boeing 727, Cessna 172	Collided in air, San Diego, California
15/11/1978	Chartered DC-8	Crashed near Colombo, Sri Lanka
25/5/1979	American Airlines DC-10	Crashed after take off at O'Hare International Airport, Chicago, USA
17/8/1979	Two Soviet Aeroflot jetliners	Collided over Ukraine
31/10/1979	Western Airlines DC-10	Skidded and crashed at Mexico City Airport
26/11/1979	Pakistani Boeing 707	Crashed near Jidda, Saudi Arabia
28/11/1979	New Zealand DC-10	Crashed into mountain in Antarctica
14/3/1980	Polish Illyushin 62	Crashed making emergency landing, Warsaw, Poland
19/8/1980	Saudi Arabian Tristar	Burned after emergency landing, Riyadh, Saudi Arabia
1/12/1981	Yugoslavian DC-9	Crashed into mountain in Corsica
13/1/1982	Air Florida Boeing 737	Crashed into Potomac River after take off
9/7/1982	Pan-Am Boeing 727	Crashed after take off in Kenner, Louisiana, USA
11/9/1982	US Army CH-47 Chinook helicopter	Crashed during air show in Mannheim, Germany
1/9/1983	South Korean Boeing 747	Shot down after violating Soviet airspace
27/11/1983	Colombian Boeing 747	Crashed near Barajas Airport, Madrid, Spain
19/2/1985	Spanish Boeing 727	Crashed into Mount Oiz, Spain
23/6/1985	Air India Boeing 747	Crashed into Atlantic Ocean, south of Ireland
2/8/1985	Delta Airlines jumbo jet	Crashed at Dallas Fort Worth International Airport, USA
12/8/1985	Japan Airlines Boeing 747	Crashed into Mount Ogura, Japan
12/12/1985	Arrow Air DC-8	Crashed after take off in Gander, Newfoundland
31/3/1986	Mexican Boeing 727	Crashed northwest of Mexico City
31/8/1986	Aeromexico DC-9	Collided with Piper PA-28 over Cerritos, California, USA

lists information on notable aircraft disasters. Table 1.6 summarizes accident data for various aircraft until the end of 1988.

If the aircraft does not break up in the air, it generally hits the ground at a certain angle. A typical example of a tangled wreckage of a Gulfstream Turbo Commander impacting the ground is shown in Fig. 1.12. Table 1.7 is an extensive list of aircraft disasters due to ground impact.

At Ramstein Airshow in Germany (12 August 1988), during the ill-fated Freccie tri-colour display, the two sets of aircraft interlocked as they formed the heart-shaped loop in a manner shown in Fig. 1.13. The letter scheme gives the step-by-step revelation of the disaster. Such mid-air collisions have occurred in the past. A total of 1476 aircraft were involved in such collisions from 19 December 1946 to September 1989. Table 1.8 gives data on mid-air collisions of some of the well known aircraft.

Bird strikes of aircraft, according to the research carried out at Rutgers

Table 1.6 Accident data for a number of aircraft prior to 1988.

Type	Total number
Airbus	24
Boeing 707/720 series	122
Boeing 727	157
Boeing 737	120
Boeing 747	100
Boeing 757	3
DHC-4	8
Douglas DC-3/C47	213
Douglas C54/DC-4	32
Douglas DC-7	7
Douglas DC-8	74
Douglas DC-9	105
Douglas DC-10	70
Fairchild	10
Fokker/Fairchild	
F-27	98
F-28	24
Gulfstream	15
Hawker Siddeley Trident	51
Hawker Siddeley BAe 748 series	71
Jetstream	13
Lockheed L-188 Electra	25
Lockheed L-382B Hercules	20
Lockheed L-18 Lodest	21
Lockheed L-1011 Tristar	33
Lockheed Volga	10
Miscellaneous	313
Soviet aircraft, Illusion	97
Vickers Vicount	40

Fig. 1.12 A Gulfstream Turbo Commander impacting the ground. (Courtesy of *Flight International*.)

University, New Jersey, USA, are related to the jet noise. Table 1.9 lists the number of aircraft subject to bird strikes over a 10-year period or so.

On 8 June 1989, a MIG-29 at the Paris airshow hit the ground 90 m (300 ft) from the crowd. A sheet of flame shot from the starboard engine, indicating compressor failure. The aircraft rolled over to its right and plunged downwards.

The British Midland Boeing 737−400 is one of the world's newest and most sophisticated passenger aircraft. It was fresh from Boeing's Seattle factory, with only 518 flying hours on the clock. It was the latest version of the world's most popular jetliner. Either of the 10 tonne CFM56-3C power plants could have held the plane aloft as long as fuel lasted. At around 400 m from the airport, while travelling at approximately 150 miles per hour, the two engines were in trouble and the plane crashed on the M1 motorway and hit trees as it limped into East Midlands airport, with a loss of 44 lives.

The engine on the crashed 737−400s, a CFM56−3, is fitted to all Boeing 737−300s, which are operated by 39 airlines or aircraft leasing companies worldwide.

Since 1984, when they first came on line, Boeing has delivered 488 737−300s with CFM56−3 engines to many prestigious airlines, including Air Europe, KLM, Lufthansa and Sabena.

The same CFM56−3 engine prototype, developed jointly by Snecma, a French company, and General Electric, USA, is also fitted to 737−400s − the crashed British Midland aircraft. Until the crash, 18 of these were flying, including two leased by British Midland, one by Air UK Leisure and one by Dan Air. All three companies grounded their 737−400s.

Such structural and other failures have occurred in the past and are listed earlier in this text.

In August 1989 a British tornado fighter bomber and two West German Air

Table 1.7 Aircraft impact at ground level.

Date	Aircraft	Location	Date	Aircraft	Location
15/1/1976	DC-4	Bogota, Colombia, South America	29/12/1977	Viscount	Cuenca, Ecuador
20/1/1976	HS748	Loja, USA	28/1/1983	DC-3	Cerro Grenada, Colombia, South America
1/6/1976	TU-154	Malabo			
1/8/1976	Boeing 707	Mehrabad, Iran	14/8/1978	C-46	Tota, Colombia, South America
7/8/1976	Falcon	Acapulco, Mexico	27/8/1978	DC-6	Muscat, Oman
19/9/1976	Boeing 727	Isparta, Italy	9/9/1978	–	Mexico City, Mexico
22/9/1976	DHC-6	Mosher, Iran	21/11/1978	DC-3	Rubio, Venezuela
26/9/1976	Gulfstream	Hot Springs, USA	4/12/1978	DHC-6	Steamboat Spring, USA
4/10/1976	DC-7C	Mount Kenya	28/1/1979	F-27	Rodez, France
23/11/1976	YS-11A	Greece	28/1/1979	DHC-6	Alaska, USA
30/11/1976	DC-3	Victoria, Australia	6/3/1979	F-28	Ngadirejo Sukapur, Indonesia
30/12/1976	DC-4	Trujillo	23/4/1979	Viscount	Ecuador
6/1/1977	Gates Lear	Palm Springs, USA	11/7/1979	F-28	Mount Sebayat, Indonesia
14/1/1977	DHC-6	Terrace airport, British Columbia, Canada	26/7/1979	Boeing 707	Rio de Janeiro, Argentina
			4/8/1979	HS748	Panvel, India
18/1/1977	Gates Lear	Sarajevo, Yugoslavia	29/8/1979	DHC-6	Frobisher Bay, North-Western Territory
29/3/1977	DHC-6	Bainaha Valley, Indonesia			
5/4/1977	DC-3	Edavli, India	14/9/1979	DC-9	Cagliari, Italy
10/4/1977	DC-3	Colombia, South America	14/9/1979	DC-7C	Klamath Falls, USA
13/5/1977	AN-12	Aramoun, Lebanon	21/11/1979	Arava	Navarino Island, Chile
20/7/1977	DC-3	Ethiopia	28/11/1979	DC-10	Mount Erebus, Antarctica
7/8/1977	DHC-6	El Bolson, Argentina	19/12/1979	DC-4	Cerro Toledo, Colombia, South America
4/9/1977	Viscount	Cuenca, Ecuador			
6/9/1977	DHC-6	Alaska, USA	19/12/1979	DHC-6	Colombia, South America
23/10/1977	DC-3	Manidar, Iran	23/12/1979	F-28	Ankara, Turkey
21/11/1977	BAC 1–11	San Carlos de Bariloche, Argentina	21/1/1980	Boeing 727	Teheran, Iran
11/12/1977	He111	Cercedilla, Spain	22/1/1980	DHC-6	Kenai, Alaska, USA
13/12/1977	DC-3	Evansville, USA	23/1/1980	CASA 212	Cemonyet, Indonesia
18/12/1977	DC-8	Salt Lake City, USA	12/4/1980	Boeing 727	Florianopolis, Brazil

Table 1.7 *Continued.*

Date	Aircraft	Location	Date	Aircraft	Location
24/4/1980	B-26 Invader	Slave Lake, Alberta	19/5/1982	Citation	Kassel
25/4/1980	Boeing 727	Tenerife	8/6/1982	Boeing 727	Fortaleza, Brazil
2/6/1980	F-27	Yacuiba, Bolivia	25/8/1982	CV440	Del Norte, Colorado
8/6/1980	Yak-40	Southern Angola	1/9/1982	DHC-4	Valladolid
12/6/1980	Yak-40	Tadzhikistan	12/10/1982	DC-3	Graskop, South Africa
1/8/1980	DC-8	Mexico City	29/11/1982	AN-26	ME Bibala, Angola
13/8/1980	Lear 35	Majorca	29/11/1982	DHC-6	Villavicencio
28/8/1980	CASA 212	Bursa, Turkey	7/12/1982	Metro	Pueblo
24/11/1980	DC-3	Medellin, Colombia	3/1/1983	Challenger	Hailey, Idaho, USA
18/1/1981	Skyvan	Guyana	11/1/1983	DC-8	Detroit, USA
20/1/1981	Beech 99	Spokane, Washington, USA	16/1/1983	DC-3	Bay City, USA
6/4/1981	DC-3	Laguna Soliz, South America	22/2/1983	Boeing 737	Manaus Airport
20/5/1981	Convair 440	Oaxaca, Mexico	10/3/1983	—	Uruzgan
24/7/1981	DHC-6	Madagascar	14/3/1983	Boeing 707	North of Sabha City, Indonesia
19/8/1981	HS748	Mangalore, India	30/3/1983	Learjet 25	Newark, New Jersey, USA
26/8/1981	Viscount	Florencia, Colombia, South America	6/4/1983	Learjet 35	Indianapolis, USA
2/9/1981	EMB110 Bandeirante	Paipa, Colombia, South America	16/4/1983	HS748	Khartoum, Sudan
			23/6/1983	Lockheed 18	Millhaven, Georgia
1/10/1981	Learjet	Felt, Oklahoma, USA	1/7/1983	IL-62	Labe, Guinea
9/11/1981	DC-9	Acapulco, Mexico	11/7/1983	Boeing 737	Cuenca, Ecuador
1/12/1981	DC-9	Ajaccio, Corsica	27/8/1983	Hercules	Dundo, Angola
18/12/1981	DHC-6	Lorica, Colombia, South America	7/10/1983	EMB-110	Uberaba
11/1/1982	Learjet	Narssarssuma	27/11/1983	Boeing 747	Madrid, Spain
9/2/1982	DC-3	Panay Island, Philippines	18/12/1983	Airbus A300	Kuala Lumpur, Malaysia
19/2/1982	DC-6	Cuginamarca, Colombia, South America	24/1/1984	CASA C212	Lokon Mountain, Indonesia
			3/4/1984	DHC-6	Quthing, Lesothe
			5/5/1984	Beech 200	Poza Rica, Mexico
26/3/1982	Viscount	Qeuate, Colombia, South America	28/6/1984	Bandeirante	Macae, Brazil
26/4/1982	Trident	Guilin, China	15/8/1984	DC-3	Pass Valley, Indonesia

continued

Table 1.7 *Continued.*

Date	Aircraft	Location
11/10/1984	DHC-6 Twin Otter	Mealy Mountain, USA
19/11/1984	Bandeirante	Inverness, Scotland
20/11/1984	DHC-6 Twin Otter	En route
22/12/1984	DHC-6	Bhojpur, India
1/1/1985	Boeing 727	La Paz, Bolivia
22/1/1985	Bandeirante	Buga, Colombia, South America
22/2/1985	DHC-6	Andes, Colombia, South America
18/2/1985	Boeing 727	Mount Oiz, Bilbao, South America
28/2/1985	F-28	Florencia, Colombia, South America
11/4/1985	HS125	Salta, Uruguay
15/4/1985	Boeing 737	Phuket, Thailand
22/4/1985	DC-6	Fitoy, France
21/5/1985	Citation 501	Harrison, Arkansas
27/5/1985	CV580	Oro Negro, Venezuela
19/6/1985	Merlin	Rock Springs, Texas
12/8/1985	Boeing 747	Mount Ogura, Japan
29/8/1985	Aeritalia G222	Sardinia
11/10/1985	DHC-6	Homer City, Pennsylvania, USA
22/10/1985	Learjet 24D	Juneau, Alaska, USA
18/1/1986	Caravelle	Santa Elena, Guatemala
20/3/1986	Casa 212	Manado, Northern Sulawesi, Indonesia
31/3/1986	Boeing 727	—
27/4/1986	DHC-6	Sarevena, Indonesia
12/6/1986	DHC-6	Port Ellen, Islay, Scottish Highlands

Date	Aircraft	Location
22/6/1986	DHC-6	Dembidollo, Ethiopia
19/7/1986	DC-6	Mont de la Plage
22/7/1986	MU-300	Sado Island
	Diamond 1A	
19/9/1986	EMB 120 Brasilia	Sao Jose dos Campos, Brazil
30/9/1986	DHC-6	Northern Sulawesi, Indonesia
2/10/1986	Falcon 10	Haenertsburg
3/10/1986	Skyvan	Manado
19/10/1986	TU-134A	Komatipoort, Swaziland
5/2/1987	Learjet 55	Jakiri, Cameroon
17/2/1987	Beech 200	Fukuoka, Japan
27/3/1987	Learjet 24	Eagle, Colorado, USA
19/5/1987	DHC-6	El Trompillo, Santa Cruz, Bolivia
21/6/1987	F-27	Pansauk, Burma
26/6/1987	HS748	Mount Ugo, Philippines
17/7/1987	Beech 200	Lake Tahoe, California, USA
11/10/1987	F-27	Turen Taung, China
15/10/1987	ATR42	Mount Crezzo
19/10/1987	Beech 200	Leeds/Bradford, UK
13/12/1987	SD3-60	Iligan, Philippines
4/1/1988	Boeing 737	Izmir, Turkey
18/1/1988	C-46	Colorado Mountain, USA
18/1/1988	IL-18	Chongging, China
26/2/1988	Boeing 727	Northern Cyprus
17/3/1988	Boeing 727	Cucuta Camilo Daza, Colombia, South America
19/4/1988	Let L-410	Bagdarin, Iran
	Turbolet	
6/5/1988	DHC-7	Bronnoysund

Table 1.7 *Continued.*

Date	Aircraft	Location	Date	Aircraft	Location
			Accidents due to tyre burst at ground level		
24/5/1988	DC-6	—	11/3/1978	L-188	Fort Myers, USA
24/5/1988	Learjet 35A	West Patterson, New Jersey, USA	22/12/1980	L-1011	State of Qatar
16/6/1988	F-27	Northern Burma	27/6/1983	L-1011	London, Gatwick, UK
17/6/1988	DHC-6	Tau, Samoa, Indonesia	16/11/1983	Boeing 727	Miaimi, Florida, USA
14/7/1988	DHC-6	Battle Creek Mountain, Oregon, USA	3/5/1987	A-300	Athens, Greece
31/8/1988	Bandeirante	Cerro de la Calera, Mexico			
18/9/92	Airbus PIA	Katmandu, Nepal			
4/10/92	Boeing 747-200F	Amsterdam, The Netherlands			

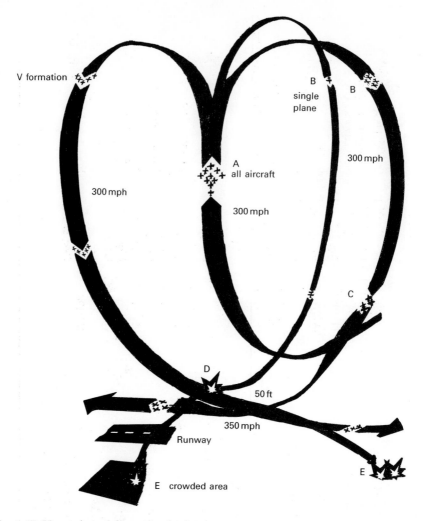

Fig. 1.13 Heart-shaped formation by jets.
C two formations aimed to complete the heart shape by passing each other in opposite
  directions.
D two groups of 5 planes just passed the group of 4 planes flying too low.
E Straight ahead, a solo plane changed direction.

Force alpha jets collided during low-flying exercises near the north German coast.
Eye-witnesses said the tornado crashed in fields only 500 m from a village school.
Similar cases of mid-air collisions are given in Table 1.8.

On 19 November 1988 a MIG-21 belonging to the Indian Air Force crashed.
The jet hit the houses near Nazamgarh in West Delhi (India), as shown in
Fig. 1.14.

On 1 December 1988, PA103, the Pan Am 747 which crashed in Scotland, was
one of the oldest jumbo jets still flying. It was fifteenth out of 710 jumbo jets built
by Boeing in Seattle and was delivered to Pan Am in February 1970. It had flown
for 72 000 hours and had completed 16 500 take-off and landing cycles, well below

Table **1.8** Mid-air collisions.

Date	Aircraft	Location	Date	Aircraft	Location
18/3/1976	DC-8 / AN24	Havana, Cuba	7/3/1984	Beech 200	Benson, Carolina, USA
10/9/1976	Trident / DC-9	Zagreb, Yugoslavia	18/4/1984	Bandeirante / Bandeirante	Imperatriz, Brazil
9/4/1977	Nord 262	Reading, USA	3/5/1985	TU-134 / AN-2	Kvov, USSR
1/3/1978	Cessna 195 / F-28 Trainer	Kano, Nigeria	12/8/1985	Beech 200 / Cessna U206F	Quinlan, Texas, USA
18/5/1978	Falcon 20 / Cessna 150	Memphis, USA	22/9/1985	Learjet 35A / Ultralight	Auburn, Alabama, USA
25/9/1978	Boeing 727 / Cessna 721	San Diego, USA	10/11/1985	Falcon 50 / Piper PA28	Cliffside Park, New Jersey, USA
17/2/1979	Boeing 707 / F-5	Over Taoyaun, Taiwan	5/3/1986	Learjet 35 / Learjet 24D	San Clemente Island
11/8/1979	TU-134 / TU-134	Dneprodzerkinsk, Ukraine	18/6/1986	DHC-6 / Bell 206	Grand Canyon, USA
6/8/1980	F-27 / Cessna 172	Bridgeport, Texas, USA	31/8/1986	DC-9 / PA28	Los Angeles, USA
17/4/1981	Jetstream / Cessna	Loveland, Colorado, USA	15/1/1987	Metro II / Mooney M20C	Salt Lake City, USA
18/7/1981	TU-206 / CL-44	Yerevan, USSR	16/6/1987	B737 / J-6	Fuzhou, China
19/1/1982	Swearingen SA226T	Rockport, Texas, USA	4/8/1987	DHC-6 / Cessna 172	Palmdale, California, USA
6/11/1982	Learjet 24 / Learjet 25	Elizabeth City, North Carolina, USA			

Table 1.9 Bird strikes.

Date	Aircraft	Location	Date	Aircraft	Location
1/1/1976	DC-10	Kastrup	20/11/1980	Falcon	Kansas City, USA
25/1/1976	Boeing 747	Istanbul, Turkey	7/4/1981	Learjet	Cincinnati, Ohio, USA
6/2/1976	Lear	Palese airport	18/4/1981	YS-11	Sand Point, Alaska
12/11/1976	Falcon	Naples, Italy	1/9/1981	Fokker F26	Ornskoldsvit, Sweden
11/7/1977	Boeing 747	Tokyo, Japan	2/2/1982	Beech 200	Nairobi, Nigeria
16/8/1977	Gates Lear	Baton Rouge, USA	15/2/1982	HS 125	Curitiba
18/1/1978	DC-9	Hamburg, Germany	17/8/1983	Lear 25	Wilmington, USA
18/2/1978	Boeing 747	Lyon, France	1/10/1983	TU-134	Krasnodar
20/2/1978	Boeing 707	Sharjah	6/11/1983	CV580	Sioux Falls, USA
4/4/1978	Boeing 737	Gosselies, Belgium	19/9/1984	Boeing 727	Kimberley, South Africa
25/7/1978	CV580	Kalamazoo, USA	6/11/1984	Boeing 737	Lasham
26/7/1978	DC-3	Guatemala	11/7/1985	Boeing 727	Bellevue, Nebraska, USA
26/5/1979	Boeing 727	Patna, India	7/12/1985	Boeing 737	Dublin, Eine
19/10/1979	Merlin	Palo Alto, California	20/7/1986	Boeing 737	Wabush, Newfoundland, Canada
8/7/1980	Airbus A300	Lyon, France			
-/10/1980	Yak-40	USSR	5/3/1987	Bandeirante	Norfolk, Nebraska, USA

Fig. 1.14 Jet aircraft (MIG-21) impact on houses in Nazamgarh, India. (Courtesy of the Indian Air Force.)

the 50 000 cycles that is considered high for a commercial jetliner. The town of Lockerbie suffered a direct hit. Forty houses were destroyed when the wreckage fell. The 747 engines, acting as missiles, landed on two rows of houses, completely destroying them. A fireball 91 m (300 ft) high lit up the sky as the aircraft blew up. Houses 10 miles east of Dumfries and 15 miles north of the Scottish border simply disappeared in the explosions. Others were set ablaze or had their roofs blown off. On impact, pieces of wreckage carved a hole in the A47 road, blocking the route through the town, and cars driving past were set on fire. Eye-witnesses reported that the doomed aircraft fell from the sky, hit, in trailing flames, a small hill east of Lockerbie and broke up, somersaulting across the main A47 London–Glasgow road before crashing into houses. One section hit a petrol station while other parts of the wreckage were scattered over 10 miles.

The Boeing corporation has manufactured 740 of the 747 aircraft since 1969, of which 13 have crashed:

(1) November 1974, a Lufthansa 747 crashed at Nairobi, 59 people died, pilot error was blamed.
(2) June 1975, an Air France 747 crashed at Bombay and burnt on the tarmac, no fatalities.
(3) May 1976, an Iranian Air Force 747 crashed at Madrid, 17 people died.
(4) March 1977, Pan Am and KLM 747s collided in fog at Tenerife Airport.
(5) January 1978, an Air India 747 overstretched on landing at Bombay and crashed into the water, 213 people died.

(6) November 1980, a Korean Air 747 overstretched on landing at Seoul and crashed into water, 14 fatalities.

(7) August 1983, a Pan Am 747 misjudged the landing at Karachi, no fatalities.

(8) September 1983, a Korean Air 747 was shot down near the Soviet Union, 269 fatalities.

(9) November 1983, an Avianca 747 landed short at Madrid, 183 people died.

(10) June 1985, an Air India 747 was blown up off the Irish coast, 329 people died.

(11) August 1985, a Japan Air Lines 747 crashed near Tokyo, 524 died.

(12) November 1987, a South African Airways 747 crashed near Mauritius, 160 died; the cause was believed to be a chemical or explosives leak in the cargo hold.

Table 1.10 gives additional information on in-flight accidents caused by bombs hidden on aircraft.

On 20 September 1989, a US Air Boeing 737−400 crashed on take off at La Guardia Airport, New York, into the East River. This was a tyical example of an impact on a water surface. The plane broke into three after hitting the water. Many such accidents have occurred. Table 1.11 lists aircraft disasters which involved water impact within the last decade or so. The total number of aircraft involved in this type of accident since 1946 is 610.

Crashes have occurred for other reasons, including failures in electrical and power plant units. The number of aircraft crashed since 1940 can be assessed from the data given in Table 1.12.

In war and in peace, an air force may intrude physically or send missiles into the airspace of other countries. Examples include the American missiles hitting the Libyan Air Force planes over the Mediterranian and the Iranian civilian plane in the Gulf, in 1989.

On 19 August 1981, a pair of US Navy F14 jet fighters shot down two attacking Soviet-built Libyan SU 22s about 60 miles from the Libyan coast. According to the Pentagon, the F14 pilots apparently saw the two Libyan jets about five miles away. Rather than turn away, one Libyan jet fired a Soviet-made Atoll air-to-air missile at the two F14s, while the other jet appeared to be moving into position to fire. The Atoll missed its target. Each US F14 then fired a Sidewinder missile, destroying both Libyan aircraft.

Since 1940, the number of missiles used has reached 300 000 plus. On 5 September 1989, a Pakistan Air Force F16 and a Norwegian F16 crashed. The Norwegian one was stolen by a Belgian technician. The plane left a crater.

Aircraft are also subject to hail impact or ice/snow accretion. Tables 1.13 and 1.14 list aircraft damaged by hail impact and ice/snow accretion, respectively. The total number of aircraft damaged by ice/snow accretion since 1946 is around 250.

A great deal of care is exercised to ensure that aircraft do not over-run or veer off the runway. Nevertheless, this has been a problem. Since 1946 there have

Table 1.10 In-flight accidents due to secretion of bombs on aircraft.

Date	Aircraft	Location	Date	Aircraft	Location
7/5/1949	DC-3	Philippines	20/11/1971	Caravelle	Formosa Strait
9/9/1949	DC-3	Quebec, Canada	26/1/1972	DC-9	Hermsdorf, Germany
13/4/1950	Viking	Hastings, UK	25/5/1972	Boeing 727	Panama/Miami
24/9/1952	DC-3	Mexico	15/6/1972	CV880	Pleiku
11/4/1955	L-749A	South China Sea	16/8/1972	Boeing 707	Rome/Tel-Aviv
1/11/1955	DC-6B	Longmont, USA	15/9/1972	F-27	Over Philippines
25/7/1957	CV240	Daggett	19/3/1973	DC-4	Ban Me Thuot, South Vietnam
19/12/1957	Armagnac	Central France			
8/9/1959	DC-3	Mexico	8/9/1974	Boeing 707	Ionian Sea
6/1/1960	DC-6B	Bolivia	15/9/1974	Boeing 727	Vietnam
28/4/1960	DC-3	Venezuela	3/6/1975	BAC 1–11	Manila, Philippines
22/5/1962	Boeing 707	Unionville, USA	1/1/1976	Boeing 720B	Al Qaysumah, Egypt
8/12/1964	C47	Bolivia	6/10/1976	DC-8	Off Barbados
8/7/1965	DC-6B	British Columbia, Canada	18/8/1978	BAC 1–11	Philippines
22/11/1966	Dakota	Meijah/Aden	7/9/1978	HS748	Colombo, Ceylon
29/6/1967	DC-6	Baranquilla/Bogota	26/4/1979	Boeing 737	Trivandrum/Madras, India
12/10/1967	Comet 4B	100 miles east of Rhodes	15/11/1979	Boeing 727	Chicago/Washington, USA
11/12/1967	Boeing 727	Chicago/San Diego, USA	9/9/1980	Boeing 727	Sacramento, California, USA
19/11/1968	Boeing 707	Over Gunnison, USA	21/12/1980	Caravelle	Riohaea, Colombia, South America
5/9/1969	HS748	Zamboanga, Africa			
22/12/1969	DC-6B	Nha Trang, Vietnam	11/8/1982	Boeing 747	Hawaii, USA
21/2/1970	CV990	Wuerenlingen, Germany	10/3/1984	DC-8	Ndjamena, Chad
21/2/1970	Caravelle	Frankfurt, Germany	23/1/1985	Boeing 727	Santa Cruz, Bolivia
14/3/1970	AN-24	Alexandria, Egypt	2/4/1986	Boeing 727	Corinth, Greece
21/4/1970	HS748	En route, France	26/10/1986	A300–600	Kochi, Shikoku
2/6/1970	F-27	En route, France	25/12/1986	Boeing 737	Arar, Saudi Arabia
26/8/1970-	AN-24	En route, France			

Table 1.11 Aircraft impact on water.

Date	Aircraft	Location	Date	Aircraft	Location
14/1/1976	Sabreliner	Recife, Brazil	8/5/1978	Boeing 727	Pensocola, USA
4/2/1976	DC-6	Santa Marta	12/5/1978	CV440	Shippingport, USA
2/4/1976	DC-3	Puerto Asis	22/7/1978	C46	Opalocka, USA
28/7/1976	IL-18	Bratislava	2/9/1978	DHC-6	Vancouver, Canada
16/9/1976	C46	Caribbean	21/9/1978	DC-3	Mantanzas, Cuba
6/10/1976	DC-8	Off Barbados	1/10/1978	DC-3	FT Walton Beach, Florida, USA
5/11/1976	DC-3	En route	5/11/1978	DC-3	Mediterranean
22/11/1976	Skyvan	Das Island	18/11/1978	DHC-6	Marie Galante
16/12/1976	DHC-6	Juan de Fuca	29/11/1978	CV240	Miami, USA
8/2/1977	C46	San Juan, Puerto Rico	23/12/1978	DC-9	Palermo, Sicily, Italy
1/3/1977	DC-3	Aden	30/1/1979	Boeing 707	Pacific
6/5/1977	C46	Off Hollywood, Florida, USA	17/2/1979	F-27	Auckland, New Zealand
28/5/1977	Yak 40	Genoa, Italy	10/3/1979	Nord 262	Los Angeles, USA
30/6/1977	L188	San Joe Maiquetia	17/3/1979	DHC-4	Barbados
17/7/1977	YS-11A	off Macton Island	17/5/1979	DC-4	Gulf of Mexico
25/7/1977	HFB320	Adjivou, Ivory Coast	11/6/1979	DC-3	Selwey River, Idaho
24/8/1977	C46	Guadeloupe	14/6/1979	DC-4	Eagle Lake, Maine
2/9/1977	CL-44	Hong Kong	7/7/1979	Lockheed	Aruba
7/11/1977	Sabre 40	New Orleans	20/7/1979	DC-6	Kingston, Jamaica
19/11/1977	Learjet	Gunabara Bay, Brazil	31/7/1979	HS748	Sumburgh
18/12/1977	Caravelle	Funchal, Madeira	11/8/1979	Lear	Athens/Jeddah
1/1/1978	Boeing 747	Bombay, India	3/9/1979	Corvette	Nice, France
2/1/1978	DC-3	Rio Grande, Brazil	11/9/1979	Boeing 707	Taoyuan, Taiwan
22/2/1978	Lear 35	Rome/Palermo, Italy	1/11/1979	DHC-6	Big Trout Lake, Canada
3/3/1978	HS748	Macuto, Venezuela	4/3/1980	Lear 25	Port au Prince
23/3/1978	DC-3	Off Grand Turk	14/3/1980	IL-62	Warsaw, Poland
1/4/1978	CV240	Unguia, Colombia, South America	13/5/1980	IL-14	Off Varadero, Cuba

Table 1.11 *Continued.*

Date	Aircraft	Location
19/5/1980	Gates Lear	Gulf of Mexico
27/6/1980	DC-9	Off Palermo, Sicily, Italy
7/8/1980	TU-154	Nouadhibou
12/9/1980	Boeing 727	Corfu airport
12/9/1980	DC-3	Freeport, Bahamas
15/9/1980	DC-6B	Haiti
3/10/1980	DC-3	En route
28/11/1980	DC-6	Bimini, Bahamas
28/3/1981	DC-4	St Croix, Virgin Islands
12/4/1981	DC-3	Mediterranean
7/5/1981	BAC 1–11	River Plate Estuary
10/6/1981	Swearingen SA226T	Cameron, La
17/6/1981	DC-3	Miraflores
26/10/1981	Constellation	St Thomas, Virgin Islands
17/1/1982	Convair 440	Honolulu, Hawaii, USA
23/1/1982	DC-10	Boston, USA
9/2/1982	DC-8	Tokyo, Japan
11/3/1982	DHC-6	East of North Cape
6/5/1982	Learjet	Savannah, Georgia, USA
9/5/1982	DHC-7	Aden
10/9/1982	Boeing 707	Khartoum, Sudan
13/2/1983	Learjet 35A	Strait of Malacca
6/6/1983	Fairchild Packet	Taiwan Strait
8/12/1983	Citation	Stornoway
28/2/1984	DC-10.30	New York, USA
15/5/1984	Learjet 35	Ushuaia, Argentina
4/8/1984	BAC 1–11	Tacloban airport, Philippines
5/8/1984	F-27	Zia, Dhaka, Bangladesh
7/8/1984	F-27	Rio de Janerio, Argentina
18/8/1984	DHC-6 Twin Otter	Tuktoyaktu, Canada
5/10/1984	Citation	Off Skiathos, Greece
13/10/1984	Catalina	Brownsville, Texas, USA
	PBT-6A	
23/10/1984	DHC-4	Sable Island, Canada
31/10/1984	DC-3	Davao/Manila, Philippines
10/11/1984	Learjet 24F	St Thomas, Virgin Islands
23/6/1985	Boeing 747	Atlantic, southwest of Ireland
27/6/1985	DC-10	San Juan, Puerto Rico
17/9/1985	Merlin III	Gulf of Mexico
10/10/1985	IAI 1124 Westwind	Sydney, Australia
25/12/1985	DC-3	Cumana, Venezuela
16/2/1986	Boeing 737	Pescadores Islands, Taiwan
22/7/1986	DC-3	San Juan, Puerto Rico
3/8/1986	DHC-6	Kingstown, St Vincent
9/9/1986	Merlin 3	McLainstow, Grand Bahamas
9/10/1986	DC-7	Dakar, Africa
28/10/1986	G73 Mallard	St Croix
23/12/1986	DC-4	Pacific Ocean
7/2/1986	Bandeirante	East coast of Papua New Guinea
31/8/1987	Boeing 737	Phuket, Africa
11/10/1987	Falcon 20	Keflavik, Africa
28/11/1987	Boeing 747	Mauritius

Table 1.12 Aircraft crashes due to electrical and power plant unit failures.

Cause of crash	Number of aircraft
Aquaplaning/hydroplaning	33
Electrical system failure or malfunction	120
Failure of power units	550
Malfunction of flying control system	210
Fuel contamination, exhaustion	95
Instrumentation misreading/malfunction	33
Airframe failure	89
Doors and windows opening or failing flight	88
Inflight smoke/fire	220
Ground fire/missiles, dog fights	18 370

Table 1.13 Aircraft damaged by hail impact.

Date	Aircraft	Location
7/9/1946	DC-4	Chicago, USA
28/6/1947	York	Central France
14/9/1947	DC-6	Monclova
25/5/1948	DC-6	Midland, UK
12/6/1948	DC-3	Harrisburg, USA
25/5/1949	DC-6	Guadalupe
29/4/1950	DC-6	En route
27/6/1951	C54A-DC	Pueblo
19/7/1951	L-749	Richmond, USA
2/6/1952	L-049	Des Moines, USA
23/5/1954	L-049	Tucumcari, Africa
19/7/1956	Viscount	Chicago, USA
8/4/1957	Viscount	Norfolk, UK
26/9/1957	DC-4	En route
27/5/1959	Viscount	En route
1/11/1963	Caravelle	En route
9/5/1965	Boeing 707	Mineral Wells, USA
5/7/1965	Boeing 707	Texarkana
17/7/1971	DC-9	Venice, Italy
19/7/1971	DC-8	Haneda
21/3/1972	Boeing 720	Riyadh, Saudi Arabia
18/6/1975	Boeing 747	Over Greece
3/8/1975	DC-9	Buffalo, USA
17/7/1976	Boeing 747	Tokyo
4/4/1977	DC-9	New Hope, USA
31/7/1981	Boeing 727	Paris, France
22/9/1983	DC-3	South Africa
6/10/1983	BAC 1−11	Ezeiza, Argentina

Table 1.14 Aircraft damaged by ice/snow accretion.

Date	Aircraft	Location
4/1/1977	Boeing 737	Frankfurt, Germany
13/1/1977	DC-8	Anchorage, USA
14/1/1977	Boeing 737	Frankfurt, Germany
15/1/1977	Viscount	Bromma airport
26/1/1977	Boeing 737	Oslo, Norway
31/1/1977	Chase C122	Anchorage, USA
21/2/1977	L-18	Truckee, USA
20/2/1978	Boeing 737	Hannover, Germany
7/3/1978	HS125	Dusseldorf, Germany
2/12/1978	DC-3	Des Moines, USA
4/12/1978	DHC-6	Steamboat Springs, Colorado, USA
4/12/1978	Gates Lear	Anchorage, USA
7/12/1978	Trident	Bovingdon
19/1/1979	Gates Lear	Detroit, USA
12/2/1979	Nord 262	Clarksburg, West Virginia, USA
19/11/1979	Citation	Castle Rock, Colorado, USA
23/11/1979	Twin Pioneer	Anchorage, USA
2/3/1980	B-26 Invader	California, USA
25/12/1980	Howard 500	Toronto, Canada
16/1/1981	DC-6B	Gambell, Alaska, USA
16/12/1981	Boeing 727	Gander, Newfoundland, Canada
13/1/1982	Boeing 737	Washington DC, USA
11/1/1983	Sabreliner	Toronto, Canada
12/3/1983	Metro II	—
21/12/1983	Beech 200	Detroit, USA
21/12/1983	Learjet 25	Kansas City, USA
13/1/1984	F-27	New York, USA
8/1/1985	SA227AC	Covington, Kentucky, USA
5/2/1985	DC-3	Charlotte, North Carolina, USA
5/2/1985	DC-9	Philadelphia, USA
11/2/1985	Jetstream 31	Macon, Georgia, USA
13/2/1985	SA226TC Metro II	Berkeley, Missouri, USA
12/3/1985	DHC-6	Bartar Island, Alaska, USA
15/12/1985	DC-3	Dillingham, Alaska, USA
15/12/1986	AN24	Lanzhou, China
18/1/1987	F-27	Castle Donington race track

been around 1100 accidents of this type. Table 1.15 gives information on aircraft impact due to over-running or veering off the runway.

Other types of aircraft accidents include aircraft hitting ground vehicles, sea-going vehicles and sea birds, but these data are extremely difficult to find and consequently have not been included here. Insufficient information is available on crashes and missile/target interaction in some former communist countries.

The Lockheed C-130 has been in continuous production since 1954 and an average of three Hercules are produced every month. More than 1800 have so far

Table 1.15 Aircraft impact due to over-running or veering off the runway.

Date	Aircraft	Location	Date	Aircraft	Location
17/11/1977	Boeing 747	JFK airport, New York, USA	15/10/1978	DC-3	Soddo, Ethiopia
19/11/1977	Boeing 727	Funchal, Maderia, Spain	24/10/1978	Lear 24D	Las Vegas, USA
19/11/1977	Gates Lear	Rio de Janeiro, Argentina	25/10/1978	DC-3	Degahabour, Ethiopia
5/1/1978	DHC-6	Leadville, USA	3/11/1978	Lear 24	Dutch Harbour, Alaska
9/1/1978	Falcon	Riviere du Loup, Canada	14/11/1978	TU154	Stockholm, Sweden
24/1/1978	CV440	San Remon, Bolivia	17/12/1978	Boeing 737	Hyderabad, India
24/1/1978	Boeing 737	Miri, Malaysia	26/12/1978	Lear 26	Sao Paulo, Brazil
12/2/1978	Boeing 737	Cranbrook, Canada	15/1/1979	Jetstream	Missoula, Montana, USA
15/2/1978	HS748	Mirgan, Canada	21/7/1979	DC-3	Williamson, USA
28/2/1978	Sabreliner	International Falls, USA	22/1/1979	Jetstar	Concord, USA
1/3/1978	DC-10	Los Angeles, USA	23/1/1979	Boeing 707	Stansted, UK
3/3/1978	DC-8	Santiago de Compostela, Spain	26/1/1979	HS125	Taos, New Mexico, USA
30/3/1978	Gates Lear	Burbank, USA	29/1/1979	DHC-6	Eastmain airport, Canada
4/4/1978	Boeing 737	Gosselies, Belgium	9/2/1979	DC-3	Hau Hau, Indonesia
20/4/1978	CV580	Cleveland, USA	15/2/1979	Boeing 747	Chicago, USA
20/5/1978	Sabre 60	Sao Paulo, Brazil	3/3/1979	IAI 1124	Aspen, Colorado, USA
25/5/1978	CV880	Miami, USA	18/3/1979	DHC-6	Akiachak, Alaska, USA
8/6/1978	DC-3	Nashville, USA	26/3/1979	IL-18	Luanda, Angola
26/6/1978	DC-9	Toronto, Canada	1/4/1979	DHC-4	Bethel, Alaska
28/6/1978	IAI 1121	Aspen, Colorado	11/4/1979	Lockheed Vega 37	Belle Glade, Florida, USA
9/7/1978	BAC 1–11	Rochester, New York, USA			
25/7/1978	DC-3	Pikangikum, Canada	16/4/1979	Boeing 747	Frankfurt, Germany
31/7/1978	FH227	Chub Bay, Bahamas	23/4/1979	Boeing 727	Tunis
29/8/1978	Boeing 747	Delhi, India	26/4/1979	Boeing 737	Madras, India
6/9/1978	CV880	Managua, Nicaragua	7/5/1979	DC-3	Sta Elena Peten, Guatemala
18/9/1978	DC-3	Komakuk, North-Western Territory	15/5/1979	DC-4	Mesa, Arizona, USA
			6/6/1979	DC-6	Charleston, West Virginia, USA
20/9/1978	DC-10	Monrovia, Liberia	21/6/1979	Constellation	Riviere Loup, Canada
7/10/1978	DC-3	Belo Horizonte, Brazil	21/6/1979	HS748	Mangalore, India

Table 1.15 *Continued.*

Date	Aircraft	Location	Date	Aircraft	Location
27/7/1979	DC-3	Bettles, Alaska, USA	4/10/1980	Learjet	Aspen, Colorado, USA
31/7/1979	HS748	Sumburgh, Germany	28/10/1980	DC-10	Waukegan, Illinois, USA
16/9/1979	DHC-6	Resolute, North-Western Territory	11/11/1980	Boeing 727	Newark, New Jersey, USA
			21/11/1980	Boeing 727	Yap Island, USA
7/10/1979	DC-8	Athens, Greece	5/12/1980	Metro	Moncton, Canada
25/10/1979	Viscount	Kirkwall, Orkney Islands	12/1/1981	DC-10	Ujung, Pedang, China
4/11/1979	Boeing 720	Multan, Pakistan	17/1/1981	Boeing 737	Antwerp, Belgium
10/12/1979	HS125	Sassandra, Spain	21/1/1981	Citation	Bluefield, West Virginia, USA
5/1/1980	Lodestar	Palmyra Island	13/2/1981	Boeing 737	Madras, India
27/1/1980	Boeing 720	Quito, Ecuador	17/2/1981	Boeing 737	Santa Ana, California, USA
30/1/1980	FH227	Shingle Point, Yukon, China	29/3/1981	Jetstar	Luton Airport, UK
31/3/1980	AN-24	Bessau	13/5/1981	HS125	Semerang, Indonesia
22/4/1980	DHC-6	Koartac, Canada	15/5/1981	IL-18	Gdansk, Poland
19/5/1980	Boeing 737	Dar-es-Salaam, Tanzania	23/5/1981	Short SC7 Skyvan	Alexander Lake, Alaska, USA
4/6/1980	Boeing 707	Bangkok, Thailand			
9/6/1980	Caravelle	Atlanta, USA	5/6/1981	F-27	Gilgit airport, Pakistan
16/6/1980	Metro	Birmingham, Alaska, USA	1/8/1981	DHC-6	Sugluk, Canada
21/6/1980	HS748	Chiang Rai, Thailand	17/8/1981	Boeing 727	Fort Lauderdale, USA
30/6/1980	DC-3	Robb Lake, Canada	19/8/1981	HS748	Mangalore, India
1/8/1980	DC-3	Smyrna Beach, Florida, USA	27/8/1981	DC-3	Dire Dawa, Arabian Desert
7/8/1980	Viscount	Leeds/Bradford, UK	7/9/1981	Packet	Dahi Creek, Alaska, USA
2/9/1980	IAI 1124	Iowa City, USA	15/9/1981	Boeing 747	Manila, Philippines
3/9/1980	Packet	Goodnews, Alaska, USA	19/9/1981	HS748	Ndola
6/9/1980	DHC-6	Seal River, Canada	23/10/1981	Boeing 707	Tokyo, Japan
10/9/1980	Learjet	Atlanta, USA	23/10/1981	C46	El Tiboy, Bolivia
12/9/1980	Boeing 727	Corfu airport, Greece	31/10/1981	DHC-6	Bafoussam
16/9/1980	Learjet	Waukegan, Illinois, USA	7/12/1981	Airbus A300	Porto Alegre, Brazil
22/9/1980	CV240	Okeechobee, Florida, USA	30/12/1981	C-46	San Juan, Puerto Rico

continued

Table 1.15 Continued.

Date	Aircraft	Location	Date	Aircraft	Location
7/1/1982	IAI 1121	—	10/12/1982	HS748	Mande Sulawesi, Indonesia
23/1/1982	DC-10	Boston, USA	17/12/1982	Metro	Montreal, Canada
3/2/1982	DC-10	Philadelphia, USA	20/12/1982	Metro III	Gillette, USA
12/2/1982	Howard 500	Fort Lauderdale, Florida, USA	7/1/1983	Boeing 727	Teheran, Iran
15/2/1982	HS125	Curitiba, Spain	9/1/1983	Convair 580	Brainerd, France
17/2/1982	Boeing 727	Miami, USA	10/1/1983	HS125	Paris, France
17/2/1982	Boeing 737	Los Angeles, USA	19/1/1983	IAI 123	Harrisburg, USA
12/3/1982	Citation	Chino, California, USA	6/2/1983	Learjet 24	St Paul Island, Alaska, USA
24/3/1982	Boeing 707	Marana, Arizona, USA	11/2/1983	Merlin	Houston, USA
5/4/1982	Swearingen SA226TC	Fort Wayne, Texas, USA	27/2/1983	Trident	Fuzhou, China
			8/3/1983	Convair 580	Canada
4/5/1982	L-1011	Santo Domingo	13/3/1983	F-27	Sao Jose de Rio Preto, Brazil
5/5/1982	Metro	Fort Wayne, Texas, USA	19/4/1983	HS125	Gaspe, USA
7/5/1982	DC-3	Calgary, Canada	2/6/1983	F-28	Branti, Italy
12/5/1982	Boeing 727	Barcelona, Spain	26/6/1983	HS125	Houston, Texas, USA
18/5/1982	Gulfstream 1	Gillette, Wyoming, USA	4/7/1983	HS748	Kasama
24/5/1982	Learjet	Uberaba, Brazil	12/7/1983	Sabreliner	Flushing
29/5/1982	DC3A	Sandwich, Illinois, USA	16/7/1983	Gulfstream	Blountville, Tennessee, USA
9/6/1982	F-27	Brisbane, Australia	4/8/1983	Boeing 747	Karachi, Pakistan
22/6/1982	Boeing 707	Bombay, India	13/8/1983	DC-8	Sanaa, Yemen
11/7/1982	HS748	Jolo, Brazil	24/8/1983	F-27	Calcutta, India
1/8/1982	F-27	Kasese, Uganda	20/9/1983	Learjet	Massena
26/8/1982	Boeing 737	Ishigaki Island	23/9/1983	Guarani II	Argentina
6/9/1982	Boeing 737	Luxor, Egypt	11/10/1983	Boeing 747	Frankfurt, Germany
13/9/1982	DC-10	Malaga, Spain	18/10/1983	Boeing 747	Hong Kong
17/9/1982	DC-8	Shanghai, China	25/10/1983	DC-8	Norfolk, Virginia, USA
29/9/1982	IL-62	Luxembourg	9/12/1983	HS125	Portland, USA
13/10/1982	HS125	Atlanta, USA	17/12/1983	EMB-110	Sao Pedro de Xingu, China
6/12/1982	Learjet	Paris, France	22/12/1983	Learjet	Eagle, Colorado, USA

Table 1.15 *Continued.*

Date	Aircraft	Location	Date	Aircraft	Location
16/1/1984	DC-3	Kissidouglas	18/8/1984	HS748	Surabaya, Indonesia
27/1/1984	F-28	Pangkal, Indonesia	24/8/1984	AN-12	Addis Ababa, Ethiopia
30/1/1984	Learjet 24	Santa Catalina, California, USA	5/9/1984	Twin Otter	Newcastle, UK
4/2/1984	Canadair CL600	Little Rock, Arkansas, USA	7/5/1986	Learjet 24	Hollywood, Florida, USA
			20/5/1986	Metro II	Hutchinson, Kansas, USA
6/2/1984	Sabreliner	St Hubert, Canada	21/5/1986	F-28	Puerto Asis, Colombia, USA
11/2/1984	Boeing 737	Tegal airport	8/6/1986	Hercules	Dondo, Angola
22/2/1984	Bandierante	Cordova, Spain	20/7/1986	Boeing 737	Wabush, Newfoundland
28/2/1984	DC-10-30	New York, USA	25/7/1986	F-27	Tabou, Ivory Coast
28/2/1984	Citation I	Fitchburg, USA	26/7/1986	F-27	Sumbe, Angola
2/3/1984	Beech 200	Vermil, USA	27/7/1986	Boeing 747	Changi, China
5/3/1984	HS748	Hyderabad, India	27/7/1986	Beech 200	Farmingdale, New York, USA
2/4/1984	Challenger	—	2/8/1986	HS 125	Bedford, Indiana, USA
28/4/1984	DC-6	Arizona, USA	6/8/1986	Learjet 55	Rutland, Vermont, USA
16/5/1984	Lockheed L382 Hercules	Palmerala, Honduras	16/8/1986	Caravelle	Calabar, Nigeria
			11/9/1986	Riley Heron	Vanua Mbalvu
			29/9/1986	Sabreliner	Liberal, Kansas, USA
16/6/1984	IL-18	Sanaa, Yemen	29/9/1986	A300	Madras, India
17/6/1984	Boeing 727	Pearson airport, Canada	19/10/1986	DC-9	Copenhagen, Denmark
24/6/1984	Boeing 707	Chicago, USA	21/10/1986	Skyvan	Nightmute, Alaska
7/7/1984	Citation I	Gualala, California, USA	25/10/1986	Boeing 737	Charlotte, North Carolina, USA
21/7/1984	DHC-6 Twin Otter	Tau	30/10/1986	Boeing 727	Aeroparque, Buenos Aires
			1/11/1986	Learjet 24	Lake Tahoe, California, USA
28/7/1984	Learjet 25B	Waterville, USA	27/11/1986	Caravelle	Arauca
4/8/1984	BAC 1-11	Tacloban	29/11/1986	DHC-6	San Juan, Puerto Rico
7/8/1984	F-27	Rio de Janerio, Argentina	17/12/1986	SA226TC	Lidkoeping/Hovby
12/8/1984	DHC-6	Back Bay, USA	26/12/1986	B727	Istanbul, Turkey

continued

Table 1.15 *Continued.*

Date	Aircraft	Location	Date	Aircraft	Location
3/1/1987	Metro II	Lidkoeping/Hovby	16/7/1987	Jet Commander	Jackson, Mississippi, USA
10/1/1987	DC-10	Ilorin, Nigeria			
10/1/1987	Metro III	Yuma, Arizona, USA	24/7/1987	BAe748	Jakarta
14/1/1987	Learjet 35A	Lugano, Italy	25/7/1987	Bandeirante	Santo Angelo
27/1/1987	F-27	Varginha, Belgian Congo	22/8/1987	Boeing 767	Scott AFB
25/2/1987	Sabreliner	East Alton, Illinois, USA	24/8/1987	Merlin IIA	Riveire Madeline
26/2/1987	Learjet 35	Centennial airport, Englewood, Colorado, USA	8/9/1987	A310	Port Harcourt
			19/9/1987	A300	Manila
23/3/1987	CV580	Dallas, USA	21/9/1987	A300	Luxor, Egypt
11/4/1987	Boeing 707	Manaus	23/9/1987	TU154	Moscow, Domodedovo
6/5/1987	Catalina	Gander	6/10/1987	Jetstream 31	Kennewick
26/5/1987	Jetstream	New Orleans, USA	28/10/1987	CV640	Bartlesville, Oklahoma, USA
20/6/1987	Falcon 20	Seletar, Singapore	13/12/1987	B737–300	Belo Horizonte
21/6/1987	Blenheim	Denham, Germany			

been produced. It is a needle-nosed, thin-winged aircraft with a huge belly and has a superb web of control systems redundancy such that they simply do not fail. The power levers, engines and utility systems have never let down the aircraft in great sweeping rises and long cruising flights. On 17 August 1988, a C-130 crashed and exploded near the Bahawalpur desert in Pakistan. The President and many Pakistani generals and US diplomats were killed. Normal weather conditions were reported. It is estimated that the angle of impact was around 30° to the vertical. The US/Pakistani report gives the height of fall as around 3000 m. According to eye-witnesses, an explosion did occur and the report suggested the crash could have occurred due to the sudden death of the crew and passengers. The theory of the crash rests on the poisonous gas PTN (pentaerythrite tetra-nitrate), a gas explosive contained in a special flask which opens under pressure at a certain altitude, generally around 3000 m.

1.7 Explosions with and without impact

Explosions have occurred in the past and are occurring at the present and there is no guarantee that they will not occur in the future. Chemical explosives, bombs, grenades, rockets, missiles, fire and gases are generally involved. Cars, lorries, vans, trains, aircraft, sea-going vessels, houses, buildings, pylons, towers, bridges, dams, ammunition dumps and many other structures including bandstands have been damaged by explosions. Some typical examples of bandstands targeted by terrorists in the UK and elsewhere are the Regiment Band in the Grande Palace, Brussels (August 1979), the Royal Airforce Central Band in Uxbridge (January 1981) and the Royal Green Jacket Bandstand in Regent's Park, London (July 1982). Barracks have been targeted too, such as the Chelsea Barracks, London (October 1981), the Parachute Regiment Officers Mess, Aldershot, Surrey (February 1972) and the Duke of York Barracks in Chelsea, London (October 1974). The devastation at Inglis Barracks at Mill Hill in North London, UK (1989), is shown in Fig. 1.15. The structure was a two-storey, gaunt, Victorian redbrick dormitory and about a quarter of it was destroyed. The roof and the first floor of this section were blown up. Upstairs the steel joists protruded out from the walls of the dormitory, entangled together. Across the parade ground, a missile fired by the blast had made a neat hole in a window in the new gymnasium block. Also bearing witness to the blast load was the steel flag-pole knocked off vertical. A similar example is given by the buildings devastated by the terrorist bomb attack at the Royal Marines School of Music in Deal, UK. Table 1.16 lists explosions reported in Great Britain from 1966 onwards. Table 1.17 lists major explosions which have occurred in other countries.

The Ojheri disaster in Pakistan of the twin cities Islamabad and Rawalpindi can be considered this century's greatest disaster. It occurred on 10 April 1988 when an ammunition dump exploded. Rockets, shells and shrapnel caused widespread destruction to life and property in a 20 km radius. Streets were rapidly

Fig. 1.15 An explosion in a dormitory at Inglis Barracks, UK.

Table 1.16 List of explosions in Great Britain.

Date	Details
23/7/1966	Bomb found at Flawith, 23 July 1966
23/2/1967	Gelignite found attached to car
17/7/1967	Unexploded mortar bomb found in golf course bunker
27/7/1967	Missile found at Milford Haven
1/8/1967	Mystery bomb found at Sand Bay
2/9/1967	Unexploded bomb was found in scrap yard
7/12/1967	Home-made bomb thrown into art school
1/1968–2/1968	Unexploded mortars and bombs were found in Weston-Super-Mare, Dagenham and Falforth Farm
21/5/1968	Petrol bombs were found
23/4/1969	Home-made bombs damaged houses in Norbury, London
31/7/1969	Unexploded bomb found in Bryanston Square, London
5/5/1970	Anti-tank bombs found in Colchester
2/6/1970	Bomb exploded in Buckingham, killing a boy
1/7/1970	Petrol bombs thrown at Army Publications office, London
7/7/1970	Bomb thrown into a recruiting office in London
10/7/1970	Bomb thrown into a retired policeman's home in London
17/8/1970	Explosion in car in Charing Cross Road, London
18/8/1970	London offices of Iberia Airways damaged by explosion
28/8/1970	Bomb exploded in bus station in Gloucester
1/1971	Four bombs exploded in London
9/2/1971	Petrol bomb thrown at house of a company manager, Jersey
13/7/1971	Petrol bombs thrown into a lounge of a public house in Bedford
1/11/1971	London Territorial Army headquarters were damaged

1/12/1971	Fire bombs damaged Town Hall at Broadstairs
4/12/1971	Army explodes old bomb on M5 motorway
3/1/1972	Man rode 8 miles on a motor bike with a live bomb, Grimsby
3/1/1972	A home in Lincoln wrecked by explosion
5/1/1972	Blast damaged a public house in Tring
2/2/1972	Explosion at gasworks in Croydon
8/2/1972	A house in Hull damaged by blast
22/2/1972	A home in Roehampton wrecked by explosion
4/4/1973	Parcels exploded at Kilburn and Paddington sorting offices, London
3/7/1973	Blast from a gas cylinder in a house in Hackney, London
30/8/1973	Solihull town centre rocked by two explosions
3/9/1973	Fire bombs thrown at four houses in Hockley
10/9/1973	Three incendiary devices exploded in Manchester
15/9/1973	Two petrol bombs thrown into a house in Nottingham
3/1/1974	Two bombs damage a building in Birmingham
22/1/1974	Paint store blast stops a train at Brixton, London
26/1/1974	Bomb blast at gasworks, St Helens
4/2/1974	Explosion wrecks a coach on M62 motorway
15/2/1974	Factory explosion in Ardeer
14/3/1974	Blast in chemical factory, Gosport
3/4/1974	Explosion at soda siphon factory, Tottenham, London
2/5/1974	Factory explosion at Isleworth
10/5/1974	Gas blast in home in Clements End
11/10/1974	Chemical plant explosion at Flixborough
16/10/1974	Dental workshop wrecked in Thetford
24/10/1974	Bomb damages a house near a school in Harrow
29/10/1974	A bomb exploded under a minister's car in Birmingham
30/10/1974	Blast demolishes telephone exchange at Sunderland
9/11/1974	Blast at dockyard, Chatham
11/11/1974	Blast wrecks a public lavatory near a tank range in Castlemartin
14/11/1974	Explosions outside a post office in Shepherds Bush, London
23/11/1974	Bombs blasted two Irish-owned premises in London
13/12/1974	Pillar-box exploded due to electrical fault in Marylebone, London
1/5/1977	Oxygen explosion shakes British Rail engineering workshop at Crewe
24/1/1977	Explosion wrecks flat in Blantyre
30/1/1977	Two explosions at chemical laboratory in Erith
3/1977	Blast in a school chemistry laboratory at Burgess Hill
5/4/1977	Gas cylinder explosion in Chessington
30/5/1977	Gas bottle explosion at Blackfriars Railway Bridge, London
21/9/1977	Explosion in a restaurant in Bristol
3/12/1977	Offshore explosion, North Yorkshire
6/1/1978	Bomb left outside charity offices in Horsham
14/3/1978	Caravan explodes in Saffron Walden
20/4/1978	Explosion in the basement of a sex shop in Soho, London
9/5/1978	Hackney flat severely damaged by a blast
12/5/1978	Blast destroys two homes in Glasgow
12/5/1978	Underground mine explosion at Ammenford
1/8/1978	Blast caused fire at a shop in Forest Gate, London
12/9/1978	Letter bomb arrived at the Iraqi Embassy in Queens Gate, London
7/10/1978	Bomb exploded in a car outside Orange Lodge Hall, Liverpool
7/10/1978	Parcel bomb delivered to a flat in Blackpool

continued

Table 1.16 *Continued.*

Date	Details
22/10/1978	Explosion from compressor on board a Navy diving vessel at Falmouth
25/10/1978	Explosion on a housing estate in Kirby
22/11/1978	Home-made bomb exploded at Ashtead
14/1/1979	A row of Essex seaside houses wrecked by a gas explosion
25/2/1979	Brentwood scout camp partially destroyed by a blast
3/3/1979	Tower block wrecked by an explosion in Battersea, London
15/3/1979	Explosion in a house at Billingham-on-Tees
1/4/1979	Gas cylinder exploded in the basement of the Carlyle Hotel, Bayswater, London
6/4/1979	Council house destroyed by an explosion in South Norwood, London
6/4/−20/6/1979	Ten explosions at various places in London
21/5/1979	Explosion at an ammunitions factory in Birmingham
9/6/1979	Four letter bombs exploded in two sorting offices in Birmingham
30/8/1979	Molotov cocktail thrown at Irish Embassy social club in Nottingham
18/12/1979	Explosion at a West End office, London
20/12/1979	Package exploded in a betting shop in Glasgow
19/2/1980	Explosion at a Boulby potash mine
20/8/1980	Oil tank exploded at Warrington
11/5/1981	Explosions and fire at an ethylene plant at Grangemouth
22/6/1981	Explosion at the Iraqi Embassy, London
22/7/1981	Fire bomb attack on an Asian family home in Middlesborough
1/10/1981	Device exploded in an Irish diplomat's car in Orpington
1/11/1981	Explosions in ice cream vendor's back garden, Luton
19/3/1982	Explosions in a Marine's camp at Otterburn
1/8/1982	Bomb exploded outside Ashraq-AR-Awsat newspaper offices, London
23/12/1982	Explosion at a Labour club in Hornsey, London
7/1/1983	Army blows open Irishwoman's van, Hull
1/−12/1983	Letter bombs delivered to 25 places in London
11/7/1983	London flat extensively damaged by an incendiary device
11/12/1983	Woolwich barracks, London, blasted
14/12/1983	Device explodes in a telephone kiosk in Oxford
14/12/1983	Device blown up in Kensington, London, in a controlled explosion
15/12/1983	Briefcase blown open outside the Hilton Hotel, London
24/5/1984	Explosion wrecks the underground water treatment plant a water station in Lancashire
16/6/1984	Three blasts and a fireball in an empty oil tanker at Milford Haven
3/12/1984	Explosion in an electricity substation on Merseyside
20/4/1985	A luggage bomb exploded at Heathrow airport, London
21/10/1985	Blast in a block of flats in Edinburgh
26/11/1985	Bomb explosion at the Iranian Embassy, London
7/1/1986	Blast in a sewer under construction in Glasgow
22/2/1986	Blast at a fireworks factory in Salisbury
16/3/1986	Blast and fire at a chemical plant in Peterlee, Durham
4/4/−9/1986	Nine blasts in streets in London
22/7/1986	Blast in an old people's home in Berkshire
4/11/1986	Electronically-detonated bombs wrecked a public lavatory cubicle in Oxford
1/1987−7/1987	Six bomb blasts in London

21/7/1987	Army defused a fire bomb outside a police station in Wolverhampton
23/12/1987	Petrol bomb attack on council houses in Manchester
24/9/1988	RAF practice bomb safely detonated in Humberside
17/1/1989	A letter bomb was defused at the Israeli Embassy, London
24/2/1989	A power bomb was set off at Bristol University
23/3/1989	Lorry carrying detonators and explosives exploded at Fengate

Table 1.17 Notable explosions in the world.

Date	Location	Deaths
31/10/1963	State Fair Coliseum, Indianapolis, USA	73
23/7/1964	Harbour munitions, Bone, Algeria	100
4/3/1965	Gas pipeline, Natchitoches, Louisiana, USA	17
9/8/1965	Missile silo, Searcy, Arkansas	53
21/10/1965	Bridge, Tila Bund, Pakistan	80
30/10/1965	Cartagena, Colombia	48
24/11/1965	Armory, Keokuk, Louisiana, USA	20
13/10/1966	Chemical plant, La Salle, Quebec, Canada	11
17/2/1967	Chemical plant, Hawthorne, New Jersey, USA	11
25/12/1967	Apartment building, Moscow	20
6/4/1968	Sports store, Richmond, Indiana, USA	43
8/4/1970	Subway construction, Osaka, Japan	73
24/6/1971	Tunnel, Sylmar, California, USA	17
28/6/1971	School, fireworks, Pueblo, Mexico	13
21/10/1971	Shopping centre, Glasgow, Scotland	20
10/2/1973	Liquefied gas tank, Staten Island, New York, USA	40
27/12/1975	Mine, Chasnala, India	431
13/4/1976	Munitions works, Lapua, Finland	45
11/11/1976	Freight train, Iri, South Korea	57
22/12/1977	Grain elevator, Westwego, Louisiana, USA	35
24/2/1978	Derailed tank car, Waverly Tennessee, USA	12
11/7/1978	Propylene tank truck, Spanish coastal campsite	150
23/10/1980	School, Ortuella, Spain	64
13/2/1981	Sewer system, Louisville, Kentucky, USA	0
7/4/1982	Tanker truck, tunnel, Oakland, California, USA	7
25/4/1982	Antiques exhibition, Todi, Italy	33
2/11/1982	Salang Tunnel, Afghanistan	1000–3000
25/2/1984	Oil pipeline, Cubatao, Brazil	508
21/6/1984	Naval supply depot, Severomorsk, USSR	200+
19/11/1984	Gas storage area, northeast Mexico City	334
5/12/1984	Coal mine, Taipei, Taiwan	94
25/6/1985	Fireworks factory, Hallett, Oklahoma, USA	21
6/7/1986	Oil rig, North Sea	166
1/6/1988	Coalmine, Brocken, West Germany	—
4/6/1988	Freight train, Arzamar, USSR	—
27/6/1988	Commuter trains, Paris, France	—
3/7/1988	Commercial Iranian airline, Persian gulf	—

littered with exploded and unexploded ammunition, including anti-tank mortars, rockets, RPG7s, anti-tank wire-guided missiles, Bazooka warheads and white phosphorous filled smoke shells and many others. Dozens of live and dead rockets and bombs caused vast damage to property and vehicles and over 150 houses were raised to the ground. Thousands died as a result of panic and mayhem as rockets and missiles showered the cities.

A very high intensity of explosion occurred in the Ojheri ammunition depot. It is widely believed that this explosion involved premature functioning of the large volume of different types of ammunitions described earlier. From the very sketchy news reports that have appeared in the media, it appears that white phosphorous shells were stacked inside sheds in combination with other shells, bombs, rockets, etc. Some of the white phosphorous shells started leaking in the sheds, generating copious fumes of oxides which combined with atmospheric oxygen. The heat generated in the process increased the interior temperature to a level sufficient to ignite the propellant fuel of the rocket motors, which developed high forward thrust and behaved like jet-propelled projectiles flying in all directions with great acceleration. Concurrently, the intense pressure produced inside was instrumental in ejecting a large number of unexploded rounds that were thrown in the air in the form of missiles with no explosive potential.

Thousands of lives were lost and many people were injured. Eye-witnesses reported that shells used to land continuously for almost a month. Many adjacent areas were littered with unexploded land mines, thousands of rounds of bullets and some RPG launchers were even sighted. The Pakistani's claimed that the tragedy was purely due to an accident.

Another major explosion in a missile storage area occurred in the military industrial complex near Al-Hillah, 40 miles south of Baghdad, and was apparently heard in the Iraqi capital. The blast occurred on 17 August 1989 and it was reported that 700 bodies were recovered. The casualties appear to have included Egyptians and Iraqi military personnel and civilians. The explosions occurred in an area where longer-range versions of the Soviet-made Scud B missile are produced, with a North Korean-manufactured supplementary fuel tank. In this field the team also developed the Badr-2000, which is the improved version of the Argentine Condor-2, ranging over 500 miles. In this area also it is believed that North Korean experts developed a technique of dismantling the warhead to rearrange the explosive charge. It is widely understood that this might be responsible for the blast which occurred at Al-Hillah, at the place chemical weapons are also produced, such as those widely used in the Gulf war. The place is in the striking range from other countries carrying different type of missiles. Table 1.18 gives such ranges.

Another disaster occurred in July 1988 at sea — on the Piper Alpha oil platform in the North Sea. The layout of this platform is shown in Fig. 1.16. Such platforms are designed to withstand the worst wave likely to hit them in 100 years. This may sound favourable, until compared with nuclear power stations or hazardous chemical plants which must be designed so that there is only a one in a million

Table 1.18 Missiles in the Gulf countries.

Country	Missile type	Range
Libya	M9 missile	410 miles
Egypt	Scud B2	380 miles
Israel	Jericho II	470 miles
Syria	Spider SS23	320 miles
Iran	Silkworm	280 miles
Saudi Arabia	CSS-2 (DF-3)	1550 to 1875 miles

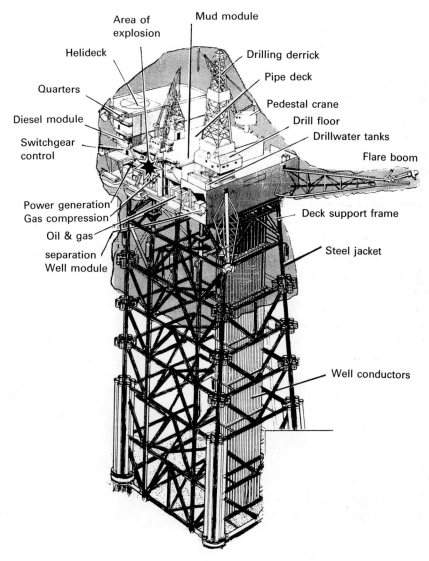

Fig. 1.16 The layout of the Piper Alpha platform.

chance of catastrophic accident from all causes. Oil platforms are inherently dangerous and it appears that, in cases such as Piper Alpha and a few others, they were designed to far lower safety standards than would be permitted on land. A gas explosion caused the platform disaster. It was due to compressor failure and involved the shattering of a compressor casing by a broken piston. Figure 1.17 shows the aftermath of the platform.

Since the top was totally integral with the jacket, side blasts could have literally torn out supporting members. The temperature in such circumstances could easily reach 800°C. Steel designed to a typical $340 \, N/mm^2$ yield strength would be reduced to just $30 \, N/mm^2$ at the height of the fire at a temperature of 800°C. Collapse of deck equipment, such as the 100 tonne drilling derrick, might have caused additional damage. The blast may have been so large that it destroyed the local gas-detection systems triggering automatic shut down. Since there was a major pipe rupture too, it might have been caused by metal fatigue.

Since this accident, the platform top sides design has inevitably found itself under close scrutiny. Among many issues are the top sides layout location of accommodation modules, detection and shut-down systems, evacuation procedures and the effectiveness of inspection, regulations and guidelines. Overall, it was the worst disaster caused by gas explosions.

Flats, houses, restaurants and public places where gas is abundantly used have been subject to explosions. In October 1971, the gas explosions in the town of Clarkston, Glasgow, Scotland, which killed 20 people, initiated a fundamental re-thinking of building methods in the basements of large buildings. The disaster

Fig. 1.17 The aftermath of Piper Alpha.

occurred because of the way shop basements were constructed in relation to the nearby gas main. The 20 shops in the terrace were built in 1965. As shown in Fig. 1.18, with the exception of shop 13, there was no access to the front basement of any shop, built in prestressed concrete, which had neither individual side walls nor a front wall — the only access, indeed, was through manholes at either end of the block. The front basements were half filled with loose clay at an angle of 45° running back from the footpath. At 1 m down and about 1 m from the building line was a 100 mm gas pipe running under the pavement. Owing to dead and imposed loads on the roadway and footpath and insufficient support from loose soils underneath, the pipe must have buckled and finally cracked. The explosions in the first few shops were minor and their floors were less severely damaged. The gas seeped backwards into what was a kind of 'tunnel' running along the terrace formed by the front basements. The gas lay in huge pockets between concrete girders. A mixture of gas and air caused an explosion when it reached the pockets.

While the possibility of gas explosions in domestic buildings cannot be ruled out, attention was focused on this hazard only in 1968 when the Ronan Point disaster, London, occurred. As shown in Fig. 1.19, Ronan Point was a 22-storey block, 192 m high and 24 m × 18 m on plan. It was built using the Larssen and Nielsen system of pre-cast concrete panel construction. On 16 May 1968 at 5.45 am there was an explosion in flat 90, a one-bedroom flat on the southeast corner of the eighteenth floor. The explosion blew out the cladding which acted as non-load bearing walls of the kitchen and living room. As the floor slab collapsed, the flank panel walls and floors above fell, causing progressive collapse of the floor and wall panels in the corner of the block right down to the podium. Gas is believed to have leaked into the flat from a broken connection at the gas cooker. It is believed that the gas could have risen to the ceiling, mixed with air and thus formed a gas/air layer at the kitchen ceiling which gradually extended downwards until it flowed under the door lintel and accumulated in the flat.

Similar explosions occurred in Putney houses and in a three-storey house in Balham Market in London, UK. The former was caused by a gas leak and the latter by a cigarette lit too close to a gas cylinder. A number of people were killed and injured.

At about 7.30 am on 4 October 1989 an explosion occurred in Guthrie Street in the centre of the old part of Edinburgh, Scotland, crumbling six flats to the ground and leaving part of their roof hanging precariously. All flanking buildings remained standing, although one was later pulled down. The scale of damage caused by the explosion suggested a large build-up of gas, probably in the stairwwell of the building. Only the basement ceiling stayed in place. A fracture was found in the 200 mm cast iron pipe located above the main sewer which led in the direction of the flats in Guthrie Street. It is probable that the gas from this fracture reached the building and caused an explosion.

Table 1.19 records gas explosions in selected countries.

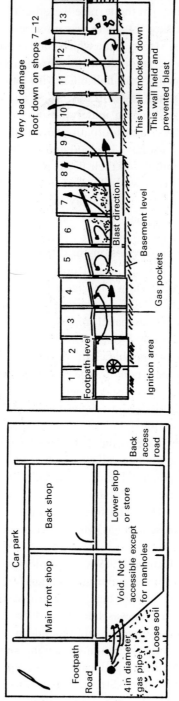

Fig. 1.18 Gas explosions in basements in Clarkston, Glasgow.

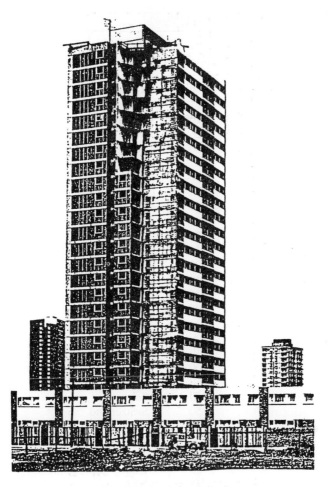

Fig. 1.19 Gas explosion at Ronan Point. (Courtesy of the British Ceramic Society.)

1.8 Nuclear explosions and loss-of-coolant accidents

Nuclear explosions result from the very rapid release of a large amount of energy within a limited space. Nuclear explosions can be many thousands of times more powerful than the largest conventional detonations. In a nuclear explosion the temperatures are comparatively greater and the energy is emitted in the form of light and heat, referred to as *thermal radiation*. The measure of the amount of explosive energy is known as the *yield* and is generally stated in terms of the equivalent quantity of TNT. For example, a one kiloton (1000 tonne) nuclear weapon is one which produces the same amount of energy in an explosion as does one kiloton of TNT. These effects are directly responsible for structural damage.

On 6 August 1945, at about 8.15 am, a new era was born amidst death and

Table 1.19 Gas explosions in selected countries.

Country	No of explosions	Country	No of explosions
Afghanistan	*50	Korea (North)	†§131
Albania	71	Korea (South)	†§310
Algeria	*310	Kuwait	*375
Argentina	†536	Libya	*†§30
Australia	†590	Malaysia	*†§51
Austria	115	Morocco	*330
Bahrain	*400	Netherlands	†150
Bangladesh	*58	New Zealand	—
Brazil	†350	Nigeria	*550
Bulgaria	†§31	Norway	†150
Burma	†370	Pakistan	*†88
Canada	†318	Philippines	§85
Chili	200	Poland	§61
China	†710	Portugal	†§220
Colombia	35	Qatar	†90
Cuba	*†20	Saudia Arabia	*210
Cyprus	†30	Singapore	*†110
Egypt	*†500	South Africa	*†§310
Fiji	†10	Spain	†§610
Finland	†200	Sri Lanka	†69
France	†250	Sudan	*†75
Germany	†150	Sweden	†§95
Greece	†350	Switzerland	†50
Hong Kong	†150	Syria	*†161
India	†§1509	Tunisia	*175
Indonesia	†170	Turkey	*†310
Iran	†§10	United Arab Emirates	*400
Iraq	*†§15	USA	†§670
Irish Republic	†150	USSR	†§1100
Israel	†40	Yugoslavia	†§210
Italy	†§450	Zaire	§89
Japan	†§310	Zambia	†120
Kenya	†§70	Zimbabwe	†110

* mostly gas cylinders; † mostly domestic; § mostly in mines or industry

destruction as the US Air Force B-29 bomber banked away after dropping a 13.5 kt bomb on Hiroshima. On 9 August 1945, at about 12.01 pm, a similar bomb was dropped over Nagasaki, killing and maiming many hundred thousands of people. Figure 1.20 shows the nuclear fireball at Hiroshima. In these two cities the nuclear bomb damage was extensive. Everywhere debris was evidence to how a disaster occurred in Japan in split seconds. People from other areas witnessed the buckling of structural frames, collapsed roofs, caved-in walls, shattered panels and damage to windows, doors, vehicles, trains, etc. In this city light structures and residences were totally demolished by the blast. Industrial buildings were

Fig. 1.20 Nuclear fireball at Hiroshima.

denuded of roofing and siding. Some robust buildings leaned away from ground zero. All masonry buildings were engulfed by the blast wave pressure and collapsed. Buildings and other structures at a distance away suffered damage. Underground pipes burst and much of the area was gutted by the sudden release of fire. A similar scene was witnessed in Nagasaki.

Other types of nuclear accident involve the nuclear reactors of commercial and military establishments. The loss-of-coolant accident at Three Mile Island (USA) and a similar one with fire at Chernobyl (USSR) received widespread media coverage.

A list of some notable nuclear accidents is given below.

- 7 October 1957   A fire in the Windscale plutonium production reactor north of Liverpool, UK, spread radioactive material throughout the countryside.
- 3 November 1957   A chemical explosion in Kasli, USSR, in tanks containing nuclear waste, spread radioactive material.
- 3 January 1961   An experimental reactor at a federal installation near Idaho Falls, USA, killed three workers. The plant had high radiation levels but damage was contained.
- 5 October 1966   A sodium cooling system malfunction caused a partial core meltdown at the Enrico Fermi demonstration breeder reactor near Detroit, USA. Radiation was contained.

- 21 January 1969   A coolant malfunction from an experimental underground reactor at Lucens Vad, Switzerland, resulted in the release of a large amount of radiation into a cavern which was then sealed.
- 19 November 1971   The water storage space at the Northern States Power Company's reactor in Monticello, USA, filled to capacity and spilled over, dumping about 50 000 gallons of radioactive waste water into the Mississippi River.
- 22 March 1975   A technician checking for air leaks with a lighted candle caused a fire at the Brown's Ferry reactor in Decatur, USA. The fire burned out electrical controls. The cost was $100 million.
- 28 March 1979   The worst commercial nuclear accident in the USA occurred as equipment failures and human mistakes led to a loss of coolant and partial core meltdown at the Three Mile Island reactor in Middletown, USA.
- 7 August 1979   Highly enriched uranium was released from a top-secret nuclear fuel plant near Erwin, USA.
- 11 February 1981   Eight workers were contaminated when over 100 000 gallons of radioactive coolant leaked into the containment building of the TVA's Suquohay 1 plant in Tennessee, USA.
- 25 April 1981   Some 100 workers were exposed to radioactive material during repairs of a nuclear plant at Tsuruga, Japan.
- 25 January 1982   A steam-generator pipe broke at the Rochester Gas & Electric Company's Ginna plant near Rochester, New York, USA. Small amounts of radioactive steam escaped into the air.
- 6 January 1986   A cylinder of nuclear material burst after being improperly heated at a Kerr-McGee plant at Gore, USA.
- April 1986   A serious accident at the Chernobyl nuclear plant about 60 miles from Kiev in the Soviet Union caused the emission of clouds of radiation that spread over several other countries.

In 1986 and 1987, the number of accidents in US commercial nuclear power plants were 2836 and 2810 respectively. In the years 1984 to 1988, the accidents which occurred outside nuclear islands are listed in Table 1.20.

Table 1.20 Nuclear accidents in various countries in the years 1984 to 1988.

Country	Number	Country	Number
Canada	1500	Netherlands	150
China	1200	Pakistan	5
France	2800	Sweden	110
Germany	3100	Switzerland	—
India	1700	UK	1750
Israel	1100	USA	4150
Italy	800	USSR	—
Japan	350		

1.9 The Gulf war

The Gulf war between Iraq and the USA and its allies was a perfect example of the kind of precise, high-technology air war one would expect in this age. The full-scale war began on 16 January 1990. For weeks the dominant image of the battle was a grainy video clip. A tiny bomb headed for a tiny building or a tiny puff of smoke exploded across the screen. The scene made the war seem remote and bloodless. Looking at the aftermath, the world discovered a total destruction of the important zones of the country and a great tragedy to its people. The initial operation of the war was purely from the air. The USA-led coalition flew more than 10 000 sorties, targetting command-and-control centres, airfields and scud missile launchers.

The long-range attack was carried out from ships, submarines and aircraft. Cruise missiles, Stealth fighter bombers, electronic jamming, the Patriot system, 'smart bombs' and night division devices all took part. The Tomahawk cruise missiles (range 780 miles), launched from naval ships and flying at the speed of a commercial airline, used digital mapping technology to penetrate beneath Iraqi radar and strike within 18 metres of their targets. The US Air Force F-117A Stealth fighters led the aircraft strike. The F-4G aircraft Wild Weasel launched missiles that homed in on the signals to knock out the emitting facility and so kept the Iraqis from co-ordinating their surface-to-air missiles (SAMs).

Some of the F-15E Eagle and F-16 Fighting Falcon attackers released their ordinance from as high as 600 metres, well above the light calibre Iraqi flak. These fighters, as well as the US Navy's F/A-18 Hornets, also delivered laser-guided or other smart bombs to their targets. The B-52 bombers created havoc and with their carpet bombing destroyed much of the enemy's military arsenal and personnel. Tornado GR-1S aircraft with F-4G escorts took out airfields and radar equipment. AWACs were used to monitor attacks. Impacts/explosions could be witnessed between Patriot and Scud missiles (also known as SS-1 missiles). GBU-15(V)2 glide bombs (weight 2450 lb, length 12.75 ft (3.9 m), also known as smart bombs) were focused on various targets.

1.10 Recent air crashes: aircraft impact at ground level

On 18 September 1992, a Pakistan International Airways Airbus crashed near Katmandu, Nepal. The jagged ridges were clouded. The aircraft passed only 2500 feet (761.9 m) above the mountains. All 167 people on board were killed when the aircraft hit the mountains.

On 4 October 1992, an El Al jet cargo 747−220F plane from Schiphol airport crashed and impacted a block of flats in Amsterdam. One engine caught fire and both starboard engines fell off thus destabilizing the plane and preventing it from flying.

Both crashes are being investigated at the time of printing.

2

DATA ON MISSILES, IMPACTORS, AIRCRAFT AND EXPLOSIONS

2.1 Introduction

This chapter introduces various types of missiles, impactors, aircraft and explosions. They include tornado-generated and windborne missiles, plant-generated missiles, missiles from jet fluid, snow/ice and rocks/boulders which have disintegrated under environmental conditions. Dropped weights are discussed together with other impactors such as trucks, lorries, cranes, tanks and naval vessels. Data exist on all popular types of available military missiles, rockets, civilian and military aircraft and helicopters. A brief introduction is given to types of explosions. The chemistry of bombs, shells, grenades, shrapnel and explosives is discussed. A full account is given of gas explosions, nuclear detonations, dust explosions and underwater explosions. Wherever possible, the reader is given full access to data which can act as an input for solving problems in specific areas. Great care is taken to ensure that, under a specific topic, sufficient data are available to compare known results. In addition, a comprehensive bibliography is provided for further in-depth studies.

2.2 Types of conventional missiles and impactors

For impact analysis and design, missiles generated by tornadoes, hurricanes and wind can be anything from roof tiles and planks to cars, lorries, boats, etc. Because of mechanical faults or for other reasons, components have been ejected from parent structures with greater velocities and, acting as missiles, have had devastating effects on the workforce and on structures.

In combat situations, military missiles are always in action to destroy sensitive targets. Aircraft and helicopter crashes produce missile effects on vital installations on the ground or in the sea. The breakaway rotors, engines, wings and tails themselves act as high-speed missiles. Vehicles, ships, tankers and high-speed boats may collide with each other or with vital installations, and consequently are a major hazard. Such impactors have caused damages worth millions of pounds.

On the environmental side, falling trees, high-speed water jets, ejecting stones from rocks during the penetration process, water waves, snow/ice loads impacting on structures and missiles generated by blasts and explosions due to gas leaks and nuclear detonations are part of a wider aspect of impact problems (covered in all sections of the bibliography). This section is, therefore, entirely devoted to data about missiles and impactors which can be used as an input for problem-solving exercises and for case studies.

2.2.1 Tornado- and wind-generated missiles

Tornadoes occur frequently in certain parts of the world. They vary considerably in their width, length and maximum speed. Owing to their small path area, the chance of recording a tornado wind for a specific zone is remote. The length, width and area are generally considered to be those bounding the area of potential damage. A great deal of research has been carried out on tornadoes. If one includes all the reported tornadoes, the average area is between 1.0 and 1.1 square miles (2.6 and 2.8 km^2). The estimated maximum wind speeds of tornadoes also vary widely, from 70 miles/h (112 km/h) (hurricanes damaging roofs and trees), to an estimated 400 miles/h (640 km/h), causing complete destruction and generating land-based missiles with greater speeds. Various codes exist for estimating wind loads for different terrains and, where necessary, they are generally referred to in estimating the wind speed and its relation to the speed at which a missile is generated. The details of this topic are beyond the scope of this book. However, a comprehensive bibliography is given for those who wish to study this in greater depth.

The characteristics of tornado-generated missiles must be studied carefully. It is important to identify objects in the path of a tornado prior to them becoming airborne. These objects can range from small debris to full-scale structural components and vehicles. Tables 2.1 and 2.2 list data on tornado- and wind-generated missiles and their characteristics.

2.2.2 Plant-generated missiles

Plant-generated missiles, according to their origin, vary in size and weight and have a wide range of impact velocities. They are generated as a result of high-energy system rupture. Rotating machinery, on disintegration, generates potentially dangerous missiles. Several references are given on the general characteristics of potential turbine missiles and broken pipes from sections of pressurized piping in nuclear power plants[2.73–2.80] Failure of large steam turbines in both nuclear- and fossil-fuelled power plants has occurred occasionally in the past due to metallurgical and/or design inadequacies, environmental and corrosion effects and failure of the overspeed protection systems. This has resulted in the loss of many mechanical items including blades, disks and rotors or their respective fragments. These can act as either high-trajectory missiles, which are ejected upward through the turbine casing, or low-trajectory or direct missiles, ejected from the turbine casing, any of which may strike an essential industrial system. The generation of the latter missiles is the more probable. Table 2.3 lists characteristics of plant-generated missiles.

Table 2.1 Tornado- and wind-generated missiles and their characteristics: wood, steel and concrete building components.

| Missile type | Geometry | | | Velocity | Weight |
	Diameter (mm)	Length (m)	Impact area (m²)	(m/s)	(kg)
Wooden plank	—	3.67	0.03	41.5	56.7
Wooden pole	200	3.67	0.03	5.73	94.8
Circular hollow sections in steel (average)	168.3	4.00	0.000026	70.2	60
Sign boards (average)	—	—	6.0	57.0	56
Steel I-beam light sections (average)	—	4.0	0.000032	40.5	100
Steel members channel sections (average)	—	3.0	0.000013	50.5	30
Steel members L-sections (average)	—	3.0	0.000015	45.5	36
Steel rafters T-sections (average)	—	3.0	0.000018	45.5	42
Steel rod	25	0.92	0.00049	75.6	3.63
Concrete lintels	—	3.0	0.025	60.5	1.80
Concrete sleepers	—	2.70	0.0031	75.0	0.20
Precast concrete beams or piles at delivery stage	—	9.0	0.09	60.5	19.44
Precast concrete wall panels	—	5.0	11.5	2.5	1380
Prestressed concrete pipes	400	—	—	—	1.100
	500	—	—	—	1.375
	600	—	—	—	1.650
	700	—	—	—	1.920
	800	—	—	—	2.200
	900	—	—	—	2.474
	1676	6.0	0.032	—	4.608
Prestressed concrete poles	—	17.0	0.0019	30.5	65.7
		12.0	0.00080	50.1	14.46
		9.0	0.000025	65.2	9.65

Table 2.2 Tornado- and wind-generated missiles and their characteristics: basic data on cars and other road vehicles.

Manufacturer Vehicle	Length (m)	Width (m)	Height (m)	Wheel base (m)	Laden weight (kg)	Max. speed (miles/h)
Alfa Romeo						
33 1.7 Sport Wagon Veloce	4.142	1.612	1.345	2.465	925	115
75 2.0i Veloce	4.330	1.630	1.350	2.510	1147	124
164 3.0 V6	4.555	1.760	1.4	2.660	1300	142
American Motors (USA)						
Jeep Wagoner limited	4.198	1.790	1.615	2.576	2074	90
Aston Martin						
Lagonda	5.820	1.790	1.3	2.91	2023	143
V8 Vantage Volante	4.39	1.86	0.1295	2.610	1650	160
Vantage Zagato	4.39	1.86	0.1295	2.610	1650	186
Audi (D)						
80 1.85	4.393	1.695	1.397	2.544	1020	113
90 Quattro	4.393	1.695	1.397	2.546	1270	125
100	4.792	1.814	1.422	2.687	1250	118
100 Turbo Diesel	4.793	1.814	1.422	2.687	1250	108
200 Avant Quattro	4.793	1.814	1.422	2.687	1410	139
Austin						
Maestro 1.60 Mayfair	4.049	1.687	1.429	2.507	946	102
Metro 1.0 Mayfair 3-door	3.405	1.549	1.361	2.251	771	86
Montego Vanden Plas EFi Estate	4.468	1.710	1.445	2.570	1111	110
Bentley (GB)						
Mulsanne	5.268	1.887	1.485	3.061	2245	119
Mulsanne Turbo R	5.268	1.887	1.486	3.061	2221	143
Bitler (D)						
Type III	4.450	1.765	1.395	—	1300	140
BMW (D)						
320i Convertible	4.325	1.645	1.380	2.57	1125	123
325i Touring	4.325	1.645	1.380	2.57	1270	132
M 3	4.325	1.645	1.380	2.57	1150	139
520i	4.72	1.751	1.412	2.761	1400	126
735i	4.91	1.845	1.411	2.832	1590	145
750i L	5.024	1.845	1.401	2.832	—	155
Z 1	3.921	—	—	2.45	110	140
Bristol (GB)						
Brigand Turbo	4.902	0.1765	1.4535	2.895	1746	150
Buick (USA)						
Lesabre T-type Coupé	4.991	1.838	1.389	2.814	1458	115
Cadillac (USA)						
Allanté Convertible	4.537	1.864	1.325	2.525	1585	110

continued

Table 2.2 *Continued.*

Manufacturer Vehicle	Length (m)	Width (m)	Height (m)	Wheel base (m)	Laden weight (kg)	Max. speed (miles/h)
Cadillac (USA)						
Allanté Convertible	4.537	1.864	1.325	2.525	1585	110
Chevrolet (USA)						
Camaro IROC-2	4.775	1.850	1.270	2.565	1525	130
Corvette Convertible	4.483	1.805	1.185	2.438	1414	142
Chrysler (USA)						
le Baron Convertible	4.697	1.738	1.326	2.546	1474	110
GS Turbo 2	4.555	1.760	1.302	2.465	1194	125
Portofino	—	—	—	—	—	150
Citroën (F)						
Ax 14 TRS	3.495	1.56	1.35	2.285	695	99
Bx 19 GTi 16v	4.229	1.657	1.365	2.655	1093	130
Bx 25 GTi Turbo	4.660	1.77	1.36	2.845	1385	126
Coleman Milne (GB)						
Grosvenor limousine	5.563	1.964	1.575	3.661	2100	115
Dacia (R)						
Duster 4 × 4 GLX	3.777	1.6	1.74	2.4	1180	70
Daihatsu (J)						
Charade LX Diesel Turbo	3.61	1.615	1.385	2.34	810	87
Charade GT ti	3.61	1.615	1.385	2.34	816	114
Fourtrak Estate EL TD	4.065	1.580	1.915	2.53	1660	83
Daimler						
3.6	4.988	2.005	1.358	2.87	1770	137
Dodge (USA)						
Daytonna Shelby 2	4.545	1.76	1.279	2.464	1220	120
Ferrari						
F40	4.43	1.981	1.13	2.451	1100	201
Mondial 3.2 Quattro valvole	4.58	1.79	1.26	2.65	1430	143
Fiat						
Croma Turbo ie	4.495	1.76	1.433	2.66	1180	131
Panda 4 × 4	3.378	1.485	1.46	2.159	761	83
Ford						
AC (GB)	3.962	1.816	1.168	2.477	907	140
Ford (D)						
Escort RS Turbo	4.046	1.588	1.348	2.4	1017	124
Granada 2.4i GL	4.669	1.760	1.41	2.761	1265	120
Scorpio 4 × 4 2.9i	4.669	1.766	1.453	2.765	1385	126
XR3i Cabriolet	4.049	1.64	1.336	2.398	925	115
Ford (GB, B)						
Sierra Sapphire GLS 2.0EFi	4.468	1.699	1.359	2.609	1060	115

Table 2.2 *Continued.*

Manufacturer Vehicle	Length (m)	Width (m)	Height (m)	Wheel base (m)	Laden weight (kg)	Max. speed (miles/h)
Ford (B)						
Sierra Ghia 4 × 4 Estate	4.511	1.694	1.359	2.612	1315	119
Ford (USA)						
Taurus	4.785	1.796	1.795	2.692	1299	105
Ginetta (GB)						
G32	3.758	1.651	1.168	2.21	753	135
Honda (J)						
Accord Aerodeck 2.0 EXL	4.335	1.651	1.335	2.6	1147	110
Legend Coupé	4.755	1.745	1.37	2.705	1395	132
Prelude 2.0L-16	4.460	1.695	1.295	2.565	1145	128
Hyundai (J)						
Pony 1.5 GLS	3.985	1.595	1.38	2.38	890	96
Stellar 1.6 GSL	4.427	1.72	1.372	2.579	1034	98
Isuzu (J)						
Trooper Turbo Diesel	4.38	1.65	1.8	2.3	1655	78
Jaguar (GB)						
Sovereign 3.6	4.988	2.005	1.358	2.87	1770	137
XJ6 2.9	4.988	2.005	1.38	3.87	1720	117
Lada (Su)						
Riva Cossack	3.708	1.676	1.638	2.197	1150	77
Samara 1300 SL	4.006	1.62	1.335	2.46	900	92
Lamborghini (I)						
Countach 5000s Quattro valvole	4.14	2.0	1.07	2.45	1446	178
Lancia (I)						
Delta 1.6 GTie	3.895	1.62	1.38	2.475	995	115
Delta HF vitegrade	3.9	1.7	1.38	1.38	1200	134
Thema 2.0ie Turbo Estate	4.59	1.755	1.433	2.66	1150	139
Thema 8.32	4.59	1.755	1.433	2.66	1400	139
Y10 Turbo	3.392	1.507	1.425	2.159	790	111
Land Rover (GB)						
One Ten Diesel Turbo	4.445	1.79	2.035	2.795	1931	73
Range Rover Vogue Turbo D	4.47	1.718	1.778	2.591	2061	90
Lincoln (USA)						
Continental	5.21	1.847	1.412	2.769	1645	112
Lotus (GB)						
Esprit Turbo	4.331	1.859	1.138	2.459	1268	152
Maserati (I)						
Bi Turbo 228	4.46	1.865	1.33	2.6	1240	151

continued

Table 2.2 *Continued.*

Manufacturer Vehicle	Length (m)	Width (m)	Height (m)	Wheel base (m)	Laden weight (kg)	Max. speed (miles/h)
Mazda (J)						
121 1.3LX Sun Top	3.475	1.605	1.565	2.295	775	99
626 2.0 GLX Hatchback	4.515	1.69	1.375	2.575	1196	111
626 2.0i GT Coupé	4.45	1.69	1.36	2.515	1230	130
RX7	4.29	1.69	1.265	2.43	1221	134
Mercedes Benz (D)						
190E 2.6	4.427	1.678	1.39	2.665	1209	130
300CE	4.655	1.682	1.41	2.715	1390	126
560 SEL	5.16	2.006	1.446	1.555	1780	147
Mercury (USA)						
Topa 3 XR5	4.468	1.747	1.339	2.537	1135	92
MG (GB)						
Maestro 2.0 EFi	4.05	1.69	1.42	2.51	975	114
Metro Turbo	3.403	1.563	1.359	2.251	840	110
Montego Turbo	4.468	1.71	1.42	2.565	1079	125
Mitsubishi (J)						
Galant Sapporo	4.66	1.695	1.375	2.6	1230	114
Starion 2000 Turbo	4.43	1.745	1.315	2.435	1308	133
Mitsubishi Colt (J)						
Lancer 1500 GLX Estate	4.135	1.635	1.42	2.38	950	95
Morgan (GB)						
Plus 8	3.96	1.575	1.32	2.49	830	122
Nissan (J)						
Bluebird 1.6 LS	4.405	4.365	1.69	1.395	1120	103
Prairie Anniversary II	4.09	1.655	1.6	2.51	1070	95
Silvia Turbo ZX	4.351	1.661	1.33	2.425	1136	124
Oldsmobile (USA)						
Trofeo	4.763	1.798	1.346	2.741	1526	115
Panther (GB)						
Kallista 2.9i	3.905	1.712	1.245	2.549	1020	112
Solo 2	4.344	1.78	2.18	2.53	1100	150
Peugeot (F)						
205 GTi Cabriolet	3.706	1.572	1.354	2.421	884	116
205 GRD	3.706	1.572	1.369	2.421	895	96
Peugeot (GB)						
309 SRD	4.051	1.628	1.379	2.469	950	99
Plymouth (USA)						
Sundance	2.463	1.71	1.339	2.463	1131	105
Pontiac (USA)						
Bonneville SSE	5.046	2.838	1.409	2.814	1504	112
Porsche (D)						
911 Speedster	4.291	1.65	1.283	2.273	1140	152
944 S	4.2	1.735	1.275	2.4	1280	140

Table 2.2 *Continued.*

Manufacturer Vehicle	Length (m)	Width (m)	Height (m)	Wheel base (m)	Laden weight (kg)	Max. speed (miles/h)
Reliant (GB)						
Scimitar 1800 Ti	3.886	1.582	1.24	2.133	889	124
Renault (F)						
Espace 2000−1	4.25	1.277	1.66	2.58	1177	105
5 GTD	3.65	1.585	1.397	2.466	830	94
21 GTS	4.46	1.714	1.415	2.659	976	113
21 Turbo	4.498	1.714	1.375	2.597	1095	141
Rolls-Royce (GB)						
Silver Spirit	5.27	1.887	1.495	3.06	2245	119
Rover (GB)						
216 Vanden Plas	4.16	1.62	1.39	2.45	945	107
820i	4.694	1.946	1.398	2.759	1270	126
Saab (S)						
900 Turbo 16S Convertible	4.739	1.69	1.42	2.525	1185	124
9000i	4.62	1.765	1.43	2.672	1311	118
Seat (E)						
Ibiza 1.5 GLX 5-door	3.638	1.609	1.394	2.448	928	107
Malaga 1.5 GLX	4.273	1.65	1.4	2.448	975	103
Skoda (CS)						
130 Cabriolet LUX	4.2	1.61	1.4	2.4	890	95
Subaru (J)						
Justy 4 × 4	3.535	1.535	1.42	2.85	770	90
XT Turbo Coupé	4.49	1.69	1.335 1.370	2.465	1139	119
Suzuki (J)						
Santana	3.43	1.46	1.69	2.03	830	68
Swift 1.3 GTi	3.67	1.545	1.35	2.245	750	109
Toyota (J)						
Celica 2.0 GTi Convertible	4.365	1.71	1.29	2.525	1195	125
Corolla GTi	4.215	1.655	1.365	2.43	945	122
Space Cruiser	4.285	1.67	1.815	2.235	1320	87
Supra 3.0i	4.62	1.745	1.31	2.595	1550	135
TVR						
S Convertible	4.0	1.45	1.117	2.286	900	128
TVR (GB)						
9205 EAC Convertible	4.051	1.628	1.379	2.469	950	99
Vauxhall						
Astra Cabriolet	2.463	1.71	1.339	2.463	1131	105
Astra GTE 2.0ie 16v	5.046	2.838	1.409	2.814	1504	112
Carlton CD 2.0i	4.291	1.65	1.283	2.273	1140	152
Carlton GSi 3000	4.2	1.735	1.275	2.4	1280	140
Senator 3.0i CD	3.886	1.582	1.24	2.133	889	124

continued

Table 2.2 *Continued.*

Manufacturer Vehicle	Length (m)	Width (m)	Height (m)	Wheel base (m)	Laden weight (kg)	Max. speed (miles/h)
Volkswagon (D)						
Golf GTi 16v	4.25	1.277	1.66	2.58	1177	105
Jetta GTi 16v	3.65	1.585	1.397	2.466	830	94
Scirocco GTX	4.46	1.714	1.415	2.659	976	113
Volvo (NL)						
360 GLY	4.498	1.714	1.375	2.597	1095	141
480 ES	5.27	1.887	1.495	3.06	2245	119
Volvo (S)						
760 GLE	4.16	1.62	1.39	2.45	945	107
Yugo (YU)						
65A GLX	4.694	1.946	1.398	2.759	1270	126

2.2.3 Impact due to jet fluid and rock blasting

Shaped-charge jets, jet fluids and rock blasting create rock fragments which are ejected with great velocities, known as ejecta velocities. Nearby structures can be subject to intense dynamic loading from such missiles. Sometimes it is difficult to assess individual break-up and to mitigate or control ejecta velocities. The work focuses primarily on the following two broad topics:

(1) Intensity of fragmentation or average fragment size resulting from the impulsive failure event and its relationship to material properties and loading conditions.
(2) Fragmentation size distribution, geometry and wave propagation.

   Research has shown that the two topics are diverse and complex, especially when related to catastrophic failure conditions. After the examination of many cases studies;[2.147–2.162] Table 2.4 has been drawn up for soil/rock. Table 2.5 gives thrust force versus penetration for a number of rocks. From these two tables, the size of the fragmental rock and its velocity can easily be estimated. If the fragment size and velocity are known, impact analysis can then be successfully carried out using various dynamic models given in this text.

2.2.4 Snow load as an impactor

Several researchers[2.279–2.340] have investigated snow/ice impact on structures. Several graphs have been plotted on failure pressure versus aspect ratio. The aspect ratio is equal to the chord length or width D_i of the interaction zone divided by the average snow/ice thickness t_i in the zone. The impact forces are calculated using the contact factor and the shape factor. The modes of ice impact

Table 2.3 Plant-generated missiles and their characteristics.

Type of missile	Weight (kg)	Impact area (cm^2)	Impact velocity (m/s)
Control rod mechanism or fuel rod	53	15.5	91.5
Disc 90° Sector	1288	4975	125
Disc 120° Sector	1600	6573	156
Hexagon head bolts			
1.4 cm dia	0.20	1.54	250
2.0 cm dia	0.30	2.30	230
2.4 cm dia	0.37	2.84	189
3.3 cm dia	0.42	3.22	150
6.8 cm dia	0.97	7.44	100
Turbine rotor fragments			
High trajectory			
Heavy	3649	5805	198
Moderate	1825	3638	235
Light	89	420	300
Low trajectory			
Heavy	3649	5805	128
Moderate	1825	3638	162
Light	89	420	244
Valve bonnets			
Heavy	445	851	79
Moderate	178	181	43
Light	33	129	37
Valve stems			
Heavy	23	25.0	27.5
Moderate	14	9.7	20.0
Other			
30 cm pipe	337.0	260.00	68
12 cm hard steel disc	1.6	113.0	140
Steel washers	0.0005	3.0	250
Winfrith test missile	15.6	176.0	240

Table 2.4 Soil/rock characteristics and ejecta velocities.

Soil/rock type	$V_L \times 10^3$ (m/s)	$V_c \times 10^3$ (m/s)	ν	E (kPa/cm² × 10⁵)	G (kPa/cm² × 10⁵)	K (kPa/cm² × 10⁵)	σ_c (kPa/cm² × 10⁵)	τ (kPa/cm²)	σ_t (kPa/cm²)	ρ (g/cm²)
Sand	0.3–1.4	0.4–2.7	—	0.0032	—	—	50–60	—	—	1.4–2.0
Clay	0.8–3.5	1.0–8.30	—	0.0032	—	—	65–100	2.0(average)	3.8	1.4–2.0
Limestone	6.2	1.00	0.26	2.16	0.85	1.70	448	105	70.0	2.41
Slate	7.0	3.30	0.24	10.20	3.85	6.50	455	70	50.0	2.45
Granite	5.30	3.31	0.21	6.10	2.55	3.80	1560	—	180.0	260.00
Shale	9.2	1.00								
Quartzite	7.3	3.70	0.25	9.30	3.80	7.90	1500	240	—	2.60
White marble	5.5	3.00	0.20	3.85	1.60	3.40	750	—	150.0	2.70
Black marble	6.0	3.28	0.33	5.75	2.20	7.00	755	350	—	2.80
Red marble	6.10	3.10	0.26	6.75	2.70	4.75	1200	215	—	2.74
Gneiss	6.10	3.40	0.28	8.30	3.40	6.40	1180	340	—	2.85
Dolomite	12.6	2.30	0.28	9.80	3.80	7.60	1880	1200	350.0	2.85
Coal	1.3	3.35	0.35	0.18	0.07	0.09	80	30	5.5	1.30

ρ = density; V_L, V_c = velocities in longitudinal and transverse directions; E, G, K = Young's modulus, modulus of rigidity and bulk modulus; σ_c, σ_t and τ = compressive stress, tensile stress and shear stress; ν = Poisson's ratio.

Table 2.5 Thrust force of a fragment versus penetration.

Thrust force (kN)	Penetration (cm/min $\times 10^{-1}$)									
	Granite	Limestone	Basalt	Charcoal	Taconite	Shale	Sandstone	Quartzite	Gneiss	Dolomite
1.50	11	17	19	15	11	16	15	17	13	11
1.75	13	19	24	32	25	31	30	20	17	14
2.00	15	21	29	38	31	36	33	25	20	19
2.25	20	25	35	43	39	39	38	27	—	—
2.50	23	29	41	45	42	40	39	31	35	33
3.00	29	35	45	55	53	45	42	35	39	37
3.25	31	39	49	63	59	50	50	45	—	—
4.00	41	55	61	69	63	55	53	52	59	57
4.25	45	63	65	73	71	60	60	59	69	65
10.00	60	139	165	178	161	120	110	80	—	—
20.00	120	239	270	310	210	160	171	100	150	147
30.00	220	300	328	400	270	230	240	150	179	171
50.00	250	349	420	450	310	310	305	210	283	270
75.00	270	410	570	610	459	459	460	310	350	339
100.00	300	500	630	670	560	550	500	358	410	400
125.00	350	535	710	750	620	610	559	452	490	480

Table 2.6 A comparative study of three theories.

Type of analysis	Breaking force (kN)	Riding-up force (kN)	Total force (kN)
Bercha and Danys[2.280]	1558	—	1558
Ralston[2.323–2.325]	1964	1196	3160
Edwards and Croasdale[2.297]	922	900	1822

load on a floating or fixed structure are intrinsically the same, but with a difference in the time-history variations. Figures 2.1 to 2.5 show ice-failure pressure curves.

The total forces computed by various theories vary considerably. For a typical example of an 18.3 m diameter conical tower, Table 2.6 compares the results calculated in three different ways. The relevant data were $D_i = 18.3$ m; $t_i = 0.91$ m; Poisson's ratio $= 0.33$; $\sigma_c =$ crushing strength $= 0.7$ MPa; $E_{ice} = 7$ GPa; cone angle $\alpha = 45°$; free board $= 6.1$ m; coefficient of friction $= 0.15$.

Using Sinha theory,[2.323–2.334] Fig. 2.4 shows the stress at the first crack and the strain rate of the snow/ice. Figure 2.5 gives a useful relationship between the unconfined compressive strength and the strain rate of the snow/ice.

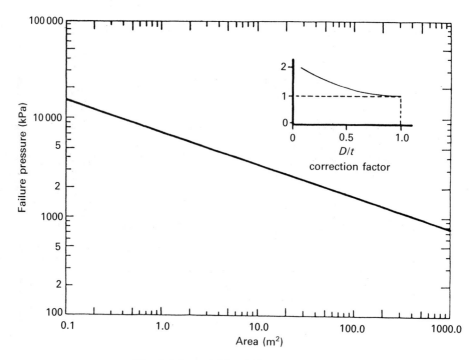

Fig. 2.1 Imperial ice-failure pressure curve.

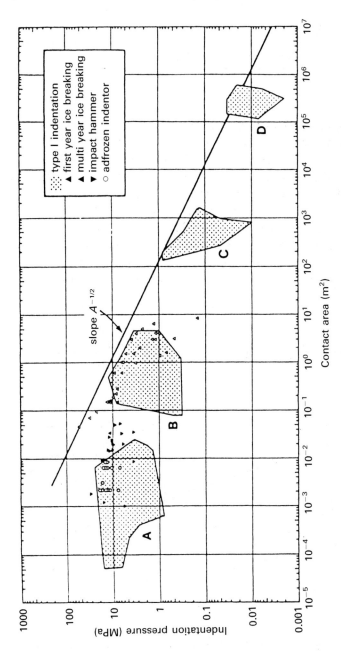

Fig. 2.2 Imperial ice-failure pressure curve. A: laboratory tests; B: medium-scale *in situ* tests, lighthouses and bridge piers; C: full-scale Arctic islands and structures; D: meso-scale models.[2.329]

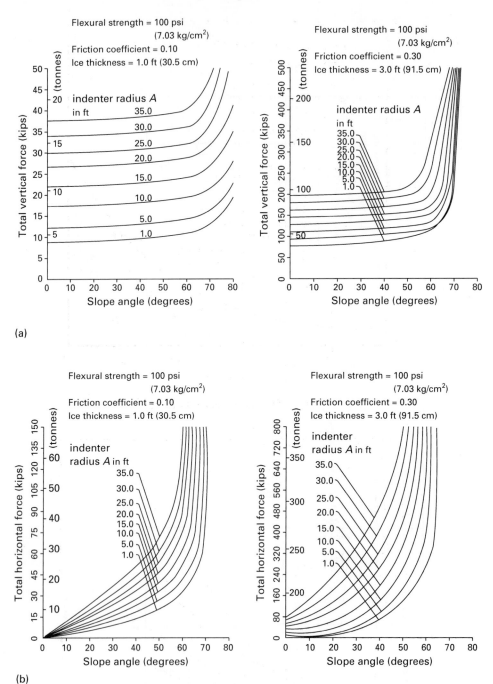

Fig. 2.3 Vertical force (a) and horizontal force (b) versus slope angle and indenter radius *A* for constant flexural strength, friction and ice thickness (kip = 1000 lb).

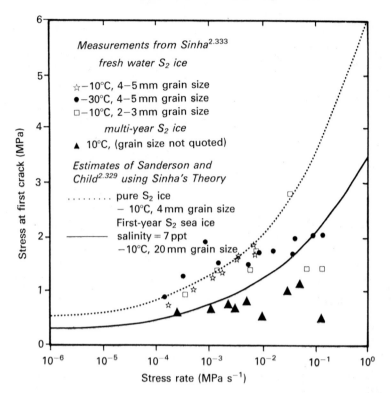

Fig. 2.4 Stress at first crack versus strain rate.[2.333]

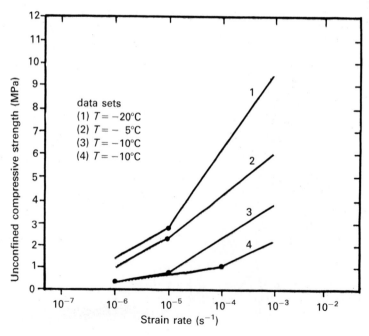

Fig. 2.5 Unconfined compressive strength versus strain rate (*API*, 1987).

2.2.5 Falling or dropped weights as impactors

Falling or dropped weights can be anything from rock falls, trees, sign boards, cars, lorries, trucks, freight containers to sensitive heavy objects such as aircraft wings and engines or nuclear waste casks. Details about cars, lorries and trucks are given in this chapter. Rock falls have tragic effects due to their appalling speed. Figure 2.6 illustrates the gravity-induced descent of rocks which are lying in a state of physical separation. Their free fall may be accompanied by rolling, bouncing or sliding or any combination thereof. The weights of these rocks vary from initially at least 1 kg to 50 000 kg over speeds in the range of 30 m/s to 100 m/s. Again the definitions and constitutive details of various rocks are given in this chapter. With size and speed any category of rocks can cause a large impact load on buildings, railway lines and on vehicles passing on a nearby motorway.

The transport of goods in large re-usable boxes (containers) can be achieved using various cranes. These cranes vary in their capacity and in their operating speeds. The weights, heights and speeds at which cranes operate can certainly give an accurate assessment of the impact loads. On construction sites, different cranes operate for lifting weights. They are shear legs, derrick cranes, crawler-mounted cranes, self-propelled, rubber-tyred wheeled cranes, self-propelled telescopic-jib cranes and their truck-mounted versions, hoists, tower and gantry cranes. Figure 2.7 shows two types of derrick cranes. For impact analysis, the data for 7 tonne and 10 tonne capacity derrick cranes are given in Table 2.7.

The *short and long crawler cranes* are of 30 and 80 tonne capacity. The hoisting speed is 40–50 m/min with a slowing speed of two revolutions per minute. The dragline bucket data and grabbing crane weights are given as 9200 kg and 2975 kg respectively. The *self-propelled telescopic-jib cranes* are of 4 to 10 tonne capacity with a travelling speed of 30 km per hour. The boom length is 6 to

Fig. 2.6 Rock falls.

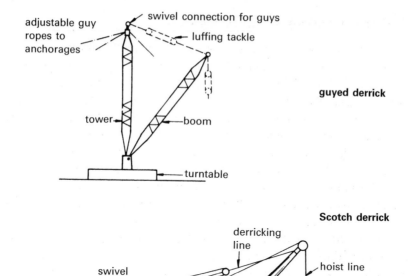

Fig. 2.7 Derrick cranes for dropped weights. (Courtesy of Crane Manufacturers' Association, UK.)

Table 2.7 Comparative data for 7 tonne and 10 tonne capacity derrick cranes.

	Up to 7 tonne capacity	10 tonne capacity and over
Hoisting speed–lifting design capacity (m/min)	30–35	10–15
Hoisting speed–lifting light load (m/min)	70	20–30
Derricking (luffing/speed) (m/min)	30	12–15
Slewing speed (rev/min)	1	0.3
Hoist motor (kW)	50	50
Slewing motor (kW)	10	30
Derricking motor (kW)	40	50
Travelling speed (m/min)	40	10
Travelling motor (kW)	20	60

Table 2.8 Data for cranes of 40 tonne capacity.

Engine size	150–200 hp
	(112–150 kW)
Machine weight	20–40 tonnes
Max hoisting speed (single fall line)	50–60 m/min
Derricking (max to min)	25 s
Slewing speed	3 rev/m
Travelling speed	30 km/h
Turning radius	10 m
Road gradient: unladen	1 in 25
Boom length: four part	20 m
three part	14 m
Overall height	3 m
Overall width	2.5 m
Overall length	8 m

8 m and width and height are approximately 2.75 m. The *truck-mounted jib crane* has a capacity of between 30 and 50 tonnes and its travelling speed is up to 75 km per hour with a maximum hoisting speed and slowing speed of 120 m/min and 0.3 revolutions per minute respectively. For larger cranes of 40 tonne capacity, Table 2.8 lists useful data. Figure 2.8 illustrates a tower crane and various jib radii,

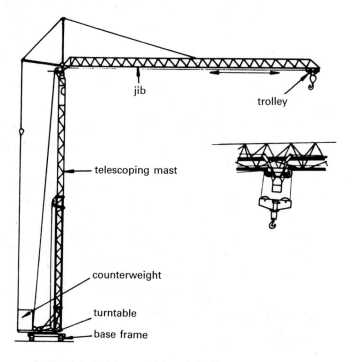

Fig. 2.8 Quick-assembly saddle-jib tower crane.

loading capacities and other major parameters required for impact analysis are given in Table 2.9.

The containers vary in sizes and have either a closed or a removable top (Table 2.10). The most common sizes of containers are 6.10 m × 2.43 m × 2.43 m and 12.20 m × 2.43 m × 2.58 m. They may be loaded or unloaded at the dockside by either shipboard cranes or dockside gantries. They are provided with lifting facilities. Figure 2.9 shows the structural arrangement of a typical dry freight container.

Spent fuel, low-level wastes, decommissioning waste, etc., are transported by rail, road and ship. The spent-fuel containers (casks or flasks) have safety problems which are similar to those of reactors. In order to contain high levels of spent-fuel radiation and decay heat emissions, the cask or flask must be safe against accidents. As shown in Fig. 2.10, the casks are usually made of carbon steel with stainless steel linings. They are usually cylindrical in shape, fitted with a lid. The lid is bolted to the body of the cylinder and sealed with O-rings. The external surface

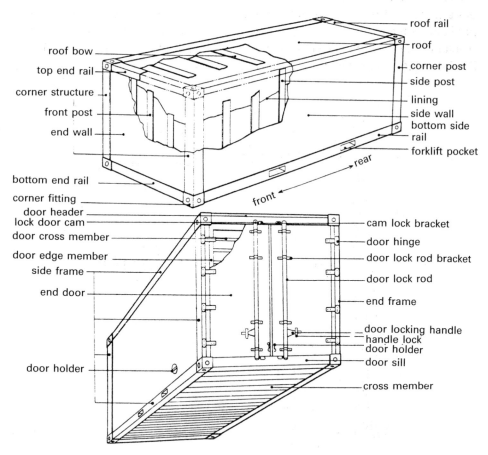

Fig. 2.9 A layout for a typical dry freight container.[2.385]

Table 2.9 Tower cranes: relevant parameters for impact analysis. (Courtesy of the Crane Manufacturers' Association, UK.)

	11	16	18	20	22	25	27	30	35	35	40	55	60	65	80
Max jib radius (m)	11	16	18	20	22	25	27	30	35	35	40	55	60	65	80
Max capacity (tonnes)	0.45	1.5	1.5	1.7	2.0	3.0	3.0	4.0	6.0	8.0	8.0	12.0	20.0	20.0	63.0
Radius (m)	7.8	8.2	10.3	11.42	9.8	10.3	11.0	9.4	8.8	14.9	10.7	14.6	14.0	19.5	23.5
Min radius (m)	3.0	3.0	3.0	3.0	3.0	3.0	3.0	3.0	3.0	3.0	3.0	3.0	3.0	3.0	3.5
Track/wheel gauge (m)	2.0	2.32	2.8	3.2	2.8	2.8	3.2	3.8	4.5	6.0	5.0	66.0	66.0	62.0	96.0
Power supply (kW)	16.0	16.0	16.0	22.0	20.0	20.0	20.0	25.0	40.0	35.0	50.0	6.0	8.0	8.0	12.5
Max hook height (m)	10.6	16.0	18.0	18.0	20.0	20.0	23.5	20.0	32.8	29.3	32.8	100.0	160.0	200.0	360.0

Table 2.10 Data for freight containers.

	40 ft Dry cargo	40 ft Dry cargo hi-cube	40 ft Reefer	40 ft Tank	40 ft Open top	40 ft Half-high open top	20 ft Dry cargo	20 ft Dry cargo
Outside dimensions								
Length	40'	40'	40'	40'	40'	40'	20'	20'
Width	8'	8'	8'	8'	8'	8'	8'	8'
Height	8'	8'6"	8'6"	4'3"	8'6"	4'3"	8'	8'
Inside dimensions								
Length	39'7"	39'6"	37'10"		39'6"	39'6"	19'6"	19'7"
Width	7'9"	7'9"	7'4"		6'8"	7'9"	7'8"	7'10"
Height	7'4"	7'10"	$7'\frac{1}{2}''$		7'6"	3'4"	7'4"	7'6"
Door opening								
Width	$7'5\frac{3}{4}''$	7'6"	7'6"		7'6"	7'6"	$7'5\frac{3}{4}''$	$7'5\frac{3}{4}''$
Height	$6'11\frac{3}{4}''$	$7'5\frac{1}{2}''$	$7'5\frac{3}{4}''$		6'11"	3'4"	$6'11\frac{3}{4}''$	$6'11\frac{3}{4}''$
Construction	Aluminium	Aluminium	Aluminium	Stainless steel AISI 304	Aluminium	Aluminium	Aluminium	Plastic plywood
Internal cubic capacity	2250 ft³	2398 ft³	1988 ft³	6020 gal	2296 ft³	1020 ft³	1096 ft³	1151 ft³
Max load capacity	60260 lb	60950 lb	50000 lb	51600 lb	59280 lb	45000 lb	44800 lb	40650 lb
Tare weight	6400 lb	6940 lb	10700 lb	9260 lb	7750 lb	6660 lb	4500 lb	4500 lb
Special features	End door opening	End door opening	Temp control meat rails	Bulk liquid; loading equipment supplied	End door opening; open top loading; swinging roof bows	End door opening; open top loading; swinging roof bows	End door opening	End door opening

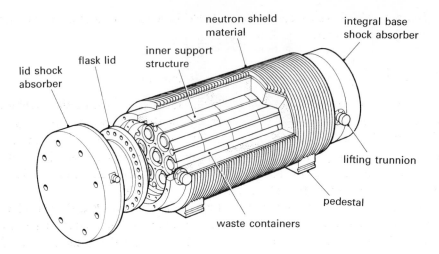

Fig. 2.10 A transport flask for vitrified waste. (Courtesy of the UK Atomic Energy Authority and CEGB.)

has fins for dissipating heat. During transportation, shock absorbers (made of stainless steel encased in balsa wood collars) are fixed to the ends. Sometimes additional lead linings are provided to support or strengthen radiological shielding. Inside the container, the fuel elements are positioned in a basket. The CEGB (now Nuclear Electric, UK) has carried out drop tests on a full-scale magnox fuel flask. This flask, according to the International Atomic Energy Agency, Vienna (IAEA) regulations, must withstand the impact of a free fall from a height of 9 m on to a hard unyielding target followed by a 30 minute totally enveloping fire at 800°C without leaking. CEGB also tested a similar container or flask in a high-speed impact test using a railway train. Data for the flask shown in Fig. 2.10 are listed in Table 2.11.

Table 2.11 Data for the transport flask illustrated in Fig. 2.10.

Description	Max external dimensions (m)	Max gross weight (t)
200 litre drum	0.61 O/D × 0.863	0.75*
3 m³ box	2.15 × 1.5 × 1.3*	15*
12 m³ box for LLW	4.0 × 2.4 × 1.85*	60*
Freight container	6.0 × 2.44 × 2.59	24
500 litre drum	0.8 O/D × 1.2	2*
3 m³ box	1.72 × 1.72 × 1.20	16
12 m³ box for ILW	4.0 × 2.4 × 1.85*	60
Transport container	2.46 × 2.46 × 2.48	70

* approximated to the nearest dimension or weight.

The data for aircraft engines are included in a study of aircraft impact data later on in this chapter.

2.2.6 Heavy lorries, trucks and bulldozers as impactors

Heavy lorries and trucks are in frequent use on many constructions sites, highways and in industrial areas. They are frequently involved in various incidents and accidents. It is, therefore, essential for the structural engineer to know the basic parameters such as payloads, masses/weights, speeds and dimensions. The standard lorries and trucks used in Britain and in the rest of Europe are given in Table 2.12 with their horsepower (HP) (kW), weights and payloads.

Trucks are described in terms of the total number of wheels and drive wheels. Thus a 4×4 truck has four wheels and four-wheel drive. An 8×4 truck has eight wheels, of which only four provide drive, thus leaving the other four as free wheels. The modern 300 tonnes rear dump truck is heavily used on site in quarrying operations. The standard truck load used for the bridge impact analysis in North America is given in Fig. 2.11. The average actual dimensions of tyres and trucks/containers and the rectangular wheel-load areas at the level surface are given in Tables 2.13 and 2.14.

Bulldozers are versatile machines and are used frequently for stripping soils, shallow excavation, maintenance of haul roads, opening up pilot roads, spreading, grading and ripping. The machine is assembled in two separate sections, comprising

(1) a base frame welded and attached with mountings for the dozer blade, the drive sprockets and rollers for the tracks;
(2) the superstructure carrying the engine, transmission, hydraulics cab and controls.

The attached blades can be U-blades, angled blades or push blades. A typical bulldozer is shown in Fig. 2.12. Data on engine sizes, weights and blade lengths and heights are given in Table 2.15. Since these bulldozers are brought on site via roads and are involved on sites for the construction of roads, runways, power stations, etc., they are treated as potential hazards. Many accidents have occurred, particularly during demolition work. Consequently it was thought necessary to list selected data on these machines.

2.2.7 Railway trains

Railway trains are frequently involved in crashes with other trains, with vehicles at crossings or with other structural objects. The structural layouts of trains vary from country to country. Only the British and French systems will be discussed here. Comprehensive data are given in Tables 2.16 to 2.23. Where crash worthiness is to be checked for other engines and railway buggies, similar data are to act as an input for preliminary impact/crash analyses, typical views and cross-sections are given in Figs 2.13 to 2.15.

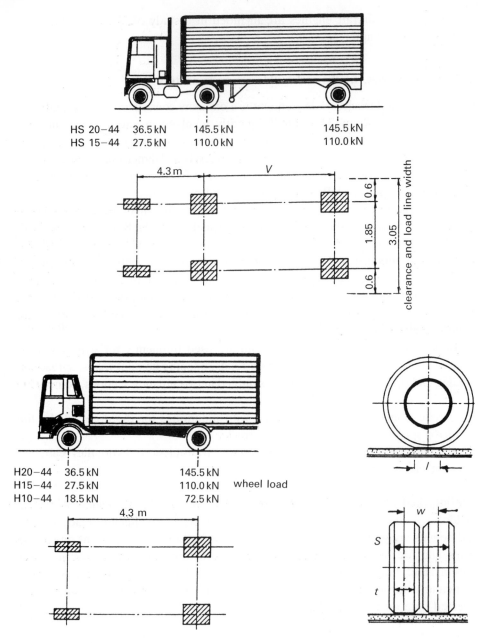

Fig. 2.11 Typical North American trucks. (Courtesy of the North American Trucks Association, New York.)

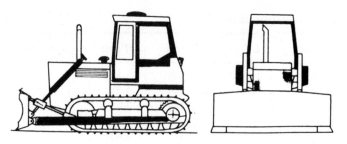

Fig. 2.12 A typical bulldozer.

Table 2.12 Vehicle weights and payloads.

	Gross vehicle HP (kW)	Capacity (m^3)	Net weight empty (tonnes)	Payload (tonnes)
	60−350 (45−261)	30+	10	up to 16.26
	200−350 (150−261)	50+	20	up to 26
	156−1600 (112−1194)	6−70	6−120	up to 150
	200−3000 (150−2235)	6−120	10−250	up to 300
	100−400 (75−298)	70−130	4−10	up to 16.26
	200−400 (150−298)	120	20	up to 32.5
	100−200 (75−150)	10−15	20	up to 35

Table 2.13 Truck/container data.

Type	L (m)	S (m)	H (m)	w_a (tonnes)
1A	12.190	2.435	2.435	30
1B	9.125	2.435	2.435	25
1C	6.055	2.435	2.435	20
1D	2.990	2.435	2.435	10
1E	1.965	2.435	2.435	7
1F	1.460	2.435	2.435	5

L = length, S = width, H = height, w_a = maximum weight.

Table 2.14 Wheel dimensions versus wheel loads.

Nominal wheel loads (kN)	Wheel dimensions (mm)				
	t	l	a^c	W	S
36.5	178	226	40×10^3	290	467
54.5	201	274	55×10^3	325	526
73.0	254	279	71×10^3	412	666

a^c = actual contact area of one tyre at the top of the wearing surface (mm^2); t and l are given in Fig. 2.11.

Table 2.15 Data on bulldozers. (Courtesy of the Federation of Master Builders, UK.)

Engine size	hp	400	300	200	100–150
	kW	298	224	149	75–112
Machine weight	tonnes	35	25	16	10–12
Blade length	m	5	4.5	4.0	3.5
Blade height	m	1.8	1.5	1.2	1.0

Recommended operating speeds km/h.

grading site roads	4–9
scarifying (eg soil stabilization)	8–18
forming ditches	4–8
spreading	4–10
trimming and levelling	9–40
snow ploughing	8–20
self-transporting	10–40

Table 2.16 Data on British railway systems.

	Class 37		
	37/0	37/3	37/9
Subclass			
Former class codes	D17/1, Later 17/3	—	—
Number range	37001–37326	37350–37381	37905–37906
Former number range	D6600–D6999	From main fleet	From main fleet
Built by	EE & RSH Ltd	EE & RSH Ltd	EE Ltd
Introduced	1960–65	As 37/3/1988	As 37/9/1987
Wheel arrangement	Co–Co	Co–Co	Co–Co
Weight (operational)	102–108 tonnes	106 tonnes	120 tonnes
Height	3.89 m	3.89 m	3.89 m
Width	2.70 m	2.70 m	2.70 m
Length	18.74 m	18.74 m	18.74 m
Minimum curve negotiable	80.46 m	80.46 m	80.46 m
Maximum speed	129 km/h	129 km/h	129 km/h
Wheelbase	15.44 m	15.44 m	15.44 m
Bogie wheelbase	4.11 m	4.11 m	4.11 m
Bogie pivot centres	11.32 m	11.32 m	11.32 m
Wheel diameter (driving)	1.14 m	1.14 m	1.14 m
Brake type	Dual	Dual	Dual
Sanding equipment	Pneumatic	Pneumatic	Pneumatic
Route availability	5	5	7
Coupling restriction	Blue Star	Blue Star	Blue Star
Brake force	50 tonnes	50 tonnes	50 tonnes
Engine type	EE 12 CSVT	EE 12 CSVT	Ruston RK 270T
Engine horsepower	1750 hp (1304 kW)	1750 hp (1304 kW)	1800 hp (1340 kW)
Power at rail	1250 hp (932 kW)	1250 hp (932 kW)	—
Tractive effort	55000 lb (247 kN)	55000 lb (247 kN)	—
Cylinder bore	10 in (0.25 m)	10 in (0.25 m)	—
Cylinder stroke	12 in (0.30 m)	12 in (0.30 m)	—

continued

Table 2.16 *Continued.*

	Class 37		
Main generator type	EE 822/10G	EE 822/10G	Not fitted
Traction alternator type	Not fitted	Not fitted	GEC 564
Auxiliary generator type	EE 911/5C	EE 911/5C	Not fitted
Number of traction motors	6	6	6
Traction motor type	EE 538/1A	EE 538/1A	EE 538/1A
Gear ratio	53:18	53:18	53:18
Fuel tank capacity	890 gal (4046 litres)	1690 gal (7682 litres)	1690 gal (7682 litres)
Cooling water capacity	160 gal (727 litres)	160 gal (727 litres)	–
Boiler water capacity	800 gal (3637 litres)	Not fitted	Not fitted
Lubricating oil capacity	120 gal (545 litres)	120 gal (545 litres)	–
Boiler fuel capacity	From main supply	Not fitted	Not fitted
Region of allocation	Eastern, Western, Scottish	Eastern, Western	Western

	Class 45	
Subclass	45/0	45/1
Former class codes	D25/1, later 25/1	D25/1, later 25/1
Number range	45 007–45 052	45 103–45 141
Former number range	D11–137	D11–137
Built by	BR Derby & Crewe	BR Derby & Crewe
Introduced	1960–62	As 45/1/1973–75
Wheel arrangement	1Co–Co1	1Co–Co1
Weight (operational)	138 tonnes	135 tonnes
Height	3.91 m	3.91 m
Width	2.70 m	2.70 m
Length	20.70 m	2.17 m
Minimum curve negotiable	100.58 m	100.58 m

continued

Maximum speed	145 km/h	145 km/h
Wheelbase	18.18 m	18.18 m
Bogie wheelbase	6.55 m	6.55 m
Bogie pivot centres	9.95 m	9.95 m
Wheel diameter (driving)	1.14 m	1.14 m
	0.91 m	0.91 m
Brake type	Dual	Dual
Sanding equipment	Pneumatic	Pneumatic
Route availability	6	7
Coupling restriction	Not multiple fitted	Not multiple fitted
Brake force	63 tonnes	63 tonnes
Engine type	Sulzer 12LDA28B	Sulzer 12LDA28B
Engine horsepower	2500 hp (1862 kW)	2500 hp (1862 kW)
Power at rail	2000 hp (1490 kW)	2000 hp (1490 kW)
Tractive effort	55 000 lb (245 kN)	55 000 lb (245 kN)
Cylinder bore	11 in (0.27 m)	11 in (0.27 m)
Cylinder stroke	14 in (0.35 m)	14 in (0.35 m)
Main generator type	Crompton CG462A1	Crompton CG462A1
Traction alternator type	Crompton CAG252A1	Crompton CAG252A1
Auxiliary generator type	Brush BL 100–30 Mk 11	Not fitted AG252A1
Number of traction motors	6	6
Traction motor type	Crompton C172A1	Crompton C172A1
Gear ratio	62:17	62:17
Fuel tank capacity	790 gal (3591 litres)	790 gal (3591 litres)
Cooling water capacity	346 gal (1572 litres)	346 gal (1572 litres)
Boiler water capacity	190 gal (864 litres)	190 gal (864 litres)
Lubricating oil capacity	Not fitted	1040 gal (4727 litres)
Boiler fuel capacity	Not fitted	From main supply
Region of allocation	Eastern	Eastern

Table 2.16 *Continued.*

	Class 47		
Subclass	47/0	47/3	47/9
Former class codes	27/2	27/2	—
Present number range	47002–47299	47301–47381	47901
Former number range	D1521–D1998	D1782–D1900	D1628(47046)
Built by	BR Crewe, Brush Ltd	BR Crewe, Brush Ltd	BR Crewe
Introduced	1962–5	1964–5	As 47/9/1979
Wheel arrangement	Co–Co	Co–Co	Co–Co
Weight (operational)	111–121 tonnes	114 tonnes	117 tonnes
Height	3.89 m	3.89 m	3.89 m
Width	2.79 m	2.79 m	2.79 m
Length	19.38 m	19.38 m	19.38 m
Minimum curve negotiable	80.46 m	80.46 m	80.46 m
Maximum speed	153 km/h	153 km/h	129 km/h
Wheelbase	15.69 m	15.69 m	15.69 m
Bogie wheelbase	4.41 m	4.41 m	4.41 m
Bogie pivot centres	12.27 m	11.27 m	11.27 m
Wheel diameter (driving)	1.14 m	1.14 m	1.14 m
Brake type	Dual	Dual	Air
Sanding equipment	Not fitted	Not fitted	Pneumatic
Heating type	Steam	Not fitted	Not fitted
Route availability	6	6	6
Coupling restriction	Not fitted	Not fitted	Not fitted
Brake force	60 tonnes	60 tonnes	60 tonnes
Engine type	Sulzer 12LDA28C	Sulzer 12LDA28C	Ruston Paxman 12RK3CT
Engine horsepower	2580 hp (1922 kW)	2580 hp (1922 kW)	3300 hp (2455 kW)
Power at rail	2080 hp (1550 kW)	2080 hp (1550 kW)	2808 hp (2089 kW)
Tractive effort	60 000 lb (267 kN)	60 000 lb (267 kN)	57 325 lb (255 kN)
Cylinder bore	11 in (0.27 m)	11 in (0.27 m)	10 in (0.25 m)

Cylinder stroke	14 in (0.35 m)	14 in (0.35 m)	12 in (0.30 m)
Main generator type	Brush TG160–60 or TG172–50	Brush TG160–60 or TG172–50	Not fitted
Main alternator type	Not fitted	Not fitted	Brush BA 1101A
Auxiliary generator type	Brush TG69–20 or TG69–29	Brush TG69–20 or TG69–28	Not fitted
Auxiliary alternator type	Not fitted	Not fitted	Brush BAA602A
ETS generator	Not fitted	Not fitted	Not fitted
ETS alternator	Not fitted	Not fitted	Not fitted
Number of traction motors	6	6	6
Traction motor type	Brush TM64–68	Brush TM64–68	Brush TM64–68 Mk 1A
Gear ratio	66:17	66:17	66:17
Fuel tank capacity	765 gal (3477 litres)	765 gal (3477 litres)	765 gal (3477 litres)
Cooling water capacity	300 gal (1364 litres)	300 gal (1364 litres)	300 gal (1364 litres)
Boiler water capacity	1250 gal (5683 litres)	Not fitted	Not fitted
Lubricating oil capacity	190 gal (864 litres)	190 gal (864 litres)	190 gal (864 litres)
Boiler fuel capacity	From main supply	Not fitted	Not fitted
Region of allocation	Eastern, Midland, Scottish, Western	Eastern, Midland, Western	Western

Class 50

Subclass	50/0	50/1
Former class codes	27/3	—
Number range	50001–50050	50149
Former number range	D400–D449	50049
Built by	EE (Vulcan Foundry) Ltd	Rebuilt Laira
Introduced	1967–8	Originally 1968, rebuilt 1987
Wheel arrangement	Co–Co	Co–Co
Weight (operational)	117 tonnes	117 tonnes
Height	3.95 m	3.95 m
Width	2.77 m	2.17 m
Length	20.87 m	20.87 m
Minimum curve negotiable	80.46 m	80.46 m

continued

Table 2.16 *Continued.*

	Class 50	
Maximum speed	161 km/h	129 km/h
Wheelbase	17.11 m	17.11 m
Bogie wheelbase	4.11 m	4.11 m
Bogie pivot centres	13.00 m	13.00 m
Wheel diameter (driving)	1.09 m	1.09 m
Brake type	Dual	Dual
Sanding equipment	Not fitted	Not fitted
Heating type	Electric–Index 61	Not available
Route availability	6	6
Coupling restriction	Orange square	Orange square
Brake force	59 tonnes	59 tonnes
Engine type	English Electric 16CSVT	English Electric 16CSVT
Engine horsepower	2700 hp (2014 kW)	2450 hp
Power at rail	2070 hp (1540 kW)	1890 hp
Tractive effort	48 500 lb (216 kN)	48 500 lb (216 kN)
Cylinder bore	10 in (0.25 m)	10 in (0.25 m)
Cylinder stroke	12 in (0.30 m)	12 in (0.30 m)
Main generator type	EE840–4B	EE840–4B
Traction alternator type	EE911–5C	EE911–5C
Auxiliary generator type	EE915–1B	EE915–1B
Number of traction motors	6	6
Traction motor type	EE538–5A	EE538–1A
Gear ratio	53:18	53:18
Fuel tank capacity	1055 gal (4797 litres)	1055 gal (4797 litres)
Cooling water capacity	280 gal (1272 litres)	280 gal (1272 litres)
Boiler water capacity	130 gal (591 litres)	130 gal (591 litres)
Region of allocation	Western	Western

Table 2.17 Data on British railway systems.

	Class 58	Class 59
Number range	58 001−58 050	59 001−59 004 JT26SS−55
Built by	BREL Doncaster, UK	General Motors Ltd, Illinois, USA
Introduced	1983−7	1986
Wheel arrangement	Co−Co	Co−Co
Weight (operational)	130 tonnes	126 tonnes
Height	3.91 m	3.91 m
Width	2.70 m	2.65 m
Length	19.13 m	21.40 m
Minimum curve negotiable	80.46 m	
Maximum speed	129 km/h	97 km/h
Wheelbase	14.85 m	17.269 m
Bogie wheelbase	4.18 m	4.14 m
Bogie pivot centres	10.80 m	13.25 m
Wheel diameter (driving)	1.12 m	1.06 m
Brake type	Air	Air
Sanding equipment	Pneumatic	Pneumatic
Heating type	Not fitted	Not fitted
Route availability	7	7
Multiple coupling restriction	Red diamond	Within type only
Brake force	62 tonnes	69 tonnes
Engine type	Ruston Paxman 12RK3ACT	EMD 645E3C
Engine horsepower	3300 hp (2460 kW)	3300 hp (2238 kW)
Power at rail	2387 hp (1780 kW)	
Tractive effort	61 800 lb (275 kN)	122 000 lb (573 kN)
Cylinder bore	10 in (0.25 m)	9 1/16 in (0.23 m)
Cylinder stroke	12 in (0.30 m)	10 in (0.25 m)
Main generator type	Brush BA1101B	EMD AR11 MLD D14A
Number of traction motors	6	6
Traction motor type	Brush TM73−62	EMD D77B
Fuel tank capacity	985 gal (4480 litres)	919 gal (4543 litres)
Cooling water capacity	264 gal (1200 litres)	212 gal (962 litres)
Lubricating oil capacity	110 gal (416 litres)	202 gal (920 litres)
Region of allocation	Midland	

Table 2.18 Data on British railway systems.

	Class 81	Class 83
Former class code		AL3
Present number range	81 002−81 019	83 009−83 012
Former number range	E3001−E3023, E3096−E3097	E3024−E3035, E3098−E3100
Built by	BRC & W Ltd	English Electric
Introduced	1959−64	1960−62
Wheel arrangement	Bo−Bo	Bo−Bo
Weight (operational)	79 tonnes	77 tonnes
Height (pantograph lowered)	3.76 m	3.76 m
Width	2.65 m	2.65 m
Length	17.22 m	17.52 m
Minimum curve negotiable	80.46 m	80.46 m

continued

Table 2.18 *Continued*.

	Class 81	Class 83
Maximum speed	129 km/h	64 km/h
Wheelbase	12.87 m	12.19 m
Bogie wheelbase	3.27 m	3.04 m
Bogie pivot centres	9.60 m	9.14 m
Wheel diameter	1.21 m	1.21 m
Brake type	Dual	Dual
Sanding equipment	Pneumatic	Pneumatic
Heating type	Electric−Index 66	Electric−Index 66
Route availability	6	6
Coupling restriction	Not multiple fitted	Not multiple fitted
Brake force	40 tonnes	38 tonnes
Horsepower (continuous)	3200 hp (2387 kW)	2950 hp (2200 kW)
(maximum)	4200 hp (3580 kW)	4400 hp (3280 kW)
Tractive effort (maximum)	50 000 lb (222 kN)	38 000 lb (169 kN)
Number of traction motors	4	4
Traction motor type	AEI 189	EE 532A
Control system	LT Tap changing	LT Tap changing
Gear drive	Alsthom Quill, single reduction	SLM flexible, single reduction
Gear ratio	29:76	25:76
Pantograph type	Stone-Faiveley	Stone-Faiveley
Rectifier type	Mercury Arc	Mercury Arc
Nominal supply voltage	25 kV ac	25 kV ac
Region of allocation	Scottish	Midland

Table 2.19 Data on British railway systems.

	Class 85	Class 86
Subclass		86/4
Former class code	AL5	AL6
Present number range	85 002−83 040	86 401−86 439
Former number range	E3056−E3095	−
Built by	BR Doncaster	EE Ltd and BR Doncaster
Introduced	1961−4	As 86/4 1984/87
Wheel arrangement	Bo−Bo	Bo−Bo
Weight (operational)	83 tonnes	83 tonnes
Height (pantograph lowered)	3.76 m	3.97 m
Width	2.66 m	2.64 m
Length	17.19 m	17.83 m
Minimum curve negotiable	120.70 m	120.70 m
Maximum speed	129 km/h	161 km/h
Wheelbase	12.87 m	13.25 m
Bogie wheelbase	3.27 m	3.27 m
Bogie pivot centres	9.60 m	9.98 m
Wheel diameter	1.21 m	1.16 m
Brake type	Dual	Dual
Sanding equipment	Pneumatic	Pneumatic
Heating type	Electric−Index 66	Electric−Index 74
Route availability	6	6
Coupling restriction	Not multiple fitted	
Brake force	41 tonnes	40 tonnes

Table 2.19 *Continued.*

	Class 85	Class 86
Horsepower (continuous)	3200 hp (2390 kW)	4040 hp (3014 kW)
(maximum)	5100 hp (3800 kW)	5900 hp (4400 kW)
Tractive effort (maximum)	50 000 lb (222 kN)	58 000 lb (258 kN)
Number of traction motors	4	4
Traction motor type	AEI 189	AEI 282AZ
Control system	LT Tap changing	HT Tap changing
Gear ratio	29:76	22:65
Pantograph type	Stone-Faiveley	Stone-Faiveley/AEI
Rectifier type	Germanium	Silicon semi-conductor

Table 2.20 Data on British railway systems.

	Class 87		Class 91
Subclass	87/0	87/1	
Number range	87 001–87 035	87 101	91 001–91 050
Built by	BREL Crewe	BREL Crewe	BREL Crewe and GEC
Introduced	1973–4	1977	1988–90
Wheel arrangement	Bo–Bo	Bo–Bo	Bo–Bo
Weight (operational)	83 tonnes	79 tonnes	80 tonnes
Height (pantograph lowered)	3.99 m	3.99 m	3.75 m
Width	2.64 m	2.64 m	2.74 m
Length	17.83 m	17.83 m	19.40 m
Minimum curve negotiable	120.70 m	120.70 m	80.80 m
Maximum speed	177 km/h	177 km/h	225 km/h
Wheelbase	13.25 m	13.25 m	17.20 m
Bogie wheelbase	3.27 m	3.27 m	3.35 m
Bogie pivot centres	9.98 m	9.98 m	10.50 m
Wheel diameter	1.16 m	1.16 m	1.00 m
Brake type	Air	Air	Air (rheostatic)
Sanding equipment	Pneumatic	Pneumatic	Pneumatic
Route availability	6	6	
Coupling restriction	Within type and Class 86	Within type and Class 86	Fitted with TDM
Brake force	40 tonnes	40 tonnes	
Horsepower (continuous)	5000 hp (3730 kW)	4850 hp (3620 kW)	6080 hp (4530 kW)
(maximum)	7860 hp (5860 kW)	7250 hp (5401 kW)	6310 hp (4700 kW)
Tractive effort (maximum)	58 000 lb (258 kN)	58 000 lb (258 kN)	
Number of traction motors	4	4	
Traction motor type	GEC G412AZ	GEC G412BZ	GEC
Control system	HT Tap changing	Thyristor	Thyristor
Gear ratio	32:73	32:73	
Pantograph type	Brecknell Willis HS	Brecknell Willis HS	Brecknell Willis HS
Nominal supply voltage	25 kV ac	25 kV ac	25 kV ac
Region of allocation	Midland	Midland	Eastern

Table 2.21 Data on French railway systems.

Class	Transmission	Rated power hp (kW)	w_a (kg)	V mph (km/h)	V_{max} mph (km/h)	Wheel dia (mm)	Total weight (tonnes)	L (mm)	First built
68 000	Electric	2225 (1660)	30 400	19 (30.6)	81 (130)	1250	106	17 920	1963
68 500	Electric	2205 (1645)			81 (130)		105		1963
65 000	Electric	1300 (970)	25 000 17 000		75 (120)	1050	112	19 814	1956
72 000 monomotor bogies (2 gears)	Electric	3020 (2250)	36 400	34.7 (65) 21.5 (34.5)	100 (160) 53 (85)	1140	110	20 190	1967
63 000	Electric	480 (355) 585 (435)	37 000 17 000	6 (10) 8 (13)	50 (80) 50 (80)	1050	68	14 680	1953 1957
63 500	Electric	605 (450)	17 100	7.5 (12)	50 (80)	1050	68	14 680	1956

66 000	Electric	1115 (830)			75 (120)		70	14 898	1959
66 600	Electric	1195 (890)			75 (120)		71	14 898	1962
67 000 monomotor bogies (2 gears)	Electric	1930 (1440)	20 600 / 31 000	26 (42) / 17.4 (28)	56 (90) / 56 (90)	1150	80	17 090	1963
67 400	Electric	2045 (1525)	29 000	23 (37)	87 (140)	1250	83	17 090	1969
Y7100	Hydrodynamic	175	7 400		34 (54)	1050	32	8 940	1958
Y7400	Mechanical	175			37 (60)	1050	32	8 940	1963
Y8000 (2 gears)	Hydrodynamic	290	6 750	20 (32)	37 (60)	1050	36	10 140	1977

w_a = max weight; V = speed; V_{max} = max speed; L = length.

Table 2.22 Data on French railway systems.

Class	Line current	Rate output hp (kW)	w_a (kg)	V mph (km/h)	V_{max} mph (km/h)	Wheel dia (mm)	Total weight (tonnes)	L (mm)	Year built
BB-8100	1500 V dc	2815 (2100)	30 400	25.8 (41.5)	65 (105)	1400	92	12 930	1949
BB-8500 (2 gear ratios)	1500 V dc	3940 (2940)	20 100 / 33 000	51.3 (82.5) / 30.6 (49.2)	93 (140) / 56 (90)	1100	78	14 700–15 570	1963
BB-7200	1500 V dc	5845 (4360)	30 000	60 (97)	112 (180)	1250	84	17 480	1977
BB-9200	1500 V dc	5160 (3850)	26 500	58 (93)	100 (160)	1250	82	16 200	1957
BB-9300	1500 V dc								1968
BB-9400	1500 V dc	2965 (2210)	27 500	31 (50)	81 (130)	1020	59	14 400	1959
CC-6500 (2 gear ratios)	1500 V dc	7910 (5900)	29 347	38.5 (62)	62 (100) / 137 (220)	1140	115	20 190	1970
CC-7100	1500 V dc	4680 (3490)	26 500	49.5 (79.5)	93 (150)	1250	105	18 922	1951
BB-12 000	25 kV, 50 Hz	3310 (2470)	36 000	29.5 (47.5)	75 (120)	1250	83	15 200	1954
BB-13 000	25 kV, 50 Hz	2680 (2000) / 2855 (2130)	25 000	40.5 (65)	65 (105) / 75 (120)	1250	84	15 200	1954 / 1956
BB-1600	25 kV, 50 Hz	5540 (4130)	31 500	53 (85)	100 (160)	1250	84	16 200	1958

Table 2.22 *Continued.*

Class	Line current	Rate output hp (kW)	w_a (kg)	V mph (km/h)	V_{max} mph (km/h)	Wheel dia (mm)	Total weight (tonnes)	L (mm)	Year built
BB-15 000	25 kV, 50 Hz	5485 (4360)	29 000	62 (100)	112 (180)	1250	88	17 480	1971
BB-16 500 (2 gear ratios)	25 kV, 50 Hz	3460 (2580)	33 000 19 200	51 (82) 30 (90)	93 (150) 56	1100	74	14 400	1958
BB-17 000 (2 gear ratios)	25 kV, 50 Hz	3940 (2940)	20 100 33 000	51.3 (82.5) 30.6 (49.2)	87 (140) 56 (90)	1100	78	14 700–14 940	1964
CC-14 100	25 kV, 50 Hz	2495 (1860)	43 000	17.7 (28.5)	37 (60)	1100	126	18 890	1954
BB-20 200 (2 current) (2 gear ratios)	25 kV, 50 Hz 15 kV, 16 2/3 Hz	3940 (2940) 2225 (1660)			56 (90) 93 (150)		80	14 490	1969
CC-21 000 (2 current) (2 gear ratios)	25 kV, 50 Hz 1.5 kV dc	7910 (5900)			62 (100) 137 (220)		122	20 190	1969
BB-22 200 (2 current)	25 kV, 50 Hz and 1500 V dc	5845 (4360)	30 000	60 (97)	112 (180)	1250	89	17 480	1977
BB-25 100 (2 current)	25 kV, 50 Hz and 1500 V dc	5540 (4130) 4560 (3400)	37 000	52 (83.5)	81 (130)	1250	84	16 200	1963

continued

Table 2.22 *Continued.*

Class	Line current	Rate output hp (kW)	w_a (kg)	V mph (km/h)	V_{max} mph (km/h)	Wheel dia (mm)	Total weight (tonnes)	L (mm)	Year built
BB-25 200 (2 current)	25 kV, 50 Hz and 1500 V dc	5540 (4130) 4560 (3400)	31 000	62 (99.5)	99 (160)	1250	84	16 200	1964
BB-25 500 (2 current) (2 gear ratios)	25 kV, 50 Hz and 1500 V dc	3940 (2940)	20 100 33 600	51 (82) 30 (48)	93 (140) 56 (90)	1100	78	14 700–15 570	1963
CC-40 100 (4 current) (2 gear ratios)	25 kV, 50 Hz 15 kV, 16 2/3 Hz 3000 V dc 1500 V dc	6000 (4480) 6000 (4480)	14 500 20 200	95.4 (153.5) 68 (110)	149 (240) 99 (160)	1080	108	22 030	1964

w_a = max weight; V = speed; V_{max} = max speed; L = length.

Table 2.23 Data on French railway systems.

Class	Cars per unit	Line voltage frequency	Motor cars per unit	Motored axles per motor car	Rated output (kW)	Max speed (km/h)	Weight (tonnes)	Length of unit (mm)	Rate of acceleration under normal load	Year first built
Z5300	4	1.5 kV dc	1	4	1180	130	154	102 800	0.7 m/s² 0 to 50 km/h	1965
Z5600	4	1.5 kV dc	2	4	2700	140	216	98 760	0.9 m/s² 0 to 50 km/h	1982
Z7100	4	1.5 kV dc	1	2	940	130	139	94 170	0.47 m/s² 0 to 50 km/h	1960
Z7300 Z7500	2	1.5 kV dc	1	4	1275	160	103	50 200	0.5 m/s² 0 to 50 km/h	1980
Z6100	3	25 kV, 50 Hz	1	2	615	120	113	74 450	0.45 m/s² 0 to 40 km/h	1964
Z6300	3	25 kV, 50 Hz	1	2	615	120	105	60 100	0.5 m/s² 0 to 40 km/h	1965
Z6400	4	25 kV, 50 Hz	2	4	2350	120	189	92 430	1 m/s² 0 to 50 km/h	1976
Z8100	4	1.5 kV dc, 25 kV, 50 Hz	2	4	2500	140	212	104 160	0.9 m/s² 0 to 50 km/h	1985
Z8800	4	25 kV, 50 Hz	2	4	2800	140	224	98 760	0.9 m/s² 0 to 50 km/h	1985
Z9500 Z9600	2	1.5 kV dc, 25 kV, 50 Hz	1	4	1275	160	115	50 200	0.5 m/s² 0 to 50 km/h	1982
Z11500 TGV 23000	2 10	25 kV, 50 Hz, 25 kV, 50 Hz 1.5 kV dc	2 2	4 12	1275 6450 3100	160 270	115 418	50 200 200 190	NA 0.5 m/s² 0 to 50 km/h	1987 1978
TGV 3300	10	25 kV, 50 Hz 16²/3 Hz, 1.5 kV dc	2	12	6450 3100 2800	270	419	200 190	0.5 m/s² 0 to 50 km/h	1981

NA = not applicable.

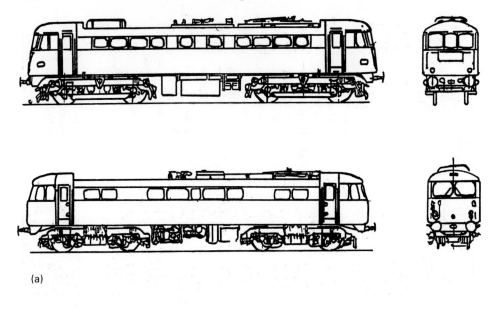

(a)

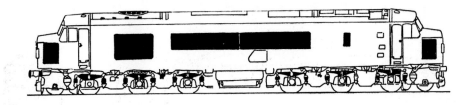

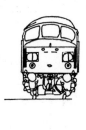

(b)

Fig. 2.13 Typical views of the British railway train. (a) classes 81 and 83; (b) class 45; (c) class 86; (d) classes 86 and 87.

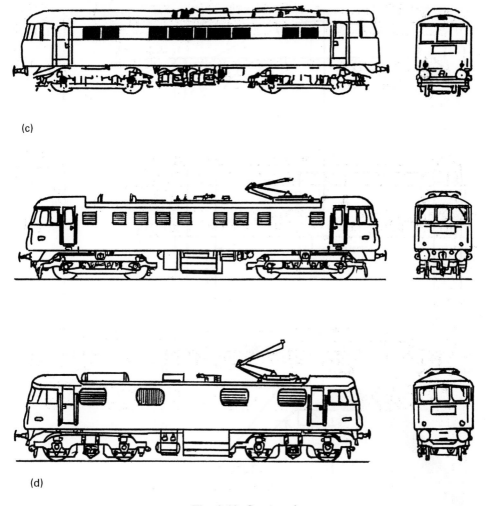

(c)

(d)

Fig. 2.13 *Continued.*

2.3 Military, air force and navy missiles and impactors

2.3.1 Introduction to bombs, rockets and missiles

Different versions of bombs, rockets and missiles are available in the defence markets. Table 2.24 summarizes the characteristics of a number of shells and bombs. Figure 2.16 shows typical cartridges, high explosive shells, shrapnel shells, hand grenades and bombs as impactors. The chemistry of such bombs, shells and cartridges is described later on in this chapter. Advances in the manufacture of bombs have reached their acme. Some of the new ones are fully described in this chapter with appropriate specifications.

Different missile systems exist in a number of countries. The range capability determines their category. Missiles with a maximum range exceeding 550 km are

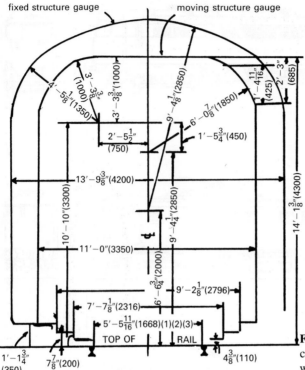

Fig. 2.14 A typical structural cross-section of a French train with circular heads.

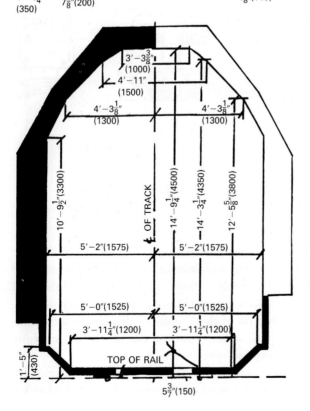

Fig. 2.15 A typical cross-section of a French train with parabolic heads.

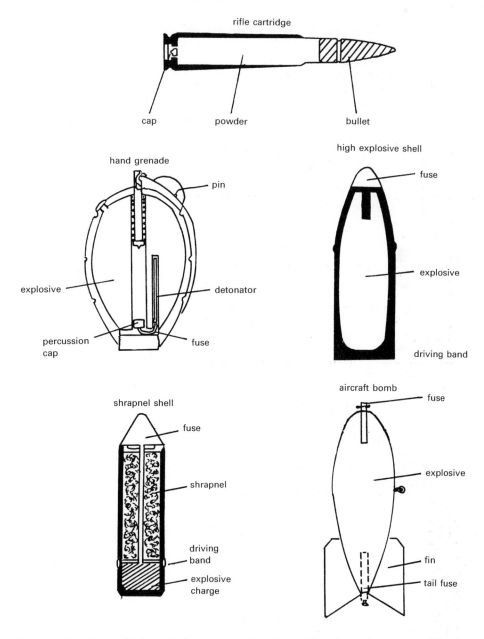

Fig. 2.16 Cartridges, high explosives, shrapnel shells and bombs as impactors. (Sketches and information from Dara Adam Khel, NWFP, Pakistan (Malik Zarak Khan).)

continued

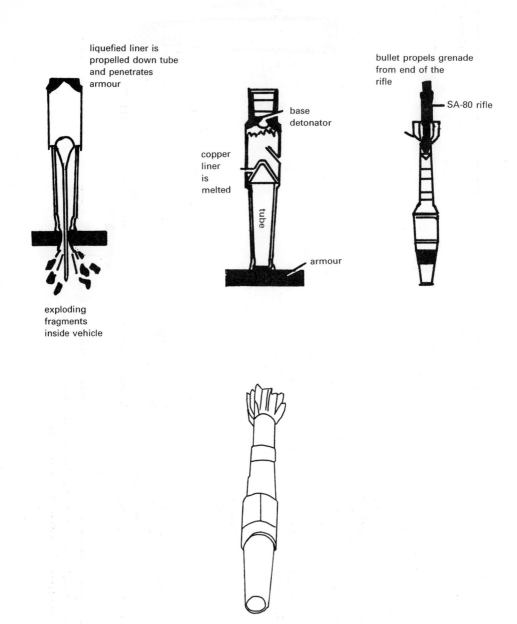

liquefied liner is propelled down tube and penetrates armour

exploding fragments inside vehicle

base detonator

copper liner is melted

tube

armour

bullet propels grenade from end of the rifle

SA-80 rifle

isometric view

Fig. 2.16 *Continued.*

Table 2.24 Guns, shells and bombs.

Types of shells and bombs	Manufacturer	Dimensions	w_L (kg)	Shape and guidance	R	V (m/s)	Warhead (kg)	Mechanisms
Double base shot gun	International	$d = 1.25$ $t = 0.150$	—	Porous disc	Short	—	—	Guns
Single or double base rim fire	International	$d = 0.875$ $t = 0.10$	—	Porous disc	Short	—	—	Guns
Double base revolver	International	$d = 1.00$ $t = 0.125$	—	Porous disc	Short	—	—	Gun
Single base rifle	International	$d = 1.25$ $d = 0.375$ $t = 0.045$	—	Tubular	Short	—	—	Gun
GBU-15 HOBOS guided bomb	Rockwell International USA	$L = 3.75$ *MK84* $S = 112$ $d = 46$ *(M118)* $S = 132$ $d = 61$	*MK84* 2240 (1016) *M118* 3404	The forward guidance system is KMU-353 or 390 electro-optical with TV camera or target seeker optics	Medium	—	—	Free fall bomb from
Bofor	Sweden	—	0.120	5500 kg guns	3.7 km	10005	0.88	Trajectory fall
ZSU-23-4	USSR	$L = 654$ $d = 29.5$ 3400 rounds	20.50	SPAA vehicles 4×3 automatic	Short	7000	—	Gun
GBU-15(V)2 glide bomb (smart bomb)	USA	$L = 388$		Imaging infra-red	Short		1111.0	

d = diameter (mm); t = web thickness (mm); L = length (cm);
w_L = weight; S = span (cm); R = distance; V = velocity.

classified *strategic* and those with ranges between 1000 and 5500 km are known as *intermediate missiles*. Missile systems with ranges less than 1000 km are called *short-range* missile systems.

Certain symbols used in the explanatory notes for the missiles are defined as follows:

- ICBM   Intercontinental ballistic missiles or strategic missile
- IRBM   Intermediate range ballistic missile
- SRM   Short-range missile
- SSM   Surface-to-surface missile
- NSA   Naval surface-to-air missile
- ADM   Air defence missile
- AAM   Air-to-air missile
- ASM   Air-to-surface missile
- RO   Rocket
- B   Bomb
- Parameters L: length; S: span; d: diameter: w_L: weight; V: speed; R: range in miles (km); SG: self-guided; P_L: payload.

It is important to mention a few of the missiles. A typical example of the ICBM/IRBM is Patriot, a SAM, and three shoulder-mounted missiles, namely Stinger, Blow Pipe and Javelin missiles. Others are described in Tables 2.25 to 2.33. Figures 2.17 and 2.18 show major components of AT2 and M77 rockets. A typical Matra bomb described in the tables is shown in Fig. 2.19.

2.3.1.1 Patriot: a SAM

In March 1972, Patriot was underway with modifications in radar, computer and guidance hardware. In July 1973, demonstration model fire control group (DMFCG) was tested. In January 1979, the programme was redesignated XMIM-104 Patriot. A full-scale development commenced in August 1979. First firing in CM electronic counter measures (ECM) was carried out in December 1976. A Patriot fire unit consists of a fire control section (FCS) and its launchers. The individual sections of the weapon from nose to tail are given below:

(1) nose radome
(2) terminal guidance system
(3) warhead section
(4) propulsion system
(5) control section.

The nose radome is fabricated from 12 mm thick slip-cast fused silica and tipped with cobalt alloy; Fig. 2.20 shows the layout. Below the silica a planar

Table 2.25 Missiles for Armed Forces (ICBM and IRBM) (China, USSR).

Types of missiles/rockets	Manufacturer/ country of manufacture	Dimensions	w_L lb (kg)	Power plant and guidance	R miles (km)	V mach	Warhead	Mechanisms
BGM-109 Tomahawk cruise missile (IRBM)	General dynamics/ Convair, San Diego, USA	$L = 5.56$ $S = 2.45$ $d = 0.53$	2200 (1000) to 4000 (1814)	Williams Research F107 Turbofan 600 lb (272 kg) Thrust Guidance MD Tercom and Inertial	1727 (2780)	0.72 (550 mph/ 885 km/h)	Thermo-nuclear 1000 lb (454 kg)	SG
CSS-2 CSS-3 (IRBM)	China	—	—	Liquid propellant Stage 2 Stage 3	(12 000)	0.8	Nuclear 2 Mt 1–5 Mt	SG
CSS-1 CSS-2 (IRBM)	China	—	—	Liquid propellant Two stages	(1200) (2700)	0.7	Nuclear 20 kt 1–3 Mt	SG
Sandal (IRBM)	—	—	—	Two stages	(2000)		1 Mt	
SSBS S-3	France	—	—	Solid propellant Two stages	(3000)	0.65	Nuclear 1.2 Mt	—
SS-N-20 Sturgeon (IRBM)	USSR	—	—	Solid propellant	(3000)	0.65	Nuclear 1.2 Mt	—
UGM-73 Poseidon missile (IRBM)	Lockheed Missiles and Space Company, California, USA	$L = 10.36$ $d = 1.88$	65 000 (29 500)	Solid motor Thiokol as first stage; Hercules as second stage; guidance: inertial	3230 (5200)	10.0	Lockheed MIRV carrying 50 kt Rvs	Submarine launch
UGM-27 Polaris (ICBM) missile	Lockheed, California, USA	$L = 9.45$ $d = 1.37$	35 000 (15 850)	Aerojet solid motor with jetavator control as first stage; Hercules motor with liquid injection as second stage	2875 (4630)	10.0	MIRV 200 kt	Submarine launch

d = diameter (m); L = length (m); S = wingspan (m); w_L = launch weight; R = range; V = speed; SG = self-guided.

Table 2.26 Missiles for Armed Forces (ICBM and ALCM) (USA, USSR).

Types of missiles/rockets	Manufacturer/country of manufacture	Dimensions	w_L lb (kg)	Power plant and guidance	R miles (km)	V mach	Warhead	Mechanisms
AGM-86A air-launched cruise missile (ALCM)	Boeing, Aerospace, Seattle, USA	$L = 4.27$ $S = 289$ $d = 64$	1900 (862)	Williams Research F107-WR-100 2 No shaft turbofans 600lb (272 kg) thrust; McDonnell Douglas Tercom with inertial system	760 (1200)	0.6 to 0.8	Thermonuclear 200 kt	B-52
LGM-30 Minuteman (ICBM) missile, located in silos	Boeing Aerospace, Ogden, USA	$L = 18.2$ $d = 183$	Model II 70116 (31800) Model III 76015 (34475)	I-Stage: Triokol TU-120 (M55E), 2×10^5 lb (9×10^3 kg) thrust II-Stage: Hercules rocket 35×40^3 lb (16×10^3 kg) thrust III-Stage: Aerojet 35×10^3 lb (16×10^3 kg) thrust	II Stage: 7000 (11250) III Stage: 8000 (12875)	1500 mph	Thermonuclear 1–5 Mt II-Stage: AVCO MK 11C III-Stage: GE MK-12 MIRV	SG
RS-12/SS13 Savage (ICBM)	USSR	–	–	Solid fuel propellant; three stages	(9400)	0.8	Nuclear 750 kt	SG
RS-16/SS17 (ICBM)	USSR	–	–	Liquid cold launch; two stages	(11000) (10000)	0.8	3.6 Mt mode 2 or MIRV 4×200 kt	SG
RS-20/SS18 (ICBM)	USSR	–	–	Liquid propellant; two stages	(10000)	1.0	5–10 Mt mode 2	
SS-23 Scalpel (ICBM)	USSR	–	–	Solid cold launch propellant; three stages	(10000)	1.0	Nuclear MIRV 8–10 × 300–500 kt	Rail; mobile
SS-25 Sickle (IRBM)	USSR	–	–	Solid cold launch propellant; three stages	(10500)	1.0	Nuclear 550 kt	

L = length (m); S = wingspan (cm); d = diameter (cm); w_L = launch weight; R = range; V = speed; SG = self-guided.

Table 2.27 Missiles for Armed Forces (SRM, SCUD, RBS, HJ) (USSR, France, UK, USA, China, Sweden).

Types of missiles/rockets	Manufacturer/ country of manufacture	Dimensions	w_L lb (kg)	Power plant and guidance	R (km)	Warhead	Mechanisms
M Type (SRM)	China	L = 910	6200	Solid propulsion; inertial	600	Nuclear	SSM
Frog-7 (SRM)	France	L = 910 d = 85	2300	Solid propulsion	70	Nuclear	SSM
SS-1C Scud B	USSR	L = 1125 d = 85	6370	Storable liquid; inertial	280	Nuclear	
SS-23 Spider (SRM)	USSR	L = 605 d = 100	3500	Solid propulsion; inertial	525 miles	Nuclear	SSM
Exocet MM 10	France	L = 578 S = 100 d = 35	850	Solid propulsion; two stages: inertial and active radar homing	70	165 kg Conventional	SSM ship
RBS 15	Sweden	L = 435 d = 50	(600)	Turbojet and boosters	150	Conventional	SSM ship
Arrow 8 (HJ-8)	China	L = 99.8 S = 47 d = 10.2	(11.3)	Solid propulsion; wire-guided	0.1 to 3	Heat	SSM anti-tank
Spandrel	USSR	L = 100 d = 16	(12–18)	Solid propulsion; semi-automatic	4	Heat	SSM anti-tank
Vigilant	UK	L = 107 S = 28 d = 11	(14)	Solid propulsion; two stages: wire manual or auto	1.375)	Heat 6 kg	
Dragon 1 M 47	USA	L = 74 S = 33 d = 13	(13.8)	Multiple solid propulsion; wire manual	1.0	Heat	SSM anti-tank

L = length (cm); d = diameter (cm); S = wingspan (cm); w_L = launch weight; R = range.

Table 2.28 Missiles for Armed Forces (BGM, M, MK, RGM, etc.) (USA, France, Israel).

Types of missiles/rockets	Manufacturer/ country of manufacture	Dimensions	w_L lb (kg)	Power plant and guidance	R (km)	Warhead	Mechanisms
TOW 2 (BGM-71 D)	USA	$L = 140$ $d = 15$	(21.5)	Solid propulsion; wire manual or auto	3.75	Heat conventional	SSM anti-tank
Copperhead (M712)	USA	$L = 137$ $d = 15$	63.5	Cannon launched; laser homing	16	Heat 6.4 kg	SSM anti-tank
Gabriel	Israel	L S d					SSM shipborne
MK I		335 135 34	430	Two stage solid propulsion; auto pilot/command	18	Conventional 100 kg	
MK II		341 135 34	430		36	100 kg	
MK III		381 135 34	560		36+	150 kg	
MK IV		381 135 34	560	Turbojet propulsion; inertial and active radar homing	200		
SS-N-3 Shaddock (SRM)	USSR	$L = 1020$ $d = 86$	4700	Solid boosters; internal turbojet radio command	460 to 735	Nuclear and conventional	SSM shipborne
Harpoon (RGM-84A)	USA	$L = 384$ $S = 830$ $d = 34$	519	Solid booster turbojet cruise; inertial and radar	90	Conventional	SSM shipborne
Deadeye 5	USA	$L = 384$ $d = 12.7$	47.5	Solid propulsion; laser homing	24	–	Guided projectile, shipborne
Crotale	France	$L = 290$ $S = 54$ $d = 15$	80	Solid propulsion; radio command	18	15 kg Conventional	NSA
Sadral	France	$L = 180$ $d = 16$	17	Solid propulsion; infra-red homing	0.3–6	3 kg Heat conventional	NSA

L = length (cm); d = diameter (m); S = wingspan (cm); w_L = launch weight; R = range.

Table 2.29 Missiles for Armed Forces (mixed) (Israel, USSR, France, Italy).

Types of missiles/rockets	Manufacturer/ country of manufacture	Dimensions	w_L lb (kg)	Power plant and guidance	R (km)	Warhead	Mechanisms
Barak missile	Israel Aircraft Industries (IAI)	$L = 217$ $d = 17$	(86)	Semi-active radar-homing missile with disposable launch canister, manually fitted to an 8-round launcher based on the MBT TCM-30 twin 30 mm anti-aircraft gun mounted	10	Conventional 2 nuclear 7 kg	NSA small patrol boats
SA-N-3 Goblet	USSR	$L = 620$ $S = 150$ $d = 33.5$	550	Solid propulsion; semi-active homing	55	80 kg Conventional	NSA
SA-N-4	USSR	$L = 320$ $d = 21$	190	Solid propulsion; semi-active homing	14.8	50 kg	NSA
Roland 3	France and Italy	$L = 260$ $S = 50$ $d = 27$	85	Solid propulsion and command	8	9 kg Conventional	ADM
SA-4 Ganef (SRM)	USSR	$L = 880$ $S = 290$	100 000	Ramjet and solid boosters; radio command	70	135 kg	ADM
SA-10 Grumble MM 10	USSR	$L = 700$ $S = 10$ $d = 45$	1500	Solid propulsion	100	Nuclear	ADM
SA-12 Gladiator	USSR	$L = 750$ $S = 350$ $d = 50$	2000	Solid propulsion; semi-active radar	80	150 kg Conventional	ADM
SA-13 Gopher (HJ-8)	USSR	$L = 220$ $S = 40$ $d = 12$	55	Solid propulsion; infra-red homing	10	4 kg Conventional	ADM

L = length (cm); d = diameter (cm); S = wingspan (cm); w_L = launch weight; R = range.

Table 2.30 Missiles for Armed Forces (medium type) (UK, USA, China, France).

Types of missiles/rockets	Manufacturer/ country of manufacture	Dimensions	w_L (lb)	Power plant and guidance	R (km)	Warhead	Mechanisms
Blood hound	UK	$L=846$ $S=283$ $d=55$	—	Ramjet and solid boosters; semi-active radar homing	80	Conventional	ADM
Blow Pipe	UK	$L=139$ $S=270$	20.7	Solid propulsion; radio command	Any short distance 3	Conventional	ADM, shoulder-fixed
Javelin	UK	$L=140$ $d=8$	—	Solid propulsion; Saclos	4+	Conventional	ADM, shoulder-fixed
Hawk MIM-23B	USA	$L=503$ $S=119$ $d=36$	627.3	Solid propulsion; semi-active homing	40	Conventional	ADM
Stinger FIM-92A	USA	$L=152$ $S=14$ $d=7$	15.8	Solid propulsion; infra-red homing	Short range 3	Conventional	ADM
MIM Hercules	USA	$L=1210$ $d=80$	4858	Solid propulsion; command	140	Nuclear and conventional	ADM
L-5B	China	$L=289$ $S=66$ $d=13$	85	Solid propulsion; infra-red	5	Conventional	AAM
L-7	China	$L=275$ $S=66$ $d=16$	90	Solid propulsion; infra-red	5	Conventional	AAM
Mistrel	France	$L=180$ $d=9$	18	Solid propulsion; infra-red	3	3 kg Conventional	AAM
Super 530	France	$L=354$	250	Solid propulsion	25	30 kg	AAM

L = length (cm); S = wingspan (cm); d = diameter (cm); w_L = launch weight; R = range.

Table 2.31 Missiles for Armed Forces (missiles and rockets) (France, USA, USSR).

Types of missiles/rockets	Manufacturer/country of manufacture	Dimensions	w_L lb (kg)	Power plant and guidance	R	Warhead	Mechanisms
550 Magic	France	$L=275$ $S=66$ $d=16$	—	Solid propulsion; infra-red	5 km	Conventional	AAM
AIM-9 Sidewinder	Naval Weapon Centre, Philco-Ford, Ford Aero-Space, General Electric, USA	$L=283$ to 2.91 $S=56$ to 64 $d=12.7$	up to 190 (86)	Rocket dyne, Thiokol, Bermite or Naval propellant; Single Grain solid MK 17, 36, 86; guidance: 9E to 9L, high power servo-system, AM/FM IR conical scan head or Raytheon track via missile	2.5+ miles cruise 6.0 miles max	XM 248 high explosive (150lb)	AAM (F-16,F-20, F-5E aircraft)
Alamo AA-10	USSR	$L=400$ $S=70$ $d=19$	200	Solid propulsion; semi-active or infra-red	30 km	Conventional	AAM
AMBAAM	USA	$L=357$ $S=63$ $d=18$	150	Solid propulsion; command and inertial	12 km	Conventional	AAM
Falcon	USA	$L=213$ $S=639$ $d=29$	115	Solid propulsion; semi-active radar	8 km	Conventional	AAM
Phoenix	USA	$L=396$ $S=92$ $d=38$	454	Solid propulsion; semi-active radar	150 km	Conventional	AAM
Sparrow AIM-7M	USA	$L=366$ $S=102$ $d=20$	227	Solid propulsion; infra-red	140 km	Conventional	AAM

L = length(cm); S = wingspan (cm); d = diameter (cm); w_L = launch weight; R = range.

Table 2.32 Missiles for Armed Forces (mixture ASM and rockets) (China, USSR, USA).

Types of missiles/rockets	Manufacturer/ country of manufacture	Dimensions	w_L lb (kg)	Power plant and guidance	R	V mach (mph)	Warhead	Mechanisms
C-601 HY-4 C-801	China	L S d 738 280 92 736 280 76 480 165 55	2440 1740 1025	Liquid propulsion Turbofan propulsion Solid propulsion All active radar	100 km 150 km 150 km	— — —	400 kg to 500 kg Conventional	ASM ASM ASM
S-9 S-10	USSR	L S d 600 150 50 350 90 30	750 300	Solid propulsion; semi-active laser	150 km	—	Conventional	ASM
AS-11	USSR	L = 350 S = 90 d = 30	300	Solid propulsion; infra-red	300 km min	—	Conventional	ASM
Sea Eagle	UK	L = 414 S = 120 d = 40	600	Turbofan inertial; active radar	110 km	—	Conventional	ASM
AGM-65 Maverick	Hughes Aircraft, Tulson, USA	L = 246 S = 71 d = 44.5	462 (210)	Thiokol TX-481 solid motor; AGM-65C laser guidance	8 miles (13 km) to 14 miles (22.5 km)	6–10	Mk 19 113 kg	ASM
AGM-69A SRAM	Boeing Aerospace, Seattle, USA	L = 4127 S = 89 d = 44.5	2230 (1010)	LPC (Lockheed Company) propulsion two-pulse solid motor; guidance: inertial	High 105 miles (170 km); low 35 miles (56 km)	(200)	Conventional Nuclear 170 kT	ASM Rocket
ASLAM Missile MK III MK IV	McDonnell Douglas, USA	L = 4127 S = 89 d = 44.5	2700 (1200)	Internal rocket/ramjet; guidance: inertial with Tercom		6–10	170 kt Nuclear and conventional	ASM

L = length (cm); S = wingspan (cm); d = diameter (cm); w_L = launch weight; R = range.

Table 2.33 Missiles for Armed Forces (ADM and NSA types) (UK, USA).

Types of missiles/rockets	Manufacturer/ country of manufacture	Dimensions	w_L lb (kg)	Power plant and guidance	R miles (km)	V mach (mph)	Warhead	Mechanisms
MGM-31A Pershing missile	Martin, Orlando, USA	$L = 10.51$ $S = 202$ $d = 101$	10150 (4600)	2 No. Thiokul solid motors; first stage – M105, second stage – M106; guidance: inertial	100 (160) to 520 (840)	8	Nuclear 400 kt	Mobile tactical system, ADM/SSM
MGM-51C Shillelagh missile	Ford, New Post Beach, USA	$L = 1.14$ $S = 29$ $d = 15.2$	60 (27)	Amoco single-stage solid jetavators; guidance: optical tracking and infra-red command link	3.0	(800)	Octal-shaped charge 15 lb (6.8 kg)	Battle tank, ADM/SSM
BGM-71 TOW missile	Hughes Aircraft, Culver City, USA	$L = 1.17$ $S = 34$ $d = 14.7$	42 (19)	Quad boost motor for recoil-less launch; guidance: with optical sighting and trailing wire	up to (3.75)	(620)	Shaped charge containing 5.3 lb (2.4 kg) high explosive	Vehicles, aircraft, ADM/SSM
Sea Dart missile	British Aerospace, UK	$L = 4.4$ $d = 42$	(550)	Marconi radar 805 SW tracker with type 909 illuminator; guidance: semi-active	(30)	—	Conventional and nuclear	NSA patrol boats and ships

continued

Table 2.33 *Continued.*

Types of missiles/rockets	Manufacturer/ country of manufacture	Dimensions	w_L lb (kg)	Power plant and guidance	R miles (km)	V mach (mph)	Warhead	Mechanisms
Sea Wolf missile	British Aerospace, UK	$L = 1.9$ $d = 18$	(82)	GWS 25 – Marconi radar (805547) search, tracking radars, Ferranti fire-control computer; guidance: radio command	(30)	–	Conventional and nuclear	NSA/SSM boats and ships
Sea Cat missile	Short Brothers, Belfast, UK	$L = 1.47$ $d = 19$	(63)	Radio command guidance with optical remote TV and radar aided tracking	(30)	–	Conventional and nuclear	NSA/SSM boats and ships
Raytheon/Martin XMIM-104A Patriot missile	Martin Orlando Division, USA	$L = 5.31$ $S = 87$ $d = 41$	3740 (1696)	Thiokol TX-486 single-thrust solid motor	30 (48)	3 to 5	Nuclear and conventional	Advanced Mobile System SAM

L = length (m); S = wingspan (cm); d = diameter (cm); w_L = launch weight; R = range; V = speed.

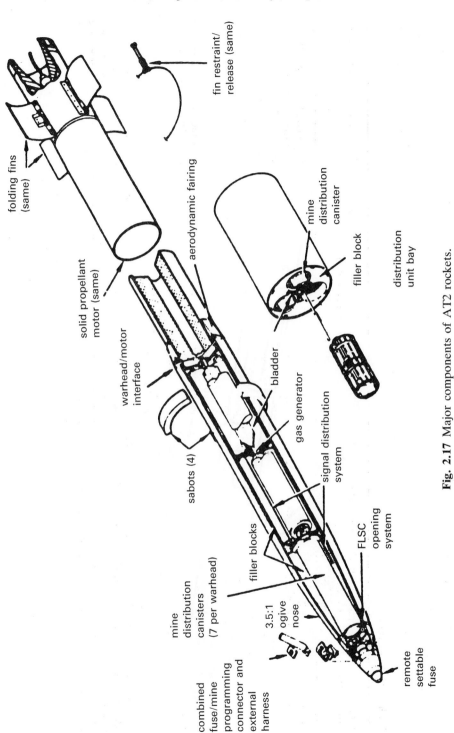

folding fins (same)

fin restraint/ release (same)

solid propellant motor (same)

aerodynamic fairing

mine distribution canister

filler block

distribution unit bay

warhead/motor interface

bladder

gas generator

sabots (4)

signal distribution system

FLSC opening system

mine distribution canisters (7 per warhead)

filler blocks

3.5:1 ogive nose

remote settable fuse

combined fuse/mine programming connector and external harness

Fig. 2.17 Major components of AT2 rockets.

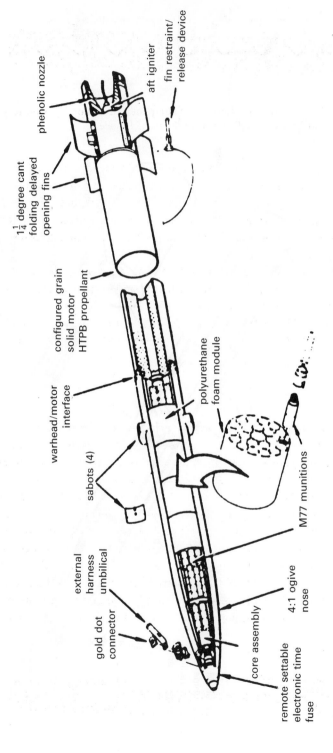

Fig. 2.18 Major components of M77 rockets.

Fig. 2.19 168 Matra Durandal, 430 lb (195 kg) penetration bomb. (Courtesy of United States Defense and Ministry of Defence Industry of France.)

seeker antenna, glimbal system and inertial platform are mounted. Behind these lie two units: the guidance system (TGS) and modular mid-course package (MMP). The latter, containing navigation electronics and computer, is located in the warhead section. Also in the warhead section, separate places are earmarked for the inertial sensor assembly, signal data converter, high explosive and safety devices.

Two external conduits on the propulsion section carry signals from the guidance electronics back to the aft-mounted control system. The Thiokol TX-486 rocket motor supports the operation of this missile.

Each Patriot XM-901 launcher carries a maximum of four rounds remotely operated from the ECS and is mounted on a wheeled semi-trailer and towed by a wheeled tractor.

2.3.1.2 *Stinger missile*

The Stinger missile system is designed to provide superior defence. This portable, shoulder-fired system can easily be deployed in any combat situation. With the introduction of the reprogrammable microprocessor (RMP), the Stinger has been able to meet the demand for sophisticated land, air or sea-based defence. Figure 2.21 shows the guidance and warhead layout of the Manpad Stinger. The system is based on a target-adopted guidance (TAG) technique which biases missile orientation toward vulnerable portions of the airframe with consequent maximum lethality. The superior lethality is derived from hit-for-kill accuracy, warhead lethality and kinetic energy. The force of impact of the Stinger on the target is equivalent to that of a medium sized vehicle travelling on a road at 60 miles per hour. Recently advanced Stinger configurations employ a rosette-pattern image scanning technique, as shown in Fig. 2.21. This capability allows

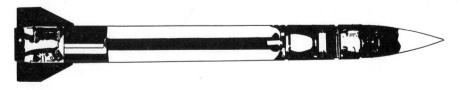

Fig. 2.20 Patriot: a SAM.

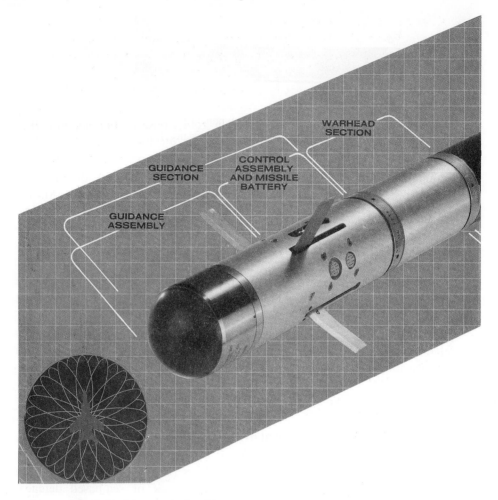

WARHEAD
SECTION

CONTROL
ASSEMBLY
AND MISSILE
BATTERY

GUIDANCE
SECTION

GUIDANCE
ASSEMBLY

Fig. 2.21 Stinger missile. (Courtesy of General Dynamics of USA.)

the missile effectively to discriminate between targets, flares and background clutter within detectable ranges, which prevents launches against false targets.

The missile has a flight motor, launch motor and an accurate propulsion system. The missile is issued as a certified round of ammunition requiring no field maintenance or associated logistical costs. The next generation of Stinger is Stinger-Post and Stinger-RMP.

2.3.1.3 Shorts' Javelin missile and Blow-Pipe missile

Javelin is a guided missile system for use as a self-defence weapon against low-flying attacking aircraft. It employs a semi-automatic command to line of sight (SACLOS) guidance system consisting of a stabilized tracking system and an auto-guidance system. A miniature television camera, aligned with the operator's stabilized sightline, detects the missile by means of its flares and computes automatically the necessary guidance demands. The system includes compensations for cross-winds, low-level targets and automatic generation of a 'lead angle' for launching the missile ahead of the crossing target. It is the improved version of the Blow-Pipe missile. As shown in Fig. 2.22, the missile structure consists of a nose section, a body tube assembly and a wing assembly.

The nose section consists of a rotary nose portion and the control surfaces actuator connected by a two-row ballrace. The rotary nose portion supports the four control surfaces and contains the fuse and cordite blast start roll position gyroscope. Four spike aerial elements on the control surface actuator form the aerial system for the missile receiver.

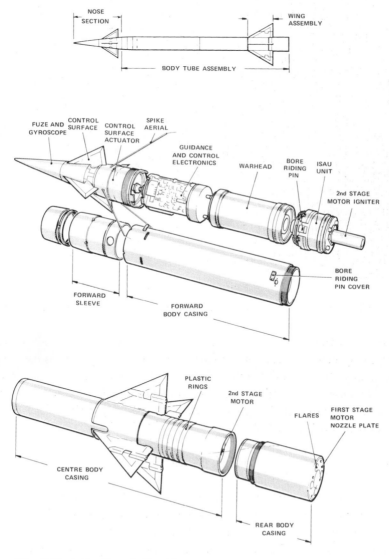

Fig. 2.22 A Javelin missile. (Courtesy of Shorts Brothers, Belfast.)

The body tube assembly consists of a forward sleeve and three tubular body casings. The sleeve joins the assembly to the nose section. The forward body casings house the guidance and control electronics, the warhead and the ignition, safety and arming units (ISAU). The centre and the rear body casings house the two-stage rocket motor and flares.

The wing assembly consists of four swept-back wings arranged in cruciform configuration on a central tube.

The control of the missile is by a twist-and-steer system using the control surfaces on the nose section. The role control surface (ailerons) rotate the nose independently of the missile body. Figure 2.22 shows the control unit.

Figure 2.23 shows the Blow-Pipe missile.

The following technical characteristics apply for the Javelin missile.

Operational use
$L = 1.394$ m, $d = 19.7$ cm, $w_L = 34$ lb (15.4 kg)
Dimensions of the aiming unit: 408 mm $\times$ 342 mm $\times$ 203 mm,
$w_L = 19.7$ lb (8.9 kg)

Field use
$L = 1.454$ m, $d = 23.5$ cm, $w_L = 41.9$ lb (19 kg)
Dimensions of the aiming unit: 482.6 mm $\times$ 431.8 mm $\times$ 271.8 mm,
$w_L = 22.3$ lb (10.1 kg)

Altitude: 1500 ft (4500 m)
Range:　　<500 m minimum
　　　　　>500 m maximum

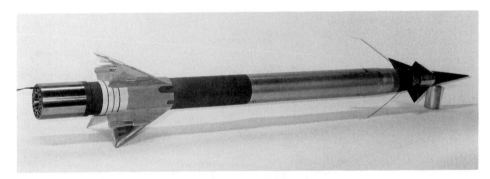

Fig. 2.23 A Blow-Pipe missile. (Courtesy of Shorts Brothers, Belfast.)

2.4 Data on civilian and military aircraft, tanks and marine vessels

2.4.1 Civilian aircraft

Civilian aircraft normally in service include Concorde, Airbus, Boeing, Antonov, BAC, Tri-Star, DC Series, Ilyushin and Tupolov.

2.4.1.1 British Aerospace/Aerospatiale Concorde

Figure 2.24(a) and 2.24(b) shows a photograph and a cut-away drawing of the aircraft Concorde 206 respectively. The aircraft has a variable geometry drooping nose linked with an A-frame and guide rails. The fuselage is situated close to the delta wing. Fuel is carried in the fin as well as in tanks which are located in the wings. This fuel can automatically be transferred between tanks at supersonic speed in order to maintain the centre of gravity of the aircraft. Table 2.34 lists basic data.

Table 2.34 Basic parameters of Concorde.

	Power plant
	$4 \times 38\,050\,lb$ (169 kN) Rolls-Royce/Sneema Olympus 593 Mk60 two-spool turbojet
S (m)	25.61
L (m)	62.10
H (m)	12.19
A_w (m^2)	358
P_L (kg)	11 340
V (km/h)	2150
w_a (kg)	186 800

S = span; L = length; H = height; A_w = wing area; P_L = payload; V = speed; w_a = weight at take-off or landing.

(a)

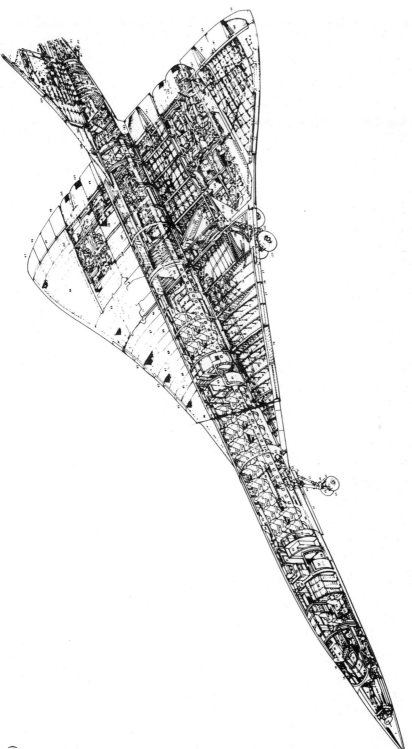

(b)

Fig. 2.24 (a) Concorde 206. (Courtesy of British Airways.) (b) A cut-away drawing of Concorde to illustrate the principal structural features.

2.4.1.2 Airbus Industrie

Figure 2.25 shows a typical configuration of the Airbus 300 B. They have standard General Electric (GE) CF-6 engines with engine pylons cantilevered to wing spars. The weather radar scanner is placed in an upward hinged radome in the nose area, supported by control columns and rubber pedals. The wing torsion box is carried through the structure near the centre of the body and the wing. The basic data of the Airbus family are summarized in Table 2.35.

2.4.1.3 Boeing

The Boeing aircraft family has a long history of research and development. Many versions have been produced; those currently used are the 727, 737, 747, 757 and 767 aircraft. Most of them have a stretched fuselage and generally use Pratt and Whitney JTD engines. Only the 757 type uses Rolls-Royce RB211 engines. The nose generally contains a radome and the main fuselage frame is set close to the wing. The rear fuselage is near the air distribution duct. Figure 2.26 represents a typical Boeing 747 aircraft. Table 2.36 gives a brief summary of the data on Boeing aircraft.

2.4.1.4 Antonov

Generally, Antonov aircraft are developed to meet requirements in the USSR for a small utility transport offering a higher performance level with comparatively short take-off and landing characteristics. Table 2.37 records the data for aircraft in service.

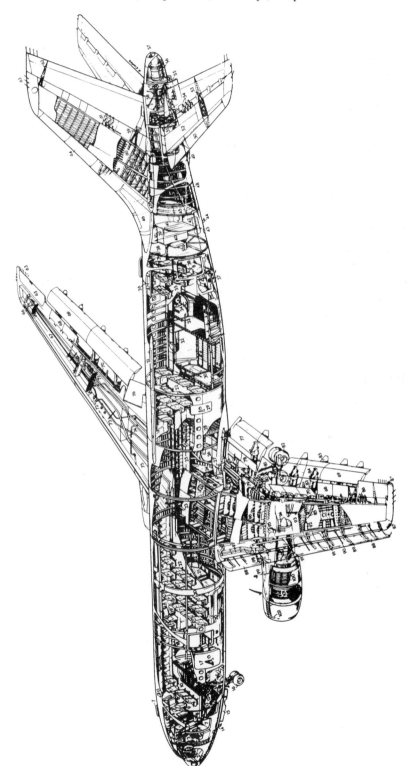

Fig. 2.25 A cut-away drawing of the structural features of a 300 B2 version of the Airbus. (Courtesy of Airbus Industrie.)

Table 2.35 Data on the Airbus family.

Type	Power plant	S (m)	L (m)	H (m)	A_w (m²)	P_L (kg)	V (km/h)	w_a (kg)
A300B2–100	2 × 51 000 lb (227 kN) GE CF6–50C turbofans	44.84	53.75	16.53	260	14 900	869	34 585
A300B2–200	2 × 51 000 lb (227 kN) GE CF6–50C turbofans	44.84	53.57	16.53	260	34 585	869	142 900
A300B2–100	2 × 51 000 lb (227 kN) GE CF6–50C turbofans	44.84	53.57	16.53	260	35 925	869	158 400
A300B4–200	2 × 52 500 lb (233.5 kN) CF6–50C1 turbofans	44.84	53.57	16.53	260	35 600	869	165 900
A310–202	2 × 48 000 lb (218 kN) GE CF6–80A turbofans	43.9	46.66	15.80	219	32 400	780	132 000

S = span; L = length; H = height; A_w = wing area; P_L = payload; V = speed; w_a = weight at take-off or landing.

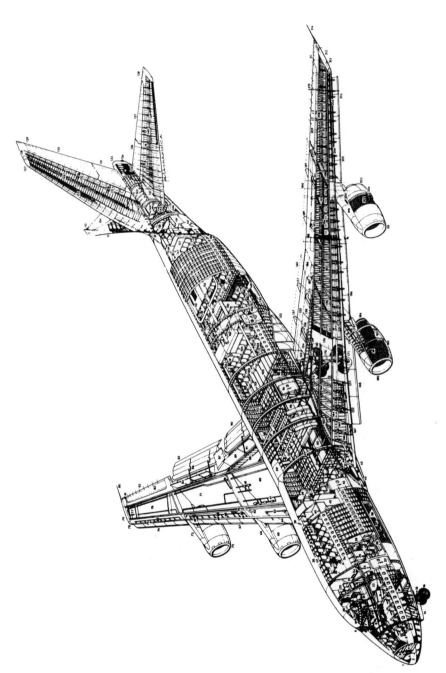

Fig. 2.26 A cut-away drawing of a Boeing 747–200B. (Courtesy of the Boeing Company, Washington, Seattle, USA.)

Table 2.36 Data on Boeing aircraft.

Type	Power plant	S (m)	L (m)	H (m)	A_w (m^2)	P_L (kg)	V (km/h)	w_a (kg)
727–200	3 × 16 000 lb (71.2 kN) Pratt and Whitney JT8D-17 turbofans	32.9	46.7	10.4	153.2	18 594	883	95 238
737–200	2 × 16 000 lb (71.2 kN) Pratt and Whitney JT8D-17 turbofans	28.3	30.5	11.4	91	15 422	775	53 297
767	2 × 44 300 lb (197 kN) Pratt and Whitney JT9D-7R 4A turbofans	47.24	48.46	15.38	200	40 224	800	128 030
757	2 × 37 400 lb (166.43 kN) Rolls-Royce RB211-535C	37.95	47.32	13.56	181.25	71 530	899	298 880
747–200B	4 × 50 000 lb (222 kN) Pratt and Whitney JT9D-7F (wet) turbofans	59.6	70.5	19.3	512	71 530	907	366 500
747–200B	4 × 53 000 lb (236 kN) Pratt and Whitney JT9D-7Q turbofans	59.6	70.5	19.3	512	69 900	907	373 300
747–200B	4 × 52 500 lb (234 kN) General Electric CF6-50E2 turbofans	59.6	70.5	19.3	512	69 080	907	373 300

S = span; L = length; H = height; A_w = wing area; P_L = payload; V = speed; w_a = weight at take-off or landing.

Table 2.37 Data for Antonov aircraft.

Type	Power plant	S (m)	L (m)	H (m)	A_w (m^2)	P_L (kg)	V (km/h)	w_a (kg)
An-12	4 × 4000 ehp Ivchenko A1-20K turboprops	38	37	9.83	119.5	10 000	550	54 000
An-22	4 × 15 000 ehp Kuznetsov NK-12MA turboprops	64.4	57.8	12.53	345	80 000	679	250 000
An-24	2 × 2500 ehp Ivenchenko A1-24 Seviiny 11 turboprops	29.2	23.53	8.32	74.98	13 300	450	21 000
An-26*	2 × 2800 ehp Ivenchenko A1–24T turboprops	29.2	23.8	8.575	74.98	5500	435	24 000
An-28	2 × 970 ehp Glushenkov TVD-10B turboprops (similar to An-14)	21.99	12.98	4.6	39.72	1550	350	6100
An-72	2 × 14 330 lb (6500 kg) Lotarev D-36 turbofans	25.83	26.58	8.24	74.98	7500	720	30 500

* An-30 and An-32 have similar status to An-26.

S = span; L = length; H = height; A_w = wing area; P_L = payload; V = speed; w_a = weight at take-off or landing.

2.4.1.5 British Aerospace (BAe)

Table 2.38 gives brief data on the BAe 146, 748 and 1−11 series of civil aircraft designed by British Aerospace.

2.4.1.6 Tri-Star

Tri-Star is a large-capacity, short-to-medium-haul transport of the Airbus type which uses turbofan engines. It generally has a three-engine layout powered by three Rolls-Royce R211 engines. Figure 2.27 shows a cut-away diagram of the Tri-Star L-1011−1 aircraft. Table 2.39 summarizes data for the Tri-Star family.

2.4.1.7 DC series

Several developed versions of the McDonnell Douglas DC series using turbojets are in service. The DC-10 Sr30 known as the 'Jumbo Twin' has now assumed its final version, and is illustrated by the cut-away diagram in Fig. 2.28. Table 2.40 summarizes data on the series.

2.4.1.8 Ilyushin

The Soviet-Union-built Ilyushin aircraft Il-18, Il-62 and Il-76T are well known in the aircraft industry. Ilyushin Il-18 is numerically the second most important airliner in the Aeroflot inventory. Ilyushin Il-76T is a specialized strategic freighter in the Soviet Union. It is marked for a secondary flight refuelling tanker role. Among the new generation of transport is the first wide-body 'airbus' type designed in the Soviet Union known as Ilyushin Il-86. The unique feature is the provision of entrance vestibules on the lower level with stowage spaces; it is equipped to permit automatic landing in bad weather conditions. Table 2.41 gives the basic data for these aircraft.

2.4.1.9 Tupolev TU

The Tupolev series consists of the TU-104, 124, 134, 144 and 154 aircraft. The TU-144 was originally designed with military-type decks with a nose raised and a vizor in place for high-speed cruising flight. The centre panel contains the engine speed/pressure ratio indicator and inlet/outlet nozzle indicators for the variable-geometry turbofan engines. The latest in the series is the TU-154. It has a three-engine T-tail layout and retains the wing pods for main undercarriage stowage. Table 2.42 briefly summarizes the relevant data for the aircraft.

Table 2.38 Data on British Aerospace aircraft.

Type	Power plant	S (m)	L (m)	H (m)	A_w (m²)	P_L (kg)	V (km/h)	w_a (kg)
146–100	4 × 6700 lb (29.8 kN) Avco Lycoming ALF502H turbofans	26.36	26.16	8.51	77.3	8300	675	33 000
146–200	2 × 6700 lb (29.8 kN) Avco Lycoming ALF502H turbofans	26.36	28.37	8.51	77.3	8845	675	39 870
748 Srs 2B	2 × 2280 eshp Rolls-Royce Dart R.Da.7 Mk 536–2	31.24	20.42	7.58	77.04	5670	448	21 090
One-Eleven 475	2 × 12 550 lb (55.8 kN) Rolls-Royce Spey 512–14DW turbofans	28.54	28.57	7.74	95.78	96 485	851	41 958
One-Eleven 500	2 × 12 550 lb (55.8 kN) Rolls-Royce Spey 512–14DW turbofans	28.54	32.61	7.47	95.78	12 000	847	45 360

S = span; L = length; H = height; A_w = wing area; P_L = payload; V = speed; w_a = weight at take-off or landing.

Table 2.39 Data on the Tri-Star family of aircraft.

Type	Power plant	S (m)	L (m)	H (m)	A_w (m²)	P_L (kg)	V (km/h)	w_a (kg)
L-1011–1 Tri-Star	3 × 42 000 lb (187 kN) Rolls-Royce RB 211–22B three-spool turbofans	47.3	54.9	16.8	321	38 810	915	195 920
L-1011–200 Tri-Star	3 × 48 000 lb (214 kN) Rolls-Royce RB 211–524 three-spool turbofans	47.34	54.15	16.86	321	33 657	913	212 281

S = span; L = length; H = height; A_w = wing area; P_L = payload; V = speed; w_a = weight at take-off or landing.

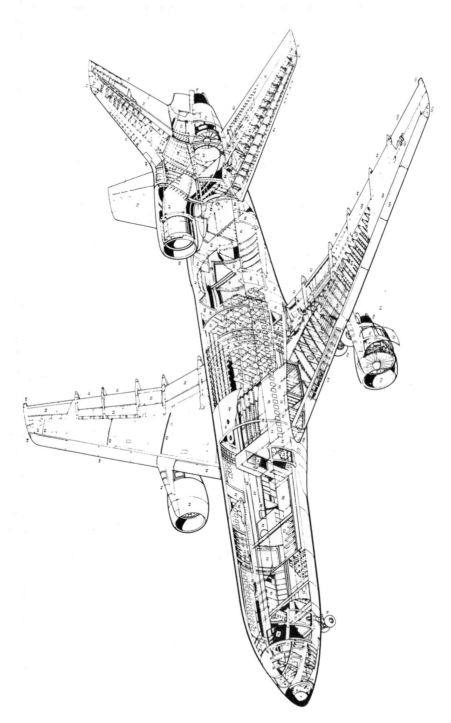

Fig. 2.27 A cut-away diagram of the Lockheed Tri-Star L-1011–1.

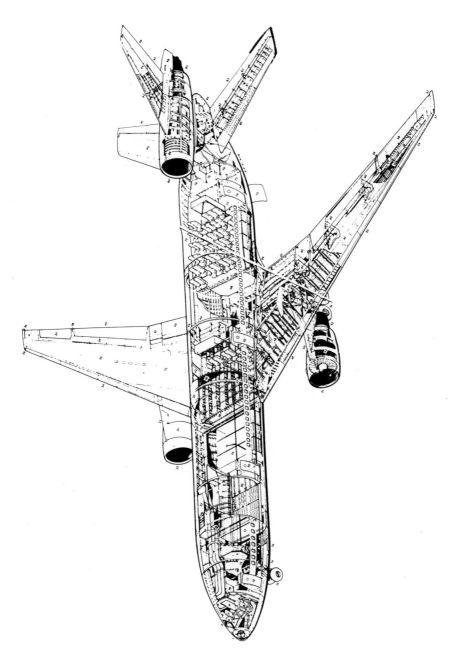

Fig. 2.28 A cut-away diagram of a DC-10 Sr30. (Courtesy of McDonnell Douglas, USA.)

Table 2.40 Data on the McDonnell Douglas DC series.

Type	Power plant	S (m)	L (m)	H (m)	A_w (m^2)	P_L (kg)	V (km/h)	w_a (kg)
DC-9-30	2 × 15 500 lb (69 kN) Pratt and Whitney T8D-15 turbofans	28.5	36.4	484	92.9	14 060	840	49 500
DC-9-40	2 × 15 500 lb (69 kN) Pratt and Whitney JT8D-15 turbofans	28.5	38.3	8.5	92.9	15 597	840	52 300
DC-9-50	2 × 16 000 lb (71.2 kN) Pratt and Whitney JT8D-17 turbofans	28.44	40.7	8.5	92.9	15 265	840	54 800
DC-9 Super 80	2 × 19 250 lb (86.64 kN) Pratt and Whitney JT8D-209 turbofans	32.9	45	9	118.8	17 842	840	63 950
DC-10-10	3 × 40 000 lb (178 kN) GE CF6-6D two-spool turbofans	47.34	55.3	17.7	329.8	46 237	925	207 747
DC-10-40	3 × 53 000 lb (236 kN) Pratt and Whitney JT9D-59A two-spool turbofans	47.34	55.3	17.7	329.8	46 237	925	207 747
DC-10-30	3 × 52 500 lb (234 kN) General Electric CF6-50C1 two-spool turbofans	50.42	55.32	17.7	338	48 327	925	250 818

S = span; L = length; H = height; A_w = wing area; P_L = payload; V = speed; w_a = weight at take-off or landing.

Table 2.41 Data on the Ilyushin aircraft.

Type	Power plant	S (m)	L (m)	H (m)	A_w (m²)	P_L (kg)	V (km/h)	w_a (kg)
Ilyushin Il-18	4 × Ivenchenko A1–20M turboprops 4250 ehp	37.4	35.9	10.17	140	14 000	625	64 000
Ilyushin Il-62	4 × 25 000 lb (113 kN) Solovier 20–30-KU turbofans	43.2	53.1	12.4	280	23 000	860	165 000
Ilyushin Il-76T	4 × Solovier D.30 KP turbofans, each with 26 455 lb St (12 000 kg)	50.5	46.59	14.76	300	40 000	850	157 000
Ilyushin Il-86	4 × Kuznetsov turbofans, each with 28 635 lb St (13 000 kg)	48.06	59.54	15.81	320	42 000	900	206 000

S = span; L = length; H = height; A_w = wing area; P_L = payload; V = speed; w_a = weight on take-off or landing.

Table 2.42 Data on the Tupolev series of aircraft.

Type	Power plant	S (m)	L (m)	H (m)	A_w (m²)	P_L (kg)	V (km/h)	w_a (kg)
TU-104	2 × 21 385 lb (97 kN) Mikulin AM 3M500 turbojet	34.54	25.85	11.9	174.4	900	800	76 000
TU-124	2 × 11 905 lb (54 kN) Soloviev D-20P turbofans	25.5	30.58	8.08	119	3500	800	26 300
TU-134	2 × 15 000 lb (66.5 kN) Soloviev D-30 turbofans	29	34.9	9	127	77 000	849	45 200
TU-144	4 × 44 000 lb St (20 000 kg) with Kuznetsov NK-144 turbofans	28.8	65.7	12.85	438	14 000	2500	180 000
TU-154	3 × 21 000 lb (93.5 kN) Kuznetsov NK-8-2 turbofans	37.5	48	11.4	202	20 000	900	91 000

S = span; L = length; H = height; A_w = wing area; P_L = payload; V = speed; w_a = weight at take-off or landing.

2.4.2 Military aircraft

2.4.2.1 British Aerospace Tornado Interdictor Strike (IDS) and Air Defence Variant (ADV)

The multi-role combat aircraft Tornado is proof that multi-national collaboration in technology at the frontiers of science is not only possible but can be successful in meeting the requirements of the world's airforces. The aircraft been produced in various batches since August 1974. Figure 2.29 shows a Tornado aircraft GR Mk 1 from 9 Squadron RAF Honington carrying four 1000 lb (454.74 kg) bombs, two full tanks and two ECM pods. Figure 2.30 shows the cut-away diagram of a Tornado IDS giving intimate details of the airframe, weaponry and internal structural, mechanical and avionic systems.

The IDS Tornado is designed primarily to fly at trans-sonic speeds, hugging the ground at a very low level and striking the targets in all weathers. With wings at the optimum sweep angle, it is highly manoeuvrable and possesses long-range capability. The wings are fully swept under speed performance, at both high altitude and low level. Relevant data are given in Table 2.43.

Fig. 2.29 Royal Airforce Tornado GR Mk 1 with four MK13/15 bombs. (Courtesy of British Aerospace, Warton Division UK.)

Panavia Tornado GR Mk 1 Cutaway Drawing Key

1 Air data probe
2 Radome
3 Lightning conductor strip
4 Terrain following radar antenna
5 Ground mapping radar antenna
6 Radar equipment bay hinged position
7 Radome hinged position
8 IFF aerial
9 Radar antenna tracking mechanism
10 Radar equipment bay
11 UHF/TACAN aerial
12 Laser Ranger and Marked Target Seeker (Ferranti), starboard side
13 Cannon muzzle
14 Ventral Doppler aerial
15 Angle of attack transmitter
16 Canopy emergency release
17 Avionics equipment bay
18 Front pressure bulkhead
19 Windscreen rain dispersal air ducts
20 Windscreen (Lucas-Rotax)
21 Retractable, telescopic, in-flight refuelling probe
22 Probe retraction link
23 Windscreen open position, instrument access
24 Head-up display, HUD (Smiths)
25 Instrument panel
26 Radar "head-down" display
27 Instrument panel shroud
28 Control column
29 Rudder pedals
30 Battery
31 Cannon barrel
32 Nosewheel doors
33 Landing/taxiing lamp
34 Nose undercarriage leg strut (Dowty-Rotol)
35 Torque scissor links
36 Twin forward-retracting nosewheels (Dunlop)
37 Nosewheel steering unit
38 Nosewheel leg door
39 Electrical equipment bay
40 Ejection seat rocket pack
41 Engine throttle levers
42 Wing sweep control lever
43 Radar hand controller
44 Side console panel
45 Pilot's Martin-Baker Mk 10 ejection seat
46 Safety harness
47 Ejection seat headrest
48 Cockpit canopy cover (Kopperschmidt)
49 Canopy centre arch

50 Navigator's radar displays
51 Navigator's instrument panel and weapons control panels
52 Foot rests
53 Canopy external latch
54 Pitot head
55 Mauser 27-mm cannon
56 Ammunition feed chute
57 Cold air unit ram air intake
58 Ammunition tank
59 Liquid oxygen converter
60 Cabin cold air unit
61 Stores management system computer
62 Port engine air intake
63 Intake lip
64 Cockpit framing
65 Navigator's Martin-Baker Mk 10 ejection seat
66 Starboard engine air intake
67 Intake spill duct
68 Canopy jack
69 Canopy hinge point
70 Rear pressure bulkhead
71 Intake ramp actuator linkage
72 Navigation light
73 Two-dimensional variable area intake ramp doors
74 Intake suction relief doors
75 Wing glove Krüger flap
76 Intake bypass air spill ducts
77 Intake ramp hydraulic actuator
78 Forward fuselage fuel tank
79 Wing sweep control screw jack (Microtecnica)
80 Flap and slat control drive shafts
81 Wing sweep, flap and slat central control unit and motor (Microtecnica)
82 Wing pivot box integral fuel tank
83 Air system ducting
84 Anti-collision light

85 UHF aerials
86 Wing pivot box carry-through, electron beam welded titanium structure
87 Starboard wing pivot bearing
88 Flap and slat telescopic drive shafts
89 Starboard wing sweep control screw jack
90 Leading-edge sealing fairing
91 Wing root glove fairing
92 External fuel tank, capacity 330 Imp gal (1 500 l)
93 AIM-9L Sidewinder air-to-air self-defence missile
94 Canopy open position
95 Canopy jettison unit
96 Pilot's rear view mirrors
97 Starboard three-segment leading-edge slat, open
98 Slat screw jacks
99 Slat drive torque shaft
100 Wing pylon swivelling control rod
101 Inboard pylon pivot bearing
102 Starboard wing integral fuel tank
103 Wing fuel system access panels
104 Outboard pylon pivot bearing
105 Marconi "Sky-Shadow" ECM pod
106 Outboard wing swivelling pylon
107 Starboard navigation and strobe lights
108 Wing tip fairing

109 Double-slotted Fowler-type flaps, down position
110 Flap guide rails
111 Starboard spoilers, open
112 Flap screw jacks
113 External fuel tank tail fins
114 Wing swept position trailing edge housing
115 Dorsal spine fairing
116 Aft fuselage fuel tank
117 Fin root antenna fairing
118 HF aerial
119 Heat exchanger ram air intake
120 Starboard wing fully swept back position
121 Airbrake, open
122 Starboard all-moving tailplane (taileron)
123 Airbrake hydraulic jack
124 Primary heat exchanger
125 Heat exchanger exhaust duct
126 Engine bleed air ducting
127 Fin attachment joint
128 Port airbrake rib construction
129 Fin heat shield
130 Vortex generators
131 Fin integral fuel tank
132 Fuel system vent piping
133 Tailfin structure

Fig. 2.30 A cut-away diagram of a Panavia Tornado GR Mk 1.

34 ILS aerial
35 Fin leading edge
36 Forward passive ECM housing
37 Fuel jettison and vent valve
38 Fin tip antenna fairing
39 VHF aerial
40 Tail navigation light
41 Aft passive ECM housing
42 Obstruction light
43 Fuel jettison
44 Rudder
45 Rudder honeycomb construction
46 Rudder hydraulic actuator (Fairey Hydraulics)
47 Dorsal spine tail fairing
48 Thrust reverser bucket doors, open
49 Variable area afterburner nozzle
50 Nozzle control jacks (four)
51 Thrust reverser door actuator
52 Honeycomb trailing edge construction
53 Port all-moving tailplane (taileron)
54 Tailplane rib construction
55 Leading-edge nose ribs

156 Tailplane pivot bearing
157 Tailplane bearing sealing plates
158 Afterburner duct
159 Airbrake hydraulic jack
160 Turbo-Union R.B.199-34R Mk 101 afterburning turbofan engine
161 Tailplane hydraulic actuator
162 Hydraulic system filters
163 Hydraulic reservoir (Dowty)
164 Airbrake hinge point
165 Intake frame/production joint
166 Engine bay ventral access panels
167 Engine oil tank
168 Rear fuselage fuel tank
169 Wing root pneumatic seal

170 Engine driven accessory gearboxes, port and starboard (KHD), airframe mounted
171 Integrated drive generator (two)
172 Hydraulic pump (two)

173 Gearbox interconnecting shaft
174 Starboard side Auxiliary Power Unit, APU (KHD)
175 Telescopic fuel pipes
176 Port wing pivot bearing
177 Flexible wing sealing plates
178 Wing skin panelling
179 Rear spar
180 Port spoiler housings
181 Spoiler hydraulic actuators
182 Flap screw jacks
183 Flap rib construction

184 Port Fowler-type double-slotted flaps, down position
185 Port wing fully swept back position
186 Wing tip construction
187 Fuel vent
188 Port navigation and strobe lights
189 Leading-edge slat rib construction
190 Marconi "Sky-Shadow" ECM pod
191 Outboard swivelling pylon
192 Pylon pivot bearing
193 Front spar
194 Port wing integral fuel tank
195 Machined wing skin/stringer panel
196 Wing rib construction
197 Swivelling pylon control rod
198 Port leading-edge slat segments, open
199 Slat guide rails
200 External fuel tank
201 Inboard swivelling pylon
202 Inboard pylon pivot bearing
203 Missile launch rail
204 AIM-9L Sidewinder air-to-air self-defence missile
205 Port mainwheel (Dunlop), forward retracting
206 Main undercarriage leg strut (Dowty-Rotol)
207 Undercarriage leg pivot bearing
208 Hydraulic retraction jack
209 Leg swivelling control link

210 Telescopic flap and slat drive torque shafts
211 Leading-edge sealing fairing
212 Krüger flap hydraulic jack
213 Main undercarriage leg breaker strut

214 Mainwheel door
215 Landing lamp
216 Hunting JP 233 Airfield Attack Weapon (two, side-by-side)
217 Submunitions compartments (30 SG357 runway penetration bombs and 215 HB876 area denial weapons in each JP 233)
218 Port shoulder pylon
219 Fuselage shoulder pylon (two)
220 ML twin stores carriers
221 Hunting BL 755 cluster bombs (eight)
222 Mk 83 high speed retarded bomb
223 Mk 13/15 1,000-lb (454-kg) HE bomb

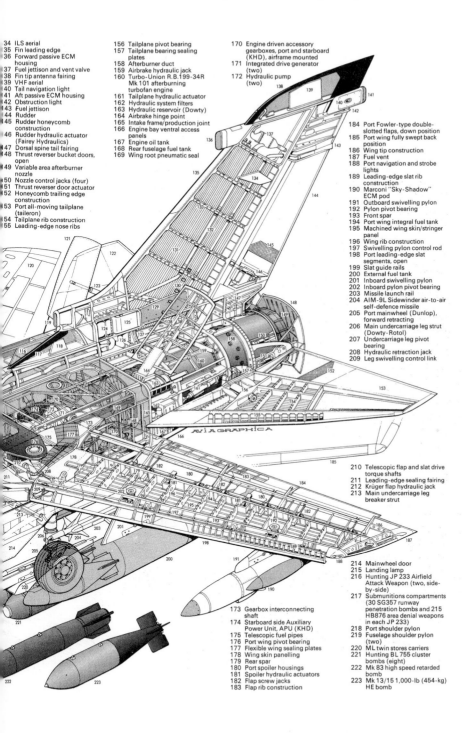

Table 2.43 Data on the Tornado IDS and ADV aircraft.

	Power plant	
	Interdictor Strike (IDS) Turbo-Union RB 199–34R (101 or 103) after burning turbofan MK 8090 lb (3670 kg) to 15 950 lb (7253 kg) after burning thrust	Air Defence Variant (ADV) As for IDS, with MK 104
S (m)	8.60 max swept 13.90 max unswept	8.60 at 67° sweep 13.90 at 25° sweep
L (m)	16.67	18.68
H (m)	5.95	5.95
A_w (m^2)	–	–
P_L (kg)	9000	9000
V (Mach)	Mach 2 at high level Mach 1 at low level	Mach 2.2
w_a (kg)	28 000	28 000
Armament	4 × MK 13/15 1000 lb (454.74 kg) bombs 2 AIM-9L missiles 8 MK 83 retarded bombs 2 CBLS-200 practice bomb containers 4 Kormoram ASM 8 × BL755 cluster bombs	

S = span; L = length; H = height; A_w = wing area; P_L = payload; V = speed; w_a = weight at take-off or landing.

2.4.2.2 Northrop F-5E and F-20 Tigershark

Both the F-5E and the F-20 are combat aircraft. The acceleration time of these aircraft varies from 900 miles/hour^2. Figures 2.31 and 2.32 illustrate the F-5E in full combat form and the layout respectively. Figures 2.33 and 2.34 show the F-20 in full combat form and the layout respectively.

Relevant data on the F-5E and F-20 aircraft are listed in Table 2.44.

Fig. 2.31 The Northrop F-5E in full combat form. (Courtesy of Northrop.)

Fig. 2.32 The layout of the Northrop F-5E. (Courtesy of Northrop.)

Fig. 2.33 The Northrop F-20 Tigershark in full combat form. (Courtesy of Northrop.)

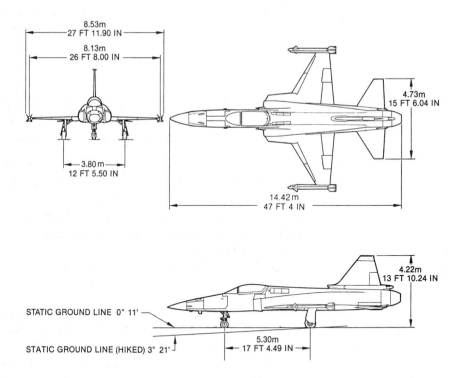

Fig. 2.34 The layout of the F-20 Tigershark. (Courtesy of Northrop.)

Table 2.44 Basic parameters for the F-5E and F-20 aircraft.

	Power plant	
	Engine 2GEJ 85−21 5000 lb (2268 kg) thrust each	Engine GEF404-GE100 1800 lb (8164 kg) thrust each
S (m)	7.98 with missiles 8.53 without missiles	8.5 with missiles
L (m)	14.45	14.42
H (m)	4.07	4.10 (4.73 with wheels)
A_w (m^2)	28.1	27.5
P_L (kg)	6350	7263
V (miles/h)	850	1300
w_a (kg)	11 213.8	12 700
Armament Air-to-air	2 No. 20 mm guns and AIM 9 Sidewinder missiles	
Air-to-ground	2 No. 20 mm guns and 9 bombs of 3020 kg	

S = span; L = length; H = height; A_w = wing area; P_L = payload; V = speed; w_a = weight at take-off or landing.

2.4.2.3 General Dynamics F-16

This is the most important combat aircraft and is known as the 'fighting falcon'. Various versions have been developed under the Multi-national Staged Improvement Programme (MSIP). This programme was accomplished in three stages:

Stage I   F-16A+/B+   Early wiring and structural provisions for the incorporation of future systems. Production deliveries began in November 1981 and ended in March 1985.

Stage II   F-16C/D   Core avionic cockpit and airframe provisions to accommodate emerging systems. Production deliveries began in July 1984.

Stage III   F-16A/B/C/D   Installation of advanced systems as these became available.

The F-16C/D versatility is increased with the common engine bay which permits the installation of either the F100-PW-220 or F110-GE-100 improved engine or future derivatives of these. The US Navy has the F-16N, a single-plate fighter which is a derivative of the F-16C multi-role fighter and is powered by the F110-GE-100 engine and has the growth potential to simulate the next generation of threat fighters. It is equipped with the AIM-9 series of Sidewinder missiles and ACMI/TACTS pods. A two-seater fighter/trainer aircraft is known as the TF-16N for the navy and the F-16D for the airforce. Figures 2.35 and 2.36 show the F-16C and F-16N versions of the aircraft with the AIM-9 series of Sidewinder missiles. Figure 2.37 shows a typical cut-away diagram of the F-16 aircraft.

Relevant data on the F-16 series of aircraft are listed in Table 2.45.

Fig. 2.35 F-16C aircraft in full combat form. (Courtesy of General Dynamics.)

Fig. 2.36 F-16N aircraft in full combat form. (Courtesy of General Dynamics.)

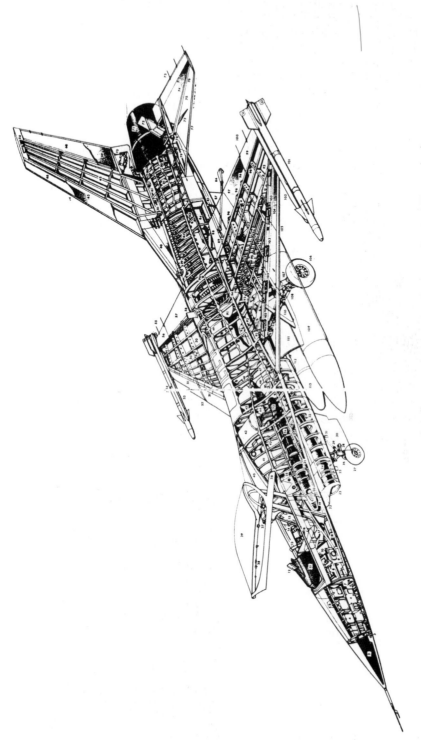

Fig. 2.37 A cut-away diagram for the F-16 aircraft. (Courtesy of General Dynamics.)

Table 2.45 Data on the F-16 series of aircraft.

	F-16A and F-16B	F-16C and F-16D	F-16N	TF-16N
Power plant	Pratt and Whitney turbofan two shaft 24 000 lb (10 885 kg) thrust F100-PW-100	F100-PW-200 F100-PW-220 F110-GE-100 25 000 lb (11 340 kg) thrust	F110-GE-100 25 000 lb (11 340 kg) thrust	F110-GE-100 25 000 lb (11 340 kg) thrust
S (m)	9.45 10.01 (with Sidewinder)	9.45 10.01 (with Sidewinder)	9.895 (without Sidewinder)	
L (m)	14.52	15.03	15.10	
H (m)	5.01	5.09	5.10	
A_w (m²)	27.87	27.87	27.87	
P_L (kg)	33 000 lb (14 969 kg)	37 500 lb (16 781 kg)	37 500 lb (16 781 kg)	
V (miles/h)	1300	1300	1300	
w_a (kg)	12 000 lb (5443 kg)	12 430 lb (5638 kg)	17 278 lb (7836 kg)	

S = span; L = length; H = height; A_w = wing area; P_L = payload; V = speed; w_a = weight at take-off or landing.

2.4.2.4 General Dynamics F-111

This is an early version of the fighter plane which is still in service and shall be phased out in the very near future. It is a two-seater, all-weather attack bomber and the well known versions are the F-111A, F-111F and EF-111A, made initially by Grumman Aerospace and then taken over by General Dynamics. Data on the aircraft are given in Table 2.46.

Table 2.46 Data on the General Dynamics F-111.

	Power plant
	Pratt and Whitney two shaft turbofans, thrust range 18 500 lb (8390 kg) to 20 350 lb (9230 kg)
S (m)	10.35±
L (m)	22.4
H (m)	5.22
A_w (m²)	35.75
P_L (kg)	20 943
V (km/h)	1450
w_a (kg)	41 400 to 54 000

S = span; L = length; H = height; A_w = wing area; P_L = payload; V = speed; w_a = weight at take-off or landing.

2.4.2.5 *British Aerospace Jaguar*

The Jaguar GR.1 and T.2, A and E were developed jointly by British Aerospace (BAe) and Dessault-Breguet of France. Figure 2.38 shows two Jaguars of the Indian Air Force in combat form. A cut-away diagram of the Jaguar showing the laser ranger on the marked target seeker behind a chisel nose is shown in Fig. 2.39. All versions have nose radar, a refuelling probe and the option of the overwing pylons for light dog fight missiles such as the air-to-air Matra 550 Magic. Data are listed in Table 2.47.

Fig. 2.38 Two Jaguar Internationals of the Indian Air Force comprising a single-seater strike aircraft and a two-seater trainer. (Courtesy of British Aerospace, Warton Division.)

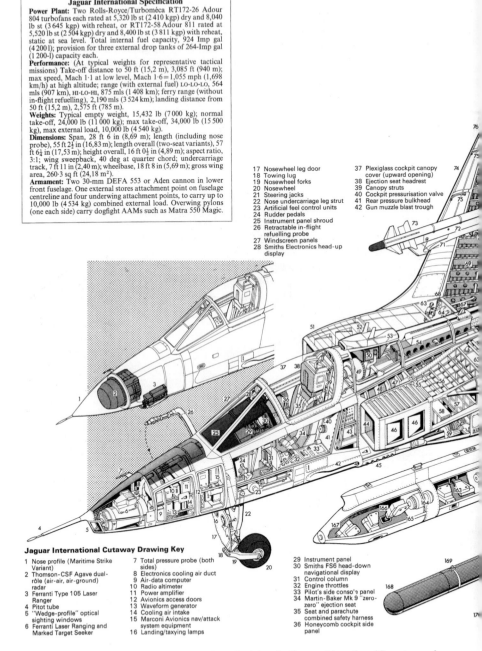

Jaguar International Specification

Power Plant: Two Rolls-Royce/Turboméca RT172-26 Adour 804 turbofans each rated at 5,320 lb st (2 410 kgp) dry and 8,040 lb st (3 645 kgp) with reheat, or RT172-58 Adour 811 rated at 5,520 lb st (2 504 kgp) dry and 8,400 lb st (3 811 kgp) with reheat, static at sea level. Total internal fuel capacity, 924 Imp gal (4 200 l); provision for three external drop tanks of 264-Imp gal (1 200-l) capacity each.

Performance: (At typical weights for representative tactical missions) Take-off distance to 50 ft (15,2 m), 3,085 ft (940 m); max speed, Mach 1·1 at low level, Mach 1·6 = 1,055 mph (1,698 km/h) at high altitude; range (with external fuel) LO-LO-LO, 564 mls (907 km), HI-LO-HI, 875 mls (1 408 km); ferry range (without in-flight refuelling), 2,190 mls (3 524 km); landing distance from 50 ft (15,2 m), 2,575 ft (785 m).

Weights: Typical empty weight, 15,432 lb (7 000 kg); normal take-off, 24,000 lb (11 000 kg); max take-off, 34,000 lb (15 500 kg), max external load, 10,000 lb (4 540 kg).

Dimensions: Span, 28 ft 6 in (8,69 m); length (including nose probe, 55 ft 2½ in (16,83 m); length overall (two-seat variants), 57 ft 6½ in (17,53 m); height overall, 16 ft 0½ in (4,89 m); aspect ratio, 3·1; wing sweepback, 40 deg at quarter chord; undercarriage track, 7 ft 11 in (2,40 m); wheelbase, 18 ft 8 in (5,69 m); gross wing area, 260·3 sq ft (24,18 m²).

Armament: Two 30-mm DEFA 553 or Aden cannon in lower front fuselage. One external stores attachment point on fuselage centreline and four underwing attachment points, to carry up to 10,000 lb (4 534 kg) combined external load. Overwing pylons (one each side) carry dogfight AAMs such as Matra 550 Magic.

17 Nosewheel leg door
18 Towing lug
19 Nosewheel forks
20 Nosewheel
21 Steering jacks
22 Nose undercarriage leg strut
23 Artificial feel control units
24 Rudder pedals
25 Instrument panel shroud
26 Retractable in-flight refuelling probe
27 Windscreen panels
28 Smiths Electronics head-up display

37 Plexiglass cockpit canopy cover (upward opening)
38 Ejection seat headrest
39 Canopy struts
40 Cockpit pressurisation valve
41 Rear pressure bulkhead
42 Gun muzzle blast trough

Jaguar International Cutaway Drawing Key

1 Nose profile (Maritime Strike Variant)
2 Thomson-CSF Agave dual-rôle (air-air, air-ground) radar
3 Ferranti Type 105 Laser Ranger
4 Pitot tube
5 "Wedge-profile" optical sighting windows
6 Ferranti Laser Ranging and Marked Target Seeker
7 Total pressure probe (both sides)
8 Electronics cooling air duct
9 Air-data computer
10 Radio altimeter
11 Power amplifier
12 Avionics access doors
13 Waveform generator
14 Cooling air intake
15 Marconi Avionics nav/attack system equipment
16 Landing/taxying lamps
29 Instrument panel
30 Smiths FS6 head-down navigational display
31 Control column
32 Engine throttles
33 Pilot's side conso's panel
34 Martin-Baker Mk 9 "zero-zero" ejection seat
35 Seat and parachute combined safety harness
36 Honeycomb cockpit side panel

Fig. 2.39 A cut-away diagram of a Jaguar equipped with missiles and bombs. (Courtesy of British Aerospace, Warton Division.)

43 Battery and electrical
   equipment bay
44 Port engine intake
45 Gun gas vents
46 Spring-loaded secondary air
   intake doors
47 Boundary layer bleed duct
48 Forward fuselage fuel tank
   (total system capacity 924
   Imp gal/1 200 l)
49 Air conditioning unit
50 Secondary heat exchanger
51 Starboard engine air intake
52 VHF homing aerials
53 Heat exchanger intake/
   exhaust duct
54 Cable and hydraulic pipe
   ducting
55 Intake/fuselage attachment
   joint

56 Duct frames
57 Integrally-stiffened machined
   fuselage frames
58 Ammunition tank
59 30-mm Aden cannon
60 Ground power supply socket
61 Mainwheel stowed position
62 Main undercarriage hydraulic
   lock strut
63 Leading-edge-slat drive
   motors and gearboxes
64 Fuel system piping
65 Wing panel centreline joint
66 Anti-collision light
67 IFF aerial
68 Wing/fuselage forward
   attachment joint
69 Starboard wing integral fuel
   tank
70 Fuel piping provision for
   pylon mounted tank
71 Overwing missile pylon
72 Missile launch rail
73 Matra 550 Magic air-to-air
   missile
74 Starboard leading-edge slat
75 Slat guide rails
76 Starboard navigation light
77 Tacan aerial
78 Flap guide rails and
   underwing fairings

79 Outboard double-slotted flap
80 Starboard spoilers
81 Inboard double-slotted flap
82 Flap honeycomb
   construction
83 Flap drive shaft and
   screwjacks

84 Spoiler control links
85 Wing/fuselage aft
   attachment joint
86 Heat exchanger air scoop
87 Control runs
88 Air conditioning supply
   ducting
89 Fuselage fuel tank access
   panels

90 Honeycomb intake duct
   construction
91 Engine intake frame
92 Hydraulic accumulator
93 Flap hydraulic motor and
   drive shaft

94 No 2 system hydraulic
   reservoir
95 Primary heat exchanger
96 No 1 system hydraulic
   reservoir
97 Heat exchanger exhaust
   ducts
98 Rear fuselage integral fuel
   tank
99 Inward/outward fuel vent
   valve
100 Dorsal spine fairing
101 Fin spar attachment joint
102 Tailfin construction
103 Starboard tailplane
104 Fin tip ECM fairing
105 VHF/UHF antenna fairing
106 Recognition light
107 Tail navigation light
108 VOR aerial
109 Rudder honeycomb
    construction

110 Fuel jettison pipe
111 Tailcone
112 Brake parachute housing
113 Rudder hydraulic jack
114 Tailplane trailing edge
    discontinuity
115 Honeycomb panel
    construction
116 Tailplane rib construction
117 Tailplane spar pivot joint
118 Differential all-moving
    tailplane hydraulic jack
119 Tailplane mounting frames
120 Fire extinguisher bottle
121 Arrester hook (extended)
122 Variable-area shrouded
    exhaust nozzle
123 Afterburner duct
124 Port ventral fin
125 Firewall
126 Engine rear suspension joint
127 Rolls-Royce/Turboméca
    Adour 804 (-26) turbofan
128 Port inboard double-slotted
    flap
129 Engine accessories
130 Hydraulic system ground
    servicing connectors
131 Airbrake hydraulic jack
132 Port airbrake (extended)
133 Wing fence (in place of
    missile pylon)
134 Spoiler hydraulic jack
135 Fixed portion of trailing edge
136 Port spoilers

137 Port outer double-slotted
    flap
138 Flap honeycomb
    construction
139 Wing tip fairing
140 Port navigation light
141 Matra Type 155 rocket
    launcher (18 SNEB rockets)
142 Outboard stores pylon
143 Port leading edge slat
144 Slat screw jacks
145 Port wing integral fuel tank
146 Machined wing skin/stringer
    panel
147 Pylon fixing
148 Inboard stores pylon
149 Twin mainwheels
150 Pivoted axle beam
151 Shock absorber strut
152 Main undercarriage leg strut
153 Undercarriage pivot
    mounting
154 Fuselage sidewall
    construction
155 Main undercarriage leg door
156 Mainwheel doors
157 Fuselage centreline pylon
158 Reconnaissance pod
159 Infra-red linescan
160 Data converter
161 Air conditioning pack
162 Rear rotating camera drum
    (rôle interchangeable)
163 Twin Vinten F95 Mk 10 high
    oblique cameras
164 Drum rotating electric motor
    and gearbox
165 Forward rotating camera
    drum
166 Twin Vinten F95 Mk 10 low
    oblique cameras
167 Forward looking Vinten F95
    Mk 7 reconnaissance camera
168 Matra Durandal, 430-lb
    (195-kg) penetration bomb
169 Pylon attachment shackles
170 264 Imp gal (1 200 l)
    auxiliary fuel tank

Table 2.47 Data on the British Aerospace Jaguar.

	Power plant
	2 No. Rolls-Royce Turboméca Adour two shaft turbofans 7305 lb (3313 kg) to 8000 lb (3630 kg) thrust
S (m)	8.69
L (m)	15.4 to 16.42
H (m)	4.92
A_w (m²)	—
P_L (kg)	6800
V (km/h)	1450
w_a (kg)	1550

Armament and other data
      2 No. 30 mm DFA 553 each with 150 rounds
      5 No. pylons with total external loads of 4536 kg with guns
      2 No. 30 mm Aden for its T-2 model
      Matra 550 Magic air-to-air missiles

Jaguar A and B and EMK 102 Adour engines ⎫
Jaguar S MK 104s ⎫
             MK 108 ⎬ Adour engines   ⎬ Using digital quadruplex fly-by-wire
Jaguar Act   control system
Jaguar FBW ⎭

S = span; L = length; H = height; A_w = wing area; P_L = payload; V = speed; w_a = weight at take-off or landing.

2.4.2.6 *Avions Marcel Dassault aircraft*

The major aircraft designed by Marcel Dassault are the Estendard, the Breguet F1 and the Mirage 3, 5, 2000 and 4000. The Mirage 2000 is designed to take full advantage of the Mirage F/3/5 and the F1 and is a multi-role, medium-size, single-seat combat aircraft with a variable camber delta wing, as shown in Fig. 2.40. The Mirage 2000B is a two-seater, designed for the same operational capabilities as the single-seater Mirage 2000. The Mirage 4000 is a twin-engine aircraft and is a logical follower of the Mirage 2000, benefitting from various research work on advanced aerodynamics.

With regard to size, the Super Mirage 4000 ranks between the Grumman F14 Tomcat and the McDonnell Douglas F18 Hornet. Its twin SNECMA 53 engines, in the 10 tonne thrust class (22 000 lb), provide a thrust/weight ratio greater than 1.0, ensuring performance figures above those of all currently known aircraft in the same class.

In contrast with the F14, F15, Tornado or MIG 23 of the same category, with either a variable-geometry wing or a separate tailplane, the Super Mirage 4000 profits from reduced drag, combined with the simple and compact design of the pure delta wing configuration.

Relevant data on the Dassault aircraft are given in Table 2.48.

Fig. 2.40 The Dassault Mirage 2000. (Courtesy of AMD, France.)

Table 2.48 Data on the Dassault aircraft.

Type and power plant	S (m)	L (m)	H (m)	A_w (m²)	P_L (kg)	V (km/h)	W_a (kg)
Dassault Breguet F1 Single-seater multi-mission fighter, 7200 kg thrust, SNECMA Atar, 9K-50 single shaft turbojet	8.4	15	4.5	—	7400	1472	14900
Estendard IVM and IVP Single-seater strike fighter, 4400 kg thrust, SNECMA Atar, 8B single shaft turbojet	9.6	14.4	4.26	—	5800	1083	10000
Super Estendard Single-seater strike fighter, 5110 kg thrust, SNECMA Atar, 8K-50 single shaft turbojet	9.6	14.31	4.26	—	6300	1200	11500
Mirage 3 and 5 Single-seater or two-seater interceptor, trainer and reconnaissance aircraft, 6000 kg thrust, SNECMA Atar, 9B single shaft turbojet	8.22	15.5	4.25	—	6156	1390	12000
Mirage 2000 Mirage 315 and F-1 improved version of these aircraft with engines SNECMA turbofans	9	15	4.5	—	7800	2200	9000
Mirage 4000 SNECMA M53, single shaft bypass turbofan 8 stage axial compressor 2 × 14500 lb (2 × 6579 kg) thrust	12	18.7	4.5	—	13000	2300	16100

Armament　Mirage 4000
　　Internal cannons 2 × 30 mm
　　4 long-range missiles
　　4 air-to-ground missiles
　　2 air-to-surface missiles

Bombs:　anti-runway Durandal up to 27
　clean or retarded (250 kg) up to 27
　laser guided (250 kg) up to 27
Rockets 68 mm

S = span; L = length; H = height; A_w = wing area; P_L = payload; V = speed; w_a = weight at take-off or landing.

2.4.2.7 McDonnell Douglas F-15 Eagle

For the past 15 years the Eagle has been the world's foremost air superiority fighter, despite continuing advances in the capabilities of threat aircraft. To meet future demands, McDonnell Douglas has supplied the US Air Force with F-15Cs and two-seater F-15Ds, as part of a multi-stage improvement programme. The first F-15 to be developed under the programme was delivered in June 1985.

The programme equips the F-15 with an improved central computer, armament control system and radar. The programme also includes modifications to the aircraft to enable it to use the advanced medium-range air-to-air missile (AMRAAM), an anti-satellite system (ASAT) and the joint tactical information distribution system (JTIDS). The changes are included in all models of the F-15 delivered to the US Air Force and will be fitted to some F-15s already in service with the Air Force.

On a deep interdiction mission, the F-15 carries a weapons load like the one shown in Fig. 2.41. In addition, six Mk 82 low-drag, general purpose bombs are attached with stub pylons to conformal fuel tanks. Above them is a GBU-10 laser-guided bomb and two AIM-120A AMRAAMs. Additional fuel is carried in the large fuel tank beneath the centre of the F-15E. Directly above it is the targeting pod for the low-altitude navigation and targeting infra-red for night (LANTIRN) system. That pod together with a navigation pod enables the F-15E to make precise attacks at night or in poor weather. Figure 2.42 shows a cut-away diagram of the aircraft.

The first F-15 designed for long-range interdiction missions, as well as air superiority, has been flight tested by McDonnell Douglas and the US Air Force. Designated the F-15E, it will supplement the Air Force's ageing fleet of F-111 fighter-bombers. The Air Force plans to buy 392 F-15s. The F-15E can carry heavy payloads farther than previous F-15s and can deliver weapons accurately in adverse weather, day or night.

Relevant data are listed in Table 2.49.

Fig. 2.41 McDonnell Douglas F-15 in a vertically accelerated position, with missiles. (Courtesy of McDonnell Douglas.)

Table 2.49 Data on the F-15.

	Power plant
	2 No. Pratt and Whitney F-100-PW-220 each with 24 000 lb thrust
S (m)	13.05
L (m)	19.45
H (m)	5.64
A_w (m²)	—
P_L (kg)	7000
V (km/h)	2500
w_a (kg)	20 000

Armament	4 AIM-9L/M infra-red-guided Sidewinder missiles; 4 AIM-7F/M radar-guided Sparrow missiles; 8 advanced medium-range air-to-air missiles (AMRAAMs); M-61 20 mm Gatling gun with 940 rounds of ammunition. Accommodates a full range of air-to-ground ordnance.

S = span; L = length; H = height; A_w = wing area; P_L = payload; V = speed; w_a = weight on take-off or landing.

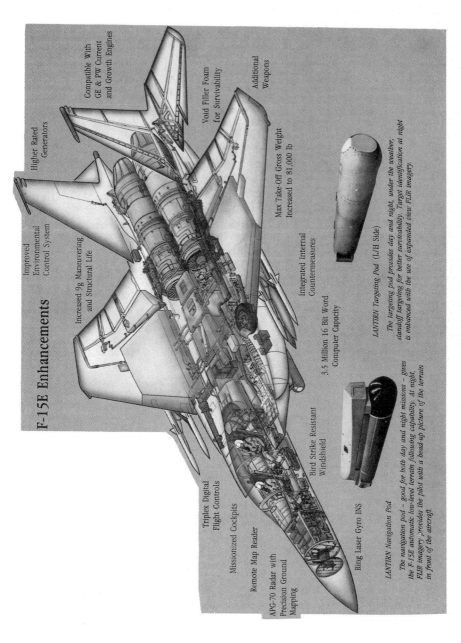

Fig. 2.42 A cut-away diagram of the F-15E (Courtesy of McDonnell Douglas.)

2.4.2.8 McDonnell Douglas F/A-18 Hornet

The F/A-18 Hornet is a multi-role, high-performance, tactical aircraft (Fig. 2.43), which can perform fighter strike or intercept missions. The twin-engine, multi-mission aircraft is capable of operating from both aircraft carriers and short bases. It is designed to replace the F-4 fighter and A-7 attack aircraft. It is a superior fighter with attack capabilities which include close-in and beyond-visual-range, all-weather and day/night strike.

Data on the F/A-18 are listed in Table 2.50.

Table 2.50 Data on the F/A-18 Hornet.

	Power plant
	2 No. F404-GF-400 low bypass turbofan engines each in 1600 lb (70.53 kN) thrust and with a thrust/weight ratio of 8:1
S (m)	11.43
L (m)	17.06
H (m)	4.7
A_w (m^2)	37.2
P_L (kg)	—
V (km/h)	2700
w_a (kg)	24 402

Armament Up to 7711 kg maximum on nine stations: two wing-tips for Sidewinder heat-seeking missiles; two outboard wings for air-to-ground ordnance; two inboard wings for Sparrow radar-guided missiles, air-to-ground, or fuel tanks; two nacelle fuselage for Sparrow missiles or sensor pods; one centreline for weapons, sensor pods or tank. Internal 20 mm cannon mounted in nose.

S = span; L = length; H = height; A_w = wing area; P_L = payload; V = speed; w_a = weight at take-off or landing.

Fig. 2.43 F/A-18 Hornet with Harpoon missiles. (Courtesy of McDonnell Douglas.)

2.4.2.9 Grumman F-14 Tomcat

The F-14A Tomcat represents the culmination of the US Navy's efforts for a total air superiority of fighters through the use of an advanced airframe with a variable sweep wing and a long-range weapon system. Some technical changes, including engine thrusts, have been introduced in the F-14B, C and D versions. Figure 2.44 shows the F-14A in full combat form armed with two AIM-54A Phoenix missiles, two AIM-7F Sparrow missiles and two AIM-9G Sidewinders while Fig. 2.45 shows a F-14D. Figure 2.46 illustrates the wing and gloves movement of the F-14. Data for the Tomcat are shown in Table 2.51.

Fig. 2.44 The F-14A in combat form. (Courtesy of Grumman Corporation, USA.)

Fig. 2.45 The F-14D in combat form. (Courtesy of Grumman Corporation, USA.)

Fig. 2.46 The F-14 with wing and glove movement. (Courtesy of Grumman Corporation, USA.)

Table 2.51 Data for the Grumman F-14 Tomcat.

Power plant		
F-14A $2 \times 20\,900$ lb (9480 kg) thrust Pratt and Whitney TF30−1412A		F-14B,C $2 \times 28\,090$ lb (12 741 kg) thrust Pratt and Whitney F401−400
Two shaft after-burning turbofans		
S (m)	11.630	(68° sweep)	safely landing
	19.54	(20° sweep)	
L (m)	18.89		
H (m)	4.88		
A_w (m²)	—		
P_L (kg)	17 010		
V	Mach 2.3 or 1564 mph maximum speed, 400−500 km/h cruise speed, 125 km/h approaching speed		
w_a (kg)	27 216		
Armament	AIM-54 Phoenix missiles AIM-7 Sparrow missiles AIM-9 Sidewinder missiles M61−A1 Vulcan 20 mm cannon		

S = span; L = length; H = height; A_w = wing area; P_L = payload; V = speed; w_a = weight at take-off or loading.

2.4.2.10 Soviet Union MIG aircraft

The MIG aircraft in operation with a number of world air forces are the MIG-17 (Hip-H), MIG 19–195, MIG-5F, MIG-19PM (NATO name 'Farmer'), MIG-23BN (Flogger H), MIG-23MF (Flogger G), MIG-23CKD (Flogger J), MIG-24 (Hind D), MIG-25R (Foxbat B) and the MIG-29CKD (Fulcrum). The MIG-23 is further accommodated with a 23 mm GSH-23 twin-barrel gun on the ventral central line plus various mixes of air-to-air missiles of 2 No. infra-red AA-7 'Apexes' and infra-red or radar-homing AA-8 'Aphids'. Table 2.52 is a comparative study of various MIG aircraft.

It is interesting to note that the MIG-25 has four underwing pylons, each carrying one AA-6 air-to-air (two radar and two infra-red) missiles with VO guns employed. On the other hand, the MIG-27 has a 23 mm barrel Gatling-type gun in a belly. Seven external pylons are provided for a wide range of ordnance including guided missiles AS-7 'Kerry' and tactical nuclear weapons. All ECM are internal. The pylons on the outer wing are piped but not fixed drop tanks. Since they do not pivot, they can only be loaded when the wings remain unswept.

2.4.2.11 Other important fighter/bomber aircraft

Data are included in this section for the well-known combat aircraft Rockwell B-1 bomber, Russian Sukhoi (SU), Chinese F-6 and F-7, Saab 37 Viggen and BAe Harrier. These well proved aircraft are operational with many air forces in the world. Their layouts and performances are similar to the combat aircraft described earlier. Slight variations exist in the engine capacities, dimensions, payloads and maximum and minimum speeds. Comparative data are listed in Table 2.53.

2.4.2.12 Helicopters

Helicopters are more vulnerable than aircraft in warfare. In peace time a helicopter may crash after losing a rotor or hitting objects such as offshore platforms, buildings, helipads or their surrounding structures. Figure 2.47 shows three types of helicopters manufactured by the Soviet Union and India. Table 2.54 gives useful data for other types of helicopters.

Table 2.52 Comparative data for MIG aircraft.

Power Plant	MIG-19 (Mikoyan)	MIG-21	MIG-23 (Flogger)	MIG-25 (Foxbat)	MIG-27	MIG-29
(Engines)	*Single-seater* 2 × 6700 lb (3040 kg) to 2 × 7165 lb (3250 kg) Kimov RD-9B turbojets	*Single-seater* Range turbojet 11 240 lb (5100 kg) to 14 500 lb (6600 kg) Tumansky single shaft	*Single-seater* 17 640 lb (8000 kg) to 25 350 lb (11 500 kg) thrust, 1 Tumansky turbofan	*Single-seater* 27 000 lb (12 250 kg) thrust, 2 Tumansky R-266 after-burning turbojets	*Single-seater* 17 640 lb (8000 kg) to 25 350 lb (11 500 kg) thrust, 1-Tumansky after-burning turbofan	—
S (m)	9.00	7.15	8.7 (72° sweep) 14.4 (16° sweep)	14.0 Foxbat A	8.7 (72° sweep) 14.4 (16° sweep)	
L (m)	13.08 (S-5F)	14.35	16.15	22.3 (Foxbat A) 22.7 (Foxbat R) 23.16 (Foxbat U)	16.15	
H (m)	4.02	4.50	3.96	5.60	4.60	
A_w (m²)	—	—	—	—	—	
P_L (kg)	3760	4600	7050	14 970	9900	
V (max)	Mach 1.3 or 1480 km/h (920 mph)	Mach 2.1 or 2070 km/h (1285 mph)	Mach 1.1 or 1350 km/h (840 mph)	Mach 3.2 or 3380 km/h (2100 mph)		
w_a (kg)	9500	9800	15 000	34 930	17 750	

S = span; L = length; H = height; A_w = wing area; P_L = payload; V = speed; w_a = weight at take-off or landing.

Table 2.53 Comparative data of some important combat aircraft.

Type	Power plant	S (m)	L (m)	H (m)	P_L (kg)	V mach (mph)	w_a (kg)
BAe Harrier GR-3 model T-Mark 4 model (AV-8A, TAV-8A) FRS-1 model Single-seater Two-seater	1 × 21 500 lb (9752 kg) thrust Rolls-Royce Pegasus 103 two shaft turbofan (US designation F-402)	7.7	13.87 17.0 (GR-3)	3.43	6260	1.2 (860)	11793
F-6 Shenyang (or NATO's name FANTAN 'A') Single-seater	2 × Axial turbojets MD manufactured	10.2	15.25	3.35	4500	1.0 (760)	10700
Rockwell B-1 Four-seater	4 × 30 000 lb (13 610 kg) General Electric F101-100 two shaft augmented turbofans	41.4	45.6	10.24	115 670	0.85 (646)	179 170
Saab 37 AJ, JA Viggen SF, SH, SK versions Single-seater	1 Svenska flying motor RMB; Pratt and Whitney two shaft 25970 lb (11790 kg) to 28086 lb (12750 kg) thrust	10.6	AJ: 10.6 JA: 16.3	5.6	4500	2.0 (1320)	16000
SU-9 Fishpot B SU-11 Fishpot C Single-seater	1 Lyulka single shaft turbojet; 19840 lb (9000 kg) thrust (SU-9) and 22040 lb (10000 kg) thrust (SU-11)	8.43	SU-9 : 18.5 SU-11: 17.4	4.9	4540	0.95 (1195)	13610
SU-15 (Flagen A to E models) Single-seater	2 × Tumonsky R-25 turbofans; 16530 lb (7500 kg) thrust after burner	9.5 D Model	21.50	5.0	10100 (D) 5 100 (A)	2.3 (1520)	21000 (D) 16000 (A)
SU-17, SU-20 and SU-22 models Single-seater	(17) 1-Lyulka AL-21 F-3 thrust single shaft 17200 lb (7800 kg) (20–22) AL-7F 22046 lb (10000 kg)	14 (28° sweep) 10.6 (62° sweep)	18.75	4.75	9000	1.05 (798) to 2.17 (1432)	19000

S = span; L = length, H = height; P_L = payload; V = speed; w_a = weight at take-off or landing.

Fig. 2.47 Helicopters in service with the Indian forces. (Courtesy of Hindustan Aeronautics, Helicopter Division.)

Table 2.54 Data on helicopters.

Type	Power plant	S (m)*	L (m)†	H (m)*	A_w (m²)*	P_L (kg)	V (km/h)	w_a (kg)	R (m) d	R (m) t_b	R (m) h_t
Augusta A 109A MK II (Italy)		1.32	13.05	1.28	4.50	2600	285	1604	11.0		3.30
Augusta-Sikoysky AS-61R (HH-3F) (Italy/USA)	2 No. 1118 kW (1500 shp) GE: T58-GE-100 turboshafts	1.98	18.90	5.51	3.54	2270	213	10000	18.90	0.46	4.90
Bell 206B Jet Ranger III (USA)	1 No. 313 kW (4205 shp) Allison 250-C20J turboshaft	1.97	11.82	2.91	4.2	1451	216	742	10.16	0.33	2.54
Bell 206L-3 Long Ranger III (USA)	1 No. 485 kW (650 shp) Allison 250-C30P engine	1.97	13.02	3.14	2.35	1882	246	998	11.08	0.33	2.90
Bell 212 Twin Two Twelve (USA)	Pratt and Whitney PT6 turboshafts coupled to a combined gearbox with a single output shaft producing 1342 kW (1800 shp)	2.86	17.46	4.53	4.5	5080	185	2720	14.69	0.5	3.91
Bell 412 (USA)	Pratt and Whitney PT6-3B-1 turbo twin pac (2 turboshafts) with a total 1044 kW (1025 shp)	2.86	17.07	4.32	7.9	5397	230	2935	14.02	0.29	3.29
HMZ-9A (China and Aerospatiale)	SA 365N Dauphin 2 twin turboshafts					4100	285	2050			

Table 2.54 Continued.

Type	Power plant	S (m)*	L (m)†	H (m)*	A_w (m²)*	P_L (kg)	V (km/h)	w_a (kg)	R (m) d	R (m) t_b	R (m) h_t
Kawasaki KV10711A-4 (Japan)	2 No. 1044 kW (1400 shp) GE: CT58−140−1 or Ishikawajima-Harima CT58−1H1−140−1 turboshaft with max continuous rating 932 kW (1250 shp)	2.01	25.40	8.83	4.24	3172	241	9706	15.24	0.43	5.13
MBB-BO 105 S (Canada)	2 No. Allison 250-C28C turboshafts giving 820 kW (1100 shp)	1.94	10.40	3.10	3.0	2600	243	1380	12.80	0.35	3.25
MIL-Mi-24 (Hind) MIL-Mi-25 (Hind) (USSR)	2 No. Isotov TV3-117 turboshafts each with max rating 1640 kW (2200 shp)	3.20	21.50	6.50		7500	310	11000	17.00		
Naras (Brantly) Model 305 (India/USA)	1 No. 227.4 kW (305 hp) Textron Lycoming IVO-540-B1A flat-6 engine	1.39	10.03	2.44	3.20	1350	193	816	8.74	0.254	4.60
SA 332 Super Puma (Aerospatiale)	2 No. Turbomeca Makila 1A1 turboshafts each with 1400 kW (1877 shp)	1.55	6.05	4.92	7.80	9300	266	4475	15.60	0.60	4.60

* overall internal dimensions.
† external rotor dimension.

S = span; L = length; H = height; A_w = wing area; P_L = payload; V = speed; w_a = weight at take-off or landing; R = rotor details; d = rotor diameter; t_b = rotor blade card; h_t = overall height.

2.4.3 Main battle tanks (MBTs) as impactors

Main battle tanks are constantly being developed. They are always involved on the front line and are subject to impact and explosion. Table 2.55 lists data on some of the important MBTs currently in service. Figures 2.48 and 2.49 give the layouts of the 80-MBT, AMX-30 and C1-MBT with brief details of armament.

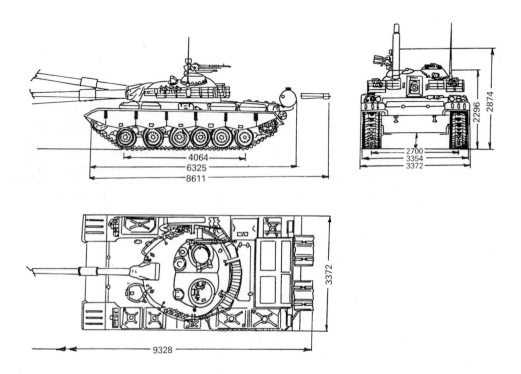

Fig. 2.48 Type 80-MBT main battle tank. (Courtesy of the China Government Arsenal, Beijing.)

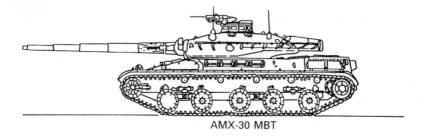

AMX-30 MBT

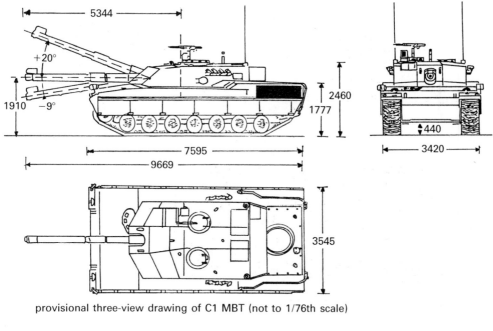

provisional three-view drawing of C1 MBT (not to 1/76th scale)

Chieftain Mk 3 MBT

Fig. 2.49 AMX-30 and C1 Chieftan Tanks. (Courtesy of the Ministries of Defence, UK and France.)

Table 2.55 Tanks as impactors.

Type of tank and power plant	S (m)	L (m) (hull)	H (m)	w_a (kg)	V (km/h)	Max range (km)	Manufacturer and country
AMX-40 Engine: Poyand V12 × 12-cylinder diesel developing 1300 hp	3.28 to 3.35	6.8	2.38	43 700	55	850	Giat, France
Engesa, EE-T1 Osorio Engine: MWMTBD 234, 4-stroke, 12-cylinder water cooled turbo-charged diesel developing 1040 hp	3.26 Track width (570 mm)	7.13	2.68	40 440	70	550	Engesa, Brazil
Leopard 2 Engine: TU MB873 Ka501, 4-stroke, 12-cylinder developing 1500 hp	3.70	7.72	2.48	55 150	72	550	Krupp, Germany
MB-30 Tamoio Engine: Saab-Scania DS1–14 diesel developing 500 hp	3.22	6.5	2.50	30 000	67	550	Bernardini, Brazil
C1 MBT Engine: Fiat V-12 MTCA turbo-charged 12-cylinder diesel developing 1200 hp	3.545	7.595	2.46	48 000	65	550	IVECO Fiat and OTTO, Italy
Challenger MBT Engine: Perkins Engines Condor 12-V 1200 12-cylinder diesel developing 1200 hp	3.518	9.80	2.5 to 2.9	62 000	56	—	Vickers, UK
Chieftan MBT Engine: Leyland L60 2-stroke 6-cylinder NO: 4MK6A or NO: 4MK8A developing 730 bhp and 750 bhp respectively	3.504	7.52	2.895	54 100 to 55 000	56	500	Vickers, UK

Table 2.55 *Continued.*

Type of tank and power plant	S (m)	L (m) (hull)	H (m)	w_a (kg)	V (km/h)	Max range (km)	Manufacturer and country
Type 80 MBT V-12 diesel developing 730 hp	3.372	6.325	2.874	38 000	60	430	North Industries, China
Type 88 MBT Engine: MTU 871 Ka-501 diesel developing 1200 hp	3.594	7.477	2.248	51 000	65	500	Hyundai, South Korea
90 MBT Engine: Mitsubishi 10-cylinder diesel developing 1500 hp	3.40	7.50	2.30	50 000	70	300	Mitsubishi, Japan
M48 MBT Engine: All Continental 12-cylinder							Variants, US Army, USA
M48 (AV-1790-5B/7/7B/7C) petrol	3.631	6.705	3.241	44 906	41.8	113	
M48A1 (AV-1790-7C) petrol	3.631	6.87	3.13	47 173	41.8	113	
M48A2 (AV-1790-8) petrol	3.631	6.87	3.089	47 173	48.2	258	
M48A3 (AVDS-1790-2A) diesel	3.631	6.882	3.124	47 173	48.2	463	
M48A5 (AVDS-1790-2D) diesel	3.631	6.419	3.086	48 987	48.2	499	
Note: all petrol 2800 hp all diesel 2400 hp							
M60 MBT, M60 A1 MBT Engine: Continental AVDS-1790-2A 12-cylinder developing 750 bhp	3.631	6.946	3.213	49 714	48.28	500	General Dynamics, USA
M60 A3 MBT Engine: Continental AVDS-1790-2C 12-cylinder developing 750 bhp	3.631	6.946	3.270	52 617	48.28	480	

continued

Table 2.55 *Continued.*

Type of tank and power plant	S (m)	L (m) (hull)	H (m)	w_a (kg)	V (km/h)	Max range (km)	Manufacturer and country
Merkava 2MBT Engine: Teledyne Continental AVDS-1790-6A V-12 diesel developing 900 hp	3.70	7.45	2.64	60 000	46	400	Sibit, Israel
MI Abrams MBT Engine: Lycoming Textron AGT 1500 gas turbine developing 1500 hp	3.657	7.918	2.886	57 154	67	500	General Dynamics/ Lima, USA
Stridsvagn 103 MBT Engine: Rolls-Royce K60 developing 240 hp or Boeing 553 gas turbine developing 490 shp	3.63	7.04	2.43	39 700	50	390	Bofors, Sweden
T80 MBT Engine: Gas turbine developing 985 hp	3.40	7.40	2.20	42 000	75	400	China and Soviet State Arsenal, USSR
Talbot MBT Engine: Continental V-12 AVDS developing 760 hp	3.39	6.36	3.35	47 000	56	600	Peugeot, Spain
Upgraded Centurian Tank Engine: Rolls-Royce MK1VB 12-cylinder developing 650 hp	3.39	7.556	2.94	50 728			
Vickers/FMC VFM 5MBT Engine: Detroit diesel model 6V-92 TA developing 552 hp	2.69	6.20	2.62	19 750	70	483	Vickers, UK
Vijayanta MBT Engine: Leyland L60 developing 535 hp	3.168	9.788	2.711	40 000	48.3	550	Department of Defence, India

S = width; L = length; H = height; w_a = combat weight; V = speed.

2.4.4 Marine vessels

2.4.4.1 Light marine vessels

Light naval vessels are classified into cargo boats, passenger boats, lightweight sailing and fishing boats, lightweight barges, ore carriers and tankers. Table 2.56 gives specific dimensions and weights of these vessels. The maximum speed for all these vessels for impact analysis is taken as 30 knots per hour. Table 2.57 gives comprehensive data on the hovercraft illustrated in Fig. 2.50.

2.4.4.2 Heavy marine vessels

The heavy vessels are classified into ships, cruisers, aircraft carriers, mine sweepers, frigates, heavy tankers and helicopter carriers. Although they are seldom involved in accidents in peace-time roles, nevertheless they are vulnerable in battle zones and are always subject to aircraft and missile attacks and terrorist attacks. Table 2.58 lists weights, basic dimensions, speeds, ranges and armament for some of the well-known naval vessels. For postulated analyses, typical layouts of vessels are given in Figs 2.51 to 2.56. For other types, the analyses will require a comprehensive geometrical layout in order to assess damage due to unwarranted attacks.

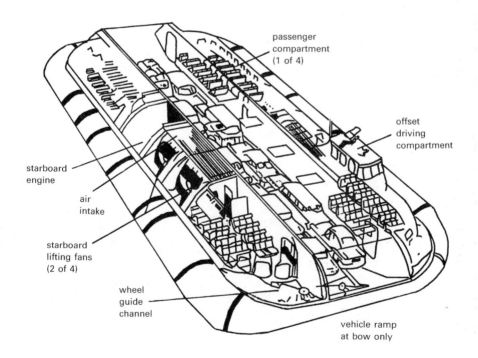

Fig. 2.50 Hovercraft layout.

Table 2.56 Lightweight vessels (weights and dimensions).

Vessel	w_a (tonnes)	L (m)	S (m)	d (m)
Barges		50	18.0	5.0
		100	20.5	5.5
		150	22.5	6.3
		200	25.0	6.6
		300	30.0	6.9
Boats	300	37	7.0	3.3
	500	43	7.8	3.8
	700	54	7.9	4.0
	1 000	61	8.9	4.5
	2 000	76	11.2	5.7
	3 000	87	12.8	6.5
	4 000	96	14.0	7.2
	5 000	103	15.1	7.8
	6 000	110	16.0	8.2
	7 000	116	16.8	8.7
	20 000	164	23.7	12.3
	25 000	176	25.5	13.3
	30 000	187	27.1	14.1
	35 000	197	28.5	14.8
	40 000	206	29.7	15.5
	50 000	222	32.0	16.7
	60 000	236	34.0	17.8
	70 000	248	35.7	18.7
	80 000	260	37.3	19.6
Cargo boats	700	52	8.3	3.8
	1 000	60	9.3	4.4
	2 000	77	11.5	5.8
	3 000	90	13.1	6.8
	4 000	100	14.3	7.7
	5 000	109	15.3	8.4
	6 000	117	16.2	9.0
	7 000	124	17.0	9.6
	8 000	130	17.7	10.1
	9 000	136	18.4	10.6
	10 000	142	19.0	11.1
	12 000	152	20.1	11.9
	15 000	165	21.6	13.0
	17 000	173	22.4	13.7
	20 000	184	23.6	14.6
Ferry boats	50	20	6.0	2.3
	100	25	7.5	2.7
	200	35	9.0	3.2
	300	42	10.0	3.5
	500	50	11.5	3.9
	1 000	64	13.0	4.4

Table 2.56 *Continued.*

Vessel	w_a (tonnes)	L (m)	S (m)	d (m)
Fishing boats	10 000	162.2	20.7	12.0
	17 000	189.5	23.6	12.7
	20 000	178.0	22.8	17.4
Motor and sailing boats				
Wooden boats	100	21.0	6.3	2.6
Steel boats	100	25.0	5.3	2.5
Wooden boats	200	29.0	7.4	3.4
Steel boats	200	33.0	6.6	3.3
Wooden boats	300	32.0	8.0	4.0
Steel boats	300	38.5	7.2	3.6
Ore carriers	1 000	61	8.9	4.8
	2 000	77	11.1	6.0
	3 000	88	12.7	6.8
	4 000	96	13.9	7.5
	5 000	104	14.9	8.1
	15 000	149	21.3	11.5
	20 000	164	23.4	12.7
	25 000	176	25.1	13.6
	30 000	187	26.6	14.4
	40 000	206	29.2	15.9
	50 000	222	31.4	17.1
Passenger boats	500	50	8.2	4.5
	1 000	65	10.0	5.3
	2 000	82	12.0	6.4
	3 000	95	13.5	7.3
	4 000	105	14.8	8.0
	5 000	113	15.8	8.8
	6 000	121	16.7	9.5
	7 000	127	17.5	10.2
	8 000	135	18.2	10.8
	10 000	145	19.2	12.0
	15 000	165	21.5	13.0
	20 000	180	23.0	13.8
	30 000	210	26.5	15.5
	50 000	245	30.5	18.0
	80 000	290	36.0	21.0
Trawl boats	400	53.8	7.9	
	800	67.2	10.2	
	1000	76.2	10.7	
	2000	87.4	13.1	
	3000	98.6	14.2	
Whaling vessels	400	53.8	8.3	
	800	62.7	9.4	
	1000	68.3	10.2	

w_a = gross weight; L = length; S = width; d = depth.

Table 2.57 Data on the principal hovercraft.

Type	Power plant	Length (ft)	Max weight (tonnes)	Max speed (mph)	General comments
					All hovercraft: (1) *Amphibious* Horsepower 228–565 with air propeller 390–538 air jet (2) *Semi-amphibious* Horsepower 150–191 (3) *Non-amphibious* Horsepower 172
Australia Hovergem G-4	Two 170–200hp aeros Two reversing aps	30	–	65	
Canada Hovergem	Three 17.5hp JLD pistons Two ducted aps	16	0.55	45	
France Aerotrain 01	Lift, two 50hp Renault auto pistons 260hp ap + rocket booster	33	2.75	215	
Aerotrain 02	Lift, Turbomeca air-bleed turbine 3000lb thrust P&W turbojet	84	22.0	185	
Aerotrain 'Orleans'	400hp Turbomeca Astazou turbine Two 1500hp Turmos, one ducted ap	47	11–12.5	112	
Aerotrain 'Cite'	Lift, 250hp auto piston Linear electric or auto wheels	32	5.5	50	
SEDAM Naviplane BC.7	Lift, 180hp Chevrolet piston Two axles, each 45hp	33	4.75	62	
SEDAM Naviplane BC.8	Lift, 880lb thrust Turbomeca jet 400hp turbine two reversing aps	79	30.0	71	
SEDAM Naviplane N-300	Two 1500hp Turbomeca Turmo turbines Two reversing aps	11	0.28	34	
SEDAM Naviplane Sports	Lift, 7hp JLO piston Two 4.3hp JLO shrouded aps	40	3.8	75	
Israel Israel American Lady Bird 2	Lift, 230hp auto 230hp aero piston ducted ap	35	2.8	52	

Table 2.57 *Continued.*

Type	Power plant	Length (ft)	Max weight (tonnes)	Max speed (mph)	General comments
Japan					
Mitsui MV-PP1	Lift, 250hp Continental 10–470	53	12.0	63	250hp 10–470 reversing ap
Mitsui MV-PP5	1050hp IHI (GE LM-100) turbine	56	12.0	—	Two reversing aps
UK					
BHC BH.8	Two 3400hp R-R Proteus turbines	96	103.0	86	Two non-swivelling aps
BHC SR.N4	Four 3400hp R-R Proteus turbines	130	194.0	81	Four swivelling reversing aps
BHC SR.N5	900hp R-R Marine Gnome turbine	39	7.5	68	Fixed reversing ap
BHC SR.N6	900hp R-R Marine Gnome turbine	48	11.5	60	Fixed reversing ap
	390hp P&W ST6B turbine	25	3.0	45	Two low-pressure air jets
Cushioncraft CC-7	Lift, two 180hp Caterpillar diesels	81	34.0	31	Two 400hp diesel ms
Denny D.2	Lift, 18hp Velocette piston	13	0.35	50	33hp Greaves shrouded ap
Express Air Rider	Lift, 16hp Velocette piston	16	0.625	34	Two 16hp Velocette shrouded aps
Hover Hawk	Lift, 185hp Cummins diesel	51	18.0	40	Two 320hp Cummins ms
Hovermarine HM.2	Two diesels or gas turbines	160	140.0	50	Waterjets or ducted ms
Hovermarine HM.4	Lift, 46hp Volkswagon piston	27	2.25	35+	90hp Porsche fixed ap

continued

Table 2.57 *Continued.*

Type	Power plant	General comments	Length (ft)	Max weight (tonnes)	Max speed (mph)
UK (Contd)					
Hovermarine Hovercat Vosper Thorneycroft VY1	Two 2000hp Lycoming TF 20 turbines	Two variable-pitch ms	96	86.0	55
USA					
Aerogem 1A	Lift, 60hp auto piston	25hp ducted fixed ap	16	0.6	60+
Beardsley Skimmer	6hp Chrysler piston	Air escape from fan duct	10	0.2	18
Bell SK-5 (7250)	1000hp GE LM-100 turbine	Fixed reversing ap	39	10.0	70
Bell SK-6 (7282)	1250hp GE LM-100 turbine	Fixed reversing ap	49	12.5	65
Bell SK-9	Two 1250hp GE LM-100 turbines	Two fixed reversing aps	56	26.5	70
Bell SK-10	Two 12000hp GE LM-1500 turbines	Two fixed reversing aps	80	148.0	92
Bell SES	140000hp (two GE 4 converted)	Waterjets or ducted ms	420	4000.0	92
Bertelsen Aeromobile 13	Two 40–49hp autos or aeros	Lift fans gimballed to rear	21.5	1.35	50–80
Cushionflight Airseat	45hp Volkswagen piston	Reversing shrouded ap	15	0.5	40
Dobson Air Car D	20hp Lloyd or similar piston	Rear vents from cushion	14	0.325	35
Gunderson Crop Sprayer	Lift, two 7hp Tecumseh pistons	15hp Hirth ap plus wheels	12	0.3	20

Table 2.58 Heavy naval vessels data.

Type of naval vessel and power plant	S (m)	L (m)	H (m)	Weight (displacement) w_a (tonnes)	V (knots)	Additional data	Range (miles)
4 × 'Agosta' class; Diesel electric, 2SEMT-piedstick, 16 PA 4185VG diesels, 3600 hp, 1 electric motor, 4600 hp (patrol submarine)	6.8	67.6	5.4	1490 to 1740	12 surface 20 diving	SSM Aerospatiale SM 39 Excocet missiles: 553 mm tube at 0.9 mach, warhead 165 kg	8500
Ex-British 'Hermes' class aircraft carrier; 2 Parsons geared turbines 76000 shp	226.9	208.8	8.7	23900 to 28700	28.0	SAM system (see Soviet)	1200
Ex-Chinese 'Huchuan' class; 3 M50 diesels 3600 hp (fast-attack hydrofoil – torpedo)	6.3	21.8	3.6		55	Guns 14.5 mm, MGS 2 No. 533 tubes, torpedoes	500
Ex-Chinese 'Shanghai II' class; 2M 50F diesel 2400 hp (fast-attack craft)	5.5	39	1.7	120 to 155	30	4–37 mm guns, shell 1.42 kg, 8 depth charges	800
Destroyers (1) Bristol; COSAG: 2 Standard Range geared, steam turbines, 30000 shp, 2 Rolls-Royce Marine Olympus TM1A gas turbines, 30000 shp, 2 shafts, 2 boilers	16.8	154.5	5.2	7100	30	Missiles: SAM: British Aerospace Sea Dart GWS 30 twin launcher, radar homing to 40km (21 nm), warhead HE; 40 missiles, limited anti-ship capability. Guns: 1 Vickers 4.5in (114mm) 55 Mk 8, 55° elevation, 25 rounds/minute to 22 km (11.9 nm) anti-surface, 6km (3.3 nm) anti-aircraft weight of shell 21 kg. 4 Oerlikon/BMARC 30 mm/75 GCM-A03 (2 twin); 80° elevation, 650 rounds/minute to 10 km (5.5 nm)	5000
(2) Birmingham, Newcastle, Glasgow, Cardiff, Exeter, Southampton, Nottingham, Liverpool	14.3	119.5	5.8	3500 to 4100	29	Missiles: SAM: British Aerospace Sea Dart GWS 30 twin launcher; radar semi-active radar homing to 40km (21 nm), warhead HE, 22 missiles limited anti-ship capability	4000

continued

Table 2.58 *Continued.*

Type of naval vessel and power plant	S (m)	L (m)	H (m)	Weight (displacement) w_a (tonnes)	V (knots)	Additional data	Range (miles)
COGOG; 2 Rolls-Royce Olympus TM3B gas turbines (full power), 50 000 shp: 2 Rolls-Royce Tyne RM1C gas turbines (cruising): 9700 shp: 4 diesel generators, 4000 kW, 2 shafts, stone manganese type XX cp propellers						Guns: 1 Vickers 4.5 in (114 nm) 55 Mk 8, 55° elevation, 25 rounds/minute to 22 km (11.9 nm) anti-surface, 6 km (3.3 nm) anti-aircraft, weight of shell 21 kg 2 Oerlikon/BMARC 20 mm GAM-BOL; 55° elevation, 1000 rounds/minute to 2 km 2 Oerlikon 20 mm MK 9; 50° elevation; 800 rounds/minute to 2 km, weight of shell 0.24 kg 2 General Electric/General Dynamics 20 mm Vulcan Phalanx Mk 15; 6 barrels per launcher, 3000 rounds/minute combined to 15 km; Still to be fitted in some Torpedoes: 6–324 mm Plessey STWS Mk 2 (2 triple) tubes; Marconi Stingray, active/passive homing to 11 km (5.9 nm) at 45 knots: warhead 35 kg Honeywell/Marconi Mk 46 Mod 2; active/passive homing to 11 km (5.9 nm) at 40 knots, warhead 45 kg	
Aircraft carriers Nimitz, Dwight D. Eisenhower, Theodore Roosevelt, Abraham Lincoln, George Washington 2 pressurized-water cooled A4W/A1G nuclear reactors: 4 geared steam turbines; 260 000 shp, 4 shafts	40.8	332.9	11.8	72 916 to 96 386	30+	Missiles: SAM 3 Raytheon Sea Sparrow Mk 29 octuple launchers; semi-active radar homing to 14.6 (8 nm) at 2.5 mach Guns: 4 General Electric/General Dynamics 20 mm Vulcan Phalanx 6-barrelled Mk 15 (3 in CVN 68 and 69), 3000 rounds/minute combined to 1.5 km Countermeasures: decoys; 4 Loral Hycor SRBOC 6-barrelled fixed Mk 36: IR flares and chaff to 4 km (2.2 nm); ESM/ECM SLQ 29 (WLR 8 radar warning and SLQ 17AV jammers) Fire control: 3 Mk 91 MFCS; NTDS action data automation; links 11 and 14; OE-82 satellite communications antenna, SSR 1 receiver	

Table 2.58 *Continued.*

Type of naval vessel and power plant	S (m)	L (m)	H (m)	Weight (displacement) w_a (tonnes)	V (knots)	Additional data	Range (miles)
French Colbert cruiser 2 sets CEM-Parsons geared turbines, 86 000 shp, 2 shafts, 4 Indrit multi-tubular boilers	20.2	180	7.7	8 500 to 11 300	31.5	SSM: 4 Aerospatiale MM38 Excocet missiles, range 42 km, warhead 165 kg, 2/100 mm guns	4000
French nuclear propelled aircraft carriers (PAN)	261.5 also deck	238	31.8				
Nuclear propulsion type K15 300 MW (PWR), 82 000 hp	2.62 / 29.4	261.5 / 138	6.1 / 7.3	34 000 to 36 000	26	Variants, SAM 7 Thomson missiles, CSF-SAAM range 17 km at 2.5 mach	
Indian Ex-British 'Majestic' class aircraft carrier 2 No. Parsons single reduction, geared turbines, 4000 shp	24.4	231.4	7.3	16 000 to 19 500	24.5	7 Bofor guns, 40 mm, range 12 km, shell weight	12 000
Light aircraft carrier (Giuseppe Garibaldi) COGAG; 4 Fiat/General Electric LM 2500 gas turbines, 80 000 hp; Tosi reduction/reversing gears, 2 shafts, 5 bladed propellers, generating capacity 6–1950 kVA diesel alternators	33.4	180	6.7	13 850	30	Missiles: SSM: 4 OTO Melera/Matra Teseo Otomat launchers, inertial cruise, active radar homing to 180 km (100 nm) at 0.9 mach, warhead 200 kg, sea-skimmer SAM: 2 Selenia Elsag Albatros octuple launchers, 48 Aspide semi-active radar homing to 13 km (7 nm) at 2.5 mach Guns: 6 Breda 40 mm/70 (3 twin) MB, 85° elevation, 300 rounds/minute to 4 km (2.2 nm) anti-aircraft, weight of shell 0.96 kg Torpedoes: 6–324 mm ILAS 3 (2 triple) tubes Whitehead A244S, anti-submarine active/passive homing to 6 km	7000
4 × Meko 360 2 Rolls-Royce 'Olympus' TM3B gas turbines 51 600 shp (destroyer)	14	125.9	5.8	2900 to 3600	30.5 / 30.5	55M Aerospatiale MM 40 Excocet missiles: range 70 km, weight of warhead 165 kg; 1 No. OTO 40 mm guns, 8 Breda/Bofor 40 mm	4500

continued

Table 2.58 *Continued.*

Type of naval vessel and power plant	S (m)	L (m)	H (m)	Weight (displacement) w_a (tonnes)	V (knots)	Additional data	Range (miles)
3 × Modified 'USDDG-2' class 2 No. GE double reduction steam turbine 70 000 shp (destroyer)	14.3	134.3	6.1	3370 to 4618	30	McDonnell Douglas harpoon; 2FMC 127 mm guns, weight of shell 32 kg; torpedo 6–324 mm tube	4 500
2 × Ex-Soviet 'Khobi' class 2 diesels 1600 bhp (tanker)	10.1	63	4.5	700 to 1500	13		2 500
2 × Ex-Soviet 'T-43' class 2 type 9D diesel 2200 hp (minesweepers)	8.4	58	2.1	500 to 580	15	2–37 mm guns, 2 depth charges	3 000
5 × Soviet 'Kashin II' class 4 sets gas turbines, 9600 shp, with sonars and helicopters	15.8	146.5	4.8	3950 to 4950	35.0	SSM: 4 SS-N-2C Styx, range 83 km at 0.9 mach, warhead 500 kg; 2–76 mm guns, 16 kg weight, range 15 km; 5 torpedoes, 533 mm tubes, 20 km range with warhead 400 kg; A/S mortar, range 6000 m, weight 31 kg	4 500
3 × Soviet 'Koni' class Codag: 1 gas turbine 1800 shp 2 diesels (frigate)	12.8	95	4.2	1700 to 1900	27 (gas) 22 (diesel)	4–76 mm guns, weight of shell 16 kg, A/S mortars (31 kg) at 600 m range; SAM range 15 kg at 2.5 mach	1 800
2 × Soviet 'Kronshtadt' class 3 type 4-D diesels, 3300 bhp (large petrol craft)	6.5	52.1	2.1	303 to 305	24 surface	Guns 1 No. 85 mm; shell weight 9.5 kg	1 500
2 × Soviet 'Whisky' class Type 37-D diesel, 4000 bhp (submarine)	6.5	76	4.9	1080	18 surface	Torpedoes 533 tube; Sonar	8 500

S = width; L = length; H = height.

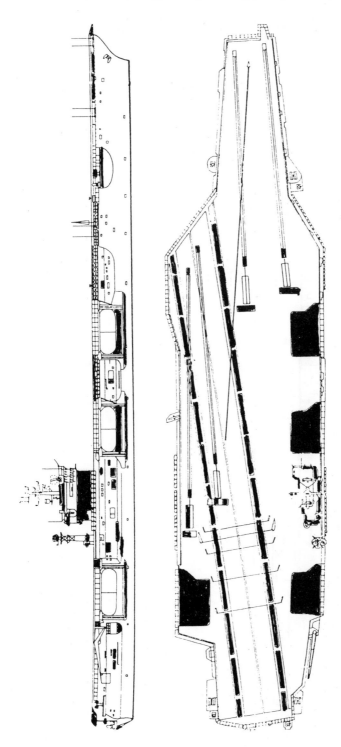

Fig. 2.51 Nimitz class (CVN) carrier. (An abbreviated detail from the original US Defense layout.)

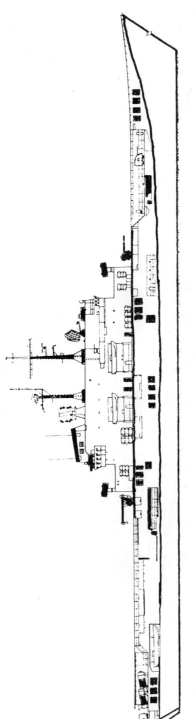

Giuseppe Garibaldi

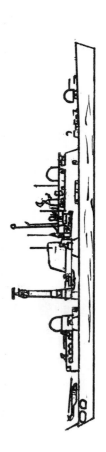

Exeter (Batch 2)

Fig. 2.52 Garibaldi and Exeter. (Courtesy of the Italian and British Defence Ministries.)

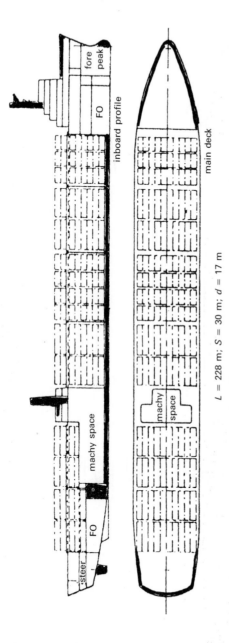

Fig. 2.53 Container ship (L = length; S = span; d = depth).[2.386]

L = 228 m; S = 30 m; d = 17 m

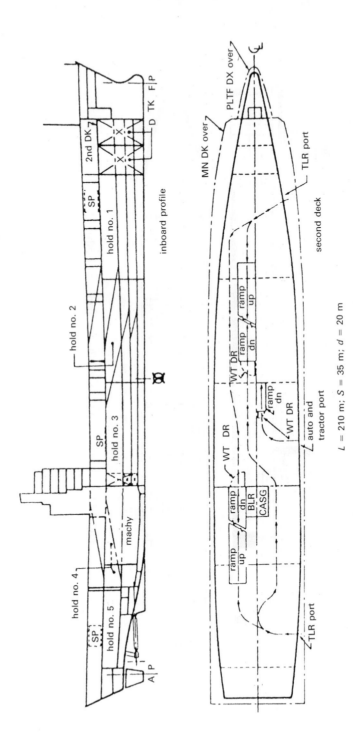

Fig. 2.54 Roll-on/roll-off ship (L = length; S = span; d = depth).[2.386]

L = 210 m; S = 35 m; d = 20 m

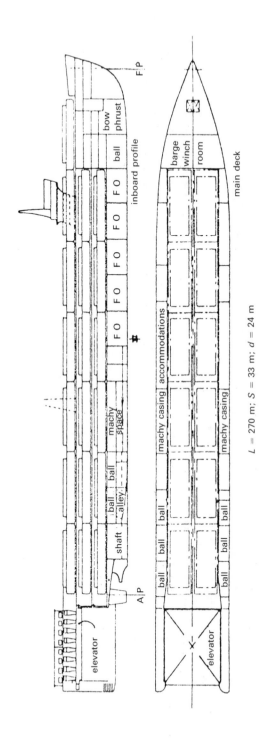

Fig. 2.55 Barge-carrying ship (platform elevator type) (*L* = length; *S* = span; *d* = depth).[2.386]:

$L = 270$ m; $S = 33$ m; $d = 24$ m

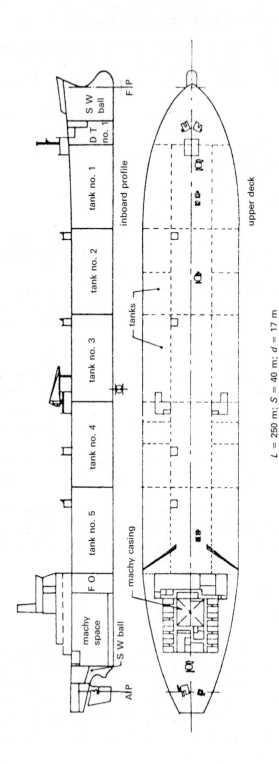

Fig. 2.56 Tanker (L = length; S = span; d = depth).[2.386]

$L = 250$ m; $S = 40$ m; $d = 17$ m

2.4.5 Offshore floating mobile and semi-submersible structures

The tension leg platform (TLP) design is a complex one. From considerations of construction and operational requirements, and because of the danger of heavy impact from sea vessels, vigorous data collection is necessary. Table 2.59 summarizes data for such a structure.

The drilling semi-submersibles can generally be characterized by the three sea conditions:

(1) Maximum drilling with significant waves of 8 m.
(2) Riser's disconnections with significant waves of 11 m.
(3) Survival conditions with significant waves of 16 m.

Based on these criteria, the basic shape and size can be directly related to the load-carrying capacity. Various drilling semi-submersibles have been designed. In Fig. 2.57, five (Aker H3, DSS 20, GVA 4000, Sonat and DSS 40) semi-submersibles are compared in terms of loadings and displacements. A typical layout of the DSS 40 is illustrated in Fig. 2.58. It is important to consider as well the wind effects during any accident conditions. Table 2.60 gives useful data on the wind velocity profile.

For flat surfaces, all the registered companies agree to take the value of 1.0. Similarly, all take a value of 1.25 for drilling derricks. The Japanese and API take 1.3 for exposed beams and girders and 1.5 for isolated shapes. For clustered deck houses or similar structures, Nippon and API take a drag coefficient of 1.1.

In addition, various dimensions are required for the semi-submersible when it is transported in a heavy lift ship. These are given in Table 2.61 for a typical case study by the Wijsmuller Transport Bureau. Taking the weights of the chains and anchors as 2.42 and 180 tonnes respectively, the total weight is assumed to be 22.20 tonnes.

The same approach is adopted when a jack-up rig is transported. The total load is generally not more than 20 tonnes. In the absence of a specific rig, the data in Table 2.62, provided by the Wijsmuller Transport Bureau, will be adopted for impact analysis.

2.5 Types of explosion

A comprehensive introduction is given to each type of explosion later on in Chapter 5. The purpose of this section is to provide chemical data under a specified classification for researchers and practising engineers/scientists who are familiar with the subject but wish to use it without extensive searching for materials for their individual problems. The author appreciates that many ordnance factories, munition and explosive plants and laboratories have accumulated a large amount of information which is confidential. He has therefore limited himself to data for which there is a reference available in the open literature.

Explosions may be due to bombs, shells, grenades and various explosives and also to gas leaks, nuclear detonation and fuel in chemical industries. Explosions occur in the air, on the surface, underground and underwater. They are classified in the text into:

(1) explosions due to bombs, shells and explosives;
(2) explosions due to gas leaks;
(3) nuclear explosions;
(4) dust explosions.

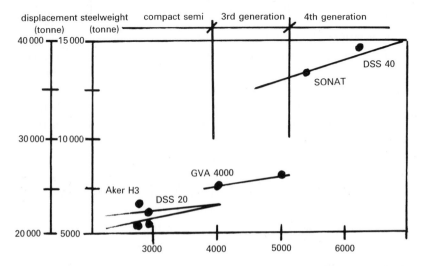

Fig. 2.57 Load-carrying capacities of drilling semi-submersibles. (Courtesy of Aker, GVA, DSS and Sonat.)

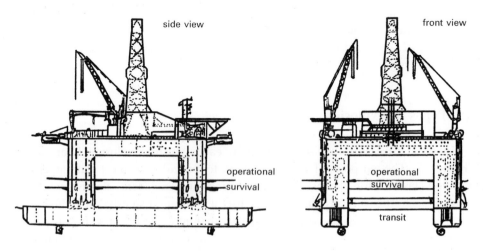

Fig. 2.58 Layout of drilling semi-submersible DSS 40. (Courtesy of DSS.)

Table 2.59 Characteristics of the tension leg platform. (Data courtesy of Conoco, Amoco, Deep Oil Technology and International Marine Development.)

Company/investigator	Hutton TLP Conoco	VMP Amoco Produce Co	Deep Oil X-1 Deep Oil Technology	Triton International Marine Development
Design environment				
Water depth (m)	150.0	264.0	61.0	182.9
Wave excitation	Irregular waves	Design wave	Design wave	Design wave
Wave height	16.6	26.0	7.6	15.2
Wave period (s)	13.9	13.0–16.0		15.0
Current velocity	0.85 m/s	2.7 m/s; lv	2.5 knots	
Wind velocity	44.0 m/s	67 m/s	60.0 knots	120.0 knots
Basic configuration				
Scale	Prototype	Model (1:60)	Model (1:3)	Model (1:6)
Deck plan	Rectangular	Square	Triangular	Rectangular
Length (m)	78.0	61.0	31.7	7.3
Breadth (m)	74.0	61.0	36.6	7.3
Depth (m)	12.0	9.1		3.2
Keel to deck (m)	69.0	64.9	20.1	13.8
Columns/cassions				
No total	6.0	4.0	6.0	6.0
No corner	4.0	4 w/cb	3.0	4.0
No interior	2.0	0	3 w/cb	2 w/cb
Corner (m)	17.7 dia	9.1–18.3 dia	3.2 dia	0.38 dia
Interior (m)	14.5 dia		0.91 dia	
Lower hull				Triangular
Length (m)	95.7	67.1	36.6	13.2
Breadth (m)	8.0	67.1	1.68 dia	1.6 dia
Height (m)	10.8		1.68 dia	1.6 dia
Overall width (m)	91.7	67.1	36.6	15.2

continued

Table 2.59 *Continued.*

Company/investigator	Hutton TLP Conoco	VMP Amoco Produce Co	Deep Oil X-1 Deep Oil Technology	Triton International Marine Development
Tendons				
Type and no	Pipe, 16	Risers, 24	Wire rope, 6	Wire rope, 6
Outside diameter (m)		0.47	0.6	0.25
Pre-tension (t)	15 000			53.0
Displacement (t)	63 000	36.3	635.0	124.0
Operating draft (m)	33.2	20.0	12.2	10.8
No. wells	32		13 risers	
Accommodation (people)	239			
Natural response periods				
Surge and sway (s)	60.0	60.0	66.0	60.0
Heave (s)	1.9	2.0	0.5	1.9
Roll and pitch (s)	1.9	2.0	2.0	1.9
Yaw (s)	43.0	60.0		43.0
Dynamic response behaviour				
Wave excitation	Irregular waves	Design wave	Random	Random
Wave height (m)	16.6	16.0	18.1	15.8
Wave period (s)	13.9	14.0	16.7	
Motion response				0.34–3.06 rms
Surge (m)	9.5	11.2	2.7 ha	
Sway (m)				
Heave (m)			0.2 da	
Roll (degrees)				
Pitch (degrees)			0.11	0.12–0.45
Yaw (degrees)	1.7			
Tension, max (t)	300.0			

Table 2.60 Wind velocity profile.

| Height (m) | Coefficients for wind | | Drag coefficients | | | |
	LLOYDS*	DNV†	LLOYDS	API§	NIPPON**	DNV
10	1.0	1.0	0.4	0.4	0.4	—
20	1.05	1.06	0.5	0.5	0.5	0.5
30	1.105	1.10				
50	1.14	1.16		spherical parts		
80	1.20	1.20		cylindrical parts		
100	1.22	1.23				
150	1.26	1.28				

* Lloyds Register, UK; § API, American Petroleum Institute, USA; † DNV, Norway; ** Nippon, Japan.

Table 2.61 Rigs and floaters.

	Rigs	Floaters
L (m)	97.5	97.5
S (m)	72.5	72.5
H (m)	38.80	
x (m)	39.5	
w_a (tonnes)	19.6	
D (m)		8.5

L = length; S = width, H = height; x = distance between floaters; w_a = weight of rig; D = depth of floaters; number of columns = 8; outer diameter of columns = 11.8 m; inner diameter of columns = 9.0 m.

Table 2.62 Platform dimensions.

Platform
$L = 84.00$ m
$S = 90.00$ m
$D = 9.50$ m
At loading/unloading = 4.05 m

Legs	
Type	Triangular lattice
Number	3
Longitudinal centres	56.80 m
Transverse centres	66.00 m
Length	156.80 m
Chord diameter	1.00 m
Chord centres	12.00 m

L = length; S = width; D = depth.

2.5.1 Bombs, shells and explosives

2.5.1.1 Bombs and shells

Bombs and shells manufactured for military purposes are illustrated in section 2.3.1 with the respective aircraft.

2.5.1.2 Explosives

During most of the period from the fourteenth to the early nineteenth century, chemistry played only a small part in the development of science. The history of industrial explosives began with the use of black powder for blasting in 1600. Black powder was originally used by the Arabs, Hindus and Chinese. It was a low-power explosive (LE). Its role was deflagration rather than detonation. In 1346 it was used as a gun powder in the battle of Crécy, as a propellant in wooden cannon. A safety fuse was invented in 1831.

Since an explosive is composed of a fuel and an oxidizer, a large amount of energy is given off on reaction. In the case of black powder, the sodium/potassium nitrate furnishes oxygen for the reaction which eventually combines with sulphur (S), and carbon (C). The equations of reaction are as follows:

(a) $2NaNO_3 + S + 2C = Na_2SO_4 + 2CO \uparrow N_2$ 195 kcal
$\quad Q = $ specific heat $= 920$ kcal/kg
$\quad n = $ number of moles of gas $= 4.7$ mole/kg
(b) $2NaNO_3 + 2S + \frac{3}{2}CO = 2NaSO_3 + \frac{3}{2}CO + N_2$
$\quad Q = 620$ kcal/kg; $n = 6.9$
(c) $2NaNO_3 + 2S + 3C = 2NaSO_3 + 3CO + N_2$
$\quad Q = 620$ kcal/kg; $n = 12.6$
(d) $2NaNO_3 + S + 2C = Na_2SO_4 + 2CO + N_2 \uparrow$
$\quad Q = 680$; $n = 8.9$

The energy depends upon the oxygen balance and the strength depends upon the value of n.

The same principal holds for various dynamites and other high explosives (HE).

Single-compound explosives are substances such as nitroglycerin, alpha-trinitrotoluene (TNT), pentaerythrite tetra-nitrate (PETN) and others. Nitro-glycerin (NG) and nitrocellulose (NC) are other substances which, when combined in a 92/8 ratio, produced a stiff gel called blasting gelantin (BG), which is a powerful explosive. The chemical formula for NG is $C_3H_5(NO_3)_3$ and it has a natural density of 1.6 g/cm³. Since it is oxygen positive, during detonation, the equation for the reaction is:

(e) $C_3H_5O_9N_3 = 3CO_2 + 2\frac{1}{2}H_2O + \frac{1}{2}O + 1\frac{1}{2}N_2 \uparrow$

The heat of reaction is about 1503 cal/g (298 K) and has the following mechanical properties: detonation velocity = 7926 m/s; pressure in the detonation wave = 25×10^4 atm.

Explosives may be divided according to use:

(1) *Commercial*   Dynamites, ammonium nitrate−fuel oil (ANFO) explosives and water-based explosives.
(2) *Military*   TNT, PETN, RDX.

Tables 2.63 and 2.64 give a useful summary of all these explosives with their individual characteristics and important parameters. Additional properties of ingredients of explosives are given in Tables 2.65 and 2.66.

2.5.1.3 *Explosive types versus projectile velocity of soil/rock deposits and crater volume*

A chemical charge exploding on the Earth's surface creates a shallow depression or cavity, thereby compacting the underlying medium. Most of the energy is lost into the air. When the explosive is buried, depending upon its depth a crater is produced, the size of which is directly related to the depth of the charge. When the charge depth is increased beyond the *optimum depth*, the material is ejected to a certain height only and most of it falls back in the crater. In certain cases an explosion cavity is formed due to the expansion of explosive gases. The 'roof surface' is ejected and soil/rock falls into the cavity. At a certain stage a *subsidence crater* is formed. The space of *soil slip* is called the *rubble chimney*. In loose and water-saturated soils, the explosive charges produce such chimneys of the volume of the crater and if the volumes of explosive gases in the space between rubble particles are combined they will give the volume of the original explosion cavity.

Tables 2.67 to 2.70 show data on the performance of various explosives in creating particle and fly rock initial velocities and crater dimensions in difficult soils/rocks. Table 2.71 shows the relationship of crater diameter and depth for mass and reinforced concrete. The data collected in these tables are from various sources given in the relevant references[2.144−2.185]

2.5.2 Gas explosions

2.5.2.1 *Introduction*

Gas explosion incidents involving domestic or commercial premises do occur from time to time. It is therefore necessary to have a proper understanding of the development of the pressure-pulse relationship and the way structures respond to pressure loading. Most hydrocarbon gases, when mixed with air at atmospheric

Table 2.63 Data on major explosives.

Explosive	Composition (%)	V_{sp} (l/kg)	Q_E (kcal/kg)	Temperature at explosion (°C or K)	V_E (m/s)	Properties
Abbcite (ammonia dynamite)	Ammonium nitrate (58); nitroglycerine (8); dinitrotoluene (2); sodium chloride (23); wood meal (9)					As a blasting explosive in coal mining
AG dynamite	Ammonium nitrate (45); nitroglycerine (45); ammonia gelatine wood flour dynamite (7.5); collodian cotton (2.5)	635	1200	4050	6600	Mining operations, underground quarrying, tunnelling in hard rock; submarine work
Amatol (mixture of TNT and ammonium nitrate)	Ammonium nitrate (80); TNT (20); or ammonium nitrate (50); TNT (50)				5000 7000	As a bursting charge for high explosive shells/bombs
Ammonal	Ammonium nitrate (65); TNT (15); charcoal (33); aluminium powder (17)					Shell filler; Blasting, tunnelling
Ammonite	Potassium nitrate (10); ammonium nitrate (77); solid hydrocarbon (1); napthalene (10); gelatinized nitroglycerine (2)	860 to 915	938 to 1020	2310 to 2600	3500 to 5100	Ammonites are the main explosive used in the USSR
Belgian permite	Ammonium nitrate (78); TNT (8); calcium silicide (14)					Dark-grey powder set off by service detonators
Black powder	Potassium nitrate (70–75); sulphur (10–14); charcoal (14–16)	260	588	2615	400	Class A (pellet), class B (powder); bombs used in USA

Table 2.63 *Continued.*

Explosive	Composition (%)	V_{sp} (l/kg)	Q_E (kcal/kg)	Temperature at explosion (°C or K)	V_E (m/s)	Properties
DD 60/40 (LE)	Picric acid (60); dinitrophenol (40)					Bursting charge for shells used by the French
Gelignite	Ammonium nitrate (70); nitroglycerine (29.3); nitro cotton (0.7)	See individual components (Table 2.64)				Used for blasting
MDPC (HE)	Picric acid (55); dinitrophenol (35); trinitrocresol (10)					Bursting charge for shells used by the French
PETN	PETN (70); TNT (30)	780	1410	4010	8200	High explosive charge for sea mines, drop bombs and torpedoes
Wetter-astralit	Gelatinized nitroglycerine (12); ammonium nitrate (57); wood meal (2); coal powder (2); sodium chloride (27)	See individual components (Table 2.64)				A German explosive used in many countries

V_{sp} = specific volume for explosive gases; Q_E = specific heat of explosion; V_E = velocity at explosion.

Table 2.64 Data on major explosive elements.

Explosive element	Composition	Density (kg/m^3)	V_{sp} (l/kg)	Q_E (kcal/kg)	Temperature at explosion (°C or K)	V_E (m/s)	3 kg Load dropped at which explosions occurred
							Impact sensitivity (mm)
Ammonium nitrate	NH$_4$NO$_3$	1600	—	1090	—	—	—
Charcoal	Black powder or lump coal (95%) and ash (5%)	1600	—	—	—	—	—
Dinitroglycol	(CH$_2$)$_2$(NO$_3$)$_2$	1400	738	1690	4230	8300	104
Dinitrophenol	C$_6$H$_3$OH(NO$_2$)$_2$	1400	—	—	—	—	—
Gun cotton	Cellulose nitrate (high nitration)	300	936	810	2640	6300	270
Hexogen	—	1700	908	1500	3850	8300	435
Nitroglycerine	C$_3$H$_5$(NO$_3$)$_3$	1600	717	1470	4110	8000	60
PETN	C(CH$_2$ONO$_2$)$_4$	1750	See Table 2.63				420
Picric acid	—	900	685	920	3620	7250	900
Potassium nitrate	KNO$_3$	—	—	—	—	—	—
Sodium nitrate	NaNO$_3$	—	—	—	—	—	—
Trinitrocresol	C$_6$HCH$_3$(OH)(NO$_2$)$_3$CH$_3$	—	—	—	—	—	—
Trinitrotoluene (TNT)	C$_6$H$_2$(CH$_3$)(NO$_2$)$_3$	1600	728	1000	2950	6800	2200

V_{sp} = specific volume for explosive gases; Q_E = specific heat of explosion; V_E = velocity of explosion.

Table 2.65 Ingredients of explosives and their properties.

Formula	Name	(g/mol)	Density (g/cm^3)	Explosive strength (%TNT)	Gurney constant (m/s)
$C_6H_6N_4O_7$	Ammonium picrate (explosive D)	246	1.72	100t, 99m, 84s, 87b, 91p, 85a	2137
$(C_6H_7N_{2.25}O_{9.5})_n$	Nitrocellulose 12% N (NC)	263(n)	—	82b	2189
$(C_6H_7N_{2.5}O_{10})_n$	Nitrocellulose 13.35% N (NC)	274(n)	1.67	102b, 140t	2473
$(C_6H_7N_3O_{11})_n$	Cellulose trinitrate 14.1% N (NC)	297(n)	1.66	95s, 138b	2966
$C_6H_8N_2O_6$	2.2 Dinitropropylacrylate (DNPA)	204	1.47	50b	1478
$(C_6H_8N_4O_9)_n$	Cellulose dinitrate 11.1% N (NC)	252(n)	1.66	87b	2170
$C_6H_8N_6O_{18}$	Mannitol hexanitrate	452	1.73	170t, 156b	3451
$(C_6H_9N_3)_{11}$	Glycerol monolactate trinitrate (GLTN)	299	1.47	115m, 148b	2952
$C_7H_3N_3O_7$	2,4,6 Trinitrobenzaldehyde	241	—	125t, 94b	2307
$C_7H_5N_3O_6$	2,4,6 Trinitrotoluene (TNT)	227	1.65	100t, 100m, 100s, 100b	2315
$C_7H_5N_3O_6$	2,4,6 Trinitrotoluene (liquid TNT)	227	1.447	103b	2350
$C_7H_5N_3O_7$	2,4,6 Trinitroanisole	243	1.41	112t, 93b	2202
$C_7H_5N_3O_7$	2,4,6 Trinitro-m	243	1.68	98t, 87b	2131
$C_7H_5N_5O_8$	N-Methyl-N, 2,4,6 tetranitroaniline (TETRYL)cresol	287	1.73	131t, 131m, 123s, 130b, 116p	2710

continued

Table 2.65 *Continued.*

Formula	Name	(g/mol)	Density (g/cm^3)	Explosive strength (%TNT)	Gurney constant (m/s)
$C_{10}H_{16}N_6O_{19}$	Dipentaerythritolhexanitrate (DPEHN)	524	1.63	128t, 142m, 123s, 182b	3268
$C_{12}H_4N_6O_{12}$	2,2',4,4',6,6' Hexanitrobiphenyl (HNBP)	424	1.74	114b, 116t	2588
$C_{12}H_4N_8O_8$	1,3,8,10 Tetranitrobenzotriazolo, (1,2a) benzotriazole (T-TACOT)	388	1.81	108b	2655
$C_{12}H_4N_8O_8$	1,3,7,9 Tetranitrobenzotriazolo (2,1a) benzotriazole (Z-TACOT)	388	1.85	108b	2656
$C_{12}H_4N_8O_{10}$	5,7 Dinitro-1-picrylbenzotriazole (BTX)	420	1.74	110b	2611
$C_{12}H_4N_8O_{12}$	2,2',4,4',6,6' Hexanitroazobenzene (HNAB)	452	1.78	122b	2693
$C_{12}H_6N_8O_{12}$	Diaminohexanitrobiphenyl (dipicramide) (DIPAM)	454	1.79	105b	2431
$C_{13}H_5N_5O_{11}$	2,2',4,4',6 Pentanitrobenzophenone (PENCO)	407	1.86	92b	2352
$C_{14}H_6N_6O_{12}$	2,2',4,4',6,6' Hexanitrostilbene (HNS)	450	1.74	108b	2524
$C_{17}H_5N_{13}O_{16}$	2,6 Bis (picrylazo)-3,5-dinitro-pyridine (PADP)	647	1.86	122b	2738

Table 2.65 *Continued.*

Formula	Name	(g/mol)	Density (g/cm^3)	Explosive strength (% TNT)	Gurney constant (m/s)
$C_{18}H_5N_9O_{18}$	2,2',2",4,4',4",6,6',6" Nonanitroterphenyl, (NONA)	635	1.78	113b	2608
$C_{18}H_6N_8O_{16}$	2,2',2",4,4',4",6,6',6" Octanitroterphenyl (ONT)	590	1.80	106b	2537
$C_{18}H_7N_{11}O_{18}$	1,3 Bis (picrylamino) 2,4,6 trinitrobenzene	665	1.79	101b	2427
$C_{21}H_9N_{15}O_{18}$	2,4,6 Tris (picrylamino), s-triazine	759	1.75	86b	2267
$C_{22}H_9N_{15}O_{20}$	2,4,6 Tris (picrylamino) s-nitropyrimidine	803	1.88	82b	2210
$C_{23}H_9N_{15}O_{22}$	2,4,6 Tris (picrylamino) 3,5 dinitropyridine	847	1.80	98b	2400
$C_{24}H_6N_{12}O_{24}$	2,2',2"',4,4',4"',6,6',6",6"' Dodecanitro-m,m'-quatraphenyl	846	1.81	114b	2625
$C_{24}H_6N_{14}O_{24}$	Azobis (2,2',4,4',6,6') hexanitrobiphenyl (AHB)	874	1.78	120b	2699
$C_{24}H_6N_{16}O_{24}$	Dodecanitro-3,3-bis (phenylazo)-biphenyl (BisHNAB)	902	1.81	127b	2785
$C_{24}H_9N_9O_{18}$	1,3,5 Tripicrylbenzene (TPB)	711	1.67	81b	2242
NO	Nitric oxide	30	1.30	69b	2212

Table 2.66 Non-aluminized explosive mixtures.

Name	Apparent formula	Composition	(g/mol)	Density (g/cm³)	Explosive Strength (%TNT)	Gurney constant (m/s)
Amatol	$C_{0.62}H_{4.44}N_{2.26}O_{3.53}$	80/20 Ammonium nitrate/TNT	100	1.60	143b, 130m, 123t	2908
Anfo	$C_{0.365}H_{4.713}N_{2.000}O_{3.000}$	96/6 Ammonium nitrate/No. 2 diesel oil	85	1.63	142b	2769
Black powder	$C_{1.25}K_{1.92}N_{5.36}O_{14.1}S_{0.31}$	75/15/10 Potassium nitrate/carbon/sulphur	—	—	17s, 50m, 10t	—
Comp A-3	$C_{1.87}H_{3.74}N_{2.46}O_{2.46}$	91/9 RDX/WAX	100	1.65	157b, 135m, 126p, 109a	2727
Comp B-3	$C_{6.851}H_{8.740}N_{7.650}O_{9.300}$	64/36 RDX/TNT	347	1.713	133m, 130t, 132p, 149b	2843
Comp C-4	$C_{1.82}H_{3.54}N_{2.46}O_{2.51}$	91/5.3/2.1/1.6 RDX/Di-(2-ethylhexyl) sebacate/polysiobutylene/motor oil	100	1.66	160b, 130m, 115p	2801

Table 2.66 *Continued.*

Name	Apparent formula	Composition	(g/mol)	Density (g/cm³)	Explosive Strength (%TNT)	Gurney constant (m/s)
Cyclotol	$C_{5.045}H_{7.461}N_{6.876}O_{7.753}$	77/23 RDX/TNT	288	1.743	159b, 112u, 133p	2979
Dynamite (MVD)	—	75/15/10 RDX/TNT/plasticizers	—	—	105s, 122m	—
LX-14	$C_{1.52}H_{2.92}N_{2.59}O_{2.66}$	95.5/4.5 HMX/estane 5702-F1	100	1.83	167b	3033
Octol	$C_{6.80}H_{10.00}N_{9.20}O_{10.40}$	76.3/23.7 HMX/TNT	388	1.809	115m, 158b, 160p	2965
PBX-9011	$C_{5.696}H_{10.476}N_{8.062}O_{8.589}$	90/10 HMX/estane	329	1.767	153b, 140p	2815
PBXC-116	$C_{1.968}H_{3.746}N_{2.356}O_{2.474}$	86/14 RDX/binder	100	1.65	117b	2342
Pentolite	$C_{2.332}H_{2.366}N_{1.293}O_{3.219}$	50/50 TNT/PETN	100	1.65	126m, 122t, 156b	2970
Smokeless powder	$C_6H_7N_{2.5}O_{10}$	99/1 Nitrocellulose/diphenylamine	275	—	116a, 108p, 93s	—

Table 2.67 Explosive and crater formation data.

Type of soil/rock	Type of explosives	Charge data				Crater data		
		Weight (kg)	Length (m)	Diameter (m)	Radius (m)	Depth (m)	Angle (degrees)	Volume (m³)
Sandstone	*Type A* Semi-gelatine	3.632	0.22	0.132	1.25	0.400	134	0.454
		3.632	0.25	0.132	2.44	0.700	148	4.26
		3.632	0.30	0.132		no crater		
	Type A Gelatine	3.632	0.18	0.132	1.83	0.850	132	3.41
		3.632	0.2	0.132		no crater		
	Ammonia-gelatine	3.632	0.12	0.152	0.750	0.310	130	0.176
		3.632	0.152	0.152	1.83	0.920	130	3.30
		3.632	0.152	0.152		no crater		
	Ammonia-dynamite	3.632	0.23	0.132	1.02	0.310	142	0.937
		3.632	0.23	0.132	1.0	1.700	124	0.5
Granite	*Type A* Semi-gelatine	2.80	0.25	0.10	1.25	0.310	150	0.46
		2.80	0.25	0.10	1.83	0.310	158	0.937
		2.80	0.3048	0.10	0.75	0.1524	154	0.08
		14.50	0.62	0.1524	1.83	0.457	105	1.306
	Type B Semi-gelatine	1.48	0.25	0.16	0.10	1.230	152	2.90
		4.355	0.60	0.10	1.20	0.620	152	0.90
		14.50	0.62	0.1524	3.0	0.62	154	3.55
	Type B Gelatine	8.62	0.47	0.128	3.20	0.40	164	5.26
Marlstone	*Type A* Semi-gelatine	1.544	0.308	0.10	1.0	0.308	138	0.44
		1.544	0.308	0.10	1.80	0.62	118	1.58
		1.544	0.308	0.10	0.16	0.150	112	0.0044
		3.620	1.37	0.13	1.82	1.37	108	5.210
		6.130	0.32	0.13	1.82	1.22	114	5.210
Chalk	*Type A* Semi-gelatine	0.908	0.22	0.08	1.01	0.4	138	0.510
		0.908	0.20	0.08	1.24	0.77	114	2.08
		2.040	0.22	0.13	2.10	1.524	106	7.39
		3.632	0.20	0.152	3.45	1.53	112	11.36
	Ammonia-gelatine	3.632	0.125	0.152	1.70	0.9	130	2.70
		3.632	0.10	0.152	1.70	1.02	118	7.73
		3.632	0.152	0.152	2.10	1.52	106	8.86
		3.632	0.13	0.152		no crater		

Table 2.67 *Continued.*

Charge depth (m)					Particle (a) or flyrock (b) velocity (m/s)					Additional data and velocity V_E of detonation (m/s) of explosive
0.1	0.2	0.3	0.4	0.5	0.6		0.9	1.5		Density = 1153.44 kg/m³ V_E = 3657.6
	b	b			b		b	b		
	60.0	60.0			30.5		15.0	2.0		
								1.5		
								b		
								2.0		
					0.6					Depth 3.05 m and greater Density = 1361.7 kg/m³ V_E = 6095.7
					b					
					30.5					
								1.5		Just broke surface
								b		
								2.0		
			0.5							Density = 1410 kg/m³ V_E = 2591
					b					
					25.0					
							0.9			
							b			
							29.0			
	0.2									Depth 3.15 m and greater Density = 1121.4 kg/m³ V_E = 1981
	b									
	90.0									
		0.3								
		b								
		75.0								
		0.3								
							0.9			
					0.6					Charge depth 2 m
								1.5		
0.1	0.2	0.3	0.4	0.5	0.6	0.7	0.9	1.5	3.0	Charge depth 1.8 m Density = 1153.44 kg/m³ V_E = 3657.6
b										
60.0										
								1.5		
									3.0	
								1.5		
								1.5		
			0.4							Density = 1121.4 kg/m³ V_E = 1981
								1.5		
								1.5		
0.1	0.2	0.3	0.4	0.5	0.6	0.7	0.9	1.5	3.0	Density = 1153.44 kg/m³ V_E = 3657.6
					0.6					
								1.5		
									3.0	

Table 2.68 Projectile weight/charge weight, horizontal distance from explosion or crater radius in soils.

Ratio	Shallow soft rock	Dry clay shale	Wet clay shale	Dense rock
Projectile weight / Charge weight	10^1	$10^{0.3}$	10^1	10^1
	1	3	3	10
	10^0	10^0	10^0	10^0
	2	3	8	30
	10^{-1}	10^{-1}	10^{-1}	10^{-1}
	5	9	10	80
	10^{-2}	10^{-1}	10^{-2}	10^{-2}
	2	10.8	20	300
	10^{-3}	10^{-3}	10^{-3}	10^{-3}
	8	25	40	–
Distance or crater radius (m)	10^{-4}	10^{-4}	10^{-4}	10^{-4}
	20	50	65	350

pressure, can produce on ignition a maximum pressure of more than $8\,\text{N/m}^2$ — a pressure unlikely to be resisted by conventional structures where explosion relief vents are not provided.

In the event of a gas explosion, the response of structures or their components must be assessed and the overpressure loadings generated as a result should be accurately evaluated. In all cases, the gas explosion loadings must not be confused with blast loadings caused by explosives. Considerable data on the full or partial venting of explosions can be gathered from the references in the text.[4.176–4.180]

However, there is still uncertainty about flammable gas or vapour likely to be encountered in domestic or commercial buildings. Whether the supply to such buildings comes from natural or manufactured gas plants, the possible hazards associated with the misuse or accidental leakage are predictable. When leakage occurs, a gas layer is likely to build up near the ceiling. Eventually, with time, this layer extend downwards and explosion occurs when it reaches a source of ignition such as a pilot light, electric contact, heater, etc. The rise and rate of explosion pressure will depend on whether the volume of the room is partially or fully filled. A further complication in the domestic environment arises when there are several inter-communicating rooms or passages full of gas layers which are capable of transmitting the explosion from one compartment to the other. The passage of gas through doorways increases its turbulence and burning rate and can thus cause a violent explosion in a compartment remote from that containing the ignition source. The burning rates of some gases are given in Table 2.72.

Table 2.69 Particle velocity for various soils.

Ratio $\dfrac{\text{Distance (m)}}{\sqrt[3]{W}\ \text{(kg)}}$	Saturated clay	Wet sandy clay	Dense sands	Sandy loam or medium sand	Dry loose sand
1	110	80	—	—	—
2	45	40	30	20	15
3	25	20	15	8	3
4	20	12	16	3	3
5	15	8	3	1.5	1.5
6	12.5	5	2	0.8	7
7	10	4	1.5	0.5	0.35
8	8	3	1	0.3	—
9	6	2.5	0.8	—	—
10	5	1.5	0.5	—	—
V_E (m/s)	5	1.5	0.5	—	—

W = charge weight; V_E = particle velocity.

Table 2.70 Crater dimensions in soils under the effects of high explosives.

Depth of CG of charge (m)/$\sqrt[3]{W}$ (kg)	Clay				Clay and mixed soil				Mixed soil				Sand and mixed soil				Sand			
	d_w	d_d	h_w	h_d	d_w	d_d	h_w	h_d	d_w	d_d	h_w	h_d	d_w	d_d	h_w	h_d	d_w	d_d	h_w	h_d
0	4.40	3.20	1.40	1.20	3.20	2.80	1.20	0.80	2.80	2.60	0.80	0.70	2.60	2.40	0.90	0.70	2.40	2.10	0.70	0.60
0.5	5.40	4.25	2.00	1.70	4.25	3.38	1.70	1.50	3.58	3.50	1.50	1.40	3.50	3.20	1.40	1.20	3.20	2.90	1.20	1.00
1.0	5.80	5.00	2.40	1.90	5.00	4.50	1.90	1.70	4.50	4.20	1.70	1.60	4.20	3.55	1.60	1.35	3.55	3.40	1.35	1.25
1.5	6.25	5.35	2.60	2.20	5.35	4.60	2.20	1.80	4.60	4.25	1.80	1.55	4.25	4.00	1.55	1.40	4.00	3.60	1.40	1.20

2.0	6.50	2.50	5.25	2.00	4.65	2.00	4.30	1.70	4.30	1.60	4.30	3.55	1.60	3.55	3.55	1.30	3.49	1.30	1.10
2.25	6.25	2.40	5.10	1.70	4.70	1.70	4.70	1.50	4.40	1.25	4.40	3.50	1.25	3.50	3.50	0.55	3.10	0.55	0.25
3.0	5.80	1.90	4.50	1.25	4.00	1.25	4.00	0.70	3.50	0.40	3.50	2.80	—	2.80	2.60	—	—	—	—
3.5	5.50	1.40	3.50	0.48	3.00	—	—	—	—	—	—	—	—	—	—	—	—	—	—
4.0	4.50	0.55	3.25	0.50	—	—	—	—	—	—	—	—	—	—	—	—	—	—	—

d and h denote crater diameter and crater height, respectively; subscripts d and w denote dry and wet soils respectively; d_d, d_w, h_d and h_w are calculated from (crater diameter or height/$\sqrt[3]{W}$), where W is the charge weight.

Table 2.71 Crater diameter and depth for mass and reinforced concrete.

Crater diameter or height (m)/ $\sqrt[3]{W}$ (kg)	Mass concrete		Reinforced concrete	
	Dia (m)	Depth (m)	Dia (m)	Depth (m)
0	0.61	0.08	0.18	0.06
0.25				
0.5	0.13	0.25	0.15	0.12
1.0	0.53	0.43	1.34	0.26
1.50	1.86	0.46	1.41	0.21
2.0	1.92	0.44	1.39	0.15
2.50	2.06	0.31	0.46	0.09
3.0	1.62	—	0.76	—
3.25	1.52	—	0.69	—

$f'_c = 0.87f_{cu}$ = concrete cylindrical strength = $26\,N/mm^2$; $f_{cu} = 30\,N/mm^2$; W = charge weight (kg).

Table 2.72 Burning velocities of some gases.

Gas (or vapour)	Burning velocity (m/s)
Methane (natural gas)	0.37
Propane	0.46
Butane	0.40
Hexane	0.40
Ethylene	0.70
Town gas	1.2
Acetylene	1.8
Hydrogen	3.4

2.5.2.2 Gas and wave pressure versus time

An ideal gas has no frictional forces or bonds between its particles. The internal energy is treated as zero. The entire internal energy is then equal to the kinetic energy and thus depends only on the absolute temperature. Where the internal energy is proportional to temperature, the gas in such an ideal case is referred to as *polytropic gas*. The entropy of this gas is a function of temperature and volume or pressure and volume.

After the instantaneous explosion, the gases generally begin to expand and, owing to the speed of this phenomenon, the process is adiabatic, i.e., there is no transfer of heat to the surrounding medium. The detonation wave from the explosion propagates in all directions and its front creates an impact on the surrounding medium by propagating a shock wave in it. At the same time, a reflected wave expands towards the centre of the source. At the centre of the source, the front of the wave is contracted and the new reflected waves propagate away from the centre. The procedure of creating new waves and repeating the process, causes

(1) the explosive gases decaying pulsation of reflected waves;
(2) the fading of the waves in the medium;
(3) the volume of the gases during reflection to increase until it reaches a maximum, thus making the explosive gas pressure low compared with that in the surrounding medium;
(4) an overpressure of the medium, compelling the gases and the medium to move in the opposite direction towards the centre of the source;
(5) a further increase in the overpressure and a new expansion, resulting in a damped pulsation of these explosive gases.

Since the explosion is characterized by two parameters, namely the maximum pressure and the rate of pressure rise, the pressure pulse generated becomes a complex phenomenon which depends on a number of factors. These are the type of gas and its medium, the internal/external blast and the boundary conditions such as the unvented and vented medium of the surrounding structures. A typical pressure/time forcing function is given in Fig. 2.59. A typical pressure/time history of a vented and an unvented explosion is given in Fig. 2.60. A vented or partially confined vented explosion and the geometry and materials used in the surrounding structures have various boundary conditions. Despite a great deal of research carried out so far on gas explosions in vented and partially vented structures, knowledge about pressure-pulse generation is still in its infancy. More comprehensive data are needed for gas explosions and the generation of the pressure pulse in various environments.

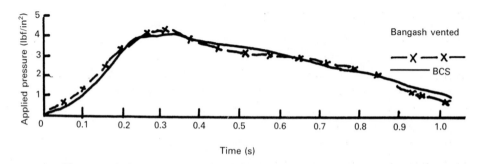

Fig. 2.59 Typical pressure/time forcing function from a gas explosion. (Courtesy of the British Ceramics Society.)

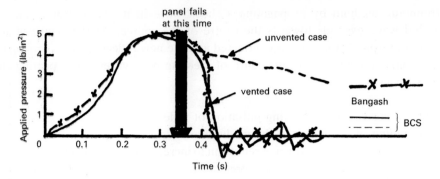

Fig. 2.60 Pressure versus time for a vented and an unvented explosion. (Courtesy of the British Ceramics Society.) (Note: $1\,lbf/in^2 = 6.895\,kN/m^2$.)

2.5.2.3 Gas concentration and pressure due to explosion

Experiments have been carried out on gas explosiohs in load-bearing brick structures. Tables 2.73 to 2.75 and Figs 2.59 to 2.61 show some data from experiments carried out on models.

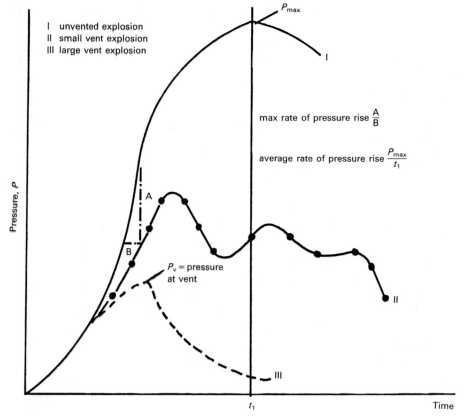

Fig. 2.61 Typical pressure/time history during an explosion — a comparative study. (Courtesy of the British Ceramics Society.)

Table 2.73 Theoretical and experimental results: a comparative study (Group 1).

Reference	Description of experiment	Rasbash[4.88] Theory I	Theory II	Cubbage and Simmonds[4.87] (1st peak)		Dragosovic[4.81] (1st peak)	Harris and Briscoe[4.79]	Cubbage and Marshall[4.91, 4.92] (1st peak)		Burgoyne and Wilson[4.88]	Runes and Decker[4.95] (1st peak)	
Cubbage and Simmonds	Explosion of town gas-air and methane-air mixtures in cubical drying ovens: $V = 0.23\,m^3$; $A = 0.297\,m^2$ $W = 21.4\,kg/m^2$ or $W = 7.8\,kg/m^2$									Decker		
	Town gas	36.8 22.3	11.5 11.5	26.9	14.1	— —	— —	151.5	54.7	Not applicable when the explosion pressure is less than $35\,kN/m^2$	15.8	—
	Methane	11.5 7.0	3.6 3.6	8.40	4.90	— —	— —	14.6	5.3		—	—
	CS_2	18.3 —	5.7 —	11.2	—	— —	— —	37.0	—		—	—
	Acetone	— 33.8	4.3 17.5	7.0	26.6	— —	— —	20.7	125.4		8.62	50.5
	Ether	— —	4.6 —	9.1	—	— —	— —	24.6	—		13.0	15.8
Burgoyne and Wilson	Explosion of pentane-air mixtures in cylindrical chambers										Decker	
	(1) $P_v = 35\,kN/m^2$; 3.25% C_5H_{12}; $V = 17\,m^3$	49.68	46.6	—		26.46	—	—		394.0	168	
	(2) $P_v = 28\,kN/m^2$; 3.5% C_5H_{12}; $V = 17\,m^3$	61.2	47.0	—		21.3	—	—		222.0	158	
	(3) $P_v = 16.8\,kN/m^2$; 2.7% C_5H_{12}; $V = 17\,m^3$ $W = 0$ for all	45.5	35.9	—		33.1	—	—		394.0	158	
Harris and Briscoe	Explosion of pentane-air mixture in a $1.7\,m^3$ vessel											
	3.0% C_3H_{12} $K = 5.1$	13.0	11.9	—		4.4	14.0	—		—	42.0	
	$K = 20.2$	22.3	21.10	—		7.3	196.0	—		258.0	74.2	
	$K = 75.8$	183.0	178.0	—		30.7	—	—		698.7	625.0	
	$W = 0$ for all											

Table 2.74 Theoretical and experimental results: a comparative study (Group 2).

Reference	Description of experiment	Maximum explosion pressure (kN/m²) obtained from experiment and theory								Reference results
		Rasbash[4.88] Theory I	Theory II	Cubbage and Simmonds[4.87] (1st Peak)	Dragosovic[4.81] (1st Peak)	Harris and Briscoe[4.79]	Cubbage and Marshall[4.91, 4.92] (1st Peak)	Burgoyne and Wilson[4.88]	Decker[4.95] (1st Peak)	
Bromma[4.38]	Explosion of propane-air mixtures and acetylene-air mixtures									
	(1) 4.5% C_3H_8; $V = 11.25 \text{ m}^3$; $K = 1.11$	8.438	2.75	0.565	8.28	—	—	—	13.3	—
	(2) 6% C_3H_8; $V = 70 \text{ m}^3$; $K = 1.38$	—	3.45	0.566	3.0	—	—	—	16.8	1.275
	(3) 6% C_3H_8; $V = 200 \text{ m}^3$; $K = 1.11$ $P_v = 5.49 \text{ kN/m}^2$ (1) and (2) open vent	8.435	8.20	2.8	8.49	—	5.49	—	13.3	5.88
Howard and Karabinis[4.88]	Explosion of propane-air mixtures of 5% in a rectangular building of volume 81 m³ in a vessel wall of least area fitted with vent panel									
	(1) $P_v = 1.4 \text{ kN/m}^2$; $K = 1$	4.87	3.9	—	4.4	—	1.4	—	2.2	4.9
	(2) $P_v = 7.35 \text{ kN/m}^2$; $K = 1$	11.3	9.83	—	10.3	—	7.35	—	3.17	10.08
	(3) Two walls vented $P_v = 1.4 \text{ kN/m}^2$	—	5.62	—	4.4	—	2.69	—	—	2.1
	(4) As (1) $P_v = 3.15 \text{ kN/m}^2$	5.0	2.72	—	6.15	—	3.15	—	3.15	4.69

Table 2.74 *Continued.*

Reference	Description of experiment	Maximum explosion pressure (kN/m²) obtained from experiment and theory							Reference results
		Rasbash[4.88] Theory I Theory II	Cubbage and Simmonds[4.87] (1st Peak)	Dragosovic[4.81] (1st Peak)	Harris and Briscoe[4.79]	Cubbage and Marshall[4.91, 4.92] (1st Peak)	Burgoyne and Wilson[4.88]	Decker[4.95] (1st Peak)	
Buckland[4.88]	Explosion of layers of natural gas-air. Enclosure: $3.7\,m \times 3\,m \times 2.4\,m$. Vent: $3\,m \times 2.4\,m$ wall								
	(1) 10% CH_4; $P_v = 0.7\,kN/m^2$; $K = 8$ 1.2 m layer	1.37 17.0	16.44	3.7	—	0.4	—	118.1	8.3
	(2) 10% CH_4-air; $P_v = 1.7\,kN/m^2$; $K = 2$ 1.5 m layer	2.87 5.77	4.11	4.7	—	1.7	—	29.5	6.6
	(3) 10% CH_2-air; $P_v = 7.2\,kN/m^2$; $K = 2.5$ 1.8 m layer	11.2 12.29	5.14	10.2	—	7.2	—	36.9	11.0
	(4) same as (1); $P_v = 0.7\,kN/m^2$; $K = 4$ 0.9 m layer	13.82 8.22	8.22	12.0	—	9.0	—	59.1	6.86

Table 2.75 Theoretical and experimental results: a comparative study (Group 3).

Reference	Description of experiment	Explosion pressure (kN/m²) obtained from experiment and theory							Reference results
		Rasbash[4.88] Theory I	Theory II	Cubbage and Simmonds[4.87] (1st Peak)	Dragosovic[4.81] (1st Peak)	Cubbage and Marshall[4.91, 4.92] (1st Peak)	Burgoyne and Wilson[4.88]	Decker[4.95] (1st Peak)	
Dragosovic	Explosion of methane-air mixtures Dimensions of building: 4 m × 2 m × 2.6 m, volume 20.8 m³ Vent covers of hardboard, glass:								
	(1) $P_v = 0$; $K = 1$ } Hardboard	0.35	2.04	0.35	3.0	—	—	81.0	—
	(2) $P_v = 0$; $K = 1.593$ } Hardboard	0.35	3.24	0.35	1.5–3.0	—	—	67.0	—
	(3) $P_v = 2.99$ kN/m²; $K = 1.593$ (glass cover)	4.70	6.14	0.35	5.9	—	—	67.0	—
	(4) $P_v = 4.75$ kN/m²; $K = 5.09$	7.48	15.10	0.35	7.75	—	—	–	—
	(5) $P_v = 21.6$ kN/m²; $K = 1.7$	21.95	25.06	0.35	24.6	—	—	1.43	—
Solberg[4.88]	Explosion of propane-air mixtures Dimensions: 4.0 m × 3.5 m × 2.5 m, volume 35 m³								
	(1) $K = 4.375$	0.35	10.83	0.35	3.0	—	—	45.3	8.98
	(2) $K = 8.75$	0.35	21.65	0.35	3.0	—	—	90.6	39.4
	(3) $K = 17.5$	0.35	43.3	0.35	3.0	—	—	181.0	60.9
Yao[4.88]	Explosion of propane-air and hydrogen-air, volume 0.765 m³			2nd peak	2nd peak				
	(1) $K = 5.1$ C_3H_8-air	17.79	23.6	12.75	9.35	—	—	43.2	—
	(2) $K = 1.44$ C_3H_8-air	5.6	20.7	3.6	7.14	—	—	13.23	—
	(3) $K = 1.61$ C_3H_8-air (free vents)	1.28	3.99	4.38	3.09	—	—	–	—

2.5.3 Nuclear explosions

2.5.3.1 Introduction

A nuclear explosion, in general, results from the very rapid release of a large amount of energy associated with high temperatures and pressures. Several basic differences exist between nuclear explosions and explosions caused by high explosives. Some of the major differences are listed below:

(1) Nuclear explosions are many thousands of times more powerful than the conventional type of explosions.
(2) For the release of a given amount of energy, the mass of the nuclear residue is comparatively smaller and is immediately converted into hot and compressed gases.
(3) The temperatures reached in a nuclear explosion are a very much higher, thus assisting the emission of a large proportion of energy in the form of light and heat. This is known as *thermal radiation*. The remaining substances, unlike conventional explosives, emit radiation for a certain period of time.

Owing to these and many other differences, the effects of the nuclear detonations require special consideration, including their dependence on the type of burst, i.e., air, surface and subsurface. An accurate assessment is required for blast loadings in all such cases with target responses.

2.5.3.2 Air blast loading

It is desirable to consider in some detail the phenomena associated with waves due to air blast. A difference in the air pressure acting on different surfaces of a structure produces a force on that structure. The destructive effect will be felt due to overpressure — the maximum value of pressure at the blast wave or shock front. This maximum value is sometimes called *peak overpressure*. The other phenomena are dynamic pressure, duration and time of arrival.

A typical 'fireball' associated with air burst occurs. Immediately after the formation of the 'fireball' it grows in size, thus engulfing the surrounding air and decreasing its own temperature. The 'fireball' rises like a hot-air balloon. The rate of rise of the radio-active cloud (from a 1-megatonne air burst) ranges from 330 miles/hour at 0.3 minutes to 27 miles/hour at 3.8 minutes. The expansion of the intensely hot gases at high pressures in the fireball causes a shock wave, moving at high velocity. The pressure rises very sharply at the moving front and falls off toward the interior region of the explosion. As the blast wave travels, the overpressure decreases and the pressure behind the front falls off. After a short interval, when the shock waves have travelled a certain distance, the pressure behind the front drops below atmospheric and is known as a *negative phase* of the blast wave forms. During the negative phase, a partial vacuum is produced and the air is sucked in. The negative phase is comparatively longer, the pressure will essentially return to ambient. The peak values of the underpressure are generally small compared with the peak positive overpressures.

Since the degree of blast damage depends largely on the drag force associated with strong winds and is influenced by the shape and size of the structure, the net pressure acting on the structure is called the *dynamic pressure* and is proportional to the square of the wind velocity and to the density of the air behind the shock front. Figure 2.62(a) and (b) shows comparisons between the overpressure and the dynamic pressure with distance and time. For very strong shocks the dynamic pressure is larger than the overpressure, but below $480\,kN/m^2$ (4.7 atmospheres) overpressure at sea level, the dynamic pressure is smaller. As shown in Fig. 2.62, the dynamic pressure decreases with increasing distance from the explosion centre. Figure 2.63 and Table 2.76 show the peak 'free-air' overpressure of shock, range versus duration.

2.5.3.3 Blast loads from a surface burst

When the incident blast wave from an explosion in air strikes a more dense medium such as land or water, it is reflected. The front of the blast wave in the air will assume a hemispherical shape, as shown in Fig. 2.64. Since there is a region of regular reflection, all structures on the surface, even close to ground zero, are subjected to air blast. Some of the blast wave energy is transferred into the ground. A minor oscillation of the surface is experienced and a ground shock is produced. For large overpressures with a long positive-phase duration, the shock will penetrate some distance into the ground and will damage buried structures.

When the front of the air blast wave strikes the face of the structure, reflection occurs. As the wave front moves forward, the reflected overpressure on the face drops rapidly to that produced by the blast wave without reflection, plus an added drag force due to wind. At the same time, the air-pressure wave diffracts around the structure and is entirely engulfed by the blast. The damage caused by diffraction will be determined by the magnitude of the loading and by its duration. If the structure has openings, there will be a rapid equalization of pressure between the inside and outside of that structure. The diffraction loading of the structure as a whole will be decreased. Since large structures have openings, diffraction and drag must not be ignored.

The loads computed for the surface explosion shall be P_{so} and $2.3\,P_{so}$ for roof/floors and walls respectively, where P_{so} is the peak incident wave pressure. Figure 2.63 and Table 2.76 show the damage-distance relationship and peak overpressure failure effects on structural components, respectively.

2.5.3.4 Shallow and deep underground explosion loadings

Shock damage to structures placed underground can be evaluated using computed codes and experiments. The degree of damage for a shallow explosion can be

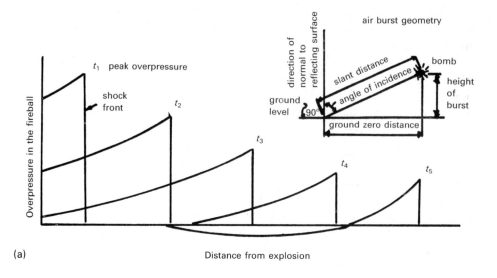

(a)

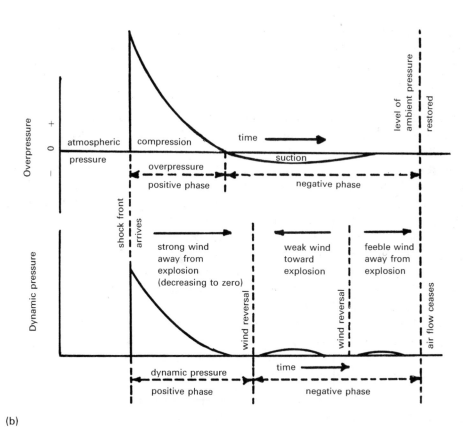

(b)

Fig. 2.62 (a) Time-dependent variation of overpressure with distance; (b) variation of overpressure and dynamic pressure with time.

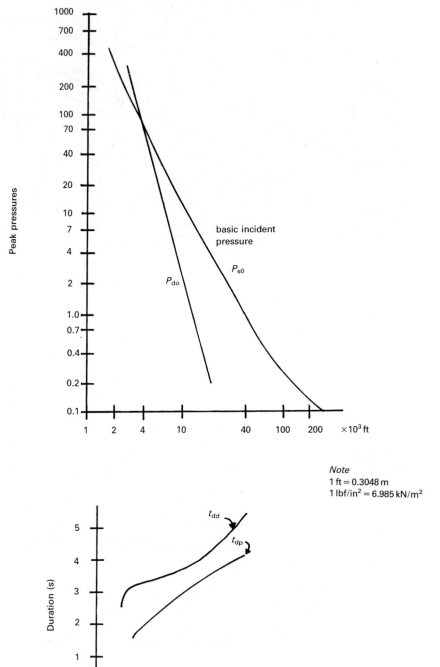

Fig. 2.63 Peak pressure, duration and range.

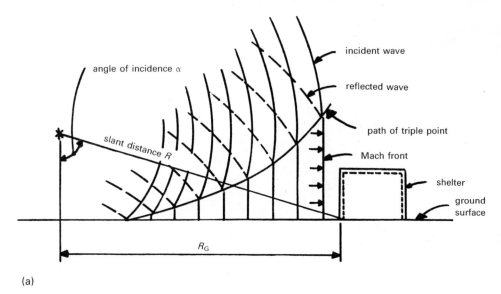

(a)

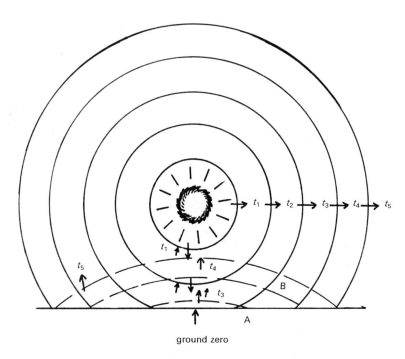

(b)

Fig. 2.64 (a) Air and (b) surface burst: incident and reflected waves meet.

Table 2.76 Peak overpressure, arrival time and duration.

Distance	Peak overpressure kPa (psi)		Wind speed m/s (mph)		Arrival time (s)		Duration (s)	
km (miles)	1 Mt	10 Mt	1 Mt	10 Mt	1 Mt	10 Mt	1 Mt	10 Mt
1 (0.6)	980 (140)	4200 (660)	980 (2200)	—	0.9	—	0.9	1.0
2 (1.25)	200 (29)	1050 (150)	400 (900)	980 (2200)	4.0	1.8	1.8	2.0
3 (1.9)	130 (18)	380 (54)	180 (400)	805 (1800)	6.0	4.7	2.3	2.4
4 (2.5)	63 (12)	220 (31)	128 (289)	490 (1100)	9.0	8.0	2.6	3.7
5 (3.1)	56 (8)	170 (24)	106 (238)	277 (620)	12	11	2.8	4.2
6 (3.8)	42 (6)	135 (19)	81 (182)	200 (450)	14	12.25	3.0	4.7
7 (4.4)	33 (4.7)	110 (16)	69 (155)	170 (380)	18	16	3.2	5.0
8 (5)	27 (3.8)	90 (13)	56 (125)	143 (320)	22	18	3.4	5.3
9 (5.6)	24 (3.4)	77 (11)	48 (108)	125 (280)	26	20	3.5	5.5
10 (6.2)	21 (3)	63 (9)	40 (90)	114 (255)	28	22	3.6	5.9
12 (7.5)	15 (2.2)	46 (6.5)	31 (70)	96 (215)	31	30	3.7	6.3
15 (9.4)	10 (1.5)	34 (4.8)	24 (53)	72 (160)	45	38	3.8	6.7
20 (12.5)	7 (1)	22 (3.3)	17 (37)	46 (103)	59	53	4.0	7.5
25 (16)	5 (0.75)	15 (2.2)	9 (20)	32 (72)	70	67	4.2	7.8

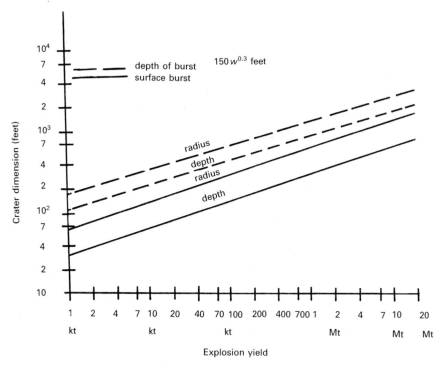

Fig. 2.65 Explosion yield versus crater dimensions for specific yields.

related to the apparent crater radius. The dependence of the crater radius and crater depth upon the depth of burst is shown in Fig. 2.65. For shallow buried structures for steel and concrete arches, the peak pressures are given in Table 2.77.

The Home Office gives the value of P_{so} (overpressure) as a full value for dry and high water level cases when roofs and floors are considered. For walls in dry ground, the value of P_{so} is 0.5 P_{so}. For sensitive structures above ground, the structural elements must fail under the minimum overpressures given in Table 2.78.

2.6 Dust explosions

2.6.1 Introduction

There are similarities between dust and gas explosions when the particle size in particular is small and the turbulence level is low. The effects on the explosion pressure of a venting dust explosion are the same as with a gas explosion. The vented gas explosion has a number of empirical formulae for computing the explosion pressures or vent areas. In the case of dust explosions not many such formulae exist and, instead, scaling laws must be relied on. The classification 'St' for the potential of dust explosion under conditions of moderate turbulence is given in Table 2.79.

Table 2.77 Peak pressures for shallow buried structures for steel and concrete arches.

Damage type	Steel structures (N/m^2)	Concrete structures (N/m^2)
Severe*	750–900	3100–4500
Moderate[†]	600–700	2000–3100
Light[§]	450–600	1800–2000

* collapsed condition.
[†] cracking, yielding, spalling.
[§] slight cracking and yielding.

Table 2.78 Structural elements versus failure pressures.

Sensitive structural element	Failure type	Overpressure (N/m^2)
Large and small glass windows	Shattering usually, occasional frame failure	7.5–15
Corrugated asbestos siding	Shattering failure	15–30
Corrugated steel or aluminium panelling	Connection failure with buckling	15–30
Brick wall panel	Shear and flexure failures	105–125
Wood siding panels	Failure occurs at the main connections	15–30
Concrete wall panels, unreinforced	Shattering of the wall	30–45

Table 2.79 Classification of dust explosion under conditions of moderate turbulence.

Dust explosion class	Explosion rate constant $K = (dP/dt)V_{max}^{1/3}$
St0	does not explode
St1	$K \leq 2 \times 10^4$ kPa m/s
St2	$2 \times 10^4 < K \leq 3 \times 10^4$ kPa m/s
St3	$K > 3 \times 10^4$ kPa m/s

The maximum overpressure in a vented zone develops because of:

(1) burning of the dust, causing release of heat and increasing pressure;
(2) flow of the unburnt and burning dust cloud, causing a decrease in the pressure.

The main factors that influence the rate of heat increase are:

(1) the chemical composition of the dust, with some moisture;
(2) the chemical composition and initial pressure and temperature of the gas;
(3) the distribution of particle sizes and shapes in the dust;
(4) the degree of dust dispersion that determines the effective specific surface available to the combustion process in the dust cloud;
(5) the distribution of dust concentration in the actual cloud;
(6) the distribution of initial turbulence in the actual cloud;
(7) the possibility of *in situ* generation of high turbulence levels by the rapid flow of the still unburnt dust cloud;
(8) the possibility of flame front distortion by acoustic wave interaction.

2.6.2 Overpressure and vent pressure versus burning rate

Figure 2.66 shows a general diagram giving a relationship between overpressure and vent opening pressure versus ignition rate. Table 2.80 gives dust explosion results and Table 2.81 gives data on coal dust/air explosions. Figure 2.67 gives explosion pressures versus vent areas.

Explosions in silos do occur, Bartknecht[4.76] has applied the cube root law to large-volume silos by allowing high flame speeds due to turbulence and the outflow of the material at vent. It was assumed that the entire cross-section of the

Table 2.80 Dust explosion results: corn starch/air explosion in vessels of volume (V) 0.0285 and 1.81 m³; maximum rate of pressure rise (Maisey[4.77]): 2200 psi/s; burning velocity: 0.26 m/s.

Description of experiment	Maximum pressure (kN/m²)	Cubbage and Simmonds[4.87] 2nd Peak (6)	Rasbash equations[4.88] (8) and (9)	Decker[4.95] cube root law
$V = 0.0285\,m^3$				
$K =$ 7.17	9.5	10.35	14.45	26.1
8.36	12.39	12.08	16.87	31.3
10.04	17.25	14.51	20.26	37.5
12.55	25.76	18.31	25.32	47.0
16.74	45.69	24.17	33.77	62.6
$V = 1.81\,m^3$				
$K = 1.784$	9.5	2.58	3.60	9.9
2.08	12.39	3.01	4.20	11.5
2.50	17.25	3.61	5.04	13.9
3.12	25.76	4.51	6.29	17.3
4.40	45.69	6.36	8.88	24.4

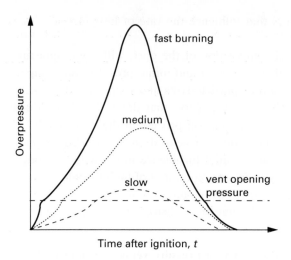

Fig. 2.66 Overpressure and vent opening pressure versus ignition rate.

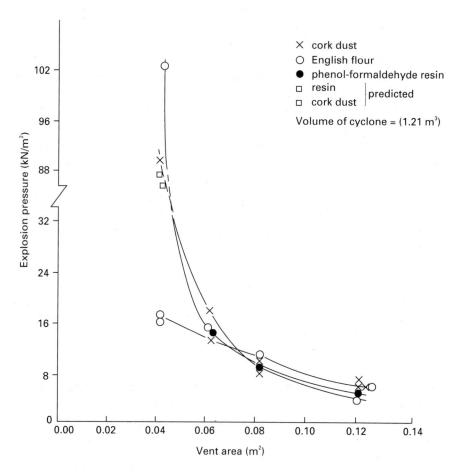

Fig. 2.67 The effect of vent area on explosion pressure in a cyclone.

Table 2.81 Coal dust/air explosion measurements from Donat[4.74] in vessels of volume (V) 1 and 30 m^3; maximum rate of pressure rise (Maisey[4.77]): 2300 psi/s; burning velocity: 0.27 m/s.

Description of experiment	Maximum pressure (kN/m^2)	Rasbash equation[4.88] (4)	Rasbash equations (8) and (9)		Decker[4.95] cube root law
$V = 1\,m^3$					
$K = 2.5$	21.0	20.2	15.74	8.07	9.8
3.33	22.05	21.45	17.45	9.81	13.0
5.0	25.2	23.95	21.0	13.31	19.6
10.0	55.6	31.45	31.45	23.8	39.2
20.0	110.2	46.45	52.4	44.7	78.4
$V = 30\,m^3$					
$K = 2.41$	28.0	19.62	15.5	7.9	9.42
3.22	31.5	20.8	17.24	9.58	12.6
4.82	35.7	23.2	20.6	12.9	18.84
9.65	57.8	30.5	30.72	23.1	37.7
19.3	147.0	45.0	50.9	43.3	75.4

silo roof could be used for relief venting, regardless of the volume. Figures 2.68 and 2.69 give guidelines for the protection of silos.

Dust explosions in ducts have been investigated by Brown[4.65] The results are no different from those produced for gas explosions in ducts. In both cases, vents near to the point of the source of ignition significantly lower the explosion pressure of the vented explosions. If the vented ducts are fitted outside a vent, then the pressure inside the container will be increased relative to a vent exhausting into an open space. The German Standard VDI gives a graph, Fig. 2.70, which estimates the effect of duct length on the internal pressures for dust explosion pressures or vented areas in vented dust explosions. All methods given below require experimental data.

(1) The K_{st} method[4.70–4.75] which is mostly used to size vent areas.
(2) The Schwab and Othmer method gives excellent predictions at both low and high explosion pressures[4.58, 4.67–4.75] This method is best used in conjunction with other methods.
(3) Maisey's equivalence coefficient and equivalence burning velocities method tends to underestimate experimental explosion pressures[4.77]
(4) Heinrich's method has some success in predicting experimental explosion pressures[4.75] With St1 and St2 dusts, this method can provide upper estimates of pressures and vented areas. This method must not be used with St2 dust.
(5) Palmer's method accurately predicts experimental explosion pressures for St1 and St2 dusts[4.83]
(6) The Rust method is more accurate for St1 than for St2 and St3 dusts[4.97]

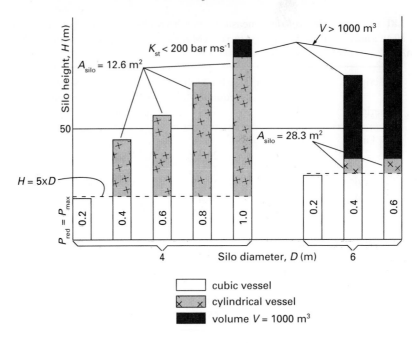

Fig. 2.68 Dusts: influence of pressure resistance (i.e. P_{red}) of silos on the allowable height, by application of the nomograms.

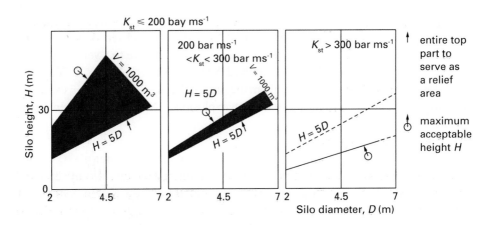

Fig. 2.69 Influence of dust explosion class on the maximum acceptable height of silos, when the nomograms are used ($P_v = 0.1$ bar, $P_{max} = 0.4$ bar).

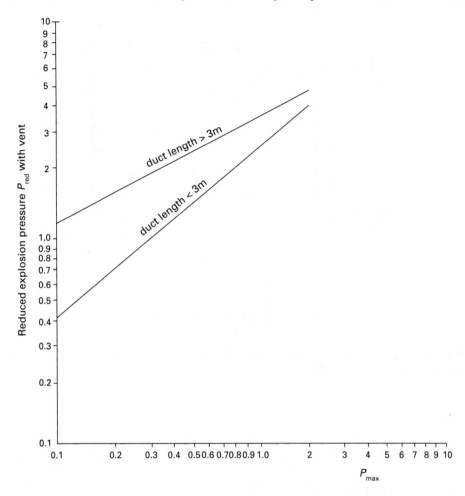

Fig. 2.70 Effect of vent ducts on vented explosion pressures.

2.7 Underwater explosions

Underwater explosion phenomena are subject to a number of physical laws and properties, including the physical conditions at the boundary of the explosive or element of detonation and the surrounding water. A physical relationship is necessary between the detonation and the propagation of disturbances. Owing to the dynamic properties of water (in the regions surrounding an explosion) the pressures are generally large and the wave velocities are not independent of such pressures. After detonation, secondary pressure pulses are generated. Such shock waves are dominant to certain distances and their character may be affected by factors such as viscosity and refraction by velocity gradients in the water.

Experimental pressure-time results as given by Hilliar[4.250] are shown in Fig. 2.71, together with an idealized curve. This curve was also examined at 14 feet

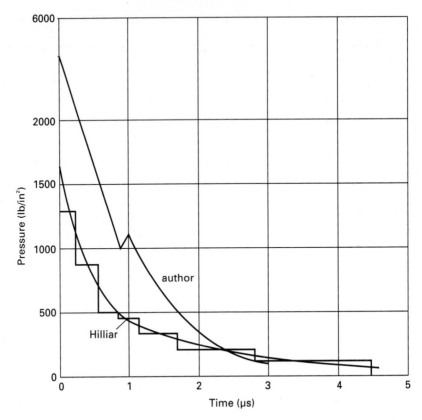

Fig. 2.71 Hilliar pressure-time curve.[4.250] ($1\,\text{lb/in}^2 = 15.444 \times 10^6 \text{ N/m}^2$.)

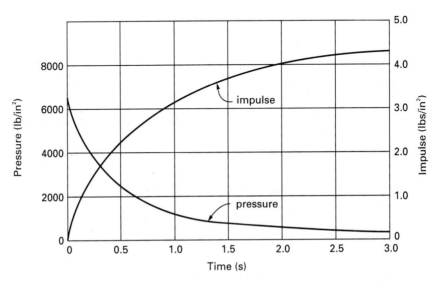

Fig. 2.72 Pressure-pulse-time relationship.[4.266] ($1\,\text{lb/in}^2 = 15.444 \times 10^6 \text{ N/m}^2$.)

Table 2.82 Pressure−reduced time relationship of the shock wave caused by 50/50 PETN−TNT.

Charge weight W (lb)	80.0	51.0	3.8
Distance R (ft)	14.0	11.9	5.0
$W^{1/3}/R$ (lb$^{1/3}$/ft)	0.308	0.312	0.312
Peak pressure P_m(lb/in^2)	5910.0	6060.0	6040.0
Reduced time constant $\theta/W^{1/3}$ (μs/lb$^{1/3}$)	69.6	72.8	69.7
Reduced impulse $I/W^{1/3}$ (lbs/in^2/lb$^{1/3}$)	0.604	0.604	0.558
Reduced energy density $E/W^{1/3}$ (inlb/in^2/lb$^{1/3}$)	287.0	263.0	273.0

1 lb = 0.4536 kg; 1 lb/in^2 = 15.444 × 10^6 N/m^2; 1 ft = 0.3048 m.

Table 2.83 Kirkwood and Motroll results.

Distance (charge radii)	Velocity U (V_s) (m/s)	Peak pressure (kilobars)
1.00	5710	>70
	4885	>70
1.86	2675	17.3
2.73	2290	9.5
3.60	1975	5.0
4.46	1875	3.8
5.33	1865	3.7

Table 2.84 Peak pressures (in lb/in^2 × 10^3, underwater of 50 lb TNT at various distances of detonation.

Distance	Pressure (lb/in^2 × 10^3)
Far end (ft)	
2	5.8
4	2.5
8	1.8
20	0.65
40	0.35
Near end (ft)	
2	5.0
4	2.3
8	1.4
20	0.5
40	0.25
Offside (ft)	
2	15
4	12
8	8
20	4
40	1.6

(4.27 m) from an 80 lb (36.3 kg) 50/50 pentolite charge by the author. Despite limitations, based on a number of factors, reasonable figures were achieved for the pressure−reduced time relationship of the shock wave and they are given in Table 2.82. These results are viewed in the light of calculations produced by Kirkwood and Motroll.[4.266] Their results are given in Table 2.83 for fresh water.

Typical curves for a pressure-time impulse are given in Fig. 2.72. Table 2.84 shows peak pressures for 50 lb TNT, at detonation distances measured by the author.

3

BASIC STRUCTURAL DYNAMICS

3.1 General introduction

Most loads acting on structures are dynamic in origin. These loads can be suddenly applied or allowed to reach full magnitude after a considerable delay. On the other hand, the structures will have various degrees of freedom with unclamped or clamped free or forced vibrations. These need to be discussed prior to the introduction of impact and explosion analysis and design.

3.2 Single-degree-of-freedom system

If a system is constrained such that it can vibrate in only one mode with a single co-ordinate system (geometric location of the masses within the system), then it is a single-degree-of-freedom system.

3.2.1 Unclamped free vibrations

A mass m is suspended by a spring with a stiffness k (force necessary to cause unit change of length). Let the mass m be displaced vertically as shown in Fig. 3.1. Then, with given restraints,

$$F - k\delta_{ST} = 0 \tag{3.1}$$

where F or $W = mg$, $\mathbf{g} =$ acceleration due to gravity and $\delta_{ST} =$ static deflection.

The mass is released and displaced from the equilibrium position. The co-ordinate x then defines the position of the mass m at any time and is taken to be positive when moving in a downward direction. Figure 3.2 shows the new positions.

3.2.1.1 Method 1: using Newton's second law of motion

This law states that the magnitude of the acceleration of a mass is proportional to the resultant force acting upon it and has the same direction and sense as this force. The following equations are obtained:

$$m\frac{\mathrm{d}^2x}{\mathrm{d}t^2} = -k(\delta_{ST} + x) + F \tag{3.2}$$

$$m\ddot{x} = -kx \text{ or } m\ddot{x} + kx = 0 \tag{3.3}$$

235

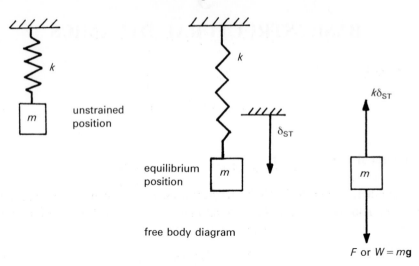

Fig. 3.1 The mass and its equilibrium position.

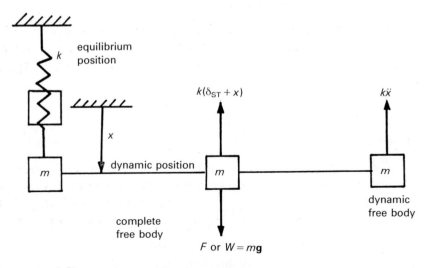

Fig. 3.2 The mass displaced from the equilibrium position.

3.2.1.2 Method 2: energy method

For a conservative system, the total energy of the system (potential energy (PE) plus kinetic energy (KE)) is unchanged at all times. Thus

$$KE + PE = \text{constant}; \quad \text{or} \quad \frac{d}{dt}(KE + PE) = 0 \tag{3.4}$$

$$KE = \tfrac{1}{2}m\dot{x}^2 \tag{3.5}$$

$$\text{PE} = \int_x^0 [F - k(\delta_{ST} + x)]dx = -\int_x^0 kxdx = \tfrac{1}{2}kx^2 \qquad (3.6)$$

Using equation (3.5)

$$\frac{d}{dt}(m\dot{x}^2 + kx^2/2) = 0 \qquad (3.7)$$

Hence

$$(m\ddot{x} + kx)\dot{x} = 0 \text{ or } m\ddot{x} + kx = 0 \qquad (3.8)$$

Since the energy balance holds all the time, including the beginning stage, it is easy to stipulate values for x and $\dot{x}$. Assuming $x = x_0$ and $\dot{x} = 0$ at time $t = 0$ for the prescribed initial conditions, then

$$\text{KE} + \text{PE} = \tfrac{1}{2}kx_0^2 \qquad (3.9)$$

$$\tfrac{1}{2}m\dot{x}^2 + \tfrac{1}{2}kx^2 = \tfrac{1}{2}kx_0^2 \qquad (3.9a)$$

Dividing both sides by $\tfrac{1}{2}kx_0^2$, the following dimensionless equation results:

$$[\dot{x}/\sqrt{(k/m)}x_0]^2 + (x/x_0)^2 = 1 \qquad (3.10)$$

Equation (3.10) can be plotted as a circle of radius unity. The loss or gain of the potential and kinetic energy can be estimated from this circle at various positions. This circle will also show the conversion of potential energy to kinetic energy, etc.

3.2.2 Solution of the equation

3.2.2.1 Method 1

The value of x must be a function such that its second derivative with respect to time will be proportional to the negative value of the function itself. Cosine and sine functions have just this property. Since the equation is of order 2, the solution must contain two arbitrary constants. Hence the value of x is written as

$$x = A \sin\omega t + B \cos\omega t \qquad (3.11)$$

where $\omega^2 = k/m$. Substitution of equation (3.11) into equation (3.3) shows that the differential equation is satisfied.

The set of arbitrary constants A, B can be replaced by another set of arbitrary constants so that

$$A = C\cos\phi$$

$$B = C\sin\phi$$

where ϕ is the *phase angle* or *phase*. Equation (3.11) then becomes

$$x = C(\sin\omega t\cos\phi + \cos\omega t\sin\phi)$$

$$= C \sin(\omega t + \phi) \qquad (3.12)$$

where C and ϕ are the new arbitrary constants defined as

$$(C\cos\phi)^2 + (C\sin\phi)^2 = A^2 + B^2 \tag{3.12a}$$
$$C = \sqrt{(A^2 + B^2)}$$

$$\tan\phi = C\sin\phi/C\cos\phi = B/A \tag{3.12b}$$

The following solution forms can also be obtained, which all satisfy the differential equation:

$$
\begin{aligned}
x &= C_1\sin(\omega t - \alpha) \\
&= C_2\cos(\omega t + \beta) \\
&= C_3\cos(\omega t - \gamma)
\end{aligned}
\tag{3.13}
$$

The evaluation of the two arbitrary constants in equation (3.11) requires information such as initial conditions and that m has both initial velocity and displacement:

$$x = x_0, \ \dot{x} = \dot{x}_0 \tag{3.14}$$

both at $t = 0$. Equation (3.11) will give

$$A = \dot{x}_0/\omega \text{ and } B = x_0 \tag{3.15}$$

The solution becomes

$$x = x_0/\omega \ \sin\omega t + x_0 \ \cos\omega t \tag{3.16}$$

Substituting equation (3.15) into equations (3.12a and b) and then into equation (3.12) gives

$$x = X\sin(\omega t + \phi) \tag{3.17}$$

where the *amplitude* of the displacement is given by

$$X = \sqrt{[(\dot{x}_0/\omega)^2 + x_0^2]} \tag{3.18}$$

and the corresponding phase angle can be written as

$$\tan\phi = x_0/(\dot{x}_0/\omega) \tag{3.18a}$$

The motion defined by equations (3.16) and (3.17) is harmonic owing to its sinusoidal form. The period T, circular frequency ω and natural frequency are computed as follows:

$$T = 2\pi/\omega$$
$$\omega = \sqrt{(k/m)} = \sqrt{(kg/W)} = \sqrt{(g/\delta_{ST})} \tag{3.19}$$
$$f = \omega/2\pi = \tfrac{1}{2}\pi\sqrt{(k/m)}$$

The velocity $\dot{x}$ and the acceleration $\ddot{x}$ are expressed by time derivatives of equations (3.16) and (3.17):

Velocity

$$\dot{x} = \dot{x}_0 \cos\omega t - x_0\omega\sin\omega t \tag{3.20}$$
$$= X\omega\cos(\omega t + \phi)$$
$$= X\omega\sin(\omega t + \phi + \pi/2)$$
$$= \dot{X}\cos(\omega t + \phi)$$

amplitude of velocity: $\dot{X} = X\omega$

Acceleration

$$\ddot{x} = -\dot{x}_0\omega\sin\omega t - x_0\omega^2 \cos\omega t \tag{3.21}$$
$$= -X\omega^2\sin(\omega t + \phi)$$
$$= X\omega^2\sin(\omega t + \phi + \pi)$$
$$= -\omega^2 x = -\ddot{X}\sin(\omega t + \phi)$$

amplitude of acceleration: $\ddot{X} = X\omega^2 = \dot{X}\omega$ (3.21a)

The velocity is ω multiplied by the displacement, and leads it by 90°. The acceleration is ω^2 multiplied by the displacement and leads it by 180°.

Diagrams of displacement, velocity and acceleration against ωt are shown in Fig. 3.3.

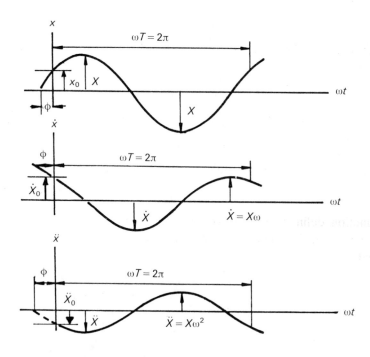

Fig. 3.3 Displacement, velocity and acceleration, with phase angles.

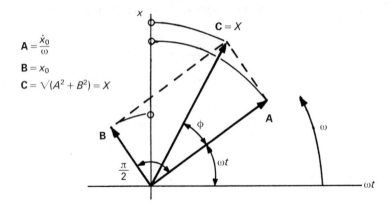

Fig. 3.4 Rotating vectors.

The phase angle ϕ indicates the amount by which each curve is shifted ahead, with respect to an ordinary sine curve.

The rotating vector concept is evolved by looking at the displacement x from equations (3.12) and (3.16). Three vectors **A**, **B** and **C**, whose relative positions are fixed, rotate with angular velocity ω. Their angular position at any time t is ωt. The vectors **A** and **B** are at right angles and the vector **C** leads **A** by the phase angle ϕ. A graphical representation is obtained by vertically projecting these vectors onto the graph of x against ωt shown in Fig. 3.4.

The displacement, velocity and acceleration curves can be generated in this manner, using rotating vectors **X**, **Ẋ** and **Ẍ**. The velocity and acceleration vectors lead the displacement vector by 90° and 180°, respectively. Their relative position is fixed and they rotate with angular velocity ω. These are shown in Fig. 3.5.

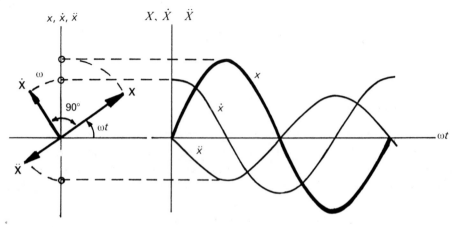

Fig. 3.5 Rotating vectors versus X, $\dot{X}$ and $\ddot{X}$.

The expressions for displacement, velocity and acceleration are given in equation (3.22).

$$x = X\sin\omega t = x_0/\omega\sin\omega t; \quad x_0 = 0$$

$$\dot{x} = X\omega\cos\omega t = \dot{X}\cos\omega t \tag{3.22}$$

$$\ddot{x} = X\omega^2(-\sin\omega t) = \ddot{X}(-\sin\omega t)$$

3.2.2.2 Method 2

Assume an exponential function

$$x = Ce^{st} \tag{3.23}$$

where C is an arbitrary constant, but s is to be determined so that the differential equation will be satisfied.

$$(s^2 + \omega^2)\, Ce^{st} = 0 \tag{3.24}$$

$$s^2 + \omega^2 = 0$$

$$s = \pm i\omega \tag{3.25}$$

$$x = C_1 e^{i\omega t} + C_2 e^{-i\omega t} \tag{3.26}$$

The Euler relation is expressed as

$$e^{i\omega} = \cos\omega + i\sin\omega \tag{3.27}$$

$$x = C_1(\cos\omega t + i\sin\omega t) + C_2(\cos\omega t - i\sin\omega t)$$
$$= (C_1 - C_2)i\sin\omega t + (C_1 + C_2)\cos\omega t$$
$$= A\sin\omega t + B\cos\omega t$$

C_1 and C_2 are conjugate complex numbers.

$$C_1 = a + ib, \quad C_2 = a - ib \tag{3.28}$$

$$i(C_1 - C_2) = i(2ib) = -2b = A$$

$$C_1 + C_2 = 2a = B$$

3.2.2.3 Torsional vibrations

Simple torsional vibrations are dependent on the torsional stiffness k_T, the angle of twist θ and the second moment of area I. Figure 3.6 shows a simple system. k_T is the torsional spring constant for the shaft, measured by the twisting moment per unit angle of twist. I represents the mass moment of inertia for the disk relative to its axis of rotation.

For an assumed positive angular displacement of the disk, the restoring torque acting on the disk would be

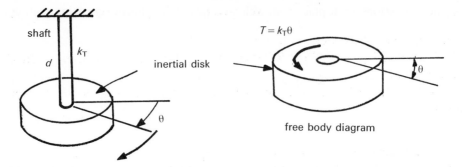

Fig. 3.6 Simple torsional vibration.

$$T = k_T\theta \tag{3.29}$$

in the direction shown in the free body diagram. The Newtonian relation for rotation about a fixed axis gives

$$I\ddot\theta = -k_T\theta \tag{3.30}$$

hence $$\ddot\theta = -k_T\theta/I = 0 \tag{3.31}$$

This is the same form as given in equation (3.8) for the rectilinear case. The solution will, by analogy, be written as

$$\theta = A\sin\omega t + B\cos\omega t = C\sin(\omega t + \phi) \tag{3.32}$$

where $\omega = \surd(k_T/I)$ depends on the physical constants of the system.

A and B or C and ϕ are determined from the initial conditions of motion.

Using the material from the previous section, and replacing x by θ, the values of θ, $\dot\theta$ and $\ddot\theta$ may be computed:

Angle of twist of shaft: $\theta = TL/GI_0$

hence the torsional spring constant $k_T = T = GI_0/L$

where I_0 is the polar second moment of area and is equal to $\pi d^2/32$, G is the modulus of rigidity and L is the length of the shaft.

Using the energy method, the kinetic energy and potential energy of the system are given by

$$KE = \tfrac{1}{2}I\dot\theta^2$$
$$PE = \tfrac{1}{2}k\theta^2 \tag{3.33}$$

Moreover, $d/dt\ (KE + PE) = 0$, which gives

$$(I\ddot\theta\dot\theta + k\theta\dot\theta) = 0 \tag{3.34}$$

Since $\dot\theta$ is always zero by virtue of

$$\dot\theta(I\ddot\theta + k\theta) = 0 \tag{3.34a}$$

hence $$I\ddot{\theta} + k\theta = 0 \tag{3.35}$$

which is the same as equation (3.31).

The remaining part of the solution for amplitudes is the same as given in earlier sections.

Tables 3.1 to 3.6 give a limited application to the above-mentioned dynamic analysis.

Table 3.1 Undamped free vibration (single-degree system): springs.

Equivalent spring stiffnesses
(1) Springs in parallel, with equal deflections of all springs.

$$F = K_1\delta + K_2\delta + \ldots K_n\delta = \delta(\Sigma k)$$
$$F = K_{eq}\delta$$
$$K_{eq} = \sum_{j=1}^{n} k_j = k_1 + k_2 + \ldots k_n$$

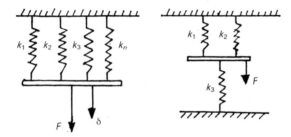

(2) Springs in series, with all springs carrying the same force.

$$F = k_1x_1 + k_2x_2 \ldots + k_nx_n$$
$$F = k_{eq}\delta$$
$$\delta = x_1 + x_2 + \ldots + x_n = \Sigma x$$
$$= F_1/k_1 + F_2/k_2 + \ldots + F_n/k_n = F/\sum_{i=1}^{n}(1/k_i)$$
$$k_{eq} = F/\delta = 1/\sum_{i=1}^{n}(1/k_i)$$
$$K = k_{eq} = 1/\sum_{i=1}^{n}(1/k_i)$$

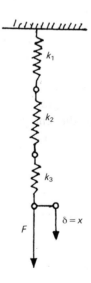

continued

Table 3.1 *Continued.*

(3) Spring mass pulley system.
   Energy method

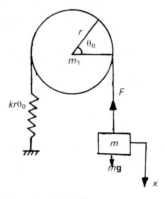

m_1 = mass of pulley

m = mass under consideration

$\text{KE} = (\text{KE})_m + (\text{KE})_{m_1}$

$\qquad = \frac{1}{2}m\dot{x}^2 + \frac{1}{2}I_0\dot{\theta}^2 = \frac{1}{2}mr^2\dot{\theta}^2 + \frac{1}{2}I_0\dot{\theta}^2$

$\text{PE} = \frac{1}{2}kx^2 = \frac{1}{2}kr^2\theta^2$

Using equation (3.4), the following equation of motion is derived:

$(mr^2\ddot{\theta} + I_0\ddot{\theta} + kr^2\theta) = 0$ or $\ddot{\theta} + (kr^2/I_0 + mr^2)\theta = 0$

but $I_0 = \frac{1}{2}m_1r^2$; $I_0\ddot{\theta} = Fr - kr^2(\theta + \theta_0)$

$\omega = \sqrt{[k/(\frac{1}{2}m_1 + m)]}; \qquad f = \omega/2\pi$

Table 3.2 Undamped free vibrations: beams.

(1) Simply supported beams with a single concentrated mass and a distributed mass.

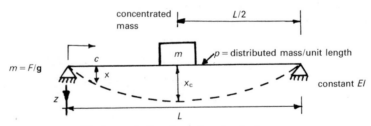

Undamped vibration of a simple beam.

Assumption: The dynamic deflection curve is the same as that due to the concentrated load acting statically on the beam.

The vertical displacement $x = \dfrac{x_c}{L^3}(3zL^2 - 4x^3)$

$\bar{F} = pL$

$\text{KE (distributed mass)} = 2\displaystyle\int_0^{L/2} \frac{\bar{F}}{2g}\left(\dot{x}_c \frac{3zL^2 - 4x^3}{L^3}\right)^2 dx$

$\qquad\qquad = \dfrac{17}{35}\bar{F}L\dfrac{\dot{x}_c^2}{2g}$

$\text{KE (concentrated mass } m) = \frac{1}{2}m\dot{x}_c^2$

$\text{PE} = \frac{1}{2}kx_c^2$

The total energy $E = KE + PE$ and, using equation (3.4), the natural frequency is given as

$$f = \omega/2\pi = \frac{1}{2\pi}\sqrt{\left(\frac{k\omega}{\overline{F} + \frac{17}{35}\overline{F}L}\right)}$$

The total mass of the beam is put with the concentrated mass.

If the mass of the beam is negligible compared to the mass acting on it, then the maximum displacement

$$x_c = \overline{F}L^3/48EI \qquad \text{since } k = \overline{F}/x_c = 48EI/L^3$$
$$\omega = \sqrt{(k/m)} = \sqrt{(48EI/mL^3)} \text{ rad/s}$$

(2) Simply supported beams with a continuous mass distribution and with constant EI.

$$EId^2x/dz^2 = M_c = pz^2/2 - pLz/2$$

$$EIx = \frac{pz^4}{24} - \frac{pLz^3}{12} + c_1z + c_2$$

When $z = 0 \quad x = 0 \quad c_2 = 0$
$$z = L \quad x = 0 \quad c_1 = pL^3/24$$

Hence $x = \dfrac{p}{24EI}(z^4 - 2Lz^3 + L^3z)$

$$\omega = \sqrt{\left[g\int_0^L\left(xdz/\int_0^L x^2dz\right)\right]} = \sqrt{(k/m)} = \sqrt{\left(\frac{15\,120}{155}\frac{EIg}{pL^4}\right)}$$
$$= 9.87\sqrt{(EI/mL^4)} = \alpha_B\sqrt{(EI/mL^4)}$$

where $\alpha_B = 9.87$
$$f = \omega/2\pi$$

(3) A cantilever beam with a single concentrated mass. The static deflection x_c of the cantilever due to the mass m is given by:

$$x_c = FL^3/3EI$$
$$k = F/x_c = 3EI/L^3$$
$$\omega = \sqrt{(k/m)} = \sqrt{(3EI/mL^3)}$$
$$f = \omega/2\pi$$

constant EI

Frequency of a simple cantilever beam.

(4) Beams fixed at both ends with continuous mass distribution.

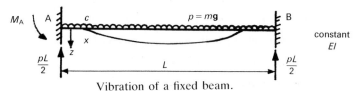

Vibration of a fixed beam.

continued

Table 3.2 *Continued.*

$$EI\frac{d^2x}{dz^2} = M_C = M_A + \frac{pz^2}{2} - \frac{pLz}{2}$$

$$EIx = \frac{1}{2}M_A z^2 = \frac{pz^4}{24} - \frac{pLz^3}{12} + c_1 z + c_2$$

$$x = 0 \qquad \frac{dx}{dz} = 0 \qquad c_1 = 0$$

$$x = L \qquad \frac{dx}{dz} = 0 \qquad M_A = pL^2/12$$

$$z = L \qquad x = 0 \qquad c_2 = 0$$

$$x_c = \frac{p}{24EI}(L^2 z^2 + z^4 - 2Lz^3)$$

$$\omega = \sqrt{\left(g\int_0^L x_c dz / \int_0^L x_c^2 dz\right)} = 22.4\sqrt{(EI/mL^4)} = \alpha_B \sqrt{(EI/mL^4)}$$

where $\alpha_B = 22.4$

$$f = \omega/2\pi$$

(5) Additional cases for the natural frequency of transverse vibration of beams with end conditions with continuous mass distribution.

$$\omega = \alpha_B \sqrt{(EI/mL^4)}$$
$$f = \omega/2\pi$$

Type	Normal mode shapes			
Cantilever	$\alpha_B = 3.52$	0.774 $\alpha_B = 22.4$	0.5 0.868	0.356 0.644 0.906
Simply supported	$\alpha_B = 9.87$	0.5 $\alpha_B = 39.5$	0.333 0.667 $\alpha_B = 88.9$	0.25 0.5 0.75 $\alpha_B = 158$
Fixed ends	$\alpha_B = 22.4$	0.5 $\alpha_B = 61.7$	0.359 0.641 $\alpha_B = 121$	0.5 0.238 0.722 $\alpha_B = 200$
Propped support	$\alpha_B = 15.4$	0.560 $\alpha_B = 50$	0.384 0.692 $\alpha_B = 104$	0.294 0.529 0.762 $\alpha_B = 178$
One end hinged other free	$\alpha_B = 15.4$ 0.736	0.446 0.853 $\alpha_B = 50$	0.308 0.616 0.898 $\alpha_B = 104$	0.235 0.471 0.707 0.922 $\alpha_B = 178$

(6) A beam vibrating vertically by both linear and torsional springs.

$$\omega = \sqrt{(6kL^2 + 3k_T)/mL^2}\ \text{rad/s}$$

$$f = \omega/2\pi \quad \omega = \theta$$

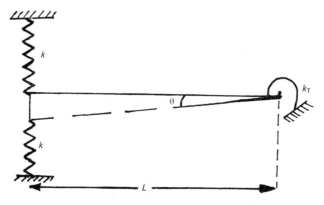

A beam with linear and torsional springs.

(7) A beam/spring type of structure.

(i) Equation of motion.

$$mL^2\ddot{\theta} + (kh^2 + mgL)\theta = 0$$

$$\omega = \sqrt{(kh^2 + mgL)/mL^2}$$

$$f = \omega/2\pi \qquad \omega = \theta$$

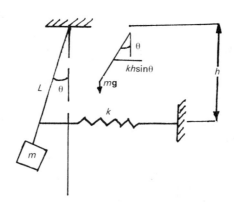

Rotating beam/mass system with a mass down.

(ii) $x = L(1 - \cos\theta)$

$$\approx \tfrac{1}{2}L\theta^2$$

$$(\text{PE})_{\max} = \tfrac{1}{2}kh^2\theta^2 - \tfrac{1}{2}mgL\theta^2$$

$$\text{KE} = \tfrac{1}{2}mL^2\theta^2$$

$$\omega = \sqrt{\left[\frac{1}{L}(kh^2/mL - 1)\right]}$$

$$f = \omega/2\pi \qquad \omega = \theta$$

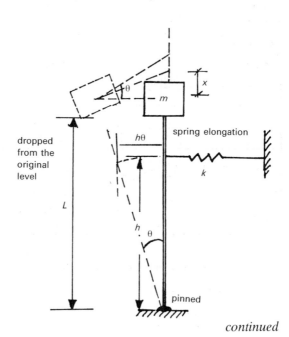

dropped from the original level

Rotating beam/mass system with a mass upward.

continued

Table 3.2 *Continued.*

(iii) m_L = mass of the lever, m_c = mass of the cylinder

$$= \rho_c(\tfrac{1}{4}\pi D^2 h)$$

Upward thrust on the cylinder = weight of water displaced

$$T_v = -\rho_w g V$$
$$= -\rho_w g(\tfrac{1}{4}\pi D^2 x_2)$$
$$= -\rho_w g(\tfrac{1}{4}\pi D^2 L\theta)$$

F_s = spring restoring force

$$= -Kx_1 = -L_1\theta x_1$$

Total restoring torque $= I_0\ddot{\theta}$

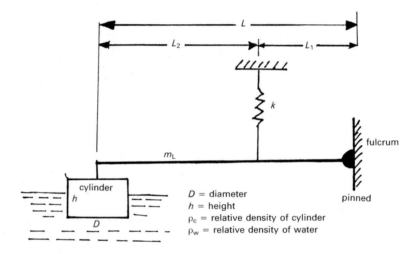

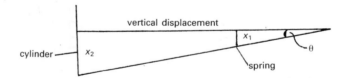

Vertical vibration of lever/cylinder system.

Equation of motion:

$$T_v L + F_s L_1 = \left(m_c L^2 + \frac{L_1}{L} m_L L^2 \right)\ddot{\theta}$$

$$\theta = (\tfrac{1}{4}\pi D^2 L^2 \rho_w + L_1^2 x)/(\tfrac{1}{4}\pi D^2 h\rho_c L^2 + LL_1 mL)$$

Hence $f = \theta/2\pi$

Table 3.3 Natural frequency of missiles with specific cross-sectional shapes and each subject to a distributed mass.

Note: circular frequency $\omega = \alpha_B \sqrt{(EI/mL^4)}$; where α_B values are as used in Table 3.2, section 5. The values of I given below are substituted in the above equation with m. The values of E and L must be known; $f = \omega/2\pi$. For unit weight all expressions are multiplied by ρ.

(1) Rectangular cross-section of a missile.

$$I_{xx} = bh^3/12$$
$$I_{yy} = hb^3/12$$

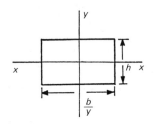

(2) Solid circular cross-section of diameter D.

$$I_{xx} = I_{yy} = \pi D^4/64$$

(3) Hollow circular cross-section of outer diameter D and inner diameter d.

$$I_{xx} = I_{yy} = \frac{\pi}{64}(D^4 - d^4)$$

(4) Thin-walled tubular cross-section of outer diameter D and thickness t.

$$I_{xx} = I_{yy} = \pi t D^3/8$$

(5) Elliptical type.

$$I_{xx} = \pi Dd^3/64$$
$$I_{yy} = \pi D^3 d/64$$

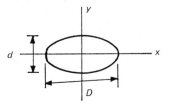

(6) Triangular cross-section with base b and height h.

$$I_{xx} \text{ (parallel to } b) = bh^3/36$$

(7) Right circular cylindrical type missile.

$$I_{zz} = \pi h R^4/2$$
$$m = \rho \pi R^2 h$$

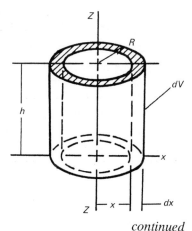

continued

Table 3.3 *Continued.*

(8) Cone-shaped missile.

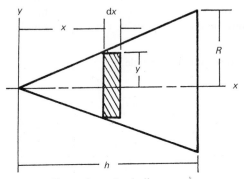

Cone-shaped missile.

$$I_{xx} = \pi R^4 h / 10$$

$$\rho = m/(\pi R^2 h/3)$$

$$I_{xx} = \frac{3}{20} W(R^2 + 4h^2)$$

$$I_{zz} = \frac{3}{10} WR^2; \quad W = \pi R^2 h/3$$

$$I_{xx}, I_{yy} = \frac{3W}{20}(R^2 + h^2/4)$$

$$I_{x_1 x_1}, I_{y_1 y_1} = \frac{W}{20}(3R^2 + 2h^2)$$

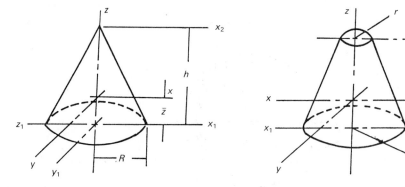

Right circular cone-shaped missile.

Table 3.3 *Continued.*

$$I_{zz} = \frac{3}{10}W(R^5 - r^5)/(R^3 - r^3)$$

$$W = \frac{\rho\pi h}{3}(R^2 + Rr + r^2)$$

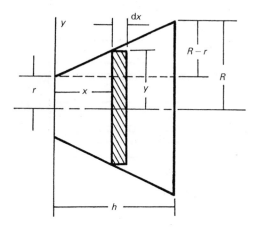

Frustrum of a cone-shaped missile.

(9) Solid spherical missile.

$$I_{xx} = \frac{8}{15}\pi r^5$$

$$x = \sqrt{(r^2 - y^2)}$$

$$y = \sqrt{(r^2 - x^2)}$$

where $\rho = m/\frac{4}{3}\pi r^3$

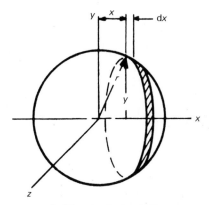

Solid spherical missile.

(10) Hollow spherical missile.

$$I_{xx} = I_{yy} = I_{zz} = \frac{2}{3}Wr^2$$

$$W = 4\rho\pi r^2$$

continued

Table 3.3 *Continued.*

(11) Wedge-shaped semi-cylindrical missile.

$$I_{zz} = \rho h \pi R^4 / 4$$

$$\text{where } \rho = m / \pi^2 h$$

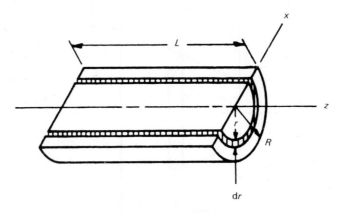

Wedge-shaped semi-cylindrical missile.

(12) Pyramid-shaped missile.
    (i) Right rectangular pyramid

$$I_{xx} = \frac{W}{20}(b^2 + 3h^2/4)$$

$$I_{x_1 x_1} = \frac{W}{20}(b^2 + 2h^2)$$

$$I_{yy} = \frac{W}{20}\left(a^2 + \frac{3}{4}h^2\right)$$

$$I_{y_1 y_1} = \frac{W}{20}(a^2 + 2h^2)$$

$$I_{zz} = \frac{W}{20}(a^2 + b^2)$$

$$W = \rho a b h / 2$$

    (ii) Regular triangular prism

$$I_{xx} = I_{yy} = \frac{W}{24}(a^2 + 2h^2)$$

$$I_{zz} = W a^2 / 12$$

$$W = \frac{\sqrt{3}}{4}a^2 h$$

Table 3.3 *Continued.*

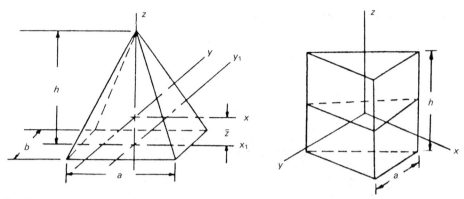

(i) Right rectangular pyramid-type missile.   (ii) Regular triangular prism-type missile.

(13) Rod-shaped missiles.
    (i) Segment of a circular rod.

$$I_{xx} = WR^2[\tfrac{1}{2} - (\sin\theta_1 \; \cos\theta_1)/2\theta_1]$$

$$I_{x_1x_1} = WR^2[\tfrac{1}{2} - (\sin\theta_1 \; \cos\theta_1/2\theta_1) + \sin^2\theta]$$

$$I_{yy} = WR^2[\tfrac{1}{2} + (\sin\theta_1 \; \cos\theta_1/2\theta_1) - \sin^2\theta/\pi^2]$$

$$I_{y_1y_1} = WR^2 \; (\tfrac{1}{2} + \sin\theta_1 \; \cos\theta_1/2\theta_1)$$

$$\bar{x} = R\sin\theta_1/\theta_1$$

$$I_c = WR^2(1 - \sin^2\theta_1/\theta_1^2)$$

$$\bar{y} = R\sin\theta_1$$

$$W = \text{density} \times \text{volume}$$

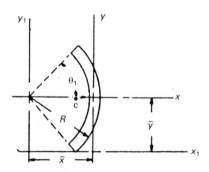

Rod (segment-shaped) missile.

    (ii) Elliptical rod.

$$I_{xx} = Wb^2(55a^4 + 10a^2b^2 - b^4)/2(45a^4 + 22a^2b^2 - 3b^4)$$

$$I_{yy} = Wa^2(35a^4 + 34a^2b^2 - 5b^4)/2(45a^4 + 22a^2b^2 - 3b^4)$$

$$I_c = I_{xx} + I_{yy}$$

continued

Table 3.3 *Continued.*

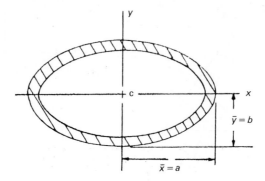

Elliptical rod as a missile.

(iii) Parabolic rod.

$$\bar{x} = \frac{\sqrt{(4a^2 + b^2)^3}}{8aL} - \frac{b^2}{16a}, \; \bar{y} = 1$$

$$I_{xx} = \frac{Wb^2}{8a^2} \left(\frac{\sqrt{(4a^2 + b^2)^3}}{L} - \frac{b^2}{2} \right)$$

$$I_{x_1 x_1} = \frac{Wb^2}{8a^2} \left(\frac{\sqrt{(4a^2 + b^2)^3}}{L} - \frac{b^2}{2} \right)$$

$$I_{yy} = \frac{W\sqrt{(4a^2 + b^2)^3}}{12L} - \frac{I_{xx}}{8} - Wx^2$$

$$I_{y_1 y_1} = \frac{W \sqrt{(4a^2 + b^2)^3}}{12L} - \frac{I_{xx}}{8}$$

$$I_c = I_{xx} + I_{yy}$$

$$W = \rho \times \text{volume}$$

$$L = \text{length}$$

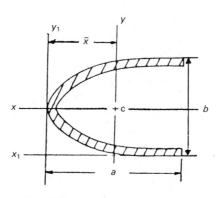

Parabolic-shaped rod as a missile.

Table 3.3 *Continued.*

(iv) U-rod-shaped missile.

$$\bar{x} = L_2^2/(L_1 + 2L_2); \qquad \bar{y} = A_1/2$$

$$I_{xx} = WA_1^2(A_1 + 6B_2)/12(A_1 + 2B_2)$$

$$I_{yy} = WB_2^3(2A_1 + B_2)/3(A_1 + 2B_2)^2$$

$$I_c = I_{xx} + I_{yy}$$

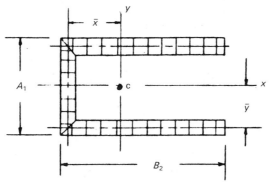

U-rod-shaped missile.

(v) V-rod-shaped missile.

$$\bar{x} = \tfrac{1}{2}L\sin\theta_1; \qquad \bar{y} = L\cos\theta_1$$

$$I_{xx} = \tfrac{1}{3}WL^2\cos^2\theta_1$$

$$I_{x_1x_1} = \tfrac{4}{3}WL^2\cos^2\theta_1$$

$$I_{yy} = \frac{WL^2}{12}\sin^2\theta_1$$

$$I_{y_1y_1} = \tfrac{1}{3}WL^2\sin^2\theta_1$$

$$I_c = I_{xx} + I_{yy}$$

$$I_c = I_{x_1x_1} + I_{y_1y_1}$$

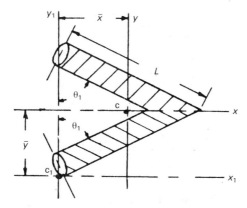

V-rod-shaped missile. *continued*

Table 3.3 *Continued.*

(vi) L-rod-shaped missile.

$$\bar{x} = \tfrac{1}{2}B_2; \quad \bar{y} = \tfrac{1}{2}A_1$$

$$I_{xx} = W A_1^2 (A_1 + 3B_2)/12(A_1 + B_2)$$

$$I_{yy} = W B_2^2 (3A_1 + B_2)/12(A_1 + B_2)$$

$$I_c = I_{xx} + I_{yy}$$

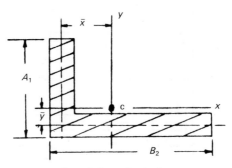

L-rod-shaped missile.

(vii) Hollow rectangular missile.

$$I_{xx} = \frac{W}{12}A_1^2(A_1 + 3B_2)/(A_1 + B_2)^2$$

$$I_{yy} = \frac{W}{12}B_2^2(3A_1 + B_2)/(A_1 + B_2)^2$$

$$I_c = I_{xx} + I_{yy}$$

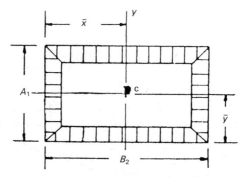

Hollow rectangular missile.

(viii) Inclined rod-shaped missile flight (not through centroidal axis).

$$I_{xx} = \tfrac{1}{3}W\sin^2\theta_1(A_1^2 - A_1B_2 + B_2^2)$$

$$I_{yy} = \tfrac{1}{3}W\cos^2\theta_1(A_1^2 - A_1B_2 + B_2^2)$$

$$I_c = I_{xx} + I_{yy}$$

Table 3.3 *Continued.*

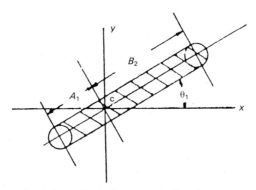

Rod-shaped missile at an inclination.

(14) Ogive-shaped missile.

$$A_s\bar{x} = \text{surface area developed}; \qquad \bar{x} = \text{centroid}$$

$$A_s\bar{x} = 2\pi \int_0^h xyds$$

$$= 2\pi R\left\{\frac{h^2}{2} - br + bh_1[\sin^{-1}(h + h_1)/R - \sin^{-1}h_1/R]\right\}$$

$$y = [(R^2 - h_1^2) - 2h_1x - x^2]^{1/2} - b$$

$$A_s = \text{surface area developed}; \qquad \bar{x} = \text{centroid}$$

$$A_s = 2\pi R\{h - b[\sin^{-1}(h + h_1)/R - \sin^{-1}h_1/R]\}$$

I_{xx} and I_{yy} can thus be easily calculated for a typical curve of radius R.

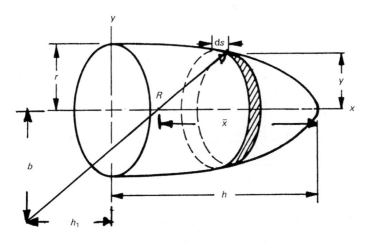

Ogive-shaped missile.

continued

Table 3.3 *Continued.*

(15) Torus and spherical sector missiles.
    (i) Torus

$$I_{xx} = I_{zz} = \frac{W}{8}\,(4R^2 + 5r^2)$$

$$I_{yy} = \frac{W}{4}\,(4R^2 + 3r^2)$$

$$W = 2\pi^2 r^2 \rho$$

$$\bar{x} = \bar{z} = R + r$$

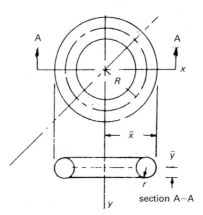

Torus-shaped missile.

(ii) Spherical sector

$$I_{zz} = \frac{Wh}{5}\,(3R - h)$$

$$W = \tfrac{2}{3}\pi\rho R^2 h$$

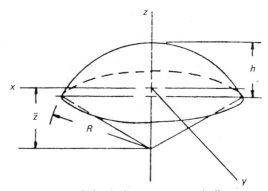

Spherical sector type missile.

Table 3.4 Natural frequency of beam/column/wall system (horizontal members are infinitely stiff).

(1) Single-bay frame.
  Stiffnesses of vertical members:

$$k_{AB} = \frac{12EI}{L^3}; \qquad k_{CD} = 12E(nI)/(nL)^3$$

$$\Sigma k = K = \frac{12EI}{L^3}(1 + 1/n^2)$$

$$\omega = \sqrt{\left[\frac{12EI}{L^3}\left(1 + \frac{1}{n^2}\right)\right]/m}$$

$$f = \omega/2\pi$$

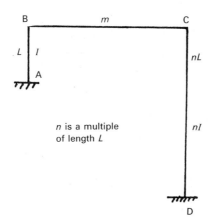

A single-bay frame.

(2) Two-bay frame.
  Stiffnesses of vertical members:

$$k_{AB} = 12En_1I/L_1$$

$$k_{CD} = 12En_2I/L_2$$

$$k_{EF} = 12En_3I/L_3$$

$$\Sigma k = K = 12EI\left(\frac{n_1}{L_1} + \frac{n_2}{L_2} + \frac{n_3}{L_3}\right)$$

$$\omega = \sqrt{(K/m)}; \qquad f = \omega/2\pi$$

continued

Table 3.4 *Continued.*

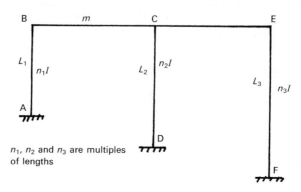

A two-bay frame.

(3) Building of height h and base area $a \times b$.
Equation of motion:

$$I_0\ddot{\theta} - (\tfrac{1}{2}mgh)\theta = -M$$

or $I_0\ddot{\theta} - (\tfrac{1}{2}mgh)\theta = -2\int_0^{1/2} kax^2\theta dx$

$$= kb^3a\theta/12$$

$$\omega = \sqrt{\left[\left(\frac{kb^3a}{12} - \frac{mgh}{2}\right)\Big/ I_0\right]}$$

$$f = \omega/2\pi$$

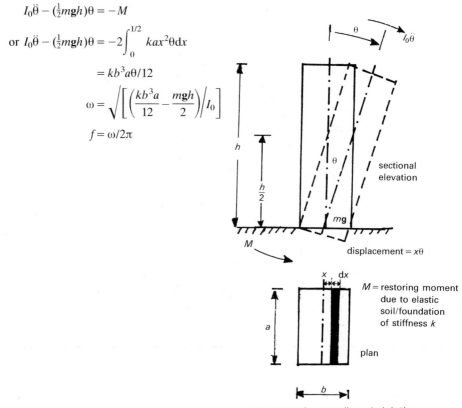

A single wall.

Table 3.5 Vibration of plates of uniform thickness.

$$f = \frac{1}{2\pi} \alpha_B \sqrt{[Et^2/\rho D^4(1-v^2)]}$$

where

f = natural frequency
ρ = density
D = diameter or length
v = Poisson's ratio
t = thickness
α_B = vibration factor

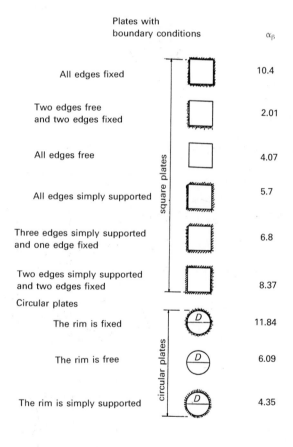

Plates with boundary conditions	α_β
All edges fixed	10.4
Two edges free and two edges fixed	2.01
All edges free	4.07
All edges simply supported	5.7
Three edges simply supported and one edge fixed	6.8
Two edges simply supported and two edges fixed	8.37
Circular plates	
The rim is fixed	11.84
The rim is free	6.09
The rim is simply supported	4.35

Table 3.6 Helical springs and springs to simulate other conditions.

---

(1) Helical springs.

The following spring stiffnesses are used where springs are substituted for structure–structure and structure–foundation interactions.

Vertical direction $k_a = Gd^4/8nD^3 \times N$

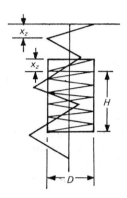

Helical springs.

Lateral direction $k_\ell = k_a \left[\dfrac{1}{0.385\, \alpha_\beta} \left(1 + \dfrac{0.77 H^2}{D^2} \right) \right] \times N$

Bending direction $k_m = I_\ell k_a$ or $I_m k_s$

Torsional $k_\theta = k_T = \dfrac{I_n k_\ell}{0.385\, \alpha_\beta \left(1 + \dfrac{0.77\ H^2}{D^2} \right)}$

Table of α_β values.

x_z/H H/D	0.10	0.20	0.30	0.40	0.50
0.10	1.00	1.00	1.00	1.00	1.00
0.25	1.00	1.03	1.03	1.03	1.03
0.50	1.08	1.08	1.08	1.08	1.08
0.55	1.08	1.10	1.10	1.17	1.20
1.00	1.08	1.10	1.15	1.25	1.34
1.25	1.08	1.15	1.30	1.45	1.60
1.50	1.09	1.20	1.43	1.70	2.03
1.55	1.10	1.40	1.50	2.10	3.00
1.75	1.15	1.45	1.60	2.50	4.40
2.00	1.20	1.55	2.00	3.04	6.00

Table 3.6 *Continued*.

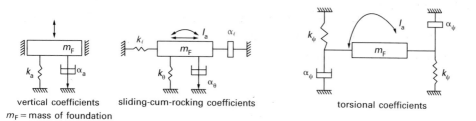

vertical coefficients sliding-cum-rocking coefficients torsional coefficients

m_F = mass of foundation

Springs representing vertical, sliding-cum-rocking and torsional effects.

(2) Foundations on springs.

The linear stress-strain relation demands that subgrade reaction coefficients k_a, k_ℓ, k_θ and k_T must be determined (see table): r_a, r_ℓ and r_θ are radii of equivalent circular bases for translation, rotation and rocking modes. α_a, α_ℓ and α_ψ are given in this same table for rigid foundations, where α_ℓ, α_a = uniform compression and shear in vertical and horizontal modes, respectively; $\alpha_\psi, \alpha_\theta$ = non-uniform rocking or twisting modes, respectively.

Determination of the coefficients k_a, k_ℓ, k_θ and k_ψ.

	Circular foundation	
Motion	Spring constant	Reference
Vertical	$k_a = \dfrac{4Gr_a}{1-\nu}$	Timoshenko and Goodier[6.73]
Lateral	$k_\ell = \dfrac{32(1-\nu)Gr_a}{7-8\nu}$	Bycroft[6.75]
Rocking	$k_T = \dfrac{8Gr_\psi^3}{3(1-\nu)}$	Borowicka[6.77]
Torsion	$k_\theta = \dfrac{16}{3}\,Gr_\theta^3$	Reissner[6.76]

	Rectangular foundation	
Motion	Spring constant	Reference
Vertical	$k_a = \dfrac{G}{1-\nu}\,\alpha_a\sqrt{(4cd)}$	Barkan[6.74]
Lateral	$k_\ell = 4(1+\nu)G\alpha_\ell\sqrt{(cd)}$	Barkan[6.74]
Rocking	$k_\psi = \dfrac{G}{1-\nu}\,\alpha_\psi\,8cd^2$	Gorbunov-Possador[6.77]

$G = E/2(1+\nu)$

continued

Table 3.6 *Continued.*

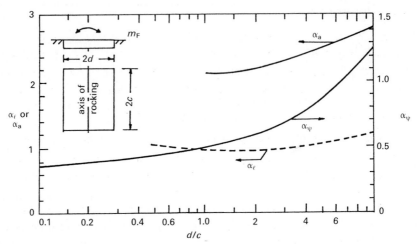

Coefficients α_a, α_ℓ and α_ψ for rectangular footings (after Richart *et al.*).[6.78]

$$r_a = \sqrt{(A_F/\pi)} \text{ for translation}$$

$$r = \sqrt[4]{4I_a/\pi} \text{ for rocking}$$

$$r_\theta = \sqrt{[2(I_x + I_y)/\pi]} \text{ for torsion}$$

Motion	Mass ratio α	Damping ratio
Vertical	$\alpha_a = \dfrac{(1-v)}{4}\dfrac{m_F}{\rho r_a^3}$	$\tau_a = \dfrac{0.425}{\sqrt{\alpha_a}}$
Horizontal (sliding)	$\alpha_\ell = \dfrac{7-8v}{32(1-v)}\dfrac{m_F}{\rho r_a^3}$	$\gamma_\ell = \dfrac{0.288}{\sqrt{\alpha_\ell}}$
Rocking	$\alpha_\psi = \dfrac{3(1-v)}{8}\dfrac{I_a'}{\rho r_\psi^5}$	$\gamma_\psi = \dfrac{0.15}{(1+\alpha_\psi)\sqrt{\alpha_\psi}}$
Torsional	$\alpha_\theta = \dfrac{I_{xx}+I_{yy}}{\rho r_\theta^5} = \dfrac{I_a}{\rho r_\theta^5}$	$\gamma_\theta = \dfrac{0.5}{1+2\alpha_\psi}$

ρ = mass density of soil;
I_a' = mass moment of inertia about a parallel axis passing through the centroid of the foundation;
$v = 0.25$ to 0.35 for cohesionless soils;
$v = 0.35$ to 0.45 for cohesive soils;
γ = engineering strain;
τ = shear.

3.2.3 Free damped vibrations

3.2.3.1 Viscous damping force

Observation shows that the amplitude of free vibrations dies away slowly. A force is introduced into the mathematical model to simulate the effects of damping. A common way to do this is to postulate that the damping force is proportional to the velocity at any instant and acting in the opposite way to the displacement, i.e.

$$F = -c \times v = -c \times \dot{x} \tag{3.36}$$

where c is the damping force per unit of velocity (damping coefficient).

The viscous type of damping is a good approximation for bodies moving at low velocities, or sliding on lubricated surfaces, or where hydraulic dashpots and shock absorbers are used.

Other types of damping may be more appropriate for structural materials, structural connections and for the entire structure. Viscous damping has mathematical advantages because it produces a viscous force which gives the same rate of energy dissipation as the actual damping force.

Other types of damping occurring are discussed in the text later on. A typical system is shown in Fig. 3.7.

From Newton's law, the equation of motion is given by

$$m\ddot{x} = -c\dot{x} - kx \tag{3.37}$$

$$\text{or } m\ddot{x} + c\dot{x} + kx = 0 \tag{3.38}$$

Use the trial solution $x = Ce^{\lambda t}$ (3.39)

$$(m\lambda^2 + c\lambda + k)Ce^{\lambda t} = 0 \tag{3.40}$$

The characteristic or auxiliary equation of the system

$$m\lambda^2 + c\lambda + k = 0 \tag{3.41}$$

Therefore $\lambda_{1,2} = -c/2m \pm \sqrt{[(c/2m)^2 - k/m]}$ (3.42)

general solution $x = C_1 e^{\lambda_1 t} + C_2 e^{\lambda_2 t}$ (3.43)

provided that $c/2m \neq \sqrt{(k/m)}$ (3.44)

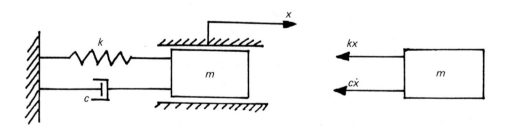

Fig. 3.7 Simple system with damping: free body.

3.2.3.2 Critical damping

The value of the damping coefficient which causes the radical part of the exponent of equation (3.42) to vanish is called the *critical damping coefficient*, c_c, defined by

$$c_c/2m = \sqrt{(k/m)} = \omega$$
$$c_c = 2\sqrt{(mk)} = 2m\omega \qquad (3.45)$$

The dimensionless parameter $\gamma = c/c_c$, called the *damping ratio*, gives a meaningful measure of the damping present in the system.

$$c/2m = \frac{c}{c_c} \times \frac{c_c}{2m} = \gamma \times \frac{2m\omega}{2m} = \gamma\omega \qquad (3.46)$$

$$\lambda_{1,2} = [-\gamma \pm \sqrt{(\gamma^2 - 1)}]\omega \qquad (3.47)$$

$$x = C_1 e^{[-\gamma + \sqrt{(\gamma^2 - 1)}]\omega t} + C_2 e^{[-\gamma - \sqrt{(\gamma^2 - 1)}]\omega t} \qquad (3.48)$$

for $\gamma \neq 1$.

First case

$\gamma > 1$ overdamping. Since $\sqrt{(\gamma^2 - 1)} < \gamma$ both exponents of equation (3.48) are real and negative, and the solution can be left in the form shown. This represents the sum of two decaying exponentials. The motion is called *overdamped*, it is non-periodic or aperiodic.

$$x = C_1 e^{-\alpha t} + C_2 e^{-\beta t} \qquad (3.49)$$

$$\alpha = [\gamma - \sqrt{(\gamma^2 - 1)}]\omega > 0 \qquad (3.50)$$
$$\beta = [\gamma + \sqrt{(\gamma^2 - 1)}]\omega > 0$$

for $\qquad \beta \gg \alpha$

C_1 and C_2 may be positive or negative, depending on the initial conditions. Five basic types of the displacement versus time curve are possible (Fig. 3.8).

Second case

$\gamma = 1$ critical damping solution is

$$x = (C_1 + C_2 t)e^{-(c/2m)t} \qquad (3.51)$$
$$= (C_1 + C_2 t)e^{-\omega t}$$

This is the product of a linear function, $C_1 + C_2 t$, and the decaying exponential $e^{-\omega t}$, as shown in Fig. 3.9. Five different basic shapes can result, similar to those shown in Fig. 3.8.

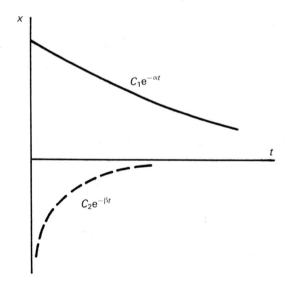

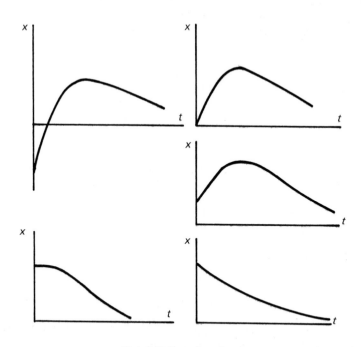

Fig. 3.8 Overdamping.

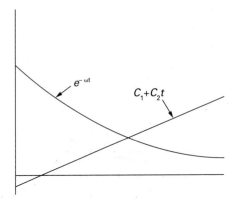

Fig. 3.9 Critical damping.

Third case

$\gamma < 1$ underdamping. For this condition $(\gamma^2 - 1)$ is negative, and the exponential multipliers of ωt in equation (3.48) are conjugate complex numbers. It is therefore desirable to write

$$x = C_1 e^{[-\gamma + i\sqrt{(1-\gamma^2)}]\omega t} + C_2 e^{(-\gamma - i\sqrt{(1-\gamma^2)})\omega t} \tag{3.52}$$

where $i = \sqrt{-1}$. With Euler's formula:

$$e^{i\theta} = \cos\theta + i\sin\theta \tag{3.53}$$

where $\theta = \omega$, the displacement x is given as

$$x = e^{-\gamma\omega t}[A\sin\sqrt{(1-\gamma^2)}\omega t + B\cos\sqrt{(1-\gamma^2)}\omega t]$$

or $\qquad x = X e^{-\gamma\omega t}\sin[\sqrt{(1-\gamma^2)}\omega t + \phi] \tag{3.54}$

$$= X e^{-\gamma\omega t}\sin(\omega_d t + \phi)$$

$$\omega_d = \sqrt{(1-\gamma^2)}\omega \tag{3.55}$$

where ω_d is the damped circular frequency

The circular frequency is reduced by viscous damping. However, for the usual case of small damping, this effect is very small and it is appropriate to assume that the frequency is unaffected.

The successive maxima X_1 and X_2 occur with the period T_d, as shown in Fig. 3.10.

$$\omega_d T_d = 2\pi; \qquad T_d = 2\pi/\omega_d = 2\pi/\sqrt{(1-\gamma^2)}\omega \tag{3.56}$$

The ratio of successive maximum amplitudes is equal to the ratio of the values of the exponential term $e^{-\gamma\omega t}$ at the corresponding times, leading to

$$X_1/X_2 = e^{-\gamma\omega t_1}/e^{-\gamma\omega(t_1 + T_d)} = e^{\gamma\omega T_d}$$

$$X_1/X_2 = e^{2\pi\gamma/\sqrt{(1-\gamma^2)}} \tag{3.57}$$

The *logarithmic decrement*, $\bar{\delta}$, which gives the rate of attenuation, is

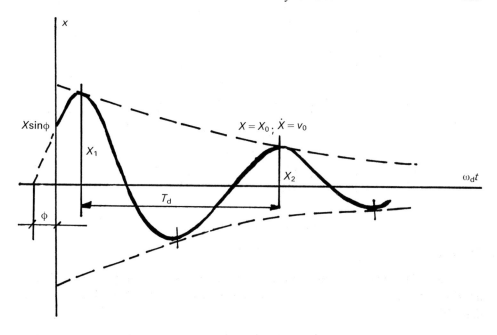

Fig. 3.10 The motion with a decaying amplitude.

$$\bar{\delta} = \log X_1/X_2 = 2\pi\gamma/\sqrt{(1-\gamma^2)} \tag{3.58}$$

For small damping this becomes

$$\bar{\delta} \doteq 2\pi\delta \tag{3.59}$$

The logarithmic decrement $\bar{\delta}$ can be calculated from the ratio of the amplitudes several cycles apart, since for viscous damping the logarithmic decrement can be related to any pair of successive amplitudes.

$$X_0/X_n = X_0/X_1, \; X_1/X_2, \; X_2/X_3 \; \ldots \; X_{n-1}/X_n \tag{3.60}$$
$$= (X_j/X_{j+1})^n$$

$$\log (X_0/X_n) = n\log(X_j/X_{j+1}) = n\bar{\delta}$$
$$\bar{\delta} = (1/n)\log(X_0/X_n) \tag{3.61}$$

The above is the basis of the experimental determination of the damping ratio for a system.

The amplitudes are obtained by measurement of an experimental record. This gives the logarithmic decrement $\bar{\delta}$ when equation (3.61) is used. The damping ratio follows from equation (3.58) when $\bar{\delta}$ is substituted.

If the damping force is truly viscous (damping force $= c \times$ velocity), the logarithmic decrement will have a value independent of the amplitude of motion; i.e. the ratio of successive peaks in the free-vibration curve will be constant. Table 3.7 gives typical examples of damping and logarithmic decrement.

Table 3.7 Example of free damped vibration.

(1) A platform is treated as a single-degree system and is deflected at mid-span by lowering the deck. The release is a sudden one. The vibration is found to decay exponentially from an amplitude of X_1 to X_2 in S cycles, such that the frequency is f_1. A vehicle with a mass m_1 is then placed at mid-span and the frequency of vibration is noted as f_2. Calculate the effective mass m_e, the effective stiffness k_e and γ for the platform.

$$f_1 = \omega/2\pi = \frac{1}{2\pi}\sqrt{\frac{k}{m}}; \quad f_2 = \frac{1}{2\pi}\sqrt{\left(\frac{k}{m+m_e}\right)}$$

$$(f_1/f_2)^2 = (m_1 + m_e)/m_1$$

$$m_1 = \frac{m_e}{(f_1/f_2)^2 - 1}$$

$$k_e = (2\pi f_1)^2 m_1$$

The logarithmic decrement $\bar{\delta} = (1/S)\log(X_1/X_2)$

$$\bar{\delta} = 2\pi\gamma/\sqrt{(1-\gamma^2)} = 1/S \ \log \ (X_1/X_2)$$

$$\gamma = \sqrt{[\tfrac{1}{4}\pi^2(1/S \ \log \ X_1/X_2)^2]/[1 + \tfrac{1}{4}\pi^2(1/S \ \log \ X_1/X_2)^2]}$$

$\gamma > 1$: overdamping, $\gamma = 1$: critical damping and $\gamma < 1$: underdamping.

(2) A structure vibrating with viscous damping makes δ_1 oscillations per second and in S_n cycles its amplitude diminishes by η per cent. Determine $\bar{\delta}$, c and γ. Assuming damping is removed, determine the decrease in proportion in the period of vibration.

$$\delta = 1/S_n \ \log[\eta/S_n \times S_1/(\eta S_1/S_n - \eta)]$$

$$= 1/S_n \ \log[\eta S_1/\eta(S_1 - S_n)]$$

$$\text{the period } T_d = 1/S_1; \quad \omega = 2\pi/T_d = 2\pi S_1$$

$$c = \bar{\delta}/S_1$$

$$\gamma = \bar{\delta}/\omega = 1/(2\pi S_1)S_n \ \log \ [S_1/(S_1 - S_n)]$$

$$T/T_d = \sqrt{(1-\gamma^2)} \ \text{decrease in proportion}$$

Coulomb damping or dry friction damping

Coulomb damping occurs when bodies slide on dry surfaces. The damping force is approximately constant provided the surfaces are uniform and the differences between the starting and moving conditions are small.

$$F_f = \mu N \tag{3.62}$$

where μ is the coefficient of kinetic friction of the materials and N is a normal force.

The equation of motion is

$$m\ddot{x} + kx \pm F_f = 0 \tag{3.63}$$

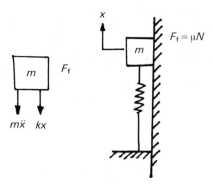

Fig. 3.11 Coulomb damping.

If the velocity $\dot{x}$ is positive, a positive sign is normally applied to F_f. The general solutions for the displacement x and velocity $\dot{x}$ are written as

$$x = x_0\cos\omega t + F_f/k\ (1 - \cos\omega t)$$
$$\dot{x} = -\omega(x_0 - F_f/k)\sin\omega t \qquad (3.64)$$

A diagram similar to Fig. 3.10 is drawn. The first negative peak for $t = \pi/\omega$ is

$$x = -(x_0 - 2F_f/k) \qquad (3.65)$$

The amplitude is reduced by $2F_f/k$. In the second half-cycle the velocity is positive and the equation of motion is written as

$$m\ddot{x} + kx = -F_f \qquad (3.65a)$$

The amplitude at the end of the next half-cycle can easily be determined by inspection. For an initial displacement of $-\bar{x}_0$ for time $t = T_d/2$, all terms in equation (3.65) are to be reversed. Where $\bar{x}_0 = x_0 - 2F_f/k$, equation (3.64) is written as

$$x = -(\bar{x}_0 - F_f/k)\ \cos(\omega t - \pi) - (F_f/k)[1 - \cos(\omega t - \pi)] \qquad (3.66)$$
$$= -(x_0 - 3F_f/k)\cos(\omega t - \pi) - F_f/k \text{ for } T_d/2 \le t \le T_d$$

In the second half-cycle the velocity is positive and the equation of motion becomes

$$m\ddot{x} + kx = -F_f \qquad (3.67)$$

At the next peak value, that is at the end of the first complete cycle with $\omega = 2\pi$, the value of x becomes

$$x_1 = x_0 - 4F_f/k \qquad (3.68)$$

The dotted line shown in Fig. 3.12 is represented by

$$x = x_0 - 4F_f/k \times t/T_d \qquad (3.68a)$$

After n such cycles the amplitude is

$$x_n = x_0 - 4nF_f/k \qquad (3.69)$$

For small damping at the nth cycle, the value of x_n will be larger. Eventually, with a large value of F_f or f, the vibration will cease at N cycles such that $F_f = k'x_N$. Equation (3.69) becomes

$$x_N = x_0 - 4NF_f/k' \qquad (3.70)$$

This phenomenon is demonstrated in Fig. 3.12 and a typical case is examined in Table 3.8.

Table 3.8 Coulomb or dry friction damping.

A bascule bridge is modelled as a single-degree of freedom system of vertical structure tied by a rotating deck with a bottom guide imposing a friction. Using the following data find the frequency of oscillation and the number of cycles executed prior to the ceasing of the deck motion.

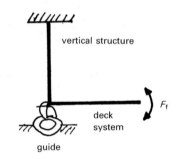

A line diagram for a bascule bridge.

Stiffness of the vertical structure = K_T
Moment of inertia of the deck system = I_0
Displacement from the original position of the deck system (in radians) = x

$$\omega = \sqrt{(K_T/I_0)}$$

$$f = \frac{1}{2\pi} \sqrt{(K_T/I_0)}$$

Loss in amplitude per cycle = $4F_f/K_T$ radians, hence the number of cycles executed prior to ceasing

$$= \frac{x}{4\ F_f/K_T} = x \times K_T/4F_f$$

The displacement–time curve can be drawn starting at time zero for x radians to a reduced value corresponding to $x \times K_T/4F_f$.

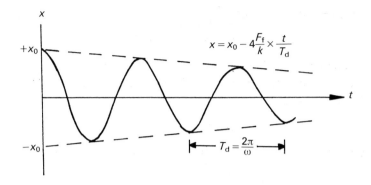

Fig. 3.12 Decay of free vibrations under Coulomb damping.

Structural damping

This is due to the internal friction of the material. This type of resistance is approximately proportional to the displacement amplitude and is independent of the frequency. Suggested values for damping for steel, reinforced concrete and prestressed concrete are 2% to 3% without cracks and 3% to 5% with well disposed cracks. Table 3.9 gives damping properties of known structural materials.

Table 3.9 Material damping.

Material	Damping values
Aluminium	0.002
Aluminium alloy	0.0039−0.001
Cast iron	0.003−0.03
Steel	0.001−0.009
Glass	0.0006−0.002
Concrete without cracks	0.01−0.03
Concrete with cracks	0.03−0.05
Rubber	1.0

3.2.4 Undamped forced vibrations (harmonic disturbing force)

The general form of a harmonic driving or forcing function is written as

$$F = F_0\sin(\omega_f t + \phi) \qquad (3.71)$$

where F_0 is a constant presenting the amplitude of the force, ω_f is the circular frequency for the harmonic driving function and ϕ is the phase angle, which depends on the initial conditions for the force. The force is defined by

$$F = F_0\sin\omega_f t \quad \text{or} \quad F = F_0\cos\omega_f t \qquad (3.72)$$

Chapter 3

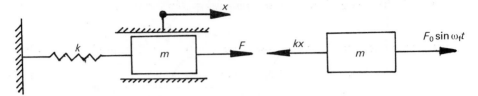

Fig. 3.13 Harmonic force excitation: free body diagram.

The model of the undamped system is shown in Fig. 3.13. Newton's law is applied and the equation of motion becomes

$$m\ddot{x} = -kx + F \tag{3.73}$$

$$\text{or } m\ddot{x} + kx = F_0\sin\omega_f t \tag{3.74}$$

The solution will be $x = x_a + x_b$ $\tag{3.75}$

where $x_a = A\sin\omega_f t + B\cos\omega_f t$ $\tag{3.76}$
     = a complementary function
     = the solution of the homogeneous equation

   x_b = the particular solution
     = a solution which satisfies the complete differential equation

   x = the complete solution
     = the sum of the free-vibration and the forced-vibration components

  The particular solution for x_b is now considered. A particular solution of the form

$$x_b = X\sin\omega_f t \tag{3.77}$$

is adopted and, upon substitution, the following equation is obtained:

$$-m\omega_f^2 X\sin\omega_f t + kX\sin\omega_f t = F_0\sin\omega_f t \tag{3.78}$$

where $X = F_0/k - m\omega_f^2$

$$\begin{aligned}
x_b &= [(F_0/(k - m\omega_f^2)]\sin\omega_f t \\
&= (F_0/k)\sin\omega_f t/[1 - \omega_f^2]/(k/m) \\
&= (X_{ST}/1 - \gamma_f^2)\sin\omega_f t
\end{aligned} \tag{3.79}$$

$$\text{and } x_b = X\sin\omega_f t \tag{3.80}$$

where $\gamma_f = \omega_f/\omega$

     = the frequency ratio = $\dfrac{\text{forced frequency}}{\text{natural (free) frequency}}$

$X_{ST} = F_0/k$ = the static displacement of the spring due to a constant force F_0

$X = X_{ST}/(1 - \gamma_f^2)$ = the amplitude of x_b $\tag{3.81}$

Equation (3.80) is considered for the particular solution of x_b.

(1) $\gamma_f < 1$; $1 - \gamma_f^2 > 0$

x_b is in phase with the force as shown in Fig. 3.14. The amplitude is given by

$$X = X_{ST}/(1 - \gamma_f^2); \qquad \gamma_f < 1 \qquad (3.82)$$

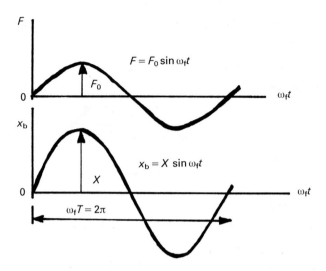

Fig. 3.14 Force and amplitude diagrams when $\gamma_f < 1$.

(2) $\gamma_f > 1$; $1 - \gamma_f^2 < 0$

$$x_b = [X_{ST}/(\gamma_f^2 - 1)](-\sin\omega_f t) \qquad (3.83)$$
$$= X(-\sin\omega_f t)$$

The positive amplitude X is given by

$$X = X_{ST}/(\gamma_f^2 - 1); \qquad \gamma_f > 1 \qquad (3.84)$$

The motion is of opposite phase to the force as shown in Fig. 3.15.

(3) $\gamma_f = 1$; $\omega_f = \omega$

The resonant amplitude $X = X_{ST}/(1 - \gamma_f^2) \to \infty$ and resonance occurs. The solution for this case can be shown as

$$x_b = (-X_{ST}\omega_f t/2)\cos\omega_f t; \qquad \gamma_f = 1 \qquad (3.85)$$

or
$$= (X_{ST}\omega_f t/2)\sin(\omega_f t - \pi/2) \qquad (3.85a)$$

The motion is oscillating with an amplitude which increases with time, as shown in Fig. 3.16. The amplitude does not become large instantaneously but requires time to build up. The motion lags the force by 90°.

The discussion has been of the forced-motion part only of the complete solution defined by equation (3.75). The free-vibration part of the motion was covered in an earlier section.

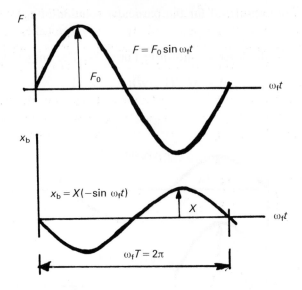

Fig. 3.15 Force and amplitude diagrams when $\gamma_f > 1$.

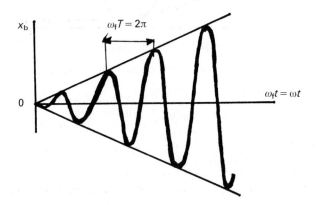

Fig. 3.16 Increase of amplitude with time.

(4) Forced-amplitude and magnification factor (MF)

For some purposes the amplitude of the forced motion is of great importance. The variation of the amplitude X with the frequency ratio γ_f can be studied, using the magnification factor MF (also called the dynamic load factor or DLF) defined as

$$\text{MF} = X/X_{ST} \tag{3.86}$$

$$= \text{forced amplitude/static deflection of spring}$$

$$\text{MF} = X/X_{ST} = 1/(1 - \gamma_f^2) \qquad \gamma_f < 1$$

or   (3.87)

$$\text{MF} = X/X_{ST} = 1/(\gamma_f^2 - 1) \qquad \gamma_f > 1$$

The diagram of MF against γ_f (Fig. 3.17) indicates that MF is greater than 1 in the range $\gamma_f = 0$ to $\gamma_f = 1$, approaching infinity as γ_f approaches 1. The resonant amplitude requires, however, some time to build up. The eventual large amplitude resulting from resonance is of great concern, since it may lead to the destruction of the system.

Since $\gamma_f = \omega_f/\omega$ and $\omega = \sqrt{(k/m)}$, γ_f can be changed by altering ω_f or m or k. It is proper to consider the effect on the amplitude if ω_f or m is changed; if k is altered, one should remember that not only does MF change but also X_{ST}.

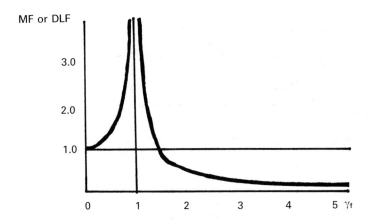

Fig. 3.17 Magnification factor (MF) versus γ_f.

(5) The complete solution and motion. The total motion is defined by $x = x_a + x_b$

$$x = X_a\sin(\omega_t + \phi) + [X_{ST}/(1 - \gamma_f^2)]\sin\omega_f t \quad \text{for } \gamma_f < 1 \tag{3.88}$$

$$x = X_a\sin(\omega_t + \phi) - [X_{ST}/\gamma_f^2 - 1)]\sin\omega_f t \quad \text{for } \gamma_f > 1 \tag{3.89}$$

The complete solution is the sum of two sinusoidal curves of different frequency.

When the forced frequency is smaller than the natural frequency, the forced-motion part can be used as a basis for plotting the free-vibration part as shown in Fig. 3.18. When the forced frequency is greater than the natural frequency, the free-vibration part serves as the axis for the forced portion and is shown in Fig. 3.19.

The complete solution for the case of resonance is:

$$x = X_a\sin(\omega t + \phi) - (X_{ST}\omega_f t/2)\cos\omega_f t \qquad \omega_f = \omega \tag{3.90}$$

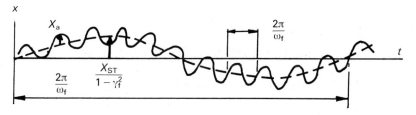

Fig. 3.18 Forced frequency smaller than natural frequency for x versus t.

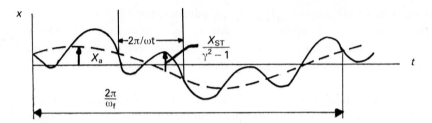

Fig. 3.19 Forced frequency greater than natural frequency for x versus t.

This is the sum of a sine wave of constant amplitude and an oscillating curve having an increasing amplitude. In the early stages, the first part may be significant, but later the forced motion part becomes predominant.

3.2.4.1 Types of pulse load

The MFs for some common simple loading cases are discussed in this section.

Rectangular pulse load

A suddenly applied load F occurs with a constant duration t_{cd} and with no damping effects. The value of the displacement x is written as

$$x = F/k \ [\cos\omega(t - t_{cd}) - \cos\omega t] \tag{3.91}$$

F/k is the static deflection and x/δ_{ST} will define the value of MF.

$$\begin{aligned}
\text{MF} &= 1 - \cos\omega t = 1 - \cos 2\pi(t/T) \text{ for } t \le t_{cd} \\
\text{MF} &= \cos 2\pi(t/T - t_{cd}/T) - \cos 2\pi(t/T) \text{ for } t \ge t_{cd}
\end{aligned} \tag{3.92}$$

where T is the natural period as before.

Triangular pulse load

Here the system is initially at rest when a load F is suddenly applied which decreases linearly to zero at time t_{cd}. The response is in two stages.

Stage 1: response before t_{cd}
$t \le t_{cd}$

$$x = \frac{F}{k} (1 - \cos\omega t) + \frac{F_1}{kt_{cd}} [(\sin\omega t/\omega) - 1] \tag{3.93}$$

$$\text{MF} = 1 - \cos\omega t + (\sin\omega t/\omega t_{cd}) - t/t_{cd}$$

Stage 2: response after t_{cd}
$t \ge t_{cd}$

$$x = \frac{F}{k\omega t_{cd}}[\sin\omega t_{cd} - \sin\omega(t - t_{cd})] - \frac{F}{k}\cos\omega t$$

$$MF = \frac{1}{\omega t_{cd}}[\sin\omega t_{cd} - \sin\omega(t - t_{cd})] - \cos\omega t$$

(3.94)

Figures 3.20 and 3.21 show design charts for rectangular and triangular pulses. The MF value is read once for a given load and the value of T is known. Table 3.10 gives a typical example with a finite time.

Table 3.10 Constant load with a finite time.

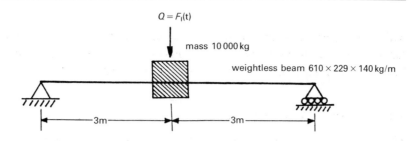

A steel beam, for which

$$L = 6\,\text{m}$$
$$t_r = 0.075\,\text{s}$$
$$I = 111\,844\,\text{cm}^4$$
$$Z_e = 3626\,\text{cm}^3$$
$$E = 200 \times 10^6\,\text{kN/m}^2 = 200 \times 10^3\,\text{N/mm}^2$$
$$F_1 = 200\,\text{kN}$$

The dynamic bending stress, σ_{dy}, is required.

$$k = \text{spring stiffness} = 48EI/L^3$$
$$= 48 \times 200 \times 10^3 \times 111\,844 \times 10^4/(6000)^3$$
$$= 4\,970\,844.4\,\text{N/mm}$$

$$T = 2\pi\ \sqrt{(10\,000/4\,970\,844.4)} = 0.28$$

Using Fig. 3.22:

$$t_r/T = 0.075/0.28 = 0.268$$

hence

$$t_{max}/t_r = 2.02$$

$$(MF)_{max} = 1.85$$

$$\sigma_{dy} = \sigma_{ST} \times (MF)_{max}$$
$$= M/Z(MF)_{max} = [200 \times 600/(4 \times 3626)]\ (1.85)$$
$$= 15.306\,\text{kN/cm}^2$$
$$= 0.001\,530\,6\,\text{kN/m}^2$$

The time at which σ_{dy} occurs is

$$t_{max} = (t_{max}/t_r)t_r = 2.02 \times 0.075 = 1.515\,\text{s} \approx 1.5\,\text{s}$$

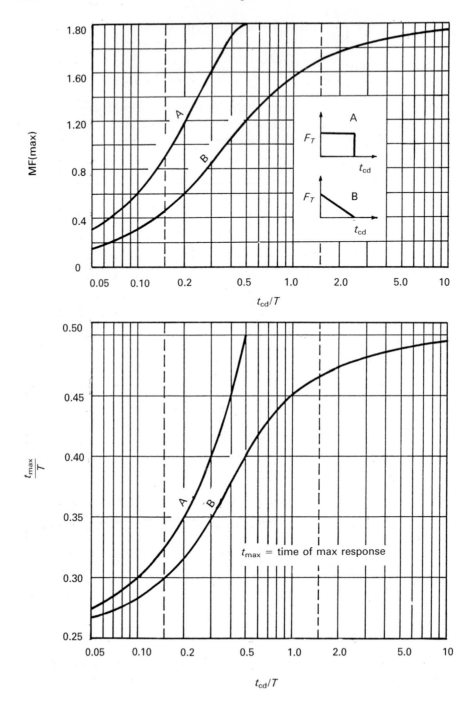

Fig. 3.20 Maximum response of single-degree elastic systems (undamped) subjected to rectangular and triangular load pulses having zero rise time. (Courtesy of the US Army Corps of Engineers.)

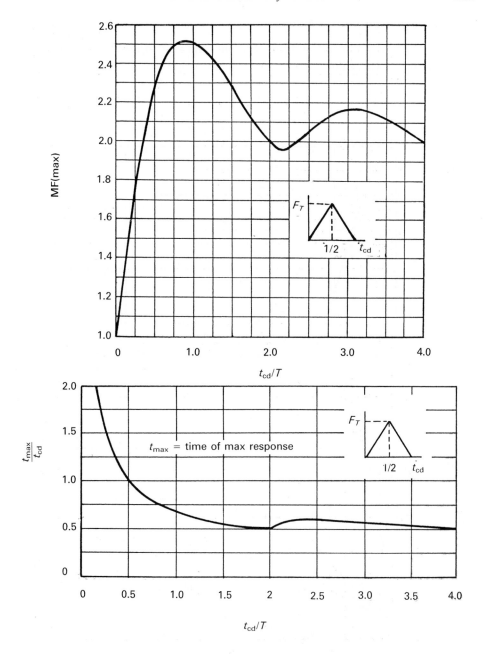

Fig. 3.21 Maximum response of single-degree elastic systems (undamped) subjected to an equilateral triangular load pulse. (Courtesy of US Army Corps of Engineers.)

Constant load with finite rise time

Here is a load which has a finite rise time but remains constant thereafter. Let the rise time be t_r. The values of MF for two stages are given opposite.

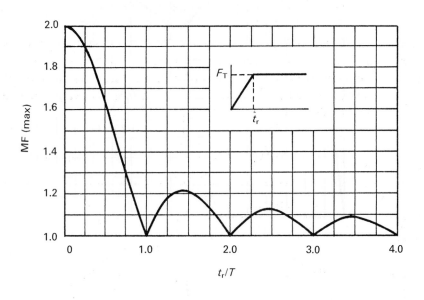

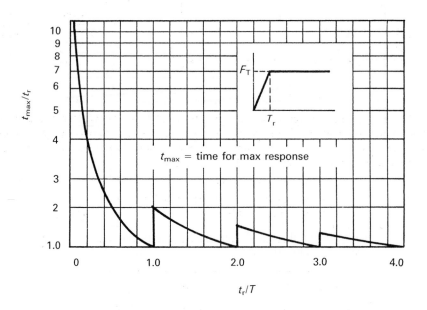

Fig. 3.22 Maximum response of single-degree elastic systems (undamped) subjected to a constant force with a finite rise time. (Courtesy of the US Army Corps of Engineers.)

Stage 1: response before T_r

$$\text{MF} = \frac{1}{t_r}\,(t - \sin\omega t/\omega) \text{ for } t \leq t_r \tag{3.95}$$

Stage 2: response after T_r

$$\text{MF} = 1 + \frac{1}{\omega t_r}\,[\sin\omega(t - t_r) - \sin\omega t] \text{ for } t \geq t_r \tag{3.96}$$

Figure 3.22 shows that when $t_r < 0.25T$, the reduction in the MF value below 2 is not significant. The loading is thus assumed to come on instantaneously. From the typical responses shown in Fig. 3.22, it can be deduced that T_r is large relative to t. Another example is that of an impulsive load of a triangular variation acting with a finite time, as shown in Table 3.11.

Table 3.11 Impulsive load with triangular variation. (The table gives an analysis independent of the charts.)

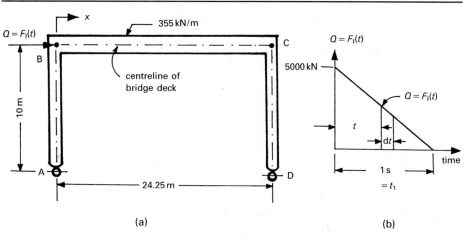

(a)   (b)

Figure (a) represents a cross-section of a concrete bridge. The bridge is subject to a time-dependent impulsive load $F(t)$ from a nearby explosion. If the impulsive load varies linearly in time, as shown in Fig. (b), determine the maximum displacement of the bridge and the dynamic shear in each pier. Use the following data:

I for each pier = $1.09\,\text{m}^4$
E concrete = $21\,\text{GN/m}^2$; $\mathbf{g} = 9.807\,\text{m/s}^2$
Deck load = $335\,\text{kN/m}$

Designation	t_0	t_1
Finite time t (s)	0.02	0.02
$F_1(t)$ (kN) load	4950	4850
V (kN) shear resistance	100	350

continued

Table 3.11 *Continued.*

$$\delta x = 1/m \int_{t=0}^{t=t_1} F(t)(t_1 - t)dt$$

The moment impulse relationship is given by

$$\delta \dot{x} = F_1(t)\frac{dt}{m} - \frac{kxdt}{m} = \frac{F_1(t) - V}{m} \times \delta t$$

$$S = kx$$

Incremental impulse $[F_1(t) - V]\delta t = I_p/m$

$$\dot{x} = \dot{x}_0 + \Sigma_i \delta \dot{x}$$

$$\partial x = \check{x}\delta t$$

$$W = 355 \times 24.25 = 8532 \, kN$$

$$\delta = FL^3/2(3EI)$$

$$\delta = 1; \; k = F = 6EI/L^3 = (6 \times 21 \times 10^6/10^3) \; (1.09) = 137\,000 \, kN/m$$

$$m = 8532/9.872 \, m^2/s = 870 \, kN\text{-}s^2/m$$

Shear resistance V at $t = 0.02$
$$= 100 \, kN$$

$$T = 2\pi \; \sqrt{(870/137\,000)} = 0.5 \, s$$

$$F_1(t) - S = 4950 - 100 = 4850; \; \text{factor} = F \times t = 4850 \times 0.02 = 97$$

$$\delta \dot{x} = 0.112; \; \dot{x} = 0.215; \; \delta x = 0.055 \times 0.02 = 0.0011$$

$$V = 137\,000 \times 0.0011 = 150 \, kN$$

3.2.5 Forced vibrations with viscous damping (harmonic force)

Figure 3.23 shows a typical forced vibration system with viscous damping represented by a spring/dashpot combination. The equation of motion is written as

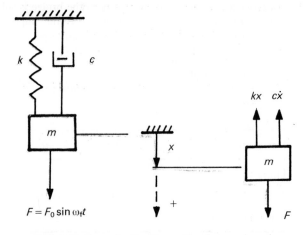

Fig. 3.23 Forced vibration with viscous damping.

$$m\ddot{x} = -kx - c\dot{x} + F \tag{3.97}$$

After rearrangement of terms, the equation of motion becomes

$$m\ddot{x} + c\dot{x} + kx = F_0\sin\omega_f t \tag{3.97a}$$

The solution is:

$$x = x_a + x_b$$

where x_a is the complementary function for small damping:

$$x_a = X'e^{-\gamma\omega t}\sin(\omega_d t + \phi)$$

decays with time to zero: *transient state*; x_b is the particular solution; the part of motion which will occur continuously, while the forcing is present; ω_f is the driving frequency: *steady state*.

Assume

$$x_b = M'\sin\omega_f t + N'\cos\omega_f t \tag{3.98}$$

After substitution into equation (3.97) the following equation is derived:

$$-m\omega_f^2(M'\sin\omega_f t + N'\cos\omega_f t) + c\omega_f(M'\cos\omega_f t - N'\sin\omega_f t) +$$
$$k(M'\sin\omega_f t + N'\cos\omega_f t) = F_0\sin\omega_f t \tag{3.99}$$

Equating the coefficients of sine and cosine on the two sides:

$$(k - m\omega_f^2)M' - c\omega_f N' = F_0$$
$$c\omega_f M' + (k - m\omega_f^2)N' = 0 \tag{3.100}$$

Solving for M' and N':

$$M' = (k - m\omega_f^2)F_0/[(k - m\omega_f^2)^2 + (c\omega_f)^2]$$
$$N' = -c\omega_f F_0/[(k - m\omega_f^2)^2 + (c\omega_f)^2] \tag{3.101}$$

Equation (3.98) can be written with trigonometric substitutions:

$$x_b = \sqrt{(M'^2 + N'^2)}\,\sin(\omega_f t - \phi)$$
$$\tan\phi = -N'/M' \tag{3.102}$$

By substitution of equation (3.101) into equation (3.102), the values of x_b and m are obtained:

$$x_b = \{F_0/\sqrt{[(k - m\omega_f^2)^2 + (c\omega_f)^2]}\}\,\sin(\omega_f t - \phi)$$
$$\tan\phi = c\omega_f/(k - m\omega_f^2) \tag{3.103}$$

The following substitutions are made:

$$\frac{1}{k}(k - m\omega_f^2) = 1 - \frac{\omega_f^2}{k/m}$$

$$= 1 - \omega_f^2/\omega^2 = 1 - r_f^2$$

(3.104)

$$\frac{1}{k}(c\omega_f) = [2c/2\sqrt{(mk)}][\omega_f/\sqrt{(k/m)}] = 2(c/c_c)(\omega_f/\omega)$$

$$= 2\gamma r_f$$

Equation (3.103) can now be written as

$$x_b = X\sin(\omega_f t - \phi)$$

(3.105)

where

$$X = F_0/\sqrt{[(k - m\omega_f^2)^2 + (c\omega_f)^2]}$$

(3.106)

$$= (F_0/k)/\sqrt{[(k - m\omega_f^2/k)^2 + (c\omega_f/k)^2]}$$

$$= X_{ST}/\sqrt{[(1 - r_f^2)^2 + (2\gamma r_f)^2]}$$

and

$$\tan \phi = c\omega_f/(k - m\omega_f^2)$$

(3.107)

$$= (c\omega_f/k)/(k - m\omega_f^2/k)$$

$$= 2\gamma r_f/(1 - r_f^2)$$

The particular solution of equation (3.105) is a steady-state motion of amplitude X, with the same frequency as the forcing condition, but it lags behind the force F by the phase angle ϕ, or by the time t' defined by:

$$t' = \phi/\omega_f$$

(3.108)

Table 3.12 gives a typical example of a simple beam. Both the steady-state amplitude X and the phase angle m are dependent upon the damping factor γ_f and the frequency ratio r_f. The complete solution is:

$$x = X'e^{-\gamma\omega t}\sin(\omega_d t + \phi) + X\sin(\omega_f t - \phi)$$

(3.109)

where X and ϕ depend on the forcing condition as well as the physical constants of the system. The constants X' and ϕ can be determined from the initial displacement and velocity.

If the force ceases after steady-state motion has been attained, a free damped vibration will follow. Whether an immediate gain, a loss, or no change in amplitude occurs depends on when the force stops. Figure 3.24 shows force/time and displacement/time relationships.

Table 3.12 Forced vibrations with damping (single degree).

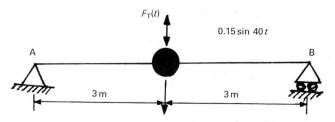

Determine the amplitude of vibration for the system shown in the diagram, for which the data are:

$$\text{length } L = 6\,\text{m}$$
$$\text{static load at mid-span} = 20\,\text{kN}$$
$$F = \text{vertical excited force} = 0.15\sin 40t\,\text{kN}$$
$$c = \text{damping coefficient} = 0.15$$
$$EI = 0.02 \times 10^6\,\text{kNm}^2$$
$$\delta_{ST} = WL^3/48EI = 20 \times (6)^3/48 \times 0.02 \times 10^6 = 0.0045\,\text{m}$$
$$f_v = \text{vertical eigen frequency} = \frac{1}{2\pi}\,\sqrt{(k/m)} = \frac{\sqrt{g}}{2\pi} \times \frac{1}{\sqrt{\delta_{ST}}}$$
$$g = 981\,\text{cm/s}^2 = 9.81\,\text{m/s}^2$$
$$f_v = 5/\sqrt{\delta_{ST}} \text{ if } \delta_{ST} \text{ in cm}$$
$$= 50/\sqrt{\delta_{ST}} \text{ if } \delta_{ST} \text{ in m}$$
$$= 50/\sqrt{0.0045} = 745.36 \text{ cycles/s}$$
$$k = 20/0.0045 = 4444.45\,\text{kN/m}$$

Substituting in equation (3.106):

$$\omega_f = 2\pi\,(745.36) = 4685.12 \text{ cycles/s}$$
$$X = 20/4444.45 \times [1/\sqrt{\{[1 - (40)^2/(4685.12)^2]^2 + (0.15)^2/(4444.45)^2 \times (40)^2\}}$$
$$= 0.0045\,\text{m}$$

Where the resonance $\omega_f = \omega$, for the maximum vibration case

$$X = 1.948\,\text{m}$$

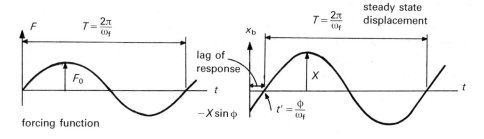

Fig. 3.24 Force/time and displacement/time diagrams.

3.2.5.1 Magnification factor MF and steady-state amplitude X

The amplitude X of the steady-state motion is important in practice. The magnification factor MF is defined by:

$$\text{MF} = X/X_{ST} = 1/\sqrt{[(1 - r_f^2)^2 + (2\gamma r_f)^2]} \tag{3.110}$$

MF plotted against $r_f = \omega_f/\omega$ for various damping ratios is shown in Fig. 3.25.

The maximum and minimum for MF are obtained when the derivative is set to zero:

$$d(\text{MF})/dr_f = 0$$

This results in

$$\frac{r_f (1 - r_f^2 - 2\gamma^2)}{[(1 - r_f^2)^2 + (2\gamma r_f)^2]^{3/2}} = 0 \tag{3.111}$$

which is satisfied by the following conditions:

(1) When $r_f = 0$, this defines the starting point of the curves. This will be a minimum point provided that $\gamma < 0.707$. For $\gamma \geq 0.707$ it will be a maximum point.
(2) When $r = \infty$, this defines the final minimum point on each curve.
(3) For $(1 - r_f^2 - 2\gamma^2) = 0$, which gives

$$r_f = \sqrt{(1 - 2\gamma^2)} \tag{3.112}$$

for $1 - 2\gamma^2 > 0$, i.e. $\leq \sqrt{2}/2 = 0.707$. This last expression defines the maximum point in the *resonant* region. Since

$$r_f = \sqrt{(1 - 2\gamma^2)} < 1 \tag{3.113}$$

the peak of the curve occurs to the left of the resonant value of $r_f = 1$.

The maximum amplitude can be determined by substituting equation (3.106) into the amplitude expression, resulting in:

$$X_{max}/X_{ST} = 1/[2\gamma \sqrt{(1 - \gamma^2)}] \tag{3.114}$$

$$\doteq \tfrac{1}{2}\gamma \quad \text{for } \gamma \ll 1 \tag{3.115}$$

When damping occurs the maximum amplitude is limited to a finite value. For $\gamma \geq 0.707$ the maximum point occurs at $r_f = 0$, and the curves drop continuously as r_f increases.

The family of curves indicates that a reduction in MF — and hence the amplitude — is obtained only in the region where r_f is large due to a high forcing frequency relative to the natural frequency of the system.

Figure 3.25 shows the effect of varying r_f upon the maximum displacement amplitude; the effect of varying r_f by changing ω_f only should be considered here. The conditions resulting from changing r_f by altering k and m (that is ω) require a separate discussion.

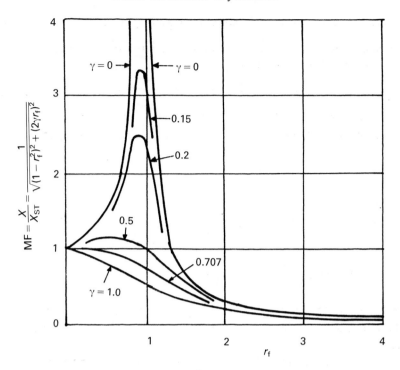

Fig. 3.25 MF versus r_f.

3.2.5.2 Phase angle ϕ

The phase angle ϕ which is defined by equation (3.107), which is also related to $2Xr_f/(1 - \gamma_f^2)$, depends on the damping factor γ and the frequency ratio r_f. Plotting ϕ against r_f for various values of γ, families of curves are obtained, as shown in Fig. 3.26.
For no damping

$$\phi = \quad 0° \text{ for } r_f = 0 \text{ to } r_f < 1$$
$$\phi = \quad 90° \text{ for } r_f = 1 \qquad\qquad (3.116)$$
$$\phi = 180° \text{ for } r_f > 1$$

For small values of γ, these conditions are approximated and the curves for small γ approach the curve for the zero-damping case.

3.2.5.3 Influence of mass and stiffness on amplitude

In studying the effect of varying r_f upon the steady-state amplitude, recall that

$$r_f = \omega_f/\omega = \omega \, \sqrt{(m/k)}$$

and hence r_f can be varied by changing k or m as well as the driving frequency ω_f.

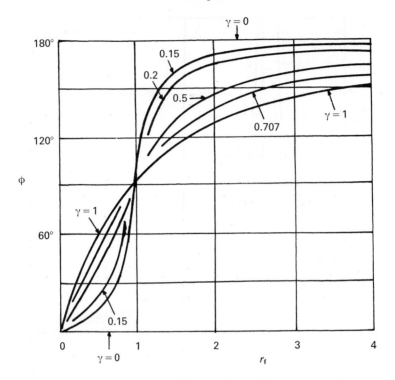

Fig. 3.26 ϕ versus r_f.

However, if either k or m is changed, based on equation (3.45), this will alter γ, as

$$\gamma = c/2\sqrt{(mk)} = c/c_c$$

and distort the interpretation of Fig. 3.25, since a different γ-curve would then have to be used. In addition, altering k will change the value of the 'static' deflection X_{ST}.

While studying the effect of varying k, the amplitude relation given by equation (3.116) can be written in the form:

$$X = F_0/\sqrt{[(k - m\omega_f^2)^2 + (c\omega_f)^2]} \tag{3.117}$$

and X can then be plotted against k for different values of the damping coefficient c, as shown in Fig. 3.27. In this case F_0, m and ω_f are constant.

Maximum and minimum points on the curves can be obtained by setting $dX/dk = 0$. From this, it is found that the maximum point occurs for $k = m\omega_f^2$ and is defined by

$$X_{max} = F_0/c\omega_f \tag{3.118}$$

In addition, all the curves approach zero as k becomes large. The initial point (for $k = 0$) is given by

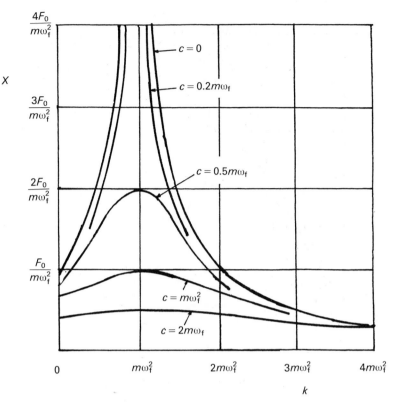

Fig. 3.27 X versus k.

$$X = F_0 / \sqrt{[(m\omega_f^2)^2 + (c\omega_f)^2]} \tag{3.119}$$

A reduction in amplitude is achieved only as k becomes large. This means that stiff springs will result in a small amplitude of motion for a given system.

The amplitude relation given by equation (3.117) can also be used to observe the effect of varying m upon the amplitude. In this case, X is plotted against m for various values of c, with F_0, k and ω_f being taken as constant. The resulting family of curves is shown in Fig. 3.28.

Maximum and minimum points on the curves can be determined by setting $dX/dm = 0$. The maximum point occurs at $m = k/\omega_f^2$ and is given by

$$X_{\max} = F_0/c\omega_f \tag{3.120}$$

All the curves approach zero as m becomes large. The initial point (for $m = 0$) is given by

$$X = F_0 / \sqrt{[k^2 + (c\omega_f)^2]} \tag{3.121}$$

It is clearly indicated that large values of m result in a reduction in amplitude.

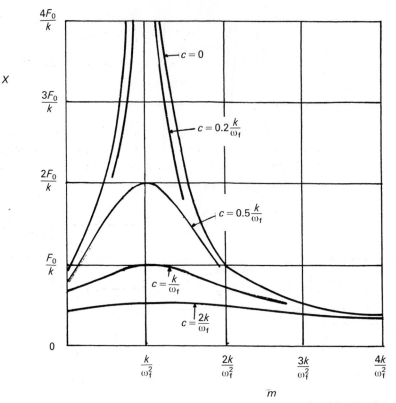

Fig. 3.28 X versus m.

3.2.5.4 Mechanical impedance method

The equation of motion is given by equation (3.97). Since

$$F = F_0 e^{i\omega t} = F_0(\cos\omega_f t) \tag{3.122}$$

then $\mathrm{Im}(F_0 e^{i\omega t})$ is the imaginary part of the solution. Let the displacement vector $X = X e^{i(\omega t - \phi)}$, the velocity and acceleration vectors will be $i\omega_f X$ and $-\omega^2 X$ respectively. Substituting these into equation (3.97) gives

$$(k - m\omega_f^2 + ic\omega_f)X e^{-i\phi} = F_0$$

or

$$X e^{-i\phi} = F_0/(k - m\omega_f^2 + ic\omega_f) \tag{3.123}$$

However

$$X e^{-i\phi} = X(\cos\phi - i\sin\phi) \tag{3.124}$$

where $\phi = \tan^{-1}\omega_f c/(k - m\omega_f^2)$. Hence equation (3.117) is derived, and

$$x_b = \text{Im}(X) \tag{3.125}$$

$$= F_0 \sin(\omega_f t - \phi) / \sqrt{[(k - m\omega_f^2)^2 + (c\omega_f)^2]}$$

The x_a value based on steady-state vibration is given earlier in equation (3.97).

3.2.5.5 Rotating unbalance

The rotating unbalance is a common source of forced vibrations. Figure 3.29 shows such a system. The arm rotates with angular velocity ω_f radians/s. The angular position of the arm is defined by $\omega_f t$ with respect to the indicated horizontal datum.

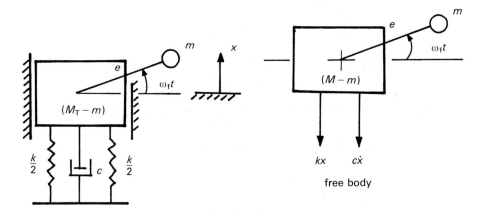

Fig. 3.29 Rotating unbalance and forced vibrations (M_T = total mass, m = eccentric mass and e = eccentricity of m).

Positive displacements x are assumed upwards. The horizontal motion of $(M_T - m)$ is prevented by guides. The vertical displacement of m is $(x + e\sin\omega_f t)$. The differential equation of motion is written:

$$(M_T - m)\frac{d^2 x}{dt^2} + m\frac{d^2}{dt^2}(x + e\sin\omega_f t)$$

$$= -kx - c\frac{dx}{dt} \tag{3.126}$$

and rearranged

$$M_T \ddot{x} + c\dot{x} + kx = me\omega_f^2 \sin\omega_f t \tag{3.127}$$

Comparison with equation (3.97) for motion forced by $F = F_0 \sin\omega_f t$ enables the steady-state solution to be set down by analogy with equation (3.105) as

$$x = X\sin(\omega_f t - \phi) \tag{3.128}$$

where

$$X = \frac{me\omega_f^2}{\sqrt{[(k - M_T\omega_f^2)^2 + (c\omega_f)^2}} \tag{3.129}$$

$$= \frac{\dfrac{me}{M}\,\omega_f^2 \times \dfrac{M_T}{k}}{\sqrt{[(k - M_T\omega_f^2/k)^2 + (c\omega_f/k)^2]}} \tag{3.130}$$

$$= \frac{\dfrac{me}{M_T} \times r_f^2}{\sqrt{[(1 - r_f^2)^2 + (2\gamma r_f)^2}} $$

$$\frac{X}{me/M_T} = \frac{r_f^2}{\sqrt{[(1 - r_f^2)^2 + (2\gamma r_f)^2]}} \tag{3.131}$$

$$\tan \phi = \frac{2\gamma r_f}{1 - r_f^2} \tag{3.132}$$

Here $\omega = \sqrt{(k/M_T)}$ represents the natural circular frequency of the undamped system (including the mass m), but x defines the forced motion of the main mass $(M_T - m)$. For this case, ϕ will be represented physically by the angle of the eccentric arm relative to the horizontal datum of $\omega_f t$. For a value of ϕ determined from equation (3.132) the arm would be at this angle when the main body is at its neutral position, moving upwards. Since the motion lags behind the driving force, the arm then leads the motion by the angle ϕ determined.

The steady-state amplitude is important, and it can be studied by plotting $X/(me/M_T)$ against the frequency ratio r_f for various values of the damping ratio γ, as shown in Fig. 3.30. Maximum and minimum points can be determined by setting

$$\frac{d}{dr_f}[X/(me/M)] = 0 \tag{3.133}$$

$$r_f = 1/\sqrt{(1 - 2\gamma^2)} > 1$$

Accordingly, the peaks occur to the right of the resonance value of $r_f = 1$.

Figure 3.30 is adequate, provided that the variation in r_f is limited to changing ω_f. Note that small amplitude occurs only at low driving frequencies, as would be expected.

Since γ (as well as r_f) is dependent on k and M_T, Fig. 3.30 does not show properly the effect of varying k or M_T. The effect of varying these can be observed by writing the amplitude relation in the form given by equation (3.133). The amplitude X can then be plotted against either k or M_T for various values of the damping coefficient c. The resulting families of curves will be identical to those of Figs 3.27 and 3.28, provided that m is replaced by M_T, and F_0 is replaced by $me\omega_f^2$. At resonance, equation (3.129) becomes:

$$X = me\omega_f/c \tag{3.134}$$

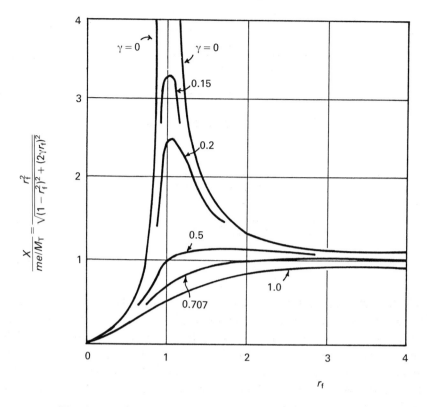

The y-axis is labelled $\dfrac{X}{me/M_\mathrm{T}} = \dfrac{r_\mathrm{f}^2}{\sqrt{(1 - r_\mathrm{f}^2)^2 + (2\gamma r_\mathrm{f})^2}}$ and the x-axis is labelled r_f. Curves are labelled $\gamma = 0$, $\gamma = 0$, 0.15, 0.2, 0.5, 0.707, 1.0.

Fig. 3.30 $X/(me/M_\mathrm{T})$ versus r_f values for the various values of γ.

It should be noted that the amplitude is dependent on the quantity '*em*' and that if either *m* or *e* is small, the amplitude will become small. The eccentric condition should be reduced as far as possible.

3.2.5.6 Force transmission and isolation

This section deals with the force transmitted to the support of the system. The force carried by the support is idealized by the spring and dashpot (Fig. 3.31) which are connected to it. The dynamic force F exerted on the system by the support is thus written in simple form as

$$F = kx + c\dot{x} \tag{3.135}$$

For steady-state displacement, x and velocity $\dot{x}$ can be substituted from equation (3.105). This gives:

$$F = kX\sin(\omega_\mathrm{f}t - \phi) + c\omega_\mathrm{f}X\cos(\omega_\mathrm{f}t - \phi) \tag{3.136}$$

where X and ϕ are defined by equations (3.106) and (3.107) respectively. Equation (3.136) can thus be rewritten, involving damping and frequency ratios, as

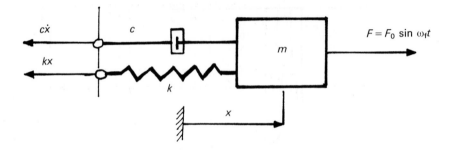

Fig. 3.31 Free-body diagram for force transmission.

$$F = \sqrt{[kX^2 + (c\omega_f X)^2]}\ \sin(\omega_f t - \phi - \beta)$$
$$= \sqrt{[k^2 + (c\omega_f)^2]}\,X\sin(\omega_f t - \alpha) \tag{3.137}$$

$$\alpha = \phi + \beta \quad \text{and} \quad \tan\beta = -c\omega_f/k = -2\gamma r_f \tag{3.138}$$

The maximum force F_T, or *force amplitude*, transmitted will be

$$F_T = X\ \sqrt{[k^2 + (c\omega_f)^2]}$$
$$= F_0\ \frac{\sqrt{[k^2 + (c\omega_f)^2]}}{\sqrt{[(k - m\omega_f^2)^2 + (c\omega_f)^2]}} \tag{3.139}$$

Dividing numerator and denominator by k gives

$$F_T = F_0\ \frac{\sqrt{[1 + (2\gamma r_f)^2]}}{\sqrt{[(1 - r_f^2)^2 + (2\gamma r_f)^2]}} \tag{3.140}$$

The ratio F_T/F_0 is defined as the *transmissibility* or *TR*. Thus

$$TR = F_T/F_0 = \frac{\sqrt{[1 + (2\gamma r_f)^2]}}{\sqrt{[(1 - r_f^2)^2 + (2\gamma r_f)^2]}} \tag{3.141}$$

Table 3.13 gives a typical example on a simple frame. The manner in which the transmitted force is influenced by the physical parameters of the system can be shown by plotting TR against r_f for various values of the damping ratio γ, as shown in Fig. 3.32. All curves start at TR = 1 for $r_f = 0$. There is a common point, or cross-over, at $r_f = \sqrt{2}$. While damping cuts down the peak force transmitted in the resonant region, it results in greater force transmission for $r_f > \sqrt{2}$. This latter effect is opposed to the influence which the damping increase has in reducing the displacement amplitude, for larger values of r_f, as indicated by Fig. 3.25. It can be shown that the peak of the curve occurs at

$$r_f = \sqrt{[-1 + \sqrt{(1 + 8\gamma^2)}]/2\gamma} < 1 \tag{3.142}$$

Figure 3.32 should be used only in conditions in which F_0 is constant and r_f is varied by changing ω_f only.

If it is required to observe the effect of varying the spring stiffness k, equation

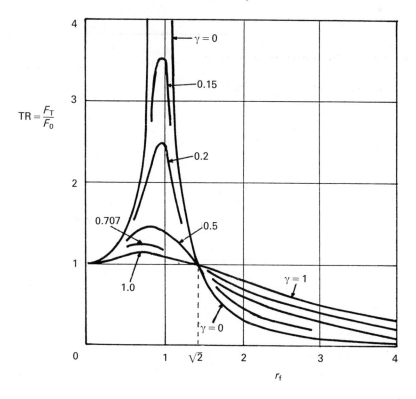

Fig. 3.32 Relationship between TR and r_f.

Table 3.13 Transmissibility.

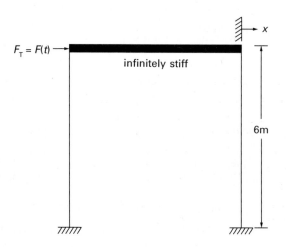

A frame shown in the figure is subjected to a sinusoidal ground motion of $250\sin5t$ at girder level. Calculate (1) the transmissibility of motion, (2) the maximum shear force in the columns of the frame and (3) the maximum bending stresses in the columns. Use the following data:

continued

Table 3.13 *Continued.*

$$K \text{ (columns)} = 5000 \text{ N/mm}$$
$$\gamma = 0.05$$
$$\omega = 7.38 \text{ rad/s}$$
$$\omega_f = 5.32 \text{ rad/s}$$
$$r_f = 5.32/7.38 = 0.721$$
$$F_0/k = 250/5000 = 0.05 \text{ mm}$$

$X =$ steady-state amplitude

$$= \frac{0.05}{\sqrt{\{[1 - (0.721)^2]^2 + (2 \times 0.05 \times 0.721)^2\}}}$$

$$= 0.103 \text{ mm}$$

(1)
$$\text{TR} = \frac{\sqrt{[1 + (2\gamma r_f)^2]}}{\sqrt{[(1 - r_f^2)^2 + (2\gamma r_f)^2]}}$$

$$= \frac{\sqrt{[1 + (2 \times 0.05 \times 0.721)]}}{\sqrt{\{[1 - (0.721)^2]^2 + (2 \times 0.05 \times 0.721)^2\}}}$$

$$= \frac{\sqrt{(1 + 0.005\,198\,4)}}{\sqrt{0.235\,751}} = 2.065$$

The relative displacement $= 0.25(0.721)^2 = 0.13 \text{ mm}$
(2) Maximum shear force $= 5000 \times 0.13/2 = 325 \text{ N}$
(3) Maximum bending stress $= 325 \times 6 \times 1000$
$$= 195 \times 10^4 \text{ Nmm}$$
$$= 1.95 \text{ kNm}$$

(3.139) can be used to plot TR against k for various values of the damping coefficient c. The transmissibility is then expressed by

$$\text{TR} = F_T/F_0 \qquad (3.143)$$

$$= \frac{\sqrt{[k^2 + (c\omega_f)^2]}}{\sqrt{[(k - m\omega_f^2)^2 + (c\omega_f)^2]}}$$

Several curves are drawn in Fig. 3.33 relating k and TR. The initial value of TR and cross-over point can now easily be determined, together with the trend of the curves in approaching $\text{TR} = 1$ for large values of k. The peak value from equation (3.145) gives the value of k:

$$k = m\omega_f^2/2 + \sqrt{[(m\omega_f^2/2)^2 + (c\omega_f)^2]} \qquad (3.144)$$

This will be greater than $m\omega_f^2$ if $c \neq 0$. Therefore, the peak will occur to the right of the $k = m\omega_f^2$ value. An appreciable reduction in the transmitted force is achieved only in the region for which $k < m\omega_f^2/2$ with light damping and small spring stiffness.

The effect of varying the mass m can also be determined from equation

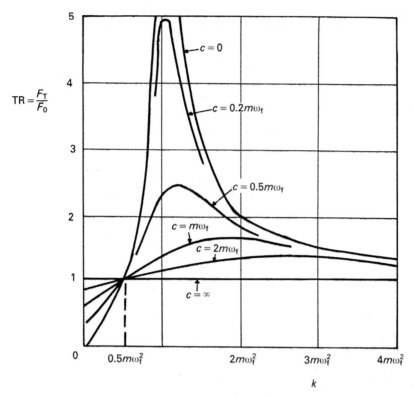

Fig. 3.33 Relationship between TR and k.

(3.143) by plotting TR against m for various values of c. This is shown in Fig. 3.34. Here, the peak value occurs at $m = k/\omega_f^2$ and is written as

$$\text{TR}_{max} = \sqrt{[k^2 + (c\omega_f)^2]}/c\omega_f \tag{3.145}$$

A reduction in the transmitted force occurs only in the region where m is large. This might not be expected. It is important to note that an increase in the mass m will, however, also result in an increase in the static force carried by the support.

The force transmitted for the case of a rotating eccentric mass m

The relation for F_T can be obtained by substituting $me\omega_f^2$ for F_0 in equations (3.139) and (3.140).

$$F(t) = F_T$$

$$= me\omega_f^2 \frac{\sqrt{[k^2 + (c\omega_f)^2]}}{\sqrt{[(k - M_T\omega_f^2)^2 + (c\omega_f)^2]}} \tag{3.146}$$

$$= me\omega_f^2 \frac{\sqrt{[1 + (2\gamma r_f)^2]}}{\sqrt{[(1 - r_f^2)^2 + (2\gamma r_f)^2]}} \tag{3.147}$$

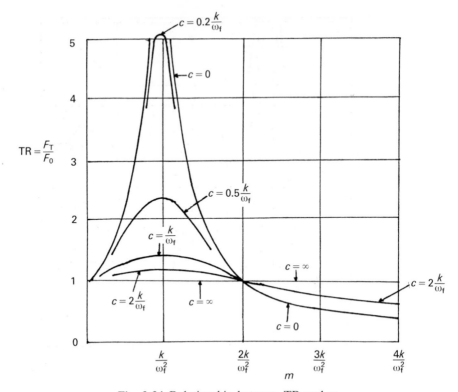

Fig. 3.34 Relationship between TR and m.

After multiplying the numerator and denominator by k/M_T, followed by re-arrangement, F_T becomes

$$F(t) = F_T = (mek/M_T) \times r_f^2 \frac{\sqrt{[1+(2\gamma r_f)^2]}}{\sqrt{[(1-r_f^2)^2 + (2\gamma r_f)^2]}} \qquad (3.148)$$

hence

$$F_T/(mek/M_T) = r_f^2 \frac{\sqrt{[1+(2\gamma r_f)^2]}}{\sqrt{[(1-r_f^2)^2 + (2\gamma r_f)^2]}} \qquad (3.149)$$

The effect of varying ω_f upon the transmitted force can be shown by plotting $F_T/(mek/M_T)$ against r_f for various values of the damping ratio γ. In so doing, k and M_T are taken as constant. The reference mek/M_T is then fixed. The curves obtained are shown in Fig. 3.35.

Damping serves to limit the transmitted force in the region of resonance. A cross-over point occurs at $r_f = \sqrt{2}$, for which $F_T/(mek/M_T)$ has a value of 2. For no damping the curve approaches a value of 1 as r_f approaches infinity. When damping is present, the force becomes very large as r_f increases, and the greater the damping is, the more rapidly this occurs. Even for small damping, the increase in transmitted force is significant. Since frequency ratios of 10 or more are quite common, the importance of considering damping is evident. The maximum and minimum points for the family of curves are determined by setting

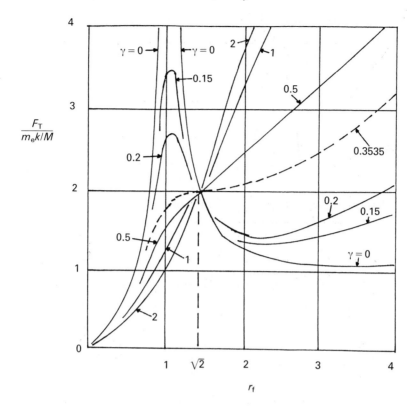

Fig. 3.35 Relationship between $F_T/(m_e k/M)$ and r_f.

$$\frac{d}{dr_f}[F_T/(m_e k/M_T)] = 0$$

The following boundary conditions are noted:

(1) For $r_f = 0$; this defines the initial point of $F_T/(m_e k/M_T) = 0$ for all curves.
(2) By the roots of the relation

$$2\gamma^2 r_f^6 + (16\gamma^4 - 8\gamma^2)r_f^4 + (8\gamma^2 - 1)r_f^2 + 1 = 0 \qquad (3.150)$$

if $0 < \gamma < \sqrt{2}/4$ there are two positive real roots of this relation. One of these will be between $r_f = 0$ and $r_f = \sqrt{2}$, and will define a maximum point on the curve. The other will be $r_f > \sqrt{2}$ and will define a minimum point on the curve. If $\gamma > \sqrt{2}/4$ there is no maximum point on the curve.

If it is required to determine the effect of varying k and M_T upon the transmitted force, this can be done by using equation (3.147). After rearrangement, the following relations are established:

$$F_T/me\omega_f^2 = \frac{\sqrt{[k^2 + (c\omega_f)^2]}}{\sqrt{[(k - M_T\omega_f^2)^2 + (c\omega_f)^2]}} \qquad (3.151)$$

Since the forcing frequency ω_f, the mass m and the eccentricity e are to be held constant, this case becomes identical to those shown in Figs 3.33 and 3.34, provided that m is replaced by M_T.

At resonance, equation (3.147) reduces to

$$F(t) = F_T = \frac{me\omega_f}{c} \times \sqrt{(k^2 + \omega_f^2)} \tag{3.152}$$

$$= X_{res} \sqrt{[k^2 + (c\omega_f)^2]}$$

Oscillating support

The oscillating support is a common cause of forced motion of a system. The following steps are taken using Fig. 3.36.

(1) Displacement of the support $y = y_0 \sin\omega_f t$. $\tag{3.153}$
(2) Displacement of the mass (assume $x > y$ and $\dot{x} > \dot{y}$).
(3) The equation of motion is written as

$$m\ddot{x} = -k(x - y) - c(x - y) \tag{3.153a}$$

or

$$m\ddot{x} + c\dot{x} + kx = ky + c\dot{y}$$
$$= ky_0\sin\omega_f t + c\omega_f y_0 \cos\omega_f t \tag{3.154}$$
$$= y_0\sqrt{[k^2 + (c\omega_f)^2]}\sin(\omega_f t - \beta)$$

as before

$$\tan\beta = -c\omega_f/k = -2\gamma r_f \tag{3.155}$$

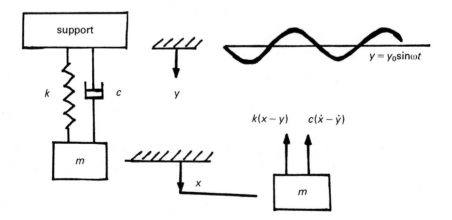

Fig. 3.36 Oscillating support.

This type of force is represented by the right-hand side of equation (3.154); the forcing amplitude would be

$$y_0\sqrt{[k^2 + (c\omega_f)^2]}$$

From equations (3.105), (3.106) and (3.107) the steady-state solution is

$$
\begin{aligned}
x &= \frac{y_0\sqrt{[k^2 + (c\omega_f)^2]}}{\sqrt{[(k - m\omega_f^2)^2 + (c\omega_f)^2]}} \times \sin(\omega_f t - \phi) \\
&= \frac{y_0\sqrt{[1 + (2\gamma r_f)^2]}}{\sqrt{[(1 - r_f^2)^2 + (2\gamma r_f)^2]}} \times \sin(\omega_f t - \phi) \\
&= X\sin(\omega_f t - \phi)
\end{aligned}
\tag{3.156}
$$

where X = amplitude

$$
= \frac{y_0\sqrt{[1 + (2\gamma r_f)^2]}}{\sqrt{[(1 - r_f^2)^2 + (2\gamma r_f)^2]}}
\tag{3.157}
$$

$$
\tan\phi = \frac{2\gamma r_f}{1 - r_f^2}
\tag{3.158}
$$

(4) The force carried by the support is defined as

$$F = k(k - y) + c(\dot{x} - \dot{y})$$

and $F = -m\ddot{x}$

Using the value of x of equation (3.156), the force carried by the support becomes

$$
F = m\omega_f^2 X\sin[\omega_f^2 - (\phi + \beta)]
\tag{3.159}
$$

$$
= m\omega_f^2 y_0 \times \frac{\sqrt{[1 + (2\gamma r_f)^2]}}{\sqrt{[(1 - r_f^2)^2 + (2\gamma r_f)^2]}} \times \sin[\omega_f t - (\phi + \beta)]
$$

$$
= \frac{y_0 k m\omega_f^2}{k} \times \frac{\sqrt{[1 + (2\gamma r_f)^2]}}{\sqrt{[(1 - r_f^2)^2 + (2\gamma r_f)^2]}} \times \sin[\omega_f t - (\phi + \beta)]
$$

$$
= y_0 k \times \frac{r_f^2\sqrt{[1 + (2\gamma r_f)^2]}}{\sqrt{[(1 - r_f^2)^2 + (2\gamma r_f)^2]}} \times \sin[\omega_f t - (\phi + \beta)]
$$

$$
= F_T\sin[\omega_f t - (\phi + \beta)]
\tag{3.160}
$$

$$
F_T = y_0 k \times \frac{\sqrt{[1 + (2\gamma r_f)^2]}}{\sqrt{[(1 - r_f^2)^2 + (2\gamma r_f)^2]}}
\tag{3.161}
$$

This value of F_T is the amplitude of the force with a maximum value. It is in phase with m. The displacement amplitude can be studied from equation (3.161) arranged as

$$
X/y_0 = \sqrt{[1 + (2\gamma r_f)^2]}/\sqrt{[(1 - r_f^2)^2 + (2\gamma r_f)^2]}
\tag{3.162}
$$

The transmitted force can be studied by the following arrangement:

$$F_T/y_0 k = \frac{r_f^2 \sqrt{[1 + (2\gamma r_f)^2]}}{\sqrt{[(1 - r_f^2)^2 + (2\gamma r_f)^2]}} \qquad (3.163)$$

This is the same expression as equation (3.149) plotted in Fig. 3.35. At resonance, equation (3.157) reduces to

$$X = y_0 \sqrt{[1 + (2\gamma)^2]}/2\gamma \qquad (3.164)$$

3.2.5.7 Energy considerations for forced motion

In the case of forced motion, energy is introduced by the positive work done by the driving force during each cycle. In the case of steady-state motion, this energy is equal to the energy dissipated by damping in each cycle.

energy input = energy dissipated by viscous damping

Energy input for harmonic force per cycle: E_{er}

The following steps are taken:

(1) Harmonic force $F = F_0 \sin\omega_f t$
(2) Steady-state motion for viscous damping

$$x = X\sin(\omega_f t - \phi)$$

hence

$$dx = X\omega_f \cos(\omega_f t - \phi)dt$$

(3) Energy input/cycle $= E_{er} = \int F_T dx$

$$E_{er} = \int F dx$$

$$= \int_0^T F_0 \sin\omega_f t \, X\omega_f \cos(\omega_f t - \phi)dt$$

$$= F_0 X\omega_f \int_0^{2\pi/\omega} \sin\omega_f t (\cos\omega_f t \cos\phi + \sin\phi)dt$$

$$= F_0 X\omega_f \int [\cos\phi(\sin\omega_f t \cos\omega_f t + \sin\phi(\sin^2\omega_f t)]dt$$

$$= F_0 X\omega_f [\cos\phi(\sin^2\omega_f t/2\omega_f) + \sin\phi t/2 - (\sin\omega_f t \cos\omega_f t/2\omega_f)]_0^{2\pi/\omega}$$

$$E_{er} = \pi F_0 X \sin\phi \qquad (3.165)$$

The maximum energy input occurs at resonance, for $\phi = 90°$

$$E_{er} = \pi F_0 X \qquad (3.165a)$$

Energy dissipated by viscous damping per cycle: E''_{VD}

The following steps are taken:

(1) Damping force $\qquad F_d = -c\dot{x}$

(2) Steady-state motion $x = X\sin(\omega_f t - \phi)$

$$\text{hence } dx = X\omega_f\cos(\omega_f t - \phi)dt$$
$$\text{and } \dot{x} = X\omega_f\cos(\omega_f t - \phi)$$
$$E''_{VD} = \int F_1(t)dx = \int (c\dot{x})dx$$
$$= c(\omega_f X)^2 \int_0^{2\pi/\omega} \cos^2(\omega_f t - \phi)dt$$
$$= c(\omega_f X)^2 \{(t/2) + [\sin(\omega_f t - \phi)\,\cos(\omega_f t - \phi)/2\omega_f]\}^{2\pi/\omega_f}$$
$$= \pi c\omega_f X^2 \tag{3.166}$$

Hence the energy dissipated per cycle is proportional to the square of the amplitude.

Boundary conditions

The vector force diagrams give an understanding of the behaviour of the system in the three regions of the frequency ratio r_f, where

$$r_f = \omega_f/\omega = \frac{\text{forcing frequency}}{\text{natural free frequency}}$$

(1) When $\omega_f \ll 1$, damping and inertial forces are small and the elastic forces are predominant. The elastic force is nearly equal to the disturbing force and the disturbing force is nearly in phase with the displacement. There is a small phase angle, as shown in Fig. 3.37.
(2) When $r_f = 1$, at resonance the phase angle is 90° and the inertial force is now larger. The elastic and inertial forces are balanced. The applied force is

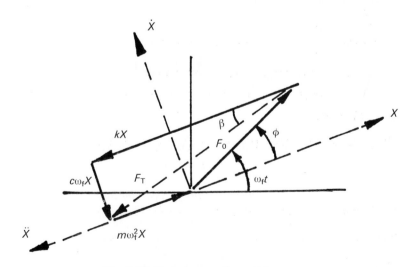

Fig. 3.37 Frequency ratio $r_f \ll 1$.

balanced by the damping force. For small damping coefficients c, the amplitude may be very large, as shown in Fig. 3.38.

(3) When $r_f \gg 1$, (Fig. 3.39) at high impressed frequencies the inertial force predominates and becomes nearly equal to the disturbing force. The latter is nearly in phase with the acceleration and the phase angle tends to be close to 180°.

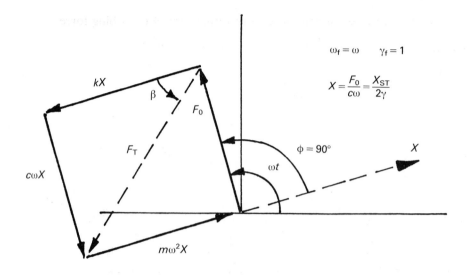

$$\omega_f = \omega \qquad \gamma_f = 1$$

$$X = \frac{F_0}{c\omega} = \frac{X_{ST}}{2\gamma}$$

$$\phi = 90°$$

Fig. 3.38 Frequency ratio $r_f = 1$.

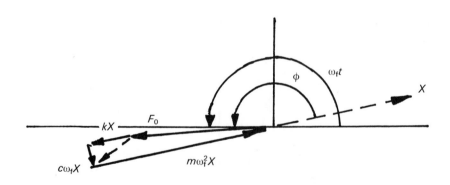

Fig. 3.39 Frequency ratio $r_f \gg 1$.

3.2.5.8 Rotating vectors and harmonically forced vibrations

The differential equation of the motion equation (3.97) can be presented in the form given below when the steady-state solution and its derivatives are substituted:

$$x = X\sin(\omega_f t - \phi)$$

$$\dot{x} = \omega_f X \cos(\omega_f t - \phi) = \omega_f X \sin(\omega_f t - \phi + \pi/2) \qquad (3.167)$$

$$\ddot{x} = -\omega_f^2 X \sin(\omega_f t - \phi) = \omega_f^2 X \sin(\omega_f t - \phi + \pi)$$

$$m\omega_f^2 X \sin(\omega_f t - \phi) - c\omega_f X \sin(\omega_f t - \phi + \pi/2) - kX \sin(\omega_f t - \phi) + F_0 \sin\omega_f t = 0$$
$$(3.168)$$

This vector relation can be interpreted as follows:

Inertial force + damping force + restoring force + disturbing force = 0

and the vector relation is shown graphically in Fig. 3.40.

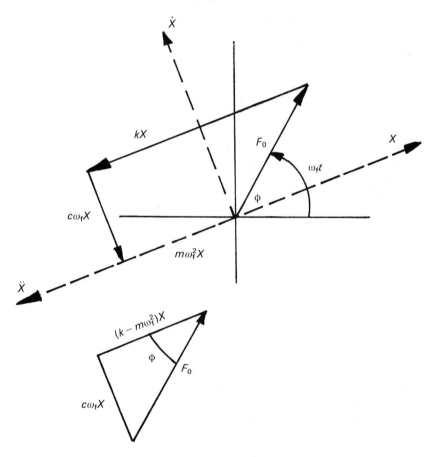

Fig. 3.40 Rotating vectors for harmonically forced vibrations.

Resulting cases

(1) The displacement lags the disturbing force by the angle ϕ, which can vary between 0° and 180°.
(2) The restoration of the elastic force is always opposite in direction to the displacement.

(3) The damping force lags the displacement by 90° and hence is opposite in direction to the velocity.

(4) The inertial force is in phase with the displacement and opposite in direction to the acceleration.

(5) The vectors remain fixed with respect to each other and rotate together with angular velocity ω_f.

For example, the right-angled triangle in Fig. 3.40 shows:

$$X = F_0 / \sqrt{[(k - m\omega_f^2)^2 + (c\omega_f)^2]}$$

$$\tan\phi = c\omega_f / (k - m\omega_f^2) \tag{3.169}$$

This relation was put into non-dimensional form in equations (3.106) and (3.113) and yielded the concept of the MF given by equation (3.110).

3.2.6 Single-degree undamped elasto-plastic system

A single-degree undamped elasto-plastic system is assumed to have the bilinear resistance function shown in Fig. 3.41. The response solution due to a suddenly applied constant load is divided into three stages:

(1) the response up to the elastic limit x_{el};
(2) the plastic response between x_{el} and the maximum displacement;
(3) the rebound after (2), when the displacement begins to decrease.

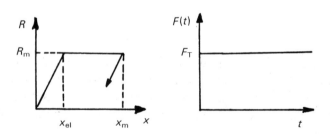

Fig. 3.41 Elasto-plastic system (R_m = maximum resistance, x_m = maximum displacement, x_{el} = elastic limit).

Stage 1

First response $x \leq x_{el}$; the initial velocity and displacements are zero.

$$x = \delta_{ST}(1 - \cos\omega t)$$

$$\dot{x} = \delta_{ST}\omega\sin\omega t \tag{3.170}$$

where $\delta_{ST} = F_T/k$ and $\omega = \sqrt{(k/m)}$. The time t_{el} at displacement x_{el} is obtained from equation (3.170).

$$\cos\omega t_{el} = 1 - x_{el}/\delta_{ST}$$

$$(3.171)$$

and
$$\sin\omega t_{el} = \sqrt{(1 - \cos^2\omega t_{el})}$$

Stage 2

Second response $x_{el} \leq x \leq x_m$

$$x_0 = x_{el}; \quad t_2 = t - t_{el}$$

$$(3.172)$$

$$\dot{x}_0 = \delta_{ST}\omega\sin\omega t_{el}$$

The equation of motion is then written as

$$m\ddot{x} + R_m = F_T$$

$$(3.173)$$

The final solution becomes

$$x = x_{el} + \frac{1}{2m}(F_T - R_m)t_2^2 + x_{el}\omega t_2\sin\omega t_{el}$$

$$(3.174)$$

Differentiating equation (3.174) and setting the result to zero, the maximum time response t_{2m} is given by

$$t_{2m} = [m\omega\delta_{ST}/(R_m - F_T)]\sin\omega t_{el}$$

$$(3.175)$$

Equation (3.175) is substituted into equation (3.174) for computing the maximum displacement x_m:

$$x_m = x_{el} + (\omega\delta_{ST} - \tfrac{1}{2})[m\omega\delta_{ST}/(R_m - F_T)]\sin^2\omega t_{el}$$

$$(3.176)$$

Stage 3

Rebound: a similar procedure is adopted with a suitable equation of motion with initial conditions from the second stage. The easiest method is to consider a residual vibration and compute the amount by which the displacement x must decrease below x_m, i.e. $(R_m - F_T)/k$, to reach the neutral position. The neutral position is given as:

$$x_m - (R_m - F_T)/k$$

$$(3.177)$$

The response is thus

$$x = [x_m - (R_m - F_T)/k] + [(R_m - F_T)/k]\cos(t - t_{2m} - t_{el})$$

$$(3.178)$$

The maximum response charts are given in Figs 3.42 to 3.45 for single-degree, undamped, elasto-plastic systems due to various load pulses.

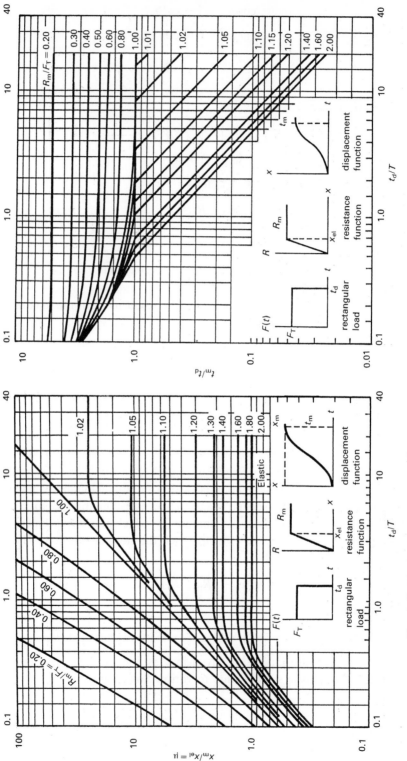

Fig. 3.42 Maximum response of elasto-plastic, single-degree, undamped systems due to rectangular load pulses. (Courtesy of the US Army Corps of Engineers.)

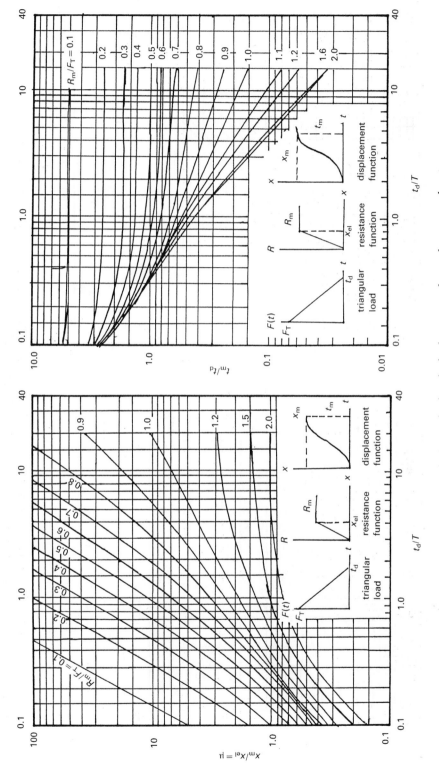

Fig. 3.43 Maximum response of elasto-plastic, single-degree, undamped systems due to triangular load pulses with zero rise time. (Courtesy of the US Army Corps of Engineers.)

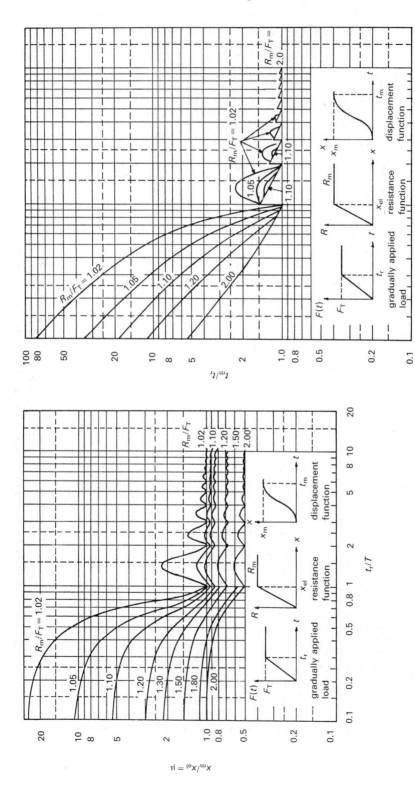

Fig. 3.44 Maximum response of elasto-plastic, single-degree, undamped systems due to a constant force with a finite rise time. (Courtesy of the US Army Corps of Engineers.)

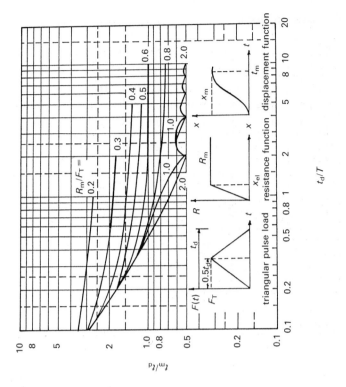

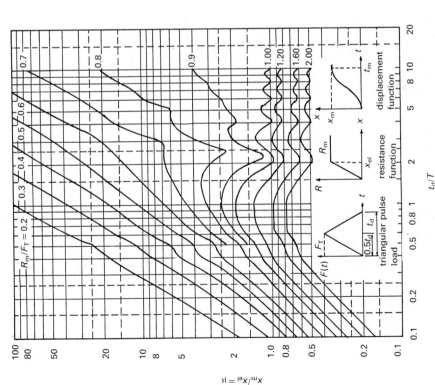

Fig. 3.45 Maximum response of elasto-plastic, single-degree, undamped systems due to equilateral triangular load pulses. (Courtesy of the US Army Corps of Engineers.)

3.3 Two-degrees-of-freedom system

Dynamic systems that require two independent co-ordinates to specify their positions are known as two-degrees-of-freedom systems. Typical examples are shown in Tables 3.13 and 3.14. There will be two natural frequencies from solution of the frequency equation of an undamped system or the characteristic equation of a damped system. A typical example of a spring-mass system is shown in Fig. 3.46(a).

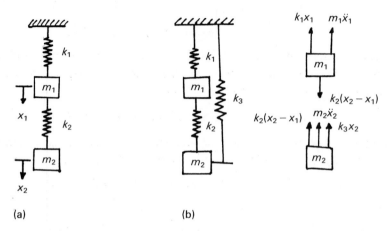

(a)   (b)

Fig. 3.46 Two-degrees-of-freedom systems.

3.3.1 Undamped free vibrations

The equations of motion are written as

$$\Sigma F = ma$$

$$m_1\ddot{x}_1 = -k_1x_1 - k_2(x_1 - x_2) \ \ \dots \ \text{mass } m_1$$

$$m_2\ddot{x}_2 = -k_2(x_2 - x_1) \ \ \dots \ \text{mass } m_2$$

(3.179)

Rearranging the above equations, the final equations of motion become

$$m_1\ddot{x}_1 + (k_1 + k_2)x_1 - k_2x_2 = 0$$

$$m_2\ddot{x}_2 - k_2x_1 + k_2x_2 = 0$$

(3.180)

Assuming the motion is periodic and is composed of harmonic motions of various amplitudes and frequencies, then one of these components as described above may be written as

$$x_1 = X_1(\sin\omega t + \phi); \qquad \ddot{x}_1 = -X_1\omega^2\sin\omega(t + \phi)$$

$$x_2 = X_2(\sin\omega t + \phi); \qquad \ddot{x}_2 = -X_2\omega^2\sin\omega(t + \phi)$$

(3.181)

When equation (3.181) is substituted into equation (3.182), which is the matrix form of equation (3.180),

$$\begin{bmatrix} m_1 & 0 \\ 0 & m_2 \end{bmatrix} \begin{Bmatrix} \ddot{x}_1 \\ \ddot{x}_2 \end{Bmatrix} + \begin{bmatrix} k_1 + k_2 & -k_2 \\ -k_2 & k_2 \end{bmatrix} \begin{Bmatrix} x_1 \\ x_2 \end{Bmatrix} = \begin{Bmatrix} 0 \\ 0 \end{Bmatrix} \tag{3.182}$$

diagonal mass   coupled stiffness
matrix   matrix

the following relationship is obtained:

$$\begin{bmatrix} (k_1 + k_2 - m_1\omega^2) & -k_2 \\ -k_2 & (k_2 - m_2\omega^2) \end{bmatrix} \begin{Bmatrix} X_1 \\ X_2 \end{Bmatrix} = \begin{Bmatrix} 0 \\ 0 \end{Bmatrix} \tag{3.183}$$

The determinant of the left-hand matrix is equal to zero since x_1 and x_2, when zero, define the equilibrium condition of the system. By expansion of the determinant the following equation is arrived at:

$$\omega^4 - [(k_1 + k_2/m_1) + k_2/m_2]\omega^2 + (k_1k_2/m_1m_2) = 0 \tag{3.184}$$

The frequency equation of the system is obtained by assuming $\omega^2 = \bar{C}$ in equation (3.184). A quadratic equation in C is obtained and hence ω.

$$\omega^2 = \bar{C} = (k_1 + k_2/2m_1) + (k_2/2m_2) \pm \sqrt{\{\tfrac{1}{4}[(k_1 + k_2/m_1)}$$
$$+ (k_2/m_2)]^2 - (k_1k_2/m_1m_2)\}} \tag{3.185}$$

If a third spring is attached with stiffness k_3, as shown in Fig. 3.46(b), the determinant of the left-hand coefficient matrix of equation (3.183) can be written as

$$\begin{vmatrix} k_1 + k_2 - m_1\omega^2 & -k_2 \\ -k_2 & k_2 + k_3 - m_2\omega^2 \end{vmatrix} = 0$$

and equation (3.184) becomes

$$\omega^4 - \{[(k_1 + k_2)/m_1] + [(k_2 + k_3)/m_2]\}\omega^2 + \{[k_2(k_1 + k_2) + k_1k_3]/m_1m_2\} = 0 \tag{3.186}$$

Other cases are similarly analysed in Table 3.13. The natural frequency is computed as $f = \omega/2\pi$.

Assuming $k_1 = k_2 = k_3 = k$ and $m_1 = m_2$, equation (3.186) becomes

$$\omega^4 - (4k/m)\omega^2 + (3k^2/m^2) = 0 \tag{3.187}$$

$\omega^2 = C$

$$C^2 - (4k/m)C + (3k^2/m^2) = 0$$

$$C_1 = \omega_1^2 = k/m$$

$$\omega_1 = \sqrt{(k/m)} \tag{3.187a}$$

The other root $\omega_2^2 = 3k/m$ or $\omega_2 = 1.73\sqrt{(k/m)}$ (3.187b)
The lowest frequency is $\omega_1 = \sqrt{(k/m)}$. Table 3.14 gives a standard case using the flexibility method.

Table 3.14 Two-degrees-of-freedom systems (free undamped).

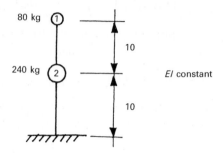

80 kg ①

240 kg ②

10

10

EI constant

Find the frequencies and modes of vibration.

$$x = X\sin\omega t$$

$$x_{10} = X_1\sin\omega t \qquad \ddot{x}_{10} = -\omega^2 X_1\sin\omega t$$

$$x_{20} = X_2\sin\omega t \qquad \ddot{x}_{20} = -\omega^2 X_2\sin\omega t$$

$$\text{Equation of motion} = -x_{10} = f_{11}m_1\ddot{x}_{10} + f_{12}m_2\ddot{x}_{20} \tag{1}$$

$$-x_{20} = f_{22}m_2\ddot{x}_{20} + f_{21}m_1\ddot{x}_{10} \tag{2}$$

Add the unit loads at points (1) and (2) and find f_{11}, f_{12}, f_{21} and f_{22}.

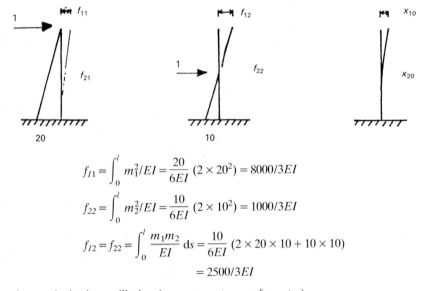

f_{11}

1

f_{21}

20

f_{12}

1

f_{22}

10

x_{10}

x_{20}

$$f_{11} = \int_0^l m_1^2/EI = \frac{20}{6EI}(2 \times 20^2) = 8000/3EI$$

$$f_{22} = \int_0^l m_2^2/EI = \frac{10}{6EI}(2 \times 10^2) = 1000/3EI$$

$$f_{12} = f_{22} = \int_0^l \frac{m_1 m_2}{EI}\,ds = \frac{10}{6EI}(2 \times 20 \times 10 + 10 \times 10)$$

$$= 2500/3EI$$

Therefore, substitutions will give (*note:* x_{10}, etc. are δ_{10}, etc.)

$$-x_{10} = \ddot{x}_{10}\,8000/3EI \times 80 + 2500/3EI \times 240\,\ddot{x}_{20}$$

$$-x_{20} = 1000/3EI \times 240\ddot{x}_{20} + 80 \times 2500/3EI \times \ddot{x}_{10}$$

$$-x_{10} = 640\,000\omega^2 x_{10}/3EI + 200\,000\;\omega^2 x_{20}/EI$$

$$-x_{20} = 80\,000\omega^2 x_{20}/EI + 200\,000\omega^2 x_{10}/3EI$$

$$x_{10}(1 + 640\,000\omega^2/3EI) + x_{20}200\,000\omega^2/EI = 0$$

$$x_{10}\,200\,000\omega^2/3EI + x_{20}(1 + 80\,000\omega^2/EI) = 0$$

Table 3.14 *Continued.*

$$\begin{bmatrix} (1 + 640\,000\omega^2/3EI) & (200\,000\omega^2/EI) \\ \left(\dfrac{200\,000\omega^2}{3EI}\right) & \left(1 + \dfrac{80\,000\omega^2}{EI}\right) \end{bmatrix} \begin{Bmatrix} x_{10} \\ x_{20} \end{Bmatrix} = 0$$

$$\begin{Bmatrix} x_{10} \\ x_{20} \end{Bmatrix} \neq 0, \qquad \text{therefore } \Delta = 0$$

$$(1 + 640\,000\omega^2/3EI)(1 + 80\,000\omega^2/EI) - (200\,000\omega^2/EI)^2 \times \tfrac{1}{3} = 0$$

$$(1 + 880\,000\omega^2/3EI) + (5.12 \times 10^{10}\omega^4/3EI) - (4 \times 10^{10}\omega^4/3E^2I^2) = 0$$

$$(1 + 880\,000k/3) + (1.12 \times 10^{10}k^2/3) = 0, \qquad \text{where } k = \omega^2/EI$$

$$k = [-b \pm \sqrt{(b^2 - 4ac)}]/2a$$

$$k = \{-88 \times 10^4/3 \pm \sqrt{[88 \times 10^4/3)^2 - 4 \times 1.12 \times 10^{10}/3]}\}/(2 \times 1.12 \times 10^{10}/3)$$

$$k = -3.571 \times 10^{-6} \text{ or } -7.5 \times 10^{-5}$$

$$\omega = \sqrt{[(3.571 \times 10^{-6})EI]} \text{ or } \sqrt{[(7.5 \times 10^{-5})EI]}$$

$$f = \omega/2\pi$$

$$\text{frequency} = 1.378 \times 10^{-3} \sqrt{(EI)} \text{ or } 3.01 \times 10^{-4} \sqrt{(EI)}$$

$$\text{frequency} = 3.01 \times 10^{-4} \sqrt{(EI)}$$

3.3.2 Free damped vibration

The spring-mass, damped, free vibration system is shown in Fig. 3.47. The equation of motion is written as

$$m_1\ddot{x}_1 + (c_1 + c_2)\,\dot{x}_1 + (k_1 + k_2)x_1 - c_2\dot{x}_2 - k_2x_2 = 0 \tag{3.188}$$

$$m_2\ddot{x}_2 + c_2\dot{x}_2 + k_2x_2 - c_2\dot{x}_1 - k_2x_1 = 0 \tag{3.188a}$$

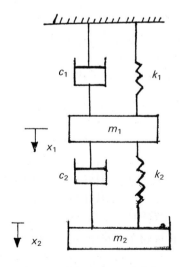

Fig. 3.47 Two-degrees, damped free system.

Using the solution procedure given earlier, the general form of the solution will be

$$x_1 = C_1 e^{st} \text{ and } x_2 = C_2 e^{st} \tag{3.189}$$

where $s = \pm i\omega t$. Substituting equation (3.189) into equation (3.188a) and dividing the two equations throughout by e^{st}, a 2×2 matrix is formed, the determinant of which is

$$\begin{vmatrix} [m_1 s^2 + (c_1 + c_2)s + (k_1 + k_2)] - (c_2 s + k_2) \\ -(c_2 s + k_2) \qquad\qquad (m_2 s^2 + c_2 s + k_2) \end{vmatrix} = 0 \tag{3.190}$$

The characteristic equation of the system after expansion of the determinant becomes

$$[m_1 s^2 + (c_1 + c_2)s + k_1 + k_2](m_2 s^2 + c_2 s + k_2) - (c_2 s + k_2)^2 = 0 \tag{3.191}$$

Since $s = \pm i\omega t$, the circular frequency ω is computed in the manner described above.

3.3.3 Forced vibration with damping

The system shown in Fig. 3.47 is now associated with the force $F_0 \sin \omega_f t$. The forces are acting parallel to the springs vertically and the equation of motion is

$$m_1 \ddot{x}_1 + (c_1 + c_2)\dot{x}_1 + (k_1 + k_2)x_1 - c_2 \dot{x}_2 - k_2 x_2 = F_0 \sin \omega t$$
$$m_2 \ddot{x}_2 + c_2 \dot{x}_2 + k_2 x_2 - c_2 \dot{x}_1 - k_2 x_1 = 0 \tag{3.192}$$

A typical example is shown in Table 3.15. If the impedance method given in section 3.2.5.4 is invoked. The following shows the replacement procedure:

$$\text{Use } F_0 e^{-\omega t} \text{ instead of } F_0 \sin \omega_f t \tag{3.193}$$

and use $X_1 e^{i\omega t}$ and $X_2 e^{i\omega t}$ instead of x_1 and x_2 respectively. By substituting equation (3.193) into equation (3.192) and dividing throughout by $e^{i\omega t}$, the following equations of motion are derived:

$$-[(k_1 + k_2) - m_1 \omega_f^2 + i(c_1 + c_2)\omega_f]X_1 - (k_2 + ic_2\omega_f)X_2 = F_0 \tag{3.194}$$
$$-(k_2 + ic_2\omega_f)X_1 + (k_2 - m_2\omega_f^2 + ic_2\omega_f)X_2 = 0$$

Using Cramer's rule:

$$X_1 = \begin{vmatrix} F_0 & -(k_2 + ic_2\omega_f) \\ 0 & k_2 - m_2(\omega_f)^2 + ic_2\omega_f \end{vmatrix} \Bigg/ \begin{matrix} [(k_1 + k_2) - m_1\omega_f^2 + \\ i(c_1 + c_2)\omega_f](k_2 - m_2\omega_f^2 + ic_2\omega_f) - \\ (k_2 + ic_2\omega_f)^2 \end{matrix} \tag{3.195}$$

Using complex variables it is of the form $(a + ib)/(c + id)$, hence

$$X_1 = (a^2 + b^2)/(c^2 + d^2)$$

Table 3.15 Forced vibrations with damping (two degrees).

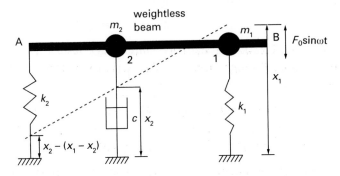

A horizontal beam with two masses is supported by springs. The location of the dashpot is shown. The force at the level of mass 1 is $F_0\sin\omega t$ and at mass 2 is zero.

$$\text{KE} = \tfrac{1}{2}m_1\dot{x}_1^2 + \tfrac{1}{2}m_2\dot{x}_2^2$$

$$\text{PE} = \tfrac{1}{2}k_1x_1^2 + \tfrac{1}{2}k_2\,[x_2 - (x_1 - x_2)]^2$$

$$\text{DE} = \text{damping energy} = \tfrac{1}{2}c\dot{x}_2^2$$

$$\frac{\text{d}}{\text{d}t}\left(\frac{\partial \text{KE}}{\partial \dot{x}_1}\right) - \frac{\partial \text{KE}}{\partial x_1} + \frac{\partial \text{PE}}{\partial x_1} + \frac{\partial \text{DE}}{\partial \dot{x}_1} = F_1 = F_0\sin\omega t$$

or

$$m_1\ddot{x}_1 - 0 + k_1x_1 - k_2[x_2 - (x_1 - x_2)] + 0 = F_0\sin\omega t$$

$$\frac{\text{d}}{\text{d}t}\left(\frac{\partial \text{KE}}{\partial \dot{x}_2}\right) - \frac{\partial \text{KE}}{\partial x_2} + \frac{\partial \text{PE}}{\partial x_2} + \frac{\partial \text{DE}}{\partial \dot{x}_2} = F_2 = 0$$

$$m\ddot{x}_2 - 0 + 2k_2\,(2x_2 - x_1) + c\dot{x}_2 = 0$$

The equations of motions are written:

$$m_1\ddot{x}_1 + (k_1 + k_2)x_1 - 2k_2x_2 = F_0\sin\omega t$$

$$m_2\ddot{x}_2 + c\dot{x}_2 + 4k_2x_2 = 0$$

They are solved for the unknown quantities using flexibility methods.

$$X_1 = F_0^2(k_2 - m_2\omega_{\text{f}}^2)^2 + c_2^2\omega_{\text{f}}^2/[(k_2 - m_1\omega_{\text{f}}^2) \times (k_2 - m_2\omega_{\text{f}}^2)$$
$$- m_2k_2\omega_{\text{f}}^2]^2 + \omega_{\text{f}}^2c_2^2(k_1 - m_1\omega_{\text{f}}^2 - m_2\omega_{\text{f}}^2)^2 \qquad (3.195\text{a})$$

The forcing function is given by

$$F_0\sin\omega_{\text{f}}t = \text{I}_{\text{m}}(F_0e^{i\omega_{\text{f}}t}) \qquad (3.196)$$

Also

$$X_1 = X_1e^{i\phi_1} = F_0(\bar{g} + i\bar{h}) = X_1(\cos\phi_1 + i\sin\phi_1)$$

where $\phi_1 = \tan^{-1}\bar{g}/\bar{h}$. Hence the steady damped vibration can be given as

$$x_1 = \text{I}_{\text{m}}(X_1e^{i\omega_{\text{f}}t}) = \text{I}_{\text{m}}(X_1e^{i(\omega_{\text{f}}t + \phi_1)}) \qquad (3.197)$$
$$= X_1\sin(\omega_{\text{f}}t + \phi_1)$$

Similarly
$$X_2 = F_0(\bar{j} + i\bar{l}) = X_2 e^{i\phi_2}$$
$$x_2 = I_m(X_2 e^{i\omega_f t}) = I_m(X_2 e^{i(\omega_f t + \phi_2)}) = X_2 \sin(\omega_f t + \phi_2)$$

where $\phi_2 = \tan \bar{l}/\bar{j}$.

$$X_2 = \sqrt{F_0^2(k_2^2 + c_2^2 \omega_f^2)/[(k_2 - m_1 \omega_f^2)(k_2 - m_2 \omega_f^2) - m_2 k_2 \omega_f^2)]^2}$$
$$+ [\omega_f^2 c_2^2(k_1 - m_1 \omega_f^2 - m_2 \omega_f^2)^2] \tag{3.198}$$

If the forcing function is $F_0 \cos\omega_f t$, then

$$x_1 = X_1 \cos(\omega_f t + \phi_1)$$
$$x_2 = X_2 \cos(\omega_f t + \phi_2) \tag{3.199}$$

3.3.4 Orthogonality principle

The orthogonality principle is expressed as

$$\sum_{j=1}^{n} m_j X_{jr'} X_{jr''} = 0 \tag{3.200}$$

where r' and r'' identify any two normal modes, j is the jth mass out of a total of n masses and m_j is the jth mass. This principle is extremely useful in the analysis of a multi-degree system. However, in the context of two degrees of freedom, this simply means that two modes are orthogonal.

Let the vibrating system given by

$$x_1 = X_1 \sin(\omega_1 t + \phi_1) + X_2 \sin(\omega_2 t + \phi_2)$$
$$x_2 = \bar{X}_1 \sin(\omega_1 t + \phi_1) + \bar{X}_2 \sin(\omega_2 t + \phi_2) \tag{3.201}$$

where X_1, X_2, $\bar{X}_1$ and $\bar{X}_2$ are the amplitudes of vibration of the two masses and ϕ_1 and ϕ_2 are the phase angles corresponding to the two circular frequencies ω_1 and ω_2.

The kinetic energy is given by

$$(\text{KE})_{max} = \tfrac{1}{2} m_1 (\dot{x}_{1_{max}})^2 + \tfrac{1}{2} m_2 (\dot{x}_{2_{max}})^2 \tag{3.202}$$

Substitution of equation (3.201) into equation (3.202) gives

$$(\text{KE})_{max} = \tfrac{1}{2} m_1 [(X_1 \omega_1)^2 + (X_2 \omega_2)^2 + (2X_1 \omega_1)(X_2 \omega_2)] + \tfrac{1}{2} m_2 [\bar{X}_1 \omega_1)^2 + (\bar{X}_2 \omega_2)^2$$
$$+ 2(\bar{X}_1 \omega_1)(\bar{X}_2 \omega_2)] \tag{3.203}$$

For the principle modes of vibration, the value of $(\text{KE})_{max}$ is given by

$$(\text{KE})_{max} = \tfrac{1}{2}[m_1(X_1 \omega_1)^2 + m_2(\bar{X}_1 \omega_1)^2] + \tfrac{1}{2}[m_1(X_2 \omega_2)^2 + m_2(\bar{X}_2 \omega_2)^2] \tag{3.204}$$

The two maximum kinetic energies must be equal, hence

$$m_1(X_1 \omega_1)(X_2 \omega_2) + m_2(\bar{X}_1 \omega_1)(\bar{X}_2 \omega_2) = 0 \tag{3.205}$$

Since ω_1 and ω_2 are not always zero

$$m_1(X_1 X_2) + m_2(\bar{X}_1 \bar{X}_2) = 0 \tag{3.206}$$

3.4 Multi-degrees-of-freedom systems

3.4.1 Undamped free vibrations

The differential equations of motions for n masses and n degrees of freedom can now be written using the stiffness method.

$$m_{11}\ddot{x}_1 + m_{12}\ddot{x}_2 + \ldots m_{1n}\ddot{x}_n + k_{11}x_1 + k_{12}x_2 \ldots k_{1n}x_n = 0$$
$$m_{21}\ddot{x}_1 + m_{22}\ddot{x}_2 + \ldots m_{2n}\ddot{x}_n + k_{21}x_1 + k_{22}x_2 \ldots k_{2n}x_n = 0$$

$$\vdots \qquad \vdots \qquad \vdots \qquad \vdots \qquad \vdots \qquad \vdots \tag{3.207}$$

$$m_{n1}\ddot{x}_1 + m_{n2}\ddot{x}_2 + \ldots m_{nn}\ddot{x}_n + k_{n1}x_1 + k_{n2}x_2 \ldots k_{nn}x_n = 0$$

In a matrix form, equation (3.207) is written as

$$\begin{bmatrix} m_{11} & m_{12} & \ldots & m_{1n} \\ m_{21} & m_{22} & \ldots & m_{2n} \\ \vdots & \vdots & & \vdots \\ \vdots & \vdots & & \vdots \\ m_{n1} & m_{n2} & \ldots & m_{nn} \end{bmatrix} \begin{Bmatrix} \ddot{x}_1 \\ \ddot{x}_2 \\ \vdots \\ \vdots \\ \ddot{x}_n \end{Bmatrix} + \begin{bmatrix} k_{11} & k_{12} & \ldots & k_{1n} \\ k_{21} & k_{22} & \ldots & k_{2n} \\ \vdots & \vdots & & \vdots \\ \vdots & \vdots & & \vdots \\ k_{n1} & k_{n2} & \ldots & k_{nn} \end{bmatrix} \begin{Bmatrix} x_1 \\ x_2 \\ \vdots \\ \vdots \\ x_n \end{Bmatrix} = \begin{Bmatrix} 0 \\ 0 \\ \vdots \\ \vdots \\ 0 \end{Bmatrix} \tag{3.208}$$

$$\text{or } [M]\{\ddot{x}\} + [K]\{x\} = 0 \tag{3.208a}$$

Where $[M]$ is the *mass* or *inertia matrix* and $[K]$ is the total or *global stiffness matrix*. Equation (3.208a) can be written as:

$$\{\ddot{x}\} + [C_D]\{x\} = 0 \tag{3.209}$$

where $[C_D]$ (the *dynamic matrix*) $= [M]^{-1}[K]$.

Sometimes the equations for multi-degree systems given by equation (3.207) are written in terms of *characteristic value problems*:

$$[K]\{X_i\} = \omega_i^2[M]\{X_i\} \tag{3.210}$$

where $[M]$ is a diagonal matrix and ω is a *characteristic number*. A more convenient form of equation (3.210) is written as

$$([K] - \omega_i^2[M])\{X_i\} = 0 \tag{3.211}$$

The determinant of the coefficients of X_i must be zero.

$$|K - \omega_i^2 M| = \begin{vmatrix} (k_{11} - \omega_1^2 M_1) & k_{12} & k_{13} & \ldots & k_{1n} \\ k_{21} & (k_{22} - \omega_1^2 M_2) & k_{23} & \ldots & k_{2n} \\ & & & & \\ k_{n1} & k_{n2} & k_{n3} & \ldots & (k_{nn} - \omega_1^2 M_n) \end{vmatrix} = 0 \tag{3.212}$$

The determinant is expanded for the nth order equations in ω_i^2 such that for each value of ω_i^2, there will be a corresponding X_i. Table 3.15 shows typical solved examples.

Table 3.16 Multi-degrees-of-freedom systems.

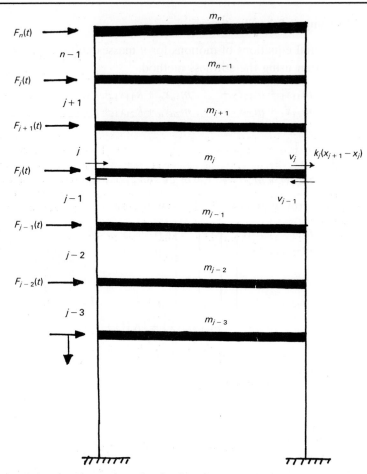

The general form of the equations of motion is given by

$$F_n(t) = m_n \ddot{x}_n + k_{n-1}(x_n - x_{n-1})$$

For the applied force at level j, the above equation is written as

$$F_j(t) = m_j \ddot{x}_j + k_{j-1}(x_j - x_{j-1}) - k_j(x_{j+1} - x_j)$$

(1) Free vibrations

$$F(t) = 0$$

The equations of motion are

$$m_2 \ddot{x}_2 + (k_2 + k_1)x_2 - k_2 x_3 = 0$$
$$m_3 \ddot{x}_3 + (k_3 + k_2)x_3 - k_3 x_4 - k_2 x_2 = 0 \qquad (1)$$
$$m_4 \ddot{x}_4 + (k_4 + k_3)x_4 - k_4 x_5 - k_3 x_3 = 0$$
$$\vdots \qquad\qquad\qquad \vdots$$
$$\vdots \qquad\qquad\qquad \vdots$$
$$\vdots \qquad\qquad\qquad \vdots$$
$$m_n \ddot{x}_n + k_{n-1}(x_{n-1}) \qquad\qquad = 0$$

Table 3.16 *Continued.*

where $x_j = X_j \sin\omega t$. The above equations are written as

$$\{[(k_2 + k_1)/m_2] - \omega^2\} X_2 - (k_2/m_2) X_3 = 0$$
$$-(k_2/m_3) X_2 + [(k_3 + k_2)/m_3] X_3 - (k_3/m_3) X_4 = 0 \qquad (2)$$
$$-(k_3/m_4) X_3 + \{[(k_4 + k_3)/m_4] - \omega^2\} X_4 - (k_4/m_4) X_5 = 0$$
$$\text{--} = 0$$
$$\text{--} = 0$$

In general, in the above equations the values of the ks are replaced by appropriate general terms. For example, $\{[(k_2 + k_1)/m_2] - \omega^2\}$ is replaced by $\{[(k_{j+1} + k_j)/m_{j+1}] - \omega^2\}$. The determinant is solved for various values of ω and hence f, the frequencies. Associated with each value of ω, one can find the characteristic vector $\mathbf{X}$ in terms of the arbitrary constant C. The mode shapes are obtained. The associated vectors $\mathbf{L_v} = \sqrt{(\mathbf{X^T X})}$ are evaluated. The matrix $\{\mathbf{e}_i\}$ is formed such that the matrix of equation (3.219) becomes the identity matrix.

For a three-storey building with the following data

j	$m_j(\text{kNs}^2/\text{m})$	$k_j(\text{MN/m})$
1	—	10
2	15 000	9
3	15 000	8
4	15 000	

$$\{\mathbf{e}_i\} = \begin{bmatrix} 0.305 & 0.688 & 0.660 \\ 0.581 & 0.416 & -0.702 \\ 0.757 & -0.597 & 0.273 \end{bmatrix} \qquad (3)$$

Equation (3.225) is solved using

$$\mathbf{x}_i = \bar{E} x_t'$$

For example

$$x_{i2}' = 0.305 x_{i2} + 0.581 x_{i3} + 0.757 x_{i4} \text{ and so on} \qquad (4)$$

The uncoupled equations of the form for a three-storey building will be

$$\ddot{x}_2' + \omega_1^2 x_2' = 0$$
$$\ddot{x}_3' + \omega_2^2 x_3' = 0 \qquad (5)$$
$$\ddot{x}_4' + \omega_3^2 x_4' = 0$$

The frequencies and periods for the three-storey building are computed as

$$f_1 = 1.768 \text{ Hz} \quad T_1 = 0.55 \text{ s}$$
$$f_2 = 4.80 \text{ Hz} \quad T_2 = 0.22 \text{ s}$$
$$f_3 = 6.96 \text{ Hz} \quad T_3 = 0.15 \text{ s}$$

(2) Forced vibrations

The time interval is $0 \leq t \leq 1$ s; $F_2(t) = 0$; $F_3(t) = 0$; $F_4(t) = 1 \times 10^5(1-t)$ kN = forcing factor in a column matrix; $F(t) = \{0, 0, 10^5(1-t)\}^T$. They are made equal to the right-hand side of equation (1). The uncoupled equations given in (5) are invoked. For example,

$$x_2' + \omega_1^2 x_2' = (0.305)(0) + (0.688)(0) + (0.757)\{10^5(1-t)\} = 7.57 \times 10^4(1-t) \qquad (6)$$

continued

Table 3.16 *Continued.*

The procedure given above is followed:

$$x'_i = X_i \sin \omega_i t$$
$$x'_2 = X_1 \sin \omega_1 t + 7.57 \times 10^4 (1 - t)$$
$$x'_3 = X_2 \sin \omega_2 t - 5.97 \times 10^4 (1 - t)$$
$$x'_4 = X_3 \sin \omega_3 t + 2.73 \times 10^4 (1 - t) \tag{7}$$

The values of x_2, x_3 and x_4 are obtained and hence the frequency f by using the text.

3.4.2 Orthogonality principle

For example, two characteristic vectors $\mathbf{X}_1$ and $\mathbf{X}_2$, each with n components, and their corresponding characteristic numbers (frequencies) ω_1 and ω_2 are considered. Two forms of equation (3.210) are written:

$$[K]\{\mathbf{X}_1\} = \omega_1^2[M]\{\mathbf{X}_1\}$$
$$[K]\{\mathbf{X}_2\} = \omega_2^2[M]\{\mathbf{X}_2\} \tag{3.213}$$

The transposed form of equations (3.213) can also be written in the following way:

$$([K]\{\mathbf{X}_1\})^{\mathrm{T}}\{\mathbf{X}_2\} = \omega_1^2([M]\{\mathbf{X}_1\})^{\mathrm{T}}\{\mathbf{X}_2\}$$
$$\{\mathbf{X}_1\}^{\mathrm{T}}([K]\{\mathbf{X}_2\}) = \omega_2^2\{\mathbf{X}_1\}^{\mathrm{T}}([M]\{\mathbf{X}_2\}) \tag{3.214}$$

Since $[M]$ is a diagonal matrix $= [M]^{\mathrm{T}}$ and $[K]$ is a symmetric matrix $= [K]^{\mathrm{T}}$, the expression of equations (3.214) gives

$$(\omega_2^2 - \omega_1^2)\{\mathbf{X}_1\}^{\mathrm{T}}[M]\{\mathbf{X}_2\} = 0 \tag{3.215}$$

Assuming $\omega_2 = \omega_1$, equation (3.215) gives

$$\{\mathbf{X}_1\}^{\mathrm{T}}[M]\{\mathbf{X}_2\} = 0 \tag{3.215a}$$

As defined earlier, the principle of orthogonality with respect to the *mass matrix* $[M]$ being unity can be represented by

$$\{\mathbf{X}_1\}^{\mathrm{T}}\{\mathbf{X}_2\} = 0 \tag{3.216}$$

The condition is satisfied regardless of the value of ω. It thus follows that:

$$\{\mathbf{X}_1\}^{\mathrm{T}}[M]\{\mathbf{X}_1\} \neq 0 \tag{3.217}$$

3.4.3 Concept of unit vectors

The length of the vector $\mathbf{L}_v$ is given as $\sqrt{(\mathbf{X}^{\mathrm{T}}\mathbf{X})}$. The unit vector is $\mathbf{e}_i$, defined by

$$\mathbf{e}_i = \{\mathbf{X}_i\}/\sqrt{(\{\mathbf{X}_i\}^{\mathrm{T}}[M]\{\mathbf{X}_i\})} \tag{3.218}$$

It is convenient to work in terms of $\mathbf{e}_i$ rather than $\mathbf{X}_i$, then

$${e_i}^T[M]{e_i} = [I] \tag{3.219}$$

and then n distinct values of ω_n^2, their characteristic vectors $\mathbf{X}_n$ and unit vectors $\mathbf{e}_n$ can be written as

$$[K][\bar{E}] = [M][E][p] \tag{3.220}$$

$$\text{where } [\bar{E}] = \begin{bmatrix} e_{11} & e_{21} & e_{31} & \cdots & e_{n1} \\ \vdots & \vdots & \vdots & & \vdots \\ \vdots & \vdots & \vdots & & \vdots \\ \vdots & \vdots & \vdots & & \vdots \\ e_{1n} & e_{2n} & e_{3n} & \cdots & e_{nn} \end{bmatrix} \tag{3.220a}$$

$$[\omega] = \begin{bmatrix} \omega_1^2 & 0 & 0 & \cdots & 0 \\ 0 & \omega_2^2 & 0 & \cdots & 0 \\ \vdots & \vdots & \vdots & & \vdots \\ \vdots & \vdots & \vdots & & \vdots \\ 0 & 0 & 0 & \cdots & \omega_n^2 \end{bmatrix} \tag{3.220b}$$

Equation (3.219) is replaced by

$$[\bar{E}]^T[M][\bar{E}] = [I] \tag{3.221}$$

where $[I]$ is the identity matrix. It is therefore easy to see that

$$[\bar{E}]^T[K][\bar{E}] = [p] \tag{3.222}$$

The basic equation of motion becomes

$$[M]\{\ddot{\bar{x}}_i\}_t + [K]\{\bar{x}_i\}_t = 0 \tag{3.223}$$

Equations (3.223) are converted into a set of uncoupled equations using the principle of transformation $\mathbf{x}_i = \bar{E}\mathbf{x}_i'$. The new set of basic uncoupled equations is written as

$$[\bar{E}]^T[M][\bar{E}]\{\ddot{x}_i'\}_t + [\bar{E}]^T[K][\bar{E}]\{x_i'\}_t = 0 \tag{3.224}$$

Using equations (3.221) and (3.222), equation (3.224) is written in n uncoupled equations

$$\{\ddot{x}_i'\}_t + \{x_i'\}_t = 0 \tag{3.225}$$

Table 3.16 illustrates the computation.

3.4.4 Undamped forced vibrations

If equation (3.223) is rewritten and equated to the forcing function $F(t)$, then following the argument given earlier, equation (3.224) assumes the form:

$$[\bar{E}]^T[M][\bar{E}]\{\ddot{x}_i'\} + [\bar{E}]^T[K][\bar{E}]\{x_i'\} = [\bar{E}]^T F(t) \tag{3.226}$$

$$\text{or } \{\ddot{x}_i\} + [\omega]\{x_i'\} = [\bar{E}]^T F(t) \tag{3.226a}$$

3.4.5 Non-linear response of multi-degrees-of-freedom systems: incremental method

The Wilson-θ method is suggested initially for the solution of the structures modelled by assuming that the acceleration varies linearly over the time interval from t to $t + \theta\delta t$, such that $\theta \geq 1$. For a value of $\theta \geq 1.38$, the Wilson-θ method becomes unconditionally stable. Consider the difference between the dynamic equilibrium conditions at time t_i and $t_i + \theta\delta t$. The following incremental equations are obtained on the lines suggested earlier

$$[M]\{\delta\ddot{x}_i'\} + [C]\{\dot{x}\}\{\delta\dot{x}_i'\} + [K]\{x\}\{\delta x_i'\} = \{\delta F_i(t)\} \qquad (3.227)$$

where δ is the increment associated with the extended time $\theta\delta t$. Thus

$$\{\delta x_i'\} = \{x(t_i + \theta\delta t)\} - \{x(t_i)\} \text{ (a)}$$
$$\{\delta\dot{x}_i'\} = \{\dot{x}(t_i + \theta\delta t)\} - \{\dot{x}(t_i)\} \text{ (b)} \qquad (3.228)$$
$$\{\delta\ddot{x}_i'\} = \{\ddot{x}(t_i + \theta\delta t)\} - \{\ddot{x}(t_i)\} \text{ (c)}$$

The incremental force is given by

$$\{\delta F_i(t)\} = \{F(t_i + \theta\delta t)\} - \{F(t_i)\} \qquad (3.229)$$

As shown in Fig. 3.48(a) and (b), both stiffness and damping are obtained for each time step as the initial values of the tangent to the corresponding curves. These coefficients are given as

$$\{k_{ij}\} = \{\delta F_{sti}/\delta x_j\}$$
$$\{c_{ij}\} = \{\delta F_{di}/\delta x_j\} \qquad (3.230)$$

During the extended time step the linear expression for the acceleration is

$$\{\ddot{x}(t)\} = \{\ddot{x}_i\} + [\delta\ddot{x}_i'/\theta\delta t \ (t - t_i)] \qquad (3.231)$$

The value of $\delta\ddot{x}_i'$ is taken from equation (3.228). Integration of equation (3.231) gives the following equations:

$$\{\dot{x}(t)\} = \{\dot{x}_i\} + \{\ddot{x}_i(t - t_i)\} + \frac{1}{2}\left\{\frac{\delta\ddot{x}_i'}{\theta\delta t} \times (t - t_i)\right\} \qquad (3.232)$$

$$\{x(t)\} = \{x_i\} + \{\dot{x}(t - t_i) + \{\tfrac{1}{2}\ddot{x}_i(t - t_i)^2\} + \left\{\frac{1}{6}\frac{\delta\ddot{x}_i'}{\theta\delta t} \times (t - t_i)^3\right\} \qquad (3.233)$$

At the end of the extended interval $t = t_i + \theta\delta t$, equations (3.232) and (3.233) are reduced to

$$\delta\dot{x}_i' = \ddot{x}_i\theta\delta t + \tfrac{1}{2}\delta\ddot{x}_i\theta\delta t \qquad (3.234)$$

$$\delta x_i' = \dot{x}_i\theta\delta t + \tfrac{1}{2}\ddot{x}_i(\theta\delta t)^2 + \tfrac{1}{6}\delta\ddot{x}_i'(\theta\delta t)^2 \qquad (3.235)$$

The values of $\delta\dot{x}_i'$ and $\delta x_i'$ are given in equation (3.228). By substituting the expression for $\delta x_i'$ from equation (3.235) into equation (3.234), the following equations are obtained.

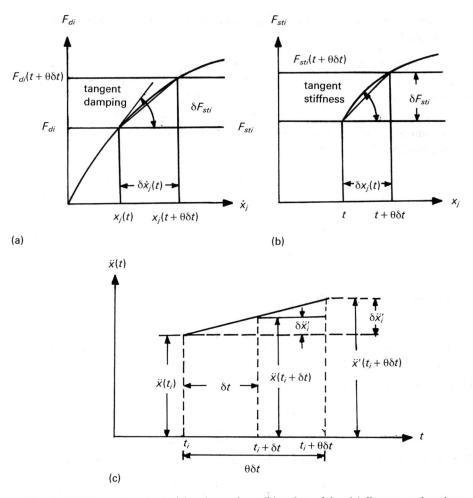

Fig. 3.48 Wilson-θ method; (a) values of c_{ij}; (b) value of k_{ij}; (c) linear acceleration.

$$\{\delta\ddot{x}_i'\} = \{[6/(\theta\delta t)^2]\delta\dot{x}_i'\} - \{[6/(\theta\delta t)^2]\dot{x}_i\} - \{3\ddot{x}_i\} \qquad (3.236)$$

$$\{\delta\dot{x}_i'\} = \{(3/\theta\delta t)\delta x_i\} - 3\{x_i\} - \tfrac{1}{2}\theta\delta t\{\ddot{x}_i\} \qquad (3.237)$$

The incremental acceleration $\delta\ddot{x}_i$ for the time interval δt can be obtained from

$$\{\delta\ddot{x}\} = \{\delta\ddot{x}_i'/\theta\} \qquad (3.238)$$

The incremental velocity and displacement for the time interval δt are given by

$$\{\delta\dot{x}_i\} = \{\ddot{x}_i\delta t\} + \tfrac{1}{2}\{\delta\ddot{x}_i\delta t\} \qquad (3.239)$$

$$\{\delta x_i\} = \{\dot{x}_i\delta t\} + \tfrac{1}{2}\{\ddot{x}_i(\delta t)^2\} + \tfrac{1}{6}\{\delta\ddot{x}_i(\delta t)^2\} \qquad (3.240)$$

$$\{x_{i+1}\} = \{x_i\} + \{\delta x_i\} \text{ at the end of the time step} \qquad (3.241)$$

$$\{\dot{x}_{i+1}\} = \{\dot{x}_i\} + \{\delta\dot{x}_i\}; \ t_{i+1} = t_i + \delta t \qquad (3.242)$$

When equations (3.236) and (3.237) are substituted into equation (3.227), the following results:

$$\{\delta \overline{F_i(t)}\} + [M]\{(6/\theta \delta t)\{\dot{x}_i\} + 3\{\ddot{x}_i\} + [C]_i\{3\{\dot{x}_i\} + \tfrac{1}{2}\theta \delta t\{\ddot{x}_i\}\} =$$
$$\{[K]_i + 6/(\theta \delta t)^2[M] + (3/\theta \delta t)[C]_i\}\{\delta x_i'\} \qquad (3.243)$$

Hence the initial acceleration for the next step is calculated at time $t + \delta t$ as

$$\{\ddot{x}_{i+1}\} = [M]^{-1}[F_{i+1}(t)] - [C]_{i+1}\{\dot{x}_{i+1}\} - [K]_{i+1}\{x_{i+1}\}] \qquad (3.244)$$

damping   stiffness
force   force
vector   vector

The procedure is repeated for t_{i+2}, etc., for the desired time.

3.4.6 Summary of the Wilson-θ method

In order to summarize the Wilson-θ integration method, the following step-by-step solution should be considered with the dynamic, impact and explosion analysis of the structures.

(1) Assemble $[K]$, $[M]$ and $[C]$.
(2) Set the initial values of x_0, $\dot{x}_0$ and $F_0(t)$.
(3) Evaluate $\ddot{x}_0$ using

$$[M][\ddot{x}_0] = [F_0(t)] - [C]\{\dot{x}_0\} - [K]\{x_0\}$$

(4) Select a time step δt (usually taken as 1.4) and evaluate

$$\theta \delta t, \; a_1 = 3/(\theta \delta t), \; a_2 = 6/(\theta \delta t), \; a_3 = \theta \delta t/3, \; a_4 = 6/(\theta \delta t)^2$$

(5) Develop the effective stiffness matrix, $[K]_{\text{eff}}$

$$[K]_{\text{eff}} = [K] + a_4[M] + a_1[C]$$

where $[K] = [\bar{K}]$ or 0 for elastic and plastic respectively.

(6) Calculate $\delta \overline{F_i(t)}$ for the time interval t_i to $t_i + \theta \delta t$

$$\{\delta F_i(t)\} = \{F(t)\}_{i+1} + [\{F(t)\}_{i+2} - \{F(t)\}_{i+1}(\theta - 1)] - \{F(t)\}_i$$

(7) Solve the incremental displacement $\{\delta x_i'\}$ (equation (3.230)) and the incremental acceleration $\{\delta \ddot{x}_i'\}$ (equation (3.232)) for the extended time interval $\theta \delta t$.
(8) Calculate $\{\delta \ddot{x}\}$ of equation (3.238).
(9) Calculate the incremental velocity and displacement of equations (3.239) and (3.240).
(10) Calculate $\{x_{i+1}\}$ and $\{\dot{x}_{i+1}\}$ for the time $t_{i+1} = t_i + \delta t$ from equations (3.241) and (3.242).
(11) Calculate $\{\ddot{x}_{i+1}\}$ at time $t_{i+1} = t_i + \delta t$ from the dynamic equilibrium equation (3.244).

A typical numerical example is shown in Table 3.17, where quadratic and cubic functions are considered instead of the linear function.

Table 3.17 Elasto-plastic analysis.

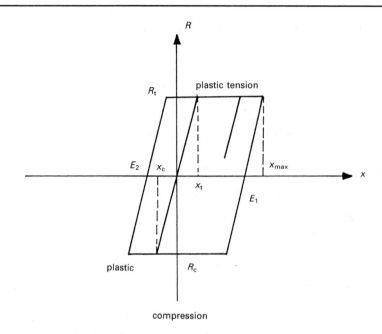

Increasing displacement $x > 0$
Decreasing displacement $x < 0$
x_t (plastic) in tension $= R_t/K$
x_c (plastic) in compression $= R_c/K$
R_t, R_c are restoring forces in tension and compression, respectively
Let $K = 3.35\,\text{kN/mm}$; $R_t = 15\,\text{kN} = R_c$; $M = 0.5\,\text{kNs}^2/\text{mm}$
$c = $ damping coefficient $= 0.28\,\text{kNs/mm}$
$x_0 = \dot{x}_0 = 0$ in the initial case
$x_0 = 0$
$x_t = 15/3.35\,\text{mm}$; $x_c = -4.48\,\text{mm}$;
$T = 2\pi\sqrt{(M/K)} = 2\pi\sqrt{(0.5/3.35)} = 2.43\,\text{s}$

For convenience, $\delta t = 0.1\,\text{s}$

$$[K]_{\text{eff}} = [\bar{K}] + a_4[M] + a_1[C]$$
$$= [\bar{K}]_p + (6/(0.1)^2)0.5 + (3/0.1)0.28$$

where $[K]_p = 0$ for plastic

$$= [\bar{K}]_p + 300 + 8.4$$
$$= [\bar{K}]_p + 308.4$$

$$\{\delta F_i(t)\} = \{\delta F(t)\} + \left(\frac{6}{\delta t}M + 3c\right)\dot{x} + \left(3M + \frac{\delta t}{2}c\right)\ddot{x}$$

$$= \{\delta F(t)\} + 3.84x + 1.514\ddot{x}$$

continued

Table 3.17 *Continued.*

The velocity increment is given by

$$\delta \dot{x} = (3/\delta t)\delta x_i - 3\dot{x}_i - (\delta t/2)\ddot{x}_i$$
$$= (3/0.1)\delta x_i - 3\dot{x}_i - (0.1/2)\ddot{x}_i$$
$$= 30\delta x_i - 3\dot{x}_i - 0.05\ddot{x}_i$$

The results are obtained on the basis of the above two equations of force and velocity increments. The step-by-step procedure is covered in section 3.4.6 and the results are tabulated as follows.

$t \quad F(t) \quad x \quad \dot{x} \quad R \quad \ddot{x} \quad [\bar{K}_p] \quad [K_{eff}] \quad \delta F(t) \quad \delta \bar{F}(t) \quad \delta x \quad \delta \dot{x}$

4

IMPACT DYNAMICS

4.1 Introduction

This chapter begins with basic impact dynamics. It includes impact effects due to vehicle/train collisions, aircraft/missile target interactions, drop weights and free-falling bodies and missiles on concrete and steel targets. An up-to-date impact simulation is included for jet fluids on soils and rocks. Brief numerical and experimental data are given on impacts and collisions on water surfaces. A special section is included on snow/ice impact.

4.2 The impactor as a projectile

An impactor in the form of a missile is first given an initial velocity and it is then possible to assume that it is moving under the action of its own weight. If the initial velocity is not vertical, the missile will move in a curve and its flight can be evaluated in terms of horizontal and vertical components of displacement, velocity and acceleration. Some typical examples are given in Table 4.1.

4.2.1 Direct impulse/impact and momentum

An impulse is defined as a force multiplied by time, such that

$$F_1(t) = \int F dt \qquad (4.1)$$

where $F_1(t)$ is the impulse, F is the force and t is the time. The momentum of a body is the product of its mass and its velocity:

$$\text{momentum} = mv \qquad (4.2)$$

where m is the mass and v is the velocity $= dx/dt$. Both velocity and momentum are vector quantities; their directions are the same. If a body is moving with a constant velocity, its momentum is constant. If velocity is to be changed, a force F must act on the body. It follows that a force F must act in order to change the momentum.

$$F = m dv/dt$$
$$\text{or} \quad F dt = m dv \qquad (4.2a)$$

331

Table 4.1 Projectile statistics.

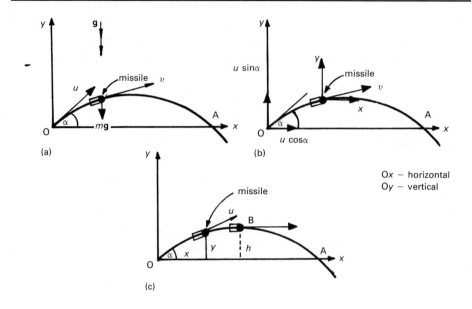

Figure (a) shows a missile projected at a velocity u from a position 0. At 0, x, $\dot{x}$ and $\ddot{x}$ are all zero. The only force on the flight is equal to mg. Hence $\ddot{y}$, the acceleration in the vertical direction, is $-\mathbf{g}$.

The general forms of the velocity and distance equations are:

$$v = u + at \tag{1}$$
$$s = ut + \tfrac{1}{2}at^2$$

$$\dot{x} = u\cos\alpha, \qquad \dot{y} = u\sin\alpha - \mathbf{g}t \ \ \text{(Fig. (b))} \tag{2}$$

$$x = (u\cos\alpha)t; \ y = (u\sin\alpha)t - \tfrac{1}{2}\mathbf{g}t^2 \tag{3}$$

$$\ddot{x} = 0; \qquad \ddot{y} = -\mathbf{g} \tag{4}$$

By elimination of t from equation (3), the trajectory equation is written in a parabolic form as

$$y = x\tan\alpha - (\mathbf{g}x^2\sec^2\alpha/2u^2) \tag{5}$$

The velocity v of the missile during flight at any instant in time is given by

$$v = \surd(\dot{x}^2 + \dot{y}^2) \ \text{with} \ \alpha = \tan^{-1}(\dot{y}/\dot{x}) \tag{6}$$

since

$$\dot{y}/\dot{x} = (\mathrm{d}y/\mathrm{d}t)/(\mathrm{d}x/\mathrm{d}t) = \mathrm{d}y/\mathrm{d}x$$

The direction of the velocity at any instant is along the tangent to the path for that particular instant. If the missile is projected from the aircraft at an angle below its level in order to hit the target at the ground level (the aircraft level is treated as horizontal), equation (5) becomes

$$y = x\tan\alpha + (\mathbf{g}x^2\sec^2\alpha/2u^2) \tag{7}$$

and all negative signs in equation (2) and equation (3) related to $\mathbf{g}$ are *positive*.

Table 4.1 *Continued.*

Case 1

If the missile is projected from the aircraft at an angle of $\alpha' = 30°$, from a distance of 700 m, and hits the target at 200 m distance, the speed and the direction are computed from Fig. (d).

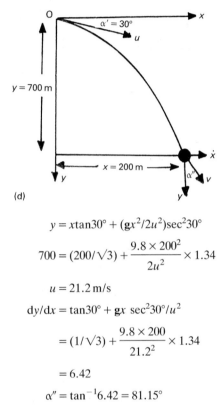

(d)

$$y = x\tan30° + (gx^2/2u^2)\sec^230°$$

$$700 = (200/\sqrt{3}) + \frac{9.8 \times 200^2}{2u^2} \times 1.34$$

$$u = 21.2 \, \text{m/s}$$

$$dy/dx = \tan30° + gx \, \sec^230°/u^2$$

$$= (1/\sqrt{3}) + \frac{9.8 \times 200}{21.2^2} \times 1.34$$

$$= 6.42$$

$$\alpha'' = \tan^{-1}6.42 = 81.15°$$

Case 2

If the missile is projected 4 m above the launch level, with a velocity of 100 m/s at an angle of 45° to the horizontal, the horizontal distance x at which it hits the ground is computed as follows:

$$y = x\tan\alpha - gx^2 \, \sec^2\alpha/2u^2$$
$$\text{or } -4 = x - (gx^2/2u^2)$$
$$\text{or } -4 = x - (gx^2/2 \times 100^2)$$

Rejecting the negative root, $x = 2045$ m.

Case 3: flight time

As shown in Fig. (a), the time taken by a missile to travel along its path from 0 to A is to be computed. At any time t:

continued

Table 4.1 *Continued.*

$$y = (u\sin\alpha)t - \tfrac{1}{2}gt^2$$

At A, $y = 0$

$$t = 2u\sin\alpha/\mathbf{g} \tag{8}$$

The other value of $t = 0$ cannot be true at A, as was assumed to be the case at 0.

Case 4: maximum height and horizontal range

Reference is made to Fig. (c). At any time t, at any point B,

$$\dot{y} = 0 = u\sin\alpha - gt$$

Hence

$$t = u\sin\alpha/\mathbf{g} \tag{9}$$

Substituting t into equation (3) of y

$$y = u\sin\alpha t - \tfrac{1}{2}gt^2$$
$$h = (u^2\sin^2\alpha/\mathbf{g}) - \tfrac{1}{2}\mathbf{g}(u\sin\alpha/\mathbf{g})^2$$
$$h = u^2\sin^2\alpha/2\mathbf{g} \tag{10}$$

The maximum range x is obtained as

$$x = ut\cos\alpha = u(2u\sin\alpha/\mathbf{g})$$
$$= 2u^2\sin\alpha\cos\alpha/\mathbf{g}$$
$$= u^2\sin 2\alpha/\mathbf{g}$$

When $\sin 2\alpha = 1$ or $\alpha = 45°$

$$x_{\max} = u^2/\mathbf{g} \tag{11}$$

Case 5

A missile hits the target at a distance x when travelling horizontally. The distance l at which the missile hits the ground after bouncing is computed below.

As shown in Fig. (d), by using the coefficient of restitution e, the speed after hitting the wall $= eu\cos\alpha$ in a horizontal direction. The maximum height reached is given by

$$h = u\sin\alpha/\mathbf{g}$$

(e)

Table 4.1 *Continued.*

The time taken to reach ground level is calculated by

$$y = h = ut + \tfrac{1}{2}\mathbf{g}t^2 = 0 + \tfrac{1}{2}\mathbf{g}t^2 \tag{12}$$

$$t = \surd(2h/\mathbf{g}) = \surd(2u^2\sin^2\alpha/2\mathbf{g}^2)$$

$$= u\sin\alpha/\mathbf{g}$$

When the missile hits the ground level at a distance *l* from the wall

$$l/x = e(u\cos\alpha)/u\cos\alpha$$

$$l = ex \tag{13}$$

where *e* is the coefficient of restitution

It can easily be proved that if the same missile hits a building floor vertically of height *h* with a velocity *v*, rebounds from there with coefficient of restitution *e* and rebounds to the floor with a coefficient of restitution *f*, then the value of *v* is given by

$$v = [2\mathbf{g}h(1 - f^2 + e^2f^2)]/e^2f \tag{14}$$

Case 6

Scene near O'Hare airport: an engine from a 10 tonne CFM56-3C power plant of an aircraft plummets at an angle α from a height *h* with a speed *v* and hits the ground at a distance *x*, as shown in Fig. (f). Assuming no air resistance is offered to the parabolic flight, determine the angle at which the engine hits the ground at a distance *x*, using the following data:

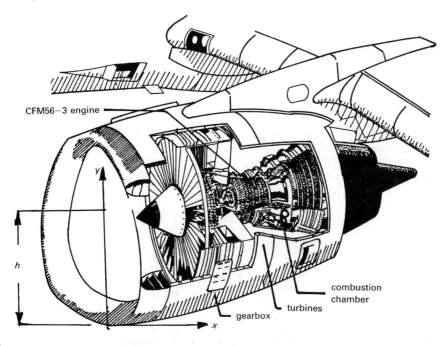

(f)

$h = 333\,\text{m} \quad x = 812\,\text{m} \quad u = 112\,\text{m/s}; \; v = 0$

continued

Table 4.1 *Continued.*

Thus

$$y = x \tan\alpha'' + (gx^2\sec^2\alpha/2u^2)$$
$$y = h = 333 \text{ m}$$
$$333 = 812\tan\alpha'' + [9.8 \times 812^2 \sec^2\alpha''/2 \times 112^2$$
$$333 = 812\tan\alpha'' + 257.56(1 + \tan^2\alpha'')$$
$$\tan\alpha'' = 0.096$$
$$\alpha'' = 5\tfrac{1}{2}° \text{ (it was reported to be almost straight)}$$

Integrating both sides:

$$\int_{t_1}^{t_2} F \mathrm{d}t = \int_u^v m\mathrm{d}v$$
$$F_1(t) = m(v - u) \tag{4.3}$$

where u and v are the velocities at times t_1 and t_2 respectively. If the initial velocity $u = 0$, equation (4.3) becomes

$$I = mv \tag{4.3a}$$

Thus the impulse of a force is equal to the change in momentum which it produces.
    Table 4.1 gives some typical examples of elastic impulse/impact phenomena.

4.2.1.1 Impacts/collisions of vehicles

When two solid bodies are in contact, they exert equal and opposite forces or impulses on each other and they are in contact for the same time. If no external force affects the motion, the total momentum in the specific direction remains constant. This is known as the *principle of conservation of linear momentum.* When two bodies, m_1 and m_2, collide (Fig. 4.1), the mass ratios are then calculated from equation (4.1):

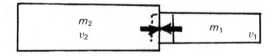

Fig. 4.1 Direct impact.

$$F_{11}(t) = m_1(v_1 - u_1) = \int F_1 \mathrm{d}t$$
$$F_{12}(t) = m_2(v_2 - u_2) = \int F_2 \mathrm{d}t \tag{4.4}$$

Since $\int F_1 \mathrm{d}t + \int F_2 \mathrm{d}t = 0$, the relationship between velocity change and mass becomes:

$$m_2/m_1 = (v_1 - u_1)/-(v_2 - u_2) \tag{4.5}$$

Table 4.2 gives some typical examples of impulse and momentum. During the collision process, although the momentum is conserved, there is a loss of energy on impact which is determined using the concept of the *coefficient of restitution*, e, which is defined as the relative velocity of the two masses after impact divided by the relative velocity of the two masses before impact. Before impact:

$$e = (v_1 - v_2)/ - (u_1 - u_2) = 0$$

when the relative velocity vanishes, and

$$e = (v_1 - v_2)/ - (u_1 - u_2) = 1 \qquad (4.5a)$$

when there is no loss of relative velocity.

Table 4.2 Direct elastic impact.

Example 1

A Fiat Panda (illustrated) with a gross laden weight of 1190 kg and travelling at 55 km/h is brought to rest in 10 seconds when it strikes a buffer. Determine the constant force exerted by the buffer.

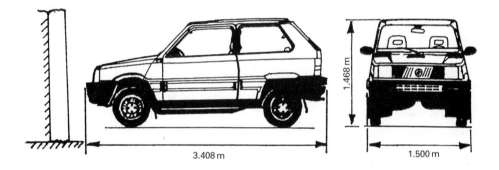

Assuming the laden weight is treated as a total mass:

$$u = 55 \, \text{km/h} = 15.28 \, \text{m/s}$$

$$F_I(t) = Ft = mv - mu$$

$$= 0 - 1190(-15.28) = 18.183 \times 10^3$$

$$Ft = 2F = 18.183 \times 10^3 \, \text{Ns}$$

$$F = 9091.6 \, \text{N} = 9.0916 \, \text{kN}$$

Example 2

A car of mass 790 kg is thrown by a hurricane and hits a wall normally at a speed of 40 m/s. Assuming the car acts as a missile and bounces away from the wall at right angles with a speed of 30 m/s, what impulsive force does the wall exert on the car? Assume no damage occurs to the wall. Approaching the wall, $u = -40 \, \text{m/s}$; leaving the wall $v = +30 \, \text{m/s}$.

$$Ft = mv - mu = (790 \times 30 - 790 \times -40)10^{-3}$$

$$= 64.39 \, \text{kN} \qquad \qquad \qquad \textit{continued}$$

Table 4.2 *Continued.*

Example 3

A windborne missile of mass m strikes a containment wall and ricochets off at 120° to its
original direction. The speed changes from $u = 40\,\text{m/s}$ to $v = 35\,\text{m/s}$. Calculate the resultant
impulse of the system. Assume no damage occurs.

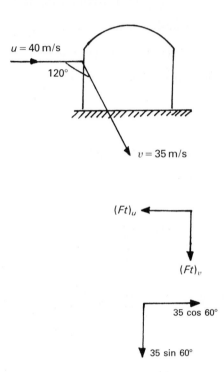

Since the direction of u and v are different before and after the impulse, the components of
the impulse and the velocities in two perpendicular directions are considered.

$$(Ft)_u = m(+35\ \cos 60°) - m(-40)$$

$$= m[40 + (35/2)] = 22.5\,m \qquad \text{where } m = \text{mass}$$

$$(Ft)_v = m(35\ \sin 60° - 0)$$

$$= m35(\sqrt{3}/2) = 30.31\,m$$

Therefore the resultant impulse or impact

$$F_I(t) = \sqrt{\{[(Ft)_v]^2 + [(Ft)_u]^2\}}$$

$$= 64.9\,m$$

Where $e < 1$, it is related to the loss in kinetic energy, and where $u_2 = 0$ (refer to equation (4.5a))

$$m_1(v_1 - u_1) + m_2(v_2) = 0$$

$$v_1 - v_2 = -eu_1$$

(4.6)

hence

$$v_1 = u_1(m_1 - em_2)/(m_1 + m_2)$$ (4.6a)

$$v_2 = u_1[(1 + e)m_1/(m_2 + m_1)]$$ (4.6b)

The original kinetic energy $(KE)' = \frac{1}{2}m_1u_1^2$

The final kinetic energy $(KE)'' = \frac{1}{2}(m_1v_1^2 + m_2v_2^2)$

$$(KE)' - (KE)'' = \frac{1}{2}m_1u_1^2 - \frac{1}{2}(m_1v_1^2 + m_2v_2^2)$$ (4.7)

Substituting the values of v_1 and v_2:

$$(KE)' - (KE)'' = (KE)' \, [m_1(1 - e^2)/(m_1 + m_2)]$$ (4.8)

The displacement resulting from a short-duration (τ) impact is given by

$$x = b(t - \tau)$$ (4.9)

where t is the time beyond τ. Details of such analysis are dealt with in Chapter 3. For dynamic analysis, the impact time is divided into n small segments and, using equation (4.3a),

$$x = \frac{1}{m} \sum_0^n v_n I_n(t - \tau_n)$$

$$= \frac{1}{m} \int_0^t F(t - \tau)d\tau$$

(4.10)

If the impact is divided into two phases such that in the first, from time t_1 to t_0, there will be compression and distortion until $(v_1 + v_2)$ are both reduced to zero (the two bodies moving together), in the second, the elastic strain energies in the bodies are restored and are separated by a negative velocity, $-V_2 = (v_1 + v_2)$. During the second phase the impulse relation between the bodies $(F_T - F_{T0})$ will be proportional to F_{T0} and the coefficient or restitution e defined above is written as

$$e = (F_T - F_{T0})/F_{T0}$$ (4.11)

where F_T is the total impulse during the impact and F_{T0} is the impulse in phase one.

At time t_0

$$V_0 = v_{10} + v_{20} = v_1 + \left(\frac{F_{T0}}{m_1} + v_2 - \frac{F_{T0}}{m_2}\right) = 0$$ (4.12)

hence

$$V = v_1 + v_2 = \left(\frac{1}{m_1} + \frac{1}{m_2}\right)F_{T0}$$ (4.13)

Similarly, at time t_2 the relationship becomes

$$V_0 - V_2 = F_T \left(\frac{1}{m_1} + \frac{1}{m_2} \right) \tag{4.14}$$

Using equation (4.11), the expression given in equation (4.5a) may be written in the form:

$$-(V_2/V) = e \tag{4.15}$$

Equations (4.6) to (4.6b) result from the above method. However, from equation (4.11) the total impulse is rewritten as

$$F_T = \left(\frac{m_1 m_2}{m_1 + m_2} \right) (1 + e)(v_1 + v_2)$$
$$= M(1 + e)V \tag{4.16}$$

where M is the equivalent combined mass of the bodies.

The changes in velocity after impact of the bodies are written as

$$\Delta V_1 = \frac{M}{m_1} (1 + e)(v_1 + v_2) = \frac{M}{m_1} (1 + e)V \tag{4.17}$$

$$\Delta V_2 = \frac{M}{m_2} (1 + e)V$$

4.2.2 Oblique impact

When two bodies collide and their axes do not coincide, the problem becomes more complex. With oblique impact, as shown in Fig. 4.2, two impulses are generated: the direct impulse, F_T, and the tangential impulse, F_T'. The latter is caused by friction between the impacting surfaces and by local interlocking of the two bodies at the common surface. Let the angular velocity of the two bodies be $\dot{\theta}_1$ and $\dot{\theta}_2$ respectively. If $F_T'/F_T = \lambda'$ and the body's centre of gravity has a co-ordinate system X and Y, the components of the vector velocity, v_1 and u_1, normal to the impact surface may be written as follows

$x_1 - y_1$ system

$$v_1 = |\bar{v}_1| \cos\theta_1 \tag{4.18}$$

$$u_1 = |\bar{v}_1| \sin\theta_1 \tag{4.18a}$$

where

$$|\bar{v}_1| = \sqrt{(v_1^2 + u_1^2)}$$

$$\alpha = \tan^{-1} (u_1/v_1)$$

Similarly, v_2 is written as

$$|\bar{v}_2| = \sqrt{(v_2^2 + u_2^2)} \tag{4.19}$$

$$\beta = \tan^{-1} (u_2/v_2) \tag{4.19a}$$

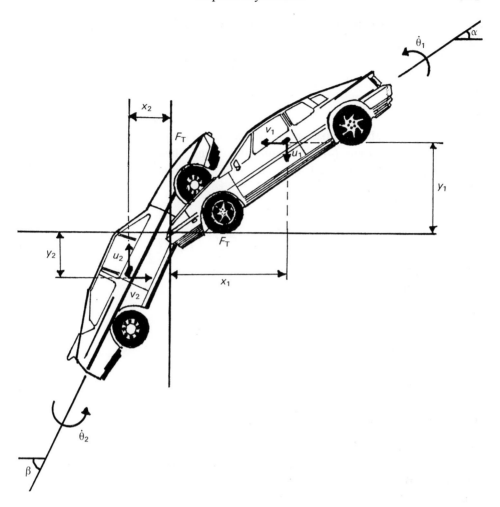

Fig. 4.2 Oblique impact.

The momentum equations for the bodies are summarized below:

$$m_1 v_1' - F_T = m_1 v_2'$$
$$m_1 u_1' - \lambda' F_T = m_1 u_2'$$
$$m_1 R_1^2 \, \theta_1' + F_T y_1 - \lambda' F_T x_1 = m_1 R_1^2 \, \dot{\theta}_2$$

body 1 (4.20)

where v_1', v_2', u_1' and u_2' are for t_1 and t_2.

$x_2 - y_2$ system

$$m_2 v_1'' - F_T = m_2 v_2''$$
$$m_2 u_1'' - F_T = m_2 u_2''$$
$$m_2 R_2^2 \, \dot{\theta}_2 + F_T y_2 - \lambda' F_T x_2 = m_2 R_2^2 \, \dot{\theta}_2'$$

body 2 (4.21)

where mR_1^2 and mR_2^2 are the second moment of inertia about the vertical axis passing through the centre of gravity. The rate of approach and the sliding of the two surfaces at the point of contact can be written as

$$\Delta V_1 = v_1 + v_2 - \dot{\theta}_1 y_1 - \dot{\theta}_2 y_2 \tag{4.22}$$

$$\Delta V_2 = u_1 + u_2 + \dot{\theta}_1 x_1 + \dot{\theta}_2 x_2 \tag{4.23}$$

The addition to these equations is the restitution given by equation (4.15) in which, when equation (4.22) is substituted and then, in the final equation, equation (4.20) is substituted, the value of F_T is evaluated as

$$F_T = \frac{V(1+e)}{c_1 - \lambda c_2} \tag{4.24}$$

where

$$c_1 = \frac{1}{m_1}\left(1 + \frac{y_1^2}{R_1^2}\right) + \frac{1}{m_2}\left(1 + \frac{y_2^2}{R_2^2}\right) \tag{4.24a}$$

$$c_2 = \left(\frac{x_1 y_1}{m_1 R_1^2} + \frac{x_2 y_2}{m_2 R_2^2}\right) \tag{4.24b}$$

Using equations (4.20) and (4.21):
$$v_2' = v_1' - (F_T/m_1)$$
$$u_2' = u_1' - (\lambda' F_T/m_1) \tag{4.25}$$

$$\dot{\theta}_2 = \dot{\theta}_1 + \frac{y_1 - \lambda' x_1}{m_1 R_1^2} F_T$$

$$v_2'' = v_1'' - \frac{F_T}{m_2}$$

$$u_2'' = u_1'' - \frac{\lambda' F_T}{m_2} \tag{4.26}$$

$$\dot{\theta}_2' = \dot{\theta}_1' + \frac{y_2 - \lambda' x_2}{m_2 R_2^2} F_T$$

Figure 4.3 shows plots for equations (4.25) and (4.26). It is interesting to note that larger values of λ' show greater interlocking of the surfaces of the two bodies and with e reaching zero, a greater plastic deformation occurs.

4.2.2.1 Case studies

(1) One body impacting a rigid barrier with no angular velocity.

$$1/m_2 = 0; \qquad v_1 = 0; \qquad u_1 = 0; \qquad \dot{\theta}_1 = 0 \tag{4.27}$$

$$c_1 = \frac{1}{m_1}\left(1 + \frac{y_1^2}{R_1^2}\right); \qquad c_2 = \frac{x_1 y_1}{m_1 R_1^2} \tag{4.27a}$$

$$v_2' = v_1' \, (y_1^2 - \lambda' x_1 y_1 - eR^2)/\bar{\lambda} \tag{4.27b}$$

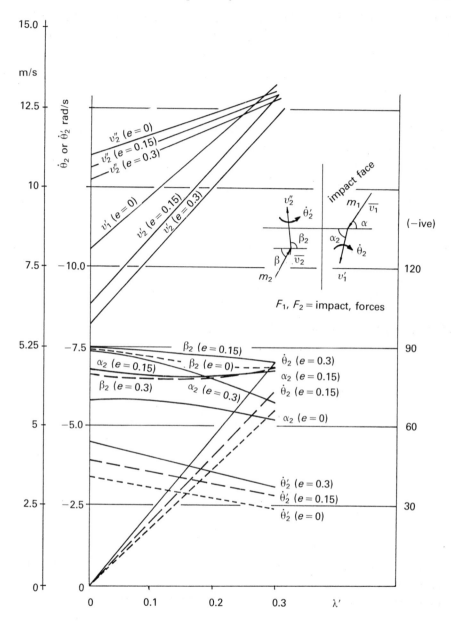

Fig. 4.3 Velocity versus λ' for oblique impact problems.

$$u_2' = u_1' - v_1' \left(\frac{\lambda'\,(1+e)R^2}{\bar{\lambda}} \right); \qquad \dot{\theta}_1 = \frac{(1+e)(y_1 - \lambda'x_1)}{\bar{\lambda}} \tag{4.27c}$$

$$\text{where } \bar{\lambda} = y_1^2 - \lambda'x_1y_1 + R^2 \tag{4.27d}$$

(2) Circular impactor with radius r_1.

$$x_1 = r_1 \text{ and } y_1 = 0 \tag{4.28}$$

$$v_2' = ev_1' \tag{4.28a}$$

$$u_2' = u_1 - \lambda'v_1' (1+e) \tag{4.28b}$$

$$\dot{\theta}_1 = -v_1'\lambda'r_1 (1+e)/R^2 \tag{4.28c}$$

For a circular impactor, $R^2 = 2r_1^2/5$

$$\dot{\theta}_1 = -v_1' (5\lambda' (1+e)/2r_1) \tag{4.28d}$$

(3) Inelastic collisions. The value of $e = 0$ in the above case studies.

Case study (1) $v_2' = v_1^2(y_1^2 - \lambda'x_1y_1)/\bar{\lambda}$

$$u_2' = u_1' - v_1' (\lambda'R^2/\bar{\lambda}) \tag{4.29}$$

$$\dot{\theta}_1 = (y_1 - \lambda'x_1)/\bar{\lambda}$$

Case study (2) $v_2' = 0; \qquad u_2' = u_1 - \lambda'v_1'$

$$\dot{\theta}_1 = -v_1'\lambda'r_1/R^2 = -2.5v_1'\lambda_1'/r_1 \tag{4.30}$$

(4) Where no interlocking exists, $\lambda' = 0$ in the above expressions.

4.3 Aircraft impact on structures — peak displacement and frequency

A great deal of work has been carried out (refer to sections 1 and 2 of the Bibliography) on the subject of missile and aircraft impact. Tall structures are more vulnerable to civilian, wide-bodied jets or multi-role combat aircraft. A great deal of work on this subject will be reported later. In this section a preliminary analysis is given for the determination of peak displacement and frequency of a tall structure when subject to an aircraft impact. As shown in Fig. 4.4, the overall dimensions of the building are given. Let A be the base area and h be the maximum height of the building. According to the principle of the conservation of momentum, if m is mass and v_1 is the velocity of the aircraft approaching the building, then using a linear deflection profile:

$$F_1 (t) = mv_1 = (\rho Ah/2\mathbf{g}) v_{20} \tag{4.31}$$

where ρ is the density or average specific weight and v_{20} is the velocity of the tip of the building.

The initial velocity, v_{20}, of the building can thus be evaluated from equation (4.31). Free vibrations studied in Chapter 3 show the time-dependent displacement $\delta(t)$ is given by

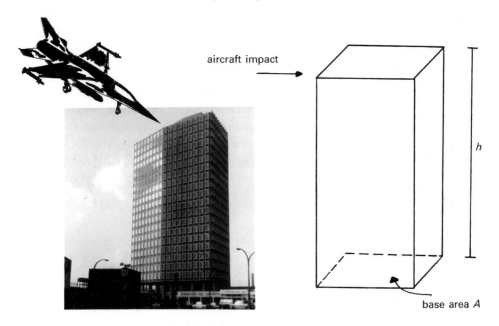

aircraft impact

h

base area *A*

Fig. 4.4

$$\delta(t) = (v_{20}/\omega)\,\sin\,\omega t \tag{4.32}$$
$$= [v_{20}/(2\pi/T)]\sin\,\omega t$$
$$= [v_{20}/\sqrt{(k_e/m_e)}]\,\sin\,\omega t$$

where ω is the circular frequency and k_e and m_e are the equivalent building stiffness and mass, respectively.

Using equation (4.31) for v_{20} and $\sin\,\omega t = 1$ for $\delta_{max}(t)$, the *peak dynamic displacement*, $\delta_{max}(t)$, is given by

$$\delta_{max}(t) = mv_1\mathbf{g}T/\pi\rho Ah \tag{4.32a}$$

The equivalent point load generated for the peak dynamic displacement is given by equation (4.32a). If that load is $F_1(t)$, then work done is equal to the energy stored and

$$F_1(t) \times \delta_{max}(t) = \tfrac{1}{2}k_e\,\delta_{max}^2(t) \tag{4.33}$$

from which
$$F_1(t) = \tfrac{1}{2}k_e\,\delta_{max}(t) \tag{4.33a}$$

While momentum is conserved, a portion of energy of the aircraft is lost on impact. The loss of energy E_1 is then written as

$$E_1 = \tfrac{1}{3}\,(\rho Ah/m\mathbf{g})(v_{20}/v_1)^2 \tag{4.34}$$

Equations in case study (1) of section 4.2.2.1 and equation (4.29) for inelastic collisions are applied with and without the interlocking parameter, λ'.

4.4 Aircraft impact: load–time functions

4.4.1 Introduction

Many sensitive installations are to be found in areas where heavy air traffic exists. Hence aircraft crashes cannot be entirely ruled out in such areas. Much effort is now being devoted to studies of aircraft impact with a clear aim of facilitating design to minimize damage to the aircraft and to the installations. Accident investigations, experiences and records are briefly discussed in Chapter 1. In this section some useful impact models are given which can be easily linked to both simplified and complex methods.

4.4.2 Stevenson's direct head-on impact model[3.167]

Work has been carried out on the remaining undamaged length of a 45 m (150 ft) long DC-8 jet which crashed into a rigid-surface, as shown in Fig. 4.5. A simplified equation of motion is written as

$$V \, (dV/dx) \, [k \, (L - x_{cr}) + m_c] = F_1(t) \qquad (4.35)$$

where V = speed of the aircraft at time t after impact
$\quad x_{cr}$ = crushed length
$\quad k$ = mass per unit length of fuselage
$\quad m_c$ = concentrated mass at wings including engines and others
$\quad F_1$ = impact force or resistance at the crash level.

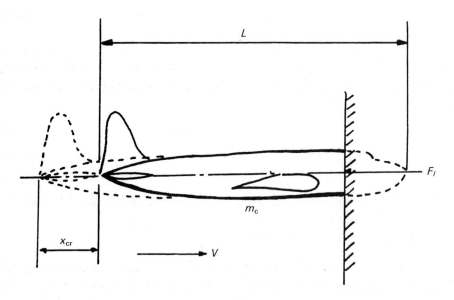

Fig. 4.5 Model aircraft impacting against a rigid surface.

Equation (4.35) is integrated:

$$F_1(t) = \tfrac{1}{2}kV_0^2 \left[\left(\frac{V}{V_0}\right)^2 - 1 \right] / \log[1 - x_{cr}/(L + m_c/k)] \tag{4.36}$$

where V_0 is the aircraft speed prior to impact.

4.4.3 Riera model[3.168]

The response of the structure was assessed by Riera. The aircraft was replaced by an equivalent force–time function. The aircraft impinges perpendicularly on a rigid target and it is assumed that it crashes only at the cross-section next to the target. The cross-sectional buckling load decelerates the remaining rigid uncrushed portion. The total impact force $F_1(t)$ is the sum of the buckling load and the force required to decelerate the mass of the impinging cross-section. Since it is a one-dimensional ideal plastic impact approach, in his model only the buckling load and the distribution of mass are needed. The equation of motion is written as:

$$F_1(t) = R_{cr}x_{cr} + m_c x_{cr}(dx_{cr}/dt)^2 \tag{4.37}$$

where m_c = mass per unit length of the uncrushed aircraft at impact
$\quad x_{cr}$ = crushed length
$\quad dx_{cr}/dt = V_{un}$ = velocity of uncrushed portion
$\quad R_{cr}$ = resistance to crushing, i.e. crushing strength.

Non-linear equations for R_{cr} and m are set up and numerical procedures are adopted for the applied forces at discrete time steps. The deceleration of the uncrushed mass m is written as:

$$G_d = \ddot{x} = -R_{cr} (x_{cr})_n / \int_{(x_{cr})_n}^{L} m_c x_{cr} dx_{cr} \tag{4.38}$$

In order to determine the current acceleration, current states of $(x_{cr})_n$ and R_{cr} at time t_n can be used. Similarly, the common kinematics relationship between acceleration, velocity, displacement and time can be used to determine conditions at time $t_{n+1} = t_n + \delta t$.

$$(\dot{x}_{cr})_{n+1} = (\dot{x}_{cr})_n + \ddot{x}_n \delta t \tag{4.39}$$

$$(x_{cr})_{n+1} = (x_{cr})_n + \dot{x}_{cr}\delta t + \tfrac{1}{2}\ddot{x}_n \delta t \tag{4.40}$$

Equation (4.37) is used to calculate the current force. The force–time history can thus be determined. A typical force–time history is given in Figs 4.6 and 4.7.

4.4.4 Model of Wolf *et al.*[3.131, 3.169]

Wolf *et al.* developed a lumped mass, elasto-plastic model, as shown in Fig. 4.8. Just prior to impact on a target of mass m_t, Fig. 4.8(a), spring stiffness k_t and damping coefficient c, the model has the mass of the fuselage which is lumped in n

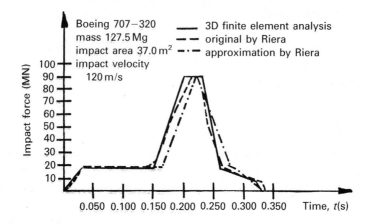

Fig. 4.6 Force as a function of time (Boeing 707−320).

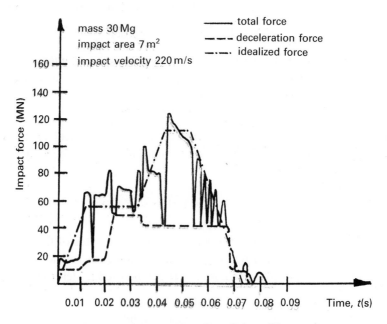

Fig. 4.7 Force as a function of time (Phantom).

nodes (Fig. 4.8(b)). The mass m_w of the part of the wing will be assumed to break away when a certain crushing length is achieved. The nodes are connected by springs k_i of length L_i. The springs work in tension and compression. For a spring next to the target, only contact in compression is allowed. In tension, after reaching the yielding force R_{yi}, the springs ideally become plastic, with a rupturing strain ε_r. At the buckling load $R_{ti}(x)$ the springs are allowed to crush completely at $\varepsilon_i = -1$. When the spring k_j, Fig. 4.8(c), reaches the value of -1, the masses m_j and m_{j+1} ideally impact plastically. Thus there is a chance for a new node to form

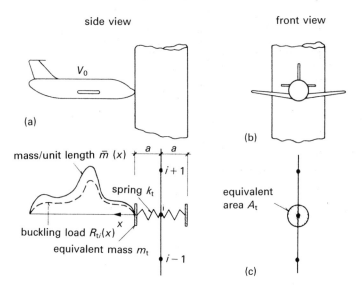

Fig. 4.8 Aircraft impact on chimney.

with mass $m_j + m_{j+1}$, Fig. 4.8(c). The spring stiffness k_j and its jth degree of freedom are deleted. Using the conservation of momentum, the velocity u_{j+1}^+ just after impact is computed:

$$u_j^+{}_{+1} = (\dot{u}_{j+1}^- \, m_{j+1} + \dot{u}_j^- \, m_j)/(m_{j+1} + m_j) \qquad (4.41)$$

where $\dot{u}_j^-$ is the displacement of the jth node, and $-$ and $+$ superscripts for before and after impact. (*Note*: the symbol u for displacement adopted here is the same as x, in this text.)

The equations of motion with discrete time steps are adopted for the force–time relationship. The total impulse $I(t)$ from the individual mass point m_1 to the target at time t_1 is given by

$$I_1(t) = -m_\ell \, [\dot{u}_t^+ - u_1^-] = (m_t + m_b)[\dot{u}_t^+ - u_t^-] \qquad (4.42)$$

Since the mass m_ℓ is distributed along the axis of the aircraft, the time for the momentum transfer δt_ℓ is given by

$$\delta t_\ell = \tfrac{1}{2} \, (t_{\ell-1} - t_{\ell+1}) \qquad (4.43)$$

where $t_{\ell-1}$ and $t_{\ell+1}$ are the times of impact of the mass points $m_{\ell-1}$ and $m_{\ell+1}$. Hence the value of $F_1(t_\ell)$ is given by

$$F_1(t_\ell) = R_\ell + I_1(t)/\delta t_\ell \qquad (4.44)$$

where R_ℓ is the force in the spring k_ℓ.

For a deformable target, as shown in Fig. 4.9,

$$F_1(t) = R_B(t) + \bar{m}(t)[\dot{u}_a(t) - \dot{u}_t(t)]^2 \qquad (4.45)$$

$$\text{where } R_B(t) = [m_a - m_b(t)]\ddot{u}_a(t) \qquad (4.46)$$

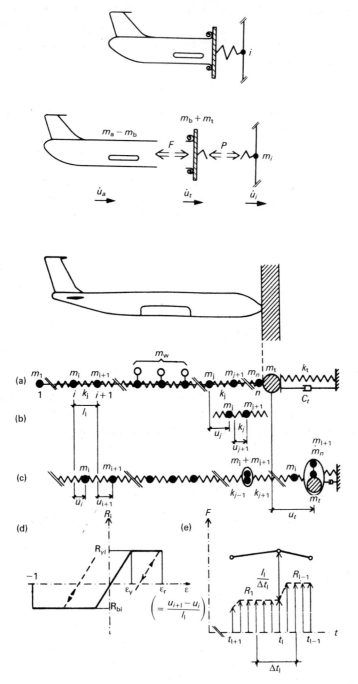

Fig. 4.9 Aircraft impact on deformable target (lumped-mass model).

where m_a, m_b and m are the total mass of the aircraft, the mass of the crushed part of it and the mass per unit length of the crushed part next to the target, respectively and $\dot{u}_a$ and $\dot{u}_t$ are the velocities of the aircraft and the target respectively. The equation of motion is written as

$$[m_b(t) + m_t]\ddot{u}(t) = P(t) - F(t) \tag{4.47}$$

where P denotes the force in the impact spring transmitted to nodes of the target. The velocity of the new target for the ideal plastic impact is given by

$$\dot{u}_t = [(m_b(t) + m_t)\dot{u}_i^- + m_t\dot{u}_i^-]/[m_b(t) + 2m_t] \tag{4.48}$$

Again the superscripts + and − indicate just after and just before impact.

Wolf et al.[3.131, 3.169] tested their work on rigid and deformable targets. Data used in their work are reproduced below:

Rigid target
Boeing 707 − 320
$m_a = 127.5\,\text{Mg}$
$m_w = 38.6\,\text{Mg}$ included in m_a
$\varepsilon_y = 2 \times 10^{-3}$; $\varepsilon_r = 5 \times 10^{-2}$

Deformable target
Impact area $= 37.2\,\text{m}^2$
$R_T = $ yielding moment/elastic moment

Figures 4.10 and 4.11 show a comparative study for two aircraft, a Boeing 707−320 and a combat aircraft FB-111, impacting on rigid targets. Figures 4.12 and 4.13 illustrate force−time relationships for deformable targets.

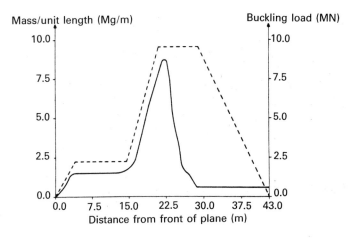

Fig. 4.10 Assumed properties of a Boeing 707−320.

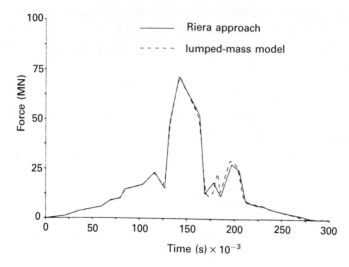

Fig. 4.11 Force−time diagrams for an FB-111.

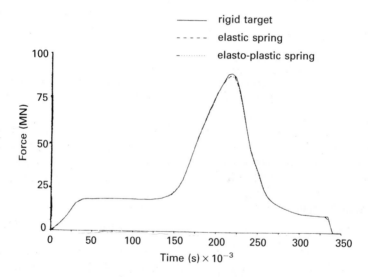

Fig. 4.12 Force−time diagram for a deformable target (frequency 50 Hz, no damping), lumped-mass model.

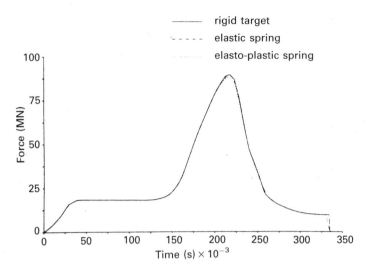

Fig. 4.13 Force–time diagram for a deformable target (frequency 50 Hz, no damping), Riera model.

4.5 Impact due to dropped weights

4.5.1 Impact on piles and foundations

4.5.1.1 Impact on piles

Piles are driven into the ground using drop hammers, single and double acting hammers, diesel and vibratory hammers. It is well understood that pile driving is not a simple case of impact which may directly be solved by Newton's law as described earlier. In fact, pile driving under impact is a case also of longitudinal compressive wave propagation and its velocity is given by

$$v_r = \sqrt{(E/\rho)} \tag{4.49}$$

where E is the Young's modulus of the pile material and ρ is the mass density.

The one-dimensional idealization for the wave equation is given which predicts the soil load resistance, stresses in piles and ultimate load capacity at the time of driving for a particular driving resistance in bpi or blows/cm. The lumped-mass model is illustrated such that the ram and pile cap are treated as masses and the hammer cushion and optimal pile cushion are considered as springs. The series of lumped masses, with attached elasto-plastic springs and dashpots for simulating soil characteristics, are connected by springs. The time-dependent analysis is carried out on the impact load of the hammer ram striking the pile cushion at an initial specified velocity. The system shown in Fig. 4.14 is associated with a time interval δt, chosen sufficiently small that the stress wave can easily travel from any top element to the next bottom element. For a length of 2.4 to 3.1 m, the value of

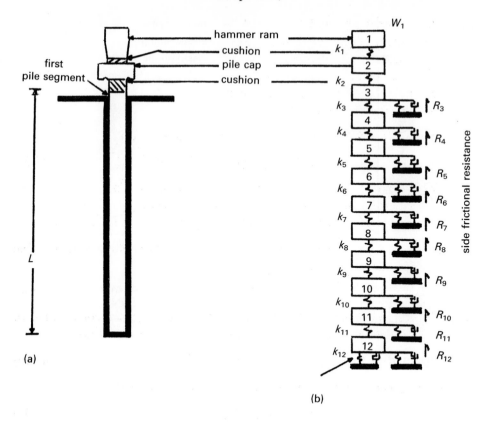

(a)

(b)

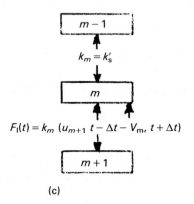

(c)

Fig. 4.14 Impact model of pile. (a) Actual pile; (b) dynamic profile of the pile; (c) forces on element m.

δt that gives a satisfactory answer is around 0.00025 s for steel and wood and 0.00033 s for concrete. For a shorter length, the actual value of δt is approximated as

$$\delta t = \alpha \sqrt{(W_m L / \bar{A} E \mathbf{g})} \qquad (4.50)$$

where $\alpha = $ constant $(0.5-0.75)$
    $W_m = $ weight of pile segment m
     $L = $ length of pile element
     $\bar{A} = $ cross-sectional area of pile
     $E = $ modulus of elasticity of pile material
     $g = $ gravitational constant

The current pile element displacement δ_m can be computed as

$$\delta_m = 2\delta'_m - \delta''_m + (F_{am} \times g/W_m)(\delta t)^2 \tag{4.51}$$

where $\delta'_m = $ element displacement in preceding time interval δt
    $\delta''_m = $ element displacement two time intervals back
    $F_{am} = $ unbalanced force in element causing acceleration $(F_{am} = m\ddot{x})$

It is not necessary to solve the equation directly. It can be done in stages. The instantaneous displacement δ_m is computed first as

$$\delta_m = \delta'_m + \dot{x}_m \delta t \tag{4.51a}$$

where $\dot{x}_m = v_m = $ velocity of the element m at δt. The relative compression or tension movement between any two adjacent elements can be written as

$$\delta_R = \delta_m - \delta_{m+1} \tag{4.51b}$$

The force F_m caused by the impact in segment m will become

$$F_m = \delta_R k_m = \delta_R (AE/L)_m \tag{4.51c}$$

where $k_m = $ element stiffness. The soil springs are computed as

$$k'_s = R'_m / \bar{K}_s \tag{4.51d}$$

where $R'_m = $ amount of estimated ultimate pile capacity P_u on each element. $\bar{K}_s = $ the soil spring properties. $\bar{K}_s$ 1.0 to 5.0 for silt/sand 1.0 to 8.0 for clay). The side and point resistance with damping may be evaluated at the side or point values j and k respectively as

$$R_m = (\delta_m - \delta_{sm})k'_s[1 + C_s \text{ or } C_p(\dot{x}_m)] \tag{4.51e}$$

where C_s is the damping for the side point and C_p, if substituted, is for the point load. Hence the accelerating force in segment m is computed by summing forces

$$F_{am} = F_{m-1} - F_m - R_m \tag{4.52}$$

The value of C_s is 0.33 to 0.66 s/m for sand and 1.3 to 3.3 for clay. The element velocity is then

$$\dot{x}_m = v_m = \dot{x}'_m + (f_{am}g/W_m)\delta t \tag{4.53}$$

where $x'_m = v'_m = $ velocity of the element m at $(\delta t - 1)$. In general, the ultimate pile capacity P_u is computed using the Bodine resonant driver (BRD) equation as

$$P_u = [\bar{A}(hp) + Br_p]/(r_p + fS_L) \tag{4.54}$$

where $\bar{A} = 550\,\text{ftlb/s}$ or $0.746\,\text{kJ/s}$

B = hammer weight, $22\,000\,\text{lb}$ or $89\,\text{kN}$ (intensity of load)

r_p = final rate of penetration in ft/s or m/s

f = frequency (Hz)

S_L = loss factor in fts/cycle or ms/cycle

    = $0.244\,\text{ms/cycle}$ for silt, sand, gravel (loose)

    = $0.762\,\text{ms/cycle}$ for silt, sand, gravel (medium)

    = $2.438\,\text{ms/cycle}$ for silt, sand, gravel (dense)

hp = horsepower delivered to the pile

Note: all the above quantities are time dependent.

4.5.2 Classical or rational pile formula

It is the basic dynamic pile capacity formula in which the coefficient of restitution e, discussed earlier, is included when the hammer impacts on the cap of the pile, as shown in Fig. 4.15. Let x_p be the penetration per blow at a point and h be the overall height of the ram prior to impact. The position of the pile y just as the hammer impacts on the cap will be equal to x_p plus the elastic compression of the parts. At impact the ram momentum is M_r:

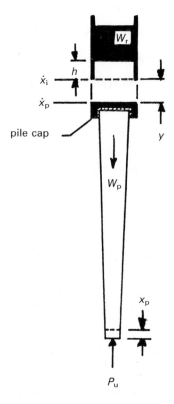

Fig. 4.15 Drop weight on piles–impact analysis.

$$M_r = W_r \dot{x}_i / g \tag{4.55}$$

At the end of the compression period, the value of M_r becomes

$$M_r = (W_r \dot{x}_i / g) - F_1(t) \tag{4.56}$$

where $\dot{x}_i$ = velocity of the ram at the impact level
W_r = weight of the ram
$F_1(t)$ = impact causing compression or a change in momentum

The corresponding velocity $\dot{x}_{pr}$ of the ram and pile at the end of compression is given by

$$\dot{x}_{pr} = [(W_r \dot{x}_i / g) - 1] g / W_r \tag{4.57}$$

If the pile momentum $M_p = F_1(t)$, then

$$\dot{x}_{pr} = F_1(t) g / W_p \tag{4.58}$$

Assuming the instantaneous velocity of the pile and ram are equal and that they have not been dislocated at the end of compression, the impact value $F_1(t)$ can be written as

$$F_1(t) = \dot{x}_i \, \frac{W_r W_p}{g \, (W_r + W_p)} \tag{4.59}$$

At the end of the period of restitution e, the momentum of the pile may be computed from

$$F_1(t) + e F_1(t) = (W_p / g) \dot{x}_p \tag{4.60}$$

Substitution of equation (4.59) into equation (4.60) gives the pile velocity $\dot{x}_p$:

$$\dot{x}_p = \frac{W_r + e W_r}{W_r + W_p} \dot{x}_i \tag{4.61}$$

At the end of the period of restitution, the momentum of the ram becomes

$$(W_r \dot{x}_i / g) - F_1(t) - e F_1(t) = W_r \dot{x}_r / g \tag{4.62}$$

After substitution of the value of $F_1(t)$, the velocity of the ram, $\dot{x}_r$, is

$$\dot{x}_r = \frac{W_r - e W_p}{W_r - W_p} \dot{x}_i \tag{4.63}$$

At the end of the period of restitution, as discussed earlier, the total energy E_p^r of the pile and ram is written as

$$E_p^r = \tfrac{1}{2} m \, \dot{x}_p^2 + \tfrac{1}{2} m \, \dot{x}_r^2 \tag{4.64}$$

Substituting the values of $\dot{x}_p$ and $\dot{x}_r$ from equations (4.61) and (4.63):

$$E_p^r = \frac{1}{2} \left[m_p \frac{W_r + e W_r}{W_r + W_p} + m_r \frac{W_r - e W_p}{W_r - W_p} \right] \dot{x}_i \tag{4.65}$$

where $m_p = W_p/g$ and $m_r = W_r/g$. The right-hand side of equation (4.65) is simplified to

$$E_p^r = \eta W_r h \frac{W_r + e^2 W_p}{W_r + W_p} \tag{4.66}$$

where η is hammer efficiency. For a 100% system, the ultimate pile capacity P_u is evaluated as

$$P_u = \eta W_r h / x_p \tag{4.67}$$

The pile top displacement will be

$$\begin{aligned}\bar{\delta} &= x_p + (P_u L / AE)_{cp} + (P_u L / AE)_{ep} + \bar{k}_s \\ &= x_p + C' + \bar{k}_s\end{aligned} \tag{4.68}$$

where the subscript cp relates to the cap block and pile cap and ep relates to elastic compression of the pile. The actual input energy to the pile system from the impact is

$$\begin{aligned}\eta W_r h &= P_u[\bar{\delta} - (P_u L / AE)_{ep}] \\ &= P_u(x_p + C')\end{aligned} \tag{4.69}$$

In terms of equation (4.66),

$$P_u = \left(\frac{\eta W_r h}{x_p + C'}\right) \left(\frac{W_r + e^2 W_p}{W_r + W_p}\right) \tag{4.70}$$

Similarly, other expressions for the pile behaviour under impact have been developed by other researchers and they are recorded in Table 4.3.

Table 4.3 Ultimate pile capacity, P_u, under impact loads.

Canadian National Building Code (use SF = 3)

$$P_u = \frac{\eta E_h C_1}{x_p + C_2 C_3} \qquad C_1 = \frac{W_r + e^2 (0.5 W_p)}{W_r + W_p}$$

$$C_2 = \frac{3 P_u}{2 \bar{A}} \qquad C_3 = \frac{L}{E} + 0.0001 \text{ (in}^3/\text{kips)}$$

Note that the product of $C_2 C_3$ gives units of e; x_p = penetration distance

Danish formula[2.395] (use SF = 3 to 6)

$$P_u = \eta E_h / (x_p + C_1) \qquad C_1 = \sqrt{(\eta E_h L / 2 \bar{A} E)} \text{ (units of } x_p)$$

where E_h = hammer energy rating.

Eytelwein formula[2.394] (use SF = 6)

$$P_u = \frac{\eta E_h}{x_p + 0.1 (W_p / W_r)}$$

Table 4.3 *Continued.*

Double-acting hammers:

$$P_u = \left[\frac{\eta W_r h}{x_p + \frac{1}{2}(c' + \bar{k}_s)}\right]\left[\frac{W + e^2 W_p}{W + W_p}\right]$$

$c' \approx C_1$; $W = W_r +$ weight of casing

Gates formula[2.395] *(Gates, 1957) (use SF = 3)*

$$P_u = a\sqrt{(\eta E_h)}(b - \log x_p)$$

where P_u is in kips or kN and E_h is in kips ft or kNm.

	x_p	a	b
F_{pS}	in	27	1.0
SI	mm	104.5	2.5

$\eta = 0.75$ for drop and 0.85 for all other hammers.

Janbu[2.395] *(use SF = 3 to 6)*

$$P_u = \eta E_h / k_u x_p \qquad C_d = 0.75 + 0.15(W_p/W_r)$$

$$k_u = C_d\{1 + \sqrt{[1 + (\lambda/C_d)]}\} \qquad \lambda = \eta E_h L / AE x_p^2$$

Modified ENR formula[2.395] *(use SF = 6)*

$$P_u = \left(\frac{1.25\eta E_h}{x_p + 0.1}\right)\left(\frac{W_r + e^2 W_p}{W_r + W_p}\right)$$

American Association of State Highway Officials (AASHTO) (SF = 6) (primarily for timber piles)

$$P_u = 2h(W_r + A_r p)/(x_p + 0.1)$$

For double-acting steam hammers take $A_r =$ ram cross-sectional area and $p =$ steam (or air) pressure. For single-acting steam hammers and gravity hammers, $A_r p = 0$. Here $\eta = 1$.

Navy-McKay formula[2.395] *(use SF = 6)*

$$P_u = \frac{\eta E_h}{x_p (1 + 0.3C_1)} \qquad C_1 = W_p/W_r$$

Pacific Coast Uniform Building Code (PCUBC)

$$P_u = \frac{\eta E_h C_1}{x_p + C_2} \qquad C_1 = \frac{W_r + k_p W_p}{W_r + W_p}$$

$$C_2 = P_u l / AE$$

where $k_p = 0.25$ for steel piles and 0.1 for all other piles.

continued

Table 4.3 *Continued.*

Coefficient of restitution e

Material	e
Broomed wood	0
Wood piles (non-deteriorated end)	0.25
Compact wood cushion on steel pile	0.32
Compact wood cushion over steel pile	0.40
Steel-on-steel anvil on either steel or concrete pile	0.50
Cast-iron hammer on concrete pile without cap	0.40

4.5.3 Impact on foundations

Useful data are given in Table 3.6 on vibration problems related to foundations. Various machines produce an impact on foundations. An additional consideration concerns energy dissipation and absorption. Typical views of the foundation for a hammer with its frame and anvil mounted are shown in Figs 4.16(a) and (b). It is treated as a two-mass-spring system, as shown in Fig. 4.16(c). Let m_1 and m_2 be the mass of the foundation and anvil respectively. The ks shown are the respective spring stiffnesses. The circular frequencies are given below:

ω_{an} = foundation of the anvil on the pad

$$= \sqrt{(k_2/m_2)} \tag{4.71}$$

ω_{ln} = the limiting circular frequency of the foundation and anvil on soil

$$= \sqrt{\left(\frac{k_1}{m_1 + m_2}\right)} \tag{4.72}$$

The equations of motion in free vibration are written as

$$m_1\ddot{x}_1 + k_1 x_1 + k_2(x_1 - x_2) = 0 \tag{4.73}$$

$$m_2\ddot{x}_2 + k_2(x_2 - x_1) = 0$$

where $x_1 = A\sin\omega_n t$ and $x_2 = B\sin\omega_n t$. The general equation can be derived:

$$\omega_n^4 - (1 + \alpha_m)(\omega_{na}^2 + \omega_{n\ell}^2)\,\omega_n^2 + (1 + \alpha_m)\,\omega_{nl}^2\omega_{na}^2 = 0 \tag{4.74}$$

or $\omega_{n1,2}^2 = \frac{1}{2}(1 + \alpha_m)(\omega_{na}^2 + \omega_{nl}^2) \pm \sqrt{\{[(1 + \alpha_m)(\omega_{na}^2 + \omega_{nl}^2)]^2 - 4(1 + \alpha_m)(\omega_{nl}^2\,\omega_{na}^2)\}}$ (4.75)

where $\alpha_m = m_2/m_1$. The values of x_1 and x_2 are

$$x_1 = \frac{(\omega_{na}^2 - \omega_{n1}^2)(\omega_{na}^2 - \omega_{n2}^2)}{\omega_{na}^2\,(\omega_{n1}^2 - \omega_{n2}^2)\,\omega_{n2}}\,\dot{x}_a \tag{4.76}$$

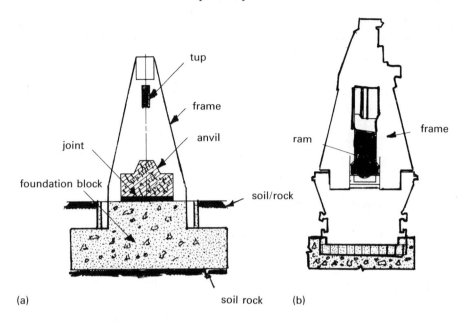

(a)   soil rock   (b)

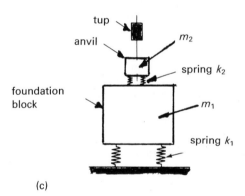

(c)

Fig. 4.16 Typical arrangement of a hammer foundation resting on soil with (a) a frame mounted on the foundation; (b) a frame mounted on the anvil; (c) two-mass-spring analogy for a hammer foundation.

$$\text{and } x_2 = \frac{(\omega_{na}^2 - \omega_{n1}^2)\,\dot{x}_a}{(\omega_{n1}^2 - \omega_{n2}^2)\,\omega_{n2}} \tag{4.77}$$

The maximum values of x_1 and x_2 occur when $\sin\omega_{n2}t = 1$. The value of $\dot{x}_a$ is the anvil velocity. The initial velocity of the anvil is computed from the impact of the tup and the anvil. If h is the drop of the tup in m, $\mathbf{g}$ is the acceleration due to gravity, $\dot{x}_{ti}$ is the initial velocity of the tup in m/s and η is the efficiency of the drop, then

$$\dot{x}_{ti} = \sqrt{2gh} \times \eta \tag{4.78}$$

if the tup is operated by a pneumatic pressure p, the area of the cylinder is A_c and the weight of the tup is W_0, then

$$x_{ti} = \sqrt{[2g(W_0 + pA_c)h/W_0]} \times \eta \tag{4.79}$$

where η ranges between 0.45 and 0.80.

The velocity of the anvil after impact, $\dot{x}_a$, is given by

$$\dot{x}_a = \dot{x}_{ti}(1 + e)/[1 + (W_2/W_0)] \tag{4.80}$$

where W_2 is the weight of the anvil and e is the coefficient of restitution.

4.5.4 Rock fall on structures

Rock fall, as described in Chapter 2, section 2.2.5, can cause severe damage to structures. Structural units can be beams, slabs or columns, or combinations thereof. In Fig. 4.17, a rock falls on a beam which has a peak mid-span deflection limited to $\delta_{max}(t)$. The rock, of weight W, falls from a height h, thus the potential energy lost during the rock fall is $W[h + \delta_{max}(t)]$ and the energy dissipated by the beam is $R_T\delta_{max}$, where R_T is the resistance when the beam fails in the development of plastic hinges. In the ultimate case, the value of R_T is

$$R_T\delta_{max}(t) = W(h + \delta_{max}) \tag{4.81}$$

However, $R_T = 4M_p/L$

and the ultimate moment M_u or M_p is given by

$$M_p = (WL/4)(1 + h/\delta_{max}) \tag{4.82}$$

where M_p is the moment for the beam at collapse.

In a similar manner, a rock fall can occur on reinforcement concrete slabs and equations can be set up for the collapse moment M_p or the ultimate moment M_u. This subject is dealt with in detail later on in the text.

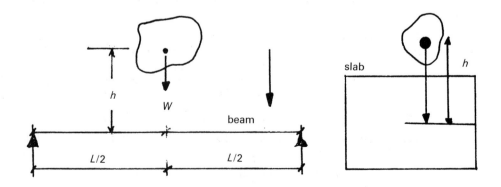

Fig. 4.17 Rock fall on a beam and slab.

4.5.5 Drop tests on nuclear flasks

4.5.5.1 Introduction

The transport of irradiated nuclear fuel is an essential part of the operation of nuclear power stations. The safety analysis of such flasks under impact conditions is on a priority list. At most nuclear stations the flask is loaded with 5 or 7 irradiated fuel elements placed in racks inside a container known as a multi-element bottle (MEB).

One drop test demonstration for different heights was conducted at the structural test centre of the Central Electricity Generating Board (CEGB) at Cheddar, UK. The altitude for the drop test demonstration was given in which the centre of gravity of the flask was over the line of impact thus ensuring that the entire energy was not dissipated in producing damage to the flask. The characteristics of such flasks are given below.

The characteristics of flasks for transporting irradiated nuclear fuel.

Type	Magnox Cuboid	AGR Cuboid	PWR Cylindrical
Total weight (tonne)	47	53	91
Wall thickness (mm)	36 steel	87.5 steel	87.5 steel
		175.0 lead	175.0 lead
External dimensions (mm)			
length	2.56	2.7	6.2
width	2.2	2.3	—
height	1.9	2.0	—
diameter	—	—	2.4
Fuel elements	200	20	7 PWR
			or 15 PWR

The heights for the side-edge and lid-corner tests were 9 m and 18 m respectively. Flask scales, apart from the prototype, were $\frac{1}{8}$, $\frac{1}{4}$ and $\frac{1}{2}$ M2a or M2b. Flask knock back results the 9 m side-edge series and the 18 m lid-corner tests are summarized below.

Summary of flask knock-back results.

Scale type		Knock back (normalized) (mm)	
		body (top/bottom)	lid (mm)
$\frac{1}{8}$ M2a	9 m	25.28/25.28	27.28
$\frac{1}{4}$ M2a	side	19.20/28.80	22.00
$\frac{1}{2}$ M2a	edge	32.00/22.20	44.80
$\frac{1}{8}$ M2a	18 m	164	
$\frac{1}{4}$ M2a	lid	192	
$\frac{1}{2}$ M2a	corner	188	

The side-edge and lid-corner arrangements are shown in Fig. 4.18(a) and (b).

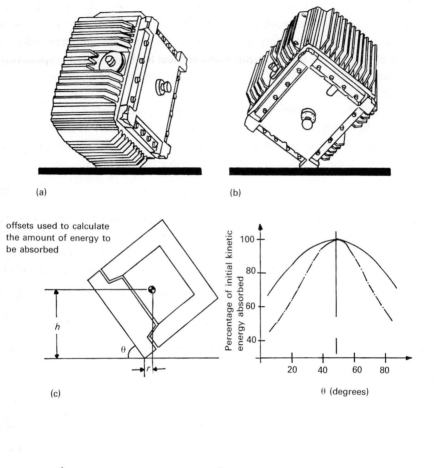

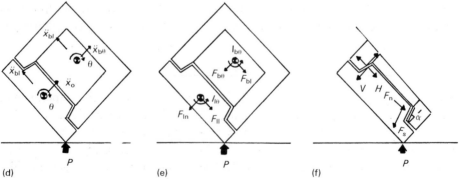

Fig. 4.18 Impact of nuclear flasks. Impact attitudes and position of the flask: (a) side edge and (b) lid corner; (c) variation of energy absorbed versus edge impacts; (d) accelerations; (e) d'Alembert forces; (f) spigot forces; b,l = body, lid; l,n = orthogonal components; I = moment of inertia.

4.5.5.2 Impact analysis and tests

Impact analysis for a number of case studies was conducted using DYNA three-dimensional finite-element analysis. Here only simple drop weight mechanics for the amount of energy absorbed is discussed. The energy absorbed for a lid-edge impact is calculated. Inertia load acceleration and spigot forces are demonstrated in Fig. 4.20(c)−(f). The spigot horizontal and vertical forces, H and V respectively, can be shown to be:

$$H = F_n \cos\alpha + F_s \sin\alpha$$
$$V = F_n \sin\alpha + F_s \cos\alpha \tag{4.83}$$

where friction is involved, $F_s/F_n = \mu$, and the relation between V and H will become

$$V = [(\tan\alpha - \mu)/(1 + \mu\tan\alpha)] \, H \tag{4.84}$$

The rigid body mechanics is given in this section for the evaluation of rotational velocity and the energy absorbed in the knock back. Assuming the flask velocity at impact is $V_0 = \dot{x}$, then using Fig. 4.20(c), the energy absorbed in knock back, E_{ha}, can be evaluated for various rotational velocities and boundary conditions as

$$E_{ha}[r_0^2/(r_0^2 + r^2)] \text{ with no friction} \tag{4.85}$$

$$E_{ha}r_0^2\{[r_0^2 + r^2 - \mu^2(r^2 + h^2)]/(r_0^2 + r^2 - \mu rh)^2\} \text{ with sliding and friction} \tag{4.86}$$

$$E_{ha}[(r_0^2 + h^2)/(r_0^2 + r^2 + h^2)] \text{ with no sliding} \tag{4.87}$$

The corresponding rotational velocities are

$$\dot{x}r/(r_0^2 - r^2) \tag{4.88}$$

$$\dot{x}(r - \mu h)/(r_0^2 + r^2 - \mu rh) \tag{4.89}$$

$$\dot{x}r/(r_0^2 + r^2 + h^2) \tag{4.90}$$

The finite element analysis will be discussed later in this text. In addition, the classical impact mechanics is given in this chapter for detailed analysis.

4.5.6 Impact/crash demonstration for trains

The train crash demonstration requires extensive analytical and experimental work. The modelling of such trains with and without nuclear flasks or other containers is complex. Figure 4.19(a) gives a mathematical model for a class 46 locomotive for which the force−time relationship is given in Fig. 4.19(b)−(d). Using detailed finite-element analysis, the force−time relationship is given in Fig. 4.20 for a class 46 locomotive impacting an unconstrained flask. The data are useful as an input to three-dimensional analysis.

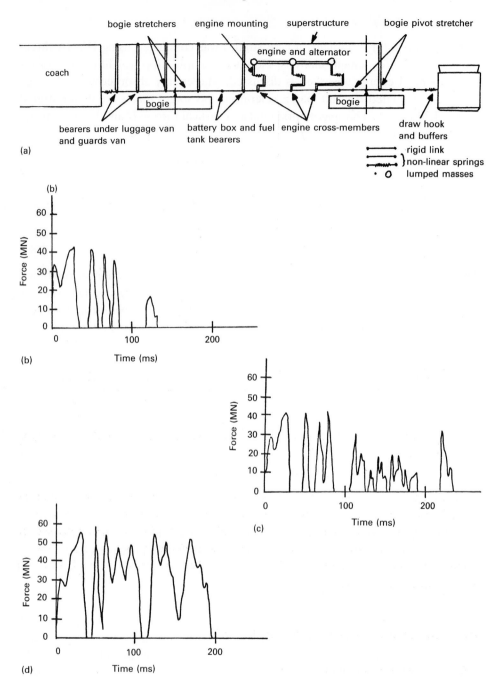

Fig. 4.19 A class 46 locomotive and the force−time relationship. (a) HST locomotive structure (original British Rail and UK Atomic Energy Authority mathematical model); (b) force−time relationship using an unconstrained flask (two-dimensional finite-element analysis); (c) force−time history for a complete HST impacting an unconstrained flask (three-dimensional finite-element analysis); (d) force−time history for an HST locomotive and one carriage impacting a rigidly fixed flask (three-dimensional finite-element analysis).

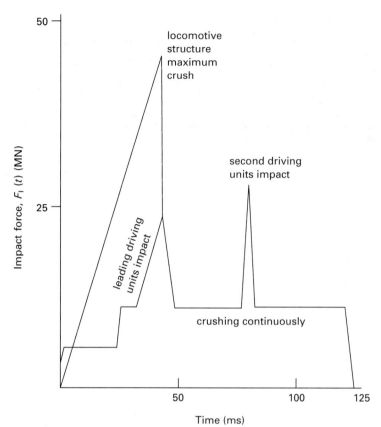

Impact force, $F_1(t)$ (MN)

Time (ms)

Fig. 4.20 Force−time histories for a class 46 locomotive impacting an unconstrained flask.

4.5.7 Load−time history: a comparative impact study

The CEGB has carried out tests of flask/train crashes on various structures including embankments, tunnels, rock and concrete structures. As mentioned, several drop tests have been demonstrated. Figure 4.21 gives a comparative study of some of the case studies. These results are a useful contribution as an input to complicated finite element analysis with a number of boundary conditions and impact load simulations.

4.6 Impact on concrete and steel

4.6.1 General introduction

When the structure receives an impact, the important consideration is to examine

(1) local damage, which includes penetration, perforation, scabbing and/or punching shear, and
(2) the overall response of the structure in terms of bending and shear, etc.

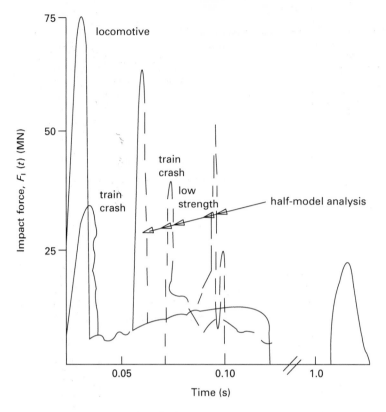

Fig. 4.21 Comparison of force–time histories.

These effects are defined below.

Penetration (x_p): the depth of the crater developed in the target at the zone of impact.

Perforation (t_p): full penetration of the target by the missile with and without exit velocity.

Scabbing (t_{sc}): the ejection of the target material from the opposite face of impact.

Spalling (t_{sp}): the ejection of the target material from the face at which impact occurred.

Figures 4.22 to 4.26 show penetration, spalling, scabbing, perforation, punching shear and overall response phenomena.

Criteria for point (1) involve the complex nature of the transient stress state. The empirical formulae developed so far, and which are described later on in this section, are based on available test data. They are valid for smaller missiles with limited deformation after impact. The impact conditions in section 4.2.1.1 are not covered by these empirical formulae since they do not penetrate the target in general. In addition to these are the interaction problems of deformable and undeformable missiles. The missile deformability lengthens the duration of impact

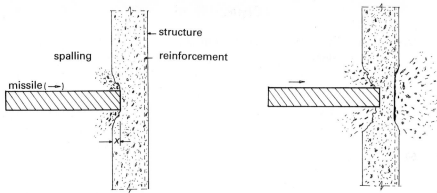

spalling

missile (→)

structure

reinforcement

Fig. 4.22 Local effect of penetration (x) and spalling.

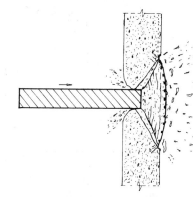

Fig. 4.23 Local effect of scabbing.

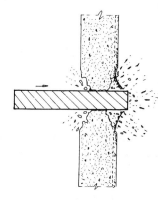

Fig. 4.24 Local effect of perforation.

Fig. 4.25 Local effect of a punch type shear failure.

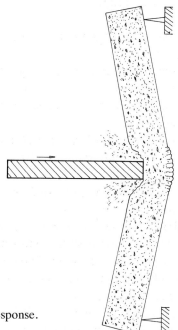

Fig. 4.26 Overall target response.

and at the same time reduces the penetration depth x_p, but has very little effect on perforation and scabbing thicknesses. A detailed list of missiles is given in Chapter 2. In the case of soft missiles, it is generally assumed that at the level of impact a significant local deformation of the missile or target occurs. In addition, when the missile is deformable, it is imperative to develop a force-time history along the lines suggested in sections 4.4 and 4.5.5, while assuming the target is rigid or flexible. As explained earlier, the graphs produced are *load−time functions*. The load imposed on a rigid target is composed of both the crushing strength and the rate of change of momentum of the missile, known as the *inertial force*. Some of this work has already been given in some detail in section 4.4. It is important to include these techniques in the local and global analyses of structures subject to missile impact. Apart from the empirical formulae given in the next section, more comprehensive non-linear analyses are required to take into consideration instantaneously formed failure mechanisms, material and geometrical non-linearities, impactor-target/structure interaction and evaluation of the overall damage to the impactors and structures.

4.6.2 Available empirical formulae

The following sections give the empirical and derived formulae for impact on targets.

4.6.2.1 Formulae for non-deformable missiles impacting on concrete targets

Petry[2.387]

Petry's formula is used for predicting the penetration depth x_p for infinitely thick concrete targets. This formula is derived from tests concerning high velocity impact on infinitely thick concrete targets. Where thickness governs the failure mode or the impact response is influenced by the size and shape of the missile and the presence of the reinforcement, this formula gives very inaccurate assessments.

$$x_p = K_m A V' \bar{R} \tag{4.91}$$

where x_p = depth of penetration in a concrete slab of thickness h
K_m = material property constant (L^3/F)
    = $4.76 \times 10^{-3}\,\text{ft}^3/\text{lb} = 2.97 \times 10^{-4}\,\text{m}^3/\text{kg}$
A = sectional mass weight of the missile per unit cross-sectional area of contact (F/L^2)
V' = velocity factor = $\log_{10}[1 + (v_0^2/v^{*2})]$
v_0 = initial velocity of missile at impact
v^{*2} = reference velocity equal to $215\,000\,\text{ft}^2/\text{s}^2$ $(19\,973\,\text{m}^2/\text{s}^2)$
$\bar{R}$ = thickness ratio

$$x_p/D_p = 1 + \exp[-4(\alpha' - 2)]$$

where $\alpha' = h/D_p = h/K_m AV'$ and D_p is the penetration depth of an infinitely thick slab. The penetration depth is restricted to less than $\frac{2}{3}h$, to satisfy the inequality below, in order to prevent penetration and spalling.

$$h \geq C_1 A \times 10^{-5} \text{ ft or an equivalent value in SI units}$$

where C_1 is taken from Fig. 4.27. The time required for penetration is derived from the *modified Petry formula*:

$$F = m\ddot{x} = mv(dv/dx) = -1.15(v^{*2}/K_m A) \exp(2.3x_p/K_m A) \qquad (4.92)$$

where $F = $ the resisting force, equal to R_m
$\quad x_p = $ penetration at any time
$\quad v = $ missile velocity at any time

Due to the non-linear nature of the equation of motion, a numerical integration is necessary to determine the velocity as a function of distance. Then:

$$\ddot{x} = -1.15(v^{*2}/K_m A) \exp[2.3(x - \delta)/K_m A] \qquad (4.93)$$

$$\text{hence } m\ddot{x} = m_t\ddot{\delta} + k\delta \qquad (4.94)$$

where $x = $ missile displacement
$\quad \delta = $ target displacement
$\quad k = $ target stiffness
$\quad m, m_t = $ mass of the missile and target respectively

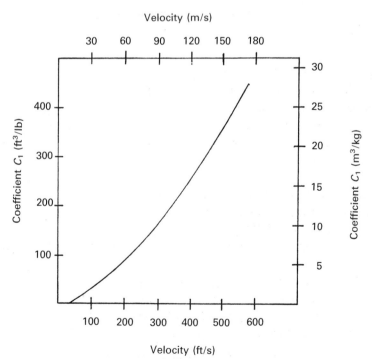

Velocity (m/s)

Velocity (ft/s)

Fig. 4.27 Minimum thickness needed to prevent penetration and spalling.

When $\delta = 0$, the acceleration $= \dfrac{m}{m_t} \times$ (missile acceleration). If $t = 0$; $x = \delta$; $\dot{\delta} = 0$; $\dot{x} = v_0$.

Army Corps of Engineers (ACE)[2.384]

This empirical formula predicts the penetration depth x_p of missiles in concrete targets. Again, it is based on high velocity impact tests on targets having infinite thickness. Since the ACE formula given below for penetration depth is not velocity dependent, it gives a non-zero value for the penetration depth when the velocity is zero.

$$x_p = 282[(Wd^{0.215})/(d^2\sqrt{f'_c})](v/1000)^{1.5} + 0.5d \qquad (4.95)$$

where W = missile weight (lb)
$\quad d$ = missile diameter (in)
$\quad f'_c$ = concrete compressive strength (psi)
$\quad v$ = velocity of the missile (ft/s)

When the penetration depth has been calculated, the perforation thickness t_p becomes

$$t_p = 1.32d + 1.24x_p \qquad (4.96)$$

and the spall thickness t_{sp} becomes

$$t_{sp} = 2.12d + 1.36x_p \qquad (4.97)$$

National Defense Research Committee (NDRC)[2.375]

Another empirical formula, proposed for non-deformable cylindrical missiles penetrating a massive reinforced concrete target, is the NDRC formula. The penetration depth of a solid missile given by this formula is

$$x_p = \sqrt{[4\bar{K}_p NWd(v/1000d)^{1.8}]} \qquad \text{for } x_p/d \leq 2.0 \qquad (4.98)$$

$$x_p = [\bar{K}_p NW(v/1000d)^{1.8}] + d \qquad \text{for } x_p/d > 2.0 \qquad (4.99)$$

where W = missile weight (lb)
$\quad x_p$ = penetration depth (in)
$\quad d$ = missile diameter (in)
$\quad v$ = impact velocity (ft/s)
$\quad f'_c$ = concrete cylinder compressive strength (psi)
$\quad \bar{K}_p = 180/\sqrt{f'_c}$
$\quad N$ = missile shape factor
$\qquad$ = 0.72 for flat-nosed bodies
$\qquad$ = 0.84 for blunt-nosed bodies
$\qquad$ = 1.00 for spherical-ended bodies

$N = 1.14$ for very sharp-nosed bodies

(*Note*: 1 psi = 6.9 kN/m^2; 1 fps = 0.305 m/s; 1 inch = 25.4 mm; 1 lb = 0.453 kg)

The formulae for scabbing and perforation thickness, t_{sc} and t_p respectively, for a solid cylindrical steel missile and infinite thickness of the target are given below.

$$t_{sc}/d = 2.12 + 1.36x_p/d \qquad \text{for } 3 \le t_{sc}/d \le 18 \tag{4.100}$$

$$t_p/d = 1.32 + 1.24x_p/d \qquad \text{for } 3 \le t_p/d \le 18 \tag{4.101}$$

$$t_{sc}/d = 7.91(x_p/d) - 5.06(x_p/d)^2 \qquad \text{for } t_{sc}/d < 3 \tag{4.102}$$

$$t_p/d = 3.19(x_p/d) - 0.718(x_p/d)^2 \qquad \text{for } t_p/d < 3 \tag{4.103}$$

Modified NDRC formula

A modification of the NDRC formula to take into account the finite thickness of a target is proposed. For large diameter missiles impacting on targets of finite thickness, the perforation and scabbing thicknesses are given below.

$$\begin{aligned} t_p/d &= 3.19(x_p/d) - 0.718(x_p/d)^2 \qquad \text{for } x_p/d \le 1.35 \\ t_{sc}/d &= 7.91(x_p/d) - 5.06(x_p/d)^2 \qquad \text{for } x_p/d \le 0.65 \end{aligned} \tag{4.104}$$

Equations (4.100) and (4.101) will have a range of $0.65 \le x_p/d \le 11.75$ and $1.35 \le x_p/d \le 13.5$, respectively. In any case, $t/d \le 3$ gives comfortable results.

Modified Ballistic Research Laboratory formula[2.200, 2.201]

The modified Ballistic Research Laboratory formula gives the perforation thickness of infinitely thick targets impacted by a non-deformable missile with high velocity as

$$t_p = (427Wd^{0.2})/(d^2 \sqrt{f_c'})(v/1000)^{1.33} \tag{4.105}$$

where t_p = perforation thickness (in)
    W = missile weight (lb)
    d = missile diameter (in)
    f_c' = concrete compressive strength (psi)
    v = missile velocity (ft/s)

and the spalling thickness $t_{sp} = 2t_p$.

Chalapathi, Kennedy and Wall (CKW)−BRL formula[6.29, 6.30, 6.54]

The penetration depth is calculated using the CKW−BRL method as

$$x_p = (6Wd^{0.2}/d^2)(v/1000)^{4/3} \tag{4.106}$$

where t_p = thickness to prevent perforation = $1.3x_p$. All units are imperial, as defined earlier.

Dietrich, Fürste (DF)−BRL formula (Dietrich F., personal communication)

The formula gives the thickness to prevent perforation as

$$t_p = (3 \times 10^{-4}/\sqrt{f_{ci}})(W/d^{1.8})(v/100)^{4/3} \qquad (4.107)$$

where d and t_p are in m, W is in kp, f_{ci} is in kp/cm^2 and v is in km/h.

Stone and Webster[2.376]

The scabbing thickness is given, as for the infinitely thick concrete targets, by

$$t_{sc} = (Wv^2/c_0') \qquad (4.108)$$

where the values of W and v are in lb and ft/s respectively and c_0' is a coefficient depending upon the ratio of t_{sc}/d. The range for which this formula is considered is 3000 psi $\leq f_c' \leq$ 4500 psi and $1.5 \leq t_s/d \leq 3$.

Chang formulae[2.268]

Chang has proposed two semi-analytical formulae for predicting perforation and scabbing thicknesses of concrete targets impacted by hard steel missiles of non-deformable type:

$$t_p = (u/v)^{0.25}(mv^2/df_c')^{0.5} \qquad (4.109)$$

$$t_{sc} = 1.84(u/v)^{0.13}(f_c')^{0.4}(d)^{-0.2}(mv^2)^{0.4} \qquad (4.110)$$

These formulae are probably validated over the following range based on random variables by Bayesian statistics:

$$16.7 \, \text{m/s} < v < 311.8 \, \text{m/s} \qquad (4.111)$$

$$0.11 \, \text{kg} < W < 343 \, \text{kg}$$

$$2.0 \, \text{cm} < d < 30.4 \, \text{cm}$$

$$232 \, \text{kg/cm}^2 < f_c' < 464.2 \, \text{kg/cm}^2$$

$$5.0 \, \text{cm} < h < 60.9 \, \text{cm}$$

where u is a reference velocity (200 ft/s or 60.96 m/s) and f_c', d, m and v are defined in other sections. The scabbing velocity, v_{sc} ft/s, is written as:

$$v_{sc} = [(1/2.469)(d^{0.2} f_c'h/W^{0.4})]^{3/2}$$

The IRS formulae[2.388]

The IRS formula for penetration is expressed as

$$t_p = 1183(f_c')^{-0.5} + 1038(f_c')^{-0.18} \exp[-0.82(f_c')^{0.18}] \qquad (4.112)$$

The IRS formula for total protection for the target against penetration, perforation and scabbing is given as

$$SVOLL = 1250(f_c')^{0.5} + 1673(f_c')^{-0.18} \times \exp[-0.82(f_c')^{0.18}] \qquad (4.113)$$

where SVOLL is the minimum wall thickness to provide complete protection. The values of t_p and f_c' are in cm and kgf/cm^2, respectively.

The Bechtel formulae[2.377]

Based on test data applicable to hard missiles impacting on nuclear power plant facilities, the two formulae for scabbing thickness are given below.

The Bechtel formula for scabbing thickness for solid steel missiles is

$$t_{sc} = (15.5/\sqrt{f_c'})(W^{0.4}v^{0.5}/d^{0.2}) \qquad (4.114)$$

The Bechtel formula for steel pipe missiles (scabbing) is

$$t_{sc} = (5.42/\sqrt{f_c'})(W^{0.4}v^{0.65}/d^{0.2}) \qquad (4.115)$$

The variables given above are in imperial units. The Bechtel formulae are also written for scabbing and perforation thickness in metric units using a reference velocity v^* (60.96 m/s):

$$t_{sc} = 1.75 \left(\frac{v^*}{v}\right)^{0.13} \frac{(mv^2)^{0.4}}{(d)^{0.3} (f_c')^{0.4}} \qquad (4.116)$$

$$t_p = 0.90 \left(\frac{v^*}{v}\right)^{0.25} \frac{(mv^2)^{0.5}}{(df_c')^{0.5}} \qquad (4.117)$$

CEA−EDF formula[2.389]

Tests were carried out for the French Atomic Energy (CEA) and Electricité de France by Berriaud, on a series of slabs subject to impactors with varying velocities (from 20 m/s to 200 m/s), thicknesses, concrete strength and reinforcement quantities. The empirical formula for a thickness to resist perforation is given by

$$t_p = 0.82(f_c')^{-3/8} (\rho_c)^{-1/8} (W/d)^{0.5} v^{3/4} \qquad (4.118)$$

where ρ is the density of concrete, and the following ranges apply:

$$30 \, \text{MPa} < f_c' < 45 \, \text{MPa}$$
$$0.3 < t_p/d < 4$$
$$75 \, \text{kg/m}^3 < p < 300 \, \text{kg/m}^3$$

where p is the reinforcement quantity. The perforation velocity, v_p, for the target thickness is given by

$$v_p = 1.7 f'_c \, (\rho_c)^{1/3} \, (dh^2/W)^{4/3} \text{ (metric)} \tag{4.119}$$

The CEA−EDF residual velocity formula, v_r, based on several tests with a correction factor K_σ is given below, with all values in imperial units:

$$v_r = \left[\frac{1}{1 + \dfrac{W_t}{W}} (v^2 - v_p^2) \right]^{1/2} \tag{4.120}$$

$$K_\sigma v_p = \left[v_p^2 - v_r^2 \left(1 + \frac{W_t}{W} \right) \right]^{1/2} \tag{4.121}$$

where v_r, v_p, K_σ and W_t are the residual velocity of the missile, perforation velocity, correction factor and the weight of the target material removed by impact, respectively. The mean and minimum values of K_σ are 1.45 and 1.225, respectively.

Haldar, Miller and Hatami method[2.378]

The impact formula of the NDRC type associated with a non-dimensional impact factor I are presented in imperial units as

$$I = WNv^2 \, \mathbf{g} d^3 f'_c \tag{4.122}$$

where $\mathbf{g} = 32.2 \, \text{ft/s}^2$ and, when substituted, the above equation becomes

$$I = 12NWv^2/32.2 d^3 f'_c$$

where N is the missile nose-shaped factor and all other notations are as defined earlier.

For various impact factors, the NDRC test results were examined using linear regression analysis for x_p/d and t_{sc}/d ratios.

$$
\begin{aligned}
t_{sc}/d &= -0.0308 + 0.2251 \, I & 0.3 \leq I \leq 4.0 & \quad \text{(a)} \\
t_{sc}/d &= 0.6740 + 0.0567 \, I & 4.0 < I \leq 21.0 & \quad \text{(b)} \\
t_{sc}/d &= 1.1875 + 0.0299 \, I & 21.0 < I \leq 455 & \quad \text{(c)} \\
t_{sc}/d &= 3.3437 + 0.0342 \, I & 21.0 \leq I \leq 385 & \quad \text{(d)}
\end{aligned} \tag{4.123}
$$

Takeda, Tachikawa and Fujimoto formula[1.70]

Takeda *et al.* proposed a formula for predicting the penetration depth into reinforced concrete slabs subject to hard missiles:

$$x_p = [\alpha/(\beta + 1)](v)^{\beta+1} \tag{4.124}$$

where $\alpha = 2^n m^{1-n}/c'\psi$

$\beta = 1 - 2n$

x_p = maximum depth of penetration (cm)

m = mass of projectile (kgs²/cm)

v = impact velocity (cm/s)

ψ = circumference of projectile (cm)

c', n = constants

Since the formula is based on the kinetic energy as input, it is valid for an energy range from 20×10^2 kgcm to 200×10^5 kgcm.

Hughes formulae[2.98]

These formulae have been developed using the dimensional analysis and test analysis results of NDRC and ACE described earlier. Front and back faces are reinforced (front face 0–0.15%; back face 0.3–1.7% each way). The penetration depth calculated for a concrete barrier, assuming no scabbing or perforation occur, is given by

$$x_p/d = 0.19NI/s' \tag{4.125}$$

where N = nose-shaped coefficient = 1; 1.12; 1.26 and 1.39 for flat, blunt, spherical and very sharp noses, respectively

I = impact parameter = $(mv^2/0.63\sqrt{f_c'})\ d^3$

s' = strain-rate factor = $1 + 12.3\ \log(1 + 0.03I)$

The thicknesses of the concrete target necessary to prevent scabbing and perforation are written as

$$t_{sc}/d = 1.74(x_p/d) + 2.3 \tag{4.126}$$

$$t_p/d = 1.58(x_p/d) + 1.4 \tag{4.127}$$

Kar formulae[2.2, 2.380]

It is claimed that most of the formulae described earlier do not include dimensions, shapes of the missiles, material properties of the missiles and targets and the size of the coarse aggregate in concrete. Kar gives the penetration depth (in inches) in the concrete targets as

$$G(x_p/d) = \alpha \bar{K}_p N(W/D)(v/1000d)^{1.8} \tag{4.128}$$

$$\text{where } G(x_p/d) = \begin{cases} (x_p/2d)^2 & \text{for } x_p/d \leq 2.0 \\ (x_p/d) - 1 & \text{for } x_p/d \geq 2.0 \end{cases} \tag{4.129}$$

where D = diameter of the actual missile in the case of a circular section, or is equal to the projectile diameter d in the case of a rectangular section or is equal to the diameter of the circles inscribed within the boundary formed by joining the extremities of impacting ends of angular I or irregular sections; the minimum value of $D = d$

α = constant equal to 1.0 in imperial units

$\quad = 0.01063$ if D, d, E, v, W and x_p are expressed in cm, cm, kN/m^2, m/s, kg and cm, respectively

$N = 0.72$ flat-nosed

$\quad = 0.72 + 0.25(n - 0.25)^{0.5} \leq 1.17$ for spherical-nosed $\qquad$ (4.130)

n = radius of nose/missile diameter

$\quad = 0.72 + [(D/d)^2 - 1]\, 0.36 \leq 1.17$ for hollow circular sections or irregular sections $\qquad$ (4.131)

$\bar{K}_p$ = penetrability factor = $(180/\sqrt{f'_c})(E/29\,000)^{1.25}$ $\qquad$ (4.132)

The depth to prevent perforation or scabbing is given by

$$(t_p - a')/d = 3.19(x_p/d) - 0.718(x_p/d)^2 \qquad \text{for } x_p/d \leq 1.35 \qquad (4.133)$$

$$(t_{sc} - a')/d = 7.91(x_p/d) - 5.06(x_p/d)^2 \qquad \text{for } x_p/d \leq 0.65 \qquad (4.134)$$

For x_p/d larger than above, the following equations are suggested:

$$(t_p - a')/d = 1.32 + 1.24(x_p/d) \qquad \text{for } 3 \leq t_p/d \leq 18 \qquad (4.135)$$

$$\beta(t_{sc} - a')/d = 2.12 + 1.36(x_p/d) \qquad \text{for } 3 \leq t_{sc}/d \leq 18 \qquad (4.136)$$

where $\beta = (29\,000/E)^{0.2}$ $\qquad$ (4.136a)

$\quad \beta = 1$ for steel missiles

$\quad a'$ = maximum aggregate size in concrete

Perry and Brown formulae[2.103, 2.104]

(1) Solid hard missile on concrete targets. The penetration depth is given (Fig. 4.28) by

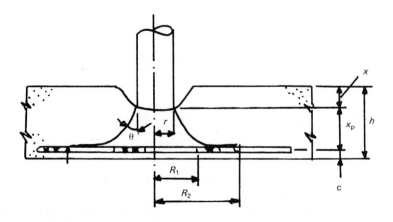

Fig. 4.28 Assumed barrier failure mechanism and derivation of hinge radius.

$$x_p/d = 9.665 \sqrt{m}[v/\sqrt{(E_c d^3)}] + 0.06 \qquad (4.137)$$

where x_p, d and v are as defined earlier and m is the missile mass. If $L < 50$, the scabbing thickness does not occur; if $L > 70$, the scabbing thickness certainly occurs.

$$L = [\sqrt{m}(v/\sqrt{d}) \, h(1 + h/d)]\sqrt{E_c}\sigma_s \qquad (4.138)$$

where h = target thickness

E_c = Young's modulus

σ_s = maximum value of nominal shear stress before damage

$\quad = \sigma_{tm}\sqrt{[1 + (\sigma_{cm}/\sigma_{tm})]}$

σ_{tm} = mean tensile stress in concrete

σ_{cm} = compressive stress

$\sigma_{cr} = 0.9\sigma'_{cr}$

$\quad = (0.9)[0.52[(1 - v^2)/(1 - v_p^2)]^{1/2} \, (E_t/E)^{1/2}E_s(t^p/r)]$

E = Young's modulus

E_t = tangent modulus

E_s = secant modulus

t^p = pipe thickness (in)

r = pipe radius (in)

v, v_p = Poisson ratios for elastic and plastic, respectively

Geometry:

$$\tan\theta = R_1 \times r/x = 2$$

$$2x = R_1 \times r$$

$$R_1 = 2(htx_p) + r$$

$$R_2 = C_L t_3$$

$$r_u = M_u D_h$$

where M_u = ultimate curvature = eu/C_h

D_h = hinge length (or $R_2 R_1$)

$(R_2 R_1)_{max} = r_u/M_u = 0.07/0.003 = 23.3$

$(R_2)_{max} = R_1 + 23$

(2) The thickness to prevent spalling is given by

$$t_{sp} = x_p + x + c \qquad (4.139)$$

$$x(x + r)^2 > \left[\frac{6W^2V_m^2 - W}{2g\pi r_u M_u}\right]\left(\frac{1728}{\rho_c\pi}\right) - R_2^2 C \qquad (4.140)$$

where $\rho_c = 0.15$ kips/ft^3

r_u = ultimate rotational capacity at hinge (rad)

C = cover (in)

M_u = ultimate moment capacity at hinge (in-kips/in) (1 kip = 1000 lbf)

$R_2 = R_1 + 23$

$$v = v_2 \quad \text{if } v_2 > v_2'$$
$$v = v_2' \quad \text{if } v_2 < v_2' < v_1$$
$$\text{if } v_2' > v_1, \text{ not applicable}$$

The spalling equations (4.139) and (4.140) are not applicable if $t_{sp} = h < 12$ inches.

(3) Pipe missile on concrete targets. The penetration depth is given by

$$x_p/d = 8[\sqrt{mv}/\sqrt{(Ed^3)}] + 0.24 \tag{4.141}$$

where d is the outside diameter of the pipe missile. The scabbing thickness can be achieved if $L > 60$ and is unlikely to occur if $L < 50$. The value of L in this case is expressed by

$$L = \sqrt{mdv}/[dh(1 + h/d)](\sqrt{E_c}/\sigma_s) \tag{4.142}$$

where the various elements of the equations have been defined previously.

Barr, Carter, Howe, Neilson[2.113, 2.114] Winfrith model

On the basis of a number of tests carried out on various target slabs, Winfrith modified the CEA−EDF formula to include the bending reinforcement quantity. The perforation velocity in conjunction with the CEA−EDF formula can now be written as

$$v_p = 1.3(\sigma_{cu})^{0.5} \; (\rho_c)^{1/6}(dh^2/m)^{2/3} \; r^{0.27} \tag{4.143}$$

where $\sigma_{cu} = f_c'$
$\quad m = W/g$
$\quad r$ = bending reinforcement quantity in % $(0.125\% \leq r \leq 0.5\%)$

All values are in metric units and $r = p$ for previous formulae.

4.6.2.2 Formulae for deformable missiles impacting on concrete targets

McMahon, Meyers, Sen model[2.390]

The model evaluates local damage including penetration and back-face spalling of reinforced concrete targets subject to the impact of deformable, tornado-generated missiles such as pipes, etc. The total penetration is given by

$$x_p = x_1 + x_2 + x_3 \tag{4.144}$$

where x_1 = penetration during time $t_1 = (Ft_1^2/6m) + v_0 t_1$
$\quad F$ = interface force = $\sigma_{cr}A$
$\quad t_1$ = rise time = $3.2 \times 10^{-6}F$
$\quad A$ = pipe area
$\quad v_0$ = missile velocity at the initial time t_0
$\quad m$ = mass of the missile

$x_2 - x_1 = x_c$ (plastic missile deformation during $t_2 - t_1$)

$$= \frac{m}{2F} (v_2 - v_1)^2 + \frac{mv_1}{F}$$

$$= \frac{m}{2F} (v_2^2 - v_1^2) \qquad (4.144a)$$

$v_1, v_2 =$ missile velocity at times t_1 and t_2

$$v_1 = (Ft_1/2m) + v_2; \qquad v_2 = (F/m)(t_2 - t_1) + v_1 \qquad (4.144b)$$

$$x_3 = x_2 \text{ (penetration during } t_3 - t_2) = -mv_2^2/2F \qquad (4.144c)$$

$$t_2 = l/\sqrt{(E_T/\rho)} + l/\sqrt{(E/\rho)} = \text{wrinkling duration} \qquad (4.144d)$$
$$\quad\text{plastic} \qquad\quad \text{elastic}$$
$$\quad\text{waves} \qquad\quad \text{waves}$$

$$t_3 = (-mv_2/F) + t_2 = \text{final time} \qquad (4.144e)$$

where ρ is material density and l is missile length.

Rotz damage model[2.381]

Rotz predicted scabbing thickness using Bayesian estimators as

$$t_{sp} = \bar{K}_p (W^{0.4} \, v^{0.65}/\sqrt{f_c'} \, d^{0.2}) \qquad (4.145)$$

where $\bar{K}_p = 5.42$ (empirical constant)
$\quad v =$ impact velocity (in)
$\quad d =$ missile diameter (in)
$\quad f_c' =$ concrete compressive strength (lb/in^3)
$\quad t_{sp} =$ scabbing thickness (in)
$\quad W =$ missile weight (lb)

Note: other soft missiles are described later on in various sections of this chapter.

PLA damage model for heavy-duty pavements[2.382]

Serviceability failure occurs in a heavy-duty pavement due to impact by either excessive vertical compressive stress in the subgrade or excessive horizontal tensile strain in the base. The allowable subgrade vertical compressive strain ε_v is given by

$$\varepsilon_v = 21600/N^{-0.28} \qquad (4.146)$$

where $\bar{N}$ is the number of repetitions of impact load.
The allowable base horizontal tensile strain, ε_h, is given by

$$\varepsilon_h = \frac{993\,500 \times f_c'}{6 \times E_b^{1.022} \, N^{-0.0502}} \qquad (4.147)$$

where E_b = Young's modulus of the base material (N/mm^2)

f'_c = characteristic compressive strength of the base material (N/mm^2)

≤ 7 N/mm^2 then $E_b = 4 \times 10^3 f'_c$

> 7 N/mm^2 then $E_b = 1.68 \times 10^4 (f'_c)^{0.25}$

If these values are not met, the pavement will be damaged. Any impact formula must be assessed against these conditions. The damage effect, D_E, given below, by Heukelom and Klomp[2.382] can then be judged in the light of cracking, spalling, etc., occurring from the impact formula:

$$D_E = W^{3.75} \, P^{1.25}$$

where W is the wheel load with impacting factor and P is the tyre pressure (N/mm^2).

4.6.2.3 Missiles on steel targets

Missiles and targets

Missiles as projectiles with non-deformable nose shape are given in Fig. 4.29. Non-deformable projectiles are assumed to be either spherical or cylindrical (refer to Chapter 2), with a nose of one of the shapes shown in Fig. 4.30. The calibre or ballistic density ρ is generally given as W/d^3, where W is the weight of the missile and d is the diameter. Owing to changes in the value of W, a longer missile is, therefore, more dense than a short one with the same diameter and material. Metal targets are generally restricted to hard missiles of non-deformable type striking the plate.

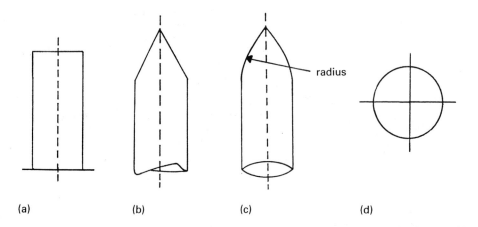

Fig. 4.29 Projectile shapes. (a) Flat; (b) conical; (c) ogival; (d) spherical.

Slow speed indentation of steel targets

(1) Conical missile. Assume a conical missile is striking a steel target with a low velocity, v_0, leaving a permanent indentation of diameter d_0 (Fig. 4.30). The yield stress of the target steel $\bar{\sigma}_e = 3\sigma_t$, where σ_t is the uni-axial stress flow of the target material. The equation of motion is written as

$$m\ddot{x}_p = mv(\mathrm{d}v/\mathrm{d}x_p) = -\bar{\sigma}_e \pi r_0^2 \qquad (4.148)$$

where v is the missile speed after penetration x_p is achieved and r_0 is the final radius of the crater $= x_p \tan\alpha$.

After substitution of the value of r_0 into equation (4.148) and integration, the crater radius and depth are written as

$$r_0 = (0.4772 m/\bar{\sigma}_e \tan^2\alpha)^{1/3} (\tan\alpha) \, v_0^{2/3} \qquad (4.149)$$

$$x_p = 2r_0 \cot\alpha/2 \qquad (4.150)$$

(2) Spherical missile. Equation (4.148) is still applicable when x_p/d is small. It is assumed that $r_0 \approx x_p d$, the equation of motion expressed in equation (4.148) is integrated and the final penetration obtained as

$$x_p = \sqrt{[(m/\sigma_e)\pi d]} v \qquad (4.151)$$

where v is the final velocity at the time of the formation of the crater.

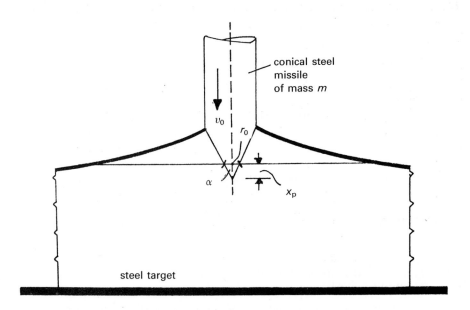

Fig. 4.30 Conical missile striking steel target.

Calder and Goldsmith velocity model (preliminary report, 1979)

The impact velocity at which the projectile penetrates a steel target completely, but comes to rest in the process, defines the ballistic limit. The formula for the residual velocity, v_r, developed is based on both impact velocity and ballistic limit and is given for a *sharp-nosed* missile as

$$v_r = \sqrt{(v_0^2 + v_B^2)} \qquad (4.152)$$

where v is the initial velocity and v_B is the ballistic limit.

In equation (4.152) it is assumed that the missile carries no material from the steel target. The residual velocity for a *blunt-nosed* missile carrying a plug of material ejected from the steel target is given by

$$v_r = \sqrt{\left[\frac{m}{m + m_p'} (v_0^2 - v_B^2) \right]} \qquad (4.153)$$

where m and m_p' are the masses of the missile and the plug, respectively.

Ballistic Research Laboratory formula (BRL)[2.200, 2.201]

This formula is based on the impact of small-diameter, high-calibre, high-density non-deformable missiles striking thin steel targets:

$$(t_p/d)^{3/2} = Dv^2/1\,120\,000\,\bar{K}_p^2 \qquad (4.154)$$

where t_p = perforation thickness (in or mm)

$\quad d = 4A_m/\pi$ = effective missile diameter (in or mm)

$\quad A_m$ = missile area (in^2 or cm^2)

$\quad v$ = impact velocity (ft/s or m/s)

$\quad D$ = missile diameter (in or mm)

$\quad W/d^3$ = calibre density of missile (lb/in^3 or kg/m^3), from which d can be evaluated

$\quad \bar{K}_p$ = steel penetrability constant depending upon the grade of the steel target; the value of $\bar{K}_p$ is generally taken as 1.0

The Stanford Research Institute (SRI) equation[2.391]

Like BRL's formula, the following equation is for small-diameter, hard missile striking a thin steel target.

$$(t_p/d)^2 + (3/128)(B/d)(t_p/d) = 0.0452D_v/\sigma_{tu} \qquad (4.155)$$

where t_p, d, D and v are as defined in the BRL formula

$\quad B$ = width of the steel target

$\quad \sigma_{tu}$ = ultimate tensile strength of the target steel (lb/in^2 or Pascals)

The formula is based upon tests with the following range of parameters:

$$0.1 \le t_p/d \le 0.8$$

$$2\,\text{lb/in}^3 \ (550 \times 10^2 \,\text{kg/m}^3) \le D \le 12\,\text{lb/in}^3 \ (3300 \times 10^2 \,\text{kg/m}^3)$$

$$0.062\,\text{in} \ (1.6\,\text{mm}) \le d \le 3.5\,\text{in} \ (89\,\text{mm})$$

$$70\,\text{ft/s} \ (21\,\text{m/s}) \le v \le 400\,\text{ft/s} \ (120\,\text{m/s})$$

$$2\,\text{in} \ (50\,\text{mm}) \le B \le 12\,\text{in} \ (300\,\text{mm})$$

$$5 \le B/d \le 8$$

$$8 \le B/t_p \le 100$$

For design purposes, the design thickness due to t_p or t_{sc} must be increased by 20%.

Kar steel target formula[2.383]

For mid-to-medium-hard homogeneous steel plates, the barrier may have a ductile failure. When steel target Brinell hardness numbers are above 350, failure by plugging may occur. For inferior quality steel, flaking may occur on the back face of the steel targets. According to Kar, for a good quality steel, back face phenomena do not generally influence the depth of penetration. The penetration or thickness to prevent perforation is given by the following:

$$t_p = \alpha(E/29\,000)(0.72 + N)\ \bar{K}_p(mv^2)^{0.667}/1067(D + d) \tag{4.156}$$

where m is the mass of the missile (lb-s²/ft) and v (ft/s), D (in) and d (in) are as defined for equations (4.128) and (4.129).

The penetrability coefficient $\bar{K}_p$ is determined from the following:

$$\bar{K}_p = (0.632\text{BHN} + 94.88)/275 \tag{4.157}$$

where BHN is the brittle hardness number of the steel target material and is limited to between 0.37 and 1.0. The above equation is still relevant if BHN < 0.37 or > 1.0.

$\alpha = 1.0$ if imperial units are used
 $= 0.0035$ for m (kg-s²/m), v (m/s), d (cm), E (kN/m²), t_p (cm)

de Marre modified formula

de Marre proposed a relationship between the specific limit energy h/d and the target penetration:

$$mv_1^2/d^3 = \bar{\alpha}(h/d)^{\bar{\beta}} \tag{4.158}$$

where m = missile mass (g)
 v_1 = limit velocity (m/s)
 d = missile diameter (cm)

h = steel target thickness (cm)
$\bar{\alpha}$ = constant between 1 and 2
$\bar{\beta}$ = constant approximately 1
h is replaced by $hf(\theta)$
θ = incidence angle
f = a function of obliquity, usually secant

Taub and Curtis model

A perceptive analysis by Taub and Curtis derived the following formula for back-face spalling or petalling type of failure:

$$mv_1^2/d^3 = \bar{\alpha}[(h/\bar{\alpha}) + \bar{\beta}] \qquad \bar{\beta} < 0 \qquad (4.159)$$

Lambert model[2.349]

The development assumes back-face thickness where petalling occurs and d to be constant and β becomes complex as a quadratic function. To overcome this, Zukas replaced $\bar{\beta}$ by $e^{-h/d} - 1$; d^3 by $d^{3-c} l^c$ and θ as stated in the case of the de Marre formula, by $h \sec^{k'} \theta$. Both c and k' are constants. Using Lambert's limit velocity database containing limit velocities for 200 cases involving:

range of mass	$\frac{1}{2}$ to $3630\,\mathrm{g}$
diameter	$\frac{1}{5}$ to $5\,\mathrm{cm}$
l/d	4 to 30
h	$\frac{3}{5}$ to $15\,\mathrm{cm}$
θ	$0°$ to $60°$
ρ_s (rod material density)	7.8 to $19\,\mathrm{g/cm^3}$

$$\bar{\alpha} = 4000^2; \ c = 0.3; \ k' = 0.75$$

the following predicted model is established for the limit velocity v_ℓ:

$$v_\ell = u(l/d)^{0.15} \sqrt{[f(z)\ d^3/m]} \ \mathrm{m/s} \qquad (4.160)$$

where $z = h(\sec\theta)^{0.75}/d$, $u = 4000$ for rolled homogeneous armour (RHA) and m depends on the density ρ_s.

$$f(z) = z + e^{-z} - 1 = \Sigma_{j=2} \ (-z)^j/j \qquad (4.161)$$

$$v = \begin{cases} 0, & 0 \le v \le v_\ell \\ \alpha(v^2 - v_\ell^2)^{1/2}, & v > v_\ell \end{cases}$$

$$v_\ell = \max\{v: v_r = 0\} = \inf\{v: v_r > 0\}$$

$$v_r = \begin{cases} 0, & 0 \le v \le v_\ell \\ \bar{\alpha}(v^p - v_\ell^p)^{1/p}, & v > v_\ell \end{cases}$$

where the value of p is generally 2 and v, v_ℓ and v_r are striking velocity, ballistic limit and missile residual velocity, respectively.

Winfrith perforation energy model[2.392, 2.393]

Using dimensional analysis, the perforation energy of the steel pipe is related to the geometric parameters and material properties by

$$E_p/(\sigma_u d^3) = A(h/d)^a(d/D)^b \qquad (4.162)$$

where E_p = perforation energy

σ_u = characteristic strength of the material = σ_e

d, D = missile and pipe diameter, respectively

h = target thickness

a, b, A = constants given in Table 4.4

Table 4.4 Permanent deformations of pipe targets.

Test	a (mm)	b (mm)	A (cm^2)	d_0' (mm)	δ (mm)	m (kg)	v (m/s)	h (mm)
255	75	115	—	—	—	1.7	93	7.4
256	70	100	170	148	21	1.7	67	7.4
257	140	215	158	115	65	7.5	75	7.3
258	170	310	99	77	112	7.3	108	7.2
259	140	250	172	114	60	7.39	69	7.1
260	165	325	113	86	104	7.29	104	7.1
264	130	220	151	111	64	4.0	105	7.3
265	135	200	159	113	67	4.0	104	7.4
266	145	280	121	93	94	4.0	136	7.1
267	135	225	—	—	—	4.0	142	7.0
268	105	—	—	—	—	4.0	117	7.2
269	130	—	—	—	—	4.0	112	7.2
270	105	—	—	—	—	4.0	108	7.2
271	100	110	180	144	35	4.0	108	7.6
272	—	—	—	—	—	4.0	108	7.7
273	120	220	—	100	83	4.0	114	7.5
274	135	250	—	100	82	4.0	114	6.9
275	140	300	135	92	93	4.0	113	7.0
276	100	120	—	—	—	4.0	108	7.4
477	80	90	—		—	1.7	130	10.5
478	45	50			11	1.7	203	18.2
479	70	80	—		—	1.7	129	10.7

continued

Table 4.4 *Continued.*

Test	a (mm)	b (mm)	A (cm²)	d_0' (mm)	δ (mm)	m (kg)	v (m/s)	h (mm)
480	55	60			17	0.6	325	18.6
481	110	—	—	—	—	4.0	180	10.6
482	105	—	—	—	—	4.0	236	18.6
483	135	200	—	—	—	4.0	136	11.0
484	90	125			36	4.0	144	18.6
485	110	165	—	—	—	4.0	143	7.2
486	110	140	—	—	—	4.0	87	7.1
487	110	145	—	—	—	4.0	113	8.1
488	105	130	—	—		4.0	67	7.2
489	90	160			36	3.1	75	7.4
490	95	180			46	3.1	84	7.2
491	100	200			55	3.1	99	7.2
492	110	220			—	3.1	143	7.2
493		oblique impact			—	4.0	180	7.3
494						4.0	120	7.0
495	185	230			231	43.0	78	7.4
496	105	180			55	34.9	49	7.5
497	155	250			95	37.5	46	7.4
498	170	380			196	44.0	78	7.3
499	170	270			350	54.2	106	10.9
500	175	385			150	29.5	70	7.1

Courtesy of A.J. Neilson, UKAEA, Winfrith, UK.

Tests have been carried out on target thicknesses of 7.1 mm and 11 mm. For a 25 mm diameter missile the perforation energy varied as the 1.8 power of the target pipe thickness and for a 60 mm diameter missile an exponent of 1.4 was obtained. At an impact energy of 41.5 kJ, the 60 mm diameter missile displaced a shear plug in the pipe wall thickness by a distance of 3 to 11 mm. Figure 4.31 shows a graph of the perforation energy plotted against pipe wall thickness and the missile diameter. In the case of E_p versus d, an exponent of 1.9 is obtained for a 7.1 mm pipe wall and 1.7 for an 11 mm pipe wall thickness, averaging both sets to 1.8. Exponents ranging from 1.5 to 1.7 have been suggested for plain steel targets. The test results based on the BRL formula are also plotted. On the basis of these tests, the perforation energy is assumed to vary as

$$E_p = Bh^{1.7}d^{1.8} \tag{4.163}$$

where B is a constant.

In another expression, the exponent is given as 1.7. If this normalized perforation energy variation is imposed on the pipe perforation, the correlation as shown in Fig. 4.31 becomes

$$E_p/\sigma_u d^3 = A(h/d)^{1.7}(d/D)^{1.5} \tag{4.164}$$

where the parameter $A\sigma_u$ has a value of 8×10^9, if SI units are chosen for E_p, d, h and D.

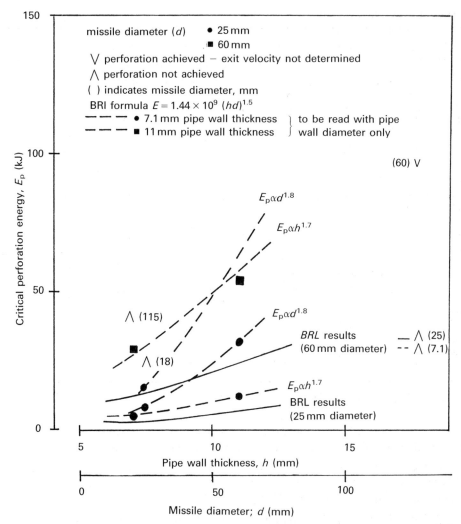

Fig. 4.31 Effect of pipe wall thickness on perforation energy of solid billet missiles. (Courtesy of A.J. Neilson, UKAEA, Winfrith, UK.)

Shot-target formula[1.31]

The maximum pressure P due to the impact force $F_1(t)$ was computed by this formula by carrying out several tests on steel targets by single and multiple shots for the shot-peening process. The shots included cast steel, cut steel wires, cast iron and glass beads. Tests were carried out using cast steel with two hardness ranges — the common one is 40–50 RC. The grade number 5240 for this research was 0.024 in (0.61 mm). The value of $F_1(t)$ when a sphere hits a flat surface is given by

$$F_1(t) = \frac{4r^{1/5}\left(\frac{15}{16}\pi\, mv^2\right)^{3/5}}{3\pi\left[\frac{(1-v_1^2)}{\pi E_1} + \frac{(1-v_2^2)}{\pi E_2}\right]^{2/5}} \tag{4.165}$$

where m = mass of the shot

r = radius of the shot

v = velocity of the shot

v_1, E_1 = Poisson's ratio and Young's modulus for the target

v_2, E_2 = Poisson's ratio and Young's modulus for the shot

The duration of impact, t, is computed as

$$t = 2.943 \left[2.5\pi\rho \, \frac{(1-v^2)}{E} \right]^{\frac{2}{5} \frac{r}{\sqrt{v}}} \tag{4.166}$$

where ρ is the density of the shot.

Table 4.5 shows the relationship between the shot radius, the impact pressure P due to $F_1(t)$ and the duration of time for velocities of 30, 40, 50, 80 and 100 m/s.

Table 4.5 Impact due to shot-peening: basic data.

Shot velocity (m/s)	Shot radius, r (mm)	Pressure, P (N/m²)	Duration time, t (s)
30	0.990	3920	8.80
	0.860	2960	7.65
	0.610	2490	5.42
	0.430	7390	3.82
	0.304	3700	2.70
	0.177	1250	1.57
40	0.990	5540	8.31
	0.860	4180	7.22
	0.610	2100	5.12
	0.430	5220	3.61
	0.304	1770	1.49
50	0.990	7240	7.95
	0.860	5460	6.91
	0.610	2750	4.90
	0.430	1370	3.45
	0.304	6820	2.44
	0.177	2310	1.42
80	0.990	1270	7.24
	0.860	9600	6.29
	0.610	4830	4.46
	0.430	2400	3.14
	0.304	1200	2.22
	0.177	4070	1.29
100	0.990	1660	6.92
	0.860	1250	6.01
	0.610	6310	4.26
	0.430	3140	3.01
	0.304	1570	2.13
	0.177	5310	1.24

4.7 Impact on soils/rocks

4.7.1 Introduction

The problem of earth penetration by a projectile continues to be tackled by the use of numerical and experimental techniques. At present, there is a growing need for a reliable evaluation of the depth of penetration, velocity and deceleration time histories. There is no doubt that the subject is highly complex as it involves the physical properties of impactors and soils/rocks, while using the acceleration time record. The study encompasses such disciplines as geology, soil mechanics, wave mechanics, dynamics of impactors and aerodynamics. An impactor can be the non-explosive type or can be generated by bombs and other detonators. Great care is taken to see that, at this stage, the reader does not confuse the two. Impact due to explosion is treated separately in this book. Figure 4.32 shows a typical earth penetrator.

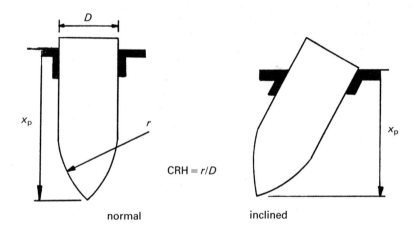

normal   inclined

$$CRH = r/D$$

Fig. 4.32 Penetration into earth (CRH = calibre radius head).

4.7.2 Empirical formulations for earth penetration

The depth of the penetration, x_p, and the soil resistance, R_T, are represented as functions of many parameters which are given below:

$$x_p = f(d, W, V_s, N, E, \sigma_{cs}) \tag{4.167}$$

$$R_T = c_0 + c_1 V_s + c_2 V_s^2 + \ldots + c_n V_s^n \tag{4.168}$$

where f = function

d, W, V_s, N, E = impactor/missile diameter, weight, velocity, nose-shape factor and Young's modulus of materials, respectively

σ_{cs} = the unconfined stress of the earth material

c_n = constants

The Robin's–Euler penetration, x_p, is computed as

$$x_p = mV^2/2c_0 \qquad (4.169)$$

Equation (4.169), in which m, the mass of the missile, has been modified to include additional constant terms of equation (4.168), is known as the Poncelet equation.

$$x_p = (m/2c_2) \log[1 + (c_2 V_s^2/c_0)] \qquad (4.170)$$

In equation (4.170) the soil resistance, R_T, is given by

$$R_T = c_0 + c_2 V_s^2 \qquad (4.171)$$

Sometimes the soil resistance is written as

$$R_T = c_1 V_s + c_2 V_s^2 \qquad (4.172)$$

The value of x_p is then written as

$$x_p = (m/c_2) \log[1 + (c_2 V/c_1)] \qquad (4.173)$$

The values of N for various shapes are given below:

Shape		N
Flat nose		0.56
Tangent ogive	(CRH* 2.2)	0.82
	(CRH 6)	1.00
	(CRH 9.25)	1.11
	(CRH 12.5)	1.22
Cone $L/d = 2$		1.08
$= 3$		1.32
Cone plus cylinder		1.28

* CRH = the calibre radius head = the radius of curvature of the nose shape/d.

For a closed impacting end, the value of N can be evaluated as follows:

$$N = 0.72 \ (CRH^{2.72}/1000) \qquad (4.174)$$

For a circular, hollow, pipe missile, N is written as

$$N = 0.72 + [(D/d)^2 - 1] \ (0.0306) \leq 1.0 \qquad (4.175)$$

where $d = \sqrt{(D^2 - D_i^2)}$
    D = outer diameter
    D_i = inner diameter

Young[2.29,2.30] carried out a thorough examination of equation (4.167) and arrived at the following expressions for x_p, the penetration distance in feet (for various velocities in ft/s) under high- and low-velocity conditions.

Low velocity

$$x_p = f_1(N)f_6(W/A)f_5(S)c_1 \log(1 + 2V_s^2 \times 10^{-5}) \qquad (4.176)$$

High velocity

$$x_p = f_1(N)f_6(W/A)f_5(S)c_2(V_s - 100) \qquad (4.177)$$

where S is a constant dependent upon soil properties averaged over a penetration distance.

The importance of the Young's equation is felt when the effects of weight and area are included.

$$f_6(W/A) = (W/A)^{1/2} \qquad (4.178)$$

The value of the soil's $f_5(S)$ is essential; however, its exact evaluation is a complicated affair. Various soil properties in this case will be related to some index of penetrability. Soil parameters are given earlier in this text. Twelve tests indicate that S lies in the range of 1 to 50, depending on the soil and its depth. The maxima for the low-velocity and high-velocity types are 220 ft/s (67 m/s) and 1200 ft/s (366 m/s), respectively. The maximum depth reached is 100 ft (30.5 m). The velocity effects are shown on the right-hand side of equations (4.176) and (4.177), respectively. The following equations give the final conclusions for equations (4.176) and (4.177) for x_p:

$$x_p = 0.53SN(W/A)^{1/2} \log(1 + 2V_s^2 \times 10^{-5})$$
$$V_s < 200\,\text{ft/s} \qquad (4.179)$$

$$x_p = 0.0031SN(W/A)^{1/2}(V_s - 100)$$
$$V_s \geq 200\,\text{ft/s} \qquad (4.180)$$

For *penetration into rock*, Maurer[2.12] and Rinehart[2.13] give the penetration as

$$x_p = K_1 (W/d)(V_s - k_2) \qquad (4.181)$$

where all notations have already been defined and K_1 and k_2 are constants defining the rock penetration resistance and constitutive materials. Tolch and Bushkovitch[2.14] have evaluated x_p for both large and small missiles striking soft rock:

$$x_p = (4.6W/d^{1.83})\bar{V} \qquad \text{small missiles}$$
$$x_p = (1.4W/d^{1.53})(\bar{V})^{1.8} \qquad \text{large missiles} \qquad (4.182)$$

where $\bar{V} = 0.001V_s$.

Rinehart[2.13,2.18,2.22,2.23] and Palmore[2.17] suggest the penetrating value of the less compact soil to be

$$x_p = (K_1W/d^2) \log_e(1 + K_2V_s^2) \qquad (4.183)$$

where K_1 and K_2 are constants. Figure 4.33 shows the experimental data given by Rinehart[2.13] and Tolch and Bushkovitch[2.14] for steel spherical missiles penetrating different soils.

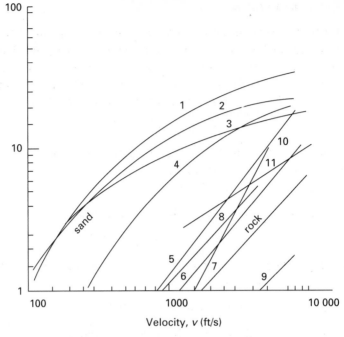

Velocity, *v* (ft/s)

Curve	Missile	Soil model
1	4.3 in cannon balls	Sand and gravel
2	Many sizes high ballistic density	Sand
3	0.356 in steel balls	Loose sand
4	0.356 in steel balls	Weakly cemented sand
5	Many	Rock—250 psi crushing strength
6	Many	Rock—2000 psi crushing strength
7	Many	Rock—10 000 psi crushing strength
8	$\frac{7}{16}$ in steel balls	Red sandstone—2780 psi
9	$\frac{9}{32}$ in and $\frac{7}{16}$ in steel balls	Granite—16 500 psi
10	High ballistic density	Soft rock
11	High ballistic density	Rock or sand—hypervelocity cratering

Fig. 4.33 Steel missiles penetrating different soils.

Where the missiles are not spherical and are not made of steel an appropriate value of the nose factor N is used and an effective value of density, ρ_{se}, should replace the steel density, ρ_s, by

$$\rho_{se} = W/(\pi/6)\rho_s d^3 \qquad (4.184)$$

A reasonable value for x_p can be derived from

$$x_p = 0.01 d\rho_{se} V_s \qquad \text{for sandy soils}$$

$$x_p = 0.001 d\rho_{se} V_s \qquad \text{for soft rocks} \qquad (4.185)$$

$$x_p = 0.004 d\rho_{se} V_s \qquad \text{for hard rocks}$$

4.7.3 Velocity and deceleration

Soils are composed of several layers. A velocity in soils depends, therefore, on the thickness and type of the constituent layers. When x_p is greater than the topmost layer, the residual velocity after this layer must be determined and used in the above equations to obtain the penetration in the second layer, and so on. Based on the x_p/d ratios, such a residual velocity is calculated as:

$$V_{R2} = (V_s^{1.25} - \tilde{V}_{s1}^{1.25})^{1/1.25} \tag{4.186}$$

where V_{R2} = the velocity at the beginning of the second layer
V_s = missile velocity at the impact level
$\tilde{V}_{s1}$ = velocity to perforate the first layer

In this way the final velocity, V_{Rn}, will be determined from

$$V_{Rn} = (V_{R(n-1)}^{1.25} - \tilde{V}_{R(n-1)}^{1.25})^{1/1.25} \tag{4.187}$$

Kar[2.2] suggested the velocity–time history of the missile in the earth material to be

$$V = V_s \exp(-\rho_s A C_s t/m) \tag{4.188}$$

where C_s is an impedance/damping constant and t is time.

Theoretical values of V_s may also be computed from other researchers' data.[2.2–2.30] The common ones are

$$V_s = 1.56\rho_{se}\, fd \qquad \text{for sand}$$

$$V_s = 0.156\rho_{se}\, fd \qquad \text{for soft rock} \tag{4.189}$$

$$V_s = 0.0624\rho_{se}\, fd \qquad \text{for hard rock}$$

where f is frequency in cycles per second and V_s is in feet per second.

The deceleration is given by

$$G_a = \dot{V} = V_s \exp(-\rho_s A C_s t/m)[-\rho_s A C_s/m](\beta) \tag{4.190}$$

Kar[2.2] proposed the value of the target flexibility to be

$$\beta = 0.5924(k_{\text{target}}/k_{\text{missile}})^{0.16055} \leq 1.0 \tag{4.191}$$

where k_{target} is the stiffness of the earth under impact and k_{missile} is the stiffness of the missile.

The deceleration is assumed to start at time $t = 0$ and increases linearly to a peak at time t given by

$$t = 4x_p/V_s^{1.25} \tag{4.192}$$

Kar[2.2] used the penetration formula (developed for a concrete target) for soil. Figure 4.34 shows the experimental results for penetration.

For low-impact velocities, an average value for the deceleration can be computed as follows:

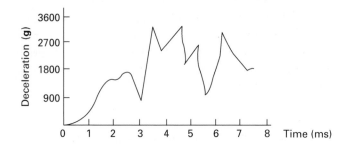

Fig. 4.34 Deceleration−time curves for penetration into limestone.

$$(G_a)^* = 50V_s/d\rho_{se} \qquad \text{for sand}$$

$$(G_a)^* = 500V_s/d\rho_{se} \qquad \text{for soft rock} \qquad (4.193)$$

$$(G_a)^* = 1250V_s/d\rho_{se} \qquad \text{for hard rock}$$

where $(G_a)^*$ is in ft/s^2.
The peak value of $(G_a)^*$ is given by

$$G_a = 1.5(\rho_{se}CV_s/\rho_s d\rho_{se}) \qquad (4.194)$$

where C is the impedance in lb/in^2/ft per second. For basalt the value of C is 750 and for pumice it is 124.

4.7.4 Impact on rock masses due to jet fluids

Section 2.2.3 gives a brief introduction to the impact problems when rock masses are subject to jet fluids. Such pressures are associated with thermal hydraulic and mechanical processes. In practice there is no control over the formation of the deep underground rock masses. The water injected into the formula is done in a relatively controlled way. Many processes can be described and simulated by mathematical modelling. The finite-element, boundary-element and finite-difference processes are just some of them. In this section the fluid motion and the heat transfer associated with it are represented by suitable transport equations covering the movement of water in fissured rock masses. To begin with, a one-dimensional flow channel subject to laminar flow is considered, and the velocity v is given by

$$v = -(TR/\mu_f)\, \partial/\partial z\, (P + \rho gx) \qquad (4.195)$$

$$= -\rho g(TR/\mu_f)\, \partial h/\partial z$$

$$\text{The momentum of the jet fluid} = \partial/\partial t\, (\rho A) \qquad (4.196)$$

$$= -\partial/\partial x\, (\rho Av) + i_f$$

where TR = transmissibility of the fissure (m^2)
μ_f = jet fluid viscosity (kN/m/s)
P = jet fluid impact (kN/m^2)
z = distance along the flow (m)
x = the vertical height (m)
A = the cross-sectional area = $x_p \times$ width
x_p = penetration depth (m)
v = velocity
i_f = injection/extraction of the fissure (kN/m/s)

Differentiation of the left-hand side of equation (4.196) gives:

$$\partial(\rho A)/\partial t = \rho(\partial A/\partial t) + A(\partial \rho/\partial t) \tag{4.197}$$
$$= \rho(\partial A/\partial P)(\partial P/\partial t) + A(\partial P/\partial \rho)(\partial \rho/\partial t)$$
$$= \rho A \beta_1(\partial P/\partial t) + \rho A \beta_2(\partial P/\partial t)$$
$$= \rho A \beta(\partial P/\partial t)$$

$$\text{where } \beta = \beta_1 + \beta_2 \tag{4.197a}$$
$$= (1/A)(\partial A/\partial P) + (1/\rho)(\partial f/\partial P)$$

β_1 and β_2 indicate the compressibility of the fluid and the compliance of the fractures which consist of shear and normal compliance terms. The value of the transmissibility (also known as transmissivity in geology) is given by

$TR = x_p^2/12$ for rocks with parallel edges and with penetration
in between (4.197b)

$TR = r^2/8$ for rock masses of cylindrical shape with hydraulic radius r

Hopkirk and Rybach[5.70] have carried out tests on several simulated holes while naming holes as RH12 and RH15. Tests were carried out for stress field directions of fissures and their spacings and geochemical information. Water was injected at the following rates:

Project I
2.55 litres/s for 2.5 hours   Total 30 hours
3.10 litres/s for a further 19.6 hours   for RH15 pressure
1.80 litres/s for a further 7.9 hours   monitored in RH12

Project II
5.4 litres/s for a period of 81 hours   Total 81 hours
for RH12 pressure
monitored in RH15

Figure 4.35 shows a comparative study of the pressure-time history for the two tests. These results may act as input data for more sophisticated studies.

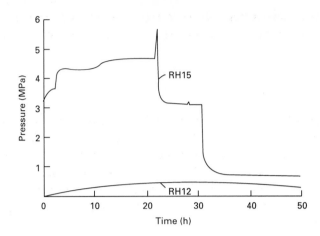

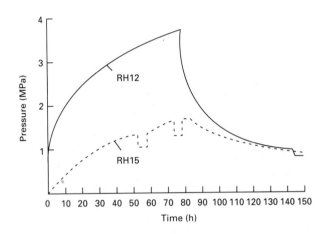

Fig. 4.35 Pressure load−time history in rock masses due to jet fluids. (a) Project I; (b) Project II.

4.8 Impact on water surfaces and waves

4.8.1 Introduction

The determination of hydrodynamic forces experienced by a body entering a water surface is an extremely difficult task since the actual water surface is not planar and is constantly changing, thus making the impact parameters statistical in nature. One has to know statistical data on wave heights, wave lengths and particle velocities. Moreover, the statistical data are influenced by whether the impacting body is smaller or larger than the wave dimensions. When the impacting body is smaller, the deceleration will be a function of small-scale wavelets. The shape of the water surface for impact or for vessel collision at sea may be selected which results in the highest deceleration, or it may be evaluated by adopting a statistical approach. A comprehensive body of work is reported in the text:[2.232–2.342]

4.8.2 Impact on water surfaces

Various theories exist on the relationship between wave heights/depths and wave lengths and the wind velocities which generate the waves. Figure 4.36 shows a relationship between sustained wind speed and significant wave height in severe storms. The water surface is not always smooth, as shown in Fig. 4.37 (a) and (b). One of the popular methods for computing the wave spectrum is the JONSWAP method. The spectral density of the water surface elevation $S_{\eta\eta}$ (f) from this spectrum is given below:

$$S_{\eta\eta} \ (f) = (\alpha g^2/(2\pi)^4 f^5) \ \exp[-\tfrac{5}{4} \ (f/f_m)^{-4}] \bar{\gamma}$$
$$\times \exp[-(f-f_m)^2/2\sigma_s^2 f_m^2] \tag{4.198}$$

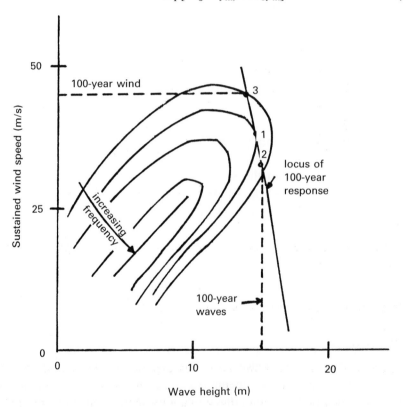

Fig. 4.36 Sustained wind speed versus significant wave heights. Joint frequency of occurrence contours for wind speed and significant wave height in severe storms are shown. Point 1 is the most likely wind/wave combination to produce the 100-year response. Point 2 is the combination of 100-year waves and associated wind producing the 100-year response. Point 3 is the combination of the 100-year wind and associated waves producing the 100-year response.

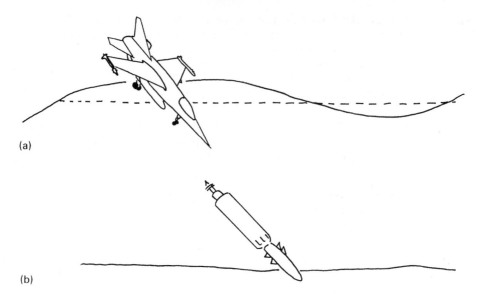

(a)

(b)

Fig. 4.37 (a) Impact of aircraft and (b) missile on sea surface.

where $\alpha = 0.046 X_f^{-0.22} / V_{w(10)}^2$ or $0.0662(f_m/2.84)^{2/3}$

       = PM constant

     $f_m = 16.04/[X_f V_{w(10)}]^{0.38}$

   $V_{w(10)}$ = wind velocity in m/s at 10 m above still water

      $\bar{\gamma} \approx 3.3$ for the North Sea

      σ_s = constant based on the scatter of the data for which the spectrum is

          derived $= \sigma_a$ for $f \le f_m = 0.07$

                $= \sigma_b$ for $f > f_m = 0.09$

    X_f = fetch in m

     f = frequency in Hz around 1.5 Hz (beyond it the energy is significant)

     g = acceleration due to gravity

Table 4.6 gives data on the JONSWAP spectrum for a specific $f = 1.5$ Hz.

$$\sigma_{\eta\eta} = \int_0^{f=1.5} S_{\eta\eta}(f)\mathrm{d}f \tag{4.199a}$$

Figure 4.38 gives the wave spectrum. Based on results from the Associated Petroleum Institute (API), typical values for the Gulf of Mexico are given in Table 4.7. The most significant finding is that the design wave shall rely on the processes such as wave motion, the probability density function of the wave height and the square of the wave period known as the Raleigh density function. Hence within each state the probability distribution functions are

$$P(H) = 1 - \exp[-2(H/H_s)]$$

$$P(T) = 1 - \exp[-0.675T/\bar{T}] \tag{4.199b}$$

Table 4.6 Data for the JONSWAP spectrum.

significant wave height, H_s	12.5 m
spectral peak period, T_p	13.5–16.5 s
wave spectrum, $S_{\eta\eta}(f)$	JONSWAP, $\bar{\gamma} = 2$
maximum crest elevation, η	13 m
wind speed at 40 m elevation μ_f	
1 h average	52 m/s
1 min average	60 m/s
3–5 s gust	75 m/s
wind spectrum, $S_\mu(f)$	'blunt'
wind speed variance, σ^2	75 m²/s²
spectral peak period, T_p	20–200 s
wind direction	aligned with waves
current profile	
above 75 m depth	1 m/s
below 75 m depth	0.1 m/s
current direction	aligned with wind
storm tide	1 m
marine growth thickness	3 cm over top 50 m
marine growth roughness height	1.5 cm over top 50 m

the JONSWAP wave spectrum is given by equation (4.198)

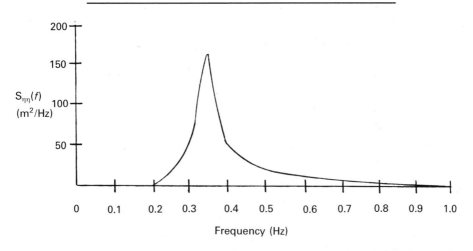

Fig. 4.38 The JONSWAP wave spectrum.

where H and H_s are the individual and significant wave heights, respectively, and T and $\bar{T}$ are the individual and mean periods, respectively.

Where the impact is on shallow water, Stokes and linear wave theories become numerically unstable since it becomes less and less sinusoidal. The crest becomes more pointed and troughs flatten. In this case, *solitary wave theory* is used and is independent of the wave length of the period of the wave.

Table 4.7 Wave frequency in the North Atlantic Ocean for a whole year.

Wave height (m)	Wave period (s)								Sum over all periods
	5	7	9	11	13	15	17		
0.75–	20.91	11.79	4.57	2.24	0.47	0.06	0.00	0.60	40.64
1.75–	72.78	131.08	63.08	17.26	2.39	0.33	0.11	0.77	287.80
2.75–	21.24	126.41	118.31	30.24	3.68	0.47	0.09	0.56	301.00
3.75–	3.28	49.60	92.69	32.99	5.46	0.68	0.12	0.27	185.09
4.75–	0.53	16.19	44.36	22.28	4.79	1.14	0.08	0.29	89.66
5.75–	0.12	4.34	17.30	12.89	3.13	0.56	0.13	0.04	38.51
6.75–	0.07	2.90	9.90	8.86	3.03	0.59	0.08	0.03	25.46
7.75–	0.03	1.39	4.47	5.22	1.93	0.38	0.04	0.04	13.50
8.75–	0.00	1.09	2.55	3.92	1.98	0.50	0.03	0.02	10.09
9.75–	0.00	0.54	1.36	2.26	1.54	0.68	0.20	0.04	6.62
10.75–	0.01	0.01	0.10	0.11	0.10	0.05	0.02	0.00	0.40
11.75–	0.00	0.00	0.03	0.08	0.17	0.06		0.00	0.34
12.75–		0.05	0.00	0.14	0.22	0.06	0.01		0.48
13.75–		0.02		0.07	0.09	0.03		0.01	0.22
14.75–				0.02	0.06	0.02	0.00	0.01	0.11
15.75–	0.00	0.02	0.00	0.02	0.02	0.02	0.01	0.01	0.08
Sum over all height	118.97	345.43	358.72	138.59	29.05	5.63	0.92	2.69	1000.00

$$\text{The surface profile} = H[\text{sech } \sqrt{(3H/4d^2)}x - ct]^2 \qquad (4.199c)$$

where H = wave height

d = water depth

c = wave celerity = $\sqrt{[g(H + d)]}$

x = distance from crest

t = time

g = acceleration due to gravity

Figure 4.39 gives a comparative study of wave heights for depths in different seas. It is important to determine the wave profile correctly prior to impact. The correct evaluation of the impact depends on whether or not the surface is planar. For example, when the impact occurs due to a projectile of any shape upon a water surface, the area of contact between body and water surface spreads faster than the speed of sound, U_0, which is the shock wave front velocity. The impact load causing pressure P for the impactor velocity V is given by

$$P = \rho_\omega U_0 V \qquad (4.200)$$

where ρ_ω is the uncompressed density of water.

By introducing a stagnating pressure of $\frac{1}{2}\rho_\omega V^2$, Kirkwood and Montroll[5.71] suggest the following interface pressure for a non-rigid body:

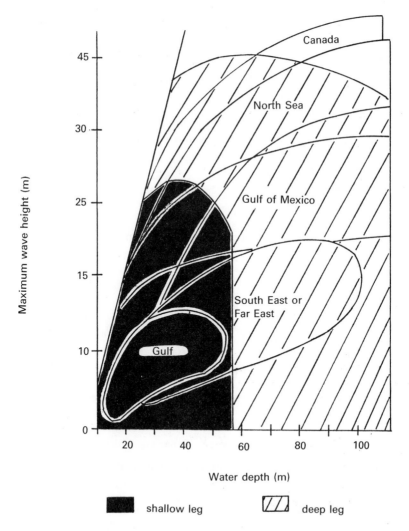

Fig. 4.39 Design conditions found in common offshore areas.

$$P = \frac{\rho_\omega U_0 V}{1 + \dfrac{\rho_\omega V}{\rho_b U_b}} \tag{4.201}$$

where $\rho_b U_b$ = the impedance
U_b = the stress wave front velocity

The particle velocity in a body and interface velocity are given as follows:

$$\text{particle velocity in body} = V_a = P/\rho_b U_b$$
$$\text{interface velocity} \qquad = V - V_a \tag{4.202}$$

For a rigid body, the pressure P (for an impact velocity) above 300 psi is given by

$$P = 2.17 \rho_\omega V^2 (U_0/V)^{0.73} \tag{4.203}$$

The shapes of the waves and of the missiles/aircraft must be taken into consideration when modifications are carried out to equations (4.201) to (4.203). The deceleration and velocity changes when the body enters the water have to be computed. It is important to include the shape factor known as the ballistic density factor, γ_{bd}, which is given by

$$\gamma_{bd} = \frac{\text{mass } m \text{ of the impacting body} \times \mathbf{g}}{\text{weight of sphere of water of the same radius } r} \tag{4.204}$$

$$= mg/\tfrac{4}{3}\pi\rho_\omega r^3$$

the initial acceleration is given by

$$\ddot{x} = 0.75 U_0 V/r\gamma_{bd} \tag{4.205}$$

When the rare faction (compression reflected) wave moves in from the outside, the value of the new $\ddot{x}_{ne}$ is given by

$$\ddot{x}_{ne} = \ddot{x}[1 - (U_0 t/r)] \tag{4.206}$$

The velocity loss ΔV is computed as

$$\Delta V = \int_{t=0}^{t=R/U_0} \frac{0.75 U_0 V}{R\gamma_{bd}} \left(1 - \frac{U_0 t}{r}\right) dt \tag{4.207}$$

$$= V(1 - e^{-\gamma_{bd}/4}) \tag{4.207a}$$

Shiffman and Spencer (pers. comm.) suggest that deceleration may be calculated as

$$\ddot{x} = \tfrac{3}{8} C_p V^2/r\gamma_{bd} \tag{4.208}$$

where C_p = the drag coefficient $\qquad$ (4.209)
$$= F_1(t)/(\pi r^2)(\tfrac{1}{2}\rho_\omega V^2)$$

Figures 4.40 and 4.41 show data for a smooth water surface subject to a spherical missile. Figures 4.42 and 4.43 illustrate data for cone-shaped missiles with the JONSWAP spectrum.

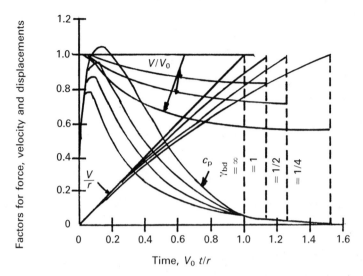

Fig. 4.40 Force, velocity and displacement versus time for spheres entering water (Shiffman & Spencer, personal communication).

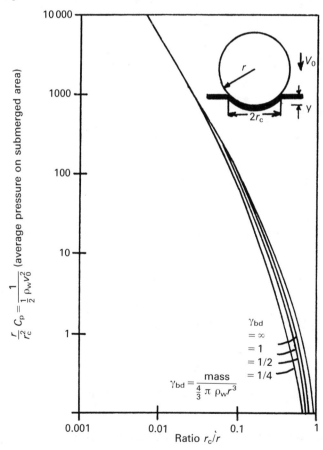

Fig. 4.41 Pressure on a sphere for normal entry into water (Shiffman & Spencer, personal communication).

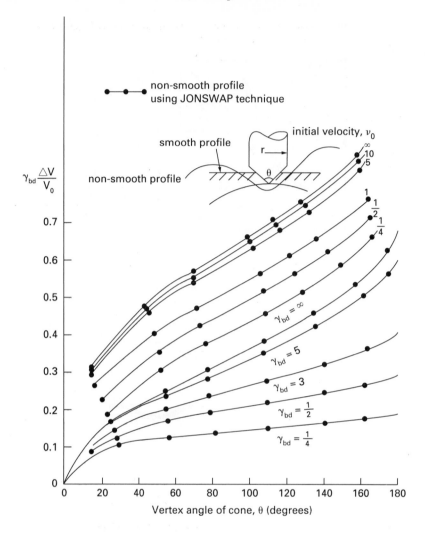

Fig. 4.42 Loss of velocity for cone-shaped missiles.

4.8.3 Impact on ocean surfaces

The high risk of collision between sea vessels themselves or between sea vessels and offshore installations has generated the need for accurate predictions of their responses to impact forces, F_I, which are transient in nature. When a sea vessel with speed V collides non-centrally with a platform at rest, the loss of kinetic energy is given by

$$E_K = \tfrac{1}{2}m_{b'}(v_{b'} \sin\theta)^2 - \tfrac{1}{2}(m_{a'} + m_{b'} + C_m)V^2 \tag{4.210}$$

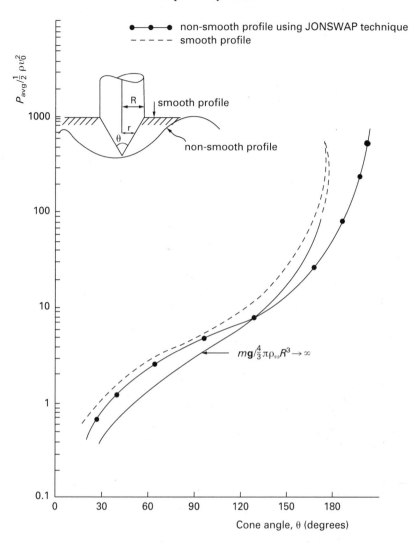

Fig. 4.43 Pressure coefficients during water entry of cones (Shiffman & Spencer, personal communication).

where C_m = added mass $0 < C_m < 1$

$m_{a'}$, $m_{b'}$ = the masses of the two ships

$v_{b'}$ = the velocity of ship b'

θ = the angle of the ship b' to the horizontal where ship a' is positioned

For the linear momentum normal to the centreline of $m_{a'}$ to be conserved

$$m_{b'}v_{b'}\sin\theta = (m_{a'} + m_{b'} + C_m)V \qquad (4.211)$$

From equations (4.210) and (4.211)

$$E_K = (m_b/2)K(v_{b'} \sin\theta)^2 \qquad (4.212)$$

$$\text{where } K = \frac{m_a + C_m}{m_{a'} + m_{b'} + C_m} = \frac{1}{1 + M/(1 + C)} \qquad (4.212a)$$

$$M = m_{b'}/m_{a'}$$

If $K \to 1$ and $m_{b'} < m_{a'}$ or $M \to 0$

$M \to \infty$; $K \to 0$

If the striking ship is of relatively small mass, the energy loss would be greater and vice versa.

In terms of vessel displacements, the kinetic energy loss is given by

$$E_K = \frac{\delta_{a'} \, \delta_{b'}}{1.43\delta_{b'} + 2\delta_{a'}} (v_b \sin\theta)^2 \qquad (4.213)$$

where $\delta_{a'}$ and $\delta_{b'}$ are the displacements of the vessels a' and b', usually in tonnes, and v_b is in knots. In terms of the *resistance factor*, R_T, measured in $m^2 - mm$, Minorsky gives

$$R_T = 1.33\Sigma d_n l_n \bar{t}_n + \Sigma d_N l_N \bar{t}_N \qquad (4.214)$$

where d_n, d_N = depth of damage in the nth member of the struck vessel and the striking vessel, respectively

l_n, l_N = length of damage in the nth member of the struck vessel and the striking vessel, respectively

$\bar{t}_n, \bar{t}_N$ = thickness of the nth member of the struck vessel and the striking vessel, respectively

The energy loss can be rewritten as

$$E_K = 233.8\bar{R}_{TS} + 175.8\bar{R}_{Tb} + 124\,000 \qquad (4.215)$$

where $\bar{R}_{TS} = \Sigma d_n l_n \bar{t}_n$ = resistance factor for the struck ship

$\bar{R}_{Tb} = \Sigma d_N l_N \bar{t}_N$ = resistance factor for the striking ship

The added mass C_m can be determined by potential flow theory. The added mass force F_{ad} is expressed as

$$F_{ad} = \rho_\omega C_m \pi a^2 \times \ddot{x} \qquad (4.216)$$

where ρ_ω = fluid density

$\ddot{x}$ = ship acceleration

a = a dimension given for various shapes by Det Norske Veritas (DNV) in Tables 4.8 and 4.9.

Where three-dimensional dynamic final element analysis is to be carried out[3.1–3.168] then wave-induced forces, the hydrodynamic reaction forces and the restoring forces[2.232–2.249] should be determined for each element. Summing up these forces and the inertia properties, the displacements, velocities and acceleration of the vessel can be determined from the equation of motion given below

$$\left[\sum_{i=1}^{n1} M_i + \sum_{j=1}^{n2} \bar{C}_\mathrm{m}\right]\ddot{x} + \left[\sum_{j=1}^{n2} C_j\right]\dot{x} + \left[\sum_{j=1}^{n2} F1_j + \sum_{i=1}^{n1} F2_i\right]\bar{x} = \sum_{j=1}^{n2} \bar{F}_j$$

(4.217)

where i = mass point number

$n1$ = total number of mass points

j = hydro element number

$n2$ = total number of hydro elements

M_i = inertia matrix of nodal point i

C_m = added mass matrix of hydro element j

C_j = damping matrix of hydro element j

$F1_j$ = restoring force matrix (due to buoyancy force contributions) of hydro element j

$F2_i$ = restoring force matrix (due to inertia force contributions) of mass point i

$\bar{F}_j$ = wave-induced forces of hydro element j

Tables 4.8 and 4.9 give the added mass coefficient, C_m.

Table 4.8 Added mass coefficients (C_m) for two-dimensional bodies. (Courtesy of Det Norske Veritas.)

Section through body	C_m	Section through body	C_m
◯ (2a)	1.0	$d/a = 0.05$ / $d/a = 0.10$ / $d/a = 0.25$ (2a)	1.61 / 1.72 / 2.19
⬭ (2a)	1.0	$a/b = 2$ / $a/b = 1$ / $a/b = 1/2$ / $a/b = 1/5$	0.89 / 0.76 / 0.67 / 0.61
⬯ (2a)	1.0	fluid, wall	2.29
▭ (2a)	1.0		
$a/b = \infty$ / $a/b = 10$ / $a/b = 5$ / $a/b = 2$ / $a/b = 1$ / $a/b = 1/2$ / $a/b = 1/5$ / $a/b = 1/10$ (2b, 2a)	1.0 / 1.14 / 1.21 / 1.36 / 1.51 / 1.70 / 1.98 / 2.23	h, $2a$	$1+\left(\dfrac{h}{2a} - \dfrac{2a}{h}\right)^2$

Table 4.9 Added mass coefficients (C_m) for three-dimensional bodies. (Courtesy of Det Norske Veritas.)

Body shape	C_m		V
1. Flat plates Circular disc 2a	0.64		$\frac{4}{3}\pi a^2$

Elliptical disc	b/a	C_m	$\frac{\pi}{6}a^2 b$
	∞	1.0	
	12.75	0.99	
	7.0	0.97	
	3.0	0.90	
	1.5	0.76	
	1.0	0.64	

Rectangular plates	b/a	C_m	$\frac{\pi}{4}a^2 b$
	1.0	0.58	
	1.5	0.69	
	2.0	0.79	
	3.0	0.83	
	∞	1.00	

Triangular plates	$\frac{1}{\pi}(\tan\theta)^{3/2}$		$\frac{a^3}{3}$

2. Bodies of revolution Spheres 2a	0.5		$\frac{4}{3}\pi a^2$

Ellipsoids		C_m		
	b/a	Axial	Lateral	$\frac{4}{3}\pi a^2 b$
lateral	1.5	0.30	0.62	
	2.0	0.21	0.70	
	2.51	0.16	0.76	
	3.99	0.08	0.86	
	6.97	0.04	0.93	
	9.97	0.02	0.96	

3. Square prisms	b/a	C_m	$a^2 b$
	1	0.68	
	2	0.36	
	3	0.24	
	4	0.19	
	5	0.15	
	6	0.13	
	7	0.11	
	10	0.08	

4.8.4 Wave impact on rock slopes and beaches

4.8.4.1 Introduction

The wave impact phenomenon is complex as it involves, apart from the influence of the wave spectrum and storm duration discussed in section 4.8.2, static and dynamic slope instability of porous and non-porous types of beaches, distortion of rubble mound revetments and the formation of a profile with damaged zones. The characteristics of rocks under impact are discussed in Chapter 2. Waves can be of plunging and surging types and can easily be influenced by parameters such as height, period and the slope angle. In this section only well known work is reported.

4.8.4.2 Wave impact on a dynamically stable profile

Prior to impact force evaluation, it is essential to define the dynamically stable profile. Powel[5.72] as shown in Fig. 4.44, defined the profile by two power curves. The upper curve starts at the crest and the profile defines the run-up and run-down area, up to the transition to the steep part (line with angle β). The lower curve starts at the transition and describes the step. The complete profile is then shown in the figure with a further definition of a crest height, a length for the upper curve and a depth for the lowest point of incipient motion. For evaluation of static stability, various formulae have been given in the report provided by the International Commission for the Study of Waves, PIANC[5.73,5.74] Most stability formulae are in agreement, but for armour units and block revetments the buoyant mass of the stone, W' (Fig. 4.45) is given by

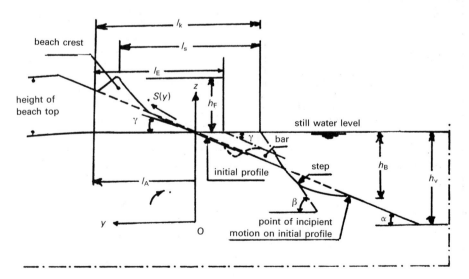

Fig. 4.44 Model for a dynamically stable profile[2.396]

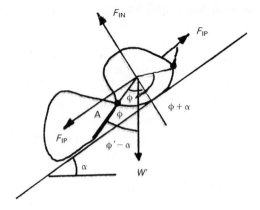

Fig. 4.45 Schematization of incipient instability.

$$W' = (\rho_a - \rho_\omega)d_{n50}^3 \qquad (4.218)$$

where ρ_ω = mass density of water

ρ_a = mass density of stone

d_{n50}^3 = nominal diameter with index n of 50% value of sieve curve

The impact load $F_1(t)$ is schematized by two forces, one parallel to the slope, $F_{IP}(t)$, and the other normal to the slope, $F_{IN}(t)$. If the value of $F_1(t)$ is given by

$$F_1(t) = \rho_\omega g C d^2 H \qquad (4.219)$$

where $F_1(t)$ = impact due to wave

C = coefficient

d = diameter of stone

then

$$F_{IN}\sin\phi\,\frac{d}{2} + F_{IP}\cos\phi\,\frac{d}{2} = gW'\sin(\phi - \alpha)\frac{d}{2} \qquad (4.220)$$

where ϕ = angle of repose

α = angle of seaward slope

Assuming a coefficient C_1 for the normal wave force, $F_{IN}(t)$, a coefficient C_2 for the parallel wave force, $F_{IP}(t)$, and assuming $d = Kd_{n50}$ (where K is a coefficient), then

$$\rho_\omega g C_1 d_{n50}^3 H \sin\phi K^3/2 + \rho_\omega g C_2 d_{n50}^3\, H\cos\phi K^3/2 = g(\rho_a - \rho_\omega)d_{n50}^4 \sin(\phi - \alpha)K^4/2 \qquad (4.221)$$

Equation (4.221) can be written as

$$H/\Delta d_{n50} = K\sin(\phi - \alpha)/(C_1\sin\phi + C_2\cos\phi) \qquad (4.222)$$

where $\Delta = (\rho_a - \rho_\omega)/\rho_\omega$

By defining the friction coefficient, μ, as $\mu = \tan\phi$, equation (4.222) can be finally rewritten as

$$H/\Delta d_{n50} = K(\mu\cos\alpha - \sin\alpha)/(\mu C_1 + C_2) \tag{4.223}$$

The wave steepness, S, using the deep water wave length $L = \mathbf{g}T^2/2\pi$, becomes

$$S = 2\pi H/\mathbf{g}T^2 \tag{4.224}$$

The value of S is then related to the slope angle of the slope of the stone beach by

$$S = (\tan\alpha/\bar{\bar{\xi}})^2 \tag{4.225}$$

where $\bar{\bar{\xi}}$ is a similarity parameter with the mean value $\bar{\bar{\xi}}_m$.

The significant wave height, H_s, and the average period, T_m, can both be calculated from the spectrum as given in section 4.8.2. For the *plunging wave* region, there is a functional relationship between $H_s/\Delta d_{n50}$ and $\bar{\xi}_n$ and they are given by[5.76-5.90]

$$\text{impermeable core} \quad H_s/\Delta d_{n50} \times \sqrt{\bar{\bar{\xi}}_m} = 4.1(S/\sqrt{N})^{0.2} \tag{4.226}$$

$$\text{permeable core} \quad H_s/\Delta d_{n50} \times \sqrt{\bar{\bar{\xi}}_m} = 5.3(S/\sqrt{N})^{0.2} \tag{4.227}$$

$$\text{homogeneous core} \quad H_s/\Delta d_{n50} \times \sqrt{\bar{\bar{\xi}}_m} = 5.7(S/\sqrt{N})^{0.2} \tag{4.228}$$

where N is the number of waves.

For surging waves, equations similar to (4.226) to (4.228) have been derived:[5.77-5.110]

$$\text{impermeable core} \quad H_s/\Delta d_{n50} = 1.35(S/\sqrt{N})^{0.2} \sqrt{\cot\alpha(\bar{\bar{\xi}}_m^{0.1})} \tag{4.229}$$

$$\text{permeable core} \quad H_s/\Delta d_{n50} = 1.07(S/\sqrt{N})^{0.2} \sqrt{\cot\alpha(\bar{\bar{\xi}}_m^{0.5})} \tag{4.230}$$

$$\text{homogenous structure} \quad H_s/\Delta d_{n50} = 1.10(S/\sqrt{N})^{0.2} \sqrt{\cot\alpha(\bar{\bar{\xi}}_m^{0.6})} \tag{4.231}$$

When the permeability coefficients are included, the final $H_s/\Delta d_{n50}$ relationships for both plunging and surging waves are written as, respectively,

$$H_s/\Delta d_{n50} \times \sqrt{\bar{\bar{\xi}}_m} = 6.2p^{0.18} (S/\sqrt{N})^{0.2} \tag{4.232}$$

$$H_s/\Delta d_{n50} = 1.0p^{-0.13} (S/\sqrt{N})^{0.2} \sqrt{\cot\alpha(\bar{\bar{\xi}}_m)^p} \tag{4.233}$$

where p is the permeability factor.

The complete range of dynamic stability, $H_s/\Delta d_{n50}$, can be covered by the combined wave height−period, H_0T_0, as

	Breakwater	Rock slopes and beaches
$H_s/\Delta d_{n50}$	1 to 4	6 to 20
HT	< 100	200 to 1500

It is important to define various heights and lengths as they influence the impact load:

(a) Run-up length, l_r:
$$H_0 T_0 = 2.9(l_r/d_{n50} \, N^{0.05})^{1.3}$$

(b) Crest length, l_c:
$$H_0 T_0 = 21(l_c/d_{n50} \, N^{0.12})^{1.2}$$

(c) Step length, l_s:
$$H_0 T_0 = 3.8(l_s/d_{n50} \, N^{0.07})^{1.3}$$

(a) Crest height, h_c:
$$h_c/H_s N^{0.5} = 0.089(s_m)^{-0.5} \tag{4.234}$$

(b) Step height, h_s:
$$h_s/H_s N^{0.07} = 0.22(s_m)^{-0.3} \tag{4.235}$$

(c) Transition height, h_t:
$$h_t/H_s N^{0.04} = 0.73(s_m)^{-0.2} \tag{4.236}$$

The slope, $\tan \gamma$, is given by

$$\tan \gamma = 0.5 \tan \alpha \tag{4.237}$$

The slope, $\tan \beta$, is given by

$$\tan \beta = 1.1 \tan \alpha^A \tag{4.238}$$

where $A = 1 - 0.45 \exp(-500/N)$.

For an *oblique wave* impact, all the above parameters are reduced by $\cos\phi$, except the crest length, l_c.

The number of waves impacting on the structure influence the damage. Thompson and Shuttler[5.110] produced results from five long duration tests with N up to 15 000. Table 4.10 gives the relationship between the number of waves, N, and damage, $S(N)/S(5000)$, which means all damage is related to the final damage after 5000 waves. The standard deviation for the ratio $S(N)/S(5000)$ in the region $N = 1000$ to 5000 is about 0.1 and is independent of the number of waves. A function that meets this requirement is given by

$$f(S) = a[1 - \exp(-bN)] \tag{4.239}$$

where a and b are curve-fitting coefficients and are found to be 1.3×10^{-4} and 3×10^{-4} respectively.

The influence of the storm duration on stability for the whole range of N is given by the following equation:

$$f(S) = S(N)/S(5000) = 1.3[1 - \exp(-3 \times 10^{-4}N)] \tag{4.240}$$

The damage due to wave impact is limited to 1.3 times the damage after $N = 5000$. Table 4.10 gives results for various waves and the corresponding damage. The influence of the wave period on the damage due to wave impact for

Table 4.10 Relationship between the number of waves (N) and the damage ($S(N)/S(5000)$).

N	$S(N)/S(5000)$
0	0
5 000	1
7 500	1.15
10 000	1.25
12 750	1.27
15 000	1.30

both plunging and surging waves can be assessed using the follow equations developed by Pilarczyk and Den Boer:[5.99]

Plunging waves

$$H_s/\Delta d_{n50} = 2.25(\bar{\bar{\xi}})^{-0.5} \, (\mu\cos\alpha + \sin\alpha) \tag{4.241}$$

Surging waves

$$H_s/\Delta d_{n50} = 0.54 \, \sqrt{\cot\alpha}[(\bar{\bar{\xi}})^{0.5} \, (H/L)^{-0.25}] \times (\mu\cos\alpha + \sin\alpha) \tag{4.242}$$

The results of these are given in Table 4.11. The influence of permeability for various values of S is also given for homogeneous, permeable and impermeable structures for damage levels $S = 3$ and $S = 8$. For wave impact analysis, other useful parameters are given in Table 4.12.

Table 4.11 Relationship between $H_s/\Delta d_{n50}$ ($\mu\cos\alpha + \sin\alpha$) and $\tan\alpha/\sqrt{(H/L)}$.

$\bar{\bar{\xi}}$	Cotα					S					
				K		3			8		
						H_0	P_e	I_m	H_0	P_e	I_m
	1.5	2.5	3.5	5.0	10.0						
0.6	2.80	2.80	2.80	2.80	2.80						
0.7	2.70	2.70	2.70	2.70	2.70						
0.8	2.50	2.50	2.50	2.50	2.50						
1	2.20	2.20	2.20	2.20	2.20						
2	1.65	1.65	1.65	1.80	2.25	2.25	2.20	1.60	2.70	2.60	2.00
3	1.65	1.70	1.80	2.20	3.00	1.80	1.70	1.30	2.25	2.20	1.70
4	1.68	1.80	2.10	2.30	—	1.75	1.65	1.40	2.40	2.20	1.50
5	1.69	1.80	2.20	2.50	2.80	2.30	2.00	1.25	2.60	2.40	1.55
6	1.70	2.20	2.30	—	—	—	—	1.25	—	—	1.50
7	1.80	2.30	—	—	—	—	—	1.30	—	—	1.70
8	1.90	—	—	—	—						
9	2.00	—	—	—	—						
10	—	—	—	—	—						

K $= H_s/\Delta d_{n50} \, (\mu\cos\alpha + \sin\alpha)$ P_e = permeable
$\bar{\bar{\xi}}$ $= \tan\alpha/\sqrt{(H/L)}$ I_m = impermeable
H_0 = homogeneous

Table 4.12 Wave impact parameters.

Expression and symbol	Value	
Wave steepness $= S$ or S_m	0.01–0.06	
Similarity parameter $= \xi$		
plunging	0.50–7.50	
surging		
collapsing	3.0	
spilling	0.2	
Damage $S/\sqrt{N}$	<0.9	
N	250–10 000	
$H_s/\Delta d_{n50}/\xi_m$	0.533–0.61	
Damage level S	$N(1000)$	$N(3000)$
2	−0.5	−0.42
3	−0.54	−0.52
5	−0.57	−0.57
8	−0.50	−0.52
12	−0.42	−0.53
Permeability factor P	0.1–0.6	
The angle of wave impact ψ	0–50°	

4.9 Snow/ice impact

4.9.1 Introduction

At present, a great deal of controversy exists as to whether the ice-structure dynamics is the cause of a quasi-static loading or a direct impact. Nevertheless, the vibrations appear to be a complex combination of ice crushing and ice-structure actions. The records[2.280–2.340] indicate that the force−time relationship is greatly dependent on the velocity of the ice floe. Some methods, mostly empirical, are available for the evaluation of impact loads.[2.280–2.333] All methods quoted in this section are based on the theory of elastic plates on elastic foundations. The plasticity theory has been developed by Ralston,[2.323–2.325] which is in good agreement with the early model tests. Some have gone further, such as Croasdale,[2.289,2.290] whose published theories included ride-up on sloping beaches. A brief introduction to snow/ice impact has been given already in Chapter 2 and in this section the work is supported by analysis and empirical formulae.

4.9.2 Empirical formulae

4.9.2.1 *General consideration*

In all circumstances the force−time relationship can be characterized in terms of low and high velocities. The snow/ice strength depends greatly on strain rate. The stress−strain rate relationship begins from the left-hand side of the curve, known as *ductile failure range*, the middle zone is a *transition range* and the right-hand portion of the curve is the *brittle failure range*. The transition between the right and the left is highly complicated since both brittle and ductile failure can exist. Figure 4.46 shows the relationship between compressive strength and the strain rate per second and the mechanical properties depend on temperature, salinity, density, grain/specimen size, loading rate and failure mode.

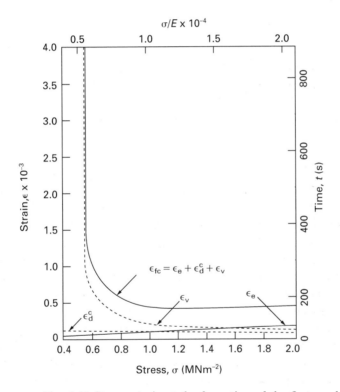

Fig. 4.46 Stress−strain at the formation of the first cracks.

4.9.2.2 *Sinha model*[2.331−2.333]

The strain is made up of three components:

$$\epsilon = \epsilon_e + \epsilon_d + \epsilon_v \tag{4.243}$$

where the portions of the total strain are
ϵ_e: elastic strain

ϵ_d: delayed elastic or time-dependent strain
ϵ_v: viscous or permanent strain

It is possible to estimate the onset of cracking of snow/ice on the basis of knowing ϵ_c. The value of ϵ_d is computed as

$$\epsilon_d = (c_1/E)(d_1/d) \sum_{i=1}^{n+1} \delta\sigma(1 - \exp\{-[a_T(n+1-i)\delta t]^b\}) \qquad (4.244)$$

where E = Young's modulus (9.5 GPa)
$\quad d$ = grain diameter of the columnar crystals of ice
$\quad d_1$ = unit of grain diameter (e.g. 1 mm for grain sizes in mm)
$\quad a_T = 2.5 \times 10^{-4}\ \text{s}^{-1}$ at $T = -10°C$
$\quad c_1 = 9$
$\quad b = 0.34$
$\quad \delta\sigma$ = stress increase per time increment
$\quad n$ = number of time increments
$\quad \delta t$ = time per increment

For a columnar crystal of size 4.5 mm and a temperature of $-10°C$, Sinha determined that first cracks occurred when the delayed elastic strain exceeded 1.04×10^{-4}. Figure 4.47 illustrates the stress–strain time-dependent relationship at the formation of the cracks. The impact load is then computed as

$$F_1(t) = \sigma A \qquad (4.245)$$

where A is the area in m^2.
    If the strain rate ranges from 1×10^{-3} to 1×10^{-5} are considered, the mean compressive strengths for $-5°C$ and $-20°C$ are 2.34 ± 1.08 MPa to 2.79 ± 0.69 MPa, respectively. For circular structures the impact load is then written as

$$F_1(t) = PdT_i' \qquad (4.246)$$

where d = structure diameter
$\quad T_i'$ = local ice thickness
$\quad P$ = ice failure pressure
For impact analysis, the shape factors for circular and nose types (60°) are 1.0, 0.90 and 0.59, respectively.

4.9.2.3 Timco model[2.334, 2.335]

In the ductile mode Timco uses the following form of the Korzhavin equation:

$$F_1(t) = cS_F IdT_i'\sigma_c \qquad (4.247)$$

where c = contact coefficient
$\quad S_F$ = shape coefficient
$\quad I$ = indentation factor
$\quad d$ = diameter of leg of submersible
$\quad T_i'$ = ice sheet thickness

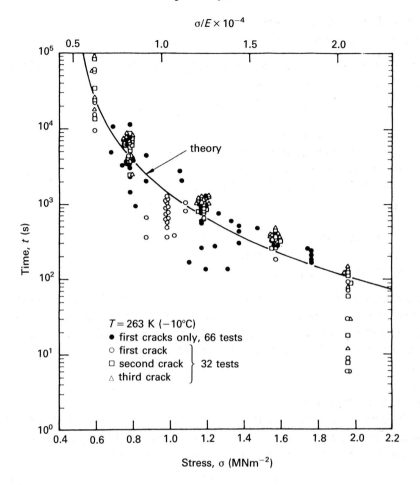

Fig. 4.47 Experimental and theoretical study of stress–time relationship at various cracks (after Sinha[2.331–2.333]).

σ_c = uni-axial unconfined compressive strength of ice at the strain rate $30(\dot{\epsilon})^{0.22}$MPa

$\dot{\epsilon} = V/2d$ = strain rate

If the width of support is considered, the following relationships should be used for the contact factor c:

Width of support (interaction width) (m)	Velocity of floe movement (m/s)		
	0.5	1.0	2.0
3–5	0.7	0.6	0.5
6–8	0.6	0.5	0.4

4.9.2.4 Sanderson model[2.329]

This theory is known as the *reference stress method*. The impact load on a structure is determined by

$$F_1(t) = IdT_i'[1 - (v/v_0)^{1/2}][v/I\psi dA \ \exp(-Q/RT)]^{1/n} \qquad (4.248)$$

where $F_1(t)$ = total load
I = indentation factor
T_i' = ice sheet thickness
v = velocity in m/s
ψ = angle of attack
Q = constant
R = radius
T = time
n = number of impacts
A = area
d = depth occupied on structures

4.9.2.5 Nevel's model[2.322]

Nevel's model produces ultimate failure of the ice plates which can be applied to forces on sloping structures. On conical structures, the impact force produced by failure of a series of ice wedges (radial cracks of the ice/snow) is given as

$$F_1(t) = [1.05 + 2.0(a/l) + 0.5(a/l)^2 + 0.5(a/l)^3][b_0\sigma_f(T_i')^2/6] \qquad (4.249)$$

where σ_f = ice flexural strength $\approx$ 700 kPa
T_i' = ice thickness
b_0 = constant defining the width of the wedge = b/x'
x' = distance along the wedge
a = loaded length
l = characteristic length of the plate = $(Et^3/12\rho g)^{1/4}$

Bercha and Danys[2.280] used this theory to present an elastic analysis for the ice-breaking component of the ice-impact force on a conical structure.

4.9.2.6 Ralston model[2.323–2.325]

An approach for ice forces on a conical structure using plastic limit analysis has been proposed by Ralston. His results can be expressed as equations for H and V — the horizontal and vertical forces.

$$H = A_4[A_1\sigma_f(T_i')^2 + A_2\rho g(T_i')^2d^2 + A_3\rho g T_i'(d^2 - d_T^2)] \qquad (4.250)$$

due to ice breaking   ice pieces sliding
on cone

$$V = B_1H + B_2\rho g T_i'(d^2 - d_T^2)$$

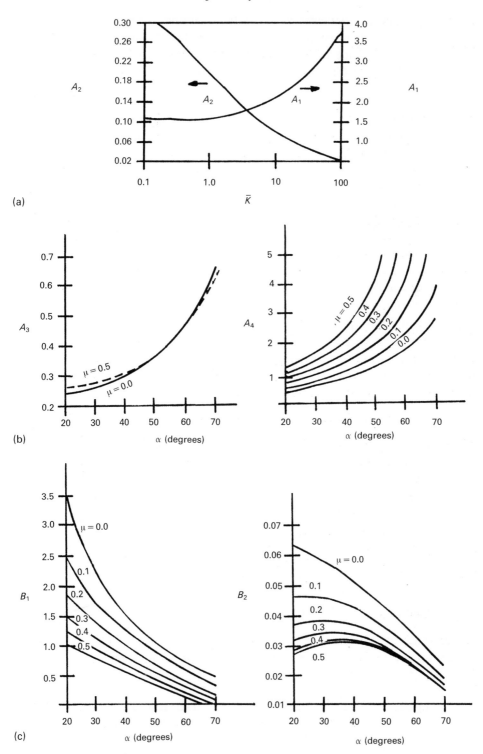

Fig. 4.48 The A and B coefficients for Ralston's plastic analysis (after Ralston[2.323–2.325]).

where d_T is the top diameter and d is the waterline diameter. A_1 and A_2 are coefficients dependent on $\rho_w g d^2 / \sigma_f T_i'$ and A_3, A_4, B_1 and B_2 are coefficients dependent on the cone angle α and friction μ. Values for the coefficients are reproduced in Fig. 4.48(a) to (c); $\rho g d^2 / \sigma_f T_i' = \bar{K}$ is taken in Fig. 4.48(a).

4.9.2.7 Watts' model[2.335]

The numerical model is similar to the one presented by Timco in equation (4.247), except diameter d is not included and the shape factor S_F is defined as

$$S_F = [5(T_i'/d) + 1]^{1/2} \tag{4.251}$$

4.9.2.8 Iyer's model[2.333]

The approach of Iyer is an empirical one and is based on the results of small indentation tests. In this approach, prior to the enveloping of the structure, contact is established locally between the ice and the structure. The local pressure will be a function of the aspect ratio d/t and the contact area. Figure 4.49 shows the curve of failure pressure versus contact area for various values of d/T_i'. An empirical relationship of the following form has been established:

$$F_1(t) = 6.8(dT_i')^{-0.3} \tag{4.252}$$

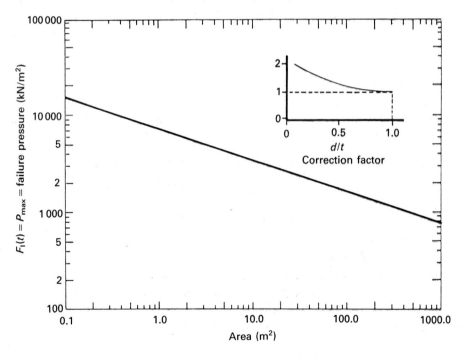

Fig. 4.49 Iyer's impact load.[2.333]

where the product of d and T_i' is the tributary area (m) and $F_1(t)$ is the impact force in MPa.

4.9.2.9 Norwegian model[2.307, 2.308]

The local impact load on the structure can be determined using the following formula:

$$F_1(t) = P_1 = 2[1 + (b_1/h)^{-0.6}][(A/A_0)^{-0.165}]\sigma_c A \qquad (4.253)$$

where b_1 = the horizontal breadth of the local load area
$\quad h$ = the extreme 'vertical' depth of the total area of the structure in contact with the ice
$\quad A$ = the local load area in cm^2
$\quad A_0 = 1.0\,cm^2$
$\quad \sigma_c$ = the maximum uni-axial crushing strength of ice

The effect on the pressure of the confinement of the ice or the multi-axial state of stress under which crushing takes place is represented by

$$K_\sigma = 2[1 + (b_1/h)^{-0.6}] \qquad (4.254)$$

4.9.2.10 Kato and Sodhi model[2.311]

Kato and Sodhi proposed an expression for the crushing force frequency in terms of the aspect ratio of ice contact and the speed of ice:

$$f = CV/T_i' \qquad (4.255)$$

where f = frequency in cycles/s
$\quad V$ = ice speed in m/s
$\quad T_i'$ = ice sheet thickness in m
$\quad C$ = coefficient as a function of column diameter, d, in m, as given in Fig. 4.50(b).

The impact forces measured on a single column are given below:

$$F_{mean} = \sigma_{mean}[5(T_i'/d) + 1]^{1/2} T_i' d$$

$$F_{max} = \sigma_{max}[5(T_i'/d) + 1]^{1/2} T_i' d$$

where T_i' = ice thickness in m
$\quad d$ = column diameter in m
$\quad \sigma_{mean}$ = mean crushing pressure in $MN/m^2 = 1.36\sigma_c$
$\quad \sigma_{max}$ = maximum crushing pressure in $MN/m^2 = 2.33\sigma_c$
$\quad \sigma_c$ = unconfined crushing strength of ice in MN/m^2
The impact force time history is given in Fig. 4.50(a).

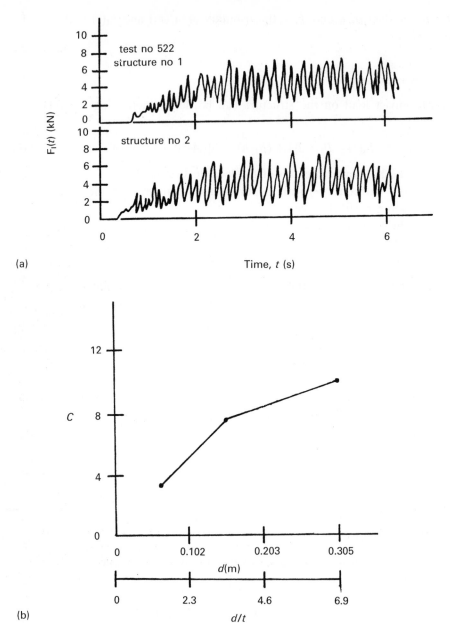

(a)

(b)

Fig. 4.50 (a) Typical ice-crushing force−time histories; (b) coefficient *C* versus column diameter *d* and aspect ratio *d/t* (after Kato & Sodhi[2.311]).

4.9.2.11 *Three-dimensional finite element method*

The finite element is fully dealt with later on in this text. The following areas are to be included:

(1) Use of proper failure criteria in the model. The most appropriate to be adopted is by Frederking and Timco[2.302] and is shown in Fig. 4.51.
(2) Development of finite elements with correct ice characterization, including strain rate dependence and brittle and ductile failure modes.
(3) Development of ice/structure interface elements which include modelling of friction and adfreeze.
(4) Incorporation of scale factor in the yield function and ice strength values.

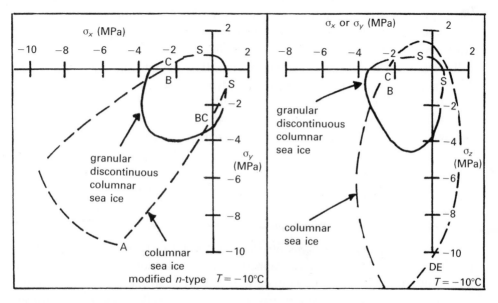

Fig. 4.51 Failure envelope for S_2 ice at two different temperatures. Outline of the failure envelope based on the present and previous tests on columnar sea ice at $\dot{\epsilon}_n = 2 \times 10^{-4}$/s and $T = -10°C$. The solid line represents the extent of the failure envelope for granular/discontinuous-columnar sea ice at the same temperature and strain rate. (After Frederking & Timco[2.302].)

5

EXPLOSION DYNAMICS

5.1 Introduction

An explosion is a rapid release of energy and can happen in air, on the Earth's surface, underground and underwater. Typical examples arise from chemical and nuclear explosives, thermal and hydroelectric sources and gas ignition. Section 2.5 of Chapter 2 gives a useful introduction to sources of explosion and also presents comprehensive data on explosions. The purpose of this chapter is to give a comprehensive treatment of explosion dynamics. The data given in section 2.5 will act as input to various equations given in this chapter. The reader must now be familiar with the basis of structural dynamics given in Chapter 3. Since explosions may generate missiles, the reader is advised to have a thorough knowledge of Chapter 4. Section 4 of the Bibliography provides comprehensive coverage of in-depth treatment of specific areas.

5.2 Fundamental analyses related to an explosion

5.2.1 Stress waves and blast waves

Stress waves represent the basis of explosion in the surrounding medium, be it gaseous, liquid or solid. They are defined as moving parts of the medium, being in a state of stress such that the boundaries are *waves* and the rest of the medium consists of *wave fronts*. Stress waves are sometimes called *deformation waves*. These stress waves are divided into normal and tangential waves representing stresses in those directions. Normal waves may be divided into *pressure waves* and *tensile waves* or *rarefaction waves*. The normal waves are also known as P (longitudinal) waves and the S (tangential) waves as *transverse waves*. In geological media, the surface is the interface of individual layers of rock and soil or air and soil/rock. The interface between them is known as the *free surface*. The surface waves are then classified according to the shape and sense of the trajectories followed by movements in a medium. The surface waves are given below.

(1) *The Rayleigh wave (R wave)*. This wave exhibits a planar elliptic motion in the medium such that its semi-axes decrease rapidly at a certain distance from a given depth.
(2) *The Love wave (Q wave)*. These waves exhibit a spatial motion in a medium and their components are parallel and normal to the plane of propagation.

(3) *The hydrodynamic wave (H wave)*. These waves are similar to R waves and propagate along the surface of the liquid and are free from shear stresses.
(4) *The composite wave (C wave)*. The surface particles exhibit a complex phenomenon when compared with others.

The R wave propagation is expressed as:

$$\delta_x = W_{E(t)} \, L'\{\exp(-qz) - 2q\bar{s}[(\bar{s})^2 + (L')^2\exp(-\bar{s}z)]\} \times \sin(\omega t - L'x) \quad (5.1)$$

$$\delta_z = W_{E(t)} \, L'\{\exp(-qz) - 2(L')^2[(\bar{s})^2 + (L')^2\exp(-\bar{s}z)]\} \times \cos(\omega t - L'x) \quad (5.2)$$

where x, y and z are the co-ordinate axes

$L' = \omega/v_{SR}$
ω = circular frequency
v_{SR} = velocity of sine wave in the Y-direction
t = time
δ = particle displacement
$q = \omega^2(v_{SZ}^2 - v_{RZ}^2)/v_{PZ}^2 \, v_{SZ}^2$
$\bar{s} = \omega^2(v_{SZ}^2 - v_{RZ}^2)/v_{SZ}^2 \, v_{RZ}^2$
$W_{E(t)}$ = work done
v_{SZ} = propagation velocities of transverse waves
v_{PZ} = propagation velocities of longitudinal waves
v_{RZ} = propagation velocities of Rayleigh waves

The Q wave is generally propagating in the Y-direction if the x, y plane lies on the interface of the layer and on the mass medium. The displacement δ_y is a function of time t and the co-ordinates x and z. The Q wave is expressed as

$$\delta_y(x,z,t) = (A\cos\alpha'z + B\sin\alpha'z)\exp[L'(v_z t - x)] \quad (5.3)$$

where A,B = constants
L' = a wave number
v_z = phase velocity
$\alpha' = L'\sqrt{[v_z^2/(v_{sz}^2 - 1)]}$

The phase velocity v_z is determined from

$$v_z = \frac{G_m}{G_s L' d} \times v_{szs} \times \sqrt{\left\{\left[\tan^{-1}\sqrt{\left(1 - \frac{v_z^2}{v_{szsn}^2}\right)\Big/\left(\frac{v_z^2}{v_{szs}^2} - 1\right)}\right]^2 + 1\right\}} \quad (5.4)$$

where v_{szs} = propagation velocity of the explosion
v_{szsn} = propagation velocity in the mass half space
G_m, G_s = moduli of elasticity in shear and mass half space, respectively

5.2.1.1 Explosions and surface waves

Explosions can occur in different forms. The ones of note are explosions of a spherical charge and of a cylindrical charge. In blast dynamics, slab charges are popular. All these are discussed later in specific sections.

Explosion of a spherical charge

When a charge is located at a sufficient depth below the surface, the explosion causes vibrations to propagate in the form of spherical wave fronts in a longitudinal direction, with mass particles displaced radially. This displacement is generally calculated [4.40–4.106,4.250] as

$$
\begin{aligned}
\delta_r &= \text{radial displacement} \\
&= \frac{r_s^2 v_{zp}\alpha}{r_s^2\,(\beta r_s - v_{zp})} \times \left\{ \frac{r_s}{v_{zp}} \left[1 - \left(t - \frac{r_s}{v_{zp}} - \frac{r_s}{\beta r_s - v_{zp}} \right)\beta \right] \right. \\
&\quad \left. + t - \frac{r_s}{\beta r_s - v_{zp}} \right\} e^{-\beta(t - r_s/v_{zp})}
\end{aligned}
\tag{5.5}
$$

where v_{zp} = velocity in m/s
r_s = radius of the cavity of the charge (sometimes known as R_w)
t = time
α, β = constants for the particular type of charge

Explosion of a cylindrical charge

The following gives displacements δ_R, δ_θ and δ_ϕ in cylindrical co-ordinates for a cylindrical cavity:

$$
\delta_R = \frac{R_0^2\, hp}{4\, m_R v_{zp} R} \left(1 - \frac{2v_{zs}^2}{v_{zp}^2} \cos^2\phi \right) \frac{\partial F(t - R/v_{zp})}{\partial t}
\tag{5.6}
$$

$$
\delta_\theta = 0; \quad \delta_\phi = \frac{R_0^2\, hp \sin 2\phi}{4\, \nu v_{zs} R} \left[\frac{\partial F\,(t - R/v_{zs})}{\partial t} \right]
\tag{5.7}
$$

where R_0 = base of a cylinder with height h
m_R = relative mass
$\delta_R, \delta_\theta, \delta_\phi$ = displacements in cylindrical co-ordinates R, θ and ϕ, respectively
ν = Poisson's ratio
t = time

5.2.1.2 Damping of stress waves

When a stress wave caused by an explosion propagates in a material it is damped at a certain x and its amplitude $X(x)$ is given by

$$
X(x) = X_0 e^{(-\gamma x)}
\tag{5.8}
$$

where $X(x)$ = amplitude at a distance x
X_0 = amplitude of the wave at a source
γ = damping ratio

5.2.1.3 *Formation of blast waves from explosives and their scaling laws*

Section 2.5 gives a useful introduction to blast wave formation. As stated, when the blast wave from an explosion travels outward, the volume of air included behind the shock front becomes so great that the initial volume of the explosion becomes unimportant. Positive and negative phases are predicted. In this section, blast waves occur due to explosives. The finite size of the explosive charge from chemical explosives has a different effect from that of nuclear explosives. As seen in Fig. 5.1, the effect is initially to spread out the energy with a reduction in peak overpressure over an appreciable distance from the charge source. The influence of the charge diameter on the blast wave is thus related to two volume displacements — one from a point source with no initial atmospheric change and the other from a point source in a nominal standard pressure. This is given by

$$c_f = (\rho V / \rho_{TN} V_{TN})^{1/3} \tag{5.9}$$

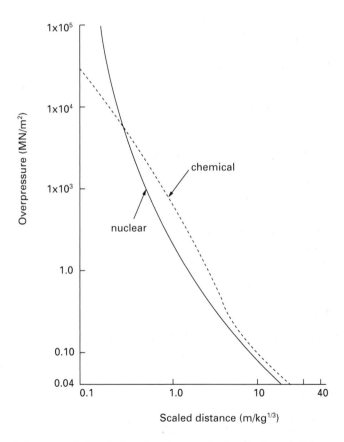

Fig. 5.1 The behaviour of chemical and nuclear explosions in terms of overpressure versus scaled distance.

where c_f = charge size factor

V, V_{TN} = actual and required TNT volumes, respectively

ρ, ρ_{TN} = actual and standard TNT atmosphere densities, respectively

For a chemical explosion, the relationships between explosion overpressure and ambient atmosphere, duration time and impact load are compared with those for a nuclear explosion in Table 5.1. Impulse or impact depends on the peak overpressure, p_{so}, in the shock front and on the duration of the wave. In some cases, the rate of decay of the overpressure can influence the impulse/impact load. The scaling principle to explosion has also been given in section 2.5. The scaling law for explosion is defined for the distance for *uniform atmosphere* as

$$\text{scaled distance} = x \times \rho_{TN}^{1/3}/E_R^{1/3} \qquad (5.10)$$

For a non-uniform atmosphere the scaled distance will be different. The energy release, E_R, is almost equal to the weapon yield.

The *scaling of the overpressure* is given by

$$\text{actual pressure} = \text{overpressure} \times p_a \qquad (5.11)$$

where p_a is the atmospheric pressure.

The *scaling time*, t_{sc}, is given by

$$t_{sc} = t_a \times f_t / Y^{1/3} \qquad (5.12)$$

where f_t, the transmission factor for time, is given by

$$f_t = \left(\frac{\rho}{\rho_{TN}}\right)^{1/3} \left(\frac{T}{T_{TN}}\right)^{1/6} \qquad (5.13)$$

and is expressed in terms of atmospheric pressure and temperature

where T, T_{TN} = actual and TNT temperatures, respectively

Y = weapon yield, W

The *direct impulse/impact scaling* per unit area is written as

$$F_1(t)/\text{unit area} = (\text{scaled impulse/area})(p_{so}/p_s')\,(t_a/t_{sc}) \qquad (5.14)$$

where p_s' = standard overpressure for reference explosion

Transmission factors for distance and time for large explosions, with large path distances and variations in atmospheric pressure and temperature, can be written in integral form:

$$\text{scaled distance} = \frac{1}{Y^{1/3}} \int_{x_1}^{x_2} \left(\frac{\rho}{\rho_{TN}}\right)^{1/3} dx \qquad (5.15)$$

A *transmission factor to conform to variations of pressures and temperatures* will be given by

$$\bar{f}_d = \text{transmission factor for distance}$$

Table 5.1 A comparison of overpressures, time duration and impact loads for chemical and nuclear explosions.

Chemical	Nuclear
$$\frac{p_{so}}{p_a} = \frac{40.4x^2 + 810}{\sqrt{[(1+434x^2)(9.77x^2)(1+0.55x^2)]}}$$	$$\frac{p_{so}}{p_a} = \frac{5.12}{x^2} \times \sqrt{(1+0.00013x^2)}$$

Chemical:

$$\frac{t_d}{Y^{1/3}} = \frac{990 + 4.65\times10^5 x^{10}}{(1+125\times10^3 x^3)(1+6.1x^6)} \times \frac{1}{\sqrt{(1+0.02x^2)}}$$

$$F_I(t)/\text{unit area} = \frac{0.7}{x^2} \times \frac{\sqrt[2]{(1+360x^4)}}{\sqrt[3]{(1+0.27x^3)}}$$

t_d = duration time in seconds for 1 kilotonne TNT
x = distance in m

Nuclear:

$$t_d/Y^{1/3} = (200\sqrt{1} + 1\times10^{-6}x^3)(1/Z)$$

where

$$Z = \sqrt{(1+0.025x)}[\sqrt[6]{(1+5.25\times10^{-13}x^5)} \times \sqrt{(1+2\times10^{-5}x)}]$$

$$F_I(t)/\text{unit area} = \frac{0.92}{x^3} \times \frac{\sqrt[2]{(1+1050x^4)}}{\sqrt[3]{(1+0.135x^3)}}$$

Example
Transmission factor for a distance = 0.7
Energy release = 100 kg TNT at a height of 12 000 m
Atmospheric pressure = 260 mbar
A point is 15 m away
Scaled distance = $0.7(15/100^{1/3}) \approx 2.30$
Peak overpressure ratio = 1.51
Transmission factor = 0.615
Scaled duration = 1.32 ms
Actual duration = $100^{1/3}(1.32/0.615) = 9.97$ ms
Overpressure = $1.52 \times 260 = 390$ mbars = 0.39 bars

$$\bar{f}_d = \frac{1}{x} \int_{x_1}^{x_2} \left(\frac{\rho}{\rho_{TN}} \right)^{1/3} dx \tag{5.16}$$

where x represents the actual distance. Similarly, the transmission for time is given by

$$f_t'' = \int_{x_1}^{x_2} \left(\frac{\rho}{\rho_{TN}} \right)^{1/3} \left(\frac{T}{T_{TN}} \right)^{1/2} \tag{5.17}$$

The subjects of open air, underground and underwater explosions are fully discussed under different headings in this chapter. The above elements are common to all of them.

5.3 Explosions in air

Explosion characteristics, including duration, are based on the sudden release of energy. An explosion may be due to nuclear detonation, explosives, gas or dust. These have been discussed in section 2.5. As stated, the magnitude of an explosion in relative values is known as the explosive yield. One generally accepted standard is the energy released in an explosion of TNT (symmetrical 2,4,6-trinitrotoluene). As stated in section 2.5, the front of the shock wave is quite steep and, as a result, the pressure may be treated as instantaneous. The dynamic load is then characterized by a rapidly reached peak value which decreases as the blast wave decays. The net effect of the load depends on the structure of the blast wave and on the geometry and construction of the structure. The basic relationship for such a blast wave having a steep front is given by Rankine-Hugonist,[4.251] and is based on the conservation of mass, energy and momentum at the shock front. Using Fig. 5.2

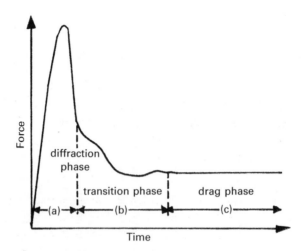

Fig. 5.2 Force−time history.

and the above conditions together with the equation of state for air, the blast wave pressure is written as

$$\frac{U}{v_s} = \left(1 + \frac{6}{7}\frac{p_{so}}{p_a}\right)^{1/2} \tag{5.18}$$

$$\frac{u}{v_s} = \frac{5}{7}\frac{p_{so}}{p_a}\left(1 + \frac{6}{7}\frac{p_{so}}{p_a}\right)^{-1/2} \tag{5.19}$$

$$\frac{\rho}{\rho_a} = \frac{7 + 6\,p_{so}/p_a}{7 + p_{so}/p_a} \tag{5.20}$$

$$\frac{p_r}{p_{so}} = 2\,\frac{7p_a + 4p_{so}}{7p_a + p_{so}} \tag{5.21}$$

$$\frac{q}{p_{so}} = \frac{5}{2}\,\frac{p_{so}}{7p_o + p_{so}} \tag{5.22}$$

$$\left(\frac{v_{so}}{v_s}\right)^2 = \frac{(p_{so} + p_a)\,(p_{so} + 7p_a)}{6p_{so} + 7p_a} \tag{5.23}$$

$$M^2 = \left(\frac{u}{v_{so}}\right)^2 = \frac{25}{7}\,\frac{p_{so}^2}{(p_{so} + p_a)\,(p_{so} + 7p_a)} \tag{5.24}$$

$$\frac{\bar{R}}{\bar{R}_a} = 0.727 p_{so}/p_a \tag{5.25}$$

$$\frac{p_r}{q_{do}} = \frac{4}{5}\,(4 + 7p_a/p_{so}) \tag{5.26}$$

where v_{so} = speed of sound in the air behind the shock front
v_s = speed of sound in ambient air
M = Mach number
p_a = pressure of ambient air
p_{so} = overpressure
p_r = reflected pressure
q_{do} = dynamic pressure = $\frac{1}{2}\,\rho u^2$
$\bar{R}$ = Reynolds number per foot (flow behind shock front)
$\bar{R}_a$ = Reynolds number per foot for ambient air sea level (6.89×10^6)
u = particle velocity of flow behind the shock front
U = shock front velocity
ρ = density of air behind shock front
ρ_a = density of ambient air

Figure 5.3 shows reflected pressure, dynamic pressure and Mach number for side-on overpressure. When the blast wave is vertical and strikes the front face of the structure, normal reflection occurs and the entire facade of the building or structure is instantly subjected to the reflected overpressure, p_{ro}, which is greater than p_{so}, the overpressure in the immediate surroundings. As a result, the blast

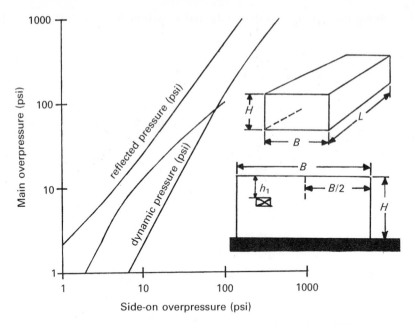

Fig. 5.3 Overpressure p_{so} versus p_r, M and q_{do} for a building.

air flows from the region of high pressure to the one of low pressure, forming a rarefaction wave with a velocity u_{rr} over the front of the structure. It then progresses inward from the edges of the structure, moving with a velocity v_s in the reflected medium. This speed varies with time as the blast wave decays. For example, as shown in Fig. 5.3, if one takes a small panel of a structure, the wave varies at this panel with a corresponding time h_1/u_{rr}, where h_1 is the distance from the top to that panel and u_{rr} is the rarefaction velocity. Assuming this time is t_1, the relieving time, t_r, is about twice that required for the sound wave, which is $t_2 - t_1 = 2x/u_{rr}$, where t_2 is the forward time and x is the distance through which the pressure relief is obtained for the length of the structure, L, width, B, and height, H. Figure 5.4 gives load distributions on various faces against time. The peak diffraction pressure, p_{df}, and peak drag load, p_d, are given by

$$p_{df} = p_r A \tag{5.27}$$

$$p_d = p_D C_D A \tag{5.28}$$

where A is the projected area, C_D is the drag coefficient, p_r is the reflected pressure and p_D is the dynamic pressure $\approx p_{so}$.

The blast wind from an explosion exerts loads on structures which are quite similar to those developed by natural winds. Nevertheless, these wind load surfaces are transient in nature and of considerably greater magnitude than those developed by conventional winds. The drag coefficient is also defined as

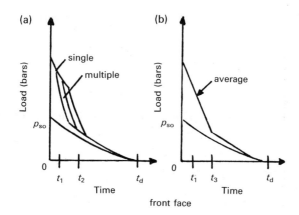

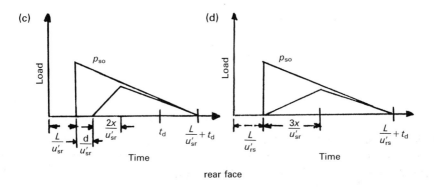

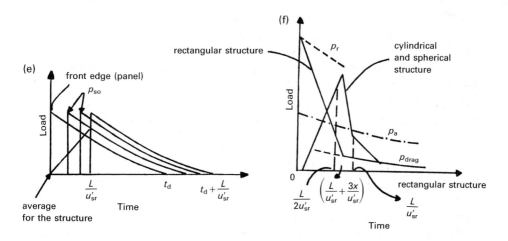

Fig. 5.4 Load–time relationships for blasts in open air. (a) Several panels; (b) average for the structure; (c) one panel; (d) average for the rear face; (e) side or top face; (f) drag type loads.

$$C_D = \text{drag energy/kinetic energy} = (p_d/\rho)/\tfrac{1}{2}\, u^2 \qquad (5.29)$$
$$= 2p_d/\rho u^2$$

For an ideal gas explosion in air

$$\rho = \text{density} = p_a/RT \qquad (5.30)$$

where R is the gas constant, T is temperature, p_d is the drag load and p_a is the pressure. The value of R is equal to 287 J/kg-K.

The value of C_D is also written in terms of Mach numbers as

$$C_D = 2p_d/k_r M^2 p_a \qquad (5.31)$$

The value of p_d is then written as

$$p_d = \tfrac{1}{2} C_D \rho u^2 = \tfrac{1}{2} k_r M^2 C_D p_a \qquad (5.32)$$

where $k_r = $ heat capacity ratio $= C_p/C_v \approx 1.4$
$\quad C_p = $ specific heat capacity at constant pressure
$\quad C_v = $ specific heat capacity at constant volume

The Mach number may be written as

$$M = u/(kRT)^{1/2} \qquad (5.33)$$

An individual small panel experiences load from an explosion when the shock front has traversed the distance L, the entire length of the structure, and a compression wave has travelled a distance h_1 from the near edge into the panel. The time for the shock will be L/u'_{sr}, where u'_{sr} is the speed of the shock, a value close to U. After these times the pressure on the panel increases and becomes instantaneous pressure on the rear face equal to $p_{stag} - p_{drag}$. Figure 5.4(c) and (d) show the load–time function for the rear face for a small panel and an average for the entire rear face. Similarly, for the panels along the side or top of a structure, ignoring reflection and p_{stag} (stagnation), the overpressure diagram is as shown in Fig. 5.4(e). Figure 5.4(f) illustrates dynamic drag type loading. To summarize, the load–time function will assume the generalized form shown in Fig. 2.72.

The stagnation pressure is an important consideration. The value of p_{stag} is computed as

$$p_{stag}/p_a = 1 + (k/2)M^2 + (k/8)M^4 + K/48(z-k)M^6 + \ldots \qquad (5.34)$$

$$(p_{stag} - p_a) - p_d(\text{or } q) = p_a(k/8)M^4 + p_a[k/48\ (z-k)]M^6 + \ldots \qquad (5.35)$$

In the case of a gas explosion in air, equation (5.35) shows a non-compressible fluid flow and $p_d = q$ represents the blast impact, but for a non-compressible fluid, with the speed of sound being infinite, all flows are at zero Mach number M. Equation (5.35) becomes

$$(p_{stag} - p_a) = p_d = q \qquad (5.36)$$

5.3.1 Thickness of the shock front

The thickness of the shock wave is the ratio of the velocity jump between two points u_1 and u_2 divided by the maximum velocity gradient $(du/dx)_{max}$ in a specific zone. In terms of Mach number, the thickness t_{sh} of the shock front using the Rankine-Hugonist equation is given by

$$t_{sh} = [(11 + 7M)/\rho(M - 1)] \ 10^{-8} \tag{5.37}$$

5.3.2 Evaluation of stagnation pressure, stagnation and post-shock temperatures

The stagnation pressure, p_{stag}, is given in equation (5.35) in terms of ambient pressure and is now defined in terms of the velocity of sound, v_s, and v_{so}, which is the speed of sound after the shock front. The value of p_{stag} is given by

$$p_{stag} = p_2 \left[1 + \frac{(k - 1)(v_{so}/v_s)^2}{2(T_2/T_a)} \right]^{k/k-1} \tag{5.38}$$

where p_2 = shock-generated pressure
T_2 = shock-generated temperature
T_a = ambient temperature
v_{so} = blast-generated velocity

The temperature known as the blast stagnation temperature obeys the relationship

$$\frac{T_{stag}}{T_o} = \frac{T_2}{T_a} + \tfrac{1}{2}(k - 1) \left(\frac{v_{so}}{v_s} \right)^2 \tag{5.39}$$

The value of k is generally taken as 1.4. The post-shock temperature T_{ps} is given for $k = 1.4$ by

$$\frac{T_{ps}}{T_a} = \left(\frac{v_{so}}{v_s} \right)^2 = \frac{(p_2/p_a + 6) \ (p_2/p_a)^{5/7}}{6(p_2/p_a) + 1} \tag{5.40}$$

where T_{ps} is the temperature under post-shock.

5.3.3 Oblique shock

A shock wave may occur in a plane that is oriented at an angle θ to the direction of the blast wind flow. Let that velocity be v_{so}; its components are v_1 and v_2, as shown in Fig. 5.5. In this oblique shock phenomenon, the velocity vectors are related in terms of the angle θ and the angle α of the shock plane with respect to the on-coming stream by

$$v_{so2}/v_2 = \tan(\alpha - \theta) \tag{5.41}$$

where v_{so2} and v_2 are the velocities after shock of the normal and parallel components, respectively. The Mach number is given by

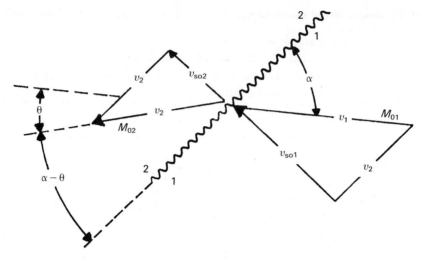

Fig. 5.5 Oblique shock.

$$M_1 \text{ (normal to the shock plane)} = v_{so1} \sin\alpha/v_{s1} \tag{5.42}$$
$$= M_{o1} \sin\alpha$$

For the downstream component v_{so2}:

$$M_2 \text{ (normal to the shock plane)} = v_{so2}/v_s \tag{5.43}$$
$$= v_{so2} \sin(\alpha - \theta)/v_{s2}$$
$$= M_{o2} \sin(\alpha - \theta)$$

where v_{s1} and v_{s2} are shock velocities in planes 1 and 2.

The following equations may be derived along the same lines as equations (5.38) to (5.40):

$$p_2/p_a = \frac{kM_{o1}^2 \sin^2\alpha - \dfrac{k-1}{2}}{(k+1)/2} \tag{5.44}$$

For the temperature and the speed-of-sound effects, T_2/T_a is given by

$$T_2/T_a = (v_{s2}/v_{s1})^2 = \frac{1 + \dfrac{k-1}{2}(M_{o1}^2 \sin^2\alpha)\left(kM_{o1}^2 \sin^2\alpha - \dfrac{k-1}{2}\right)}{\left(\dfrac{k+1}{2}\right)^2 M_{o1}^2 \sin^2\alpha} \tag{5.45}$$

The Mach numbers of the upstream and downstream velocities, before and after the shock, are related by

$$[M_{o2} \sin(\beta - \theta)]^2 = \frac{2 + (k-1)M_{o1}^2 \sin^2\alpha}{2kM_{o1}^2 \sin^2\alpha - (k-1)} \tag{5.46}$$

The values of p_a and T_a are in plane 1, i.e. $p_a = p_2$ and $T_a = T_1$. The value of k is generally taken as 1.4.

The relation between the angle of the shock plane, α, and the angle of deflection, θ, may be found from:

$$\frac{\tan (\alpha - \theta)}{\tan \alpha} = \frac{2 + (k - 1)M_{o1}^2 \sin^2\alpha}{(k + 1)M_{o1}^2 \sin^2\alpha} \tag{5.47}$$

The *maximum deflection* can easily be obtained from Fig. 5.6 for a given Mach

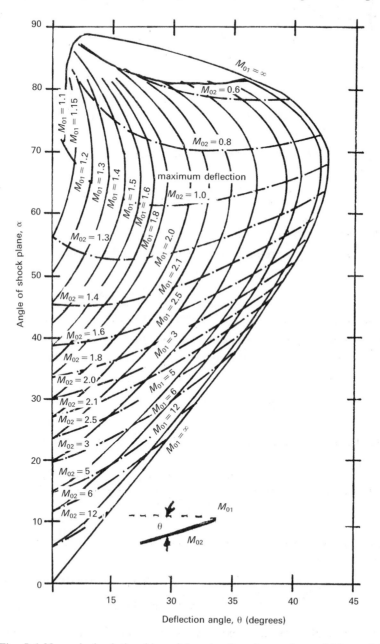

Fig. 5.6 Numerical relationships of θ and α for given values of M_{o1} and M_{o2}.

number. Using the manipulated version of equation (5.47), the value of α is computed as

$$\sin^2\alpha = \frac{1}{4kM_{o1}^2} \times (k+1) \, M_{o1}^2 - 4 + \sqrt{\{(k+1)[(k+1)M_{o1}^4 + 8 \, (k-1)M_{o1}^2 + 16]\}}$$

$$(5.48)$$

5.4 Shock reflection

5.4.1 Normal shock reflection

The reflected shock front exhibits the same particle velocity as that of the incident shock. Nevertheless, the characteristics of the two shocks moving through different media are different. In a similar manner, the particle velocity ratio v_{so}/v_{so1} is related to $p_2/p_a = p_r/p_2$, using section 5.3.2, by

$$(v_{so}/v_{so1})^2 = \frac{\left(\frac{2}{k}\right)\left(\frac{p_r}{p_2} - 1\right)^2}{(k+1)\left(\frac{p_r}{p_2}\right) + (k-1)} = \left(\frac{v_{so}}{v_{so2}}\right)\left(\frac{T_2}{T_a}\right) \qquad (5.49)$$

where p_r is the absolute pressure generated in the reflected shock.

Equation (5.49) can easily be written in terms of p_r/p_2 as

$$p_r/p_2 = \frac{(3k-1)(p_2/p_1) - (k-1)}{(k-1)(p_2/p_1) + (k+1)} \qquad (5.50)$$

where p_1 is the pressure of the unshocked air $\approx p_a$.

The Mach number, M_r, for the reflected shock can similarly be related to the Mach number for the incident shock, M_{o1}, by

$$M_r^2 = \frac{2kM_{o1}^2 - (k-1)}{(k-1)M_{o1}^2 + 2} \qquad (5.51)$$

The relationship between p_r and p_1, the incident shock and the Mach number can be written in the following form:

$$p_r/p_1 = \frac{(p_2/p_1)[(3k-1)(p_2/p_1) - (k-1)]}{(k-1)(p_2/p_1) + (k+1)} \qquad (5.52)$$

$$= \frac{[(3k-1)M_1^2 - 2(k-1)][2kM_1^2 - (k-1)]}{(k^2-1)M_{o1}^2 + 2(k+1)}$$

Using $k = 1.4$, equations (5.50) to (5.52) can be modified. The resulting equations are similar to equations (5.18) to (5.26). For example, equation (5.51) becomes

$$M_r^2 = \frac{7M_{o1}^2 - 1}{M_{o1}^2 + 5} \qquad (5.53)$$

and equation (5.52) becomes

$$p_r/p_1 = \frac{(4M_{o1}^2 - 1)(7M_{o1}^2 - 1)}{3(M_{o1}^2 + 5)} \tag{5.54}$$

The temperature after the shock is greater than the ambient temperature $T_1 = T_a$. The reflected value T_r can easily be derived after algebraic manipulation similar to equation (5.40).

$$T_r/T_1 = T_a = \left(\frac{v_{sr}}{T_a}\right)^2 = \frac{\left[(k-1)\left(\frac{p_2}{p_1}\right) + 1\right]\left[3(k-1)\frac{p_2}{p_1} - (k-1)\right]}{k\left[(k+1)\frac{p_2}{p_1} + (k-1)\right]} \tag{5.55}$$

The *reflection coefficient* is defined by

$$C_r = \frac{p_r - p_1}{p_2 - p_1} \tag{5.56}$$

$$= \frac{\text{reflected overpressure}}{\text{overpressure in the incident shock}}$$

$$= \frac{(3k-1)(p_2/p_1) + (k+1)}{(k-1)(p_2/p_1) + (k+1)}$$

$$= \frac{(3k-1)M_{o1}^2 + (3-k)}{(k-1)M_{o1}^2 + 2}$$

For $k = 1.4$

$$C_r = \frac{8M_{o1}^2 + 4}{M_{o1}^2 + 5} \tag{5.57}$$

5.4.2 Oblique reflection

Figure 5.7 illustrates the basic concept of oblique reflection. An incidental shock at M_{o1} with an incident angle of α causes a corresponding reflected shock in Mach number:

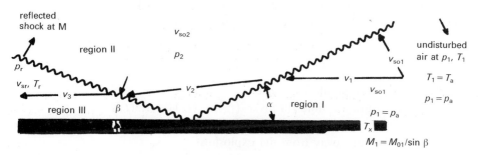

Fig. 5.7 Oblique reflection of shock waves.

$$M_r = M_{o1} \sin \alpha \qquad (5.58)$$

The angle β of this reflected shock is given by

$$\beta = (\alpha - \theta) \qquad (5.59)$$

where θ is the deflection angle. From equation (5.54), the value of p_r/p_1 becomes

$$p_r/p_1 = (7M_r^2 - 1)(7M_{o1}^2 - 1)/36 \qquad (5.60)$$

The reflection coefficient is derived using equation (5.56) with $k = 1.4$:

$$C_r = \frac{p_r - p_1}{p_2 - p_1}$$

$$= \frac{(7M_r^2 - 1)(7M_{o1}^2 - 1) - 36}{42(M_{o1}^2 - 1)} \qquad (5.61)$$

The Mach number M_{o2} in region II for $k = 1.4$ will be

$$[M_{o2} \sin(\alpha - \theta)]^2 = \frac{5 + M_{o1}^2}{7M_{o1}^2 - 1} \qquad (5.62)$$

5.5 Gas explosions

A general introduction to explosions due to explosives and gas leaks is given in sections 2.5.1.2 and 2.5.2. As explained earlier in the text, the gas blast effects are in the form of a shock wave composed of a high-pressure shock front which expands outwards from the centre of the detonation. As these waves impinge on the structure, the whole structure might be engulfed by the shock pressures. The explosion can be one of the following types:

(1) Open-air or unconfined (air-burst and surface-burst loads).
(2) Partially-confined (exterior or leakage pressure loads).
(3) Fully-vented (interior or high-pressure loads).

The open-air explosion is fully discussed in section 5.4. In the case of a gas explosion, the duration of the pressure is short in comparison to its response time, the impulse rather than the pressure pulse governs. The fictitious peak pressure P_{gf} is given by

$$P_{gf} = 2F_1(b)/t_o \qquad (5.63)$$

where $F_1(b)$ = average blast impulse
t_o = duration time (in ms) = $(t_A)_F - (t_A)_A + 1.5(t_o)_F$
$(t_A)_F$ = time of arrival of the blast wave at structures defined by the largest slant (away from the explosion)
$(t_A)_A$ = time of arrival of the blast wave at structures defined by the normal distance (nearest to the explosion)

$(t_o)_F$ = duration of the blast pressure at a structure further from the explosion (for multiple reflection of the blast waves this is increased by 50%)

In partially-confined or vented cubicles (chambers), the mean pressure p_{mo} generated by a spark or charge is given by

$$p_{mo} = 2410(Q/V)^{0.72} \qquad (5.64)$$

where Q/V is the charge/volume ratio (lb/ft^3).

Where the structure has a small A/V ratio and the blast pressure is less than 150 psi (2316.6 MN/m^2), the interior pressure is calculated as

$$\Delta p_i = C_L(A_o/V)\delta t \qquad (5.65)$$

where C_L = leakage pressure coefficient and is a function of $p - p_i$, the pressure difference at the opening
Δp_i = interior pressure increment
A_o/V = area of the opening/volume of the structure
δt = time increment

The variation of C_L versus $p - p_i$ (in psi) is given in Fig. 5.8. Table 5.2 gives a procedure for gas concentration affecting Δp_i. In the case of *multiple explosions*

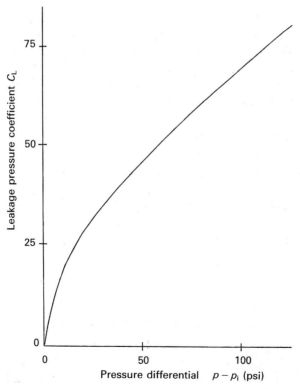

Fig. 5.8 Leakage pressure coefficient C_L versus pressure differential.

Table 5.2 Gas concentration and gas leakage time.

Gas flow out $= (V_c/100)(V_a + V_g)dt$	(1)
Gas flow in $= V_g dt$	(2)

Increase in gas concentration

$(dV_{con}/100) = [V_g - (dV_{con}/100)(V_a + V_g)dt/V]$ .... (3)

or $dV_{con} = [100V_g/(V_A + V_g)]\{1 - \exp[-(V_a + V_g)t/V]\}$ .... (4)

where dV_{con} = percentage of gas in gas/air mixture
$\qquad\qquad$ = volume percentage of gas concentration
$\qquad\qquad t$ = time
$\qquad V_a, V_g$ = volume flow rate of air and flow rate of leaked gas rate, respectively

Equation (4) can be used to relate dV_{con} at any time from the onset of leakage with the factors of gas leakage rate, ventilation rate and room volume.

dV_{con} (%)	Time t (h)	dV_{con} (%)	Time t (h)
0	0	4.00	0.42
0.50	0.10	5.00	0.50
1.00	0.15	5.50	0.53
2.0	0.2	6.00	0.55
2.50	0.21	7.00	1.25
3.00	0.25	8.00	1.50
3.50	0.30		

Gas leakage rate $= 5\,\text{m}^3/\text{h}$; vent rate $= 55\,\text{m}^3/\text{h}$
cubicle volume $= 30\,\text{m}^3$ (3 m high $\times$ 10 m^2 in area)

occurring several milliseconds apart, the first blast wave shall be assumed to be ahead of others. If the time delay between explosions is not too large, the subsequent blast waves will merge with the first blast wave at a distance which depends on:

(1) The magnitude of the individual explosion.
(2) The time delays between the initiations of explosions.
(3) The interference of obstructions.

The fully-vented types of explosions are discussed later on in this chapter. Reference may be made to the dynamic analysis given in sections 5.2 and 5.3. For a fully-confined gas explosion, the overpressure is given by

$$p_{so} = p_o \text{ (final temperature } t_f/\text{initial temperature } t_a) \qquad (5.66)$$

In terms of the burning velocity v_b and the expansion ratio α_E, the pressure generated in a spherical medium of radius R is given for a fully confined case by

$$p_{so} = p_o \exp[\alpha_E^2(\alpha_E - 1)(v_b t/R)^3] \qquad (5.67)$$

The rate of pressure rise is computed by differentiating equation (5.67) with respect to time:

$$dp_{so}/dt = 3p_o/R^3\alpha_E^3 \ (\alpha_E - 1)v_b^3 t \tag{5.68}$$

If the spherical assumption is excluded and $\Delta p = p_{so} - p_o$, then

$$\alpha_E^2(\alpha_E - 1) \approx \alpha_E^3$$

$$dp_{so}/dt = \Delta p \approx (v_b\alpha_E)^3 t^3/V \tag{5.69}$$

In a vented, confined gas explosion, the pressure generated depends upon the characteristics of the gas/air mixture and the type of enclosure. From the area of explosion relief, the overpressures generated in a vented explosion may be predicted. A number of experimental and theoretical studies on vented explosions have been carried out[4.100-4.202] There are a number of empirically-based methods of predicting explosion overpressures. Table 5.3 gives a summary of some of the well known ones. Where volume scaling is necessary, results obtained at one scale may be transformed to another by

$$A_{v2} = A_{v1}(V_2/V_1)^{2/3} \tag{5.70}$$

where A_{v1} is the necessary vent area of the test compartment with a volume V_1 and A_{v2} is the necessary vent area for a second compartment of volume V_2. The idea is that, in both cases, an explosion pressure exceeding a given value should be prevented. The value of v_b given in Table 5.3 is affected by the gas turbulent factor β and temperature. The value of v_b may be modified as follows:

$$v_{bt} = \text{velocity affected by turbulence} = \beta' v_b \tag{5.71}$$

$$v_{bT} = \text{velocity affected by temperature} = v_b(T_f/T_a) \tag{5.72}$$

where T_f/T_a = final temperature/ambient temperature
= expansion ratio resulting from an increase in ambient temperature

On the basis of experiments using two inter-connected rooms, each having a volume of $28\,m^3$, Cubbage and Marshall[4.92] have produced an empirical equation which can predict an overpressure p_{so2} developed in the second compartment, following ignition in the first, as

$$p_2 = (ap_1 + bp_2^2)^{0.5} \tag{5.73}$$

$$\text{where } a = (V_2/V_1)[46(KW)_{2AV} \ v_b/V_2^{0.33}] \tag{5.73a}$$

$$\text{and } b = (V_2/V_1)(K_2/K_{1,2}) \tag{5.73b}$$

where V_1 is the volume of room 1 in m^3, V_2 is the volume of room 2 in m^3, $(KW)_{2AV}$ is the average value of the term KW for room 2 in kg/m^2, $K_{1,2}$ is the vent coefficient between rooms 1 and 2 and K_2 is the vent coefficient for room 2; p_1 and p_2 are in mbar.

The Runes equation adopted by the US National Fire Protection Association[4.93] gives an explosion relief area for explosion venting:

Table 5.3 A summary of empirical formulae and vented parameters.

Author	Formulae for pressures (mbar)	Range of application
Cubbage and Simmonds[4.87]	$p_1 = v_b(4.3KW + 28)/V^{1/3}$ $p_2 = 58v_bK$ v_b (m/s) = the burning velocity Vent cladding can be of any material, provided that no restraining force (other than the minimum of friction) is used to maintain the vent in position	$L_{max}:L_{min} \leq 3:1$ $K \leq 5$ $W \leq 24\,kg/m^2$
Cubbage and Marshall[4.91,4.92]	$p_m = p_v + 23(v_b^2 KW/V^{1/3})\{\ldots \ldots [f(\lambda,\lambda_o)]\}$ v_b (m/s) Predicts maximum pressure generated, irrespective of whether this is p_1 or p_2. For $\lambda \leq 750\,kJ/m^3$, and $p_v < 350\,mbar$, use $f(\lambda,\lambda_o) = 1 - exp[-(\lambda - \lambda_o)/(\lambda + \lambda_o)]$. For $\lambda \leq 750\,kJ/m^3$ and $p_v > 350\,mbar$, use $f(\lambda,\lambda_o) = (\lambda - \lambda_o)/\lambda$. For hazard assessment use $f(\lambda,\lambda_o) = 1$ to calculate the maximum possible pressure rise.	$L_{max}:L_{min} \leq 3:1$ $K \leq 4 = A_s/A_v$ $2.4\,kg/m^2 \leq W \leq 24\,kg/m^2$ $p_v \leq 490\,mbar$ A_s = area of the enclosure in the plane of the vent
Rasbash[4.88]	$p_m = 1.5p_v + 77.7v_bK$ v_b (m/s) Formula essentially predicts the second peak pressure, p_2; vent cladding can be any material held in place by a positive force	$L_{max}:L_{min} \leq 3:1$ $K \leq 5$ $W \leq 24\,kg/m^2$ venting overpressure = $p_v \leq 70\,mbar$
Rasbash et al.[4.89]	$p_m = 1.5p_v + \ldots v_b\{[(4.3KW + 28)/V^{1/3}] + 77.7K\}$ v_b (m/s)	$L_{max}:L_{min} \leq 3:1$ $K \leq 5$ $W \leq 24\,kg/m^2$ $p_v \leq 70\,mbar$

Table 5.3 *Continued.*

Author	Formulae for pressures (mbar)	Range of application
	$K = A_s'/A_v$	
	A_s' = minimum area of the smallest side of the enclosure	
	A_v = area of the enclosure	
	or $\quad K = V^{2/3}/A_v \not> 5$	
	V = volume of the non-cubical vessel	
Bradley and Mitcheson[4.85,4.86]	Safe vent areas $\bar{A}$, the maximum $p_m \not> P_v$, the vent opening pressures	
	(1) Initially uncovered vents	
	$\quad \bar{A}/\bar{v}_b \geq \exp[(0.64 - p_m)/2] \quad$ for $p_m > 1$ atmosphere	
	$\quad$ and	
	$\quad \bar{A}/\bar{v}_b \geq (0.7/p_m)^{0.5} \quad$ for $p_m < 1$ atmosphere	
	(2) Initially covered vents	
	$\quad \bar{A}/\bar{v}_b \geq (2.4/p_v)^{1.43} \quad$ for $p_v > 1$ atmosphere	
	$\quad$ and	
	$\quad \bar{A}/\bar{v}_b \geq (12.3/p_v)^{0.5} \quad$ for $p_v < 1$ atmosphere	
	$\quad \bar{A}$ = normalized vent area = $C_d(A_v/A_{ST})$	
	$\quad A_{ST}$ = surface area of the enclosure	
	$\quad \bar{v}_b$ = normalized burning velocity = $(v_b/v_{so})[(\rho_{uo}/\rho_{bo}) - 1]$	
	$\quad v_{so} = (\gamma_u P_o/\rho_{uo})^{0.5}$	
	$\quad v_b$ = initial laminar burning velocity	
	$\quad v_{so}$ = speed of sound in the unburnt gas	
	$\quad \rho_{uo}$ = initial density of unburned gas	
	$\quad \rho_{bo}$ = density of combustion products	
	$\quad \gamma_u = $ constant $= c_d'/c_v = \dfrac{\text{discharge coefficient}}{\text{volume coefficient}}$	

$$A_v = KL_1L_2/(p_v)^{1/2} \qquad (5.74)$$

where A_v = vent area (m^2) to resist the overpressure p_{so} to p_v (mbar)

L_1, L_2 = smallest dimensions (m) in rectangular enclosures of rooms 1 and 2, respectively.

5.6 Dust explosions

Section 2.6 gives an introduction to dust explosions. The well known methods introduced are the K_{st}, Schwal and Othmer, Maisey, Heinrich, Palmer and the Rust ones. There are similarities between gas and dust explosions, especially if the particle size of the dust is small and the turbulence level is low. Moreover, dust contains more volatiles. The qualitative effects on the explosion pressure of venting a dust explosion are the same as for a gas explosion. Several methods exist for estimating vent areas or explosion pressures in dust explosions. Vented gas explosions have a number of empirical relationships for calculating vent areas and explosions, as indicated in section 5.5. There are not many empirical equations for dust explosions. Many just consider simple scaling methods by extrapolating results from small vessels to large vessels. There are three experimentally-based methods which are discussed below:

(1) The vent ratio method.
(2) The vent coefficient method.
(3) The K_{st} factor method.

The vent ratio method is based on the vent area divided by the volume of the vessel. For vessels greater than 1000 ft^3 (283.68 m^3), the NFPA code 68[4.93,4.259] recommends the following:

Volume V	Vent ratio X	
1000–25 000 ft^3 (28.368–709.2 m^3)	1 ft^2/30–50 ft^3	
>25 000 ft^3	1 ft^2/80 ft^3	For heavy reinforced concrete walls
(> 709.2 m^3)	1 ft^2/60–80 ft^3	For light reinforced concrete, brick and wood wall construction
	1 ft^2/50–60 ft^3	For lightweight construction
	1 ft^2/10–50 ft^3	Large part of the volume with equipment

To convert to SI units, take 1 ft^3 = 0.028368 m^3

The vent coefficient method is the same as used for gas. The vent coefficient K is given by

$$K = L_1 L_2 / A_v \qquad (5.75)$$

where L_1, L_2 = the two smallest dimensions of the enclosure
A_v = the vent area

The K_{st} factor method[4.70−4.75] is the most widely used method for estimating explosion pressures and vent areas. Donat and BartKnecht's research work gives

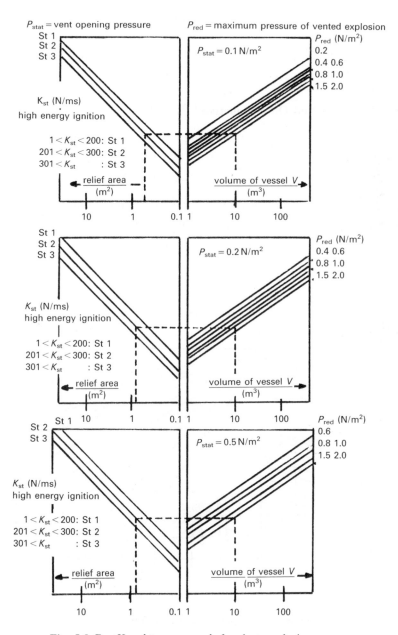

Fig. 5.9 BartKnecht nomograph for dust explosions.

data such as shown in fig. 2.78. The Hartmann bomb[4.58–4.67,4.75] and the 20 litre sphere method are the two techniques. The Hartmann bomb is a cylindrical tube of volume 1.2 litres, filled with dust. The dust is dispersed by an air blast using an ignitor. In the 20 litre sphere method, the dust is injected from a pressurized container. A standard time delay between dust injection and ignition is prescribed. From the maximum rate for pressure rise, the value of K_{st} based on the cube root law is computed as

$$K_{st} = (dp/dt)V^{1/3} \text{ bar m/s} \tag{5.76}$$

Various nomographs produced by BartKnecht for dust explosions are shown in Fig. 5.9.

5.6.1 The Schwal and Othmer method[4.58,4.67,4.75]

These authors used the Hartmann type apparatus with a volume of 1.3 litres. The vented explosion pressure p_v in lb/in² is given by

$$p_v = p_{max} \text{ (closed)}/10^{SX} \tag{5.77}$$

where S is the slope of the semi-log plot relating the measured p_v and the vent ratio X (ft²/100 ft²). Figure 5.10 shows the nomograph. The pressure is given by

$$\begin{aligned}(dp/dt)_{max} &= 5500/10^{2.443S}\\(dp/dt)_{av} &= 2700/10^{2.443S}\end{aligned} \tag{5.78}$$

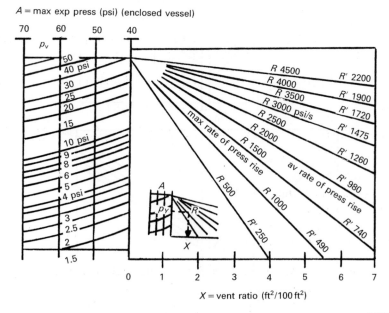

Fig. 5.10 Nomograph for pressure vent area ration (after Schwal & Othmer[4.58,4.67,4.75]).

By rearrangement of the terms of equation (5.78), the values of S may be derived:

$$\text{maximum } S = (1/2.443)\log[5500/(dp/dt)_{max}]$$

or

$$\text{average } S = (1/2.443)\log[2700/(dp/dt)_{av}]$$

(5.79)

5.6.2 Maisey method[4.77]

Here similarities between gas and dust explosions are investigated by use of the equivalence coefficient method. The method relates dust explosion pressures to gas explosion pressures under similar conditions. The average rates of pressure rise and maximum explosion pressures are measured in an enclosed Hartmann bomb. The standard gas explosion medium is hexane/air. Maisey gives the following empirical relation:

$$p_v G/p_v D = \left[(dp/dt)^G_{av} \times \frac{p^D_{max}}{p^G_{max}(dp/dt)^D_{av}} \right] \times f_P f_S f_T$$

(5.80)

where G and D represent gas in Maisey's Hartmann apparatus and f_P, f_S and f_T are correction factors for the initial pressure, fuel concentration and turbulence, respectively

Table 5.4 gives an extract of the results from Maisey's Hartmann apparatus which are relevant to structural products.

5.6.3 Heinrich method[4.75]

This is a theoretical method to predict the vented dust explosion pressure and vent areas. The effective pressure relief is given by

$$p_v - p_a = (dp/dt)_v = (dp_{ex}/dt)_v - \text{rate of pressure rise}$$

(5.81)

Equation (5.81) is transformed to the following equation:

$$A_v = (V_L)^{1/3} V^{2/3} (dp_{ex}/dt)_{v,v_L} / C'_d (2RT/\bar{M})^{1/2} [p_{v_{max}}(p_{v_{max}} - p_a)]^{1/2}$$

(5.82)

where C'_d = discharge coefficient
T = temperature
$\bar{M}$ = mean molecular weight
R = gas constant
p_a = ambient pressure
V_L = volume of the tested vessel

For an adiabatic process, the equation for A_v, the vent area, is given by

$$A_v = \frac{V_L^{1/3}}{\gamma'} \left(\frac{\gamma' + 1}{2} \right)^{1/(\gamma' - 1)} \frac{\bar{M}(\gamma' + 1)}{2RT\gamma'} \frac{V^{2/3}}{C'_d p_v} \left(\frac{dp_{ex}}{dt} \right)_{max} V_L$$

(5.83)

Table 5.4 Maisey's results.[4.77]

Material	Explosibility index	Ignition temp.(°C) cloud	Most explosive mixture				Rate of pressure rise	
			Minimum ignition energy cloud (J)	Minimum explosive concentration (oz/ft³)	Maximum pressure (psi g)		Average	Maximum
Thermoplastic resins								
Acetal								
Acetal, linear (polyformaldehyde)	>10.0	440	0.02	0.035	89		1600	4100
Acrylic								
Methyl methacrylate polymer	6.3	480	0.02	0.03	84		900	2000
Methyl methacrylate moulding compound, cyclone fines	>10.0	440	0.015	0.02	101		450	1800
Methyl methacrylate moulding compound, drier fines	6.1	440	0.02	0.03	80		800	1800
Thermosetting resins								
Alkyd								
Alkyd moulding compound, mineral filter, not self-extinguishing	>0.1	500	0.12	0.155	40		150	300
Allyl								
Allyl alcohol derivative CR-39, from dust collector	>10.00	510	0.02	0.035	91		3000	7500
Amino								
Melamine formaldehyde, unfilled laminating type, no plasticizer	>0.1	810	0.32	0.085	81		350	800
Melamine formaldehyde, unfilled laminating type, with plasticizer	0.7	790	0.05	0.065	91		800	1800
Urea formaldehyde, spray-dried	>0.1	530	1.28	0.135	52		200	500
Urea formaldehyde moulding compound, grade II fine	1.0	460	0.08	0.085	89		1300	3600
Epoxy								
Epoxy, no catalyst, modifier or additives	>10.0	540	0.015	0.02	74		2600	8500
Epoxy-bisphenol A mixture	1.9	510	0.035	0.03	85		1000	2200

Table 5.4 *Continued.*

Material	Explosibility index	Ignition temp.(°C) cloud	Minimum ignition energy cloud (J)	Most explosive mixture			Rate of pressure rise	
				Minimum explosive concentration (oz/ft³)	Maximum pressure (psi g)		Average	Maximum
Special resins and moulding compounds								
Cold-moulded								
Gilsonite, from Michigan	>10.0	560	0.025	0.02	89		1200	3800
Petroleum resin (blown asphalt), regular	>10.0	510	0.025	0.025	94		2200	4600
Coumarone-indene								
Coumarone-indene, hard	>10.0	550	0.01	0.015	93		2800	11000
Natural								
Cashew oil phenolic, hard	>10.0	490	0.025	0.025	81		1500	4000
Lignin, pure	0.8	510	0.16	0.065	80		1700	4700
Rosin, DK	>10.0	390	0.01	0.015	87		2800	12000
Rosin, pine	–	440	–	0.055	82		1900	7500
Shellac	>10.0	400	0.01	0.02	73		1400	3600
Sodium resinate, dry size, plasticized	1.0	440	0.1	0.015	81		800	2000
Rubber								
Rubber, crude, hard	7.4	350	0.05	0.025	80		1200	3800
Rubber, synthetic, hard, 33% sulphur	>10.0	320	0.03	0.03	93		1100	3100
Fillers								
Cellulose	2.8	480	0.08	0.055	119		1800	4500
Cellulose flock, chemical cotton, fine cut	1.4	460	0.06	0.065	87		1300	2900
Cork	9.7	470	0.045	0.035	80		2300	7000
Cotton flock, ground, filler for phenolic moulding compound	>10.0	470	0.025	0.05	94		1800	6000
Wood flour, filler	>10.0	450	0.04	0.04	97		2300	7500
Coal, Pittsburgh	1.0	610	0.06	0.055	83		800	2300
Hexane, quiescent	1880.0	235	0.0002	0.05	100		2000	4250

Heinrich suggests that $\bar{M}$ has the value 29 g mole^{-1} for dust/air mixtures, and C'_d has a value of 0.8. The values of γ' and R for an ideal gas are considered to be satisfactory for dust/air mixtures. The value of γ' is C'_d/C_v.

5.6.4 Palmer's equation[4.83]

Palmer suggests that, in enlarged containers with a large ignition source, combustion takes place throughout the entire volume of the container. Palmer gives an expression relating the mass rate, $\bar{M}_b$, of combustion products to maximum pressure, p_{max}, as

$$\frac{d\bar{M}_b}{dt} = \frac{p_{max}}{p_a} \times \frac{\rho_a}{\gamma' p_1} \left(\frac{dp}{dt}\right)_{max \ (enclosed)} \tag{5.84}$$

where p_1 = pressure in a closed vessel
    p_a = ambient pressure
    $\gamma' = C'_d/C_v$
    ρ_a = density

The following equations have been derived for low- and high-pressure cases:

Low pressure

$$p_{max} - p_a = 2.3\rho_c P_o[(V/A_v)(dP/dt)_{max}]2/C_d^2\gamma'^2 p_{max(enclosed)}^3 \tag{5.85}$$

where ρ_c is the density of an unburnt dust suspension at pressure p_a and of the combustion products at pressure $p_1 \approx 0.6 p_{max}$.

High pressure

$$\frac{1}{p_{max} - p_a} = \frac{1}{p_{max(enclosed)} - p_a} + \frac{K_p A_v p_{max(enclosed)}}{0.8 V \rho_c (dp/dt)_{max}} \tag{5.86}$$

where K_p is a constant and is defined by

$$K_p = C'_d\{\gamma'(\rho_a/p_a)[2/(\gamma' + 1)]^{(\gamma' + 1)/(\gamma' - 1)}\}^{1/2} \tag{5.86a}$$

The derivation of equation (5.86) utilizes the approximation

$$(dp/dt)_{av} = 0.4(dp/dt)_{max} \tag{5.86b}$$

Figure 5.11 shows a comparison of Palmer's results with those of Donat.

5.6.5 Rust method

This method relies on data from closed vessels. It includes a moving flame front travelling spherically and vent panel dynamics for a pressure–time profile. Table 5.5 gives a summary of the method. The values of E_x and C_2 are functions given in Fig. 5.12.

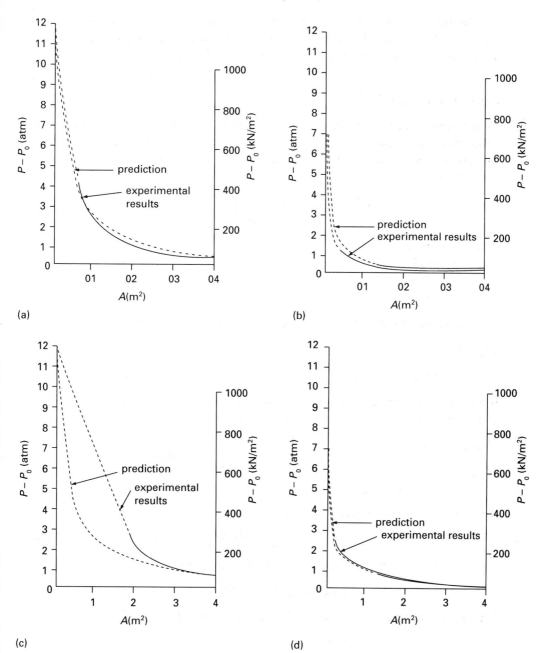

Fig. 5.11 A comparison of Palmer's[4.83] and Donat's[4.74] results. (a) Aluminium dust explosion pressures in a $1\,m^3$ vessel; (b) coal dust explosion pressures in a $1\,m^3$ vessel; (c) aluminium dust explosion pressures in a $30\,m^3$ vessel; (d) coal dust explosion pressures in a $30\,m^3$ vessel.

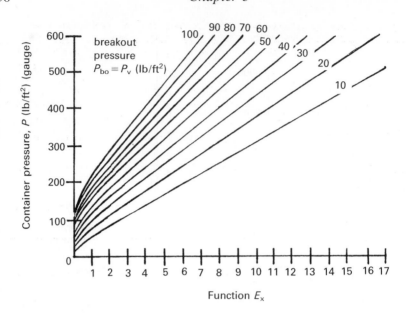

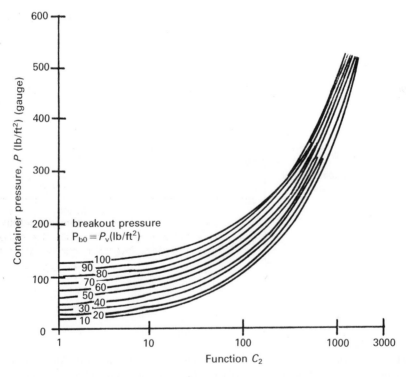

Fig. 5.12 E_x and C_2 functions.

Table 5.5 A summary of the Rust method.

Steps of calculation

(1) The explosion pressure in an enclosed vessel is given by

$$P = K_D t^3/V, \text{ where } K_D = (4/3)\pi S_u^3 P_{\text{max(enclosed)}} \tag{1}$$

(2) Differentiating:

$$dP/dt = 3K_D t^3/V \tag{2}$$

Eliminating t:

$$(3K_D/V)(PV/K_D)^3 = 3K_D^{1/3} P^{2/3}/V^{1/3}$$

where $K_D = [(dP/dt)/16]^{1/3}$ from Hartman data

(3) A_v = vent area in ft^2

$$A_v = (8.354 \times 10^{-5}) F_s (P_{\text{max(enclosed)}} V)^{2/3} K_D^{1/3}/P_{\text{max}}^{1/2}$$

where F_s = shape factor = A_F/A_{SP}
A_F = large spherical surface of the container
A_{SP} = surface area of the sphere which has a volume equal to that of the container

The distance x or rotation θ can be calculated for the achievement of full venting by checking the release panel dynamics

Values of x and θ

Horizontal translation:

$$x_i = \frac{g}{w}\left(\frac{V}{K_D}\right)^{2/3} C_2$$

Horizontal rotation around vertical shaft:

$$\theta_1 = \frac{3g}{2wL}\left(\frac{V}{K_D}\right)^{2/3} C_2$$

Vertical translation upwards:

$$X_i = g\left(\frac{V}{K_D}\right)^{2/3}\left(\frac{C_2}{w} - E_x\right)$$

w = weight of vent panel, lb/ft^2

Vertical rotation around horizontal shaft, against gravity:

$$\theta_1 = \frac{3g}{2L}\left(\frac{V}{K_D}\right)^{2/3}\left(\frac{C_2}{w} - E_x\right)$$

Vertical rotation around horizontal shaft, with gravity:

$$\theta_1 = \frac{3g}{2L}\left(\frac{V}{K_D}\right)^{2/3}\left(\frac{C_2}{w} + E_x\right)$$

In the above equations, C_2 and E_x are defined as

$$C_2 = \frac{p^{5/3}}{20} - \frac{(P_{bo})^{4/3} p^{1/3}}{4} + \frac{(P_{bo})^{5/3}}{5}$$

$$E_x = \frac{p^{2/3}}{2} - (P_{bo})^{1/3} p^{1/3} + \frac{(P_{bo})^{2/3}}{2}$$

$P_{bo} = P_v$ = vent opening pressure

continued

Table 5.5 *Continued.*

Distance of panel movement to achieve full venting:
Horizontal translation:

$$x = \frac{WL}{2(W+L)}$$

L = length of vent panel
W = width of vent panel

$$\theta = \frac{W}{W+L}$$

p = explosion pressure

Dust	Vessel volume (m³)	p_{max} dp/dt (pressure rise)			
		atm (gauge)	kN/m²	atm/s	kN/m²s
Coal	1	69	690	—	—
	30	—	—	27	2700
Aluminium	1	12.0	1200	—	—
	30	—	—	195	19500
Organic pigment	1	10.4	1040	—	—
	30	—	—	92	9200

5.7 Explosions in soils

The energy of explosion is used in the fields of civil/structural engineering, mining, agriculture and forestry. Explosions in underground installations arise from explosives, plant failure or from external rockets/missiles detonating at great depths. Nuclear explosions are adopted for finding underground resources. In mining, explosions have been adopted for tunnelling, coal mining and canal works. In agriculture and forestry, explosive methods are used to sink pits for tree planting, stump extraction and irrigation system construction. The effects of an explosion depend on the characteristics of the soils, i.e. whether hard or semi-rocks, cohesive soils or cohesionless or loose soils. Section 2.5.1.3 gives useful relevant information. In addition, Table 5.6 presents a soil classification on the basis of explosions occurring underground.

The pressure of the explosive gases immediately after detonation causes the soil adjacent to the charge surface to be crushed and eventually to change into a liquid state. The zone is *highly deformed*. The size of the deformation changes as the distance from the centre of the source of detonation increases. Due to high tensile stresses, cracks/fissures are developed. From the source, the soil which looks like a brittle one is surrounded by a chamber of three zones in the order: *crushing zone*, *rupture zone* and *elastic zone*. Various waves, their reflections and rarefactions occur which have been described earlier. In some cases swelling occurs.

Figure 5.13 shows a typical phenomenon of the soil subject to internal explosion. Chemical and nuclear charges exploding on the Earth's surface (contact explosions) create depressions by compacting the soil beneath. Most of the explosion energy is dissipated into the air. As discussed in section 2.5.1.3, various craters are formed. This area is discussed in detail later on in this chapter.

Table 5.6 Classification of soils under explosion effects.

Soil	Density, ρ (g/cm³)	Propagation velocity of a longitudinal wave in a mass v_{sl} (ms⁻¹ × 10³)	Poisson's ratio, ν	Modulus of elasticity, in tension and compression, E (kNcm⁻² × 10⁵)	Modulus of elasticity in shear, G (kNcm⁻² × 10⁵)	Bulk modulus, K (kNcm⁻² × 10⁵)
Sand	1.4–2.0	0.3–1.3	—	0.003	—	—
Clay	1.4–2.5	0.8–3.3	—	0.003	—	—
Limestone	2.42	3.43	0.26	2.17	0.85	1.71
Slate	2.45	6.92	0.24	10.22	4.13	6.50
Granite	2.60	5.20	0.22	6.20	2.54	3.77
Quartzite	2.65	6.42	0.25	9.26	3.70	7.89
Limestone	2.70	6.33	0.33	7.31	2.74	4.36
Granite ore	2.71	6.41	0.33	7.57	2.84	7.59
Slate	2.71	5.75	0.25	7.60	3.04	5.09
Marble white	2.73	4.42	0.20	3.84	1.60	3.32
Marble red	2.73	5.47	0.26	6.75	2.68	4.74
Marble black	2.82	5.90	0.32	5.74	2.18	7.09
Gneiss	2.85	6.08	0.28	8.35	3.26	6.38
Crystalline dolomite	2.85	6.60	0.28	9.83	3.83	7.59
Gabbrodiabase	2.85	5.40	0.26	7.40	2.88	5.58
Coal	1.25	1.20	0.36	0.18	0.07	0.09

continued

Table 5.6 *Continued.*

Soil	Lame's constant (kNcm^{-2} × 10^5)	Propagation velocity of transverse wave, v_{st} (ms^{-1} × 10^3)	Compression strength, σ_c (kNcm^{-2})	Shear strength, τ (kNcm^{-2})	Tensile strength, σ_t (kNcm^{-2})
Sand	—	—	50–60	—	—
Clay	—	—	65–105	1–2	4
Limestone	0.91	1.86	450	110	70
Slate	3.75	4.06	460	70	50
Granite	2.06	3.10	1550	—	180
Quartzite	3.70	3.70	1480	240	—
Limestone	5.56	3.70	1620	210	—
Granite ore	5.71	3.20	1520	180	—
Slate	3.07	3.32	1760	—	40
Marble white	1.06	2.80	750	—	150
Marble red	2.94	3.10	1200	—	250
Marble black	3.85	3.28	750	—	210
Gneiss	4.18	3.37	1175	340	—
Crystalline dolomite	5.03	3.63	1885	1125	340
Gabbrodiabase	3.12	3.14	1500	—	230
Coal	0.05	0.72	80	30	5

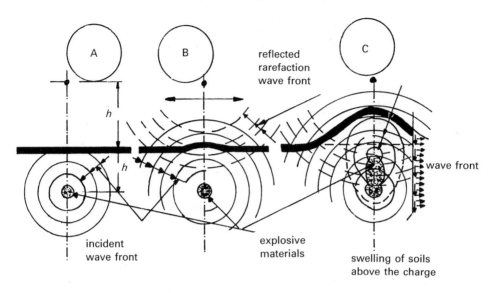

Fig. 5.13 Explosions in soils.

5.7.1 Explosion parameters for soils/rocks

The wave profile in soils is dependent upon soil type, time and distance of the source. It is widely believed that up to a distance of 100 to 120 γ_s (γ_s being equal to R_w, the radius of the charge), the charge weight W follows the relation:

$$W = \tfrac{4}{3}\pi \, \rho_w R_w^3 \tag{5.87}$$

where ρ_w is the density of the charge. Here the maximum overpressure is higher than the absolute value of the underpressure due to rarefaction. At a distance of 400 to 500 R_w, the overpressure is low and is of the same order as the absolute value of the underpressure of the rarefaction wave. The zone of the fissures does not exceed 5 to 6 R_w. Just before the fissure zone, the crushing zone generally extends from 2 to 3 R_w. The explosive wave in an unbounded rock medium obeys the law of model similarity; the overpressure duration $(t/R_w) \times 10^3$ on the relative distance $\bar{x} = x/R_w$ is written for a *spherical PETN* charge as

$$t/R_w = 10^{-3}(\alpha_0 + \alpha_1 \bar{x}) \tag{5.88}$$

where $\alpha_0 = 2.51$ for diabase with $\alpha_1 = 4.56 \times 10^{-3}$
$\alpha_0 = 3.18$ for marble with $\alpha_1 = 5.30 \times 10^{-2}$
$\alpha_0 = 4.38$ for granite with $\alpha_1 = 0.135$
$\alpha_0 = 4.10$ for water-saturated limestone with $\alpha_1 = 0.192$

The corresponding overpressure can be written as

$$p_{so} = \left(\frac{a_1 \times 10^3}{\bar{x}^3} + \frac{a_2 \times 10^3}{\bar{x}^2} + \frac{a_3}{\bar{x}} \right) \times 10^2 \tag{5.89}$$

The value of a_1, a_2 and a_3 are given below:

Rock	a_1	a_2	a_3
Diabase	18.60	88.80	202.00
Marble	1.68	4.70	46.67
Granite	1.28	20.00	38.6
Saturated limestone	−1.50	21.30	−3.91

Under the same PETN charge, the maximum mass velocity, v_m, is computed as

$$v_m = \frac{\bar{\alpha}_1}{\bar{x}^3} + \frac{\bar{\alpha}_2}{\bar{x}^2} + \frac{\bar{\alpha}_3}{\bar{x}} \text{ cm/s} \qquad (5.90)$$

For granite, $\bar{\alpha}_1 = 33\,100$ cm/s; $\bar{\alpha}_2 = -398$ cm/s; $\bar{\alpha}_3 = 36.25$ cm/s. For other soils where $\bar{x} > 50$ all constants become equal for all kinds of explosives. The corresponding p_{so} is given by

$$p_{so} = \rho v_{pz} \times 10^2 (v_m) \text{ kN/cm}^2 \qquad (5.91)$$

where ρ = rock density, kNcm^{-4}/s^2

$\quad v_{pz}$ = longitudinal wave velocity, cm/s

The impulse $F_1(t)$ of the pressure wave on soils is written as

$$F_1(t) = \int_0^t p_{so}(t)dt \qquad (5.92)$$
$$= A_2 \sqrt[3]{W}(1/\bar{x})a_2 \text{ kNs/cm}^2$$

where values of A_2 and a_2 for water-saturated sands, in particular, vary as $A_2 = 0.08$ to 0.03 and $a_2 = 1.05$ to 1.50; $\bar{x}$ is the distance from the explosion.

The time duration $t = \tau_m$ for a large number of soils from the moment of explosion to the instant at which p_{so} develops at a distance $\bar{x}$ is given by

$$t = \tau_m = 4.35 \times 10^{-3} \sqrt[3]{W}(\bar{x})^{1.6} \text{ seconds} \qquad (5.93)$$

For contained and contact explosions in sandy loams, the values of p_{so} are written as

$$p_{so} = 11.1(\bar{x}/\sqrt[3]{W})^{-2.7} \text{ kN/cm}^2 \text{ (contained explosion)} \qquad (5.94)$$

$$p_{so} = 11.1[\bar{x}/\sqrt[3]{(K_w \times W)}]^{-2.7} \text{ kN/cm}^2 \text{ (contact explosion)} \qquad (5.95)$$

where K_w is the reduction coefficient of the charge ≈ 0.28. The maximum mass velocity v_m for both contained and contact explosions can be written as

$$v_m \text{ (contained explosion)} = 4.72[\bar{x}/(\sqrt[3]{W})^{2.06}] \qquad (5.96)$$

$$v_m \text{ (contact explosion)} = 1.08[\bar{x}/(\sqrt[3]{W})^{-1.65}] \qquad (5.97)$$

where v_m is in m/s and $\bar{x}$ is the distance in m.

Cylindrical charges are used in earth work. Assuming, for comparison, that the cylindrical charge is W_c, then the arrival time of the wave front is computed as

$$t/\sqrt{W_c} = 0.335 \ \bar{x}^{10.061} \ \text{sm}^{1/2}/\text{kg}^{1/2} \tag{5.98}$$

The mass velocity is written as

$$v_m = 263(\bar{x})^{-0.965} \ \text{m/s} \tag{5.99}$$

The following values are computed for $t/\sqrt{W_c}$, $\bar{x}$ and v_m:

$\bar{x}$	v_m	$t/\sqrt{W_c}$
20	9.0	4.0
40	7.0	12.5
60	5.7	22.0
80	3.9	30.0
100	3.0	40.0
120	2.5	48.0
140	2.0	58.0

5.7.1.1 Parameters for wave types

An underground explosion, as stated earlier, causes a surface motion of the earth medium. Some of these waves have been described in section 5.2. The *P wave* is produced by the pressure wave transmitted by the first source. The value of the velocity v_{sz} (vertical) is given by

$$v_{sz} = K^P(W^{1/3}/\bar{x})$$
$$= K^P(\sqrt[3]{W})^2 \ \text{cm/s} \tag{5.100}$$

The time t^P for the maximum growth of the motion is given by

$$t^P = K^t(\sqrt[6]{W}) \tag{5.101}$$

The amplitude X^P is given by

$$X^P = K_x[\sqrt[6]{W}(\sqrt[3]{W}/\bar{x})^2] \ \text{cm} \tag{5.102}$$
$$= 2v_{sz}(t/\pi)$$

The values of constants K^P, K^t and K_x are given below:

	K^P	K^t	K_x
Loam	1100	0.01	75
Saturated sand	700	0.015	60
Granite	700	0.0032	15
Limestone	700	0.0032	15

The maximum soil velocity v_s^n for the N wave, which is transmitted by the second source — the cupola-shaped swelling of the soil surface in the epicentre region, is given by

$$v_s^n = K^n(\sqrt[3]{W}/\bar{x})^{1.7} \text{ cm/s} \tag{5.103}$$

The time t^n is evaluated as

$$
\begin{aligned}
t^n &= 0.0065 W^{1/6} \text{ s (for rocks)} \\
&= 0.06 W^{0.21} \text{ s (for clays)}
\end{aligned}
\tag{5.104}
$$

The maximum amplitude X^n is given by

$$X^n = v_s^n(t^n/2\pi) \tag{5.105}$$

For saturated sands, loams, limestones and granite, the values of v_s^n are, respectively, $300-900$, $1750-3000$ and $4200-4500$ m/s. In general form, v_s^n is given as in equation (5.103).

Section 5.2.1 gives a brief introduction on the formation of R waves and their effects on soils. The R wave is a long-period wave which is transmitted by the second explosion source. It expands conically with moving particles departing from the radial direction towards the axis of symmetry. The maximum vertical and horizontal motion velocities, v_{szv} and v_{szh}, respectively, are given by

$$v_{szv} = 2\pi X_{RZ}/t_R \tag{5.106}$$

$$v_{szh} = v_{szv}(X_{RX}/X_{RZ}) \tag{5.107}$$

where X_{RZ} and X_{RX} are the amplitudes of motion in the vertical and horizontal directions, respectively.

$$X_{RZ} = 1900 \ (x_e/v_{sz})\theta(d_0/x_e)f(\bar{x}) \text{ mm} \tag{5.108}$$

where x_e = the rupture zone = $4.5\sqrt[3]{W}$ for granite
$$= 3\sqrt[3]{W} \text{ for limestone}$$
$$= (6/8)\sqrt[3]{W} \text{ for clay}$$
$$= 2.5\sqrt[3]{W} \text{ for sandy clay}$$

d_0 = depth of charge

$$f(\bar{x}) = (\bar{x})^{-0.5}e^{-1.75(\bar{x})^{0.2}} \tag{5.109}$$

$$\bar{x} = x/\sqrt[3]{W} \text{ m/kg}^{1/3}$$

x	$f(\bar{x})$
100	1.24×10^{-3}
200	4.59×10^{-4}
500	1.025×10^{-4}
1000	2.99×10^{-5}
2000	7.5×10^{-6}
3000	3.18×10^{-6}

5.7.2 Explosion cavity

When an explosion occurs in soil, cavities are created which become filled with explosive gases of certain temperatures and pressures. These gases penetrate into the voids, expel the contained water and make the cavity dry. After a certain time, these explosive gases penetrate further into voids and fissures of the soil, the walls of the cavity eventually slide down and the soil is deformed. In water-bearing sands, the explosion cavity is distorted in a few days. In loamy-water (unsaturated) soils, the cavity happens to be stable for some years. However, in sandy-water (unsaturated) soils, the deformation occurs immediately after the detonation. The shape and size depend on the characteristics of the soil and the type of charge.

For a *spherical charge*, the cavity radius R_{vd} is given by

$$R_{vd} = \bar{K}_{vd}R_w$$
$$= \bar{K}^*_{vd} \sqrt[3]{W} \qquad (5.110)$$

where R_w = charge radius of a spherical charge
$\quad\quad W$ = charge weight
$\bar{K}^*_{vd} \approx 0.053 K_{vd}$
where K_{vd} = coefficient of proportionality m/kg$^{1/3}$

For a *cylindrical charge*, the explosive cavity will have a cylindrical shape, except at the terminal regions.

$$R_{vd} = \bar{K}_{vd}R_w = \bar{K}^*_{vd}\sqrt{W_c} \qquad (5.111)$$

	K_{vd}	$\bar{K}_{vd}$ $(m^{3/2}\,kg^{-1/2})$	$\bar{K}^*_{vd}$
Loam	11.3/13.1	28.3	0.4
Sandy soil	5.6/7.4	24.8	0.35
Limestone	3.8/4.7	19.5	0.27
Shale	1.7/2.9	10.2	0.19

For water-bearing soils, $R_{vys} = (1.2/1.3)\,R_{vd}$ of the above.

5.7.2.1 Dynamic analysis for explosion cavity formation

A great number of theories exist on the laws of deformation of soils. This is due to the fact that the constituent materials of soils and their characteristics vary. General reference to a crater normally means the visible crater or hole left after an explosion. The true crater is the hole actually excavated after the external body has penetrated and exploded within the soil. Debris actually falls back into the true crater. If the explosion occurs deep enough, the true cavity or crater is called a *camouflet*. This and other types are given in Fig. 5.14. Data on crater prediction in concrete and soil/rock, etc., are given in Tables 2.67 to 2.70. For the dynamic analysis of cratering, two Lagrange equations of motion for any medium are given:

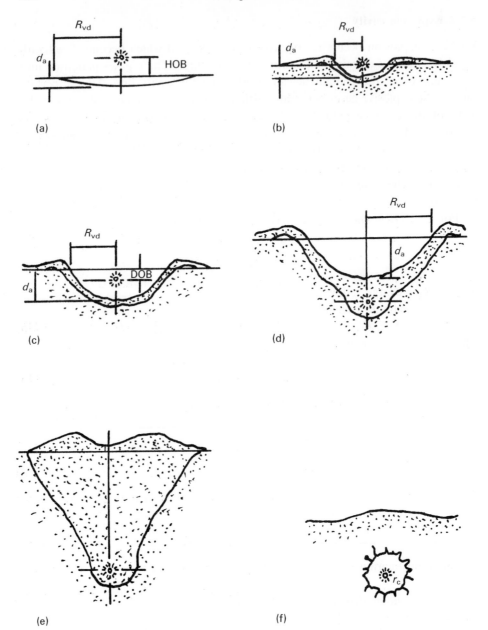

Fig. 5.14 Crater formations. (a) Low airburst (HOB = height of burst); (b) surface burst; (c) shallow depth of burst (DOB); (d) optimum DOB; (e) deeply buried; (f) camouflet.

$$\partial p/\partial x = -\rho_{1s}[(\partial v_m/\partial t) + v_m(\partial v_m/\partial x)] \tag{5.112}$$

$$\partial/\partial x[\bar{x}^2 v_m(\bar{x}, t)] = 0 \tag{5.112a}$$

The boundary conditions for the shock wave front are formed according to the law of conservation of mass and momentum.

$$\rho_a \dot{x}_m(x,t) = \rho_{1s}(\dot{x}_m - v_m) \quad \text{conservation of mass} \tag{5.113}$$

$$p_m - p_a = \rho_a \dot{x}_m v_m \quad \text{conservation of momentum} \tag{5.114}$$

where v_m = mass velocity of soil depending on (x,t)
x_m = distance
t = time
p_m = pressure dependent on (x,t)
ρ_{1s} = density of the soil medium behind the shock front
ρ_a = normal density at ambient temperature

On the interface of the gases and the medium, the above pressure and velocity are written as

$$p_{pa} = p(x,t) = \rho_{1s}[(f(t)/x) - (f^2(t)/2x^4)] + B(t) \tag{5.115}$$

where x is a general value of $x_m = R_{vp}$.

$$v_m = v_{mp} = f(t)/x^2 \tag{5.116}$$

From equations (5.116) and (5.112a), the functional values of $f(t)$ and $\dot{f}(t)$ are evaluated as

$$f(t) = [1 - (\rho_a/\rho_{1s})]x_m^2 \dot{x}_m \tag{5.117}$$

$$\dot{f}(t) = [1 - (\rho_a/\rho_{1s})]x_m(2\dot{x}_m^2 + x_m \ddot{x}_m) \tag{5.118}$$

From equations (5.115), (5.117) and (5.118), the following becomes the volume deformation phenomenon:

$$dV = \Delta V_{vd}/V_{vd} = (x)(x^2 \sin\phi \, d\phi \, d\psi \, dx) \tag{5.119}$$

where ϕ and ψ are geometrical parameters. The following pressure p is computed:

$$p = p_a + \rho_a \bar{D}\dot{x}_m^2 + \bar{D}_1(2\dot{x}_m^2 + x_m \ddot{x}_m)[(x_m/x) - 1] - \tfrac{1}{2}\bar{D}_1 \dot{x}_m^2[(x_m/x)^2 - 1] \tag{5.120}$$

where $\bar{D} = [1 - (\rho_a/\rho_{1s})]$

$$\bar{D}_1 = \rho_{1s}\bar{D}$$

The medium between the shock wave front and the explosives dictates the law of conservation of mass. Equation (5.113) dictates

$$\tfrac{4}{3}\pi(x_m^3 - x_w)\rho_a = \tfrac{4}{3}\pi(x_m^3 - R_{vp}^3)\rho_{1s} \tag{5.121}$$

Neglecting x_w^3 in comparison with x_m^3

$$R_{vp} = \bar{D}^{1/3}v_m \tag{5.122}$$

where R_{vp} is the radius of the sphere of the explosive gases. By substituting x_m into equation (5.122) and taking $p = p_{pa}$ and $x = R_{vp}$, the following equation is derived:

$$\bar{P} = 2(p_{pa} - p_a)/\rho_{1s}(1 - \bar{D}^{1/3}) = 2R_{vp}\ddot{R}_{vp} + a\dot{R}_{vp}^2 \tag{5.123}$$

$$\text{where } a = 3 + \bar{D}^{1/3}(1 + \bar{D}^{1/3} + \bar{D}^{2/3}) \tag{5.124}$$

Knowing that

$$d/dR_{vp}(\dot{R}_{vp}^2 R_{vp}^a) = PR_{vp}R_{vp}^{a-1} \tag{5.125}$$

$$v_m^2 = v_{pv}^2 = \frac{1}{R_{vp}^a}\int_{x=R_w}^{x=R} PR_{vp}R_{vp}^{a-1}dR_{vp} \tag{5.126}$$

$$p_{pa} = p_w^*(V_w/V_{vp})^9 \tag{5.127}$$

the crater radius R_{vd} may be computed from

$$m_{vp} = \left\{\frac{2p_a}{\rho_{1s}(1 - \bar{D}^{1/3})a}\left[\left(\frac{R_{vd}}{R_{vp}}\right)^2 - 1\right]\right\}^{1/2} \tag{5.127a}$$

and in some cases

$$v_{pv}^2 = {}^*v_w^2$$

$$\text{as } R_{vd} = R_w\left[\frac{{}^*v_w^2 + \dfrac{2{}^*p_w}{(9 - a)\rho_{1s}(1 - \bar{D}^{1/3})}}{2p_a/a\rho_{1s}(1 - \bar{D}^{1/3})}\right]^{1/a} \tag{5.128}$$

Since $m_{vp} = dR_{vp}/dt$, by substituting into equation (5.127a) and integrating between R_w and R_{vd}, the time t required for the creation of the explosion crater is given by

$$t = \frac{R_{vd}}{{}^*v_w}\left[\frac{\rho_{1s}{}^*v_w^2 (1 - \bar{D}^{1/3})a}{2p_a}\right]^{1/2}\bar{F} \tag{5.129}$$

$$\text{where } \bar{F} = \bar{F}_1 + \frac{2}{a+2}\left(\frac{R_w}{R_{vd}}\right)^{(a+2)/2}$$

$$\bar{F}_1 = \frac{1}{a}\int z^{(2-a)/a}(1 - z^a)d(z^a) = \frac{\sqrt{\pi}}{a}\frac{\Gamma[(2+a)/2a]}{\Gamma[(1+a)/a]}$$

5.7.3 Ground shock coupling factor due to weapon penetration

The stress and ground motions will be greatly enhanced if a weapon penetrates more deeply into the soil or a protective burster layer before it detonates. The concept of an equivalent effect coupling factor is introduced to account for this effect on the ground shock parameters. The coupling factor, f_c^*, is defined as the ratio of the ground shock magnitude from a partially buried or shallow-buried weapon (Fig. 5.15) (near-surface burst) to that from a fully buried weapon (contained burst) in the same medium. It does not indicate the size of the charge, but

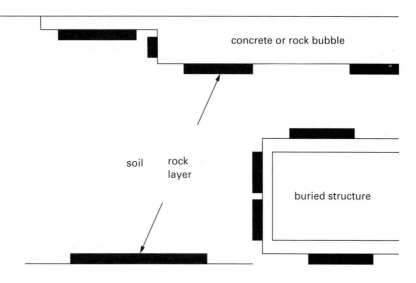

Fig. 5.15 A buried structure.

it is a reduction factor for a contained burst, as described earlier in section 5.7.1.

A single coupling factor is applicable for all ground shock parameters that depend upon the depth of burst (measured to the centre of the weapon) and the medium in which the detonation occurs, i.e. soil, concrete or a combination thereof plus air.

$$(f_c^*) = \frac{(P,V,d,I,a) \text{ near surface}}{(P,V,d,I,a) \text{ contained}}$$

The coupling factor $(f_c^*) = 0.14$ for air is practically constant.

When a weapon penetrates into more than one material (see Fig. 5.16), for example, a long bomb that passes partially through a concrete slab, steel plates, etc., and into the soil beneath them, the coupling factor is then equal to the sum of the coupling factors for each material, weighted according to the proportion of the charge contained within each medium. The coupling factor is written as

$$(f_c^*) = \Sigma f_{ci}^*(W_i/W) \tag{5.130}$$

where f_{ci}^* = coupling factor for each component
W_i = weight of the charge in contact with each component
W = total charge weight

Since most bombs are cylindrical, the coupling factor can also be written as

$$f = \Sigma f_{ci}^* (L_i/L) \tag{5.131}$$

where L_i is the length of the weapon in contact and L is the total length.

The peak pressure p_{so}, velocity v_{so} and acceleration $\dot{v}_{ps}$ are written in terms of the coupling factor f_c^* and are given below:

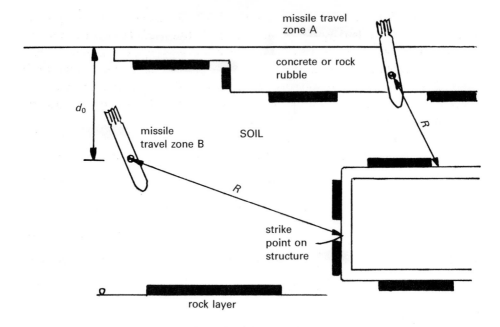

Fig. 5.16 Missile exploding on a buried structure.

$$p_{so} = f_c^* \rho \dot{v}_{ps} 160(R/W^{1/3})^{-n} \qquad (5.132)$$

$$v_{so} = f_c^* 160(R/W^{1/3})^{-n} \qquad (5.133)$$

$$\dot{v}_{sp} W^{1/3} = f_c^* 50 \dot{v}_{sp}(R/W^{1/3})^{-n-1} \qquad (5.134)$$

$$\delta/W^{1/3} = f_c^* 500(1/\dot{v}_{sp})(R/W^{1/3})^{-n+1} \qquad (5.135)$$

$$F_1(t)/W^{1/3} = f_c^* \rho[1.1(R/W^{1/3})]^{-n+1} \qquad (5.136)$$

where p_{so} = peak pressure (psi)
$\quad f_c^*$ = coupling factor for near-surface detonations
$\quad v_{so}$ = seismic velocity (ft/s)
$\quad R = x$ = distance to the explosion (ft)
$\quad W$ = charge weight (lb)
$\quad v_{ps}$ = peak particle velocity (ft/s)
$\quad \dot{v}_{sp}$ = peak acceleration (**g**s)
$\quad \delta$ = peak displacement (ft)
$\quad F_1(t)$ = peak impulse (lb s/in²)
$\quad \rho$ = mass density (lb s²/ft⁴) = 144 or equivalent
$\quad n$ = attenuation coefficient

These imperial values can be converted into SI units using standard conversions. A standard input is given in Tables 2.67 to 2.70.

As the weapon detonates near a structure, shock reflections from the ground surface or from layers such as a water table or rock layer can combine with the directly transmitted stress waves to cause a significant change in the magnitude and/or time t of the loading on the underground structure. Reflections from the ground surface will produce tensile waves which may combine with the incident wave to reduce the impact load $F_1(t)$ on the upper parts of the structure. Reflections from layers below the explosion will produce secondary compression waves, as stated earlier, which can combine with the incident stress to increase significantly the total loading on lower sections of the structure.

The path length travelled by the directly transmitted wave to any point on the structure is the straight line distance from the source to the point, which will be

$$x_i = \sqrt{[(d_0 - z)^2 + x^2]} \tag{5.137}$$

The total path length of the wave reflected from the ground surface will be

$$x_r = \sqrt{[(d_0 + z)^2 + x^2]} \tag{5.138}$$

The total path length of a wave reflected from a deeper layer is then written as

$$x_\ell = \sqrt{[(2h - d_0 - z)^2 + x^2]} \tag{5.139}$$

where h = thickness of layer
$\qquad d_0$ = depth of the bomb from the ground surface
$\qquad z$ = depth of the point on the structure from the ground surface
$\qquad x$ = horizontal distance to the point on the structure

Figures 5.17 and 5.18 show the ground shock coupling factor (f_c^*) and the propagation paths for a burst in a layered medium.

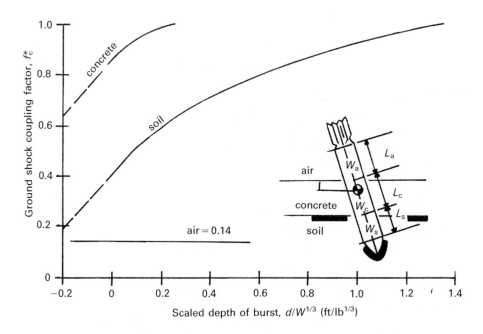

Fig. 5.17 Ground shock and scaled depth of burst.

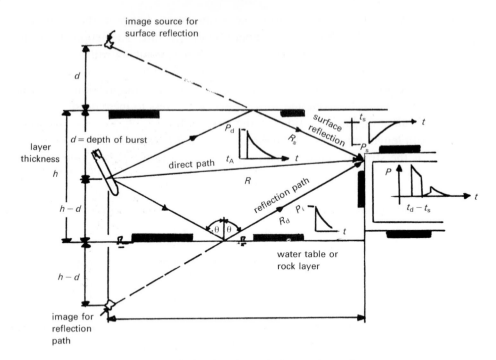

Fig. 5.18 Reflection from surface to lower layer.

Using equation (5.139), the reflected wave magnitude and the total stress–time history at the target location can be assessed from the following simple relations:

$$\sigma_d = p_{so}R_d e^{-\alpha t/t_d} \qquad t \geq t_d = R_d/v_{su} \tag{5.140}$$

$$\sigma_s = -p_{so}R_s e^{-\alpha t/t_s} \qquad t \geq t_s = R_s/v_{su} \tag{5.141}$$

$$\sigma_\ell = K p_{so}R_\ell e^{-\alpha t/t_\ell} \qquad t \geq t_\ell = R_\ell/v_{su} \tag{5.142}$$

where v_{su} = velocity at the upper layer
$\qquad \sigma_d$ = directly transmitted stress (distance R_d)
$\qquad \sigma_s$ = stress reflected from the surface (distance R_s)
$\qquad \sigma_\ell$ = stress reflected from a lower layer (distance R_ℓ)

The peak stress σ_{pi} at a distance R_i is given for each by

$$\sigma_{pi} = \rho v_s 160 (R_i/W^{1/3})^{-n} \tag{5.143}$$

The reflection coefficients C_r given in equations (5.56) and (5.61) are applied.
The shock load stress is given by

$$\sigma_{total} = \sigma_d + \sigma_s + \sigma_\ell \tag{5.144}$$

A typical half-crater profile is shown in Fig. 5.19.

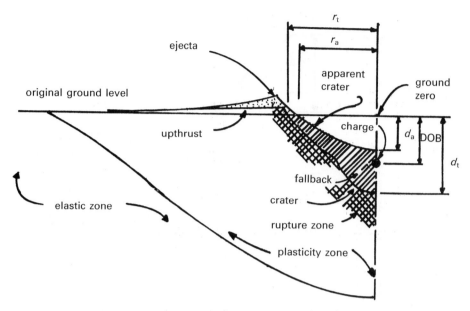

Fig. 5.19 Half-crater profile. (DOB = depth of burst; r_a = apparent radius, d_a = apparent depth; r_t = true radius; d_t = true depth.)

5.8 Rock blasting — construction and demolition

5.8.1 Rock blasting using chemical explosives of columnar shape and a shot hole

In all rock blasting work with explosives, the importance of different factors that control blasting and consequent rock fragmentation assume a priority. These factors are energy, detonation pressure, detonation velocity and explosive velocity. Although a few simple operations (drilling holes, loading of holes with explosives, detonation using a blasting cap from a distance, etc.) are needed to cause rock fragmentation, nevertheless the transient stresses and their distribution set up in the rock when an explosive detonates are complex since many variables are involved. A typical phenomenon of cracking and spalling for different types of burden is given in Fig. 5.20. The single-hole shots made in rock using PETN-based explosives produce a peak gaugehole pressure p_{gh}:[4.1–4.49]

$$p_{gh} = 764 \times \frac{2}{1 + \left(\dfrac{\rho_w v_{sd}}{\rho_r v_s}\right)\rho} \times p_D^{0.753} \times (f_{ci}^*)^{-0.715} \times \alpha_E^{0.785} \times d_g^{-1.6} \qquad (5.145)$$

where p_{gh} = peak gaugehole pressure in a water-filled hole in the rock (psi)

ρ_w = density of explosive (g/cm^3)

v_{sd} = detonation velocity of explosive (km/s)

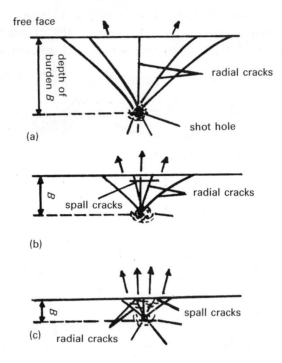

free face

Fig. 5.20 Crack patterns at different depths. (a) Large burden; (b) intermediate burden; (c) small burden.

ρ_r = density of rock

v_s = sonic velocity in rock (5.2 km/s)

p_D = detonation pressure

$$(p_D = 2.1(0.36 + \rho_w)v_{sd}^2 \text{ kilobar})\tag{5.146}$$

f_{ci}^* = volume decoupling ratio = effective volume of shot hole:volume of explosive

α_E = calculated maximum expansion work/in of charge length, kcal/in (charge weight in g/in of charge (calculated maximum expansion in kcal/g))

d_g = distance of gaugehole from shot hole (in)

Substitution of equation (5.146) into (5.145) gives the following equation for p_{gh}:

$$p_{gh} = \frac{2663}{1 + \dfrac{\rho_w v_{sd}}{\rho_r v_s}} \times (0.36 + \rho_w)^{0.573} \times v_{sd}^{1.506} \times (f_{ci}^*)^{-0.715} \times \alpha_E^{0.785} \times d_g^{-1.6}\tag{5.147}$$

In the case of the *loading density* ρ_L, which means bore hole filling, equation (5.147) is modified to

$$\rho_L = 1/\alpha_E\tag{5.148}$$

For granite, based on the examinaton of the data from the detonation velocity series, the average fragment size plotted against the logarithm of a corrected peak gaugehole/pressure[4.248] gave a reasonably straight line if the corrected peak gauge-hole pressure $p_{gh,cor}$ is

$$p_{gh,cor} = p_{gh} \times \frac{1 + \dfrac{\rho_w v_{sd}}{\rho_r v_s}}{2} \times \frac{1}{1 + \dfrac{v_{sd}^2}{v_s^2} - \dfrac{v_{sd}}{v_s}} \qquad (5.149)$$

The average size, F_s, was calculated[4.248] to be

$$F_s = -2.11 \log p_{gh,cor} + 9.02$$

$$\text{where } F_s = F_{av}\left(\frac{1}{B^3} + \frac{1}{L^3}\right)^{1/3} \qquad (5.150)$$

$$F_s = -2.11 \log \left[p_{gh} \frac{\left(1 + \dfrac{\rho_w v_{sd}}{\rho_r v_s}\right)}{2\left(1 + \dfrac{v_{sd}^2}{v_s^2} - \dfrac{v_{sd}}{v_s}\right)} \right] + 9.2$$

where the term in square brackets defines the corrected pressure $p_{gh,cor}$ and B is the burden (in) and L is the shot-hole length. Figure 5.21 illustrates a plot of F_s versus $p_{gh,cor}$ for various series.

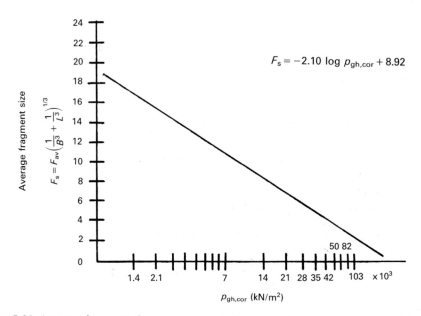

Fig. 5.21 Average fragment size versus corrected pressure for energy, pressure, detonation velocity and burden series (soils: granite; explosive: PETN).

5.8.2 Primary fragments

Sometimes the explosion of a cased donor charge results in the creation of primary fragments. They are produced when an explosive container shatters. These fragments have small sizes and very high initial velocities which, in turn, depend on the thickness of the metal container, the shape of the explosive (spherical, cylindrical or prismatic), the shape of the end or middle of the container (conical, oval), etc. On the basis of report TM55-1300[4.261] the following expressions shall be included in the overall analysis given in this chapter.

5.8.2.1 Explosives with cylindrical containers: fragment velocity

The initial velocity v_0 (ft/s) is given by

$$v_0 = (2E')^{1/2} \left(\frac{W/W_c}{1 + W/2W_c} \right)^{1/2} \tag{5.151}$$

where $(2E')^{1/2}$ = Gurney energy constant
$= 7550$ for pentolite
$= 7850$ to 8380 for RDX/TNT
$= 6940$ for TNT
$= 7460$ for tetryl
$= 7450$ for torpex
$= 7710$ for H-6

The explosion of a cased charge will produce a large number of fragments, N_f, with varying weights which can be obtained from the following equation:

$$\log_e N_f = \log_e \left(\frac{8W_c}{{}^*M_A^2} \right) - \frac{\sqrt{W_f}}{{}^*M_A} \tag{5.152}$$

where N_f = number of fragments $> W_f$
W_f = weight of primary fragment
W_c = total weight of the cylindrical portion of the metal casing
${}^*M_A$ = fragment distribution parameter
$$= {}^*B \, {}^*t_c^{5/6} d_i^{1/3} \, [1 + ({}^*t_{av}/d_i)] \tag{5.152a}$$
${}^*B$ = constant depending upon explosives and casings
$= 0.24$ to 0.35
t_c = thickness of the metal
d_i = inside diameter of the casing
${}^*t_{av}$ = average time

For a *large fragment*, $N_f = 1$ is substituted into equation (5.152) in order to compute its weight W_f. For other cases, reference is made to Table 5.7.

Table 5.7 Initial velocity of primary fragments.

Type	Cross-sectional shape	Initial fragment velocity, v_0	Maximum v_0	Remarks
Cylinder		$\sqrt{(2E')}\left[\dfrac{\dfrac{W}{W_c}}{1+\dfrac{W}{2W_c}}\right]^{1/2}$	$\sqrt{(2E')}\sqrt{2}$	
Sphere		$\sqrt{(2E')}\left[\dfrac{\dfrac{W}{W_c}}{1+\dfrac{3W}{5W_c}}\right]^{1/2}$	$\sqrt{(2E')}\sqrt{\tfrac{5}{3}}$	
Steel-cored cylinder		$\sqrt{(2E')}\left[\dfrac{\dfrac{W}{W_c}}{1+\dfrac{(3+a)\,W}{6\,(1+a)\,W_c}}\right]^{1/2}$ where $a = d_{c0}/1.6d_i$	$\sqrt{(2E')}\sqrt{\left[\dfrac{6(1+a)}{(3+a)}\right]}$	If the steel core is many times more massive than the explosive, this expression for the initial velocity should be modified by multiplying it by the expression: $$\sqrt{[1-(0.2W_{c0}/W)]}$$ where W_{c0} is the weight of the steel core (lb)
Plate		$\sqrt{(2E')}\left[\dfrac{\dfrac{3W}{5W_c}}{1+\dfrac{W}{5W_c}+\dfrac{4W_c}{5W}}\right]^{1/2}$	$\sqrt{(2E')}\sqrt{3}$	This expression applies for a rectangular explosive in contact with a metal plate having the same surface area; it is assumed that the entire system is suspended in free air and its thickness is small in comparison to its surface area, so that the resulting motions are essentially normal to the plane of the plate

continued

Table 5.7 *Continued.*

Type	Cross-sectional shape	Initial fragment velocity, v_0	Maximum v_0	Remarks
Hollow cylinder		—	—	Although an expression to predict the velocity of fragments is not available, an upper limit of the initial velocity may be obtained using the expression for a solid cylinder and a lower limit from the expression for a single plate; the ratio of the explosive weight to the casing weight (W/W_c) of the hollow cylindrical charge is used in both expressions
Sandwich plates		If $W_{c1} \neq W_{c2}$ $\sqrt{(2E')}\left[\dfrac{W}{W_{c1}+W_{c2}g+\dfrac{W}{3}(1-g+g^2)}\right]^{1/2}$ where $g=\left(W_{c1}+\dfrac{W}{2}\right)\Big/\left(W_{c2}+\dfrac{W}{2}\right)$ If $W_{c1}=W_{c2}=W_c$ $\sqrt{(2E')}\left[\dfrac{\dfrac{W}{2W_c}}{1+\dfrac{W}{6W_c}}\right]^{1/2}$	$\sqrt{(2E')}\sqrt{3}$	This expression applies to a rectangular explosive sandwiched between two metal plates having the same surface area as the explosive; it is assumed that the entire system is suspended in free air and its thickness is small in comparison to its surface area, so that the resulting motions are essentially normal to the plane of the plates

W, W_c, W_{c0}, W_{c1}, W_{c2} (lb); d_1, d_{c0} (in); v_0, $\sqrt{(2E')}$ (ft/s)

5.8.2.2 *Fragment velocity at boundary and uncased*

The minimum boundary velocity $*v_b$ at which propagation of the explosion occurs as a result of the primary fragment impact is

$$*v_b^2 = Ke^{5.37t_c/w_f^{1/3}}/W_f^{2/3}(1 + 3.3t_c/W_f^{1/3}) \qquad (5.153)$$

where K = a constant for the sensitivity of the explosive material contained in the acceptor charge

$= 2.78 \times 10^6$ for pentolite

$= 4.10 \times 10^6$ for cyclotol (60/40)

$= 3.24$ to 4.15×10^6 for RDX/TNT

$= 16.30 \times 10^6$ for TNT

$= 14.5 \times 10^6$ for amatol

$= 3.55 \times 10^6$ for torpex

Fragment velocities v_b range between 1200 and 4000 ft/s. Fragment weights W_f are no more than 3 oz. Tables 2.67 to 2.70 give useful data for certain known cases.

5.8.3 Blasting: construction and demolition

Much has been written about chamber charges in this text. Here distinctive features are given about the *slab charges*, which can afford an improvement in parameters of practical importance, such as directivity and accuracy of throw. The *directivity of throw* $\vec{f_d}$ is written as the volume of the debris projected in a given direction divided by the total volume removed to the free face. The other parameter is the *accuracy factor*, f_{ac}, which is written as the ratio of the volume of the rock thrown away to a predetermined area to the total volume of the rock moved to the free face. In the slab-charge system, the value of f_{ac} is higher than 90% under normal conditions.

Figures 5.22 and 5.23 explain clearly the chamber and slab systems. As seen, a chamber charge projects rock fragments in all directions. This implies a greater part of rock burden within the limits of the open pit. In the case of a sloping face, the chamber charge blast removes the rock burden of a volume contained by the angle BOC; this is dumped on the left-hand side, to be removed by bulldozers or other excavators. Thus it is not economical. The picture is totally different in the case of slab charges, the move is in a given direction in a compact form. This is due to the fact that the trajectory of the rock fragments is perpendicular to the baseline of the slab charge. The blasted rock may be removed to a desired distance by varying the angle of inclination α, as shown in Fig. 5.22(b). Since the fragments fly as a compact mass, air drag has much less effect on their movement than in chamber-charge blasting. The rock mass can then be easily moved to a required distance. Slab charges can be either of the layered type or the concentrated type, as shown in Fig. 5.23.

The parallel deep-hole charge is used on slopes as well. In this method, the

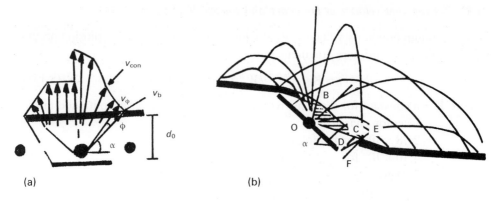

Fig. 5.22 Slab and chamber charges: initial projections. (a) Plain face; (b) sloping face.

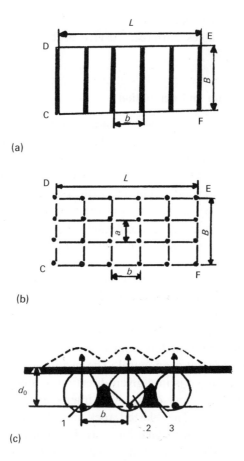

Fig. 5.23 System layout of slab charges. (a) Layered charges; (b) concentrated charges; (c) b = distance, 1 = charge, 2 = gas chamber, 3 = volume of undisturbed rock.

material to be blasted in holes moves freely from a high-pressure zone to a lower-pressure zone in the depression. Since the energy is distributed in this unevenly along the slope, it is better to charge the holes area-wise and the intermediate spaces are filled with inert material. The rock blasting by slab charges forms trenching, and will have an even surface if the distance between consecutive rows of charges does not exceed the value of b (Fig. 5.23(c)), where

$$b = d_0 e_f^{2/3} \qquad (5.154)$$

where b = spaces between charges
d_0 = depth of concentrated charges
e_f = efficiency factor = R_{vd}/d_0 (see Table 5.8)
R_{vd} = crater radius, as calculated in section 5.7

For concentrated charges, the initial velocity v_{con} along the line of least resistance is given by

$$v_{con} = A(Q^{1/3}/d_0)^m C_f \qquad (5.155)$$

where Q = mass of concentrated charges
A, m = physio-mechanical properties of the soil, examples of which are given below:

Soil type	A	m
Loess	8	3.0
Sand	9	2.4
Loam	16	2.0
Clay	22	1.8
Hard rock	40	1.5

C_f = correction factor

$$= \sqrt{\left\{ \frac{1}{2} + \frac{1}{[(q/k_e)^{0.57} + 1]^{m+1} + [0.5(q/k_e)^{0.57} + 1]^{m+1}} \right\}} \qquad (5.155a)$$

q = specific consumption of explosives
$= Q/1.047 d_0^3 \tan^2\phi_{lim}$
k_e = computed consumption of explosives

The efficiency factor is written as

$$e_f = \frac{1}{2} \sum_0^\phi v_\phi^2 d_m / Q E_{ne} \qquad (5.156)$$

where v_ϕ = velocity at an angle ϕ
E_{ne} = specific energy of explosive
$d_m = \frac{1}{3}\pi d_0^3 \tan^2\phi$ $\qquad\qquad$ (5.156a)
$\cos\phi_{lim} = 1/\sqrt{(\tan^2\phi_{lim} + 1)}$ $\qquad\qquad$ (5.156b)
$\tan\phi_{lim} = \sqrt[3]{[Q/k_e d_0^3(1 + d_0/50)]}$ $\qquad\qquad$ (5.156c)

Table 5.8 Efficiency factor e_f versus consumption of explosive q for concentrated charges in hard rock.

q (kg/m³)	e_f
0.5	0.20
1.0	0.35
1.5	0.42
2.0	0.50
3.0	0.60

5.8.3.1 Crushing of fragments due to collisions between themselves and their impact on structures

Rock fragments are further crushed by the kinetic energy of flight and the potential energy which they impact on structures. The degree of crushing depends on the value of q, size, rock hardness, density, etc. The energy balance equation has to be developed. The text gives relevant information in earlier chapters on impact and the dynamics associated with it. The energy balance equation is formed from the following constituent relationships:

$$\text{kinetic energy (KE)} = qE_{ne}e_f = E_1 l_{max}^3/k_{loc}^2 \qquad (5.157)$$

where l = fragment size; l_{max}^3/k_{loc}^2 = the volume
$\quad E_1$ = energy in fragment

$$\text{potential energy (PE)} = l^3\rho gH/k_{loc}^3 \qquad (5.158)$$

$$E_{crush} = \text{crushing energy} = K_s(\text{KE} + \text{PE})$$

where H = height of the rock
$\quad K_s$ = stiffness coefficient at impact, indicating that part of the energy is utilized for crushing
$\quad k_{loc}$ = parameter ≈ 1.3

If A_0 is the initial surface area, it is given by

$$A_0 = 6l^2/k_{loc}^2 \qquad (5.159)$$

The stiffness coefficient, k_{cr}, indicating the size reduction due to impact, is expressed as

$$k_{cr} = 1 + (l_{max}/6k_{loc} - e)(e_f qE_{ne} + \rho gH) \qquad (5.160)$$

where e is the energy in a unit fragment.
    The values given above are assessed in the following order:

$$q = 0.5\,\text{kg/m}^3;\ E_{ne} = 4 \times 10^6\,\text{J/kg};\ K_s = 0.01;\ E = 5 \times 10^{10}\,\text{N/cm}^2;$$
$$\rho = 2300\,\text{kg/m}^3;\ H = 15\,\text{m}$$

5.8.3.2 *Wedge-shaped charges (for angle 0° < α < 45°)*

Wedge-shaped charges are included with slab charges. The only difference is that their thickness is variable. In cross-section, their shape is trapezoidal, as shown in Figs 5.24 and 5.25. The efficiency is written as

$$e_f = R_{vd}/d_{03} = \cot\beta \tag{5.161}$$

$$Q_3 = k_e d_{03}^3 (H\, d_{03}/50)(0.4 + 0.6 e_f^3)$$

For spacings and explosions, the rest of the procedure is given in section 5.8.3. In the case of a multi-charge layout of a wedge shape, the specific consumption q of

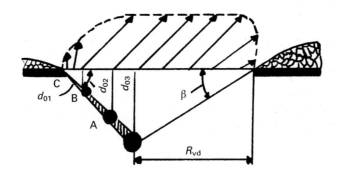

Fig. 5.24 Projection velocity for a wedge-shaped charge.

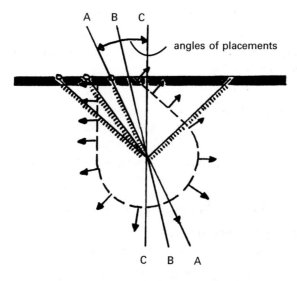

Fig. 5.25 Crater shapes for a wedge-shaped charge. (Charge placed at different angles to the vertical plane.)

the explosives must ensure that the rock is projected to a distance l_x, which is given by

$$l_x = 2d_{cr}(\cot\alpha + \cot\beta) \tag{5.162}$$

where l_x is the depth of the design pit at which the next row of charges is laid. The distance between two rows (the first and second ones) is given by

$$b_1 = d_{01} \sqrt{(q/k_e)}$$
$$\text{where } d_{01} = \tfrac{1}{4}l_x \tag{5.163}$$

In the same manner, the distance between the nth and the $(n+1)$th row of charges is given by

$$b_n = d_{0n} \sqrt{(q/k_e)} \tag{5.164}$$

$$\text{where } d_{0n} = d_{0(n-1)} + b_{i-1}\tan\beta \tag{5.165}$$

The expansion time t_{exp} between successive explosions or rows of charges is written as

$$t_{exp} = C_1 d_0 V \sqrt{(\rho q/2E_{ne} - e_f)} \tag{5.166}$$

where C_1 = a coefficient which prevents moving rock from achieving an instant
velocity of $v = \sqrt{(2E_{ne}e_f/\rho)}q \approx 1.5$
V = volume/kg $\approx 1\,m^3/kg$ as a unit

The optimum distance b between the rows of blast holes must obey the following relationship:

$$b \le H(0.4/\sin 2\alpha)$$

where H is the height of the slope being blasted.

5.9 Explosions in water

5.9.1 Introduction

The explosive wave properties and development in water are very similar to that of an air explosion (sections 2.5 and 5.3). Major differences do, however, exist. The well known one is the violent compression by the surrounding water of the expansion of the explosive gases. The high pressure, the weight of water and its inertia are responsible for this phenomenon. The rate of compression depends also on the depth of the explosion. Due to the inertia of water, the explosion pressure exceeds the hydrostatic pressure. A repeated compression generates a new expansion of the gases and, as a result, pulsating spherical gas bubbles rise towards the water surface. When the bubble appears and expands, a new pressure wave is generated which is the secondary wave; the first, being the primary wave, looks very different from this one. The pressure of the secondary is around 5 to

$$v_{xs} = \sqrt{\left\{\left(\frac{p_x}{p_a}\right)\left(1 - \frac{\rho_a}{\rho_{sw}}\right)\right\}} \tag{5.171}$$

For a water shock wave, p_a is generally neglected owing to heavy compression. However, for $p_a < p_x$, the value of the shock U (m/s) is written as

$$U = v_{xs}\left(1 - \frac{\rho_a}{\rho_{sw}}\right) \tag{5.172}$$

The corresponding energy term is written as

$$\Delta E = E_{nx} - E_{na} = \frac{p_x}{2}(V_a - V_{xs}) \tag{5.173}$$

where E_{nx} = energy for the p_x value

E_{na} = energy under ambient conditions

For PETN and TNT explosives, the following values for the above parameters have been reported:

	v_{xs} (m/s)	ρ_{sw}/ρ_a	$p_x \times 10^3$ (kN/cm^2)	U (m/s)	U/v_{xs}	ΔE (cal/g)
PETN (1.69 g/cm^3)	2725	1.636	195	7020	0.835	800
TNT (1.60 g/cm^3)	2185	1.557	136	6100	0.872	570

The pressure p_m on a structure inside water is written as

$$p_m = [p_x/(1 - t/\tau)]^{a_v - 1} \tag{5.174}$$
$$= p_w(R_w/R^*)^{a_v - 1}\cos^2\alpha$$

where $a_v = 3$ for a spherical charge

$a_v = 2$ for a cylindrical charge

$a_v = 1$ for a flat charge

R_w = radius of the charge

R^* = general radius of the gas sphere

α = angle at which the pressure acts on the structure

For chemical charges, the duration time $\tau = 10^{-4}$ s. When the charge touches the underwater structures the value of t is zero. For such short periods the effect of loading on the structure does not depend on the pressure magnitude p_m, but on the impact $F_1(t)$. The total impact is then written as

$$F_1(t) = \int_0^\tau p_x \, dt = p_m \int_0^\tau \left(1 - \frac{t}{\tau}\right)^{a_v - 1} dt \tag{5.175}$$
$$= -p_m \int_0^\tau \left(1 - \frac{t}{\tau}\right)^{a_v - 1} d\left(1 - \frac{t}{\tau}\right)$$
$$= p_m \frac{\tau}{a_v}$$

As shown in Fig. 5.27, a *nearby explosion* is effective if the distance l' (distance of the structure from the charge) is defined as $R_w < l' \leq 10R_w$ if ρ_a and p_a are both zero. For spherical and cylindrical charges, respectively, the values of $F_1(t)$ are given below:

$$F_1(t) = \frac{\bar{A}W}{(l')^2} \cos^4\alpha \tag{5.176}$$

$$F_1(t) = \left(\frac{2\bar{A}W_c}{l'}\right)\cos^3\alpha \tag{5.177}$$

where $\bar{A} = ev_{xs}$ $\hfill(5.178)$

$e =$ coefficient of restitution

$W, W_c =$ spherical and cylindrical charges (mass/unit length) respectively

For example, in an explosion occurring above a *circular structure* of radius r, assuming the explosion epicentre coincides with the centroid of the structure, the impact caused by the explosion can be written as (Fig. 5.28):

$$d(F_1'(t)) = 2\pi r dr(F_1(t)) \tag{5.179}$$

where $r = l' \tan\alpha$

$dr = (l'/\cos^2\alpha)d\alpha$ (α is the angle shown on Fig. 5.28)

Equation (5.179) can be written as

$$d(F_1'(t)) = 2\pi l' \tan\alpha(l'/\cos^2\alpha)d\alpha(\bar{A}W/l')\cos^4\alpha d\alpha$$
$$= 2\pi \bar{A}W \sin\alpha\cos\alpha d\alpha \tag{5.180}$$

$$F_1'(t) = \int_0^{\alpha_d} 2\pi \bar{A}W \sin\alpha\cos\alpha d\alpha = \pi \bar{A}W \sin^2\alpha_d \tag{5.181}$$

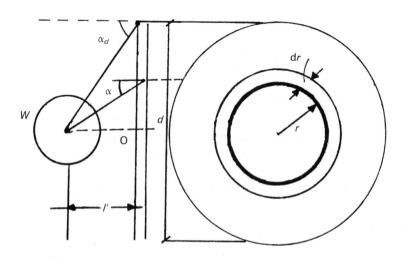

Fig. 5.28 A circular structure under an explosive load.

For structures of other shapes, refer to Chapter 3 on basic structural dynamics.
The value of r can be any dimension L (length) and B (breadth). For an infinite
circular structure, $r \rightarrow \infty$ and $\alpha_0 \approx \pi/2$, then

$$F'_l(t) = \pi \bar{A} W \tag{5.182}$$

It is important to know the *state of water*, i.e. the basic relationships of
hydrostatics. The volume $V(T,p)$ and the pressure p must have a relation. The
empirical equation of the state of water gives a useful relation for p and V, which
in turn depend on T and p values. The Bridgman equation[4.259] is written as

$$p = (109 - 93.7V)(T - 348) + 5010V^{-5.58} - 4310 \; (\text{kp/cm}^2) \tag{5.183}$$

In comparison to gaseous explosions, the reactions of *solid explosives* cause
greater problems owing to high temperatures and densities. Jones[4.252–4.254] developed an equation for the assumed state by fitting data from Bridgman as
follows:

$$p = \alpha^* A^* e^{-\alpha^*(V/N)} - B + RTf \tag{5.184}$$

where $\alpha^* = 0.263/\text{cm}^3$
$A^* = 855 \, \text{kcal/mole}$
$T = \text{temperature}$
$N = \text{number of moles}$
$B = 0.139 \, \text{kcal/mole}$
$f = 0.313/\text{cm}^3$
$R = \text{gas constant}$

On the basis of data for nitrogen, Jones[4.252–4.254] introduced higher temperatures and comparable densities. The total energy E_{NT} of the products of explosion
becomes

$$E_{NT} = \Sigma[N_a E_a + (N - N_s)(E_a + \tfrac{3}{2}RT)] \tag{5.185}$$

where $N_a = \text{number of moles for each molecular species (solid or gas)}$
$N = \Sigma N_a$
$N_s = \text{number of moles of solid products}$
$\tfrac{3}{2}RT = \text{energy of interaction for the gas molecules plus vibrational energy}$
$\tfrac{3}{2}RT/\text{mole}$

5.9.3 Major underwater shock theories

Some major underwater shock theories evaluate relations between overpressure
and the reduced distance. Some examples are given below.

5.9.3.1 Kirkwood–Bethe theory[4.255]

This theory covers initially a spherical shock wave in terms of kinetic enthalpy
$\omega(R^*,t)$. The pressure–time curve is related by

$$p(R,t) = (a_0/R')p(\tau) \tag{5.186}$$

where a_0 = initial radius of the gas sphere
　　　R' = radius of the shock front

$$p(\tau) = \rho_w(a(\tau)/a_0)\omega(a,\tau) \tag{5.186a}$$

　　　a = radius of the gas sphere at any stage

When the shock front reaches a point R, the value of τ becomes

$$\tau = \tau_0 + (\partial\tau/\partial t)R(t - t_0) = \tau_0 + (1/\gamma_{sh})(t - t_0) \tag{5.187}$$

The parameter γ_{sh}, which is the measure of the time scale behind the shock front relative to the gas sphere, increases rapidly as the shock front travels outwards. Kirkwood and Bethe finally arrived at a pressure–time relationship of

$$p(r,t - t_0) = p(0)x'(a_0/R^*)e^{-(t-t_0)/\theta} \tag{5.188}$$

where $p(0) = (0)\omega(a_0)$
　　　$\theta = \gamma_{sh}\theta_1$

The value of γ_{sh} in equation (5.188) was taken as

$$\gamma_{sh} = 1 - \frac{1}{\beta c_0(Z_a + 1)} + \frac{2a_0}{c_0\theta_1}\frac{\beta\omega(a_0)}{c_0}\left[\log\frac{Z}{Z_a} - \frac{4(Z - Z_a)}{(Z + 1)(Z_a + 1)}\right] \tag{5.188a}$$

where $c_0 = v_{s0}$ (initial), defined earlier
　　　θ_1 = parameter θ/γ_{sh}
　　　θ = incremental time = $2(a_0/c_0)(\beta\omega a_0/c_0)(R^*/a_0)$

Z is a function of xa_0/R^*. The quantity Z_a is strictly a function of $xa_0/a(\tau_0)$, and is given by the relation

$$Z_a = 1 + \frac{1}{2\omega(a_0)x_d} + \sqrt{\left\{\frac{1}{\omega(a_0)x} + \left[\frac{1}{2\omega(a_0)x}\right]^2\right\}} \tag{5.188b}$$

$$Z = (1 + \beta\sigma)/\beta\sigma \tag{5.188c}$$

$$\sigma = \sum_{\rho_0}^{\rho} (1/\rho)v_s d \tag{5.188d}$$

where $v_s = c$ = velocity of sound in water = $\sqrt{(dp/d\rho)}$

$$\beta = [(u/c_0 - 1]/\sigma$$

$\dot{x} = u$ = particle velocity
x_d = dissipation factor

　　　In terms of charge weight, the peak pressure p_m may then be written as an approximate value in terms of the power laws as

$$p_m = k(W^{1/3}/R^*)^\alpha \tag{5.189}$$

　　　The following table gives the numerical values of some of the parameters which are given in section 2.6 and below for TNT.

R^*/a_0	1.00	10	25	50	100
x	1.00	0.40	0.30	0.259	0.23
γ_{sh}	1.00	4.03	6.42	8.10	9.54
θ/a_0 $(10^{-5}\,\mathrm{s/cm})$	0.345	1.40	2.20	2.80	3.30
p_m $(\mathrm{lb/in}^2)$	537 959	20 199	6100	26 225	1169

The empirical value of α is around 1.16, based on the slope of the curves.

5.9.4 Penney and Dasgupta theory[4.258]

The finite amplitudes of the spherical waves involve many differential equations with a number of unknown parameters. Penney and Dasgupta have solved them for TNT, and derived the value of the peak pressure as a function of shock radius R_w^*:

$$p_m\ (\mathrm{lb/in}^2) = 103\,000\ (a_0/R^*)e^{2a_0/R^*} \tag{5.190}$$
$$= 14\,000\ (W^{1/3}/R^*)e^{0.274w^{1/3}/R^*}$$

where $a_0 = R_w$.

5.9.5 A comparative study of underwater shock front theories

It is vital to see how the above-mentioned theories and many others[4.250–4.266] do compare. Those of Kirkwood–Bethe, Penney–Dasgupta, Cole, Hilliar and Taylor are summarized in Fig. 5.29. The average of all these curves, using a curve-fitting approach, gives the following relationships:

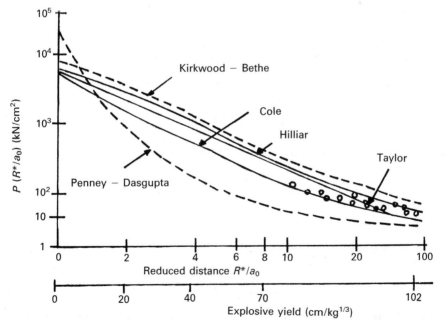

Fig. 5.29 A comparison of various underwater shock front theories.

$$p \ (\text{kN/cm}^2) = \frac{1}{R^*/a_0} \left[356 + \frac{114.5}{R^*/a_0} - \frac{2.45}{(R^*/a_0)^2} \right] \quad \frac{0.05 \le R^*}{a_0 \le 10 \, \text{m/kg}^{1/3}}$$

$$p \ (\text{kN/cm}^2) = \frac{1}{R^*/a_0} \left[293.5 + \frac{1386.5}{R^*/a_0} - \frac{1780}{(R^*/a_0)^2} \right] \quad \frac{10 \le R^*}{a_0 \le 50 \, \text{m/kg}^{1/3}} \tag{5.191}$$

where $R^*/a_0 = R/\sqrt[3]{W}$

For a shock-wave impact, the value of $F_1(t)$ is given by

$$F_1(t) = \int_0^t p(t)dt = p\theta^*(1 - e^{-t/\theta^*})$$

$$= 604 \ \sqrt[3]{(W)}(R^*/a_0)^{-0.86} \tag{5.192}$$

5.9.6 Shock wave based on a cylindrical charge explosion

Here a cylindrical column charge detonates, the wave front of which is a thin zone of chemical reaction followed by explosive gases of high pressure and temperature. If R is the distance from the charge axis, the reduced distance is adjusted in the above equations in section 5.9.5 as follows:

$$R^*/a_0 = R/\sqrt{W_c} \tag{5.193}$$

W_c is given by

$$W_c = W_{cs}Q_{sp}/Q_{WT} \tag{5.193a}$$

where W_{cs} = relative mass of the charge of the given explosive (kg/m)
$\quad Q_{sp}$ = explosive specific heat energy (kcal/kg)
$\quad Q_{WT}$ = specific energy for TNT or other explosives $\approx 1000 \, \text{k/cal per kg}$

The peak pressure is given by

$$p_m = 720(R^*/a_0)^{-0.72} \ \text{kg/cm}^2 \tag{5.194}$$

In equation (5.192), the value of θ^* is written as

$$\theta^* = 10^{-4} \ \sqrt{(W_c)}(R^*/a_0)^{0.45} \quad \text{s} \tag{5.195}$$

The numerical values of p versus R^*/a_0 are given below:

$p \ (\text{kN/cm}^2)$	$R^*/a_0 \ (\text{m/kg}^{1/3})$
10^4	0.1
10^3	1
$(10^3 + 50)$	5
10^2	10
10	10^2

Figure 5.30 shows a comparison between spherical and cylindrical shock-wave effects.

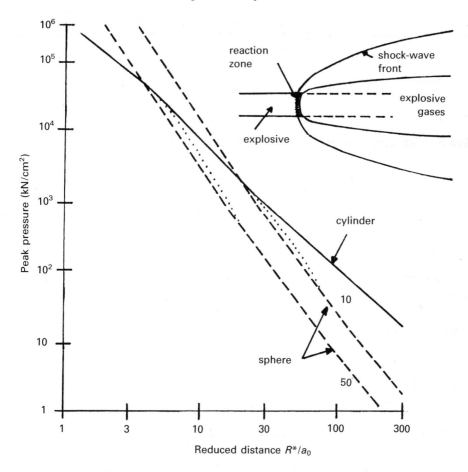

Fig. 5.30 Peak pressure versus reduced distance: a comparative study of spherical and cylindrical shock-wave effects.

5.9.7 Underwater contact explosions

For contact explosions, $2W$ is substituted for W in all relevant equations given in section 5.9.6.

5.9.8 Underwater shock-wave reflection

Much has been said on this subject in this chapter. The pressure acting on the underwater *static structure* due to this effect is given by

$$p_r(\text{kN/cm}^2) \approx 2p(t) + \frac{2.5[p(t)]^2}{p(t) + 19\,000} \qquad (5.196)$$

In an elastic medium, the value of $p_r(t) \approx 2p(t)$. Table 5.9 gives useful data on parameters relevant to underwater explosions (values other than those mentioned

Table 5.9 Relationships between pressure, velocity, density and temperature in water.

Pressure in the wave front, $p_m(\text{kN/cm}^2)$	Velocity of the wave front, $\dot{x} = v$ (m/s)	Velocity of water in the wave front, $u(t)$ (m/s)	Water density, ρ_w (g/cm^3)	Velocity of sound in the wave front, v_{s0} (m/s)	Water temperature, T (°C)
0	1460	0	1.000	1460	0
200	1490	13	1.013	1500	2.0
400	1510	26	1.024	1540	2.4
600	1540	40	1.032	1580	2.6
800	1560	58	1.040	1620	3.0
1 000	1590	67	1.044	1660	3.4
1 400	1640	93	1.058	1740	4.0
1 600	1670	106	1.065	1780	4.4
2 000	1720	133	1.075	1860	5.8
2 600	1800	173	1.090	1980	8.0
3 000	1850	200	1.100	2060	8.8
4 000	1940	240	1.120	2160	14.0
5 000	2040	280	1.140	2240	18.0
6 000	2100	320	1.160	2360	22.0
7 000	2190	360	1.175	2420	24.0
8 000	2240	400	1.200	2500	30.0
9 000	2300	420	1.210	2600	32.0
10 000	2400	450	1.220	2660	35.0
20 000	2840	680	1.325	3200	68.0
40 000	3600	1100	1.450	4040	136.0
60 000	4140	1430	1.545	4740	214
80 000	4600	1680	1.615	5162	300
100 000	5000	1940	1.665	5600	400
200 000	6460	3000	1.850	7100	870
300 000	7800	3800	1.970	8160	1390

in the table should be interpolated or extrapolated). When the structure is moving in the water with a velocity $\dot{x} = v(t)$, an overpressure will exist of the value $\rho_w v_{s0} v(t)$ in front of it. The value of the underpressure of the same value will have a negative sign. The procedure is shown in Fig. 5.31. $\rho_w v_{s0}$ is the *characteristic impedance*. The reflected pressure $p_r(t)$ acting on the moving structure can be written as

$$p_r(t) = 2p(t) - \rho_w v_{s0} u(t) \tag{5.197}$$

where $u(t) = \mathrm{d}v(t)/\mathrm{d}t$.

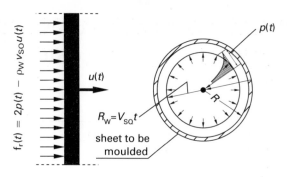

Fig. 5.31 Moving structure under shock load.

6

DYNAMIC FINITE-ELEMENT ANALYSIS OF IMPACT AND EXPLOSION

6.1 Introduction

A great deal of work has been published on finite-element techniques.[2.150,3.1−3.168] This chapter presents the dynamic finite-element analysis for impact and explosion. Plasticity and cracking models are included. Solid isoparametric elements, panel and line elements represent various materials. Solution procedures are recommended. This chapter ends with some load−time functions for selected cases using finite-element analysis. There are many publications on the topic of impact and explosion[2.1−2.384,4.1−4.231] which can be used for in-depth studies.

6.2 Finite-element equations

A three-dimensional finite-element analysis is developed in which a provision has been made for time-dependent plasticity and rupturing in steel and cracking in materials such as concrete, etc. The influence of studs, lugs and connectors is included. Concrete steel liners and studs are represented by solid isoparametric elements, shell elements and line elements with or without bond linkages. To begin with, a displacement finite element is adopted.

The displacement field within each element is defined in Fig. 6.1 as

$$\{x\} = [N]\{x\}^e = \sum_{i=1}^{n} (N_i[I]\{x\}_i) \tag{6.1}$$

The strains and stresses can then be expressed as

$$\{\epsilon\} = \sum_{i=1}^{n} ([B_i]\{x_i\}) = [D]\{\sigma\} \tag{6.2}$$

In order to maintain equilibrium with the element, a system of external nodal forces $\{F\}^e$ is applied which will reduce the virtual work (dW) to zero. In the general equilibrium equation, both equations (6.1) and (6.2) are included. The final equation becomes

$$(\{d\delta\}^e)^T\{F\}^e = (\{d\delta\}^e)^T \int_{vol} [B]^T\{\sigma\}dV \tag{6.3}$$

In terms of the local co-ordinate (ξ,η,ζ) system, equation (6.3) is written as

496

$$\{F\}^{\rm e} = \int_{\rm vol} [B]^{\rm T}[D]\{\epsilon\}{\rm d}\xi, {\rm d}\eta, {\rm d}\zeta \ \det[J]\{x\}^{\rm e} \tag{6.4}$$

The force–displacement relationship for each element is given by

$$\{F\}^{\rm e} = [K]^{\rm e}\{u\}^{\rm e} + \{F_{\rm b}\}^{\rm e} + \{F_{\rm s}\}^{\rm e} + \{F_{\sigma}\}_i^{\rm e} + \{F_{\epsilon}\}_c^{\rm e} \tag{6.5}$$

where the element stiffness matrix is

$$[K_{\rm c}] = \int_{\rm vol} [B]^{\rm T}[D][B]{\rm d}V \tag{6.5a}$$

The nodal force due to the body force is

$$\{F_{\rm b}\}^{\rm e} = -\int_{\rm vol} [N]^{\rm T}\{G\}{\rm d}V \tag{6.5b}$$

The nodal force due to the surface force is

$$\{F_{\rm s}\}^{\rm e} = \int_{s} [N]^{\rm T}\{p\}{\rm d}s \tag{6.5c}$$

The nodal force due to the initial stress is

$$\{P_{\sigma}\}_i^{\rm e} = \int_{\rm vol} [B]^{\rm T}\{\sigma_0\}{\rm d}V \tag{6.5d}$$

The nodal force due to the initial strain is

$$\{P_{\epsilon}\}_i^{\rm e} = -\int_{\rm vol} [B]^{\rm T}[D]\{\epsilon_0\}{\rm d}V \tag{6.5e}$$

Equations (6.4) and (6.5) represent the relationships of the nodal loads to the stiffness and displacement of the structure. These equations now require modification to include the influence of the liner and its studs. The material compliance matrices $[D]$ are given in Tables 6.1 and 6.2. The numerical values are given for various materials or their combinations in Tables 6.3 to 6.6. These values of the constitutive matrices are recommended in the absence of specific information.

If the stiffness matrix $[K_{\rm c}]$ for typical elements is known from equations (6.4) and (6.5) as

$$[K_{\rm c}] = \int_{\rm vol} [B]^{\rm T}[D][B]{\rm dvol} \tag{6.6}$$

the composite stiffness matrix $[K_{\rm TOT}]$, which includes the influence of liner and stud or any other material(s) in association, can be written as

$$[K_{\rm TOT}] = [K_{\rm c}] + [K_{\ell}] + [K_{\rm s}] \tag{6.7}$$

where $[K_{\ell}]$ and $[K_{\rm s}]$ are the liner and stud or connector matrices.

If the initial and total load vectors on the liner/stud assembly and others are $[F_{\rm T}]$ and $[R_{\rm T}]$, respectively, then equation (6.4) is rewritten as

$$\{F\}^{\rm e} + \{F_{\rm T}\} - \{R_{\rm T}\} = [K_{\rm TOT}]\{x\}^* \tag{6.8}$$

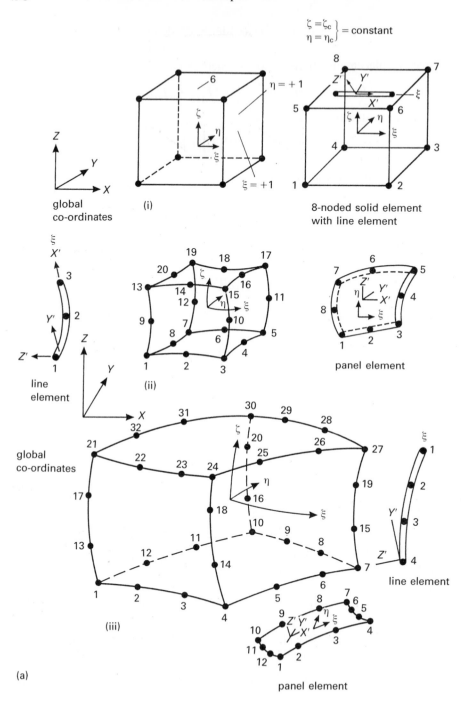

Fig. 6.1 (a) Isoparametric elements. (i) Parent element, three-dimensional isoparametric-derived element; (ii) solid element (20-noded); (iii) 32-noded solid element. (b) Line elements within the body of the solid isoparametric elements (ISS = isoparametric solid element).

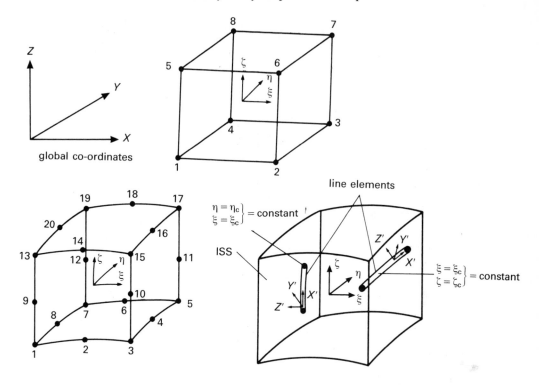

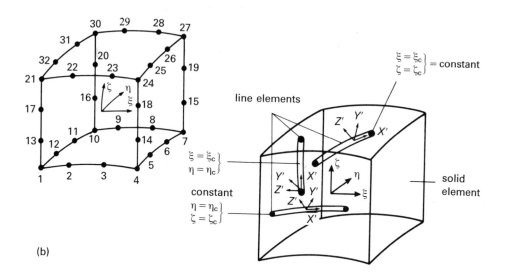

Fig. 6.1 *Continued.*

The displacement $\{x\}^*$ is different from $\{x\}$ in equation (6.4), since it now includes values for both unknown displacements and restrained linear boundaries. Hence $\{x\}^*$ is defined in matrix form as

$$\{x\}^*_{x,y,z} = \begin{Bmatrix} x_{unx} \\ x_{uny} \\ x_{unz} \\ x_{bx} \\ x_{by} \\ x_{bz} \end{Bmatrix} = \begin{Bmatrix} x_{un} \\ x_b \end{Bmatrix} \tag{6.9}$$

where x_{un} and x_b are displacement values in unrestrained or unknown conditions and restrained conditions. Similarly, the values for $\{F_T\}$ and $\{R_T\}$ can also be written as

$$\{F_T\} = \begin{Bmatrix} F_{un} \\ F_b \end{Bmatrix}_{x,y,z}$$

$$\{R_T\} = \begin{Bmatrix} R_{un} \\ R_b \end{Bmatrix}_{x,y,z} \tag{6.10}$$

The quantities for the liner corresponding to unknown displacements can be written as

$$[K_\ell]\{x_{un}\}_{x,y,z} = \{F_{un}\}_{x,y,z} \tag{6.11}$$

The shear force τ acting on studs or any other type is evaluated as

$$\{\tau\} = [K_s]\{x_{un}\}_{x,y,z} \tag{6.12}$$

Table 6.7 gives the $[K_s]$ matrix modified to include the stiffness of the liner.

Table 6.1 Material compliance matrices $[D]$ with constant Poisson's ratio.

(a) $[D]$ for steel components (isotropic)

Constant Young's modulus and Poisson's ratio

$$[D] = \frac{E}{(1+v)(1-2v)} \begin{bmatrix} 1-v & v & v & 0 & 0 & 0 \\ v & 1-v & v & 0 & 0 & 0 \\ v & v & 1-v & 0 & 0 & 0 \\ 0 & 0 & 0 & \dfrac{1-2v}{2} & 0 & 0 \\ 0 & 0 & 0 & 0 & \dfrac{1-2v}{2} & 0 \\ 0 & 0 & 0 & 0 & 0 & \dfrac{1-2v}{2} \end{bmatrix}$$

continued

Table 6.1 *Continued.*

(b) $[D]$ for concrete: variable E and constant ν

$$
\begin{aligned}
&D_{11} = \frac{E_1(E')^3 - E_{cr}}{E''} &&D_{12} = \frac{\nu E_1 E_2 (E')^2 + E_{cr}}{E''} &&D_{13} = \frac{\nu E_1 E_3 (E')^2 + E_{cr}}{E''} &&D_{14} = 0 &&D_{15} = 0 &&D_{16} = 0 \\[6pt]
& &&D_{22} = \frac{E_2 E_3 (E')^2 + E_{cr}}{E''} &&D_{23} = \frac{\nu E_2 E_3 (E')^2 + E_{cr}}{E''} &&D_{24} = 0 &&D_{25} = 0 &&D_{26} = 0 \\[6pt]
& && &&D_{33} = \frac{E_3(E')^3 - E_{cr}}{E''} &&D_{34} = 0 &&D_{35} = 0 &&D_{36} = 0 \\[6pt]
& && && &&D_{44} = G_{12} &&D_{45} = 0 &&D_{46} = 0 \\[6pt]
& && && && &&D_{55} = G_{23} &&D_{56} = 0 \\[6pt]
& && && && && &&D_{66} = G_{31}
\end{aligned}
$$

$$[D] =$$

$$
\begin{aligned}
E_{cr} &= \nu^2 E_1 E_2 E_3 E' \\
E' &= (E_1 + E_2 + E_3)/3 \\
E'' &= (E')^3 - 2E_1 E_2 E_3 \nu^2 - E'\nu^2(E_1 E_2 + E_1 E_3 + E_2 E_3) \\
G_{12} &= E_{12}/2(1+\nu) \\
E_{12} &= (E_1 + E_2)/2 \\
G_{23} &= E_{23}/2(1+\nu) \\
E_{23} &= (E_2 + E_3)/2 \\
G_{31} &= E_{31}/2(1+\nu) \\
E_{31} &= (E_3 + E_1)/2
\end{aligned}
$$

Table 6.2 [D] with variable Young's modulus and Poisson's ratio for concrete and other materials.

$$
\begin{bmatrix}
D_{11}=\dfrac{(1-v_{23}v_{32})}{\bar{E}}E_1 & D_{12}=\dfrac{(v_{12}+v_{12}v_{32})}{\bar{E}}E_2 & D_{13}=\dfrac{(v_{13}+v_{12}v_{23})}{\bar{E}}E_3 & D_{14}=0 & D_{15}=0 & D_{16}=0 \\[2mm]
D_{21}=\dfrac{(v_{21}+v_{23}v_{31})}{\bar{E}}E_1 & D_{22}=\dfrac{(1-v_{13}v_{31})}{\bar{E}}E_2 & D_{23}=\dfrac{(v_{23}+v_{13}v_{21})}{\bar{E}}E_3 & D_{24}=0 & D_{25}=0 & D_{26}=0 \\[2mm]
D_{31}=\dfrac{(v_{31}+v_{21}v_{32})}{\bar{E}}E_1 & D_{32}=\dfrac{(v_{32}+v_{12}v_{31})}{\bar{E}}E_2 & D_{33}=\dfrac{(1-v_{12}v_{21})}{\bar{E}}E_3 & D_{34}=0 & D_{35}=0 & D_{36}=0 \\[2mm]
D_{41}=0 & D_{42}=0 & D_{43}=0 & D_{44} & D_{45}=0 & D_{46}=0 \\[2mm]
D_{51}=0 & D_{52}=0 & D_{53}=0 & D_{54}=0 & D_{55} & D_{56}=0 \\[2mm]
D_{61}=0 & D_{62}=0 & D_{63}=0 & D_{64}=0 & D_{65}=0 & D_{66}
\end{bmatrix}
$$

$$\bar{E}=1-v_{12}v_{21}-v_{13}v_{31}-v_{23}v_{32}-v_{12}v_{23}v_{31}-v_{21}v_{13}v_{32}$$

$$
\begin{aligned}
E_1 v_{21} &= E_2 v_{12} & D_{44} &= G_{12} \\
E_2 v_{32} &= E_3 v_{23} & D_{55} &= G_{23} \\
E_3 v_{13} &= E_1 v_{31} & D_{66} &= G_{13}
\end{aligned}
$$

The values of G_{12}, G_{23} and G_{13} are calculated in terms of the modulus of elasticity and Poisson's ratio as follows

$$G_{12}=\frac{1}{2}\left[\frac{E_1}{2(1+v_{12})}+\frac{E_2}{2(1+v_{21})}\right]=\frac{1}{2}\left[\frac{E_1}{2(1+v_{12})}+\frac{E_1}{2\left(\dfrac{E_1}{E_2}+v_{12}\right)}\right]$$

$$G_{23}=\frac{1}{2}\left[\frac{E_2}{2(1+v_{23})}+\frac{E_3}{2(1+v_{32})}\right]=\frac{1}{2}\left[\frac{E_2}{2(1+v_{23})}+\frac{E_2}{2\left(\dfrac{E_2}{E_3}+v_{23}\right)}\right]$$

$$G_{13}=\frac{1}{2}\left[\frac{E_3}{2(1+v_{31})}+\frac{E_1}{2(1+v_{13})}\right]=\frac{1}{2}\left[\frac{E_3}{2(1+v_{31})}+\frac{E_3}{2\left(\dfrac{E_3}{E_1}+v_{31}\right)}\right]$$

For isotropic cases: $E_1 = E_2 = E_3 = E$ $v_{12}=v_{13}=v_{23}=v_{21}=v_{31}=v_{32}=v$

Table 6.3 Material properties of concrete, bovine, steel and composites.

(1) Concrete

$\sigma_1 = \sigma_2$ $\quad v_{13} = v_{23} = 0.2$ for any value of σ_3 up to 500 bar
$\sigma_1 < \sigma_2$ $\quad v_{13} > v_{23}$
$\sigma_1 = 0$ $\quad v_{13} = 0.2$ to 0.4 for any value of σ_2
$\qquad\quad = 0.4$ for up to $\sigma_3 = 500$ bar
for 80°C temperature, the above values are increased by 35–50%

E_c (kN/mm^2)	24	30	35	40
v	0.15–0.18	0.17–0.20	0.20–0.25	0.25–0.30

alternatively

$$v = 0.2 + 0.6(\sigma_2/\sigma_{cu})^4 + 0.4(\sigma_1/\sigma_{cu})^4$$

where σ_1, σ_2 and σ_3 are the pressures/stresses along the three principle axes and σ_{cu} is an ultimate compressive stress of concrete.

(2) Bovine material

$E_1 = 11$ to 18 GPa; $E_2 = 11$ to 19 GPa; $E_3 = 17$ to 20 GPa
$G_{12} = 3.6$ to 7.22 GPa; $G_{13} = 3.28$ to 8.65 GPa; $G_{23} = 8.285$ to 8.58 GPa
$v_{12} = 0.285$ to 0.58; $v_{13} = 0.119$ to 0.31; $v_{23} = 0.142$ to 0.31
$v_{21} = 0.305$ to 0.58; $v_{31} = 0.315$ to 0.46; $v_{32} = 0.283$ to 0.46

(3) Steel

$E = 200$ GN/m^2; $v = 0.3$ to 0.33

(4) Composite

Hot-pressed silicone nitride (HPSN) versus tungsten carbide
$\qquad\qquad\qquad\downarrow$ $\qquad\qquad\qquad\qquad\qquad\downarrow$
$\qquad\quad E = 320$ GPa $\qquad\qquad\qquad\quad E = 320$ GPa
$\qquad\quad v = 0.26$ $\qquad\qquad\qquad\qquad\quad v = 0.24$

Carbon fibre (reinforced epoxy with 60% fibres by volume)

	Longitudinal	Transverse
Tensile strength (σ_{tu})	1750 MPa	60 MPa
Compressive strength	1300 MPa	—
Tensile modulus (E_t)	138 GPa	9.1 GPa
Compressive modulus (E_c')	138 GPa	9.1 GPa
Failure strain in tension (ϵ_{tu})%	1.34	0.8
Failure strain in compression (ϵ_{cu})%	0.85	2.9

Chapter 6

Table 6.4 Material properties of additional composites.

Steel indenter   versus

$\downarrow$

$E = 200\,\text{GN/m}^2$

$\nu = 0.3$ to 0.33

(1) Plexiglass

$E = 3.435\,\text{GN/m}^2$

$\nu = 0.394$

(2) Laminate: thornel 300/5208 with fibres oriented $(0, +60, -60)$

$E_1 = 50\,\text{GN/m}^2$; $E_2 = 11.6\,\text{GN/m}^2$

$G_{11} = 19\,\text{GN/m}^2$; $G_{12} = 4.0\,\text{GN/m}^2$

$\nu_{11} = 0.31$ $\nu_{12} = 0.06$

(3) Aluminium and FRPs

	Aluminium	BFRP*	GFRP†	CFRP*	CFRP§
$E(\text{GN/m}^2)$	70	78.7	7.0	70	180
ν	0.3	0.32	0.30	0.30	0.28

* quasi-isotropic; † random mat; § unidirectional.

(4) Graphite/epoxy

(Web stiffened foam sandwich panels with orthotropic facing and a number of 4 equally embedded stiffeners in a polyurethane (PU) core)

$E_1 = 120.7\,\text{GN/m}^2$; $E_2 = 7.93\,\text{GN/m}^2$

$G_{12} = G_{23} = G_{13} = 5.52\ \text{GN/m}^2$

$\nu_{12} = 0.30$

Polyurethane foam

$E = 0.0431\,\text{GN/m}^2$; $G = 0.017\,\text{GN/m}^2$; $\nu = 0.267$

(5) Boron/epoxy composites

$E_1 = 219.8\,\text{GN/m}^2$; $E_2 = 21.4\,\text{GN/m}^2$; $\nu = 0.208$

$E_{p1} = 2.41\,\text{GN/m}^2$; $E_{p2} = 0.04\,\text{GN/m}^2$;

$G_p = 0.008\,\text{GN/m}^2$;   p at plastic level

$\sigma_{yt} = 1.1\,\text{GN/m}^2$

(6) Layers of woven roving and chopped strand mat

$E = 14.5\,\text{GN/m}^2$; $\sigma_{yt} = 215\,\text{N/mm}^2$; $\nu = 0.21$

Table 6.4 *Continued.*

(7) Other materials

Type	E_1 (GN/m²)	E_2 (GN/m²)	G (GN/m²)	v_{12}
CSM/polyester	8	8	3	0.32
WR/polyester	15	15	4	0.15
Glass fibre/polyester	25	25	4	0.17
UD glass/polyester	40	10	4	0.3
UD kevlar/epoxide	76	8	3	0.34
UD carbon/epoxide	148	10	4	0.31
GY70/epoxy (celion with graphite fibre)	102	7.0	4.14	0.318
MODMORE II/epoxy HMS/E (with graphite fibres)	76.8	9.6	5.83	0.305
T300/E (thornel 300/epoxy with graphite fibres)	54.86	12	5.83	0.30
GL/E (glass/epoxy)	30.3	14.9	5.84	0.32

Carbon fibre (60% volume) reinforced epoxy compound:

	Longitudinal	Transverse
E_t (GN/m²)	140.0	9.00
E_c (GN/m²)	140.0	9.00
σ_{tu} (GN/m²)	1.8	0.06
σ_{cu} (GN/m²)	1.3	0.27
v	0.3	0.02

Table 6.5 Material properties for brick and stone masonry and soil/rock.

(1) Brick masonry

Brick strength $f_b = 20-70$ N/mm² $E = 300f_b - 2000$
$f_b > 70$ N/mm² $E = 100f_b + 12\,750$

Brick strength (MN/m²)	Mortar	Mortar mean cube strength (MN/M²)	Wall thickness (mm)	Wall strength (MN/m²)
92	1:¼:3	19.30	102.5	18.40
46	1:¼:3	13.70	102.5	15.65
46	1:1:6	5.94	102.5	10.48

continued

Table 6.5 *Continued.*

(2) Stone masonry

Stone	Strength, f_b (MN/m²)	Mortar strength (MN/m²)		Failure stress (MN/m²)
Sandstone	112.0	0.78		2.78
Limestone	31.0	2.78		4.88
Whinstone	167.0	2.78	1:2:9 mix	9.86
Granite	130.6	2.78		12.32

(3) Soil/rock

	$E \times 10^2$ MN/m²	ν	Density, ρ (kg/m³)
Fine sand	57.456	0.35	
Silty clay	48.84	0.40	
Silty sand	47.88	0.35	
Plastic clay	3.56	0.40	
Silt stone	8.4	0.30	2622
Limestone	114.0	0.25	2671
Alluvial clay	5.0	0.20	
Clay (embankment fill)	20.0	0.20	1517
Saturated soil	200.0	0.30	
Jointed rock	150.0	0.25	
Sandstone	255.0	0.11	

For high plasticity, the frictional angle $\phi'_c = 18°$
For low plasticity, the frictional angle $\phi'_c = 25°$
For rocks, ϕ'_c ranges between 20° and 30°
The adhesion coefficient c is around $1\,kN/m^2$

Table 6.6 Material properties of timber.

Basic stresses and moduli.

Strength group	Bending (N/mm²)	Tension (N/mm²)	Compression to grain (N/mm²)		E_{min} (N/mm²)
	parallel to grain		parallel	perpendicular	
S_1	37.5	22.5	24.4	7.5	13 800
S_2	30.0	18.0	20.0	6.0	11 900
S_3	24.0	14.4	17.9	4.8	10 400
S_4	18.7	11.2	15.5	3.7	9 200
S_5	15.0	9.0	13.3	3.0	7 800

Table 6.6 *Continued.*

Dry grade stresses and moduli.

Grade/species	Bending (N/mm²) parallel	Tension (N/mm²) to grain	Compression to grain (N/mm²) parallel	perpendicular	E_{min} (N/mm²)
SS/Douglas fir	6.2	3.7	6.6	2.4	7000
GS/Douglas fir	4.4	2.6	5.6	2.1	6000
SS/Redwood Whitewood	7.5	4.5	7.9	2.1	7000
GS/Corsican pine	5.3	3.2	6.8	1.8	5000
GS/European pine	4.1	2.5	5.2	1.4	4500

Plywood: all stresses and moduli are multiplied by the following factors.

Grade/glued laminated	Bending (N/mm²) parallel	Tension (N/mm²) to grain	Compression to grain (N/mm²) parallel	perpendicular	E_{min} (N/mm²)
LA/4	1.85	1.85	1.15	1.33	1.0
LB/10	1.43	1.43	1.04	1.33	0.9
LB/20 or more	1.48	1.48			
LC/10	0.98	0.98	0.92	1.33	0.8
LC/20 or more	1.11	1.11			
	Permissible stresses 8 N/mm²		–	–	8700
12 mm ply thickness	5 N/mm²		–	–	7400

6.3 Steps for dynamic non-linear analysis

The solutions of equations (6.6) to (6.12) require a special treatment such as under any increment of dynamic loading, stresses, strains and plasticity are obtained in steel, concrete and composites such as the liner and its anchorages and other similar materials. An additional effort is needed to evaluate the rupture of the steel or other material when cracks develop, especially in concrete beneath the liner or its anchorages.

The dynamic coupled equations are needed to solve the impact/explosion problems and to assess the response history of the structure, using the time increment δt. If $[M]$ is the mass and $[C]$ and $[K]$ are the damping and stiffness matrices, the equation of motion may be written in incremental form as

$$[M]\{\ddot{x}(t)\} + [C_{in}]\{\dot{x}(t)\} + [K_{in}]\{\delta(t)\} = \{R(t)\} + \{F_1(t)\} \qquad (6.13)$$

Table 6.7 Stiffness matrix and load vector for stud/liner.

$$
[K_b]_{6\times 6} = \pi dL
\begin{bmatrix}
(l^2E_h + pE_v + \gamma^2E_t) & (lmE_h + pqE_v + \gamma sE_t) & (lnE_h + \gamma t\,E_v) & (-l^2E_h - p^2E_v - \gamma^2E_t) & (-lmE_h - pqE_v - \gamma sE_t) & (-lnE_h - \gamma tE_v) \\
 & (m^2E_h + q^2E_v + s^2E_t) & (mnE_h + stE_t) & (-lmE_h - pqE_v - \gamma sE_t) & (-m^2E_h - q^2E_v - s^2E_t) & (-mnE_h - stE_t) \\
 & & (n^2E_h + t^2E_v) & (-lnE_h - \gamma tE_v) & (-mnE_h - stE_t) & (-n^2E_h - t^2E_v) \\
 & & & (l^2E_h + p^2E_v + \gamma^2E_t) & (lmE_h + pqE_v + \gamma sE_t) & (lnE_h + \gamma tE_v) \\
 & \text{symmetrical} & & & (m^2E_h + q^2E_v + s^2E_t) & (mnE_h + stE_t) \\
 & & & & & (n^2E_h + t^2E_v)
\end{bmatrix}
$$

Component stiffnesses using springs:

E_h = horizontal stiffness
E_v = vertical stiffness
E_t = longitudinal stiffness

$$
\{\Delta P^e\}_{6\times 1} = \pi dL
\begin{Bmatrix}
-l\Delta\sigma_h - p\Delta\sigma_v - \gamma\Delta\sigma_l \\
-m\Delta\sigma_h - q\Delta\sigma_v - s\Delta\sigma_l \\
-n\Delta\sigma_h - t\Delta\sigma_l \\
l\Delta\sigma_h + p\Delta\sigma_v + \gamma\Delta\sigma_l \\
m\Delta\sigma_h + q\Delta\sigma_v + s\Delta\sigma_l \\
n\Delta\sigma_h + t\Delta\sigma_l
\end{Bmatrix}
$$

πdL = perimeter of the steel

$\left.\begin{array}{l} l,m,n \\ p,q,\gamma \\ s,t \end{array}\right\}$ = direction cosines

where $F_1(t)$ is the impact/explosion load. If the load increment of $F_1(t)$ is $\delta P_n(t)$, where n is the nth load increment, then

$$P_n(t) = P_{n-1}(t) + \delta P_n(t) \tag{6.13a}$$

and hence $\{R(t)\} = \{\delta P_n(t)\}$, which is the residual time-dependent load vector. The solution of equation (6.13) in terms of $t + \delta t$ for a δt increment becomes

$$[M]\{\ddot{x}(t + \delta t)\} + [C_{in}]\{\dot{x}(t + \delta t)\} + [K_{in}]\{\delta R(t + \delta t)\} + \{\delta P(t + \delta t)\} \tag{6.14}$$

where 'in' denotes initial effects by iteration using the stress approach; $\delta P(t + \delta t)$ represents the non-linearity during the time increment δt and is determined by

$$\{\sigma\} = [D]\{\epsilon\} - \{\epsilon_0\} + \{\sigma_0\} \tag{6.15}$$

The constitutive law is used with the initial stress and constant stiffness approaches throughout the non-linear and the dynamic iteration. For the iteration:

$$\{x(t + \delta t)\}_i = [K_{in}]^{-1}\{R_{TOT}(t + \delta t)\}_i \tag{6.16}$$

The strains are determined using

$$\{\epsilon(t + \delta t)\}_i = [B]\{x(t + \delta t)\}_i \tag{6.17}$$

where $[B]$ is the strain displacement. The stresses are computed as

$$\{\sigma(t + \delta t)\}_i = [D]\{\epsilon(t + \delta t)\}_i + \{\sigma_0(t + \delta t)\}_{i-1} \tag{6.18}$$

where $\{\sigma_0(t + \delta t)\}$ is the total initial stress at the end of each iteration. All calculations for stresses and strains are performed at the Gauss points of all elements.

The initial stress vector is given by

$$\{\sigma_0(t + \delta t)\}_i = f\{\epsilon(t + \delta t)\}_i - [D]\{\epsilon(t + \delta t)\}_i \tag{6.19}$$

Using the principle of virtual work, the change of equilibrium and nodal loads $\{\delta P(t + \delta t)\}_i$ is calculated as

$$F_1(t + \delta t) = \{\delta P(t + \delta t)\}_{iTOT} \tag{6.20}$$

$$= \int_{-1}^{+1} \int_{-1}^{+1} \int_{-1}^{+1} [B]^{T''}\{\delta\sigma_0(t + \delta t)\}_i d\xi d\eta d\zeta$$

$$\sigma_0(t) = \{\sigma_0(t + \delta t)\}_i = 0$$

where $d\xi$, $d\eta$ and $d\zeta$ are the local co-ordinates and T'' is the transpose. The integration is performed numerically at the Gauss points. The effect load vector $F_1(t)$ is given by

$$F_1(t + \delta t) = \{\delta P(t + \delta t)\}_{iTOT}$$
$$= -[\delta C(t)_{in}](\{x(t + \delta t)\}_i - \{x(t)\})$$
$$\quad -[\delta C(t + \delta t)]_i\{x(t + \delta t)\}_i - [\delta K(t)_{in}](\{x(t + \delta t)\}_i - \{x(t)\}_i) \tag{6.21}$$
$$\quad -[\delta K(t + \delta t)]_i\{x(t + \delta t)\}_i$$

The Von Mises criterion is used with the transitional factor f^*_{TR} to form the basis of the plastic state, such as shown in Fig. 6.2,

$$f^*_{TR} = \frac{\sigma_y(t) - \sigma_{y-1}(t)}{\sigma(t+\delta t)_i - \sigma(t+\delta t)_{i-1}} \tag{6.22}$$

The elasto-plastic stress increment will be

$$\{\delta\sigma_i\} = [D]_{ep}\{\sigma(t+\delta t)\}_{i-1}(1 - f^*_{TR})\{\delta\epsilon\} \tag{6.23}$$

If $\sigma(t+\delta t)_i < \sigma_y(t)$, it is an elastic limit and the process is repeated. The equivalent stress is calculated from the current stress state where stresses are drifted; they are corrected from the equivalent stress−strain curve.

The values of $[D]_{ep}$ and $[D]_p$ are derived using plastic stress/strain increments.

In the elasto-plastic stage, the time-dependent yield function is $f(t)$. It is assumed that the strain or stress increment is normal to the plastic potential $Q(\sigma,K)$. The plastic increment, for example, is given by

$$\delta\epsilon(t+\delta t)_p = \partial Q/\partial\sigma = \lambda b \tag{6.24}$$

where λ = proportionality constant > 0
$b \approx \partial Q/\partial\sigma(t+\delta t)$

When $f(t) = Q$

$$\delta\epsilon(t+\delta t)_p = \lambda a$$

$$a = \partial f/\partial\sigma(t+\delta t)$$

therefore, $df = [\partial f/\partial\sigma(t+\delta t)]\, d\sigma(t+\delta t) + (\partial f/dK)dK \tag{6.25}$

If A is the hardening plastic parameter, then

$$A = \frac{1}{\lambda}\,(\partial f/dK)dK$$

An expression can easily be derived for the proportionality constant λ

$$\lambda = \frac{a^{T''}D\delta\epsilon(t+\delta t)}{[A + a^T Db]} \tag{6.26}$$

hence $\delta\epsilon(t+\delta t)_p = b\lambda$

The value of the elasto-plastic matrix $[D]_{ep}$ is given by

$$[D]_{ep} = D - \frac{Dba^{T''}Db}{[A + a^{T''}Db]} \tag{6.27}$$

The value of the plastic matrix $[D]_p$ is given by

$$[D]_p = \left[\frac{Dba^{T''}D}{A + a^{T''}Db}\right] \tag{6.28}$$

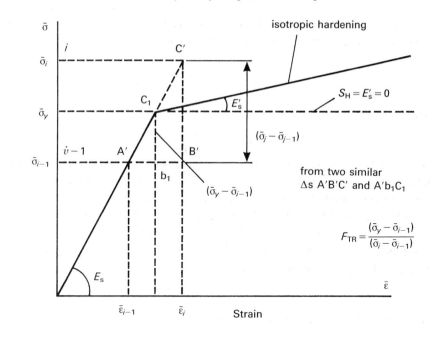

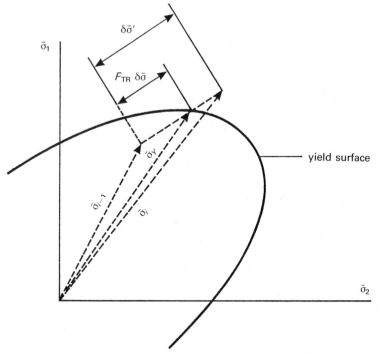

Fig. 6.2 Transitional factor and plastic point.[1.149]

where $[D]$ is the compliance matrix for the elastic case.

The elasto-plastic stress increment is given by

$$\{\delta\sigma_i\}_t = [D]_{\text{ep}}\{\sigma_i\}_t^{Y^*}(1 - f_{\text{TR}}^*)\{\delta\epsilon\}_t \tag{6.29}$$

For the sake of brevity, $\{\delta\sigma_i\}_t = \delta\sigma(t + \delta t)$ for the ith point or increment and other symbols are as given above. The total value becomes

$$\{\sigma_i\}_{\text{TOT}} = \{\sigma_i\}_t^{Y^*} + \{\delta\sigma_i\}_t \tag{6.30}$$

If $\{\sigma_i\}_t < \sigma_{yt}$ it is an elastic point and $\{\sigma_i\}_t = \{\sigma_i'\}_t$. The process is repeated. Looking at the plastic point in the previous iteration, it is necessary to check for unloading when $\sigma \geq \sigma_y$, the unloading will bring about the total stress $\{\sigma_i\}_t = \{\sigma_{i-1}\}_t + \{\delta\sigma_i'\}_t$, and set $\{\sigma_y\}_t = \{\sigma_{i-1}\}_t$. Then loading at this point gives

$$\{\delta\sigma_i\} = [D]_{\text{ep}}\{\sigma_{i-1}\}_t\{\delta\epsilon\}_t \tag{6.31}$$

The total stress is then written as

$$\{\sigma_i\}_{\text{TOT}} = \{\sigma_{i-1}\}_t\{\delta\sigma_i\} \tag{6.32}$$

Stresses are calculated using the elasto-plastic material matrix, which does not drift from the yield surfaces, as shown in Fig. 6.2. Stresses are corrected from the equivalent stress–strain curve by

$$\{\sigma_{\text{corr}}\} = \{\sigma_{i-1}\}_t + K\{\delta\epsilon_p\}_t \tag{6.33}$$

where $\{\delta\epsilon_p\}_t = \sqrt{\frac{2}{3}}\{\sqrt{(\delta\epsilon_{ij}^P \ \delta\epsilon_{ij}^P)}\}_t = $ equivalent plastic strain increment. K is the strain-hardening parameter, such that $\{\delta\epsilon_p\}_t = \lambda$. The equivalent stress is calculated from the current stress state, as shown below

$$\{\sigma_i\}_{\text{eq}} = f\{(\sigma_i)\}_t \tag{6.34}$$

$$\text{the value of } \sigma_{\text{corr}}/\sigma \text{ is a } \textit{factor} \tag{6.35}$$

Therefore the correct stress state on the yield surface is given by

$$\{\sigma_i\} = \textit{factor} \times \{\sigma_i\} \tag{6.36}$$

A reference is made to Fig. 6.2 for evaluating this factor as f_{TR}^*.

6.3.1 Buckling state and slip of layers for composite sections

Within the above stages, there can be a possibility of plastic buckling of the liner or any embedded anchors or layers. The buckling matrix is developed so that at appropriate stages the layer/liner/anchor system is checked against buckling. The plastic buckling matrix is given below

$$(K^i + \lambda_c K_G^i)F_T = 0 \tag{6.37}$$

where K^i is the elasto-plastic stiffness matrix as a function of the current state of plastic deformation using the above steps and K_G is the geometric stiffness matrix.

$$\lambda_c = 1 + E_{PS} \tag{6.37a}$$

where E_{PS} represents accuracy parameters.

Where composite layers, liner and studs are involved, the incremental slips from the nodal displacements are assessed in the following manner:

$$\{\Delta S_i\}_{x,y,z} = [T'']\{d\delta_i\} \tag{6.38}$$

where S_i is a slip at node i and T'' is the transformation matrix given in Table 6.8. The total slip at iteration is given without subscripts as

$$\{S_i\} = \{S_{i-1}\} + \{S_i\} \tag{6.39}$$

The strains are computed as

$$\{\epsilon_i\}_t = \{\epsilon_{i-1}\}_t + \{\delta\epsilon_i\}_t \tag{6.40}$$

The incremental stress $\{\sigma_B\}$ between the studs and concrete or between any composite materials for the ith node can then be computed as

$$\{\delta\sigma_{Bi}\}_t = [K_s]\{\sigma_{Bi-1}\}_t\{\delta S_i\}_t \tag{6.41}$$

The total stresses are

$$\{\sigma_{Bi}\}_{TOT} = \{\sigma_{Bi-1}\}_t + \{\delta\sigma_{Bi}\}_t \tag{6.42}$$

If $|S_i| > S_{max}$ the bond between the stud and the concrete or any composite materials is broken and the pull-out occurs, i.e. $\{\sigma_{Bi}\} = 0$ and S_{max} has a value which is maximum. If $|S_i| < S_{max}$ the value $\{S_i\}$ is calculated. The procedure is linked with the general finite element work discussed already in the non-linear dynamic cases for impact and explosion.

Table 6.8 T'' transformation matrices.

$$[T''_\epsilon] = \begin{bmatrix} l_1^2 & m_1^2 & n_1^2 & l_1m_1 & m_1n_1 & l_1n_1 \\ l_2^2 & m_2^2 & n_2^2 & l_2m_2 & m_2n_2 & l_2n_2 \\ l_3^2 & m_3^2 & n_3^2 & l_3m_3 & m_3n_3 & l_3n_3 \\ 2l_1l_2 & 2m_1m_2 & 2n_1n_2 & (l_1m_2 + l_2m_1) & (m_1n_2 + m_2n_1) & (l_1n_2 + l_2n_1) \\ 2l_2l_3 & 2m_2m_3 & 2n_2n_3 & (l_2m_3 + l_3m_2) & (m_2n_3 + n_2m_3) & (l_2n_3 + l_3n_2) \\ 2l_1l_3 & 2m_1m_3 & 2n_1n_3 & (l_1m_3 + m_1l_3) & (m_1n_3 + m_3n_1) & (l_1n_3 + n_1l_3) \end{bmatrix}$$

$$[T''_\sigma] = \begin{bmatrix} l_1^2 & m_1^2 & n_1^2 & 2l_1m_1 & 2m_1n_1 & 2l_1n_1 \\ l_2^2 & m_2^2 & n_2^2 & 2l_2m_2 & 2m_2n_2 & 2l_2n_2 \\ l_3^2 & m_3^2 & n_3^2 & 2l_3m_3 & 2m_3n_3 & 2l_3n_3 \\ l_1l_2 & m_1m_2 & n_1n_2 & (l_1m_2 + l_2m_1) & (m_1n_2 + n_1m_2) & (l_1n_2 + l_2n_1) \\ l_2l_3 & m_2m_3 & n_2n_3 & (l_2m_3 + l_3m_2) & (m_2n_3 + n_2m_3) & (l_2n_3 + l_3n_2) \\ l_1l_3 & m_1m_3 & n_1n_3 & (l_1m_3 + l_3m_1) & (m_1n_3 + m_3n_1) & (l_1n_3 + n_1l_3) \end{bmatrix}$$

6.3.2 Strain rate effects based on the elastic-viscoplastic relationship for earth materials under impact and explosion

It is assumed that for each dynamic loading increment, the strain rate ϵ_{ij} can be expressed as the sum of the elastic and viscoplastic components:

$$\{d\epsilon_{ij}\}_t = \{d\epsilon_{ij}\}_e + \{d\dot\epsilon_{ij}\}_{vp} \tag{6.43}$$

Where the subscripts e and vp denote the elastic and viscoplastic components, respectively. The elastic strains are related to the stress rate $\dot\sigma_{ij}$ by

$$\{d\dot\epsilon_{ij}\}_e = \frac{1}{9K} \times (dJ_1/dt)\delta_{ij} + \frac{1}{2G} \times dS_{ij}/dt \tag{6.44}$$

where J_1 = deviatoric stress: first invariant
$S_{ij} = \{\sigma_{ij} - \frac{1}{3}J_1\delta_{ij}\}_t$
δ_{ij} = Knonecker delta
K = elastic bulk modulus
G = elastic shear modulus

The shear modulus is expressed in terms of the invariant J'_2, where

$$J'_2 = \tfrac{1}{2}S_{ij}\,S_{ij} \tag{6.44a}$$

$$\text{then} \qquad K = \frac{K_i}{1 - K_1} \, [1 - K_1 e^{-K_2 J_1}] \tag{6.44b}$$

$$G = \frac{G_i}{1 - G_1} \, [1 - G_1 e^{-G_2\sqrt{J'_2}}] \tag{6.44c}$$

where K_i, K_1, G_i, G_1, K_2 and G_2 are material constants. The values of J'_2 and Ss are given below

$$J'_2 = \tfrac{1}{2}(S_x^2 + S_y^2 + S_z^2) + \tau_{xy}^2 + \tau_{yz}^2 + \tau_{zx}^2 \text{ is the second stress invariant} \tag{6.44d}$$

$$S_x = \sigma_x - \sigma_m \qquad S_y = \sigma_y - \sigma_m \qquad S_z = \sigma_z - \sigma_m \tag{6.44e}$$

The linear values of K and G are

$$K = \frac{E}{3(1 - 2v)}; \; G = \frac{E}{2(1 + v)} \tag{6.44f}$$

Tables 6.1 and 6.2 are used for variable properties of E and v.

The components of the viscoplastic strain rate are calculated using the above-mentioned plastic flow rule for rate-sensitive material.

$$\{\dot\epsilon_{ij}\}_{vp} = \gamma[f(\sigma_D/B) \, \delta\sigma_D/\delta\sigma_{ij}]_t \tag{6.45}$$

where γ = viscosity parameter
$f(\sigma_D/B) = f(\sigma_s - \beta/B)$
σ_s = static yield stress
B = material parameter

$$\beta = f^{-1}\left\{\frac{1}{\gamma}\left[\frac{\frac{1}{3}(d\dot\epsilon_{KK})_{vp} + (2\dot\epsilon^{-1})_{vp}^2}{3(\partial\sigma_s/\partial J_1)^2 + \frac{1}{2}(\partial\sigma_s/\partial J'_2)^2}\right]\right\} \tag{6.45a}$$

$(\dot{e}^{-1})_{vp} = [\frac{1}{2}d\dot{e}_{ij}\ d\dot{e}_{ij}]^{1/2}_{vp}$ and is the square root of the second invariant

of the viscoplastic strain rate   (6.45b)

Using this bulk modulus approach for soils, the time-dependent stress–strain relation is given in Table 6.9. With reference to *rocks*, the failure strength of the rock is defined in exactly the same way as described earlier; the values for E and v will vary. Nevertheless, the various alternative failure models given in Table 6.10 for rocks are related in terms of strain rates by

$$\bar{M} = \frac{\sigma_{dyn}}{\sigma_s} = 1 + c \ \log \frac{\dot{\epsilon}}{\dot{\epsilon}_s} \tag{6.46}$$

where σ_{dyn} = dynamic stress

σ_s = static stress

$\dot{\epsilon}$ = strain rate (dynamic)

$\dot{\epsilon}_s$ = strain rate (static)

c = constant

The range of strain rate is $\dot{\epsilon} = 10^{-5} \ \mathrm{s}^{-1}$ up to $5 \times 10^{0} \mathrm{s}^{-1}$. The dynamic failure criterion can then be written as

$$\dot{\sigma}_f = \tau_0 X_3 + \sigma \left[\frac{\sigma^2_{cu} X^2_1 - 4\tau^2_0 X^2_3}{4\sigma_{cu} X_1 \tau X_3} \right] \quad \text{for } \sigma \nless 0 \text{ compression} \tag{6.47}$$

$$\dot{\sigma}^2_f = \tau^2_0 X^2_3 + \sigma \left[\frac{\tau^2_0 X^2_3 - \sigma^2_{tu} X^2_2}{\sigma_{tu} X_2} \right] - \sigma^2 \quad \text{for } \sigma \ngtr 0 \text{ tension} \tag{6.48}$$

$$\nless A\dot{\epsilon}^{1/3}$$

where $X_1 = \frac{3}{40} \bar{M} + \frac{1}{25} \bar{M}^2$

$X_2 = \frac{3}{100} \bar{M} + \frac{7}{1000} \bar{M}^2$

$X_3 = \frac{1}{40} \bar{M} + \frac{1}{100} \bar{M}^2$ (6.49)

τ_0 = octahedral shear stress under static loads

Table 6.9 Bulk modulus model for earth materials under impact.

$$
\begin{Bmatrix} \delta\sigma_x \\ \delta\sigma_y \\ \delta\sigma_z \\ \delta\tau_{xy} \\ \delta\tau_{yz} \\ \delta\tau_{zx} \end{Bmatrix}_t
=
\begin{bmatrix}
(K+\frac{4}{3}G) & (K-\frac{2}{3}G) & (K-\frac{2}{3}G) & 0 & 0 & 0 \\
 & (K+\frac{4}{3}G) & (K-\frac{2}{3}G) & 0 & 0 & 0 \\
 & & (K+\frac{4}{3}G) & 0 & 0 & 0 \\
 & & & G_{12} & 0 & 0 \\
 & \text{sym} & & & G_{23} & 0 \\
 & & & & & G_{13}
\end{bmatrix}
\begin{Bmatrix} \delta\epsilon_x \\ \delta\epsilon_y \\ \delta\epsilon_z \\ \delta_{xy} \\ \delta_{yz} \\ \delta_{zx} \end{Bmatrix}_t
$$

or in short $\{\delta\sigma\} = [D]\{\delta\epsilon\}$

where $[D]$ is the required material matrix

$$G_{12} = G_{23} = G_{13} = G = G_e - \alpha \ \log_e \frac{J_2}{J^e_2} \quad \text{for } J_2 > J^e_2$$

$$G = G_e \quad \text{for } J_2 \le J^e_2$$

In the case where the soil/rock is orthotropic, the values of G_{12}, G_{23} and G_{13} are given as indicated in Table 6.2.

Table 6.10 Numerical models for rocks.

---

(1) Sandstone

$$\tau = 1538 + \sigma \tan\phi \tag{1}$$

where τ and σ = shear and compressive stresses, respectively
$\phi = 29°15'$

(2) Rupture of sandstone: Mohr failure envelope

$$\tau_{max}/\sigma_{cu} = 0.1 + 0.76(\sigma_m/\sigma_{cu})^{0.85} \tag{2}$$

where σ_{cu} is the uni-axial compressive stress at rupture under pure shear $\sigma_1 = -\sigma_3$

(3) Realistic rock including friction

$$\alpha(\sigma_1 + \sigma_2 + \sigma_3) + (\sigma_1 - \sigma_2)^2 + (\sigma_2 - \sigma_3)^2 + (\sigma_3 - \sigma_1)^2 = K^* \tag{3}$$

where $\alpha = \sqrt{(6 \tan\phi)}/\sqrt{(9 + 12 \tan\phi)}$ } can be obtained from
$\quad K^* = \sqrt{(6c)}/\sqrt{(9 + 12 \tan\phi)}$ } Mohr envelope
$\quad \phi$ = angle of friction
$\quad c$ = cohesion

A generalized Mohr coulomb criterion is written as

$$\tau^2 = [\sqrt{(n + 1)} - 1][\sigma_{tu}^2 - \sigma\sigma_{tu}] \tag{4}$$

where σ, τ = normal stress and shear stress on the fractured plane
$\quad \sigma_{tu}$ = uni-axial tensile strength
$\quad n = \sigma_{cu}/\sigma$ = brittleness
$\quad \sigma_{cu}$ = uni-axial compressive strength

Equation (4) can be expressed in terms of σ_m, mean stress, and σ_s, the maximum shear stress, by

$$\sigma_s = \sigma_{tu} - \sigma_m \qquad \text{for } \sigma_{tu} > \sigma_{m0} > \sigma_m$$

$$\sigma_s = \tau_0 \sqrt{[1 - (\sigma_m/\sigma_{tu}) - (\tau_0/2\sigma_{tu})^2]} \qquad \text{for } \sigma_{m0} < \sigma_m \tag{5}$$

where $\sigma_m = (\sigma_1 + \sigma_2)/2; \qquad \sigma_s = (\sigma_1 - \sigma_2)/2$
$\quad \sigma_1, \sigma_2$ = principle stresses $(\sigma_1 > \sigma_2)$

$$\sigma_{m0} = \sigma_{tu} - \tau_0^2/2\sigma_{tu} \tag{6}$$

$$\tau_0 = [\sqrt{(n + 1)} - 1]\sigma_{tu}$$

The stress state is assessed for σ_m and σ_s from the failure surface as

$$R = \sigma_s/\sigma_{s(critical)} \geq 1 \tag{7}$$

representing the failure condition. If $\sigma_s = \sigma_{s(critical)}$, equation (5) is satisfied.

In the case of *brick material*, Khoo and Hendry relationships given below are used in the above failure models and strain rate simulations. The non-linear principle stress relationship (bi-axial) is given by

$$\sigma_1/\sigma_{cu} = 1 + 2.91(\sigma_2/\sigma_c)^{0.805} \tag{6.50}$$

where σ_1 = major principle stress

$\quad\sigma_2$ = minor principle stress

$\quad\sigma_{cu}$ = uni-axial compressive strength

The brick-failure envelope with the mortar tri-axial strength curve is given by the polynomials

$$\sigma_t/\sigma_{tu} = 0.9968 - 2.0264(\sigma/\sigma_{cu}) + 1.2781(\sigma/\sigma_{cu})^2 - 0.2487(\sigma/\sigma_{cu})^2 \tag{6.51}$$

$$\sigma_3/\sigma_{cu} = -0.1620 + 0.1126(\sigma_1/\sigma_{cu}) + 0.0529(\sigma_1/\sigma_{cu})^2 - 0.0018(\sigma_1/\sigma_{cu})^3 \tag{6.52}$$

where σ/σ_{cu} = ratio of compressive strength

$\quad\sigma_t/\sigma_{tu}$ = ratio of tensile strength

$\quad\quad\sigma_t = \alpha\sigma_3$ where $\alpha = 0.15$ and 0.40 for mortars of $1:\frac{1}{4}:3$ and $1:1:6$, respectively.

6.3.3 Finite element of concrete modelling

A number of modelling methods are available for simulation into the finite-element method[1.149,3.1−3.168] On impact and explosion work, methods such as the endochronic, Ottoson and Blunt crack have been widely used. They are covered in this section. The bulk modulus model of Table 6.9 is reviewed to include cracking with and without aggregate interlocking. On the basis of the endochronic concept, which is widely reported[1.28−1.30,1.149] the following equation applies

$$\{\delta\sigma_{x,y,z}\}_t + \{\delta\sigma^p_{x,y,z}\}_t = [D^*_T]\{\epsilon^*_{x,y,z}\}_t \tag{6.53}$$

where the superscript p denotes stresses in the plastic case. Table 6.11 gives details of uncracked and cracked cases for equation (6.52). When cracks in three directions are open the concrete loses its stiffness, then

$$[D^*_T] = [0] \tag{6.54}$$

Stresses $\{\sigma_i\}_t$ are checked against the cracking criteria. For example, if there is one crack normal to the X-direction, the concrete can no longer resist any tensile stress in that direction, then

$$\delta\sigma^*_x = 0$$

Then

$$D_{11}\delta\epsilon^*_x + D_{12}\delta\epsilon^*_y + D_{13}\delta\epsilon^*_z = \delta\sigma^{p*}_x$$

$$\delta\epsilon^*_x = \frac{\delta\sigma^{p*}_x}{D_{11}} - \frac{D_{12}}{D_{11}}\delta\epsilon^*_y - \frac{D_{13}}{D_{11}}\delta\epsilon^*_y \tag{6.55}$$

In a similar manner, examples for shear terms can be written as

$$\delta\tau^*_{xy} + \delta\tau^{p*}_{xy} = \beta' D_{44}\gamma^*_{xy}$$
$$\delta\tau^*_{yz} + \delta\tau^{p*}_{yz} = D_{55}\delta\gamma^*_{yz} \tag{6.56}$$
$$\delta\tau^*_{zx} + \delta\tau^{p*}_{zx} = \beta' D_{66}\delta\gamma^*_{zx}$$

6.3.3.1 Blunt crack band propagation

The smeared crack concept, rather than the isolated sharp inter-element crack concept described above, is gaining ground. Here the element topology does change. The smeared crack band of a blunt front is that in which one can easily select cracks in any direction without paying a penalty, even if the crack direction is not truly known. Bazant *et al.*[1.28–1.30] and Bangash[1.149] introduced the equivalent strength and energy variation which are utilized for crack propagation once it is initiated within the element. The equivalent strength criterion is used for crack propagation by specifying an equivalent stress within the surrounding elements of an existing crack at which cracking should be propagated. The expression for the equivalent strength σ_{eq} is given (see Fig. 6.3) as

$$\sigma_{eq} = C[EG_f/W(1-2v\sigma^0_2/\sigma^0_1)]^{1/2} \tag{6.57}$$

where $C =$ a constant dependent on the choice of elements
$E =$ elastic modulus
$v =$ Poisson's ratio
$W = A/\delta a = A/r\cos\alpha$
$A =$ area of the element at the front

Table 6.11 Cracks using endochronic theory.

Uncracked matrix

$$\begin{Bmatrix} \delta\sigma_x + \delta\sigma^p_x \\ \delta\sigma_y + \delta\sigma^p_y \\ \delta\sigma_z + \delta\sigma^p_z \\ \delta\tau_{xy} + \delta\tau^p_{xy} \\ \delta\tau_{yz} + \delta\tau^p_{yz} \\ \delta\tau_{zx} + \delta\tau^p_{zx} \end{Bmatrix}_t = \begin{bmatrix} D_{11} & D_{12} & D_{13} & 0 & 0 & 0 \\ & D_{21} & D_{23} & 0 & 0 & 0 \\ & & D_{33} & 0 & 0 & 0 \\ & & & \beta' D_{44} & 0 & 0 \\ & & & & \beta' D_{55} & 0 \\ & & & & & \beta' D_{66} \end{bmatrix} \begin{Bmatrix} \delta\sigma_x \\ \delta\sigma_y \\ \delta\sigma_z \\ \delta\gamma_{xy} \\ \delta\gamma_{yz} \\ \delta\gamma_{zx} \end{Bmatrix}_t$$

where $D_{11} = D_{22} = D_{33} = K + \frac{4}{3}G$ $\beta =$ aggregate inter locking $\approx \frac{1}{2}$ to $\frac{3}{4}$
$D_{12} = D_{13} = D_{23} = K - \frac{2}{3}G$

$$D_{44} = G_{12} = \frac{1}{2}\left[\frac{E_1}{2(1+v_{12})} + \frac{E_2}{2(1+v_{21})}\right]$$

$$D_{55} = G_{23} = \frac{1}{2}\left[\frac{E_2}{2(1+v_{23})} + \frac{E_3}{2(1+v_{32})}\right]$$

$$D_{66} = G_{13} = \frac{1}{2}\left[\frac{E_3}{2(1+v_{31})} + \frac{E_1}{2(1+v_{13})}\right]$$

The values of E and v are given in Tables 6.3 to 6.6   *continued*

Table 6.11 *Continued.*

Cracked matrix

$$\sigma_1 - \text{direction:}[D]^* = \begin{bmatrix} 0 & 0 & 0 & 0 & 0 & 0 \\ 0 & \left(D_{22} - \dfrac{D_{12}^2}{D_{11}}\right) & \left(D_{23} - \dfrac{D_{12}D_{12}}{D_{11}}\right) & 0 & 0 & 0 \\ 0 & \left(D_{23} - \dfrac{D_{31}D_{21}}{D_{11}}\right) & \left(D_{33} - \dfrac{D_{13}D_{13}}{D_{11}}\right) & 0 & 0 & 0 \\ 0 & 0 & 0 & \beta'D_{44} & 0 & 0 \\ 0 & 0 & 0 & 0 & D_{55} & 0 \\ 0 & 0 & 0 & 0 & 0 & \beta'D_{66} \end{bmatrix}$$

$$\sigma_2 - \text{direction:}[D]^* = \begin{bmatrix} \left(D_{11} - \dfrac{D_{21}}{D_{22}}\right) & 0 & \left(D_{13} - \dfrac{D_{12}D_{23}}{D_{22}}\right) & 0 & 0 & 0 \\ 0 & 0 & 0 & 0 & 0 & 0 \\ \left(D_{31} - \dfrac{D_{21}D_{32}}{D_{22}}\right) & 0 & \left(D_{33} - \dfrac{D_{23}^2}{D_{22}}\right) & 0 & 0 & 0 \\ 0 & 0 & 0 & \beta'D_{44} & 0 & 0 \\ 0 & 0 & 0 & 0 & \beta'D_{55} & 0 \\ 0 & 0 & 0 & 0 & 0 & D_{66} \end{bmatrix}$$

$$\sigma_3 - \text{direction: }[D]^* = \begin{bmatrix} \left(D_{11} - \dfrac{D_{13}^2}{D_{33}}\right) & \left(D_{12} - \dfrac{D_{13}D_{23}}{D_{23}}\right) & 0 & 0 & 0 & 0 \\ \left(D_{21} - \dfrac{D_{31}D_{32}}{D_{33}}\right) & \left(D_{22} - \dfrac{D_{23}D_{23}}{D_{33}}\right) & 0 & 0 & 0 & 0 \\ 0 & 0 & 0 & 0 & 0 & 0 \\ 0 & 0 & 0 & D_{44} & 0 & 0 \\ 0 & 0 & 0 & 0 & \beta'D_{55} & 0 \\ 0 & 0 & 0 & 0 & 0 & \beta'D_{55} \end{bmatrix}$$

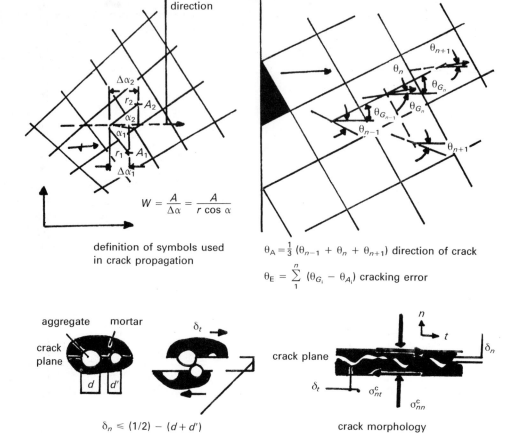

$$W = \frac{A}{\Delta\alpha} = \frac{A}{r \cos \alpha}$$

definition of symbols used
in crack propagation

$\theta_A = \frac{1}{3} (\theta_{n-1} + \theta_n + \theta_{n+1})$ direction of crack

$\theta_E = \sum_{1}^{n} (\theta_{G_i} - \theta_{A_i})$ cracking error

$\delta_n \leqslant (1/2) - (d + d')$

Fig. 6.3 Blunt crack propagation.

r = centroidal distance from the last cracked elements in the front element
α = angle measured from the established crack to the line between the centroids
G_f = energy release rate = $\delta E(P_i a)/\delta a = \Delta E(P_i a)/\delta a$
a = cracked area
P_i = loads

The band length is specified as $a + \Delta a/2$.

In the initial state prior to cracking, the strain energy U_0 is based on the principal stresses σ_2^0, where σ_1^0 is the largest tensile stress. After cracking, σ_1^1 becomes 0 and $\sigma_2^1 = 0$, which is used for the current value of U_1. The change of strain energy $\delta U = U_0 - U_1$ is equated to the crack length ($\Delta a \times G_f$), where U is the total strain energy in a cracked body and δU is the energy released by the structure into the element which cracks. The crack direction within an arbitrary grid is given by

$$\theta_A = \tfrac{1}{3}(\theta_{n-1} + \theta_n + \theta_{n+1}) \tag{6.58}$$

where θ_A is the average crack direction, θ_{n-1} and θ_n are the cracking angles of the next to last cracked element, and θ_{n-1} is the impending cracking angle of the element adjacent to the crack front. Since in the arbitrary grid the cracking direction is specified by the accumulated error of the cracked direction, the accumulated cracking error θ_E is given by

$$\theta_E = \sum_{i}^{n} (\theta_{G_i} - \theta_{Ai}) \tag{6.59}$$

where n is the number of cracked elements and θ_G is the actual average crack propagation angle within each element. A better formulation by Gamborov and Karakoc, reported by Bangash,[1.149] is given of the tangent shear modulus G^{CR} whenever cracking is initiated:

$$G^{CR} = \frac{\sigma_{nt}^c}{\epsilon_{nn}^{CR}} k \frac{1}{r(a_3 + a_4|r|^3)} \tag{6.60}$$

or

$$G^{CR} = \frac{\sigma_0}{\epsilon_{nn}^{CR}} k \{1 - [2(P/D_a)\epsilon_{nn}^{CR}]^{1/2}\}$$

where

$$k = \frac{a_3 + 4a_4|\gamma|^3 - 3a_3 a_4 \gamma^4}{(1 + a_4\gamma^4)^2} \tag{6.61}$$

and a_3, a_4 = coefficients as a function of the standard cylindrical strength f_c'

τ_0 = crack shear strength (ranging from 0.25 to $0.7f_c'$)

D_a = maximum aggregate size (up to 4 mm)

P_c = large percentage of crack asperities

$\gamma = \delta_t/\delta_n$

δ_t, δ_n = crack displacements along the normal and tangential directions (Figs 6.3 and 6.4)

Curves have been plotted showing a decrease in the value of G^{CR} when a crack opening linearly increases at increasing shear. The reference[1.149] gives a constitutive law in which a confinement stress within the rough crack model is given by

$$\sigma_{nn}^c = -a_1 a_2 \frac{\delta_t \sigma_{nt}^c}{(\delta_n^2 + \delta_t^2)^q} \tag{6.62}$$

where σ_{nn}^c = interface normal stress

σ_{nt}^c = interface shear stress

a_1, a_2 = constant ($a_1 a_2 = 0.62$)

q = a function of the crack opening; taken to be 0.25

For different types of crack dilatancy δ_n/σ_{nt}^c (Fig. 6.3) the tangent shear modulus

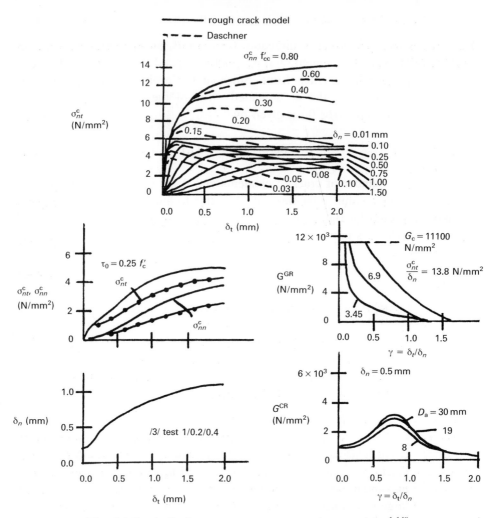

Fig. 6.4 Crack displacement versus tangent shear modulus.[1.149]

G^{CR} is plotted against the ratio r of the crack displacement. The value of σ_{nt}^c is given by

$$\sigma_{nt}^c = \tau_0 \left[1 - \sqrt{\left(\frac{2p}{D_a} \epsilon_{nn}^{CR} \right)} \right] r \frac{a_3 + a_4|r|^3}{1 + a_4 r^4} \qquad r = \gamma_{nt}^{CR} / \epsilon_{mn}^{CR} \qquad (6.63)$$

where p is the crack spacing and CR is cracked concrete,

$$\epsilon_{nn}^{CR} = \delta n/p = \text{strain against } \sigma_{nn}^c \qquad (6.64)$$

The σ_{nn}^c values have been computed using points common to the curves of crack opening and constant confinement stress.

6.3.3.2 Ottoson failure model

The Ottoson four-parameter model has a smooth but convex surface with curved meridians determined by the constants a and b.[1.150]

The analytical failure surface is defined by

$$f(I_1, J_2, J) = a\frac{J_2}{(f_c')^2} + \lambda \frac{\sqrt{J_2}}{f_c'} + b \frac{I_1}{f_c'} - 1 = 0 \qquad (6.65)$$

where $I_1 = \sigma_x + \sigma_y + \sigma_z =$ the first invariant of the stress tensor (6.65a)

$\quad J_2 =$ the second invariant of the stress deviator tensor

$\quad = \frac{1}{2}(S_x^2 + S_y^2 + S_z^2) + \tau_{xy}^2 + \tau_{yz}^2 + \tau_{zx}^2$ (6.65b)

$\quad J = \cos 3\theta = 1.5\sqrt{3}(J_3/\sqrt{J_2})$ (6.65c)

$\quad J_3 =$ the third invariant of the stress deviator tensor

$\quad = S_x S_y S_z + 2\tau_{xy}\tau_{yz}\tau_{zx} - S_x\tau_{yz}^2 - S_y\tau_{xz}^2 - S_z\tau_{xy}^2$ (6.65d)

$\quad S_x = \sigma_x - I_1/3$

$\quad S_y = \sigma_y - I_1/3$ (6.65e)

$\quad S_z = \sigma_z - I_1/3$

$\quad \lambda = \lambda\,(\cos 3\theta) > 0 \qquad$ a and b are constant

$\quad \lambda = K_1 \cos(\frac{1}{3}\cos^{-1}(K_2\cos 3\theta))$ for $\cos 3\theta > 0$

$\quad \lambda = K_1 \cos(\pi/3 - \frac{1}{3}\cos^{-1}(-K_2\cos 3\theta))$ for $\cos 3\theta \le 0$

$\qquad K_1, K_2,$ a and b are material parameters $(0 \le K_2 \le 1)$

$\quad f_c' =$ uni-axial compressive cylinder strength for concrete $= 0.87\sigma_{cu}$

$\quad \sigma_t =$ uni-axial tensile strength for concrete

Table 6.12 lists some of the relevant parameters.

The knowledge of the mechanical properties of concrete and the reinforcement (conventional and pre-stressing steel) at high strain rates is essential for rational application of materials in those constructions where impact and explosion loadings can be expected. The usual magnitude of the strain rate $(d\epsilon/dt = \dot{\epsilon})$ for all concrete structures is of the order of $5 \times 10^5/s$ in the range of the ultimate load. For reinforcement, the range is between 10^5 and $10^2/s$. Table 6.13 gives relevant data. Figure 6.5 gives experimental stress–strain relationships for reinforcement for various strain rates. The theoretical expression in equation (6.46) was used.

6.4 Ice/snow impact

Chapter 2 gives a thorough survey on data regarding the effect of ice floes. When floating ice sheets move under the influence of strong winds and currents, a sea-going vehicle or a semi-submersible will be subject to an impact given by

$$F_I(t) = F_{IO}(t) + F_I'(t) \qquad (6.66)$$

where $F_{IO}(t)$ and $F_I'(t)$ are constant and fluctuating values of the ice impact force, respectively. The value of $F_{IO}(t)$ is given by

Table 6.12 Ottoson's failure model for concrete.[1.150]

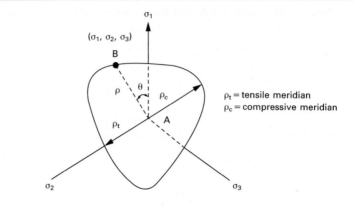

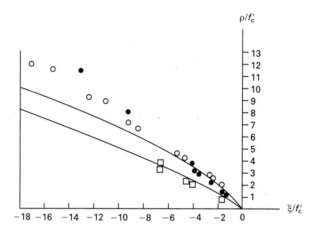

Comparison for the four-parameter model.

Four material parameters ($k = \sigma_t/\sigma_c$)

k	a	b	K_1	K_2
0.08	1.8076	4.0962	14.4863	0.9914
0.10	1.2759	3.1962	11.7365	0.9801
0.12	0.9218	2.5969	9.9110	0.9647

Values of the function ($k = \sigma_t/\sigma_c$)

k	λ_t	λ_c	$\lambda_c\lambda_t$
0.08	14.4925	7.7834	0.7378
0.1	11.7109	6.5315	0.5577
0.12	9.8720	5.6979	0.5772

Table 6.12 *Continued.*

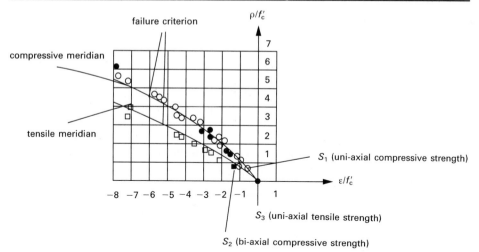

Determination of material parameters (S_1, S_2, S_3, S_4 = failure stresses).

The following three failure states were represented:

(1) Uni-axial compressive strength, f_c' ($\theta = 60°$). Uni-axial tensile strength, σ_t ($\theta = 0°$) = Kf_c'.
(2) Bi-axial compressive strength, $\sigma_1 = \sigma_2 = -1.16\sigma_c$; $\sigma_3 = 0$ ($\theta = 0°$) (test results of Kupfer)[1.16]
(3) The tri-axial state (ξ/f_c', ρ/f_c') = (-5, 4) on the compressive meridian ($\theta = 60°$).

Table 6.13 Strain rate for concrete and reinforcement.

The relationship between the fracture strain ϵ_f and the strain rate $\dot{\epsilon}$ is given by

$$\epsilon_f = \alpha\dot{\epsilon}^{1/3} \tag{1}$$

for plain concrete $\alpha = 206$; for reinforced concrete $\alpha = 220$
Another expression given by DELFT for plain concrete is

$$\epsilon_f = 100 + 109\dot{\epsilon}^{1/2} \tag{2}$$

In equation (1), for say $\dot{\epsilon}/s = 35$, the value of ϵ_f for reinforced concrete will be 740×10^{-6}. For $\dot{\epsilon}/s = 30$, the value of ϵ_f for plain concrete will be 650×10^{-6}.

For fibre-reinforced concrete, the influence of the strain rate upon the tensile strength for concrete is given by

$$\sigma_t = \alpha + \beta \log_e \dot{\epsilon} \tag{3}$$

$\alpha = 0$ for no fibres, i.e. plain concrete
$\sigma_t = 1.7 + 0.0364 \log_e \dot{\epsilon}$ for 3% fibres
$\sigma_t = 1.87 + 0.0424 \log_e \dot{\epsilon}$

For low, intermediate and high strain rate, DELFT gives an expression:

$$\sigma_t = \alpha + \beta N \tag{4}$$

continued

Table 6.13 *Continued.*

where N is the number of fibres/reinforcements

	Low	Intermediate	High
α	3.32	4.87	5.49
β	1.85×10^{-3}	2.85×10^{-3}	6.3×10^{-3}

For the fracture energy, G_f as stated in the endochronic theory will be modified as follows:

$$G_f = \alpha + \beta N$$

	Low	Intermediate	High
α	12.72	22.90	29.200
β	0.12	0.18	0.211

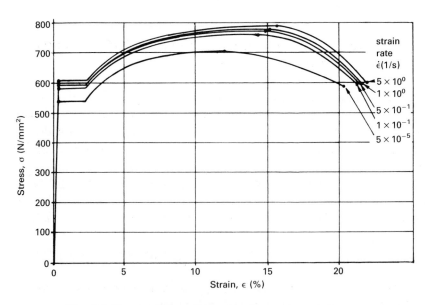

Fig. 6.5 Stress–strain relationships for various strain rates.

$$F_{IO}(t) = S_1 S_2 S_3 f(\dot{\epsilon}) TW \times h\sigma_c \ \text{(or } f_c') \tag{6.67}$$

where S_1 = contact factor around the member during crushing

S_2 = shape factor S_f of the impactor

S_3 = temperature factor which is $(1 - 0.012T)/(1 - 0.012T_s)$ $(T: 0.5°C > T > -20°C)(T_s = -10°C$ standard temperature)

σ_c = compressive strength of ice measured at a strain rate of $5 \times 10^{-4}/s$ which is $\dot{\epsilon}_0$

W = transverse width of the member

h = ice sheet thickness

$$f(\dot{\epsilon}) = (\dot{\epsilon}/\dot{\epsilon}_0)^{\gamma}; \qquad \phi = a_1 + a_2(h/W)^{1/2} \qquad (6.67a)$$

$$\dot{\epsilon} = \dot{x}/4W \qquad (6.67b)$$

where $\dot{x}$ = ice flow velocity

γ = empirical coefficient dependent on strain rate

a_1, a_2 = factors dependent on the ice thickness/diameter ratio of a member

Depending on the type of impact (direct or angular) and the stiffnesses of members and ice floes, the global equation of motion, (6.13), will be influenced by roll, pitch and yaw motions (θ_r, ϕ_p, ψ_y) and surge, sway and heave motions (θ_s, ϕ_s, ψ_h), respectively. Generally, the values of θ_r, ϕ_p, ψ_y, θ_s, ϕ_s and ψ_h range as shown below:

$$\theta_r = -0.012 \text{ to } 0.04 \times 10^{-3} \text{ radians}; \ \theta_s = -0.75 \text{ to } 6\,\text{m}$$
$$\phi_p = -0.012 \text{ to } 0.04 \times 10^3 \text{ radians}; \ \phi_s = -1.5 \text{ to } 3\,\text{m}$$
$$\psi_y = -0.012 \text{ to } 0.04 \text{ radians}; \ \psi_h \approx \theta_s$$

6.5 Impact due to missiles, impactors and explosions: contact problem solutions

Contact problems have been introduced in Chapter 3, using the spring concept. In addition, at the time of impact constraints are imposed on global equations such as (6.13) to (6.21). Hallquist *et al*.[3.40] developed a useful concept of *master* and *slave nodes* sliding on each other. As shown in Fig. 6.6, slave nodes are constrained to slide on master segments after impact occurs and must remain on a master segment until a tensile interface force develops. The zone in which a slave segment exists is called a *slave zone*. A separation between the slave and the master line is known as *void*. The following basic principles apply at the interface:

(1) Update the location of each slave node by finding its closest master node or the one on which it lies.
(2) For each master segment, find out the first slave zone that overlaps.
(3) Show the existence of the tensile interface force.

Constraints are imposed on global equations by a transformation of the nodal displacement components of the slave nodes along the contact interface. Such a transformation of the displacement components of the slave nodes will eliminate their normal degrees of freedom and distribute their normal force components to the nearby master nodes. This is done using explicit time integration, as described later under solution procedures. Thereafter impact and release conditions are

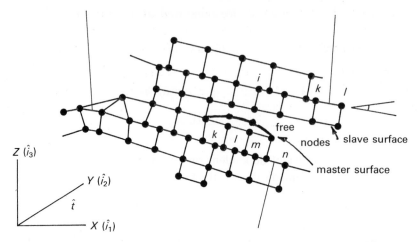

Fig. 6.6 Hallquist contact method (modified by Bangash).[1.149]

imposed. The slave and master nodes are shown in Fig. 6.6. Hallquist *et al*.[3.40] gave a useful demonstration of the identification of the 'contact point', which is the point on the master segment to the slave node n_s and which finally becomes non-trivial. As shown in Fig. 6.6, when the master segment t is given the parametric representation and $\hat{\mathbf{t}}$ is the position vector drawn to the slave node n_s, the contact point co-ordinate must satisfy the following equations:

$$\frac{\partial \hat{\mathbf{r}}}{\partial \xi} (\xi_c, \eta_c) \times [\hat{\mathbf{t}} - \hat{\mathbf{r}}(\xi_c, \eta_c)] = 0$$

$$\frac{\partial \hat{\mathbf{r}}}{\partial \eta} (\xi_c, \eta_c) \times [\hat{\mathbf{t}} - \hat{\mathbf{r}}(\xi_c, \eta_c)] = 0 \qquad (6.68)$$

where (ξ_c, η_c) are the co-ordinates on the master surface segment S_i. Where penetration through the master segment S_i occurs, the slave node n_s (containing its contact point) can be identified using the interforce vector $\mathbf{f}_s$ added, then:

$$\mathbf{f}_s = -lk_i n_i \qquad \text{if } l < 0 \qquad (6.69)$$

to the degrees-of-freedom corresponding to n_s, and

$$\mathbf{f}^i_m = N_i (\xi_c, \eta_c) \mathbf{f}_s \qquad \text{if } l < 0 \qquad (6.70)$$

$$\text{where } l = \hat{\mathbf{n}}_i \cdot [\hat{\mathbf{t}} - \hat{\mathbf{r}}(\xi_c, \eta_c)] < 0 \qquad (6.70a)$$

A unit normal $\hat{\mathbf{n}}_i = \hat{\mathbf{n}}_i(\xi_c, \eta_c);$ $\quad \hat{\mathbf{t}}_i = \hat{\mathbf{n}}_i \sum_{j=1}^{n} N_j(F_1)^j(t) \qquad (6.70b)$

$$k_i = f_{si} K_i A_i^2 / V_i \qquad (6.70c)$$

where $(F_1)^j(t) =$ impact at the jth node
$\quad k_i =$ stiffness factor for mass segment S_i

K_i, V_i, A_i = bulk modulus, volume and face area, respectively
$\quad\quad f_{si}$ = scale factor normally defaulted to 0.10
$\quad\quad N_i = \frac{1}{4}(1 + \xi\xi_i)(1 + \eta\eta_i)$ for a 4-node linear surface

Bangash extended this useful analysis for others such as 8-noded and 12-noded elements.[1.149] On the basis of this theory and owing to the non-availability of the original computer source, a new sub-program CONTACT was written in association with the program ISOPAR. CONTACT is in three-dimensions; the values of N_i for 8- and 12-noded elements are given in Table 6.14.

6.6 High explosions

The pressure P is generally defined as a function of relative volume and internal energy. Chapters 2 and 5 present useful treatments of this subject. The evaluation of the final pressure due to explosion for various case studies is dealt with in Chapter 5. Assuming $F_1(t)$ is the final surface load, the pressure P must replace $F_1(t)$ in relevant equations by taking into consideration the surface volume on which it acts. All equations defining relevant detonation pressures P must first be evaluated as shown in Chapter 5. They are then first applied as stated, in equations (6.13) to (6.21).

The cause of explosions can be nuclear (air burst or underground), gas, chemical, dust, bombs and explosives. The pressures, which are time-dependent, can then act as surface loads on the body of the element concerned or at nodal points of the element as concentrated loads derived on the basis of shape functions. It is essential to choose a proper time-aspect ratio as it will affect the type of solution procedure adopted. The interaction between the loads and the structure can be considered and the method shown in section 6.5 must be included.

6.7 Spectrum analysis

Spectrum analysis is an extension of the mode frequency analysis, with both base and force excitation options. The response spectrum table is generally used and includes displacements, velocities and accelerations. The force excitation is, in general, used for explosions and missile aircraft impact. The masses are assumed to be close to the reaction points on the finite element mesh rather than the master degrees of freedom. The base and forced excitations are given below. For the base excitation for wave

$$\gamma_i = \{\psi_i\}_R^{T'}[M]\,\{\mathbf{b}\} \tag{6.71}$$

For the impact excitation

$$\gamma_i = \{\psi_i\}_R^{T''}\,\{F_1(t)\} \tag{6.72}$$

Table 6.14 N_i for 8- and 12-noded elements.[1.149]

Eight-noded membrane element

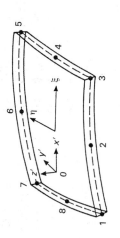

Node i	Shape functions $N_i(\xi,\eta)$	Derivatives	
		$\dfrac{\partial N_i}{\partial \xi}$	$\dfrac{\partial N_i}{\partial \xi}$
1	$\frac{1}{4}(1-\xi)(1-\eta)(-\xi-\eta-1)$	$\frac{1}{4}(1-\eta)(2\xi+\eta)$	$\frac{1}{4}(1-\xi)(2\eta+\xi)$
2	$\frac{1}{2}(1-\xi^2)(1-\eta)$	$-\xi(1-\eta)$	$-\frac{1}{2}(1-\xi^2)$
3	$\frac{1}{4}(1+\xi)(1-\eta)(\xi-\eta-1)$	$\frac{1}{4}(1-\eta)(2\xi-\eta)$	$\frac{1}{4}(1+\xi)(2\eta-\xi)$
4	$\frac{1}{2}(1-\eta^2)(1+\xi)$	$\frac{1}{2}(1-\eta^2)$	$-\eta(1+\xi)$
5	$\frac{1}{4}(1+\xi)(1+\eta)(\xi+\eta-1)$	$\frac{1}{4}(1+\eta)(2\xi+\eta)$	$\frac{1}{4}(1+\xi)(2\eta+\xi)$
6	$\frac{1}{2}(1-\xi^2)(1+\eta^2)$	$-\xi(1+\eta)$	$\frac{1}{2}(1-\xi^2)$
7	$\frac{1}{4}(1-\xi)(1+\eta)(-\xi+\eta-1)$	$\frac{1}{4}(1+\eta)(2\xi-\eta)$	$\frac{1}{4}(1-\xi)(2\eta-\xi)$
8	$\frac{1}{2}(1-\eta^2)(1-\xi)$	$-\frac{1}{2}(1-\eta^2)$	$-\eta(1-\xi)$

Table 6.14 *Continued.*

Twelve-noded membrane element

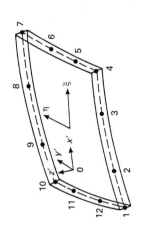

Node i	Shape functions $N_i(\xi,\eta)$	Derivatives $\dfrac{\partial N_i}{\partial \xi}$	$\dfrac{\partial N_i}{\partial \eta}$
1	$\frac{9}{32}(1-\xi)(1-\eta)[\xi^2+\eta^2-\frac{10}{9}]$	$\frac{9}{32}(1-\eta)[2\xi-3\xi^2-\eta^2+\frac{10}{9}]$	$\frac{9}{32}(1-\xi)[2\eta-3\eta^2-\xi^2+\frac{10}{9}]$
2	$\frac{9}{32}(1-\eta)(1-\xi^2)(1-\xi)$	$\frac{9}{32}(1-\eta)(3\xi^2-2\xi-1)$	$-\frac{9}{32}(1-\xi^2)(1-\xi)$
3	$\frac{9}{32}(1-\eta)(1-\xi^2)(1+\xi)$	$\frac{9}{32}(1-\eta)(1-2\xi-3\xi^2)$	$-\frac{9}{32}(1-\xi^2)(1+\xi)$
4	$\frac{9}{32}(1+\xi)(1-\eta)[\xi^2+\eta^2-\frac{10}{9}]$	$\frac{9}{32}(1-\eta)[2\xi+3\xi^2+\eta^2-\frac{10}{9}]$	$\frac{9}{32}(1+\xi)[2\eta-3\eta^2-\xi^2+\frac{10}{9}]$
5	$\frac{9}{32}(1+\xi)(1-\eta^2)(1-\eta)$	$\frac{9}{32}(1-\eta^2)(1-\eta)$	$\frac{9}{32}(1+\xi)(3\eta^2-2\eta-1)$
6	$\frac{9}{32}(1+\xi)(1-\eta^2)(1+\eta)$	$\frac{9}{32}(1-\eta^2)(1+\eta)$	$\frac{9}{32}(1+\xi)(1-2\eta-3\eta^2)$
7	$\frac{9}{32}(1+\xi)(1+\eta)[\xi^2+\eta^2-\frac{10}{9}]$	$\frac{9}{32}(1+\eta)[2\xi+3\xi^2+\eta^2-\frac{10}{9}]$	$\frac{9}{32}(1+\xi)[2\eta+3\eta^2+\xi^2-\frac{10}{9}]$
8	$\frac{9}{32}(1+\eta)(1-\xi^2)(1+\xi)$	$\frac{9}{32}(1+\eta)(1-2\xi-3\xi^2)$	$\frac{9}{32}(1-\xi^2)(1+\xi)$
9	$\frac{9}{32}(1+\eta)(1-\xi^2)(1-\xi)$	$\frac{9}{32}(1+\eta)(3\xi^2-2\xi-1)$	$\frac{9}{32}(1-\xi^2)(1-\xi)$
10	$\frac{9}{32}(1-\xi)(1+\eta)[\xi^2+\eta^2-\frac{10}{9}]$	$\frac{9}{32}(1+\eta)[2\xi-3\xi^2-\eta^2+\frac{10}{9}]$	$\frac{9}{32}(1-\xi)[2\eta+3\eta^2+\xi^2-\frac{10}{9}]$
11	$\frac{9}{32}(1-\xi)(1-\eta^2)(1+\eta)$	$-\frac{9}{32}(1-\eta^2)(1+\eta)$	$\frac{9}{32}(1-\xi)(1-2\eta-3\eta^2)$
12	$\frac{9}{32}(1-\xi)(1-\eta^2)(1-\eta)$	$-\frac{9}{32}(1-\eta^2)(1-\eta)$	$\frac{9}{32}(1-\xi)(3\eta^2-2\eta-1)$

where $\{\psi_i\}_R$ = the slave degree of freedom vector mode
$\qquad M$ = mass
$\qquad \{\mathbf{b}\}$ = unit vector of the excitation direction
$\qquad \{F_1(t)\}$ = an input force vector due to impact and explosion

The values of $\{\psi\}_R$ are normalized and the reduced displacement is calculated from the eigenvector by using a mode coefficient $\{\mathbf{M}\}$.

$$\{\mathbf{x}\}_i = [\mathbf{M}_i]\{\psi\}_i \tag{6.73}$$

where $\{\mathbf{x}\}_i$ = reduced displacement vector and $[\mathbf{M}_i]$ = mode coefficient and where
(a) for velocity spectra

$$[\mathbf{M}_i] = [\dot{x}_{si}]\{\gamma_i\}/\omega_i \tag{6.74}$$

where $\dot{x}_{si}$ = spectral velocity for the ith mode;
(b) for force spectra

$$[\mathbf{M}_i] = [F_{si}]\{\gamma\}/\omega_t^2 \tag{6.75}$$

where F_{si} = spectral force for the ith mode;
(c) caused by explosion P or impact $F_1(t)$

$$[\mathbf{M}_i] = [\ddot{x}_{si}]\{\gamma_i\}/\omega_i^2 \tag{6.76}$$

where $\ddot{x}_{si}$ = spectral acceleration for the ith mode;
(d)

$$[\mathbf{M}_i] = [x_{si}]\{\gamma_i\}/\omega_i^2 \tag{6.77}$$

$\{x\}_i$ may be expanded to compute all the displacements, as in

$$\{\mathbf{x}_{\gamma'}\}_i = [K_{\gamma'\gamma'}]^{-1}[K_{\gamma'\gamma}]\{\mathbf{x}_i\}_R \tag{6.78}$$

where $\{\mathbf{x}_{\gamma'}\}_i$ = slave degree of freedom vector of mode i
$[K_{\gamma'\gamma'}],[K_{\gamma'\gamma}]$ = sub-matrix parts
$\qquad \gamma,\gamma'$ = retained and removed degrees of freedom

The impact/explosion load is then equal to

$$[[K_{\gamma\gamma}] - [K_{\gamma'\gamma}][K_{\gamma'\gamma'}]^{-1}[K_{\gamma'\gamma}]]\{\mathbf{x}_\gamma\}$$
$$= [\{F_\gamma\} - [K_{\gamma\gamma'}][K_{\gamma\gamma'}]^{-1}\{F_{\gamma'}\}]$$
$$\text{or } [\bar{K}]\{\bar{x}\} = \{\bar{F}_1(t)\} \tag{6.79}$$

$$\text{where } [\bar{K}] = [K_{\gamma\gamma}] - [K_{\gamma\gamma'}][K_{\gamma'\gamma'}]^{-1}[K_{\gamma'\gamma}] \tag{6.80}$$

$$\{\bar{F}_1(t)\} = \{F_\gamma\} - [K_{\gamma\gamma'}][K_{\gamma'\gamma'}]^{-1}\{F_{\gamma'}\} \tag{6.81}$$

$$\{\bar{x}\} = \{\mathbf{x}_\gamma\} \tag{6.82}$$

and $[K]$ and $\{\bar{F}_1(t)\}$ are generally known as the substructure stiffness matrix and the impact load vector, respectively.

6.8 Solution procedures

Three types of solution procedure are available for impact and explosion analysis, namely, time-domain, frequency-domain and modal analysis.

6.8.1 Time-domain analysis

The following steps are adopted using a direct implicit integration procedure.

Initialization

(1) The effective stiffness matrix is

$$[K_0^*] = (6/\tau^2)[M] + (3/\tau)[C_0] + [K_0] \tag{6.83}$$

(2) Triangularize $[K_0^*]$

For each time step, calculate the displacement $\{x_{t+\tau}\}$

- Constant part of the effective load vector

$$
\begin{aligned}
\{R_{t+\tau}^*\} &= \{R_t\} + \theta(\{R_{t+\delta t}\} - \{R_t\}) + \{F_t\} + [M] \\
&\times (6/\tau^2)\{x_t\} + (6/\tau)\{\dot{x}_t\} + 2\{\ddot{x}_t\}) + [C_0]((3/\tau)\{x_t\} + 2\{\dot{x}_t\} + (\tau/2)\{x_t\})
\end{aligned}
\tag{6.84}
$$

- Initialization, $i = 0$, $\{\delta F_{t\to t+\tau}^i\} = 0$
- Iteration
  (a) $i \to i+1$
  (b) Effective load vector $\{R_{t+\tau_{TOT}}^*\} = \{R_{t+\tau}^*\} + \{\delta F_{t\to t+\tau}^{i-1}\}$
  (c) Displacement $\{x_{t+\tau}^i\}[K_0^*]\{x_{t+\tau}^i\} = R_{t\to t+\tau_{TOT}}^{*i}$
  (d) Velocity $\{\dot{x}_{t+\tau}^i\} + (3/\tau)(\{x_{t+\tau}^i\} - \{x_t\}) - 2\{\dot{x}_t\} - (\tau/2)\{\ddot{x}_t\}$
  (e) Change of initial load vector caused by non-linear behaviour of the material

$$
\begin{aligned}
\{\delta F_{t\to t+\tau}^i\} &= -[\delta C_{o\to t}]\{\dot{x}_{t+\tau}^i\} - \{\dot{x}_t\}\omega - [\delta C_{t\to t+\tau}^i]\{\dot{x}_{t+\tau}^i\} \times [\delta K_{o\to t}] \\
&\quad (\{x_{t+\tau}^i\} - \{x_t\}) - [\delta K_{t\to t+\delta t}^i]\{x_{t+\tau}^i\}
\end{aligned}
\tag{6.85}
$$

  In fact, $\{\delta F_{t\to t+\tau}^i\}$ is calculated using the initial-stress method.
  (f) Iteration convergence

$$\|\{\delta F_{t\to t+\tau}^i\} - \{\delta F_{t\to t+\tau}^{i-1}\}\|/\|\{\delta F_{t\to t+\tau}^i\}\| < \text{tol} = 0.01 \tag{6.86}$$

or, analogously, on stresses.

Calculation of velocity and acceleration

Calculate the new acceleration $\{\ddot{x}_{t+\delta t}\}$, velocity $(\dot{x}_{t+\delta t})$, displacement $\{x_{t+\delta t}\}$ and initial load $\{F_{t+\delta t}\}$:

$$\{\ddot{x}_{t+\delta t}\} = (6/\theta\tau^2)(\{x_{t+\tau}\} - \{x_t\}) - (6/\tau\theta)\{\dot{x}_t\} + (1 - (3/\theta))\{\ddot{x}_t\} \tag{6.87}$$

$$\{\dot{x}_{t+\delta t}\} = \{\dot{x}_t\} + (\tau/2\theta)\{\ddot{x}_t\} + \{\ddot{x}_{t+\delta t}\} \tag{6.88}$$

$$\{x_{t+\delta t}\} = \{x_t\} + (\tau/\theta)\{\dot{x}_t\} + (\tau^2/6\theta^2)(2\{\ddot{x}_t\} + \{\ddot{x}_{t+\delta t}\}) \tag{6.89}$$

$$\{F_{t+\delta t}\} = \{F_t\} + \{\delta F^i_{t\to t+\tau}\} \tag{6.90}$$

Calculation by quadratic integration

When the velocity varies linearly and the acceleration is constant across the time interval, appropriate substitutions are made into equation (6.13), giving

$$[f_1[M] + f_2[C_t] + [K_t']]\{x_t\} = \{F_1(t)\} + \{f_3([C_t], [M], x_{t1}, x_{t2}, \ldots)\} \tag{6.91}$$

where f_1, f_2 are functions of time. This results in an implicit time integration procedure. The only unknown is $\{x_t\}$ at each time point and this is calculated in the same way as in static analysis. Equation (6.91) is then written as

$$\left(\frac{2}{\delta t_0\,\delta t_0}[M] + \frac{\delta t_0 + \delta t_1}{\delta t_0\,\delta t_0}[C] + [K_t']\right)\{x_t\} = \{F_1(t)\} + [M]\left(\frac{2}{\delta t_0\,\delta t_1}\{x_{t-1}\} - \right.$$

$$\left. \frac{2}{\delta t_1\,\delta t_0}\{x_{t-2}\}\right) + [C_t]\left(\frac{\delta t_0}{\delta t_0\,\delta t_1}\{x_{t-1}\} - \frac{\delta t_0}{\delta t_0\,\delta t_1}\{x_{t-2}\}\right) \tag{6.92}$$

where $\delta t_0 = t_0 - t_1$ and $t_0 = $ time of current iteration

$\delta t_1 = t_1 - t_2$ and $t_1 = $ time of previous iteration

$\delta t_2 = t_2 - t_3$ and $t_2 = $ time before previous iteration

$\delta t_2 = \delta t_0 + \delta t_1 = t_0 - t_2$ and $t_3 = $ time before t_2

Calculation by cubic integration

Equation (6.91) becomes cubic and hence is written as

$$(a_1[M] + a_2[C_t] + [K_t'])\{x_t\} = \{F_1(t)\} + [M](a_3 x_{t-1}) - a_4\{x_{t-2}\} +$$

$$a_5\{x_{t-3}\} + [C](a_6\{x_{t-1}\} - a_7\{x_{t-2}\} + a_8\{x_{t-3}\}) \tag{6.93}$$

where a_1 to a_8 are functions of the time increments; these functions are derived by inverting a 4×4 matrix.

For clear-cut solutions, the size of the time step between adjacent iterations should not be more than a factor of 10 in non-linear cases and should not be reduced by more than a factor of 2 where plasticity exists.

6.8.2 Frequency-domain analysis

The original equation of motion is reproduced as

$$[M]\{\ddot{x}\} + [C]\{\dot{x}\} + [K]\{x\} = \{J_F\}\{F\} \tag{6.94}$$

where $\{J_F\}$ is a vector with all components zero except the last one, which is 1. The terms $[K]$ and $[C]$ shall be frequency dependent. The value of $\{F\} = [K_N]\{x_s\}$ can be taken for solutions of rigid rock problems. If the excitation with frequency ω assumes the form $e^{i\omega t}$, then

$$\dot{x} = i\omega x_s; \quad \ddot{x}_s = -\omega^2 x_s; \quad \{x\} = i\omega\{x\} \text{ and } \{x\} = -\omega^2\{x\} \tag{6.95}$$

Equation (6.94) can thus be written as

$$([K] + i\omega[C] - \omega^2[M])\{x\} = \{J_F\}K_n x_s \tag{6.96}$$

For a given value of ω, a set of algebraic equations is solved using any numerical scheme. The displacement of a mass can be written as

$$\{x\} = ([K] + i\omega[C] - \omega^2[M])^{-1} \{J_F\}K_n x_n \tag{6.97}$$

From displacements, accelerations, velocities, strains and stresses can be computed. The amplification function (AF) for each frequency x_1/x_s may be derived. Repeated solutions of equation (6.95) are necessary for a proper definition of this function. If the fast Fourier transform is used then AF must be tabulated at each frequency interval.

6.8.3 Modal analysis

This type of analysis is described fully in Chapter 3.

6.9 Force or load–time function

Chapters 2 and 3 give useful data on force or load–time functions for various cases. On the basis of the *contact* method given in section 6.5 along with the proposed finite-element analysis, Fig. 6.7 illustrates a comparative study of the load–time functions of a number of aircraft decelerating at different speeds and impacting flexible targets. The analytical divisions of various manufacturers have provided details of the finite-element mesh schemes, connectivity relations and material properties of various zones of aircraft. For large problems, an IBM 4381 with a CRAY-2 front-ended has been used in this text. Both crushed and uncrushed parts together with energy release or 'take-off' were examined. The flow of material across the interface was included in order to obtain the reaction load at the damaged zone for each aircraft. The engine was treated as a hard core at the appropriate level with different material properties. The hard core was assumed to represent the engine as a hard missile. Each case study consumed 7 hours cpu (central processing unit) time. It therefore becomes a useful contribution towards the analysis of structures under aircraft impact.

The United Kingdom Atomic Energy Authority at Winfrith, Dorset (UK), carries out numerous tests on impact problems. Figure 6.8 shows a missile launcher which has the following characteristics:

(1) A maximum projectile energy of 3 MJ and a mass of 2000 kg.
(2) Two barrels with diameters of 150 and 300 mm and three barrels (horizontal impact facility) of 0.5, 1.0 and 2.0 m.
(3) A projectile velocity in the range of 10 to 350 m/s.

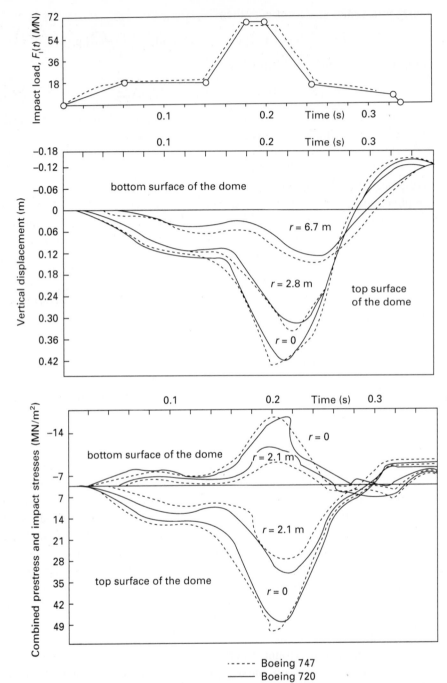

Fig. 6.7 Aircraft impact on flexible structures — a comparative study.

Using the computer program HONDO-II, reaction loads were computed for three hemispherically-ended steel missiles with different stiffnesses impacting a concrete face. Figure 6.9 gives the force or load−time functions for three missiles,

Fig. 6.8 Missile launcher. (Courtesy of A. Neilson, UK Atomic Energy Authority, Winfrith, UK.)

which were verified experimentally on concrete slabs.

Figure 6.10 gives the load−time function for a semi-submersible subject to ice-floe impact.

Figure 6.11 gives a load−time function for a pipe impacting plain concrete blocks. The pipe had a diameter of 100 mm and a wall thickness of 5 mm. The block dimensions were 500 mm × 500 mm × 100 mm of normal concrete of 50 N/mm². The velocity of the pipe was 10 m/s. Both experimental results and results from the ISOPAR program are plotted on Fig. 6.11. The bulk modulus method was used as a failure model for concrete. The block was divided into 8-noded, isoparametric, solid elements. Program contact was used along with program ISOPAR for computing the load−time function. The block was treated as rigid.

Figure 6.12 shows a pressure−time forcing function for a gas explosion for both vented and unvented cases.

6.9.1 Finite-element mesh schemes

Finite-element analysis has been carried out for a number of case studies discussed in this book. They range from beams, slabs and panels to aircraft, ships, tanks, etc. In this section a typical finite-element mesh scheme for a battle tank is given in Fig. 6.13. This mesh scheme is used for collision and impact analyses. Where tanks have different geometric features, this standard mesh is modified to include them for the accurate prediction of damage due to impact of missiles, rockets, bombs, etc., exploding near or at the tank level.

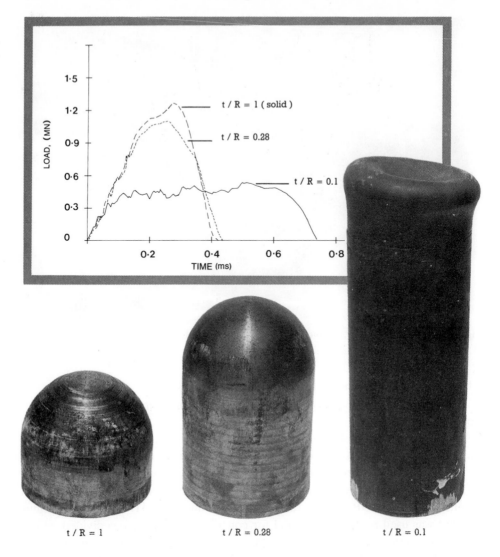

Fig. 6.9 Load−time functions for three hemispherically-ended steel missiles. t = thickness, R = radius. (Courtesy of A. Neilson, United Kingdom Atomic Energy Authority, Winfrith, UK.)

Figures 6.14 to 6.16 show finite-element schemes for PWR reactor vessels supported by ancillary structures, aircraft fuselage for dropped weight analysis and a typical chimney for the impact analysis. Figure 6.17 gives finite-element analysis of a compartment having vents and is subject to gas explosion using the pressure−time relationships given in Fig. 6.12. Figure 6.18 indicates a typical finite-element mesh scheme for a cooling tower discussed later on in Chapter 7.

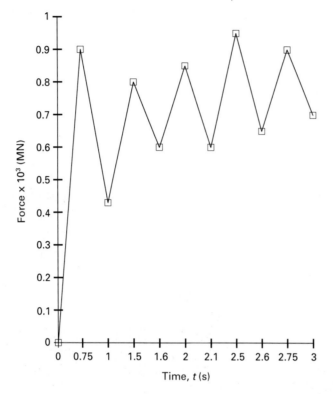

Fig. 6.10 Ice-floe impact: the load–time function for a semi-submersible.

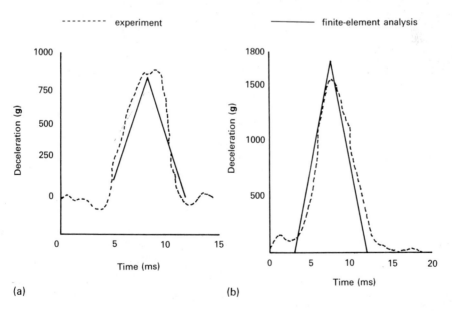

Fig. 6.11 A 100 mm diameter pipe impacting plain concrete (missile deceleration). (a) Direct fracture analysis; (b) plasticity and damage analysis.

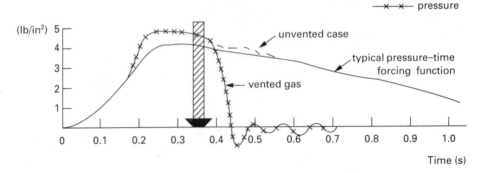

Fig. 6.12 Pressure–time relationship from a gas explosion. (Courtesy of the British Ceramic Society.)

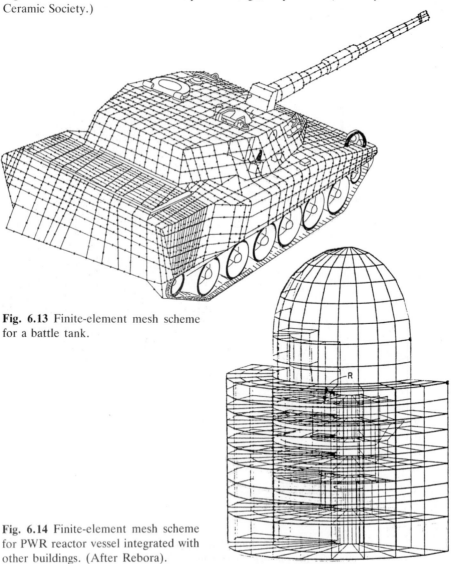

Fig. 6.13 Finite-element mesh scheme for a battle tank.

Fig. 6.14 Finite-element mesh scheme for PWR reactor vessel integrated with other buildings. (After Rebora).

Fig. 6.15 Finite-element scheme for a fuselage of an aircraft as a dropped weight.

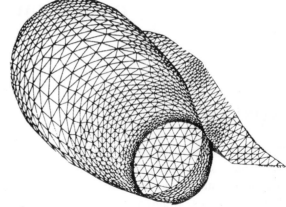

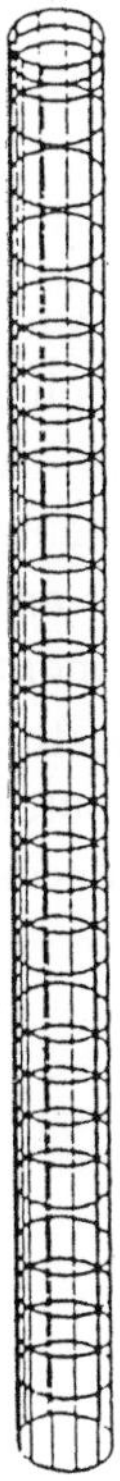

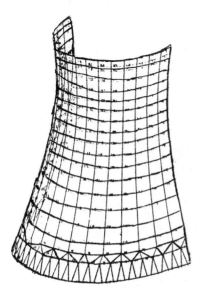

Fig. 6.16 Finite-element mesh scheme for a chimney.

Fig. 6.17 Finite-element analysis of gas explosion damage in a compartment with vents.

Fig. 6.18 Finite-element mesh scheme for a cooling tower.

7
CASE STUDIES

INTRODUCTION

This chapter covers case studies on impact/explosion in the fields of steel, concrete, composites, soils, rocks, water tunnels/chambers, brick-built structures, shells, underwater structures, offshore structures, structural ice and many others. A number of tables, graphs and design examples are included to justify the contents of this book. Conventional missiles, military missiles/bombs and their impact are analyzed. Some sections are included for important structures such as containment vessels under aircraft crashes and underground nuclear shelters, etc. For the solution of additional case studies the appendix gives the layout of the computer program and data can be found from Chapter 2 and from numerous references given in the Bibliography.

A: STEEL AND COMPOSITES

A.1 Steel structures

A.1.1 Impact on steel beams

A number of steel beams under impact are analyzed. The beams are treated as rate-sensitive and the input energy of the impactor is assumed to be high. The strain effects (rate of strain occurring at a particular solution) for a particular state of strain are included. The impactor is assumed to strike a steel beam at mid-span. Figure A1.1 shows a typical 8-node, isoparametric, finite-element mesh for a steel beam of rectangular cross-section. The material is assumed to be elastic/visco-perfectly plastic. Program ISOPAR is used to analyze the beam when

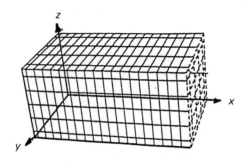

Fig. A1.1 Finite-element mesh of a steel beam.

subjected to a cone-shaped impactor. The following data are considered:

British beam	*American beam*

British beam

Span lengths, L: 3 m to 10 m
Mass, M: 30 kg to 57 kg per m width
Impactor velocity, v_s: 1 to 25 m/s
Density of steel, ρ: 7800 kg/m³ (386 kips/in³)
Young's modulus, E_s: 200 GN/m² (30×10^6 lb/in²)
Yield stress, σ_y: 250 MN/m²
Poisson's ratio, v: 0.3
Strain rate, $\dot{\epsilon}$: 40.4/s

American beam

$L = 30$ ft
Dead weight = 1000 lb
MF = DLF = 1.4
$F_1(t) = 222.4$ kN (50 000 lb)

Finite elements: 8-noded isoparametric elements; 200
Assumed impact loads: triangular loads, load−time relations vary,

$$\text{typical ones} \quad F_1\,(t)_{\text{max}} = \; 222.4\,\text{kN} \; (50\,000\,\text{lb})$$

$$t_r = 0.08\,\text{s}$$

The procedure for sizing up the beam using the US code for one case is as follows:

$F_1(t) + 20\,000$ lb deadload
$L = 30$ ft; 1000 lb/ft deadload of the beam
The beam is assumed to be fixed at both ends
DLF $\not> 1.4$; $t_r/T \not< \frac{2}{3}$ (see the response chart in Chapter 3)

$$M_{\text{max}} = \frac{w_d L^2}{12} + \frac{F_D L}{8} + \frac{F_1(t)}{8}(\text{DLF})$$

$$= (1000 \times 30^2/12) + (20 \times 30 \times 10^3/8) + (50\,000 \times 30/8)1.4$$

$$= 412\,500 \text{ in lb}$$

$$Z_p = (M \times 12)/\sigma_y = (412 \times 10^3 \times 12)/(75 \times 10^3)$$

$$= 65.92\,\text{in}^3$$

$$\text{Section adopted } Z_p = 175.4\,\text{in}^3 \rightarrow 24\text{WF76}$$

$$I = 2096\,\text{in}^4$$

$$K = 192EI/L^3 = 258\,760\,\text{lb/in}$$

Uniformly distributed load if treated separately (see tables in this section)

$$K_L = 1.0$$

Concentrated mass $K_M = 1.0$
  Distributed mass $K_M = 0.37$

$$M_e = \Sigma K_M M = (20\,000 \times 1.0 + 30\,000 \times 0.37)/(386 \times 1000)$$

$$= 0.081 \text{ kip-s}^2/\text{in}$$

$$= 0.081 \times 10^3 \text{ lb-s}^2/\text{in}$$

$$k_e = kK_L = 258\,760 \times 1.0\,\text{lb/in}$$
$$= 258\,760\,\text{lb/in}$$

$$T = 2\pi\sqrt{(M_e/K_e)} = 0.111\,\text{s}$$

$$t_r/T = 0.08/0.111 = 0.72$$

$$\text{DLF} = 1.35$$

$$M_{\text{max}}\,(\text{as above}) = \frac{1.35}{1.40} \times 412\,500 = 397\,767.86\,\text{lbft}$$

$$\sigma = M/Z_p = 412\,500/175.4 = 2351.8\,\text{psi}$$

The beam size is adequate. If the rectangular size is adopted, then the depth taken is 24 inches. The value of the height H varies from 1 m (3 ft) to 9 m (27 ft).

For the finite-element analysis, the required depth at which the impact can occur due to a falling cone-shaped impactor can influence the failure zone. Figure A1.2 gives a failure zone for the beam with a ratio of $L/H = 400$ under a triangular impact with $F_1(t) = 222.4\,\text{kN}$ and $t_r = 0.08\,\text{s}$. In addition, various falling weights have been considered. Figure A1.3 gives a relationship between δ/H and $K = M\,v_s/(500 \times 10^6)H^2$ for different ratios of L/H for simple, fixed and continuous beams.

Fig. A1.2 Plasticity and fracturing of a steel beam.

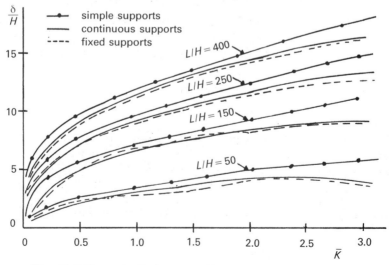

Fig. A1.3 Dynamic displacement (δ = displacement; H = height).

A.1.2 Impact on steel plates

Plates are subject to impact and blast modes. A constant finite-element mesh scheme, suggested in Fig. A1.4, has been adopted for solutions of various problems. All shapes of impactors are considered. The following summarizes the input data used in solving a number of problems:

General data
Impactors: 10 mm diameter, 35 mm length, flat-ended.
Plate thickness: rectangular and circular 20,40,60,80,100,120,140 mm.
Velocity: 3 to 300 m/s.

Material property: yield stress $220-270 \text{ N/mm}^2$
                     failure strain $30-50\%$
                     ultimate tensile stress $269-414 \text{ N/mm}^2$

A.1.2.1 Perforation of circular steel plates

A number of circular plates of cold-rolled mild steel were examined under the impact load. In the light of the current armoured plate thicknesses and armour chosen velocities, the finite-element analysis adopted gives useful relationships between the perforation thickness and the flat-ended impactor velocity giving bulge heights. These are given in Figs A1.5 and A1.6. The displacement−time relation for these circular plates is given in Fig. A1.7.

A.1.2.2 Flat steel plates (square and rectangular)

Flat plates are generally used in various armoured vehicles. Missiles and bombs are used to penetrate these plates. Finite-element analysis has been carried for the plates. Again, for the purpose of analysis, the finite-element mesh scheme shown in Fig. A1.4(b) is adopted. A number of plate thicknesses used in the analysis are plotted against various velocities for a number of sizes and yields of bombs in Fig. A1.8. The diameter, D, of the piercing bomb acts as an additional parameter

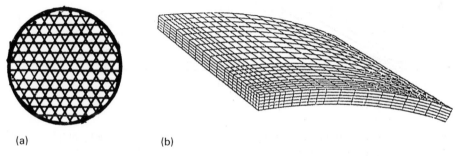

(a)   (b)

Fig. A1.4 Finite-element mesh schemes for (a) circular plates and (b) rectangular plates.

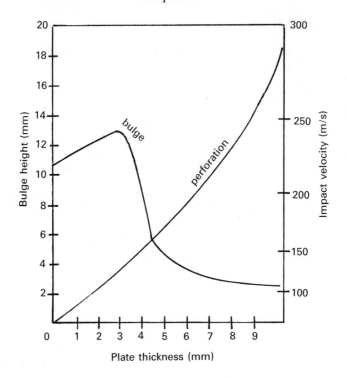

Fig. A1.5 Bulge height and perforation versus plate thickness.

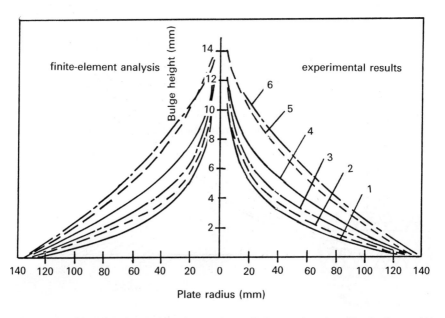

Fig. A1.6 A profile of bulge height for various thicknesses and radii of plates. (Plate thicknesses: $1 = 2$ mm; $2 = 3$ mm; $3 = 4$ mm; $4 = 5$ mm; $5 = 6$ mm; $6 = 8$ mm.)

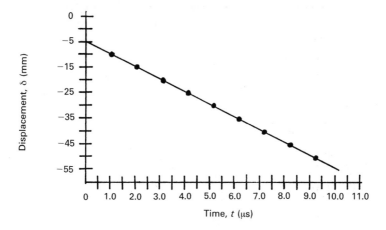

Fig. A1.7 Displacement−time relationship.

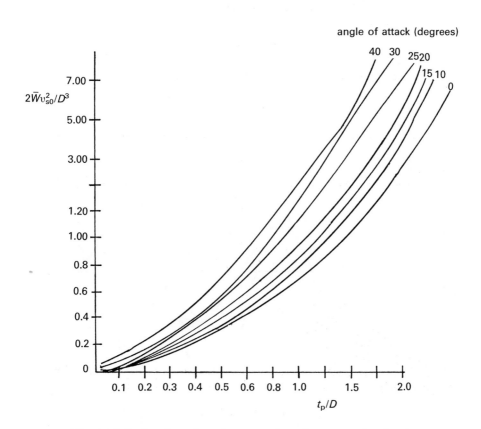

Fig. A1.8 Perforation of steel armour plates by penetrating bombs.

in the evaluation of the perforation thickness t_p. The angle of attack ranges between 0° and 40°. The pressure−time, displacement−time, velocity−time, acceleration−time and strain−time relationships are given in Figs A1.9 to A1.13. A typical three-dimensional, transient, dynamic finite-element analysis was carried out on a steel nose cone impacting a hard steel plate surface in a few milliseconds. The bomb is generally cone-shaped. The entire scenario is well demonstrated in Fig. A1.14.

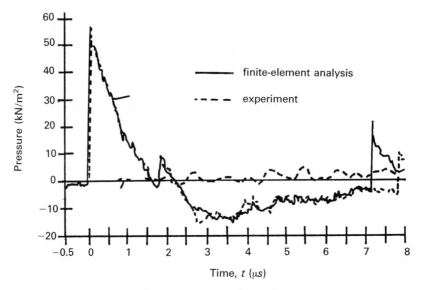

Fig. A1.9 Pressure−time relationship.

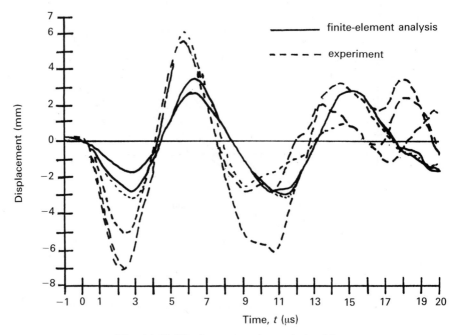

Fig. A1.10 Displacement−time relationship.

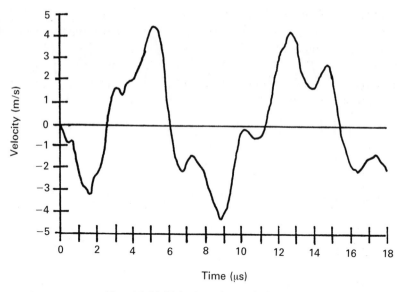

Fig. A1.11 Velocity–time relationship.

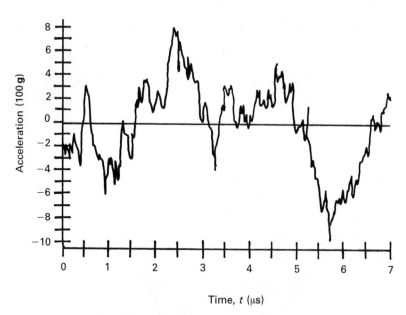

Fig. A1.12 Acceleration–time relationship.

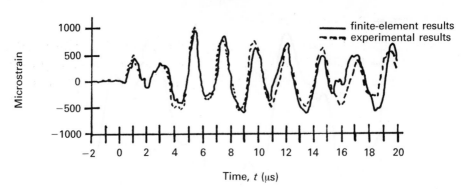

Fig. A1.13 Strain−time relationship.

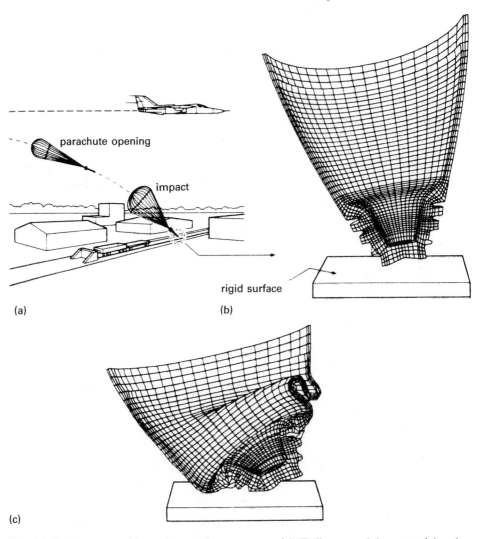

(a)

(b)

(c)

Fig. A1.14 Impact crushing of a steel nose cone. (a) Delivery and impact of bomb; (b) initial mesh of nose cone (before impact); (c) deformed mesh (15 ms after impact). (After Chiesa & Callabressi[2.397])

A.1.2.3 *Hammer drop on steel plates*

A simply supported rectangular flat steel plate $1.3\,\text{m} \times 1.3\,\text{m} \times 6\,\text{mm}$ thick was subject to a hammer drop at $5\,\text{m}$. The force−time relationship is given in Figs A1.15. The acceleration−time and the frequency responses are given in Figs A1.16 and A1.17.

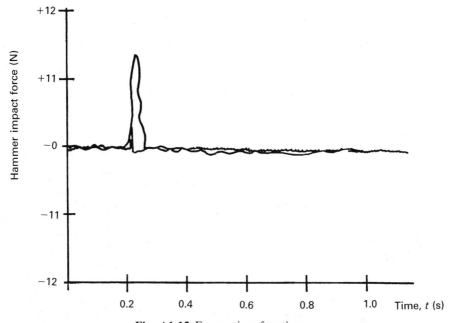

Fig. A1.15 Force−time function.

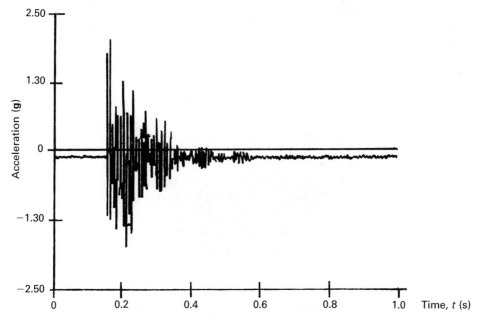

Fig. A1.16 Acceleration−time response.

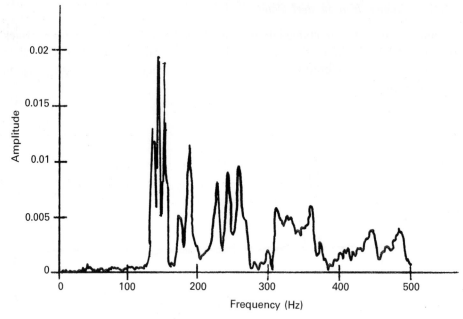

Fig. A1.17 Amplitude–frequency response.

A.2 Composite structures

A.2.1 Composite plates

Figure A2.1 is considered for the finite-element analysis of composite plates. Where specific directions of fibres are suggested, these are treated as line elements either placed on the solid element nodes or in the body of the solid elements or dispersed elements.

A.2.1.1 Thornel 300 epoxy composite plate

A tri-directional Thornel 300 (T300) epoxy plate is considered, subject to 50 mm diameter steel spheres with a velocity of 2 m/s to 20 m/s. The plate thickness is assumed to vary from 5 mm to 25 mm for a circular or rectangular plate with radius or lengths 90 mm to 250 mm. Three fibres of the tri-directional T300 are assumed to lie within 2 mm radius or square lengths. The following additional specifications are adopted:

Material properties		Composite target
Young's modulus (tension), kN/m^2	6.895×10^{-6}	48.27×10^6
(compression), kN/m^2	5.980×10^{-6}	
Shear modulus, kN/m^2	2.500×10^{-6}	18.41×10^6
Tensile stress, kN/m^2	51.000×10^{-3}	
Compressive stress, kN/m^2	$13\,960.000 \times 10^{-3}$	
Poisson's ratio	0.400	

Boundary conditions: plates simply supported, fixed and continuous. Figures A2.1 to A2.3 show the relationship between impact force and impact velocity for fixed and continuous plates for various plate thicknesses. Figures A2.4 to A2.6 show the results of the damaged plates.

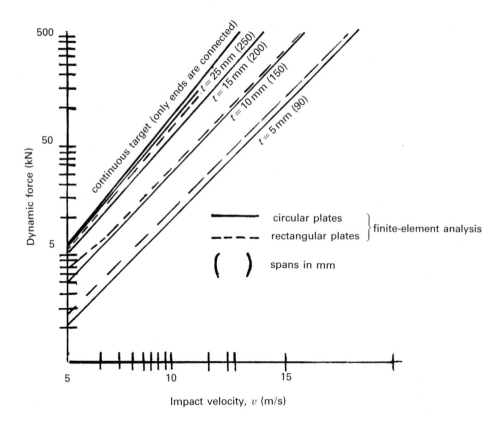

Fig. A2.1 Plates fixed at boundaries.

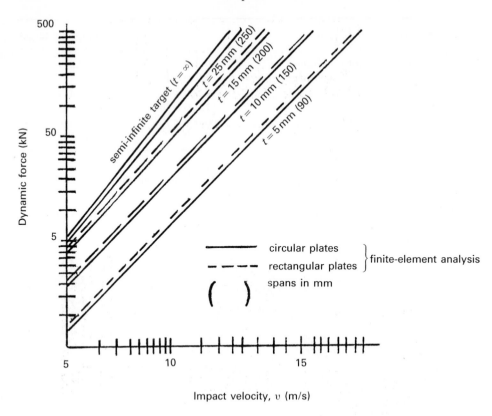

Fig. A2.2 Plates fixed at boundaries.

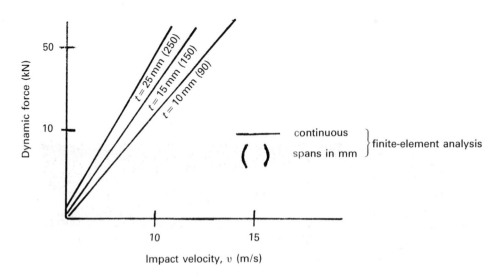

Fig. A2.3 Plates fixed at boundaries.

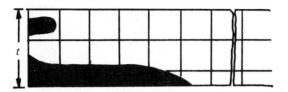

Fig. A2.4 Simply supported T300 epoxy plate.

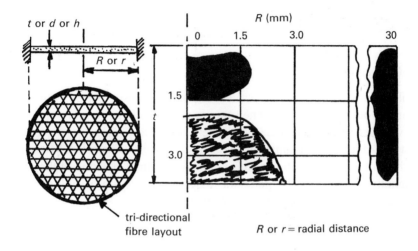

tri-directional
fibre layout

R or r = radial distance

Fig. A2.5 A T300 epoxy plate fixed at the edges.

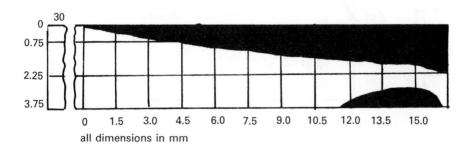

all dimensions in mm

Fig. A2.6 A T300 epoxy plate with continuous support.

A.2.1.2 Impact on a polymethyl methacrylate plate (PMMA)

Shular has developed an experiment using a velocity interferometer for observing with great precision the particle velocity history of a point within PMMA material.[2.398] The PMMA projectile nose piece was 6 mm thick. The experimental results were correlated with the finite-element analysis performed by Bangash on ISOPAR.[1.149] The additional finite-element data included in this analysis are

> 8-noded isoparametric elements: 135/spaced 800 μm
> disk size: 220 mm
> disk thickness: 3 mm
> impact velocities: 0.06, 0.15, 0.30, 0.46, 0.64 mm/μs
> shock speed: 2.834 to 3.349 mm/μs

These results are illustrated in Figs A2.7 to A2.9.

A.2.1.3 Impact on a plexiglass plate

A similar analysis to that used for a T300 epoxy composite plate (section A.2.1.1) was carried out for plexiglass. The same number of elements were adopted, only fibres were excluded. Spherical and cone-shaped steel impactors were assumed. Impact velocities were assumed to be 2 m/s to 10 m/s. Figures A2.10 and A2.11 show the impact−deformation relationship and the final cracking of the plexiglass. Throughout the analysis the plate was assumed to rest on a solid support. The following additional specifications were included:

$$\text{impactor } E_s = 200\,\text{GN/m}^2$$
$$F_I(t) = 0.2\,\text{kN to } 1.6\,\text{kN}; \ \nu_s = 0.30$$
$$\text{plexiglass } E_{pg} = 2.8\,\text{GN/m}^2; \ \nu_{pg} = 0.396$$

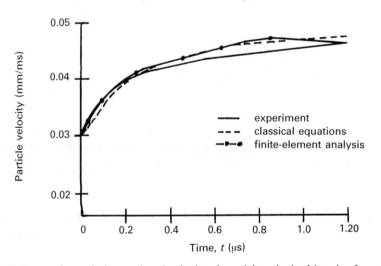

Fig. A2.7 Comparison of observed and calculated particle-velocity histories for a 0.10 mm/μs impact (t = time after arrival).

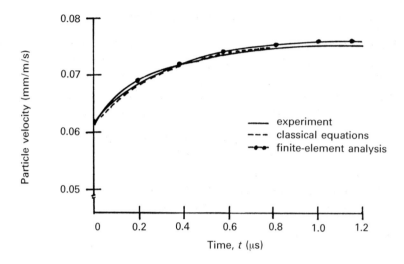

Fig. A2.8 Comparison of observed and calculated particle-velocity histories for a 0.15 mm/μs impact ($t =$ time after arrival).

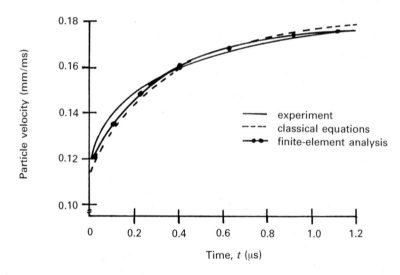

Fig. A2.9 Comparison of observed and calculated particle-velocity histories for a 0.30 mm/μs impact ($t =$ time after arrival).

A.2.1.4 Impact on a carbon fibre/epoxy plate

The effects of impact on a carbon fibre/epoxy plate depend on the type of fibre, the fibre/matrix bond strength and the fibre orientation. The purpose of this analysis is to evaluate the stress−strain behaviour of this plate and the final damage it receives under a 50 mm diameter spherical steel impactor. Carbon

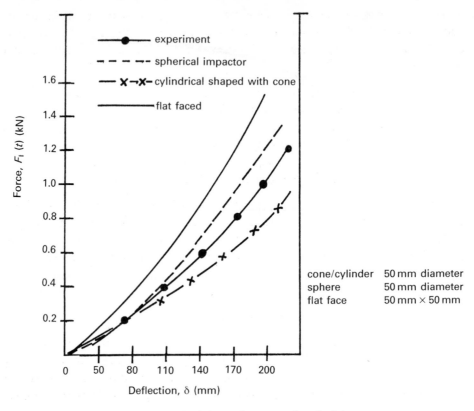

Fig. A2.10 Load–deformation curve for plexiglass.

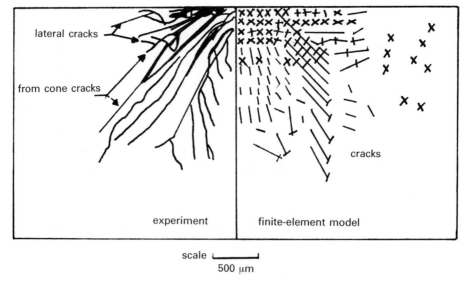

Fig. A2.11 Cracking of plexiglass.

fibres were assumed to be of 7 µm diameter, placed in small spaces with dimensions equal to twice the fibre diameter. The plate was the same size as the plexiglass one. The force–time relationship is illustrated in Fig. A2.12. The following data were considered:

Fibre modulus, E: 200 GN/m^2
Transverse modulus, E: 6.5 GN/m^2
Poisson's ratio, v: 0.32
Shear modulus, G: 4.5 GN/m^2
Tensile strength, σ_{yt}: 3.53×10^3 kN/m^2
Compressive strength, σ_c: 1.65×10^3 kN/m^2

The stress–strain relationship for the carbon fibre elements is shown in Fig. A2.13. For various orientations of the fibres, the tension/compression strengths of the plate, displacement–time and acceleration–time relationships are shown in Figs A2.14 to A2.16. The crack patterns were exactly the same as shown in Fig. A2.11 for plexiglass.

A.2.1.5 Glass-reinforced plastic (GRP) under impact

The loading conditions applied to the GRP plates are similar to those applied to the carbon fibre/epoxy plates (section A.2.1.4). On the basis of the loading conditions given in Fig. A2.12 and the impact velocity of 9 m/s, the crack pattern for a spherical impactor of 50 mm was enhanced. The general distribution of the

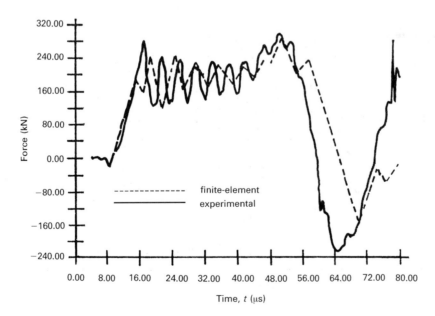

Fig. A2.12 Force–time relationship for a carbon fibre/epoxy plate.

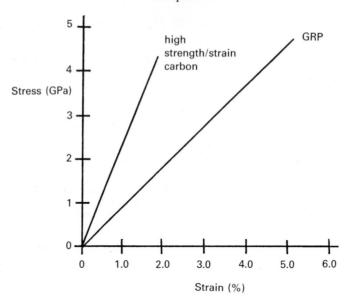

Fig. A2.13 Stress−strain curves for carbon and glass reinforced plastic (GRP) plates.

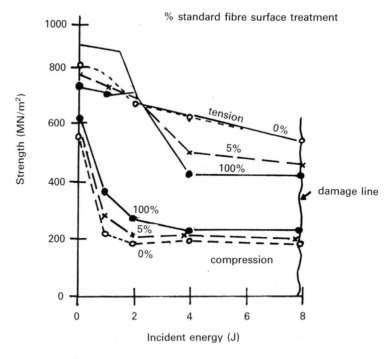

Fig. A2.14 The effect of fibre surface treatment on the residual tension and compression strengths of carbon fibre epoxy $[(O_2 \pm 45)_2]_8$ laminates.

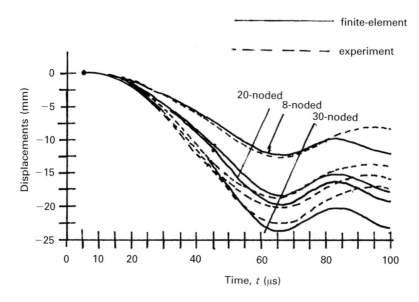

Fig. A2.15 Displacement–time relationship.

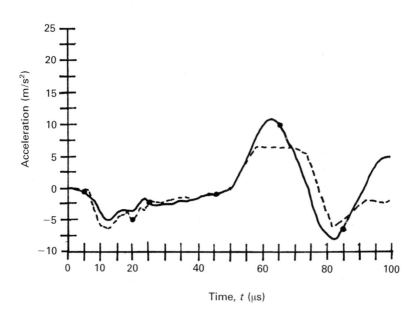

Fig. A2.16 Acceleration–time relationship.

crack was similar to that of the carbon fibre/epoxy plate given in section A.2.1.4. The material is suitable for use in concrete to counter the effects of impact; this is described later in section B.

A.3 Impact analysis of pipe rupture

A.3.1 Experimental data

The UK Atomic Energy Technology Centre (UKAEA) at Winfrith, Dorset, carried out tests on the response of pipes, typical of those used in the pressurized circuits of nuclear power plants, to impact by free-flying missiles from the disintegration of pressurized plants or rotating machinery. The horizontally placed target pipes were impacted by missiles imparting kinetic energies of up to 3 MJ. All impacts occurred at the mid-span of the targets.

A.3.1.1 Data

Target pipes: length, L: 1.8 m
                      nominal bore, d: 150 mm
                      wall thickness, d_t: 11 mm and 18.2 mm

Missiles: cylindrical mild steel billets with flat, conical and hemi-spherical ends.
            mass range: 1.75 to 54.2 kg
            impact velocity range: 46 to 325 m/s

Material property: yield stress σ_y
                      Loughborough University flow stress concept $\sigma = K\epsilon^n$
                      where $K = 1000$ MPa; $n = 0.21$; $\epsilon = $ strain
                      $\sigma_y = 330$ MPa

Cowper-Symonds relation: $\sigma_y/\sigma_{yd} = 1 + |\dot{\epsilon}/D|^{1/p}$
                                  where D and p are empirically determined constants,
                                  p = 5 and D in compression/s:

Wall	Lower yield stress	Upper yield stress
40	15 800	4 300
80	7 100	11 500

Relative perforation energies of different missile nose shapes:
flat-faced	1.0
hemi-spherically ended	0.4
90° including angle cone	0.3

A.3.1.2 Results from UKAEA tests

Table A3.1, from UKAEA, summarizes experiments on pipe targets. DYNA(3D) a three-dimensional finite-element analysis was used for pipe collapse analysis.

Table A3.1 Summary of experiments on pipe targets. (Courtesy of UKAEA.)

Test no	Missile Diameter (mm)	Missile Mass (kg)	Impact velocity (ms^{-1})	Target thickness (mm)	Boundary condition	Remarks
255	25.4 (flat end)	1.7	93	7.4	clamped	Missile perforated front surface and struck rear surface at $44\,\text{ms}^{-1}$; deformation of missile prevented withdrawal
256	25.4 (flat end)	1.7	67	7.4	clamped	Indentation of 21 mm on front surface
257	115 (flat end)	7.5	75	7.3	clamped	Indentation of 65 mm on front surface; rear surface deflection 13 mm; flange weld split
258	115 (flat end)	7.37	108	7.2	clamped	Indentation of 112 mm on front surface; rear surface deflection 26 mm; flange welds split
259	115 (flat end)	7.39	69	7.1	suspended	Indentation of 60 mm on front surface; rear surface deflection 10 mm
260	115 (flat end)	7.29	104	7.1	suspended	Indentation of 104 mm on front surface; rear surface deflection 30 mm
264	60 (flat end)	4.0	105	7.3	clamped	Indentation of 64 mm on front surface; rear surface deflection 8 mm
265	60 × 4.9 wall (axial impact)	4.0	104	7.4	clamped	Indentation of 67 mm on front surface; rear surface deflection 12 mm; missile length reduced by 13 mm to 672 mm
266	60 × 4.9 wall (axial impact)	4.0	136	7.1	clamped	Indentation of 94 mm on front surface; missile length reduced by 20 mm to 665 mm
267	60 (flat end)	4.0	142	7.0	clamped	Missile perforated front and rear surfaces
268	60 (flat end)	4.0	117	7.2	clamped	Water-filled, 1 MPa internal pressure; missile perforated front surface forming 'keyhole'
269	60 (flat end)	4.0	112	7.2	clamped	Water-filled, 1 MPa internal pressure; missile perforated front surface forming irregular hole

continued

Table A3.1 *Continued.*

Test no	Missile Diameter (mm)	Mass (kg)	Impact velocity (ms^{-1})	Target thickness (mm)	Boundary condition	Remarks
270	60 (flat end)	4.0	108	7.2	clamped	Water-filled, 6 MPa internal pressure; missile perforated front surface forming 'keyhole'
271	60 (flat end)	4.0	108	7.6	clamped	Water-filled, 13 MPa internal pressure; indentation of 35 mm on front surface and start of shear plug
272	60 (flat end)	4.0	108	7.7	clamped	Water-filled, 13 MPa internal pressure; missile perforated front surface and initiated extensive crack
273	60 (flat end)	4.0	114	7.5	clamped	Indentation of 82 mm on front surface; transient surface strains recorded
274	60 (flat end)	4.0	114	6.9	clamped	Indentation of 82 mm on front surface; transient surface strains recorded
275	60 (hemispherical end)	4.0	113	7.0	clamped	Indentation of 93 mm on front surface; lateral extent of deformation zone — 150 mm
276	60 (flat end)	4.0	113	7.0	clamped	Water-filled, 13 MPa internal pressure; missile just perforated front surface — disc wedged in hole — and was ejected from pipe at ~40 ms^{-1}; transient surface deformation recorded
477	25.4 (flat end)	1.7	130	10.5	clamped	Missile perforated front face and struck rear surface at 44 ms^{-1}; deformation of missile end ($\varnothing$ 26–27 mm) prevented withdrawal; 5 mm pimple on rear face
478	25.4 (flat end)	1.7	203	18.2	clamped	Missile did not perforate, impact end bent through 90° and diameter increased to 35 mm; indentation of 11 mm max (6 to edge of contact mark) on pipe surface

Table A3.1 *Continued.*

479	25.4 (flat end)	1.7	129	10.7	clamped	Missile perforated front face and struck rear at 51 ms^{-1}; deformation of missile end ($\varnothing$ 26 mm) prevented withdrawal; 10 mm pimple on rear face
480	25.4 (flat end)	0.6	325	18.6	clamped	Missile did not perforate, impact end bent through 15 mm and diameter increased to ~43 mm; indentation of 17 mm max (9 to edge of contact mark) on pipe surface
481	60 (flat end)	4.0	180	10.6	clamped	Missile perforated front and rear faces; residual velocity from front face ~74 ms^{-1}; impact and exit holes in line and showed clearly defined shear plug boundaries; missile diameter increased to 63 mm, residual length ~177 mm
482	60 (flat end)	4.0	236	18.6	clamped	Missile perforated front and rear faces; impact at ~6° to normal; exit velocity not determined; missile diameter increased to 70 mm, residual length 175 mm
483	60 (flat end)	4.0	136	11.5 (nominal)	clamped	Missile perforated front and rear faces; residual velocity from front face ~66 ms^{-1}; impact at ~7° to normal, entry and exit holes laterally displaced by ~60 mm; entry hole showed brittle tearing and cracking around whole perimeter, exit hole showed clearly defined shear plug boundary
484	60 (flat end)	4.0	144	18.6	clamped	Missile partially penetrated front surface displacing shear plug between 3 and 11 mm (min/max); negligible missile deformation

Table A3.1 *Continued.*

Test no	Missile Diameter (mm)	Mass (kg)	Impact velocity (ms⁻¹)	Target thickness (mm)	Boundary condition	Remarks
485	60 (hemispherical end)	4.0	143	7.2 (nominal)	clamped	Missile perforated front and rear surfaces; negligible impact end deformation; residual velocity from front face $\sim 98\,\mathrm{ms^{-1}}$
486	60 (90° conical end)	4.0	87	7.1	clamped	Missile perforated front face and emerged 80–90 mm through rear face and wedged in; impact end flattened to $\sim \varnothing\,8\,\mathrm{mm}$
487	60	4.0	113	8.1	clamped	Missile perforated front and rear surfaces; residual velocity from front face $\sim 81\,\mathrm{ms^{-1}}$; negligible impact end deformation; residual velocity from rear face $\sim 60\,\mathrm{ms^{-1}}$
488	60 (90° conical end)	4.0	67	7.2	clamped	Missile perforated front surface but total penetration by conical section not obtained; missile rebounded; impact end flattened to $\sim \varnothing\,8\,\mathrm{mm}$
489	80 × 10 (rectangle)	3.1	75	7.4	clamped	Impact parallel to pipe axis; missile did not perforate — maximum indentation 36 mm
490	80 × 10 (rectangle)	3.1	84	7.2 (nominal)	clamped	Impact $\sim 5°$ skew to pipe axis; missile did not perforate — maximum indentation 46 mm; missile deformed to $\sim 83 \times 10\,\mathrm{mm}$ max
491	80 × 10 (rectangle)	3.1	99	7.2	clamped	Impact $\sim 10°$ skew to pipe axis; missile did not perforate — maximum indentation 55 mm; missile deformed to $\sim 82 \times 11\,\mathrm{mm}$ max
492	80 × 10 (rectangle)	3.1	143	7.2 (nominal)	clamped	Impact parallel to pipe axis; missile perforated front face; missile deformed to $85 \times 12\,\mathrm{mm}$
493	60 (oblique impact)	4.0	180	7.3	clamped	Missile perforated pipe completely; a section of wall with developed surface $\sim 170 \times 60\,\mathrm{mm}$ removed completely; indentation to initiation of fracture $\sim 50\,\mathrm{mm}$; knockback of missile edge $\sim 6\,\mathrm{mm}$

Table A3.1 *Continued.*

494	60 (oblique impact)	4.0	120	7.0	clamped	Missile did not perforate, some surface cracks in skid mark; knockback of missile edge ~4 mm
495	150 × 18 (lateral impact)	43.0	78	7.4	clamped	Missile did not perforate pipe; impact zone indentation 231 mm and section almost completely flattened; pull-in forces in target pipe destroyed edge support system; missile showed no apparent deformation
496	150 × 7.2 (lateral impact)	34.9	49	7.5	clamped	Missile did not perforate pipe; impact zone indentation 55 mm; imparting pipe totally flattened and welded rear supports in good contact with it
497	150 × 11.0 (lateral impact)	37.5	46	7.4	clamped	Missile impacted at ~20° to 'vertical' axis and did not perforate pipe; impact zone indentation 95 mm; impacting pipe flattened to $\varnothing$ 125/140 mm
498	150 × 11.0 (axial impact)	44.0	78	7.3	clamped	Missile did not perforate pipe; maximum indentation 198 mm and no indication of plugging failure; negligible deformation of missile
499	150 × 18 (axial impact)	54.2	106	10.9	clamped	Missile did not perforate pipe; maximum indentation 350 mm; central section under impact zone flattened and start of plugging failure visible, split formed on one portion of impact mark; negligible deformation of missile but sabot collapsed; pull-in forces sufficient to strip prepared welds on both end flanges
500	100 × 6 (axial impact)	29.5	70	7.1	clamped	Missile did not perforate pipe; maximum indentation 150 mm; indication of plugging failure initiation; negligible deformation of missile

A typical collapse analysis of the missile pipes for various time intervals is demonstrated in Fig. A3.1. Pressure−time graphs and stresses/strains at different positions
have been evaluated.

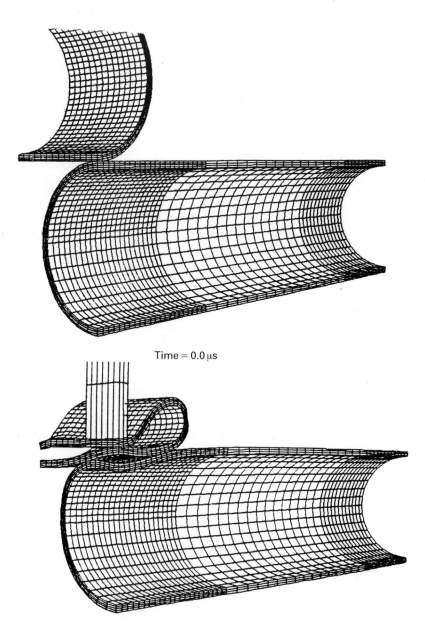

Time = 0.0 µs

Total collapse of missile pipe (time = 3.7 µs)

Fig. A3.1 Pipe collapse analysis at various time intervals (ms = microseconds). (Courtesy
of the UK Atomic Energy Technology Centre, Winfrith, Dorset.)

A.3.1.3 *Pipe whip analysis*

Pipe whip analysis is extremely important in areas of sensitive structures. In nuclear power plants, pipe rupture is a possibility. Impact forces are generated at points of pipe-to-pipe and pipe-to-wall impact, as demonstrated by the UKAEA experiment/analysis (see Fig. A3.1). The author has used a multi-level, sub-structuring technique in the ISOPAR program (also using material non-linearities, a plasticity model and time integration) to examine pipe whip. The following specifications were adopted:

> Pipe dimensions outside: 100 mm
> Thickness: 18.2 mm
> Elbow radius: 143 mm
> Temperature: 320°C
> Maximum impact force: 750 kN
> Whipping time: 10.5 μs
> Velocity of impact: 32 m/s

Figure A3.2 shows the finite-mesh scheme for the pipe rupture analysis at the elbow mid-point. The force−time relation is given in Fig. A3.3, together with the crush.

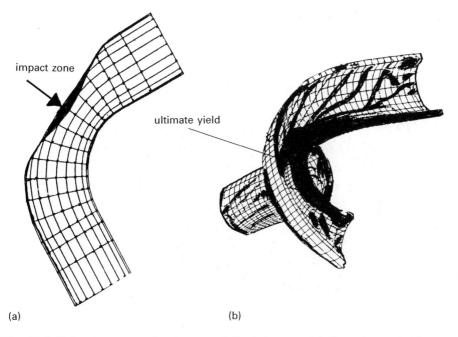

impact zone

ultimate yield

(a)   (b)

Fig. A3.2 Finite-element mesh scheme and final damage. (a) Top pipe under impact; (b) top and bottom pipes whipping.

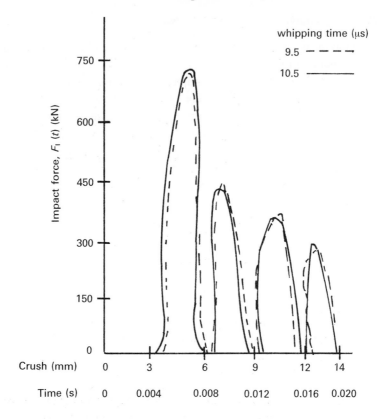

Fig. A3.3 Force−time versus pipe crush.

A.4 Explosions in hollow steel spherical cavities and domes

A.4.1 Steel spherical cavities

Five hollow steel spheres of various cavities and thicknesses of the type shown in Fig. A4.1 were examined after explosion caused by a spherical charge. The spheres were of the elastic-visco plastic type, for which the constitutive equation is:

$$\dot{\epsilon}_{ij} = (\dot{S}_{ij}/2G) + (1/2\eta)[1 - (k/I_2')]S_{ij}$$

where $\dot{\epsilon}_{ij}$ = deviator strain rate
k = strain hardening parameter
S_{ij} = deviator stress
I_2' = second invariant of stress deviator
$\dot{S}_{ij}$ = deviator stress state
G = shear modulus
η = viscosity coefficient

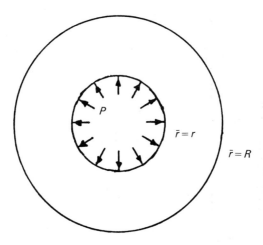

Fig. A4.1 Explosion in the cavity of a sphere.

The following data were used:

$$\lambda = \sqrt{[(4G + 3K)/3]} = 52.28 \times 10^{-5} \, \text{mm/s}$$
$$\delta t = 4 \times 10^{-5} \, \text{s}; \ E = 200 \, \text{GN/m}^2; \ v = 0.3$$
$$G/\eta = 1 \times 10^6/\text{s}; \ r/R = 1.5, 2, 3, 5, \infty$$

Fifty pounds of explosive material with a spherical shape propagation were used. The associated flow rule and the von Mises failure criteria were adopted. A time of two seconds was considered for the total explosion duration.

Figure A4.2 shows the stress−strain history from a classical approach in the radial direction at two locations based on dynamic analysis. Twenty-noded iso-parametric elements were adopted for the analysis. The total number of elements was 180. As seen, the visco-elastic strain increases along the curve 1,2,3. There is a shear drop at point 3 and the deformation seems to be elastic along the 3−4 line. The deformation is visco-plastic along the 3−6 line. A perfectly plastic material seems to be visco-plastic owing to the longer time for dynamic and static stress−strain curves when they intersect. The time-dependent tangential stresses which develop after the explosion are shown in Fig. A4.3. Figure A4.4 shows the time-history for tangential stresses at the cavity of the shell for various R/r ratios.

A.4.2 Steel domes

Finite-element analysis was carried out on explosions in a steel spherical dome. Again, a 50 lb explosive with spherical propagation was assumed. The deformed shape for two seconds is shown in Fig. A4.5, when the explosion occurs at 40 m from the centre of the dome of radius 80 m. The results of this analysis must remain classified information. The damage due to the explosion is also shown in Fig. A4.5.

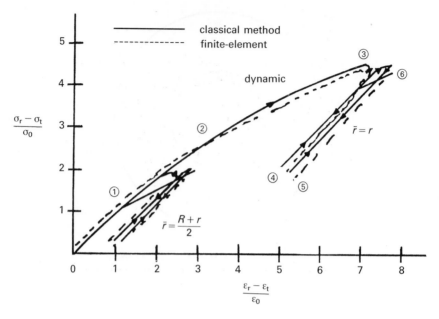

Fig. A4.2 Non-dimensional stress–strain history; σ_t, σ_r = stresses in the tangential and radial directions, respectively; ϵ_t, ϵ_r = strains in the tangential and radial directions, respectively and σ_0 = average stress.

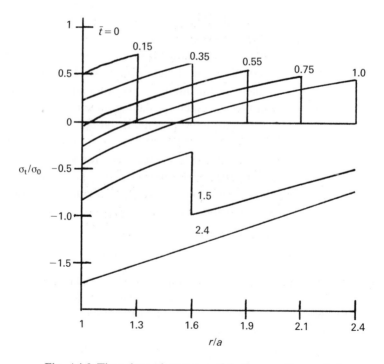

Fig. A4.3 Time-dependent tangential stresses after explosion.

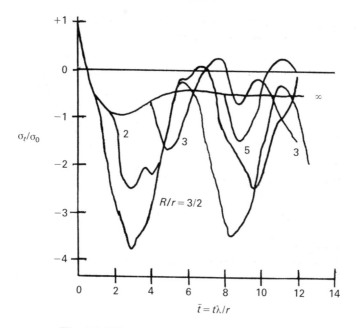

Fig. A4.4 Tangential stresses at the cavity surface.

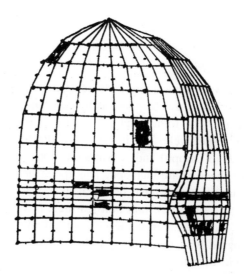

Fig. A4.5 Finite-element analysis of a steel dome with an internal explosion.

A.5 Dropped/impact analysis of a shipping container for radioactive material

This section deals with the impact resistance performance of a container for radioactive material. Several constitutive models, including the effects of strain rate on plastic deformation, have been developed. A mixture of 20-noded isoparametric solid elements and gap elements were used to analyze this container. It is understood that both low-carbon steel and stainless steel are used in the manufacture of these containers. For the low-carbon steel, the combined effects of strain rate and strain hardening (S_H) must be included. In some cases, lead is added as a cover material. The effects are summarized in the following equations:

stainless steel:

$$\sigma = \sigma_{stat} \{1 + (0.001\dot{\epsilon})^{1/8.91}\} + S_H(\epsilon_p)^{0.81}$$

low-carbon steel:

$$\sigma = \sigma_{stat} \{1 + (0.01755\dot{\epsilon})^{1/5}\} + S_H \{1 + (-0.0003\dot{\epsilon})^{1/3}\}(\epsilon_p)^{0.81}$$

lead:

$$\sigma_H = S_H \{1 + (2.33\dot{\epsilon})^{1/1.5}\}(\epsilon_p)^{0.43}$$

The vertical velocity is defined by $\sqrt{(2gH)}$. Values for the density, ρ, are given as:

$$\begin{aligned}
&\text{stainless steel:} && 7.845 \times 10^3 \, \text{kg/m}^3 \\
&\text{low-carbon steel:} && 7.855 \times 10^3 \, \text{kg/m}^3 \\
&\text{lead:} && 11.08 \times 10^3 \, \text{kg/m}^3
\end{aligned}$$

Figure A5.1 shows a typical finite-element mesh scheme for the container adopted by NIREX UK Ltd. This mesh is modified in the ISOPAR program to include gap elements. The total number of solid, gap and spring elements are 1950, 370 and 150, respectively. The impact velocities for 9 m and 20 m drops are taken to be 13.3 m/s (30.8 mph) and 19.8 m/s (46 mph). For fracture effects, both ductile tearing and shearing are included. The upper yield stress, the lower stress and the ultimate tensile stress have been calculated as 500 N/mm², 410 N/mm² and 534 N/mm², respectively. Certain bolt features, including the thread pattern, were conveniently modelled. Since the contents of the container are relatively heavy when compared with the container itself, it is necessary to include non-linear springs inside to cater for the interaction between the contents and the container.

A.5.1 Finite-element analysis and results

Non-linear dynamic finite-element analysis was carried out on the contents/container system. The 20-node isoparametric elements were associated with gap

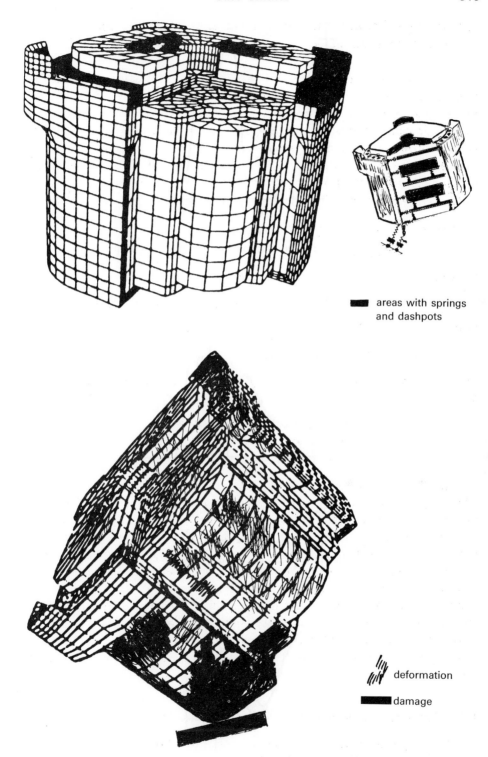

areas with springs
and dashpots

deformation

damage

Fig. A5.1 Finite-element analysis of a container for radioactive material.

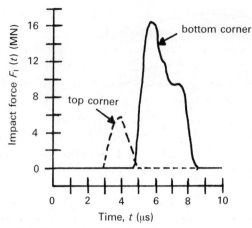

Fig. A5.2 Force−time functions for 9 m and 20 m drops.

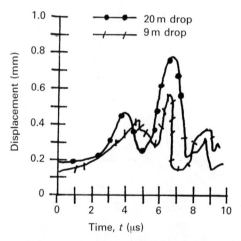

Fig. A5.3 Displacement−time relationship for the axial drop.

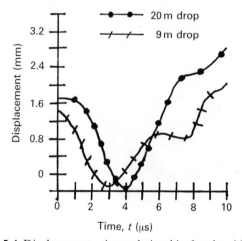

Fig. A5.4 Displacement−time relationship for the side drop.

elements involving master and slave nodes sliding over one another. The impact points were given large numbers of such nodes. The outside contact between the container and the impact area was represented by a series of non-linear springs and dashpots. The number of increments was 21 where the solution procedure was converged. The deformation was assumed to be by plastic flow rather than by fracture processes. The load–time function is given in Fig. A5.2. Figures A5.3 and A5.4 give the displacement–time relationships for the axial and side drops for the 9 m and 20 m drops. The joint separation/bolt joint relationship is illustrated in Fig. A5.5. The stress–strain relationship for various strain rates is given in Fig. A5.6. Table A5.1 summarizes the main results.

Table A5.1 Summary of results for the vertical drops of 9 m and (20 m).

Time (μs)	Acceleration (g) ($\times 10^3$)	Stresses (N/mm²)		Deformation (%)	
		Meridional	Hoop	Height	Width
0	0	0 (0)	0 (0)	0 (0)	0 (0)
200	5	35 (45)	270 (380)	−0.9 (−1.0)	+1.7 (+2.3)
500	10	101 (120)	285 (430)	−1.1 (−1.5)	+1.72 (+3.3)
1000	15	178 (210)	290 (490)	−1.15 (−1.5)	+1.75 (+3.95)
1500	20	230 (267)	310 (375)	−1.28 (−1.8)	+1.80 (+4.3)
2000	25	270 (300)	332 (401)	−1.38 (−2.1)	+2.0 (+4.9)
5000	30	350 (401)	357 (490)	−1.67 (−2.9)	+2.8 (+5.35)
10 000	35	375 (490)	385 (535)	−3.0 (−4.0)	+4.5 (+7.35)

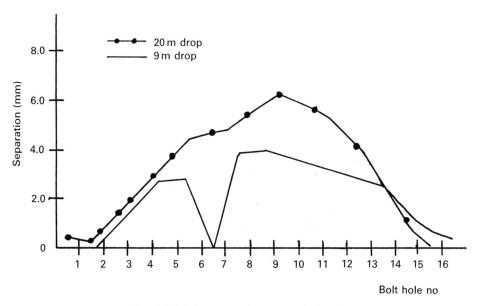

Fig. A5.5 Joint separation versus bolt joint.

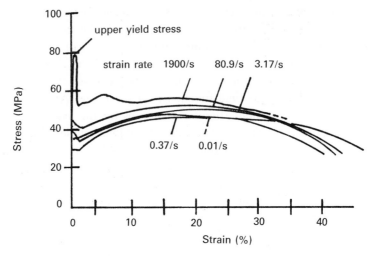

Fig. A5.6 Stress−strain relationship.

A.6 Car impact and explosion analysis

A.6.1 General data

Principles of impulse and momentum are frequently adopted in the study of various vehicle collisions. Rational velocity changes are included in the overall analysis. The energy loss ranges from 32 to 90% for a non-zero moment impulse with a range of restitution coefficients between −0.001 and 0.810. For a zero-moment impulse with the same range of energy loss the restitution coefficients will be between 0 and 0.4. The friction coefficients along the impact range are between −0.07 and 0.90. The collision can be at any angle, as shown in Fig. A6.1. Table A6.1 gives additional analytical features.

A.6.2 Finite-element analysis and results

A typical finite-element mesh scheme for a car is shown in Fig. A6.2. Where the analysis includes a human body, a seat-belted dummy is added, as shown in Fig. A6.3. Bovine materials (given in the text) and steel material properties are included in the analysis of bones and the car. Empty body and body with packings, as dampers, for different positions in the car have been examined. Table A6.1 is thus included in the global analysis. The Jernström model given in Table A6.2 is simulated in the finite-element analysis for the crash investigation. A force−time function, given in Fig. A6.4, is considered for the finite-element analysis.

The vehicles impacted the steel and aluminium barriers at velocities ranging from 30 to 50 mph. Additional input data:

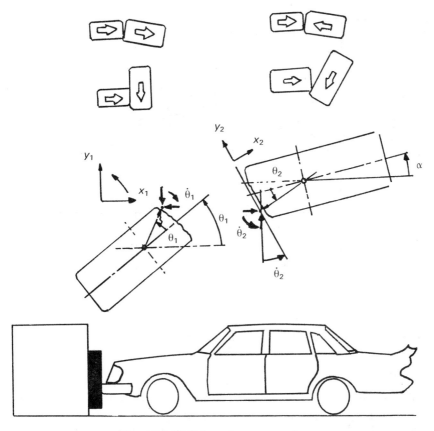

Fig. A6.1 Collision phenomena of cars.

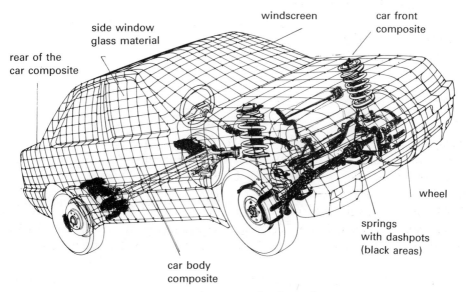

Fig. A6.2 Finite-element mesh scheme for a car.

Table A6.1 Analysis of car rotation and dummy movement.

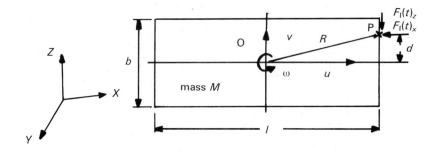

P = rotating point of a car

ω = angular velocity

u_0, v_0 and u, v = initial and final velocities, respectively

$u = \omega d$; $v = -\omega l/2$; O = centroid

$$\omega = (u_0 d)/[(I_0/M) + R^2]; \quad I_0 = [(l^2 + b^2)/2](M/12)$$
$$= u_0(12d)/(l^2 + b^2 + 12R^2)$$

Rigid body impact: $-F_1(t)_x = M(u - u_0)$
$$-F_1(t)_z = Mv$$
$$I_0\omega = F_1(t)_x d - F_1(t)_z(l/2)$$
$$\Sigma F_{xi} = M_i \, (\dot{v}_{xi} + \theta_{yi} y_{zi})$$
$$\Sigma F_{zi} = M_i \, (\dot{v}_{zi} - \theta_{yi} v_{xi})$$
$$\Sigma M_{yi} = I_{yi} \ddot{\theta}_{yi}$$

where ΣF_{xi}, ΣF_{zi} = sums of the applied forces in the X and Z directions

ΣM_{yi} = sum of the applied moments perpendicular to the XZ plane; they are components of $F_1(t)$

$\dot{v}_{xi}$, $\dot{v}_{zi}$ = transitional accelerations in the X and Z directions

$\ddot{\theta}_{yi}$ = angular acceleration

I_{yi} = moment of inertia, referred to the centre of gravity

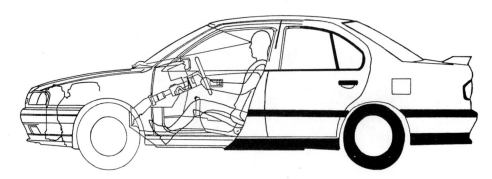

Fig. A6.3 Car with a human dummy included. (For mesh scheme for car see Fig. A6.2.)

Table A6.2 Jernström simulation data. (Courtesy of C. Jernström[2.399])

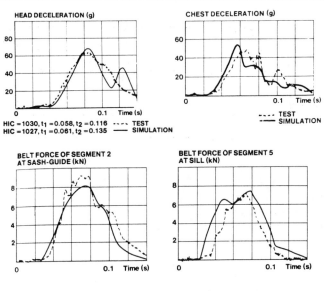

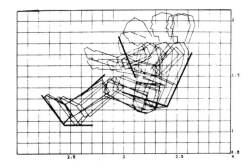

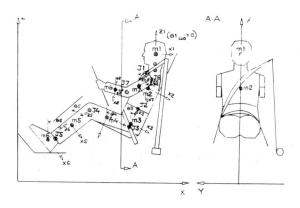

50th percentile male dummy, 10 perc, 3-point belt driver's seat position, initial velocity 15.65 m/s, simulation results.

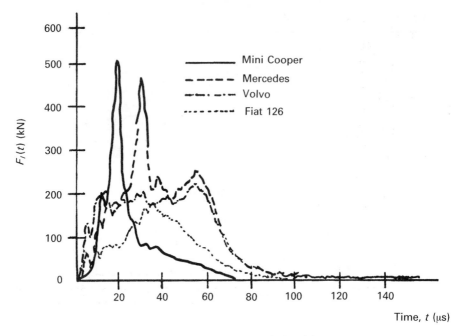

Fig. A6.4 Force−time relationship.

Cars:	solid isoparametric elements (20-noded type, quadratic)	2700
	prism isoparametric elements (20-noded type)	298
	gap elements, mixed dashpots	735
	total nodes	17 805
Barriers:	solid isoparametric elements (20-noded type, single layer)	235
	total increments	10

Total time for impact analysis	140 s/case	
t (time interval)	14 s	
Maximum impact load $F_1(t)$	500 kN	

An IBM 4381 computer with Cray-2 front ended was used for analysis.

Figure A6.5 shows the deceleration−time relationship for two types of car. The damaged car is shown in Fig. A6.6. The force−deflection and the car crash with time results are shown in Fig. A6.7. The barriers were examined for several impacts and various car velocities. Figure A6.8 illustrates the results for flat steel and aluminium barriers braced and belted to vertical posts.

During the car crash investigation the entire dummy body was examined. The seat belts were assumed to be tension elements, each carrying 20 line nodes in

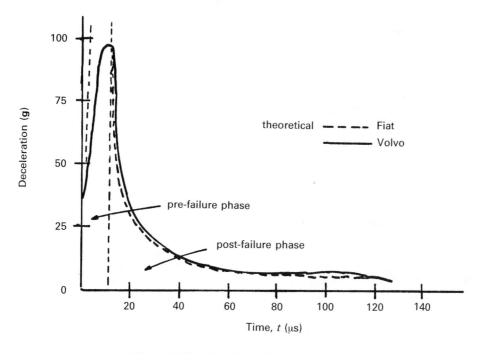

Fig. A6.5 Deceleration−time relationships.

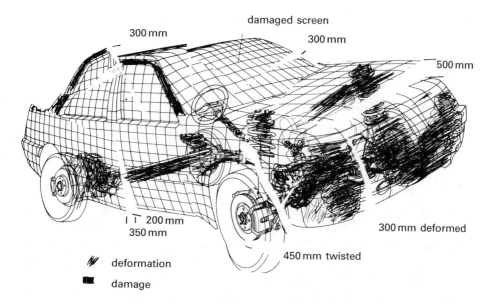

Fig. A6.6 Car: post-mortem.

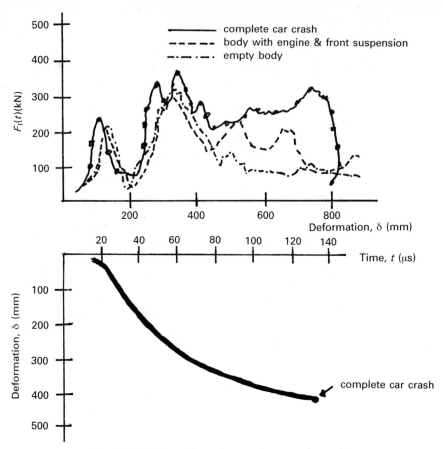

Fig. A6.7 Deformation−time and car crash results.

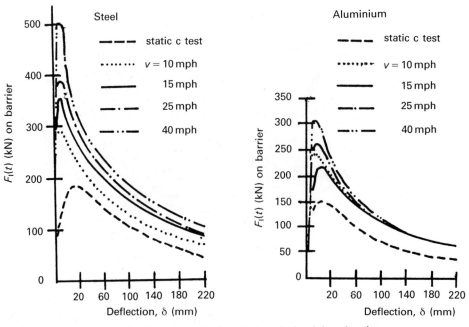

Fig. A6.8 Analytical results for steel and aluminium barriers.

total, sub-divided into 3-noded isoparametric elements. A typical damaged skull is shown in Fig. A6.9. The number of finite elements representing the skull part are 350 and 210 for bones and tissues, respectively. The brain elements numbered 105. All were 20-node isoparametric elements.

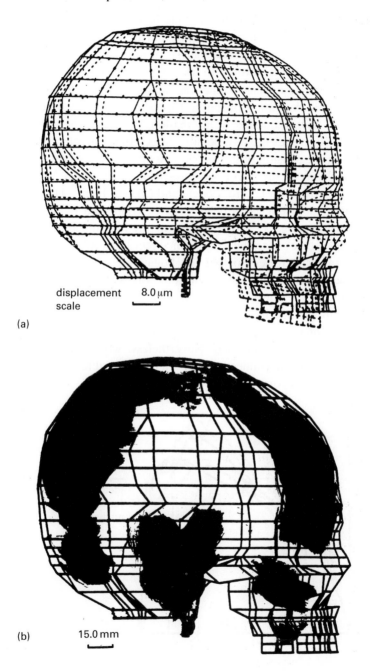

Fig. A6.9 Skull damage. (a) Finite-element mesh and displacement due to impact; (b) final skull damage.

A.7 Train crash phenomenon

A.7.1 General information

Train crashes are becoming a regular phenomenon. Extensive analytical work has been carried out to evaluate the force–time relationships and damage caused to various components. Train-to-train crashes and train-to-barrier crashes are common. The basic philosophy of crash analysis of moving vehicles was given earlier in this text. This section is devoted to the study of a train crashing a barrier. The objective is to evaluate damage to the train and the barrier due to the high-speed impact. During the derivation of the best force–time–acceleration relationships for a high-speed train a number of international railway experts were consulted.

Figure A7.1 illustrates the best possible force–time–acceleration relationship. This is based on the division of the locomotive into three regions defined by cross-members which restrain the lateral movements of the carriages. The three regions are (1) the superstructure, (2) front bogie and (3) rear bogie. They are connected by a series of middle bogies as dead weights, connected by springs and bogie bolsters, as shown in Fig. A7.2. The superstructure is assumed to sit in the bearing segments on the bogies in order to transfer direct horizontal and shear forces. A vertical restraint is provided so that the bogies are not lifted. The bogies are assumed to receive the buffing and draw hook forces. When impact occurs due to head-on collision, the inertial forces are transferred through the locomotive chassis and the bogie bearings. Many railway authorities suggest that the number of bogies does not influence the peak load for a short duration, so the model given in Fig. A7.2 can be justified. Nevertheless, when the locomotive chassis decelerates the rear bogie connections experience inertial forces. These effects are included. A class 46 locomotive was adopted.

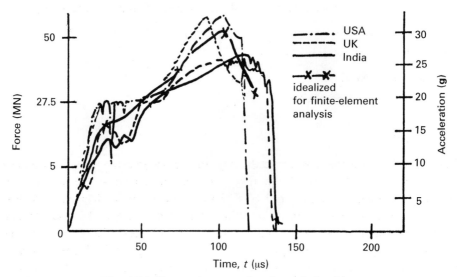

Fig. A7.1 Force–time–acceleration relationship.

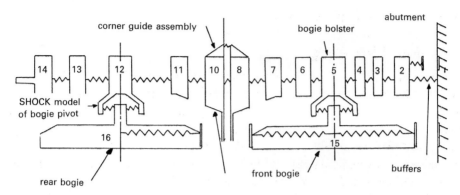

Fig. A7.2 Finite-element model.

A.7.2 Finite-element analysis

A finite-element analysis was carried out on the proposed class 46 locomotive. The data are given in full in Chapter 2. The buffers are assumed to be infinitely rigid and the locomotive has buffers mounted on it. Additional data:

Locomotive and bogies
Isoparametric solid elements (8-noded type)   1100
Spring/bolsters   300
Bogies   2100 each
Spring/bolsters (middle)   330 total
Rear or last bogie (8-noded type)   2100
Total number of nodes   33 109
Number of increments   51

Material model: von Mises with rigid plastic behaviour and associated rule

Computer: Cray 2 front-ended with Norske Data
(total time: 6 hours and 10 minutes)

Locomotive velocity: 30 mph − pre-impact
                     20 mph − post-impact

A.7.3 Results

Figure A7.3 shows the systematic damage to the locomotive at various impact time levels. At 0.10 second the first bogie crashed against the locomotive, resulting in a locomotive acceleration of 22 mph. The complete damage to the first bogie at 20 seconds was 2.3 m crush when the impact load ended at the fifty-first increment. The springs were given a rotation of 1.5 rad/s. The damage to the locomotive is also illustrated in Fig. A7.3. Figure A7.4 gives the load−deflection relationship of the locomotive for various energy absorptions for the first, second and rear bogies with a damaged locomotive.

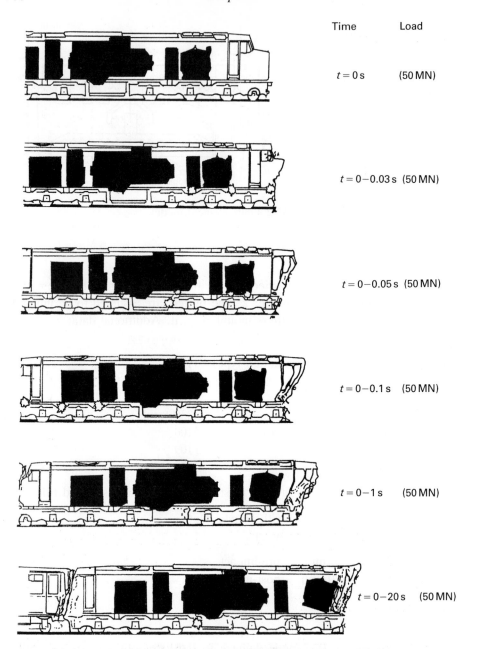

Time   Load

$t = 0\,\mathrm{s}$ (50 MN)

$t = 0-0.03\,\mathrm{s}$ (50 MN)

$t = 0-0.05\,\mathrm{s}$ (50 MN)

$t = 0-0.1\,\mathrm{s}$ (50 MN)

$t = 0-1\,\mathrm{s}$ (50 MN)

$t = 0-20\,\mathrm{s}$ (50 MN)

Fig. A7.3 A systematic damage phenomenon to the locomotive and the first bogie.

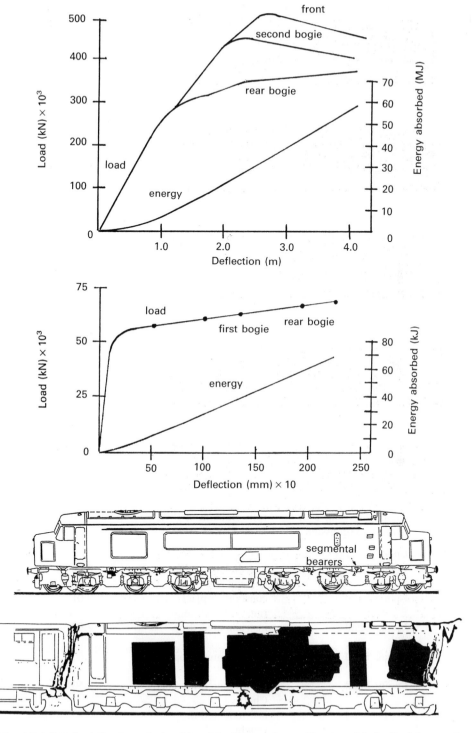

Fig. A7.4 Load–deflection relationship of the locomotive and bogies with the final damaged locomotive zones.

B: CONCRETE STRUCTURES

B.1 Introduction

In this section a number of case studies in concrete are examined under impact and explosion conditions. In most cases, the results obtained from empirical formulae and finite-element analysis are compared with those available from experiments and site monitoring. Some concrete structures have been analyzed later on in other sections as well. The choice is based simply on their usage.

Table B1.1 Strain rates for concrete under conditions of compression, tension and bending.

	Test apparatus for high strain rates	Maximum strain rate (s^{-1})
Under compression		
Any type of compression	drop weight	10.00
	drop weight	2.00
	drop weight	20.00
	hydraulic	0.50
	split Hopkinson	1000.00
	pressure (bar)	120
Under tension		
uni-axial tension	compressed air-driven loading	0.05
splitting tension	hydraulic	0.20
uni-axial tension	'pellet method' — high velocity projectile	20.00
uni-axial tension	split Hopkinson (bar)	0.75
uni-axial tension with effect of compression	constant strain rate	0.01
uni-axial tension with bi-axial compression	split Hopkinson (bar)	0.50
Under bending		
4-point bend	instrumented Charpy impact	0.20
4-point bend	constant displacement rate	0.01
3-point bend	instrumented drop weight	2.00
3-point bend	instrumented drop weight	1.00
3-point bend	instrumented modified Charpy	0.50
2-point bend	instrumented drop weight	0.50
3-point bend	instrumented drop weight	1.80

Impacts and explosions or blasts demand that the effects of the strain rate of concrete must be included in the global analysis of concrete structures. The text gives numerous data on this subject. Table B1.1 gives the values of strain rates for concrete under compression, tension and bending conditions.

B.2 Concrete beams

B.2.1 Reinforced concrete beams

A reinforced concrete beam which has already been tested in the laboratory must be analyzed so that a basis is established for the validation of the numerical tools. Ohnuma, Ito and Nomachi beams were chosen for this work; over 18 beams have been analyzed. Figure B2.1 shows the impact loading system which generates

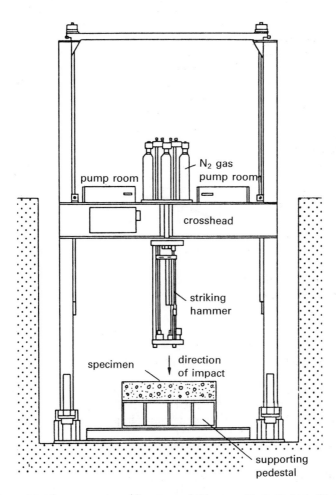

Fig. B2.1 Impact testing apparatus. (Courtesy of Ohnuma H. & C. Ltd, Civil Engineering Laboratory, CRI EPI, Abiko 1646, Abiki-Shi, Chiba-ken, Japan and S.G. Nomachi, Hokkaido University, Japan.)

impact on specimens by a high-speed striking hammer. The striking hammer is a
hard steel cylinder with a diameter of 98 mm. The weight of the hammer is 70 kg.
The maximum velocity was restricted to 50 m/s. Figure B2.2 shows the relationship
between impact force and impact velocity. The bending failure occurred at low
velocity and shearing failure occurred at high-velocity impact.

The load−time function adopted is shown in Fig. B2.3. Figure B2.4 shows the
finite-element mesh scheme adopted for A-20, A-30 and A-50 beams. Velocities
ranging from 0 to 30 m/s for the same impact were chosen. The impact forces
were calculated and are plotted on Fig. B2.2; a strain rate of $20\,s^{-1}$ was used. The
damage for the A-20 beam is shown in Fig. B2.5. For concrete, the endochronic
theory was adopted with the tension cut-off as explained in Table B2.1. The
reinforcement was assumed to be 3-noded isoparametric elements and to lie on
the main nodes of the solid isoparametric elements representing concrete.

The author has also carried out an experiment on a reinforced concrete beam
subject to an impact load of 2000 kN for a duration of 5 seconds. Figure B2.5
shows the damaged zones obtained by finite-element analysis and these compare
well with those predicted by the experiment.

The feature of high-strain loading and the propagation of stress waves from
the point of explosion have been discussed in detail in the text. Such stress waves
can cause non-uniform stress conditions, which ultimately cause cracks even
remote from the point of explosion. Mortar bars 25 mm in diameter and 1 m long
were cast vertically in 50 mm layers and compacted with a 5 mm diameter steel

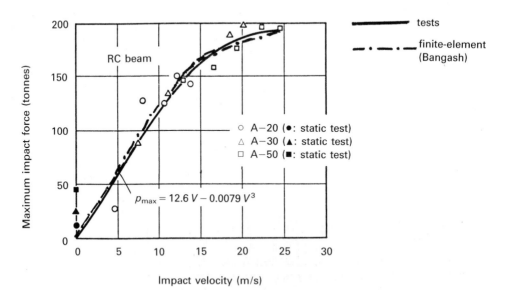

Fig. B2.2 Impact force versus impact velocity for reinforced concrete (RC) beams. Dynamic
response and local fracture of reinforced concrete beams/slabs under impact loading.
(Courtesy of Ohnuma H. & C. Ltd, Civil Engineering Laboratory, CRI EPI, Abiko 1646,
Abiki-Shi, Chiba-ken, Japan and S.G. Nomachi, Hokkaido University, Japan.)

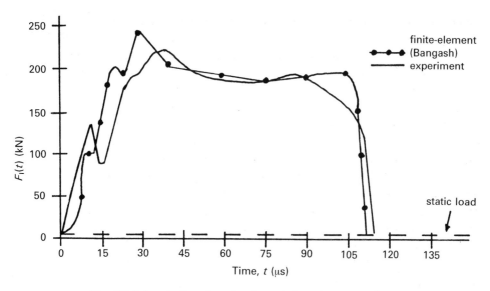

Fig. B2.3 Load−time function for reinforced concrete beams.

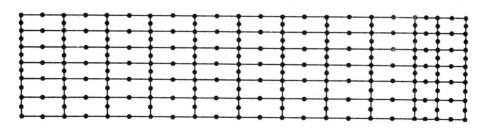

Fig. B2.4 Finite-element mesh scheme, with 20-node isoparametric elements, for reinforced concrete beams of the A-20 to A-50 types.

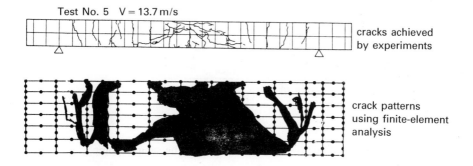

Fig. B2.5 A damaged A-20 beam under a 2000 kN impact with a 13.7 m/s velocity. (Courtesy of Ohnuma H. & C. Ltd, Civil Engineering Laboratory, CRI EPI, Abiko 1646, Abiki-Shi, Chiba-ken, Japan and S.G. Nomachi, Hokkaido University, Japan.)

Table B2.1 Tension cut-off and softening in concrete. Concrete is assumed to be elastic-brittle in tension. When a crack occurs, the stress normal to it can be immediately released and drops to zero.

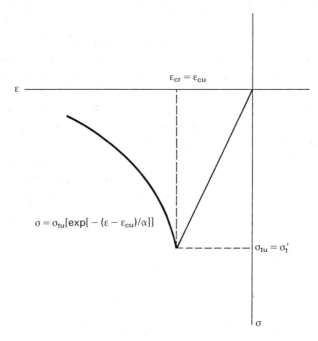

Strain softening model

α = softening parameter
$= G_r - \frac{1}{2}\sigma_{tu}\varepsilon_0(dv)^{1/3}/\sigma_{tu}(dv)^{1/3} > 0$
dv = volume
$\varepsilon_0 = \varepsilon_{cr}$ = strain at cracking
σ_{tu} = tensile strength of concrete
$(dv)^{1/3}\varepsilon_{eq}$ = crack opening
ε_{eq} = equivalent crack strain, based on smeared crack width

where ε is the nominal tensile strain in the cracked zone, ε_{cu} is the cracking and α, the softening parameter, is given by

$$\alpha = (G_f - \frac{1}{2}\sigma_{tu}\varepsilon_{cu}l_c)/\sigma_{tu}l_c > 0$$

where

$$l_c = \frac{\text{volume containing a crack}}{\text{area of crack}} = \frac{V}{A_{cs}}$$
$$\approx (dV)^{1/3}$$

and

dV = volume of concrete represented by sample point

$$G_f = \int_0^\infty \sigma(w)\, dw = \text{fracture energy needed to separate two cracks}$$

w = crack width = $l_c\,\varepsilon_{cu}$

tamping rod. The vibrator was attached to the mould. Electrical (resistance) strain gauges were positioned. Four gauges in each position were connected in series to form a Wheatstone bridge circuit. These bars were wrapped with 5 mm thick foam rubber at 250 mm centres and supported in a Dexion steel angle. The function of the foam rubber was to prevent any iteration between the Dexion and the mortar. A small PE4 explosive charge of 20 mm diameter with a total mass of 7 g was attached to one end of the mortar bar. A detonator was then inserted into the charge. This method was adopted by Dr Watson of Sheffield University, UK, for his concrete block specimens. The author adopted this method for a specially cast reinforced concrete beam which is shown in Fig. B2.6. The fracture pattern for 7 g gelignite attached to the underside of the beam is shown in Fig. B2.7. Finite-element analysis was carried out for this beam. This time the mesh was kept coarse. The deformed mesh (20-noded isoparametric elements) and cracking are shown in Fig. B2.8. A line marked XYZ (Fig. B2.7) represent areas chopped off and letters A−H show high-level cracking zones. Again, endochronic theory was used for concrete failure with a tension cut-off. The total number of increments used was 210.

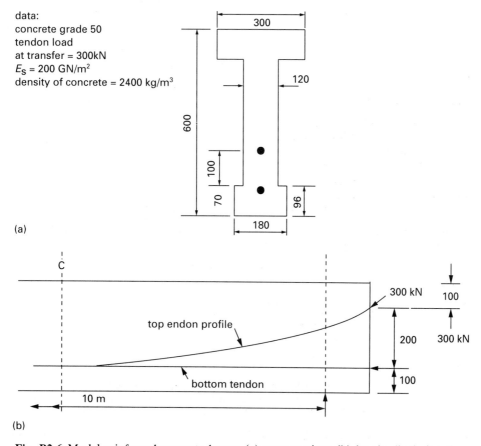

data:
concrete grade 50
tendon load
at transfer = 300kN
E_S = 200 GN/m²
density of concrete = 2400 kg/m³

(a)

(b)

Fig. B2.6 Model reinforced concrete beam: (a) cross-section; (b) longitudinal elevation.

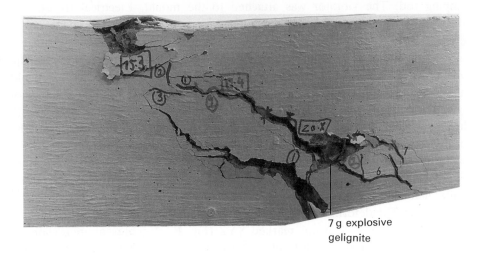

7 g explosive
gelignite

Fig. B2.7 Beam failure under an internal explosion.

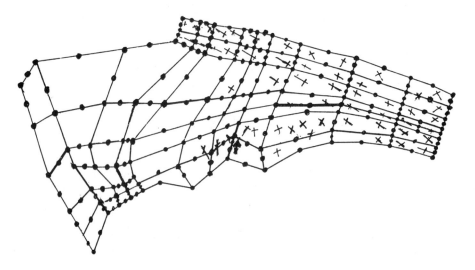

Fig. B2.8 A mesh deformed by an explosion.

B.2.2 Pre-stressed concrete beams

Twelve pre-stressed concrete beams were designed on the basis of the British Standard BS 8110. These beams were subjected to repeated impact loads of 200 kg dropped from different heights. Table B2.2 gives data on these beams. The hammer was guided by a frame and the load from each impact was transmitted through a steel bearing plate. The dynamic responses were measured, particularly displacements, using a variable displacement transformer of 0–500 Hz. Using an IBM PC with special interface software, the electrical signals were transformed

Table B2.2 Data on pre-stressed concrete beams.

P_k = characteristic pre-stressing load; e = eccentricity; h = depth; d = effective depth; L = effective length; the width of the beam is 200 mm

Beam type	L (m)	h (mm)	d (mm)	A_{sp} (mm²)	P_K (kN)	Drop heights (mm)
* REC I	1.60	200	125	60	70	
II	2.50	200	122	102	123	
III	2.50	200	122	140	172	
IV	2.50	200	122	180	220	15,30,60,90,120,160,180,250
V	2.50	200	124	102	123	300,350,400,500,600
VI	2.80	250	160	102	123	
VII	3.80	250	160	102	123	
VIII	4.80	300	160	102	123	
IX	5.80	300	160	102	123	
† BOX I	4.00	350	320	400	500	100,150,200,300,400,500,550,600
II	4.00	350	320	500	600	200,250,300,350,400,550,600,
III	4.00	400	370	600	700	250,300,350,400,550,600

* Solid rectangular beam, 200 mm wide by h deep
† Box beam 200 mm wide by h deep, with 60 mm thickness throughout

A_{sp}(top) = A_{sp}(bottom)
f_{cu}(concrete) = 40 N/mm²
E_c(concrete) = 28–30 GN/mm²
f_y(steel) = 460 N/mm²
E_s(steel) = 210 GN/m²

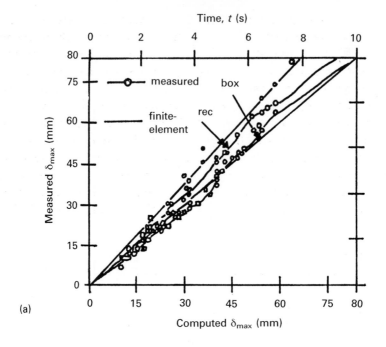

(a)

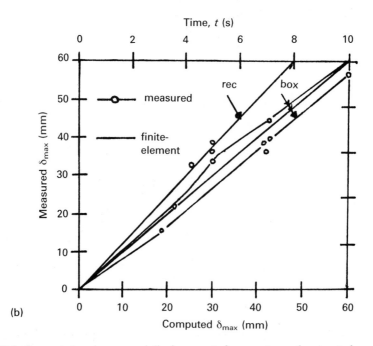

(b)

Fig. B2.9 Computed and measured displacements for pre-stressed concrete beams subjected to dropped weights. (a) Solid beams and box beams (drop heights from 50 mm to 60 mm); (b) solid beams and box beams (drop heights 15 mm to 300 mm). (Box = box beam; rec = solid rectangular beam.)

into displacements. Both programs, ISOPAR and ABACUS, were used, using 8-noded solid isoparametric elements. Pre-stressing wires/strands were represented by 2-node elements, fully bonded. The wires/strands were placed on nodes of the 8-node isoparametric elements. In some cases, the wires/strands were placed in the body of the solid elements. The finite-element mesh of a pre-stressed concrete beam is kept as shown in Fig. B2.4. The drop heights used for the experiments on these beams are given in Table B2.2. Figure B2.9 illustrates a comparative study of displacements produced from the experiments and finite-element analysis of the 12 beams. A typical damaged pattern of a pre-stressed solid concrete beam of 300 mm depth turned out to be as shown in Fig. B2.5, when subjected to multiple impacts from a drop height of 300 mm. The displacement−time relationship is given in Fig. B2.10 for the impact load−time function shown in Fig. B2.11.

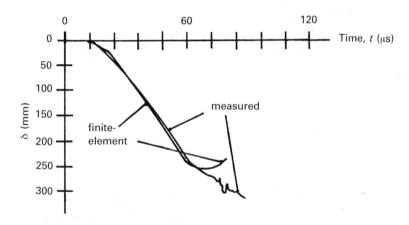

Fig. B2.10 Displacement−time relationship.

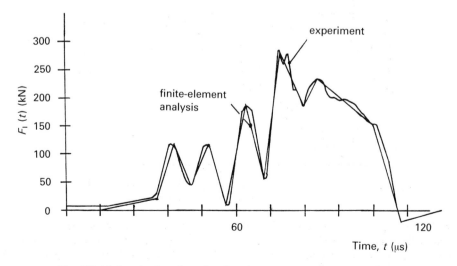

Fig. B2.11 Load−time function for the pre-stressed concrete beam.

B.2.3 Fibre-reinforced concrete beams

B.2.3.1 Introduction

References given in this text provide an in-depth study of dropped weights and projectiles impacting fibre-reinforced concrete structures. It is well known that toughness and cohesion may be greatly increased by the addition of fibres to concrete. Most of the impact work has concentrated on large masses impacting at relatively low velocities. Table B2.3 summarizes the test results on 450 mm square specimens, with various thicknesses and mix proportions, subjected to a 7.62 mm diameter, copper-sheathed, hardened steel projectile of mass 9.6 to 9.9 g, travelling at a speed of 800 m/s. In addition, Table B2.4 gives useful data on the impact tensile strength of steel-fibre concrete, much of which has already been reported in the text. Under dropped weights, the following data can be added to the existing ones: a cylindrical projectile (90 mm by 25 mm) on 200 mm cubes which are subjected to an impact load of 490 N (110.2 lb) by a hammer falling through 300 mm gives an impact energy of 60 kN/m. The general proportions for fibre-reinforced mortar were assumed to be 1:2:0.5. The generally recommended value for Young's modulus for mortar is 3.1×10^6 psi or 21.4 GPa. In most cases, the recommended impact velocity was 222 cm/s for a 60 kN impact load on fibre-reinforced specimens of dimensions 5.1 mm by 10.2 mm by 50.8 mm.

For concrete steel-fibre-reinforced (mortar and others) the compressive stresses for various mixes are as recommended for a 1 m/s impact velocity.

Composite	Mix	Fibres: length × dia. (mm)	Compressive strength, σ_{cu} (kN/m²)
Concrete steel-fibre-reinforced mortar	1:2:3,0.5	25 × 0.25	58 350
Polypropylene reinforced concrete	1:2:0:1%,0.5	6 × 0.15	58 350
Glass-fibre-reinforced concrete	1:2:0:1%,0.5	25 × 0.25	50 921

B.2.3.2 Finite-element analysis

For the current analysis, a concrete steel-fibre reinforcement of a concrete beam of a suitable size given was adopted. The following data analysis form the input for the analysis:

20-noded isoparametric elements (concrete): 84
Steel fibres 2% by volume spread impact loads falling from a 3 m height, range: 50,100,200,250,300 kN

Table B2.3 Summary of main test series results. (Courtesy of W.F. Anderson, A.J. Watson and P.J. Armstrong, Department of Civil Engineering, University of Sheffield.)

25 mm Length steel fibre type	10 mm Aggregate type	% Fibre weight by concrete weight	Penetration path length (mm)	True crater volume (mm³)
Melt extract	Limestone	0	76	492 000
		2.5	92	109 800
		5.0	80	150 500
		7.5	80	73 700
		10.0	66	353 000
	River gravel	0	56	280 500
		2.5	50	236 300
		5.0	86	135 200
		7.5	80	160 700
		10.0	68	147 700
	Basalt	0	87	276 900
		2.5	62	124 000
		5.0	102	83 400
		7.5	91	67 800
		10.0	101	67 200
Brass-coated circular indented	Limestone	0	83	339 500
		2.5	94	104 700
		5.0	91	125 300
		7.5	57	54 400
		10.0	74	59 100
	River gravel	0	74	346 100
		2.5	67	181 100
		5.0	93	280 500
		7.5	75	85 200

continued

Table B2.3 *Continued.*

25 mm Length steel fibre type	10 mm Aggregate type	% Fibre weight by concrete weight	Penetration path length (mm)	True crater volume (mm^3)
Brass-coated circular indented	Basalt	0	89	185 600
		2.5	84	147 700
		5.0	81	63 600
		7.5	84	56 800
		10.0	91	31 400
Brass-coated circular	Limestone	0	68	238 900
		2.5	87	151 300
		5.0	83	114 200
		7.5	60	62 200
		10.0	68	78 800
	River gravel	0	77	251 600
		2.5	93	149 200
		5.0	67	101 900
		7.5	71	54 100
		10.0	68	32 700
	Basalt	0	93	251 500
		2.5	85	135 500
		5.0	102	88 900
		7.5	86	48 100
		10.0	79	68 900

Table B2.4 Strain-rate, energy absorption and other data. (Courtesy of H.A. Körmeling.[2.400])

Fracture strain rate: $\epsilon_{cr} = 220\dot{\epsilon}^{1/3}$

ϵ_{cr} range: 0, 250, 500, 750 $\times 10^{-6}$

$\dot{\epsilon}(s^{-1})$ 0, 10, 20, 30, 40

Fibre volume: 0, 0.75, 1.5, 3%

Stress rate, $\dot{\sigma}$: 0.05 to 200 N/mm^2 s

$E_{n_{TOT}}$ total energy per fibre (N/m) $= G_{TOT,f} = \alpha + \beta \log_e \dot{\delta}$

$$\dot{\delta} = 100\dot{\epsilon}\,\text{mm/s}$$

	Fibre volume (m^3)		
	0	1.5	3
α	198.50	8271.8	13690.00
β	9.31	345.5	559.10
γ^2	0.71	0.8	0.97

Strain rate	Fibre content (%)			E_n
Low	0.75	0.62	0.3	33
Intermediate	0.81	0.71	35.0	35
High	0.78	0.72	40.0	40

Concrete tensile stress, σ_{tu}: 2.85 to 4.22 N/mm^2

Specimen (%)		σ_{tu} (N/mm^2)	δ (mm)	G_{TOT}	N
0		3.27 ± 0.46	0.013 ± 0.0025	—	—
1.5	low	3.75 ± 0.31	0.018 ± 0.0025	7786 ± 3426	115 ± 22.6
3		3.50 ± 0.37	0.027 ± 0.011	7560 ± 2223	190
0		4.79 ± 0.61	0.017 ± 0.001	—	—
1.5	intermediate	5.43 ± 0.52	0.022 ± 0.003	8457 ± 4235	108 ± 43.8
3		5.36 ± 0.74	0.0368 ± 0.024	12772 ± 3880	224 ± 24.7
0		5.57 ± 0.5	0.0165 ± 0.0026	—	—
1.5	high	6.49 ± 0.47	0.0224 ± 0.0061	12509 ± 3572	132 ± 35
3		6.49 ± 0.96	0.022 ± 0.004	15656 ± 5413	175 ± 46.4

δ = deformation; N = number of fibres at cross-section.

Time intervals: 0 to 3.5 µs
Strain rates/s: 10^0 to 10^{-7}
Concrete failure analysis: endochronic

All other data for concrete are assumed to be the same as given in section B2. Figure B2.12 gives the finite-element mesh scheme for solid concrete. The fibres are assumed to spread throughout the element with interlocking values of $\beta' = 0.5$. The total number of increments was 12, with a convergence factor of 0.01. The Wilson-θ method was adopted, with an implicit solution. For damage analysis, cracks occurred on the line suggested in the text.

Figure B2.13 gives the impact load–displacement relationship for two chosen strain rates. Figure B2.14 illustrates the strain–time relationship for this beam. The ratios of dynamic strength to static strength for various strain rate effects are shown in Fig. B2.15. Prior to the damage received by this beam, the ultimate deformed shape was as shown in Fig. B2.16. The final post-mortem is given in Fig. B2.17. It is interesting to note that the analytical results obtained in Fig. B2.15 compare well with those obtained by an experiment carried out on a totally different kind of beam.

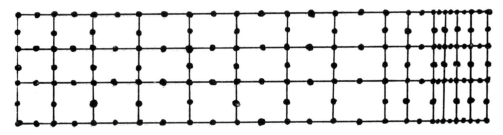

Fig. B2.12 Finite-element mesh for a steel-fibre-reinforced concrete beam.

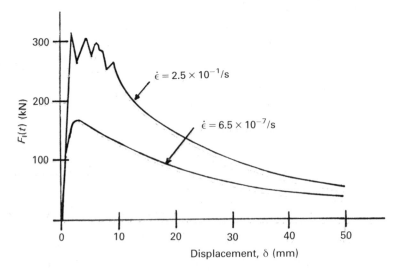

Fig. B2.13 Load–displacement relationship for two selected strain rates.

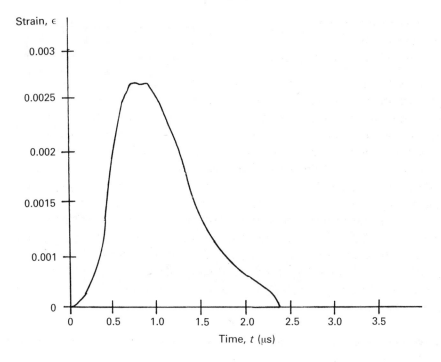

Fig. B2.14 Strain–time relationship for a fibre-reinforced beam.

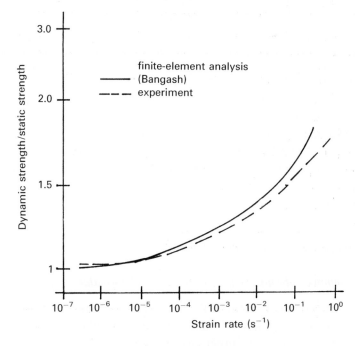

Fig. B2.15 Ratios of dynamic strength to static strength for various strain rates. (Courtesy of W. Suaris and S.P. Shah.[2.401])

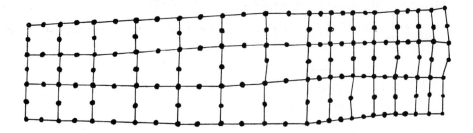

Fig. B2.16 Deformed finite-element mesh.

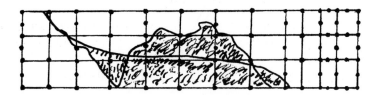

Fig. B2.17 A post-mortem for a fibre-reinforced concrete beam.

B.3 Reinforced concrete slabs and walls

B.3.1 Introduction

A number of missile formulae are given in this text for penetration, perforation and scabbing of reinforced concrete slabs and walls. On the basis of some of these empirical formulae, computer programs have been written for rapid solution of problems. Design aids are included for preliminary manual design. In addition, finite-element analyses given in the text have been carried out on slabs and their results compared with those obtained from empirical formulae and experimental tests. Similarly, slabs/walls subjected to explosion have been examined. Using design aids, slab reinforcements were calculated on the basis of both American and British practices.

B.3.2 Slabs and walls under impact loads

B.3.2.1 Computer subroutines based on empirical formulae

The Appendix shows the layout of the computer programs for impact on concrete structures based on empirical formulae already discussed in the text. The following examples are linked to these programs.

A slab with clamped edges, using the ACI/ASCE codes

The slab has clamped edges and measures 6.1 m by 0.76 m. The concrete data are:

$f'_c = 21$ MPa
$E_c = 21.5 \times 10^3$ MPa
Weight density $= 2320$ kg/m
Poisson's ratio for concrete $= 0.17$

The slab reinforcing steel data are:

$f_y = 280$ MPa
$E_s = 200 \times 10^3$ MPa
Depth to tensile reinforcing steel $= 0.70$ m

The missile is a flat-nosed cylinder with a concentrated impact load. It has a weight of intensity 1.8 N (Newton), a diameter of 178 mm and an impact velocity of 107 m/s.

The reinforced concrete section properties are given below.

(1) Average of cracked and uncracked moments of inertia:

$$I_a = \tfrac{1}{2}[(bt^3/12) + Fbd^3]$$

$b = 1$ cm, $t = 76$ cm, $d = 70$ cm. It is assumed that

$$\rho_1 = \rho'_1 = As/bd = 1.76 \times 10^{-3}$$
$$n = E_s/E_c = (200 \times 10^3)/(21.5 \times 10^3) = 9.3$$
$$\rho'_n = 1.76 \times 10^{-3} \times 9.3 = 0.0163$$

Using the chart in Table B3.1, with values of $\rho'_n = 0.0163$ and $\rho'_1/\rho_1 = 1.0$, $F = 0.012$. Hence,

$$I_a = \tfrac{1}{2}[(1 \times 76^3/12) + (0.012 \times 1 \times 70^3)]$$
$$= 20\,349 \text{ cm}^4/\text{cm}$$

(2) Elastic stiffness of the slab:

$$K_e = 12E_cI_a/[\alpha a^2(1 - v^2)]$$

The slab is square, therefore $a/b = 1.0$. Using Table B3.1, the following values are computed:

$$\alpha = 0.0671, \ v = 0.17 \text{ for concrete}$$

$$K_e = \frac{12(21.5 \times 10^3 \times 20\,349)}{0.0671 \times 610^2 \times (1 - 0.17^2)}$$
$$= 2.16 \times 10^9 \text{ N/m}$$
$$= 2.16 \text{ MN/mm}$$

Table B3.1 Slab coefficients with cracked and uncracked conditions.

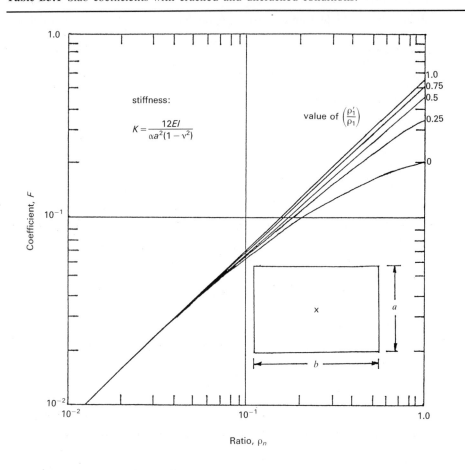

Coefficients for moment of inertia of cracked sections.

Simply supported on all four sides, with load at centre.

$\dfrac{b}{a}$	α
1.0	0.1390
1.1	0.1518
1.2	0.1624
1.4	0.1781
1.6	0.1884
1.8	0.1944
2.0	0.1981
3.0	0.2029
∞	0.2031

Fixed supports on all four sides, with load at centre.

$\dfrac{b}{a}$	α
1.0	0.0671
1.2	0.0776
1.4	0.0830
1.6	0.0854
1.8	0.0864
2.0	0.0866
∞	0.0871

(3) For a concentrated load, the load factor $K_L = 1$. The effective mass is given as

$$M_e = \int m[\phi(x)]dx$$

with a mass factor

$$K_m = M_e/M_t$$

where M_t = total mass of slab.

$$\phi = 1 - (r/R) \qquad (R = a/2)$$

$$dx = dA = 2\pi r dr$$

$$M_e = m(2\pi) \int_{r=0}^{r=R} [1 - (r/R)]r dr = m\pi R^2/6$$

The effective mass is one-sixth of the mass within the circular yield pattern.

$$m = \text{mass/unit area} = 2320 \times 0.76/9.81 = 179.7\,\text{kgs}^2/\text{m}$$

$$M_e = \frac{179.7 \times \pi \times (6.1/2)^2 \times 9.81 \times 10^{-3}}{6} = 8.56\,\text{Ns}^2/\text{mm}$$

The first natural period:

$$T = 2\pi\sqrt{(M_e/K_L K_e)} = 2\pi\sqrt{(8.56/1 \times 2.16 \times 10^6)} = 0.0125$$

The use of empirical formulae

The empirical formulae are based on empirical units and data and coefficients are based on them. The data shall be converted back in to SI units.

(1) National Defence Research Committee Formulae (NDRC) first formula
Data conversion:

Missile weight = 400 lb
Missile diameter = 7 in
Missile velocity = 350 fps
Concrete compressive strength = 3000 psi
Flat-nosed missile, therefore the nose-shaped factor $N = 0.72$

$$K = \text{concrete strength factor} = 180/\sqrt{f_c'} = 180/\sqrt{3000}$$
$$= 3.29$$

Assuming $x_p/d \leq 2.0$ and using the first formula:

$$x_p = \sqrt{[4(3.29)(0.72)(400)(7)(350/1000 \times 7)^{1.8}]}$$
$$= 10.99\,\text{in} \ (279.1\,\text{mm})$$

Check: $x_p/d = 10.99/7 = 1.57$, OK

Check perforation and scabbing with $x_p/d = 1.57$

$$t_p/d = 1.32 + 1.24(x_p/d) = 3.27, \qquad t_p = 582 \, \text{mm}$$

$$t_{sc}/d = 2.12 + 1.36(x_p/d) = 4.25, \qquad t_{sc} = 757 \, \text{mm}$$

(2) Bechtel formulae
Solid missile:

$$t_{sc} = (15.5/\sqrt{3000})(400^{0.4} \times 350^{0.5}/7^{0.2})$$
$$= 39 \, \text{in} \ (1001 \, \text{mm})$$

Hollow missile:

$$t_{sc} = (5.42/\sqrt{3000})(400^{0.4} \times 350^{0.65}/7^{0.2})$$
$$= 33 \, \text{in} \ (842 \, \text{mm})$$

(3) ACE formula

$$x_p = \frac{282}{3000} \left(\frac{400}{(7/12)^2} \right) (7/12)^{0.215} \left(\frac{350}{1000} \right)^{1.5} + 0.5(7/12)$$
$$= 20.7 \, \text{in} \ (525 \, \text{mm})$$

$$t_p = 1.23d + 1.07x_p = 1.23(7/12) + 1.07(20.7)$$
$$= 22.9 \, \text{in} \ (580 \, \text{mm})$$

(4) CKW-BRL formula
Formula for penetration:

$$x_p = (6 \times 400 \times 7^{0.2}/7^2)(350/1000)^{4/3}$$
$$= 17.8 \, \text{in} \ (453 \, \text{mm})$$

$$t_\rho = 1.3x_p = 1.3 \times 17.8 = 23.1 \, \text{in} \ (588 \, \text{mm})$$

For example, using a rectangular pulse load characteristic where the impact load $F_1(t) = W(v_s)^2/2gt_p \approx 3.69 \, \text{MN}$, the duration time t_d is given by

$$t_d = 2x_m/v_s \approx 0.00517 \, \text{s}$$

On the basis of ACI (American Concrete Institute) strength reduction analysis of concrete, the collapse load of a slab is written for a circular fan failure as

$$R_m = 4\pi\phi[(A_s - A_s')f_y(d - a/2) + A_s' f_y(d - d')](\text{DIF})$$
$$\approx 2.949 \, \text{MN}$$

where ϕ = strength reduction factor
MF or DIF = dynamic increase factor (in this case 1.19)
   A_s, A_s' = reinforcement areas in tension and compression zones

Using Fig. B3.1

$$C_T = t_d/T = 0.4155(R_m/F_1) = 0.7975$$

$$x_m/x_e = 3.795 < \text{ductility factor } \mu = 30$$

No overall failure exists.

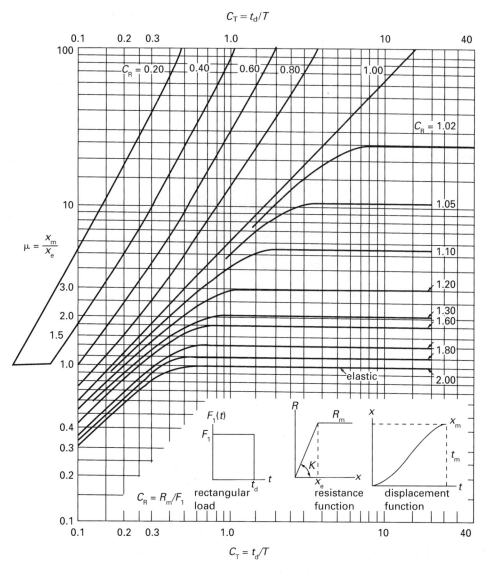

Fig. B3.1 x_m/x_e curves for an elasto-plastic system based on a rectangular impulse load. (Courtesy of J.M. Biggs.[2.402])

B.3.2.2 Finite-element analysis of reinforced concrete slabs/walls

Reinforced concrete slabs

A typical finite-element model of a reinforced concrete slab with a damaged zone is shown in Fig. B3.2. The following parameters were used:

The slab thickness varied from 100 mm to 1500 mm
For a slab thickness of 100 mm, v_s ranged from 12.1 to 21 m/s
For a slab thickness of 200 mm, v_s ranged from 20.7 to 43 m/s

The total number of 8-noded elements was 1584
The reinforcement spread was between 1% and 2%
Endochronic concrete model, $f_{cu} = 0.87 f'_c$
$$= 41.7\,\text{N/mm}^2$$
Load−time function: see Fig. B3.3

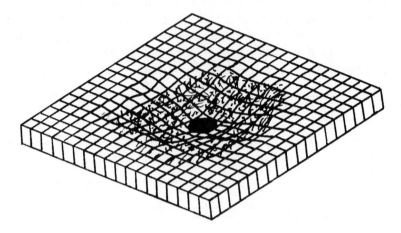

Fig. B3.2 Finite-element analysis of a damaged slab.

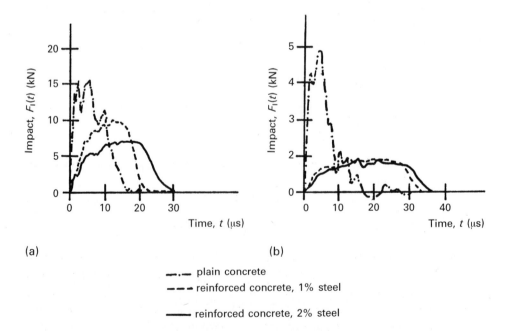

(a)   (b)

−··− plain concrete

−−− reinforced concrete, 1% steel

—— reinforced concrete, 2% steel

Fig. B3.3 Load−time function for (a) a solid impactor and (b) a pipe impactor.

Drop weight of 37.8 kg at 8 m/s
Nose shape for pipes: $N = 0.72$ and 1.0

Typical examples of the Ohnuma and Nomachi slabs were considered in order to test the finite-element program ISOPAR. Again, the numbers of elements and nodes of 8-noded isoparametric elements shown in Fig. B3.2 were employed. This step was taken to reduce the computational effort and time.

Figure B3.4 shows a typical scabbing phenomenon found by Ohnuma *et al.* (personal communication). This encouraged the author to examine their slabs when subjected to impact loads of various velocities. The loading patterns and the slab thickness were kept the same. For a 300 mm reinforced concrete slab the impact velocity was kept at 50 m/s. The penetration depths, perforations and scabbing for various impact velocities were evaluated using the ISOPAR program. Figure B3.4 illustrates a comparative study of the experimental and finite-element results for slabs of 100 mm, 200 mm and 300 mm thicknesses. The results of the finite-element analysis are within 5 to 10% of the experimental results.

Reinforced concrete walls

A reinforced concrete wall was subjected to impact loads caused by rods and pipes. The study was undertaken to determine the local response in terms of crushing, cratering, spalling, cracking and plug formation. The concrete constitutive model was assumed to be based on the bulk modulus approach with a tensile yield surface. The 8-noded isoparametric element was adopted. The optimum finite-element mesh for the wall was kept as shown in Fig. B3.2.

Additional data for this analysis using the ISOPAR program:

Target wall: diameter = 1 m
     overall depth with a haunch = 0.3 m
     main depth = 0.2 m
     concrete grade = 30
     reinforcement T-25-300

Projectile: : steel rods 100 mm long with a diameter of 20 mm and a weight of 920 g
     steel pipes 150 mm long with a diameter of 35 mm and a weight of 790 g
     velocity = 22 m/s

The effects of impact on a concrete wall of both steel rod and pipe projectiles at 50 m/s were examined. Figure B3.5 shows the penetration–time relationships for this wall at various time intervals. Table B3.2 records a comparative study of finite-element analysis and empirical formulae. The same wall was analyzed using an impact velocity of 3.3 m/s. The wall was reinforced with T25. The maximum penetration was 8 mm along the entire impact line.

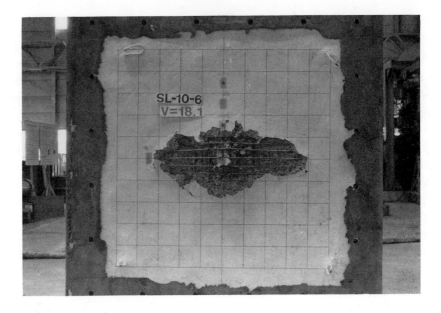

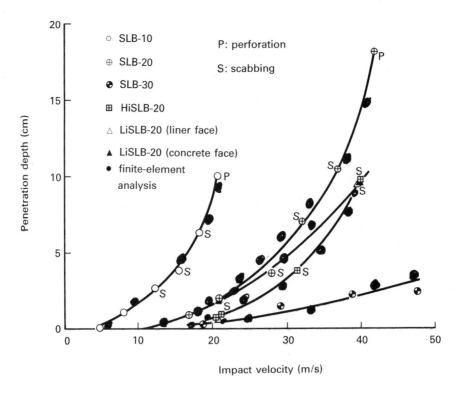

Fig. B3.4 A comparative study of experimental finite-element results. (H. Ohnuma, C. Ito and S.G. Nomachi, personal communication, August 1985.)

Table B3.2 Comparative study of finite-element analysis and empirical formulae.

Scabbing velocity/slabs	Empirical formulae			Finite-element analysis	
	Formula	(1) Penetration depth (m)	(2) Min thickness to prevent perforation (m)	Penetration (m)	Min thickness to prevent perforation (m)
12.1 m/s/760 mm slab	IRS	0.290	1.4	1.13	1.89
	HN-NDRC	0.2791	1.70		
	ACE	0.525	1.90		
Perforation velocity					
12.1 m/s/760 mm slab	DF-BRL	0.453	–	0.79	1.10
43 m/s/760 mm slab	BRL	0.8	1.02	0.83	1.15

N/d_p^*	NDRC		Finite-element analysis	
	t_{pe}/d_p^*	Velocity, v_s(m/s)	t_{pe}/d_p^*	Velocity, v_s(m/s)
0.72/300	0	–	0	–
	0.2	1.57	0.2	1.61
	0.4	3.05	0.4	2.93
0.72/600	0.6	4.20	0.6	4.10
	0.8	6.10	0.8	5.91
	1.0	7.60	1.0	7.39
1.0/300	0	–	0	–
	0.2	2.30	0.2	2.25
	0.4	3.95	0.4	3.85
	0.6	6.10	0.6	6.05
	0.8	6.30	0.8	5.15
	1.0	–	1.0	6.75

* d_p = diameter of the pipe.

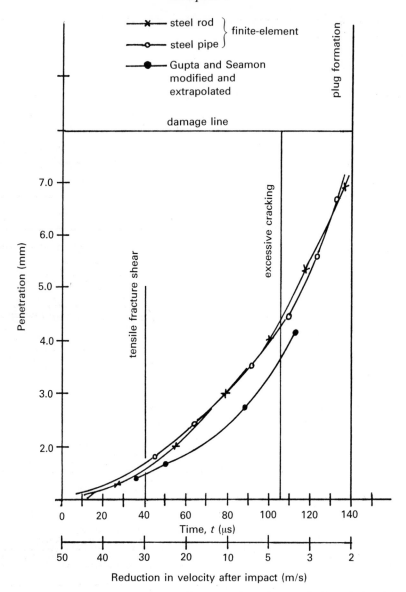

Fig. B3.5 Reinforced concrete wall penetration−time−velocity relationships.

B.3.3 Design for blast resistance

B.3.3.1 *General data and specification (BS 8110 and HMSO Guide) and TM5−1300 US Army*

The ultimate load capability of a ductile structural element subjected to blast loading can be determined by considering its capacity of sustaining an external load by relatively large plastic deformation. The design rules in the guides discussed in the following sections limit the magnitude of the plastic deformation and the level of damage to the structural elements.

B.3.3.1 General data and specifications: BS 8110 and HMSO guide on domestic shelters

Ultimate unit or resistance r_u

Depending on the mass and stiffness of the concrete element, this is written as

$$r_u = F_1\{1/[1-(1/2\,\mu)]\}$$

The moderate damage μ is taken as 3, which gives $r_u = 1.2F_1$ and the safety factor $\gamma_1 = 1.2$. The dynamic factor MF or DIF is taken as 1.1 for bending steel and 1.25 for concrete in compression.

Ultimate shear capacity, plastic moment of resistance and steel capacity

The ultimate shear stress resistance is limited to $0.4f_{cu}$ ($0.4\sigma_{cu}$).

Edge conditions	Ultimate shear stress
Cantilever	$r = (L-d)/d$
Fixed or pinned	$r = (L/2 - d)/d$

Dynamic shear stress shall not exceed $172\,\text{N/mm}^2$

Minimum area of flexural reinforcement

Reinforcement	Mild steel	High tensile steel
Main	0.25% bd	0.20% bd
Secondary	0.15% bd	0.12% bd

Connections to concrete

The allowable dynamic stresses for bolts and welds (BS 4190) are as follows:

Bolts : tension $275\,\text{N/mm}^2$
     shear $170\,\text{N/mm}^2$
     bearing $410\,\text{N/mm}^2$
Welds: tension or compression $275\,\text{N/mm}^2$
     shear $170\,\text{N/mm}^2$

Ultimate unit resistance for slabs/walls with different boundary conditions

The ultimate unit resistance of a concrete element varies according to the following:

(1) distribution of applied loads
(2) geometry of the element

(3) percentage of the reinforcement
(4) type of support.

Using the standard symbols, the ultimate unit resistance r_u for several one-way elements is given below:

Edge condition		Ultimate unit resistance, r_u
Cantilever	$\overleftarrow{\underset{L}{\rule{4cm}{0pt}}}$	$2M_{HN}/L^2$
Simple support	$\underset{L}{\rule{4cm}{0pt}}$	$8M_{HP}/L^2$
Fixed support	$\underset{L}{\rule{4cm}{0pt}}$	$8(M_{HN} + M_{HP})/L^2$
Fixed simply supported	$\underset{L}{\rule{4cm}{0pt}}$	$4(M_{HN} + 2M_{HP})/L^2$

M_{HN} = ultimate unit negative moment capacity at support.
M_{HP} = ultimate unit positive moment capacity at mid-span.
L = span length.

Two-way elements can be analyzed using yield line theory. The value of the ultimate unit resistance r_u for a two-way slab is given by:

$$R_m = r_u = [8(M_{HN} + M_{HP})(3L - x)]/H^2(3L - 4x) \text{ short span}$$

$$R_m = r_u = [5(M_{HN} + M_{HP})]/x^2 \text{ long span}$$

The ultimate unit moment capacity of a reinforced concrete element subjected to blast loading can be found by using the following equation:

$$M_u = f_y(\text{dyn}) \times A_s \times Z$$

$$M_u = 0.225 \, f_{cu}(\text{dyn}) \times bd^2$$

$$Z = (1 - 0.84 f_y(\text{dyn}) \times A_s)d/(f_{cu}(\text{dyn}) \times bd)$$

$$Z < 0.95d$$

The ultimate unit moment capacity of other shapes of slabs with respective boundary conditions against blast loads can be evaluated using tables.

B.3.3.2 Additional information from US design codes ASCE Manual 42 and TM5-1300 for blast loads

Ultimate static and dynamic moment capacity

The ultimate unit resisting moment, M_u, of a rectangular section of width b with tension reinforcement only is given by

$$\text{static } M_u = (A_s f_s / b)[d - (a/2)]$$

where A_s = area of tension reinforcement within the width b

f_s = static design stress for reinforcement

d = distance from extreme compression fibre to centroid of tension reinforcement

a = depth of equivalent rectangular stress block

$= (A_s f_s)/(0.85 f_c')$ static

b = width of compression face

f_c' = static ultimate compressive strength of concrete

Dynamic f_s is replaced by $f_y(\text{dyn}) \times \phi$. The reinforcement ratio, ρ is defined as

$$p = \rho = A_s/(bd)$$

Check on section

$$\rho_b = 0.85 K_1 \{ f_c'(\text{dyn})(87\,000)/f_y(\text{dyn})[87\,000 + f_y(\text{dyn})] \}$$

$$a(\text{dyn}) = [A_s f_y(\text{dyn})]/[0.85 f_c'(\text{dyn}) \times b]$$

Applied M varies for loadings and boundary conditions, and is generally equal to $r_u L^2/16$.

To ensure against sudden compression failures the reinforcement ratio ρ must not exceed 0.75 of the ratio ρ_b, which produces balanced conditions at ultimate strength and is given by

$$\rho_b = 0.85 K_1 [f_c' \, (87\,000)/f_s(87\,000 + f_s)]$$

where $K_1 = 0.85$ for f_c' up to 4000 psi and is reduced by 0.05 for each 1000 psi in excess of 4000 psi.

For a rectangular section of width b with compression reinforcement, the ultimate unit resistance moment is given by

$$M_u = \frac{(A_s - A_s')}{b} f_s \, (d - a/2) + \frac{A_s' \times f_s}{b} (d - d')$$

where A_s' = area of compression reinforcement within the width b

d' = distance from the extreme compression fibre to the centroid of compression reinforcement

a = depth of equivalent rectangular stress block

$= (A_s - A_s') f_s / (0.85 b f_c')$

The reinforcement ratio is given by

$$\rho' = A_s'/(bd)$$

When the compression steel reaches the value f_s at strength,

$$\rho - \rho' = 0.85 K_1 (f_c' d' / f_s d)[87\,000/(87\,000 - f_s)]$$

If $\rho - \rho'$ is less than the value given in the above equation, or when compression steel is neglected, the calculated ultimate unit resisting moment should not exceed that given by the above. The quantity $\rho - \rho'$ must not exceed 0.75 of the value of ρ_b given earlier.

Minimum flexural reinforcement

To ensure proper structural behaviour and also to prevent excessive cracking and deformation under conventional loadings, the minimum areas of flexural reinforcement should be as shown in the table below.

Pressure design range	Reinforcement	Two-way elements	One-way elements
Intermediate and low	Main	$A_s = 0.0025bd$	$A_s = 0.0025bd$
	Other	$A_s = 0.0018bd$	$A_s + A_s' = 0.0020bT_c$
High	Main	$A_s = A_s'$ $= 0.0025bd_c$	$A_s = A_s'$ $= 0.0025bd_c$
	Other	$A_s = A_s'$ $= 0.0018bd_c$	$A_s = A_s'$ $= 0.0018bd_c$

Ultimate shear capacity

The ultimate shear stress, v_u, is given by

$$v_u = V_u/(bd)$$

where V_u is the total shear.

The permitted shear stress, v_c, is given by

$$v_c = \phi[9(f_c')^{1/2} + 2500\ \rho_{max}] \leq 2.28\phi(f_c')^{1/2}$$

where ϕ is the capacity reduction factor and is equal to 0.85.

For a dynamic case, $f_c'(\text{dyn})$ replaces f_c', the static value. Wherever the ultimate shear stress exceeds the shear capacity, v_c, of the concrete, shear reinforcement must be provided to carry the excess, for example stirrups or lacing.

Minimum design stresses

Minimum static stresses are given below and are increased by dynamic increase factors.

Material	Support rotation angle, θ	Stresses (psi)		
		f_y	f_u	f_c'
Reinforcing steel A15 and A432	—	40 000 60 000	70 000 90 000	— —
Concrete	$0° < \theta \leq 2°$	—	—	≥ 2500
	$2° < \theta \leq 12°$	—	—	≥ 3000

B.3.3.3 Typical numerical examples

Typical numerical examples are given in Tables B3.3 to B3.10.

Table B3.3 Explosion resistance of reinforced concrete walls.

Data

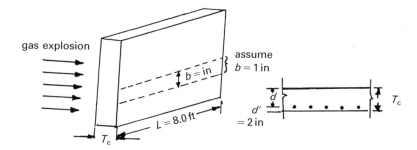

Given:
Length of span, $L = 8.0$ ft (2440 mm)
Thickness, T_c = 9.84 in (250 mm)
Effective depth, $d = 7.84$ in
Assume width, b = 1 in
f_y = 60 000 psi
f'_c = 4000 psi

Dynamic stresses
$f_y(dyn) = 1.10 \times 60\,000 = 66\,000$ psi
$f'_c(dyn) = 1.25 \times 4000 \quad = 5000$ psi

Loading
Max pressure for gas explosion $= 0.026\,89$ N/mm^2
r_u for (DL + IL) and for soil $= 0.014\,00$ N/mm^2
Total, r_u $= 0.040\,89$ N/mm^2
                              $= 0.8537$ kips/ft^2

Ultimate moment of resistance, M_u

$$M_u = (r_u \times L^2)/16 = (0.8537 \times 8^2)/16 = 3.146 \text{ ft kips}$$
$$= 40.992 \text{ in kips}$$

Checking capacity of section

$\rho = \rho_{max}$
$f'_c = 5000$ psi; $K_1 = 0.80$
$\rho_b = 0.85K_1(f'_c(dyn)/f_y(dyn)[87\,000/(87\,000 + f_y(dyn))]$

Substituting

$\rho_b = 0.02929$
$\rho_{max} = 0.75\rho_b = 0.75 \times 0.029\,29 = 0.02197$

Tension reinforcement
A_s required $= \rho_{max} \times b \times d = 0.02197(1)(7.84) = 0.172$ in^2
        $a = (A_s \times f_y(dyn)/(0.85 \times f'_c(dyn) \times b)$
          $= (0.172 \times 66\,000)/(0.85 \times 5000 \times 1) = 2.67$ in
        $\phi =$ strength reduction factor $= 0.9$ *continued*

Table B3.3 *Continued.*

$$M_{max} = \phi A_s \times f_y(\text{dyn})[d - (a/2)]$$
$$= 0.9(0.172)(66)[7.84 - (2.67/2)] = 66.46 \text{ in kips}$$
Since $M_{max} > M_u$, no compression steel needed

Tension steel
 A_s required $= 0.172 \text{ in}^2/\text{in run}$
 Use no #4 bars at 8 in $\rightarrow$ (T12−200)
 A_s provided $= 0.12 \text{ in}^2/\text{in run}$

Minimum area of flexural reinforcement
 Passive design range   = intermediate and low
 Main reinforcement, A_s $= 0.20\% \; b \times d$
                  $= 0.002(1)(7.84)$
                  $= 0.0157 \text{ in}^2 < 0.2 \text{ in}^2 \text{ OK}$
 Second reinforcement, $A_s = 0.12\% \; b \times T_c$
                  $= 0.0012(1)(9.84) = 0.0118 \text{ in}^2$
 Use no #3 bar at 8 in, (R10−200)
 A_s provided $= 0.11 \text{ in}^2 > 0.0118 \text{ in}^2$

Checking for shear
 Assume a unit area
 $V = 0.5F = 0.5(0.854) = 0.427 \text{ kN}$
 Ultimate shear stress
  $v_u = V/(b \times d) = 0.427/(1 \times 7.84)$
     $= 0.0545 \text{ kips/in}^2$
 Permissible shear stress
  $v_c = \phi[1.9f_c'(\text{dyn})^{1/2} + 2500\rho_b]$
     $= 0.85[1.9(5000)^{1/2} + 2500(0.021\,97)]$
     $= 0.17 \text{ kips/in}^2 > 0.0545 \text{ kips/in}^2$
  $v_c > v_u$, no shearing reinforcement is needed

Table B3.4 Explosion resistance design of a reinforced concrete wall/slab with boundary conditions.

Data
Required: ultimate unit resistance of a two-way element, shown below.

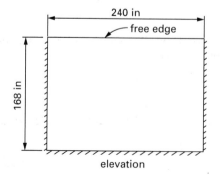

elevation

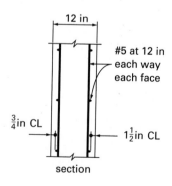

section

Table B3.4 *Continued.*

Solution
Given

   $L = 240$ in (6096 mm), $H = 168$ in (4267 mm)
   $T_c = 12$ in (305 mm)

Reinforcement: #5 at 12 in (T16−300) each way, each face
Three fixed supports capable of developing the moment capacity of the element, and one free edge
$f'_c = 3000$ psi and $f_y = 60\,000$ psi
MF = DIF→concrete = 1.25, reinforcement = 1.10

Dynamic design strengths

$f'_{dc} = \text{DIF } f'_c = 1.25(3000) = 3750$ psi
$f'_{dy} = \text{DIF } f_y = 1.10(60\,000) = 66\,000$ psi

Ultimate moment capacity, M_u
For type I sections, neglecting the small effect of the compression reinforcement, the ultimate moment capacity is given by

$$M_u = (A_s/b)f'_{dy}[d - (a/2)]$$

where $a = (A_s \times f'_{dy})/(0.85f'_{dc} \times b)$
For all sections, $A_s = 0.31\ \text{in}^2/\text{ft}$ for $b = 12$ in
   $a = (0.31 \times 66\,000)/(0.85 \times 3.75 \times 12) = 0.535$ in

(1) Horizontal direction
   • Negative moment, M_{HN}
      $d = 12 - (0.75 + 0.625/2) = 10.9375$ in
      $M_{HN} = (0.31/12)(66\,000)[10.9375 - (0.535/2)) = 18\,190$ in kips
   • Positive moment, M_{HP}
      $d = 12 - (1.50 + 0.625/2) = 10.1875$ in
      $M_{HP} = (0.31/12)(66\,000)[10.1875 - (0.535/2)) = 16\,910$ in kips
(2) Vertical direction
   • Negative movement, M_{VN}
      $d = 12 - (0.75 + 0.625 + 0.625/2) = 10.3125$ in
      $M_{VN} = (0.31/12)(66\,000)[10.3125 - (0.535/2)] = 17\,130$ in kips
   • Positive moment, M_{VP}
      $d = 12 - (1.50 + 0.625 + 0.625/2) = 9.5625$ in
      $M_{VP} = (0.31/12)(66\,000)[9.5625 - (0.535/2)] = 15\,850$ in kips

Ultimate unit resistance, r_u
The ultimate unit resistance of each sector is obtained by taking the summation of the moment about its axis of rotation (support), so that

$$M_{VN} + M_{VP} = R_c = r_u \times A_c$$

(1) Sector I

   $M_{VN} + M_{VP} = 17\,130(120) + 2(2/3)(17\,130)(60) + 15\,850(120) + (2/3)(15\,850)(60)$

   $= 6.6 \times 10^6$ in kips

continued

Table B3.4 *Continued.*

$$r_u \times A_c = r_u(240y/2)(y/3) = 40r_u \times y^2$$

$$\text{therefore,} \rightarrow r_u = (6.6 \times 10^6)/(40 \times y^2) = 164\,900/y^2$$

(2) Sector II

$$M_{HN} + M_{HP} = 18\,190(168 - y/2) = (2/3)(18\,190)(y/2)$$
$$+ 16\,190(168 - y/2) + (2/3)(16\,910)(y/2)$$
$$= 5850(1008 - y)$$

$$r_u \times A_c = 4800r_u(252 - y)$$

$$\text{therefore,} \rightarrow r_u = \frac{5850(1008 - y)}{4800(252 - y)} = \frac{1.219(1008 - y)}{(252 - y)}$$

Equating the ultimate unit resistance of the sectors:

$$164\,900/y^2 = [1.219(1008 - y)]/(252 - y)$$

simplifying, $y^3 - 1008y^2 - 135\,300y + 34\,100\,000 = 0$
and the desired root is $y = 134.8\,\text{in}$

The ultimate unit resistance (Fig. B3.6) is obtained by substituting the value of y into either equation obtained above, both of which yield:

$$r_u = 9.08\,\text{psi}$$

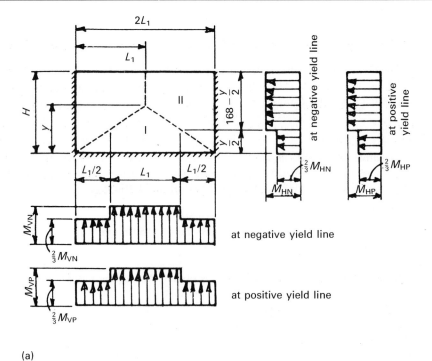

(a)

Fig. B3.6 Determination of ultimate resistance. (a) Assumed yield lines and distribution of moments. (b) free-body diagrams for individual sections.

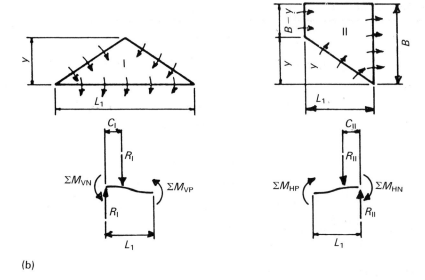

(b)

Fig. B3.6 *Continued.*

B.3.3.4 Design graphs based on the NDRC formula and the finite-element method

Using the NDRC empirical formula, the computer program developed in the Appendix has been used to calculate penetration, perforation and scabbing thicknesses, and these are given in Figs B3.7 to B3.9. A total of 300 slabs were analyzed with different percentages of reinforcement impactor sizes and shapes and velocities. The NDRC formula is compared with the finite-element method. Throughout the analysis, the finite-element mesh scheme adopted in Fig. B3.2 was used, with 8-noded isoparametric elements and with the reinforcement lying in the body as well as on the nodes of each element. A typical example based on these graphs is given below.

Scabbing thickness t_{sc} (mm)

Three limits are designed: $x_p/d < 0.648$, $x_p/d > 0.648$ and $x_p/d > 2$.

(1) $t_{sc}/d = 7.898\,(x_p/d) - 5.058\,(x_p/d)^2$
(2) $x_p/d > 0.648$
   $t_{sp}/d = 2.110 + 1.355\,(x_p/d)$

All others can be incorporated. Graphs are drawn for mass $M = 1\,\mathrm{kg}$ and nose shape $N = 0.72$; modifications may be made for other values of M and N.

Table B3.5 Transformation factors for beams and one-way slabs. (Courtesy of the US Army Design Manual against Weapon Effects.)

Loading diagram	Strain range	Load factor K_L	Mass factor K_M		Load-mass factor K_{LM}		Maximum resistance R_m	Spring constant k	Dynamic reaction V
			Concentrated mass*	Uniform mass	Concentrated mass*	Uniform mass			
simply-supported, $F = pL$	Elastic	0.64	—	0.50	—	0.78	$\dfrac{8M_P}{L}$	$\dfrac{384EI}{5L^3}$	$0.39R + 0.11F$
	Plastic	0.50	—	0.33	—	0.66	$\dfrac{8M_P}{L}$	0	$0.38R_m + 0.12F$
F at $L/2$	Elastic	1.0	1.0	0.49	1.0	0.49	$\dfrac{4M_P}{L}$	$\dfrac{48EI}{L^3}$	$0.78R - 0.28F$
	Plastic	1.0	1.0	0.33	1.0	0.33	$\dfrac{4M_P}{L}$	0	$0.75R_m - 0.25F$
$F/2$, $F/2$ at $L/3$	Elastic	0.87	0.76	0.52	0.87	0.60	$\dfrac{6M_P}{L}$	$\dfrac{56.4EI}{L^3}$	$0.525R - 0.025F$
	Plastic	1.0	1.0	0.56	1.0	0.56	$\dfrac{6M_P}{L}$	0	$0.52R_m - 0.02F$

* Concentrated mass is lumped at the concentrated load.

Table B3.6 Transformation factors for beams and one-way slabs. (Courtesy of the US Army Design Manual against Weapon Effects.)

Loading diagram	Strain range	Load factor K_L	Mass factor K_M		Load-mass factor K_{LM}		Maximum resistance R_m	Spring constant k	Effective spring constant k_E	Dynamic reaction V
			Concentrated mass*	Uniform mass	Concentrated mass*	Uniform mass				
	Elastic	0.53	—	0.41	—	0.77	$\dfrac{12M_{P_s}}{L}$	$\dfrac{384EI}{L^3}$	—	$0.36R + 0.14F$
	Elastic-plastic	0.64	—	0.50	—	0.78	$\dfrac{8}{L}\left(M_{P_s}+M_{P_m}\right)$	$\dfrac{384EI}{5L^3}$	$\dfrac{307EI}{L^3}$	$0.39R + 0.11F$
	Plastic	0.50	—	0.33	—	0.66	$\dfrac{8}{L}\left(M_{P_s}+M_{P_m}\right)$	0	—	$0.38R_m + 0.12F$
	Elastic	1.0	1.0	0.37	1.0	0.37	$\dfrac{4}{L}\left(M_{P_s}+M_{P_m}\right)$	$\dfrac{192EI}{L^3}$	—	$0.71R - 0.21F$
	Plastic	1.0	1.0	0.33	1.0	0.33	$\dfrac{4}{L}\left(M_{P_s}+M_{P_m}\right)$	0	—	$0.75R_m - 0.25F$

fixed ends

$F = pL$ L

F $\dfrac{L}{2}$ $\dfrac{L}{2}$

* Concentrated mass is lumped at the concentrated load.

M_{P_s} = ultimate moment capacity at support.

M_{P_m} = ultimate moment capacity at mid-span.

Table B3.7 Transformation factors for beams and one-way slabs. (Courtesy of the US Army Design Manual against Weapon Effects.)

simply-supported and fixed

Loading diagram	Strain range	Load factor K_L	Mass factor K_M Concentrated mass*	K_M Uniform mass	Load-mass factor K_{LM} Concentrated mass*	K_{LM} Uniform mass	Maximum resistance R_m	Spring constant k	Effective spring constant k_E	Dynamic reaction V
(diagram: $F=pL$, V_1, V_2, L)	Elastic	0.58	—	0.45	—	0.78	$\dfrac{8M_{P_x}}{L}$	$\dfrac{185EI}{L^3}$		$V_1=0.26R+0.12F$ $V_2=0.43R+0.19F$
	Elastic-plastic	0.64	—	0.50	—	0.78	$\dfrac{4}{L}(M_{P_x}+2M_{P_m})$	$\dfrac{384EI}{5L^3}$	$\dfrac{160EI}{L^3}$	$V=0.39R+$ $0.11F\pm M_{P_x}/L$
	Plastic	0.50	—	0.33	—	0.66	$\dfrac{4}{L}(M_{P_x}+2M_{P_m})$	0		$V=0.38R_m+$ $0.12F\pm M_{P_x}/L$
(diagram: F, V_1, V_2, $\frac{L}{2}$, $\frac{L}{2}$)	Elastic	1.0	1.0	0.43	1.0	0.43	$\dfrac{16M_{P_x}}{3L}$	$\dfrac{107EI}{L^3}$		$V_1=0.25R+0.07F$ $V_2=0.54R+0.14F$
	Elastic-plastic	1.0	1.0	0.49	1.0	0.49	$\dfrac{2}{L}(M_{P_x}+2M_{P_m})$	$\dfrac{48EI}{L^3}$	$\dfrac{106EI}{L^3}$	$V=0.78R-$ $0.28F\pm M_{P_x}/L$
	Plastic	1.0	1.0	0.33	1.0	0.33	$\dfrac{2}{L}(M_{P_x}+2M_{P_m})$	0		$V=0.75R_m-$ $0.25F\pm M_{P_x}/L$
(diagram: $\frac{F}{2}$, $\frac{F}{2}$, V_1, V_2, $\frac{L}{3}$, $\frac{L}{3}$, $\frac{L}{3}$)	Elastic	0.81	0.67	0.45	0.83	0.55	$\dfrac{6M_{P_x}}{L}$	$\dfrac{132EI}{L^3}$		$V_1=0.17R+0.17F$ $V_2=0.33R+0.33F$
	Elastic-plastic	0.87	0.76	0.52	0.87	0.60	$\dfrac{2}{L}(M_{P_x}+3M_{P_m})$	$\dfrac{56EI}{L^3}$	$\dfrac{122EI}{L^3}$	$V=0.525R-$ $0.025F\pm M_{P_x}/L$
	Plastic	1.0	1.0	0.56	1.0	0.56	$\dfrac{2}{L}(M_{P_x}+3M_{P_m})$	—		$V=0.52R_m-$ $0.02F\pm M_{P_x}/L$

* Equal parts of the concentrated mass are lumped at each concentrated load.
M_{P_x} = ultimate bending capacity at support. M_{P_m} = ultimate positive bending capacity.

Table B3.8 Transformation factors for two-way slabs: simple supports — four sides, uniform load. (Courtesy of the US Army Design Manual against Weapon Effects.)

Strain range	a/b	Load factor K_L	Mass factor K_M	Load-mass factor K_{LM}	Maximum resistance	Spring constant k	Dynamic reactions V_A	Dynamic reactions V_B
Elastic	1.0	0.46	0.31	0.67	$\frac{12}{a}(M_{P_{fa}} + M_{P_{fb}})$	$\dfrac{252EI_a}{a^2}$	$0.07F + 0.18R$	$0.07F + 0.18R$
	0.9	0.47	0.33	0.70	$\frac{1}{a}(12M_{P_{fa}} + 11M_{P_{fb}})$	$\dfrac{230EI_a}{a^2}$	$0.06F + 0.16R$	$0.08F + 0.20R$
	0.8	0.49	0.35	0.71	$\frac{1}{a}(12M_{P_{fa}} + 10.3M_{P_{fb}})$	$\dfrac{212EI_a}{a^2}$	$0.06F + 0.14R$	$0.08F + 0.22R$
	0.7	0.51	0.37	0.73	$\frac{1}{a}(12M_{P_{fa}} + 9.8M_{P_{fb}})$	$\dfrac{201EI_a}{a^2}$	$0.05F + 0.13R$	$0.08F + 0.24R$
	0.6	0.53	0.39	0.74	$\frac{1}{a}(12M_{P_{fa}} + 9.3M_{P_{fb}})$	$\dfrac{197EI_a}{a^2}$	$0.04F + 0.11R$	$0.09F + 0.26R$
	0.5	0.55	0.41	0.75	$\frac{1}{a}(12M_{P_{fa}} + 9.0M_{P_{fb}})$	$\dfrac{201EI_a}{a^2}$	$0.04F + 0.09R$	$0.09F + 0.28R$
Plastic	1.0	0.33	0.17	0.51	$\frac{12}{a}(M_{P_{fa}} + M_{P_{fb}})$	0	$0.09F + 0.16R_m$	$0.09F + 0.16R_m$
	0.9	0.35	0.18	0.51	$\frac{1}{a}(12M_{P_{fa}} + 11M_{P_{fb}})$	0	$0.08F + 0.15R_m$	$0.09F + 0.18R_m$
	0.8	0.37	0.20	0.54	$\frac{1}{a}(12M_{P_{fa}} + 10.3M_{P_{fb}})$	0	$0.07F + 0.13R_m$	$0.10F + 0.20R_m$
	0.7	0.38	0.22	0.58	$\frac{1}{a}(12M_{P_{fa}} + 9.8M_{P_{fb}})$	0	$0.06F + 0.12R_m$	$0.10F + 0.22R_m$
	0.6	0.40	0.23	0.58	$\frac{1}{a}(12M_{P_{fa}} + 9.3M_{P_{fb}})$	0	$0.05F + 0.10R_m$	$0.10F + 0.25R_m$
	0.5	0.42	0.25	0.59	$\frac{1}{a}(12M_{P_{fa}} + 9.0M_{P_{fb}})$	0	$0.04F + 0.08R_m$	$0.11F + 0.27R_m$

V_A = total dynamic reaction along short edge; V_B = total dynamic reaction along long edge.
$M_{P_{fa}}$ = total positive ultimate moment capacity along mid-span section parallel to short edge.
$M_{P_{fb}}$ = total positive ultimate moment capacity along mid-span section parallel to long edge.

Table B3.9 Transformation factors for two-way slabs: fixed four sides, uniform load. (Courtesy of the US Army Design Manual against Weapon Effects.)

Strain range	a/b	Load factor K_L	Mass factor K_M	Load-mass factor K_{LM}	Maximum resistance	Spring constant k	Dynamic reactions V_A	V_B
Elastic	1.0	0.33	0.21	0.63	$29.2M^s_{P_{sb}}$	$810EI_a/a^2$	$0.10F + 0.15R$	$0.10F + 0.15R$
	0.9	0.34	0.23	0.68	$27.4M^s_{P_{sb}}$	$742EI_a/a^2$	$0.09F + 0.14R$	$0.10F + 0.17R$
	0.8	0.36	0.25	0.69	$26.4M^s_{P_{sb}}$	$705EI_a/a^2$	$0.08F + 0.12R$	$0.11F + 0.19R$
	0.7	0.38	0.27	0.71	$26.2M^s_{P_{sb}}$	$692EI_a/a^2$	$0.07F + 0.11R$	$0.11F + 0.21R$
	0.6	0.41	0.29	0.71	$27.3M^s_{P_{sb}}$	$724EI_a/a^2$	$0.06F + 0.09R$	$0.12F + 0.23R$
	0.5	0.43	0.31	0.72	$30.2M^s_{P_{sb}}$	$806EI_a/a^2$	$0.05F + 0.08R$	$0.12F + 0.25R$
Elastic-plastic	1.0	0.46	0.31	0.67	$(1/a)\,[12(M_{P_{fa}} + M_{P_{sa}}) + 12(M_{P_{fb}} + M_{P_{sb}})]$	$252EI_a/a^2$	$0.07F + 0.18R$	$0.07F + 0.18R$
	0.9	0.47	0.33	0.70	$(1/a)\,[12(M_{P_{fa}} + M_{P_{sa}}) + 11(M_{P_{fb}} + M_{P_{sb}})]$	$230EI_a/a^2$	$0.06F + 0.16R$	$0.08F + 0.20R$
	0.8	0.49	0.35	0.71	$(1/a)\,[12(M_{P_{fa}} + M_{P_{sa}}) + 10.3(M_{P_{fb}} + M_{P_{sb}})]$	$212EI_a/a^2$	$0.06F + 0.14R$	$0.08F + 0.22R$
	0.7	0.51	0.37	0.73	$(1/a)\,[12(M_{P_{fa}} + M_{P_{sa}}) + 9.8(M_{P_{fb}} + M_{P_{sb}})]$	$201EI_a/a^2$	$0.05F + 0.13R$	$0.08F + 0.24R$
	0.6	0.53	0.39	0.74	$(1/a)\,[12(M_{P_{fa}} + M_{P_{sa}}) + 9.3(M_{P_{fb}} + M_{P_{sb}})]$	$197EI_a/a^2$	$0.04F + 0.11R$	$0.09F + 0.26R$
	0.5	0.55	0.41	0.75	$(1/a)\,[12(M_{P_{fa}} + M_{P_{sa}}) + 9.0(M_{P_{fb}} + M_{P_{sb}})]$	$201EI_a/a^2$	$0.04F + 0.09R$	$0.09F + 0.28R$
Plastic	1.0	0.33	0.17	0.51	$(1/a)\,[12(M_{P_{fa}} + M_{P_{sa}}) + 12(M_{P_{fb}} + M_{P_{sb}})]$	0	$0.09F + 0.16R_m$	$0.09F + 0.16R_m$
	0.9	0.35	0.18	0.51	$(1/a)\,[12(M_{P_{fa}} + M_{P_{sa}}) + 11(M_{P_{fb}} + M_{P_{sb}})]$	0	$0.08F + 0.15R_m$	$0.09F + 0.18R_m$
	0.8	0.37	0.20	0.54	$(1/a)\,[12(M_{P_{fa}} + M_{P_{sa}}) + 10.3(M_{P_{fb}} + M_{P_{sb}})]$	0	$0.07F + 0.13R_m$	$0.10F + 0.20R_m$
	0.7	0.38	0.22	0.58	$(1/a)\,[12(M_{P_{fa}} + M_{P_{sa}}) + 9.8(M_{P_{fb}} + M_{P_{sb}})]$	0	$0.06F + 0.12R_m$	$0.10F + 0.22R_m$
	0.6	0.40	0.23	0.58	$(1/a)\,[12(M_{P_{fa}} + M_{P_{sa}}) + 9.3(M_{P_{fb}} + M_{P_{sb}})]$	0	$0.05F + 0.10R_m$	$0.10F + 0.25R_m$
	0.5	0.42	0.25	0.59	$(1/a)\,[12(M_{P_{fa}} + M_{P_{sa}}) + 9.0(M_{P_{fb}} + M_{P_{sb}})]$	0	$0.04F + 0.08R_m$	$0.11F + 0.27R_m$

$M_{P_{fa}}$ = total positive ultimate moment capacity along mid-span section parallel to short edge.

$M_{P_{sa}}$ = total negative ultimate moment capacity along short edge.

$M_{P_{sb}}$ = total negative ultimate moment capacity per unit width at centre of long edge.

$M_{P_{fb}}$ = total positive ultimate moment capacity along mid-span section parallel to long edge.

V_A = total dynamic reaction along short edge; V_B = total dynamic reaction along long edge.

Table B3.10 Design aids based on yield line techniques.

Slab type	Load	Moment–load relationship
A1 simply-supported at 3 columns	n/unit area	$$m = \frac{n \times \text{area of triangle}}{\sqrt{3}} \left[\frac{7}{18} - \frac{a_v}{h}\right]$$ $$m' = (1/6)na_v^2$$
B1 simply-supported at 4 edges	n/unit area over two rectangles	$$m = \frac{nl_1l_2}{6} \times \frac{l_y - a_v}{2l_y - a_v} \times \frac{2l_y - l_2}{l_y} \times \frac{2l_x - 1}{l_x}$$
C1 simply-supported at 4 edges	n/unit area over two rectangles	$$m = \frac{F}{12} \times \frac{l_y + l_2 - a_v}{l_y + l_2} \times \frac{2l_y - l_2}{l_y} \times \frac{2l_x - 1}{l_x}$$ where $F = n(l_1l_2 + \tfrac{1}{2}l_1l_2)$

continued

Table B3.10 *Continued.*

Slab type	Load	Moment–load relationship
rectangular slab fixed at both edges supported on a column D1 $m'_1 = \mu_1 m$; $m'_2 = \mu_2 m$	n/unit area	$$m = \frac{\left[\dfrac{3nl_x l_y}{8\left(2 + \dfrac{l_y}{l_x} + \dfrac{l_x}{l_y}\right)}\right] - 0.15nc_1 d_1}{1 + 0.6\,\dfrac{\mu_1(c_1 + l_y) + \mu_2(l_x + d_1)}{(l_x + l_y + c_1 + d_1)}}$$ $= \bar{K}$ V = reaction at column $= [(0.22nl_x l_y/(1+\bar{K})] + \tfrac{1}{2}n(l_y d_1 + l_x c_1) + 2.5nc_1 d_1$
rectangular slab fixed at one edge supported on two columns E1 $m' = \mu m$	n/unit area	$$m = \frac{nl_x l_y}{18 + 12\mu}\left(\frac{5 + 2\mu}{4}\,\frac{l_x}{l_y} + 1\right)$$ where $(l_y/l_x) < 2 - (\mu/2)$ $\mu \le 1$ $m' = 0;\ m = \dfrac{nl_x l_y}{18}\left(\dfrac{5}{4}\dfrac{l_x}{l_y} + 1\right)$ simply supported where $(l_y/l_x) < 2$ for a square slab: $m' = 0;\ m = nl_x^2/8$
slab supported on four columns A2	n/unit area	$m_1 = (1/8)n(l_y^2 - 4c_1^2)$ $m_2 = (1/8)n(l_x^2 - 4d_1^2)$ $m'_1 = (1/2)nc_1^2;\ m'_2 = (1/2)nd_1^2$

B2

n/unit area

rectangular skew slabs supported at two edges and free at two edges

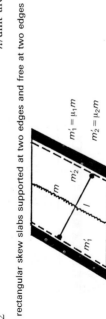

$m_1' = \mu_1 m$

$m_2' = \mu_2 m$

$$m = \frac{n}{8} \left\{ \frac{4l^2}{[\sqrt{(1+\mu_1)} + \sqrt{(1+\mu_2)}]^2} \right\}$$

C2

n/unit area

rectangular skew slabs fixed at all edges

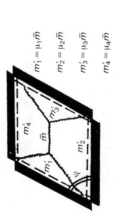

$m_1' = \mu_1 \bar{m}$

$m_2' = \mu_2 \bar{m}$

$m_3' = \mu_3 \bar{m}$

$m_4' = \mu_4 \bar{m}$

$\bar{m} = m(1.75 - \sin\psi)$ with $\psi \geq 30°$

where *m* in this case is the moment for a rectangular slab with sides at 90°

D2

n/unit area

trapezoidal slabs fixed at all supports and edges

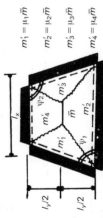

$m_1' = \mu_1 \bar{m}$

$m_2' = \mu_2 \bar{m}$

$m_3' = \mu_3 \bar{m}$

$m_4' = \mu_4 \bar{m}$

$m = m[1.75 - (1/3)(\sin\psi_1 + \sin\psi_2)]$

continued

Table B3.10 *Continued.*

Slab type	Load	Moment–load relationship
rectangular slabs fixed at two edges and free at two edges		
E2	n/unit area plus F at centre	$m = (nl^2/8) + (F/4\alpha)$
F2	n/unit area plus F at centre	$m = \dfrac{1}{\cos^2(90-\psi)}\left(\dfrac{nl_x^2}{8} + \dfrac{F}{4\alpha}\right)$

A3

circular slab supported on n_1 number of columns placed at a radius r_1

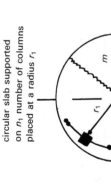

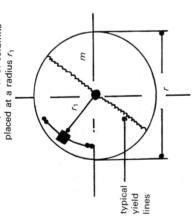

typical yield lines

n/unit area

$$m = nr^2 \left[\frac{r_1}{2r} \times \frac{1}{\dfrac{\sin\pi/n}{\pi/n}} - \frac{1}{3} \right]$$

for $r_1 = r$, i.e. columns at the rim

$$m = nr^2/6$$

B3

circular slabs rigidly supported at inner and outer circumference

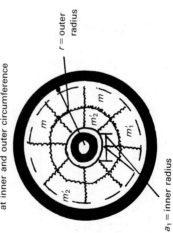

r = outer radius

a_1 = inner radius

n/unit area

$$m_1' + m_2' + 2m = n \left[\frac{(a_1 - r)^2}{15} \left(\frac{a_1}{r} + 2.75 \right) \right]$$

$a_1 = a/2$; a = inner radius

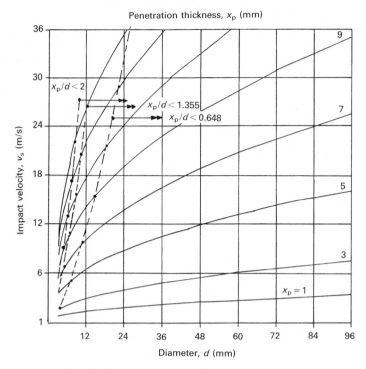

Fig. B3.7 Penetration depth versus v_s and d.

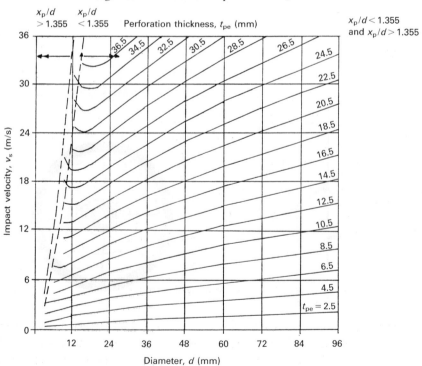

Fig. B3.8 Perforation thickness versus v_s and d.

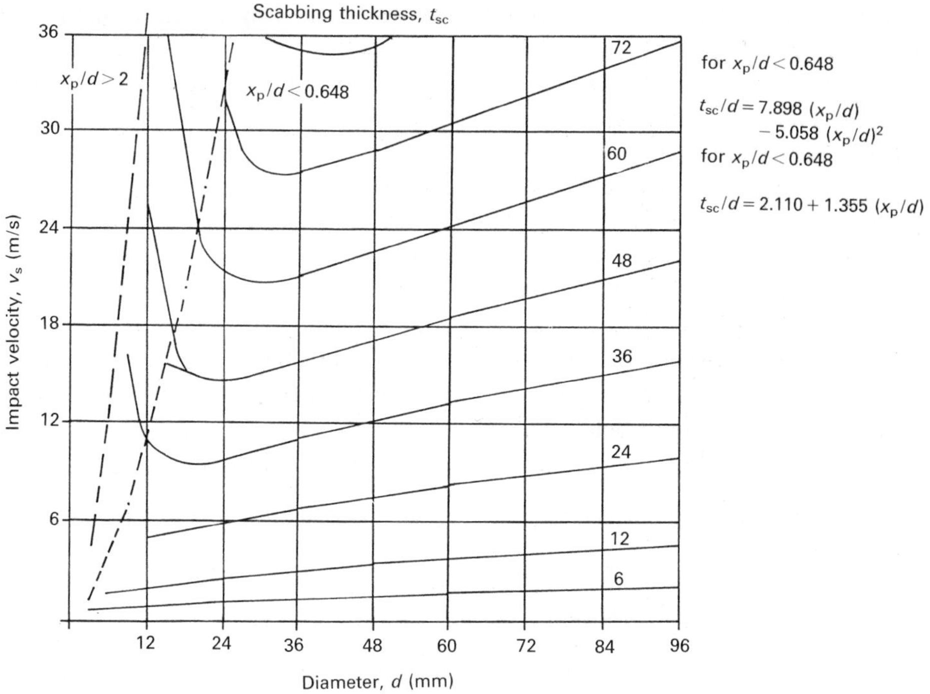

Fig. B3.9 Scabbing thickness versus v_s and d.

Penetration thickness x_p (mm)

Two limits are defined: $x_p/d < 0.648$ and $x_p/d < 2$. Graphs are drawn for $M = 1$ kg and $N = 0.72$; again, modifications are made for other values of M and N.

$$G(x_p/d) = 9.570NM \times \sqrt{(f_c'd^{-2.8}V^{1.8})}$$
$$f_c' = 35\ \text{N/mm}^2$$

Perforation thickness t_{pe} *(mm)*

Two limits are defined: $x_p/d \lesseqgtr 1.349$ and $x_p/d > 2$.

$$x_p/d < 1.349 \qquad t_{pe}/d = 3.18\,(x_p/d) + 7.18\,(x_p/d)^2$$

$$x_p/d > 1.349 \qquad t_{pe}/d = 1.28\,(x_p/d) + 1.3$$

B.4 Impact/explosion at roadways and runways

Concrete slabs resting on a specially prepared sub-grade of a roadway or runway may be subjected to an impact/explosion, caused by broken vehicles, aircraft crashes and missiles exploding on impact. In the Gulf War, runways were subjected to both direct hits by bombs and fragment impact following air bursts. It was noticed that after an explosion, the runway was broken up by the formation of front and back craters.

Figure B4.1 shows the representation of a runway by a typical finite-element mesh of 8-noded isoparametric solid elements. Three bombs were selected: 500 lb, 1000 lb and 2000 lb. The sub-grade was assumed to have a clay content in the range of 0 to 30%. Prior to explosion, the impact was treated as a repeated one. Figure B4.2 shows the load–time function for 12 repeated loads. Figure B4.3 illustrates the deformation of a portion of the runway against time. Figure B4.4 shows the damage scenario. A comparative study is shown in Fig. B4.5 for velocity and displacement of the runway. The same finite-element mesh was adopted for an explosion of a 500 lb bomb falling from 100 m. Figure B4.6 gives a summary of the results, indicating the relationship between slab thickness and penetration of the bomb below the slabs.

The following additional data formed the basis of the work:

Total number of 8-noded isoparametric elements of $\frac{1}{2}$ km by $\frac{1}{2}$ km runway area: 3000 nodes/7 layers deep
Degrees of freedom: 3/node of 23 500 nodes
Concrete strength, $f'_c = 35 \, \text{N/mm}^2$

The bulk modulus approach for soils was adopted, together with the endochronic theory for concrete. The bomb/missile velocity varied up to 3 m/s.

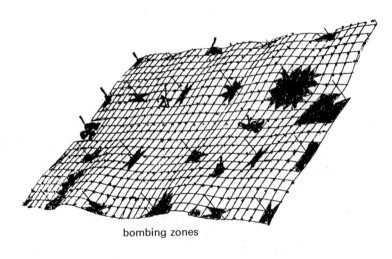

bombing zones

7 layers deep

Fig. B4.1 Finite-element mesh scheme of a runway.

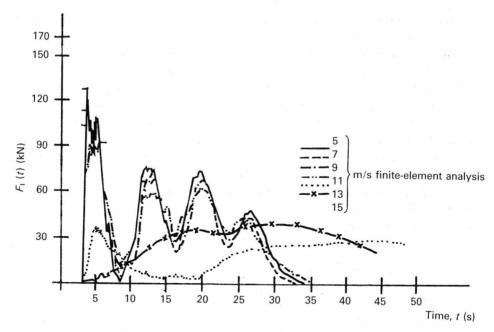

Fig. B4.2 Load−time function for a 1000 lb bomb.

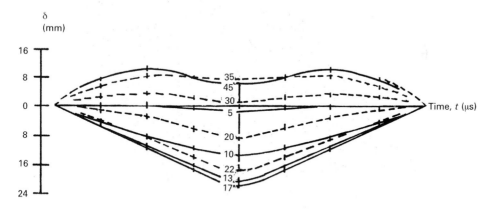

Fig. B4.3 Deformation−time relationship for a runway slab.

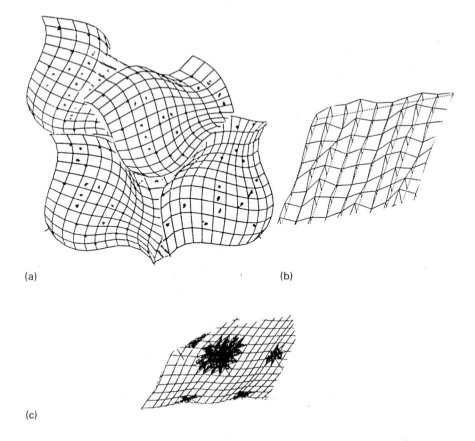

(a)   (b)

(c)

Fig. B4.4 The damage scenario: (a) destruction phenomenon; (b) deformation along the thickness; (c) key diagram.

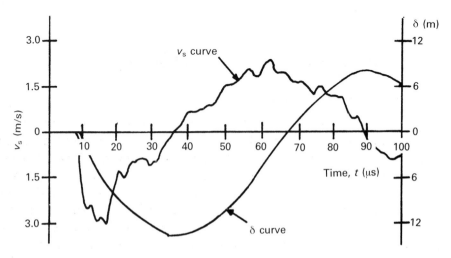

Fig. B4.5 Comparative study of velocity and displacement for a runway (v_s = velocity; δ = displacement).

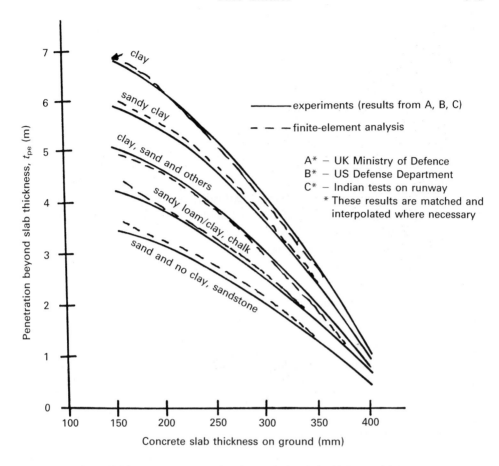

Fig. B4.6 Slab thickness and penetration beyond the slab thickness (slab concrete grade (35 N/mm²)).

B.5 Buildings and structures subject to blast loads

B.5.1 Reinforced concrete, single-storey house

A single-storey house, idealized as a box-type structure, is examined under a blast load in open air. In the ISOPAR program, the distance r of a blast load is kept variable against the weapon effect W. It is assumed that the interior of the house has enclosures and the walls have openings for doors and windows. The percentage of reinforcement varies from $1\frac{1}{2}$ to 4% of the volume of concrete. The thickness T_c and the internal height h (or H) vary. The concrete strength or grade is assumed to be 50 N/mm² — this strength is normally suitable for concrete under explosive effects. For economical reasons, the finite-element mesh schemes prepared for beams, walls and slabs discussed earlier are kept for this exercise. The following data support this exercise:

(1) Basic data

Number of beams: 6

Foundation springs with dashpots, variable soils: 300

saturated sand: $K = 438\,\text{kN/m}$

sandy clay: $K = 292\,\text{kN/m}$

gravelly soil: $K = 146\,\text{kN/m}$

Distance from the explosion, r: 3 m, 6 m, 9 m

Bomb weights: 500 lb, 1000 lb

Bomb velocity, v_{so}: 200 m/s

(2) Finite-element data

8-noded isoparametric elements: 1500/wall, 750/slab

Reinforcement placed 2-noded isoparametric on elemental nodes and the body of the element: 2100/2-node

Reinforcement has lacings: 300 lacings, 2-noded

Gap elements: 300

Increments: 20

Newmark solution implicit type

Execution time: 45 minutes 31 seconds

Figure B5.1 shows the relationship between T_c and r. The wall thickness necessary to avoid damage (perforation, spalling, etc.) for various blast loads and distances, as shown in Fig. B5.1, cannot easily be achieved if the point lies under the bottom curve. For example, a 1000 lb bomb at a distance of 1.5 m and $T_c = 1.5$ m can cause severe cracking to this wall. This is shown in Fig. B5.2 for areas close to doors and windows.

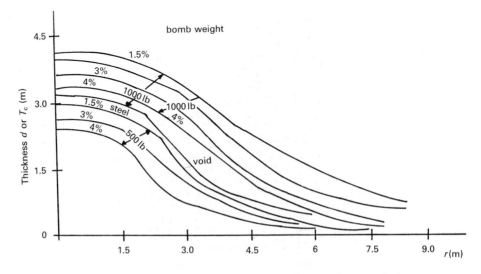

Fig. B5.1 Concrete/wall slab thickness versus distance from explosion, r.

Fig. B5.2 Cracking of walls with doors and windows.

B.5.2 Blast loads in the demolition of buildings and cooling towers

This text gives a comprehensive treatment of blast loads and their effects on structures. Figure B5.3 shows the finite-element mesh scheme for a high-rise, reinforced concrete building complex with 400 bombs, each containing 50 lb gelignite, placed along the perimeter. It is assumed that all the devices are

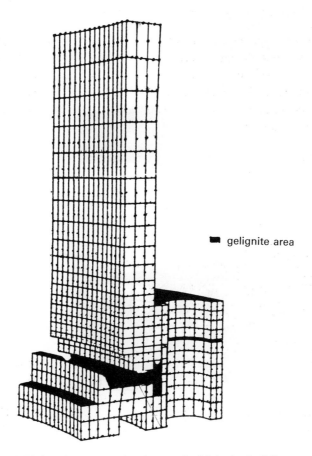

■ gelignite area

Fig. B5.3 Finite-element mesh scheme of a high-rise building.

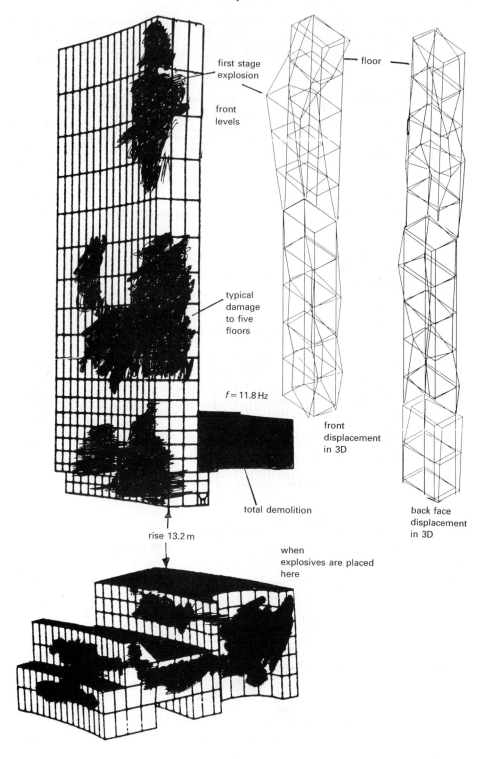

first stage
explosion

front
levels

floor

typical
damage
to five
floors

$f = 11.8\,\text{Hz}$

front
displacement
in 3D

total demolition

back face
displacement
in 3D

rise 13.2 m

when
explosives are placed
here

Fig. B5.4 Demolished state.

detonated at the same time. Figure B5.4 shows the final demolition post-mortem of this building. In practice, demolition is carried out in sequence at appropriate levels and in stages. This can easily be handled by the finite-element method.

B.5.3 Impact and explosion of cooling towers and chimneys

Table B5.1 gives the basic data for a cooling tower or a power station. Five hundred 50 lb packages of Semtex are employed in the demolition scenario. Figure B5.5 is a photograph of a cooling tower with the parameters given in Table B5.1. The finite-element prediction is very similar to the one produced by the South of Scotland Electricity Board which is shown in Fig. B5.6(a). A comparative study is given in Fig. B5.6(b).

Fig. B5.5 Balunpur cooling tower.

Table B5.1 Data for the cooling tower.

Total height = 85.3 m

Height H (m)	Diameter D (m)
0	66.4
33.7	49.2
65.4	42.6
85.3	45.4

Self-weight/unit area to be computed
Thickness: 150 mm
β values: $-15, -5, 5, 15, 25, 35, 45$ degrees
Number of 8-noded solid isoparametric elements: 1900
Number of 2-noded line elements: 1500 at nodes, 300 in the body

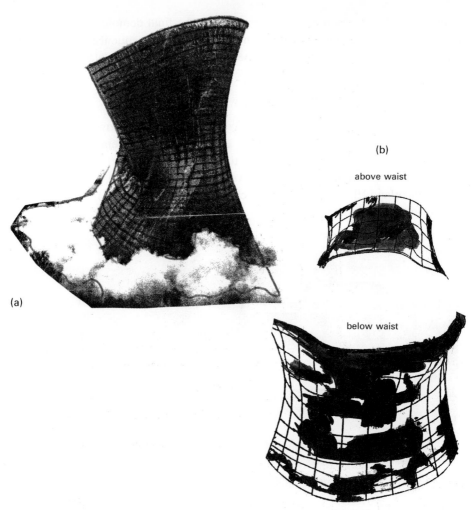

Fig. B5.6 (a) Post-mortem of a cooling tower. (b) Post-mortem of a cooling tower − finite-element approach.

Chimneys can be dealt with in a similar manner with a modification to the finite-element mesh scheme to suit the chimneys' shape and the loads acting on them. This type of work is widely reported.

B.6 Aircraft crashes on containment vessels (buildings)

Several studies have been carried out on containment vessels subjected to aircraft impact. The containments were treated both in isolation and together with surrounding buildings. The overall response with the surrounding buildings is considered. Figure B6.1 shows the load−time function for a number of aircraft prepared by the author using data given in Chapter 2.

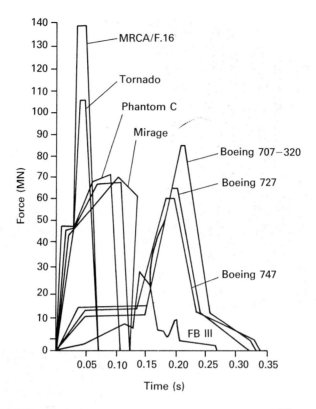

Fig. **B6.1** Force–time function. (From Bangash.[1.149])

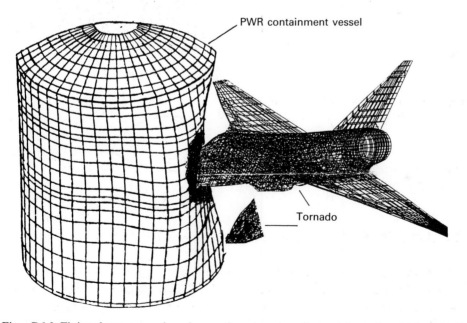

Fig. **B6.2** Finite-element mesh scheme for an aircraft–containment vessel impact phenomenon.

The author investigated the problems associated with the concrete containment vessels. Two existing reinforced and pre-stressed concrete containments were examined against extreme loads in elastic and cracking conditions. Figure B6.2 shows a three-dimensional, finite-element mesh. The analysis explored the effect of an aircraft impact on the Sizewell B vessel. Figure B6.3 shows crack propagations on both the interior and exterior surfaces of the vessel caused by the impact of a number of aircraft. The linear and non-linear displacements for impacts due to four aircraft are shown in Fig. B6.4. Displacement−time relationships using a Tornado aircraft for the linear and non-linear cases are illustrated in Fig. B6.5 for the Sizewell B vessel. Throughout the analysis, three-dimensional, 20-noded solid isoparametric elements were used for both the aircraft and the vessel concrete. The pre-stressing tendons/cables and steel liner were represented by 4- and 8-noded three-dimensional line and plate elements, respectively. Linkage

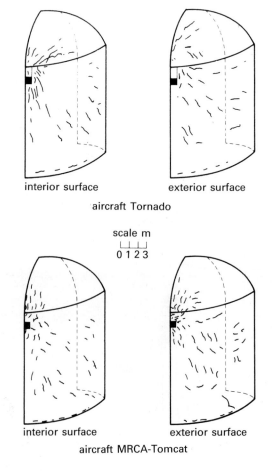

interior surface   exterior surface

aircraft Tornado

scale m
0 1 2 3

interior surface   exterior surface

aircraft MRCA-Tomcat

Fig. B6.3 A post-mortem of the containment vessel. (From Bangash[3.3])

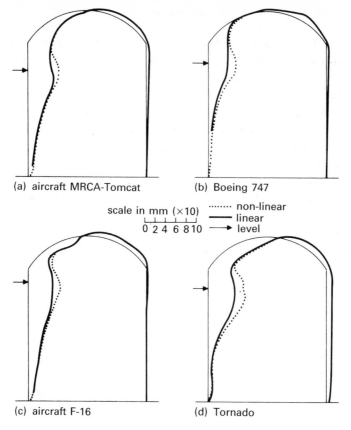

(a) aircraft MRCA-Tomcat   (b) Boeing 747

scale in mm (×10)
0 2 4 6 8 10

·········· non-linear
———— linear
———➤ level

(c) aircraft F-16   (d) Tornado

Fig. B6.4 Displacements due to impact. (From Bangash[3.3])

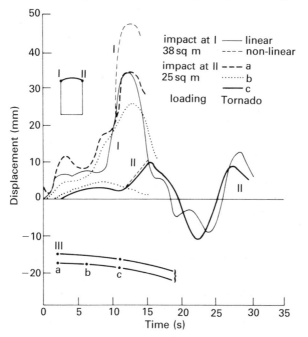

impact at I   —— linear
38 sq m   ---- non-linear

impact at II   --- a
25 sq m   ····· b
                 —— c

loading   Tornado

Fig. B6.5 Displacement−time relationship using a Tornado aircraft.

elements were included at appropriate places along the tendon layouts and the anchor positions between the concrete of the vessel and the steel liner placed on the vessel interior. The vessel boundary conditions included the general building infrastructure surrounding the reactor island. The tension criteria for cracking are given earlier, together with the general endochronic theory for concrete.

Crutzen used the semi-loof shell element to examine the damage caused to the containment vessel by a Phantom RF-4E aircraft. The original and deformed meshes are shown in Fig. B6.6. Cracking at the interior and exterior surfaces has been predicted. A similar exercise was carried out by Rebora using three-dimensional, isoparametric, 20-noded solid elements. A non-linear analysis of the vessel under the impact of a Boeing 707−320 aircraft was performed. A typical post-mortem examination of the vessel is shown in Fig. B6.7.

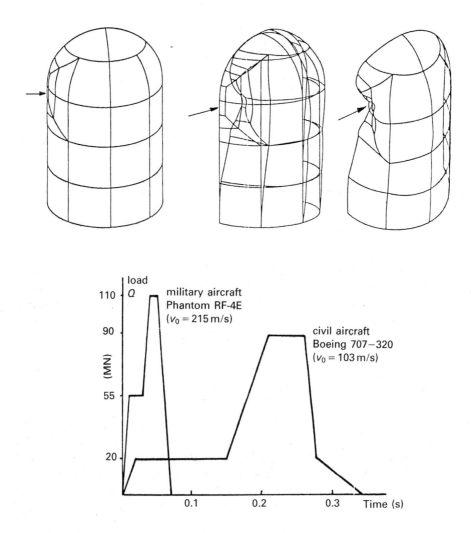

Fig. B6.6 Original and deformed finite-element meshes after impact. (Courtesy of Kraftework Union.)

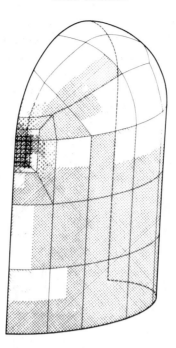

Fig. B6.7 Post-mortem examination of a shell (Rebora model). (Courtesy of Kraftework Union.)

C: BRICKWORK

C.1 General introduction

Gas explosions in buildings are a rare occurrence. Information on both the causes and effects of a gas explosion and some practical guidance on safe venting have already been discussed in this text. However, the importance of the structure's ability to withstand the blast load cannot be ignored. In this section, a brief analysis is given on impact and explosion in brickwork construction.

C.2 Finite-element analysis of explosion

Figure C2.1 shows a 7-storey building with 49 compartments. The compartments have external walls of thickness 450 mm and the thickness of the internal walls ranges between 225 and 362.5 mm. Thirteen vents are available in compartment/ flat no 32. The information on a typical 3-wall cubicle, shown in Fig. C2.2, is taken from flat no 32, where an explosion due to a gas leak occurs. A typical external wall of a cubicle is shown in Fig. C2.3. A gas leak pressure is generated in compartment/flat no 32 using the finite-element scheme given in Fig. C2.4. A pressure rise of 75.84 kN/m^2 was generated for a rise time of 0.2 second; various empirical expressions were used to calculate this pressure. Table C2.1 lists the

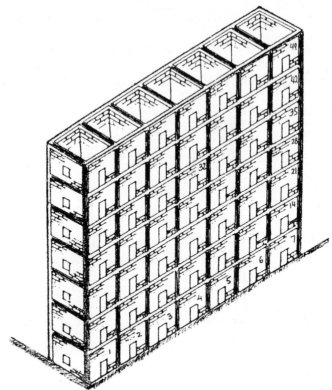

Fig. C2.1 A typical 7-storey building in brickwork.

Table C2.1 Pressure generated in a vented explosion.

```
C       THIS IS A FORTRAN PROGRAM USED FOR THE CALCULATION OF THE
C       PREDICTION OF PRESSURE GENERATED IN A VENTED CONFINED GAS
C       EXPLOSION. A FEW METHODS OF CALCULATION WERE COMPUTED AND
C       COMPARED WITH EACH OTHER, THE HIGHEST VALUE OF THE PRESSURE
C       WAS THEN CHOSEN FOR THE DESIGN PURPOSE.
C
C       K E Y S:
C
C       P = BREAKING PRESSURE OF THE RELIEF PANEL
C       S = BURNING VELOCITY OF THE GAS INVOLVED
C       W = WEIGHT PER UNIT AREA
C       A = AREA OF RELIEF PANEL
C       B = AREA OF THE SMALLEST SECTION OF THE ROOM
C       V = VOLUME OF THE ROOM
C
C     ~~~~~~~~~~~~~~~~~~~~~~~~~~~~~~~~~~~~~~~~~~~~~~~~~~~~~~~~~~~~~~~~~~~~~~
        REAL P,S,A,B,V,W
        WRITE (6,10)
   10   FORMAT (2X,'INSERT THE VALUE OF BREAKING PRESSURE,P (mbar)')
        READ (5,*)P
        WRITE (6,20)
   20   FORMAT (2X,'INSERT THE VALUE OF BURNING VELOCITY,S (m/s^2)')
        READ (5,*)S
        WRITE (6,30)
   30   FORMAT (2X,'INSERT THE AREA OF RELIEF PANEL, A (m^2)')
        READ (5,*)A
        WRITE (6,40)
```

```
40        FORMAT (2X,'INSERT THE AREA OF WALL WHERE R.P LOCATED, B (m^2)')
          READ (5,*)B
          WRITE (6,50)
50        FORMAT (2X,'INSERT THE VOLUME OF THE SECTION , V (m^3)')
          READ (5,*)V
          WRITE (6,60)
60        FORMAT (2X,'INSERT THE WEIGHT PER UNIT AREA OF R.P , W (Kg/m^2)')
          READ (5,*)W

C    TO CALCULATE THE VALUE OF VENT COVER COEFFICIENT
C    ~~~~~~~~~~~~~~~~~~~~~~~~~~~~~~~~~~~~~~~~~~~~~~~~~
          K1=B/A
          K2=V**(2/3)/A
C    TO CALCULATE THE MAXIMUM PRESSURE GENERATED BY CUBBAGE & SIMMONDS
C    ~~~~~~~~~~~~~~~~~~~~~~~~~~~~~~~~~~~~~~~~~~~~~~~~~~~~~~~~~~~~~~~~~~~
          X1=(S*4.3*K1*W)+28
          X2=V**(1/3)
          P1=X1/X2
          WRITE (6,70)P1
70        FORMAT (/,2X,'THE VALUE OF MAXIMUM PRESSURE BY CS IS : ',F9.1)
          P12=58*S*K1
          WRITE (6,70)P12
C
C    TO CALCULATE THE MAXIMUM PRESSURE GENERATED BY CUBBAGE & MARSHALL
C    ~~~~~~~~~~~~~~~~~~~~~~~~~~~~~~~~~~~~~~~~~~~~~~~~~~~~~~~~~~~~~~~~~~~
          X1=S**2*K1*W
          X2=V**(1/3)
          P2=(P+(23*(X1/X2)))
          WRITE (6,80)P2
80        FORMAT (/,2X,'THE VALUE OF MAXIMUM PRESSURE BY CM IS : ',F9.1)
C
C    TO CALCULATE THE MAXIMUM PRESSURE GENERATED BY RASBASH 1
C    ~~~~~~~~~~~~~~~~~~~~~~~~~~~~~~~~~~~~~~~~~~~~~~~~~~~~~~~~~~
          X1=(1.5*P)
          X2=(77.7*K1*S)
          P3=(X1+X2)
          WRITE (6,90)P3
90        FORMAT (/,2X,'THE VALUE OF MAXIMUM PRESSURE BY R1 IS : ',F9.1)
C
C    TO CALCULATE THE MAXIMUM PRESSURE GENERATED BY RASBASH 2
C    ~~~~~~~~~~~~~~~~~~~~~~~~~~~~~~~~~~~~~~~~~~~~~~~~~~~~~~~~~~
          X1=1.5*P
          X2=(4.3*K1*W)+28
          X3=V**(1/3)
          X4=77.7*K1
          P4=X1+S*((X2/X3)+X4)
          WRITE (6,100)P4
100       FORMAT (/,2X,'THE VALUE OF MAXIMUM PRESSURE BY R2 IS : ',F9.1)
C
C    TO COMPARE ALL THE VALUES OF MAXIMUM PRESSURE GENERATED AND CHOOSING
C    ~~~~~~~~~~~~~~~~~~~~~~~~~~~~~~~~~~~~~~~~~~~~~~~~~~~~~~~~~~~~~~~~~~~~~~
C              THE MAXIMUM VALUE FROM THE FOUR METHOD
C              ~~~~~~~~~~~~~~~~~~~~~~~~~~~~~~~~~~~~~~~
          IF (P1.GT.P2.AND.P1.GT.P3.AND.P1.GT.P4) THEN
          WRITE (6,120)P1
120       FORMAT (///,2X,'THE HIGHEST VALUE IS BY CS : ',F9.1)
          ENDIF
          IF (P2.GT.P1.AND.P2.GT.P3.AND.P2.GT.P4) THEN
          WRITE (6,130)P2
130       FORMAT (///,2X,'THE HIGHEST VALUE IS BY CM : ',F9.1)
          ENDIF
          IF (P3.GT.P1.AND.P3.GT.P2.AND.P3.GT.P4) THEN
          WRITE (6,140)P3
140       FORMAT (///,2X,'THE HIGHEST VALUE IS BY R1 : ',F9.1)
          ENDIF
          IF (P4.GT.P1.AND.P4.GT.P2.AND.P4.GT.P3) THEN
          WRITE (6,150)P4
150       FORMAT (///,2X,'THE HIGHEST VALUE IS BY R2 : ',F9.1)
          ENDIF
          STOP
          END
```

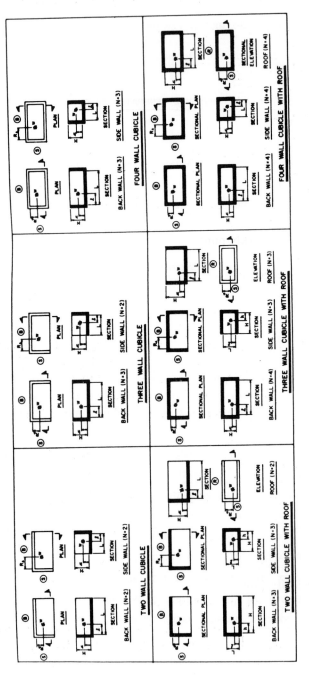

Notes

(1) B denotes back wall, S side wall and R roof

(2) Numbers in parentheses indicate number N of reflecting surfaces adjacent to surface in question.

(3) h is always measured to the nearest reflecting surface.

(4) l is always measured to the nearest reflecting surface except for the cantilever wall where it is measured to the nearest free edge.

Fig. C2.2 Barrier and cubicle configuration and parameters. (Courtesy of TM 5-1300/NAVFAC P-397/AFM 88–22.)

computer program used to evaluate such pressures. The program is linked to the finite-element ISOPAR program for the quick evaluation of pressures. The following data form part of the input for an explosion in brick buildings:

Aspect ratio of height/thickness: 1.5, 2.0, 2.5, 3.0
Openings: 1.5×4, 0.4×0.4, 1.0×2.3 m
Frame stiffness, K: 11.61, 7.3, 3.69 kN/mm
Brick size: $185.9 \times 87.3 \times 87.4$ mm
  Young's modulus, E: 6.4670 kN/mm^2
  Poisson's ratio, v: 0.096
$224.1 \times 108 \times 67.3$ mm
  Young's modulus, E: 4.2323 kN/mm^2
  Poisson's ratio, v: 0.141
Number of 8-noded isoparametric elements (main building): 900
Number of 20-noded isoparametric elements (wall): 350
Mortar (cement, lime, sand): 1:1:6
  Young's modulus, E: 2.465 kN/mm^2
  Poisson's ratio, v: 0.244
Number of gap and 3-noded elements
  Main building: 1950
  Wall: 350

Figure C2.5 illustrates the pressure pulses for four types of explosions. The corresponding deflection type relationship is given in Fig. C2.6. Figure C2.7 shows the deflections for various pressures against the distance of the explosion, and Fig. C2.8 shows the pressure pulse for the single wall of Fig. C2.3. Figures C2.9 and C2.10 illustrate the post-mortems of the building and wall, respectively.

C.3 Bomb explosion at a wall

A 250 kg GP bomb is thrown by a missile at a velocity of 100 m/s. The impact force is calculated by

$$F_1(t) = 0.55 \times 10^6 \ (mv_{im}^2/u_{im})$$

where m = weapon mass (kg)
    v_{im} = impact velocity (km/s)
    u_{im} = normal penetration (m) due to impact

The finite-element analysis is carried out on a wall (Fig. C2.4). The damaged zones for the front and back are indicated by Fig. C2.11(a) and (b).

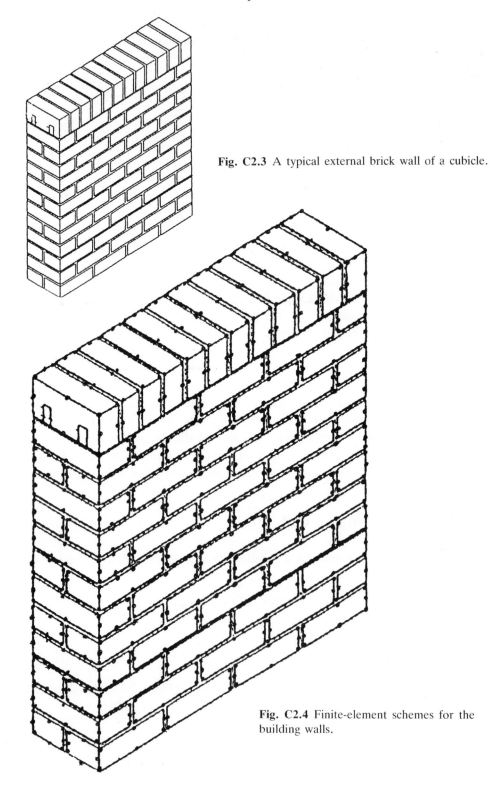

Fig. C2.3 A typical external brick wall of a cubicle.

Fig. C2.4 Finite-element schemes for the building walls.

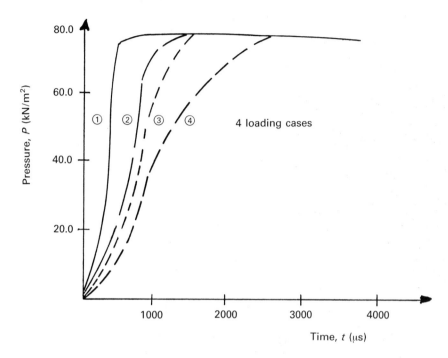

Fig. C2.5 Pressure pulses for four types of explosion.

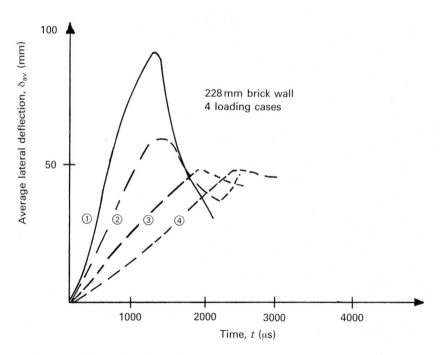

Fig. C2.6 The effect of the pressure rise time on the maximum deflection.

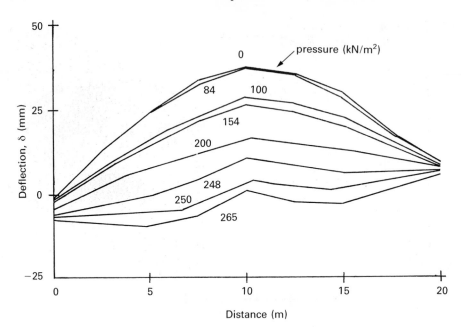

Fig. C2.7 Pressure pulse versus distance of explosion for a wall.

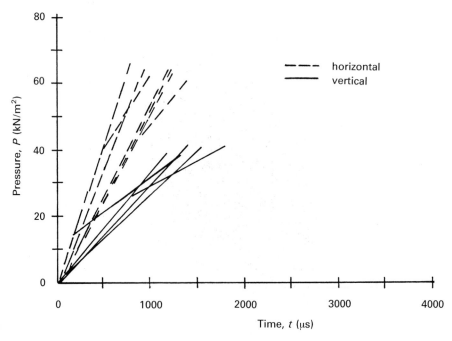

Fig. C2.8 Pressure−time relationship for a wall.

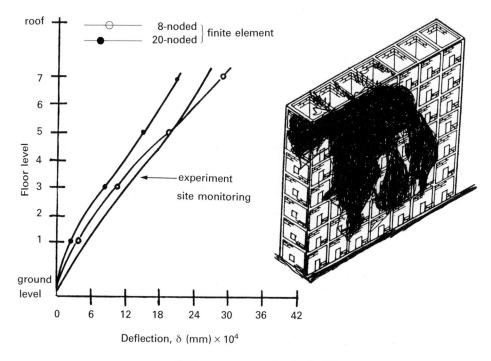

Fig. C2.9 Post-mortem for the building.

Fig. C2.10 Post-mortem for the wall.

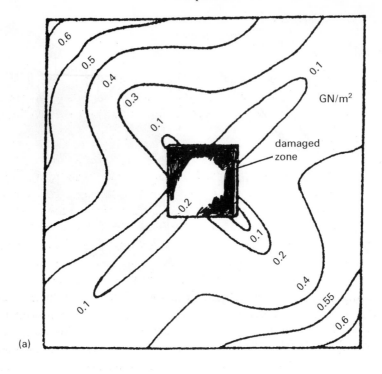

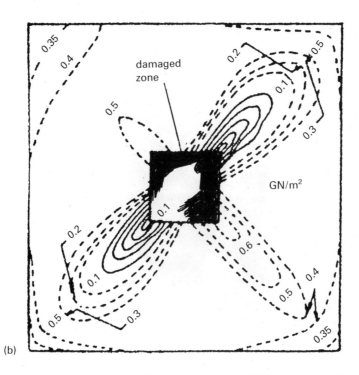

Fig. C2.11 Damage to (a) the front of the wall and (b) the back of the wall.

D: ICE/SNOW IMPACT

D.1 Introduction

A great deal of work has been carried out on ice impacting on ice or on structures. Major work has already been reported in this text regarding the prediction of impact loads before, during and after indentation and collapse conditions.

D.2 Finite-element analysis

Finite-element analysis has been carried out on a concrete platform subject to ice impact. Non-linear visco-elasticity based on the displacement method was considered. At each time step, a successive substitution-type iteration was applied to the global equations of motion, while a Newton–Raphson method combined with the α-method of numerical time integration was used in the constitutive equations at the Gauss integration points in the finite-element mesh. Variable interface conditions between the ice and the structure were simulated to limit the effects of interface adfreeze or friction. To allow the deliberate development of certain normal compressive stresses at the interface, an adaptive procedure was used to free those points at the interface. The mechanical behaviour of the sea ice was modelled as an isothropic material with creep laws. The three-dimensional yield surface was represented by Jones's tri-axial tests and is given in Fig. D2.1(a) and (b). The range of strain rates covered can be seen in Fig. D2.1(b). As an example, a Condeep-type platform (Fig. D2.2) was assumed to lie in the ice environment and to be subjected to ice impact.

The major parameters for the iceberg impacting the platform are:

Iceberg size: 50 000 tonnes
Equivalent diameter: 38.2 m
Freeboard: 3.8 m
Draft: 34.4 m
Initial velocity: 1.0 m/s

$$K_{IC} = YF_1(t)/D^{1.5} = 0.08 \text{ MPa} \sqrt{m} \qquad Y = 1.72(D/D_n) - 1.27$$

where D, D_n = original and notch root diameters, respectively.

Data for icebergs
Added mass coefficient: 0.70
Form drag area, 50 000 tonne berg: 1315.00 m^2
Drag coefficient: 0.80
Flexural strength: 700.00 kPa
Elastic modulus, E: 3.00 GPa
Friction coefficient: 0.15
Adfreeze strength: 200.00 MPa
Total horizontal iceberg load: 790.00 MN
Total vertical iceberg load: 530.00 MN

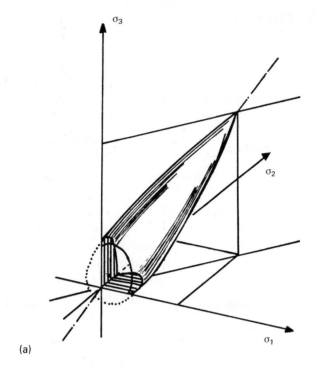

(a)

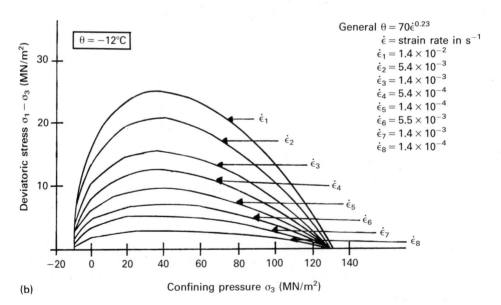

(b)

Fig. D2.1 Jones's tri-axial tests. (a) Yield surface for isotropic ice, new formulation; (b) Jones's tri-axial stress.

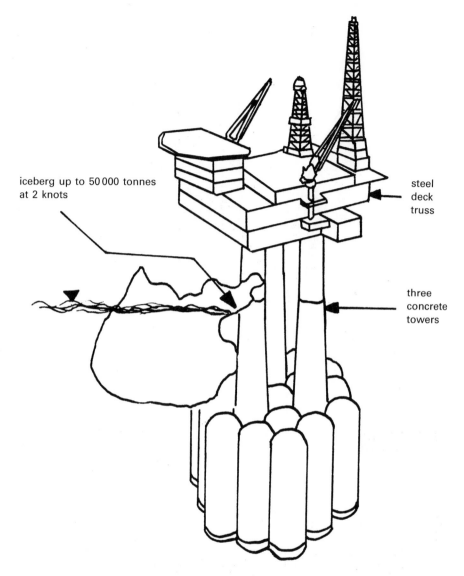

iceberg up to 50 000 tonnes
at 2 knots

steel
deck
truss

three
concrete
towers

Fig. D2.2 Concrete gravity platform subject to ice impact.

Figure D2.3 shows a typical finite-element mesh scheme for the platform. Figure D2.4 gives the force—time history of the iceberg. The displacement—time history of the platform is given in Fig. D2.5. The post-mortem of the platform is given in Fig. D2.6.

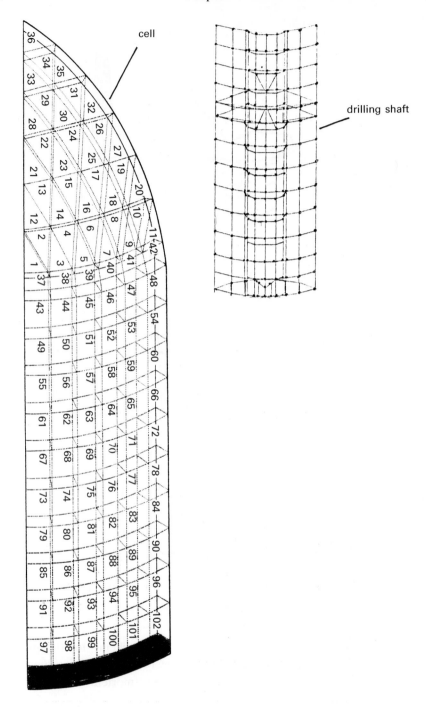

Fig. D2.3 Finite-element mesh scheme for the platform.

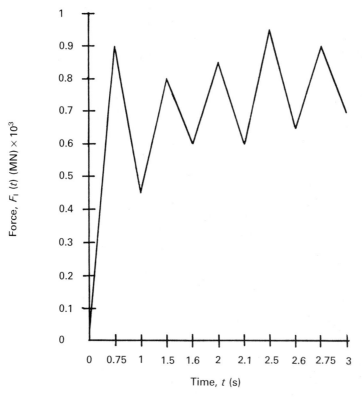

Fig. D2.4 Force−time function for the iceberg.

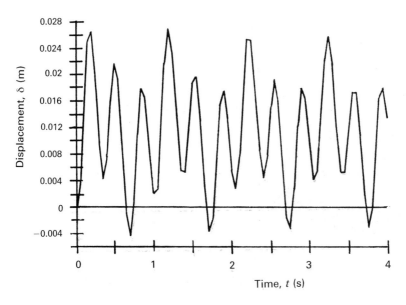

Fig. D2.5 Displacement−time function.

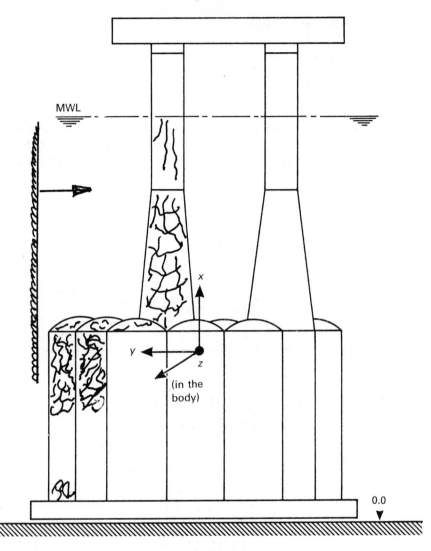

Fig. D2.6 Platform post-mortem.

E: NUCLEAR REACTORS

E.1 PWR: Loss-of-coolant accident (LOCA)

E.1.1 Introduction to LOCA

A loss-of-coolant accident (LOCA) occurs as a result of a penetration to the main coolant boundary such that the primary circuit water is released through the break to the containment area, causing a rapid decrease in the pressure and temperature

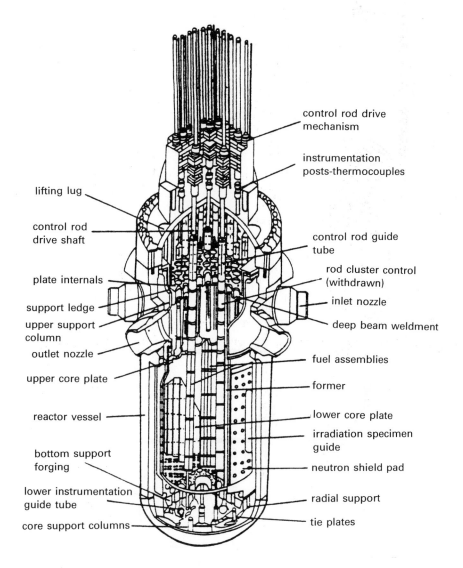

Fig. E1.1 Cut-away diagram showing the PWR steel vessel and internals (former CEGB Sizewell B inquiry). (From Bangash[3.3])

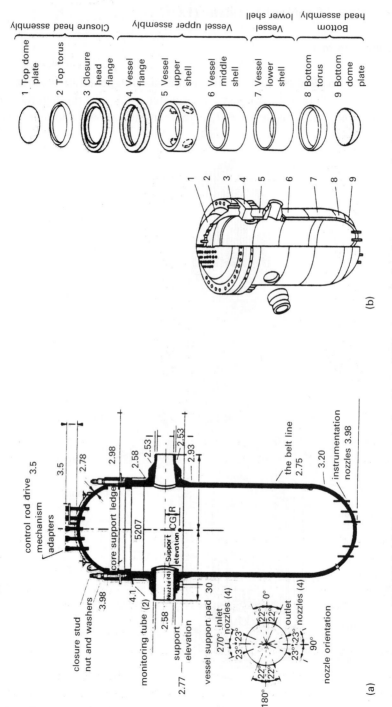

Fig. E1.2 (a) The reactor vessel of a 4-loop PWR system (safety factors computed using the finite-element method). (b) Diagram showing sections from which the PWR vessel may be fabricated when sections are either rolled plate or forged material (shaded parts are common to both). (Courtesy of Marshall[5.155]) (From Bangash[3.3])

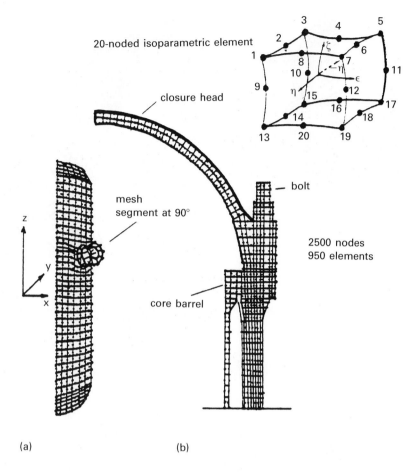

20-noded isoparametric element

closure head

mesh segment at 90°

bolt

2500 nodes
950 elements

core barrel

(a)   (b)

Fig. E1.3 Finite-element mesh generation of the vessel. (a) Generalized vessel mesh. (b) Closure head−wall flange region. (From Bangash[3.3])

of the primary coolant. This will give an impact thermal shock load. The streamline break accident (SLBA) occurs as a result of a complete and partial rupture of a steam line inside the containment vessel. A rapid cool-down and depressurization of the primary circuit normally take place. In order to restore the reactor coolant pressure, a pressure loading unconnected with LOCA is required.

E.1.2 Description of the PWR vessel and its materials

Figure E1.1 shows a typical PWR vessel for Sizewell B. The vessel steel must possess high toughness and strength coupled with adequate weldability in thick sections, with generally low-alloy steels containing manganese, nickel and molybdenum. The material grade for plates is SA-533 B alloys, and SA-508 alloys for

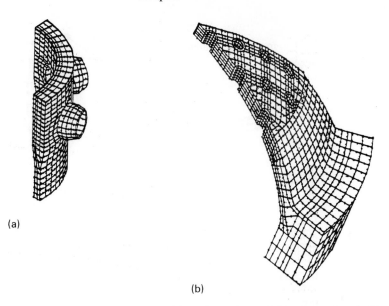

(a)

(b)

Fig. E1.4 Finite-element mesh generation of (a) wall nozzles and (b) closure head. (From Bangash[3.3])

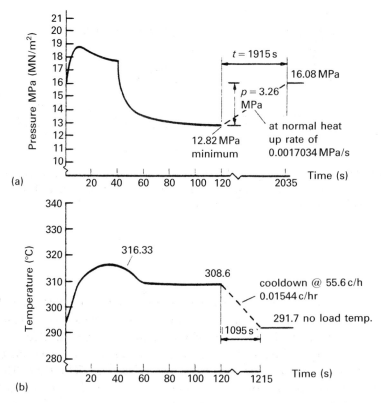

(a)

(b)

Fig. E1.5 Loss of load (without immediate reactor trip) (Westinghouse revision no 5). (a) Reactor coolant pressure and (b) temperature variations. (Courtesy of Marshall[5.155]) (From Bangash[3.3])

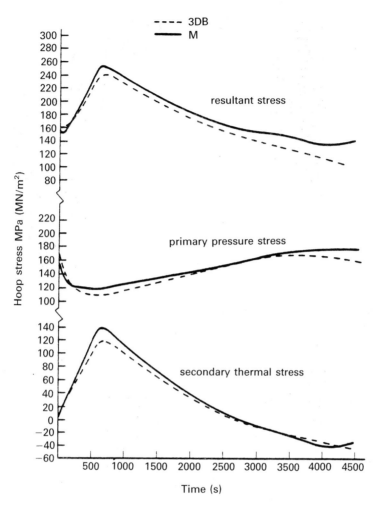

Fig. E1.6 Belt-line reactor trip from full power. Time variation of hoop stress in vessel material from clad/vessel interface. 3 DB: three-dimensional finite-element analysis (Bangash); M: Marshall[5.155] (From Bangash[3.3])

forging. Both must be in quenched and tempered conditions. The suitability of these steels rests on the mechanical properties such as yield stress, ultimate tensile strength, elongation to fracture and charpy impact energy affected by thermal aging, strain aging and neutron irradiation. The vessels are made out of thick-section plates of up to 360 mm or from ingots of over 200 000 kg. The ingots generally develop cavities of up to 3 mm in the v-segregation regions. These are healed by hot working processes. Both plates and forgings are welded. Figure E1.2 shows vessel fabrications.

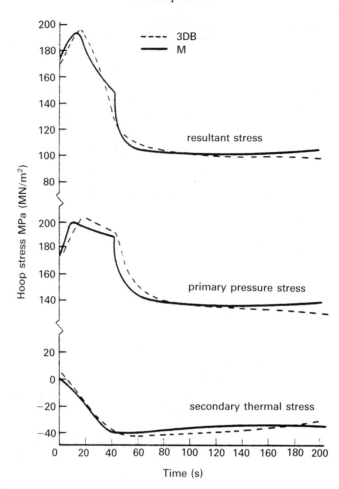

Fig. E1.7 Time variation of hoop stress in vessel material from clad/vessel interface (belt-line loss of load). 3DB: Bangash; M: Marshall[5.155] (From Bangash[3.3])

Table E1.1 shows the data used in the three-dimensional finite-element analysis. Figures E1.3 and E1.4 show the finite-element mesh generation scheme of the vessel, wall nozzles and closure heads. These are the important sections and locations in the reactor vessels from the point of view of fracture assessment. Figure E1.5(a) and (b) shows the pressure−time and temperature−time relationships. Figures E1.6 to E1.18 indicate various stresses in different zones due to a LOCA. The defect size in each region is checked using the R6 method of the former CEGB (UK) which is given in Tables E1.2 to E1.4.

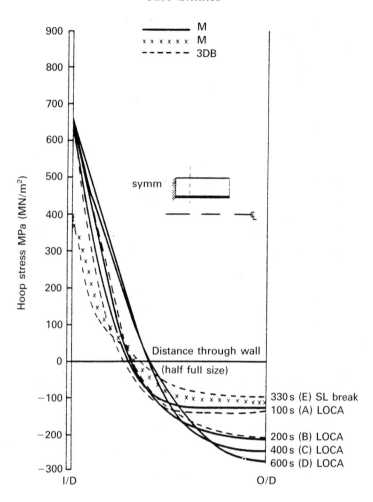

Fig. E1.8 Secondary thermal stress at various durations: belt-line region, LOCA condition. 3DB: Bangash; M: Marshall.[5.155] (From Bangash[3.3])

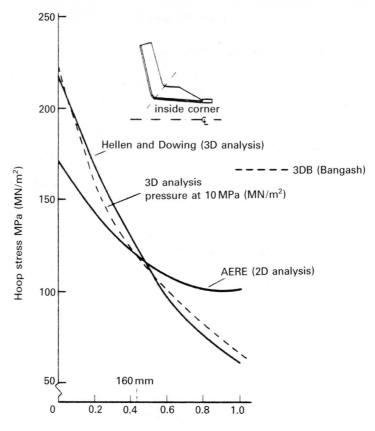

Fig. E1.9 Comparative study of results for the inside corner of an inlet nozzle. (From Bangash[3.3])

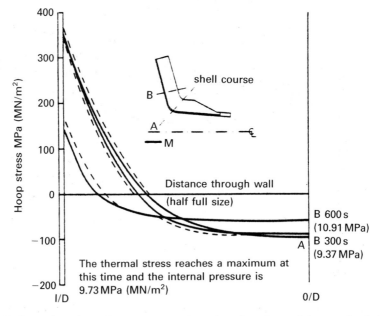

Fig. E1.10 Secondary thermal stresses at various durations for an inlet nozzle shell course large streamline break. M: Marshall[5.155] (From Bangash[3.3])

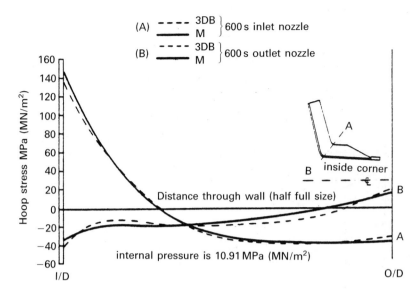

Fig. E1.11 Inlet nozzle inside corner: reactor trip from full power. 3DB: Bangash; M: Marshall[5.155] (From Bangash[3.3])

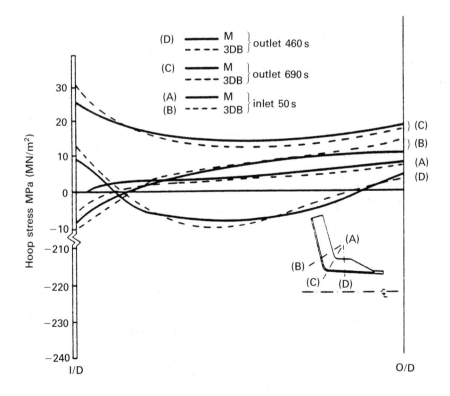

Fig. E1.12 Inlet nozzle inside corner: loss of power, secondary thermal stresses. 3DB: Bangash; M: Marshall[5.155] (From Bangash[3.3])

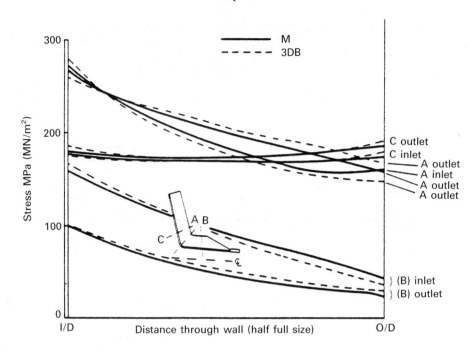

Fig. E1.13 Inlet nozzle hoop stress plotted through the wall at the sections shown. Basic pressure 17.24 MPa (MN/m^2). 3DB: Bangash; M: Marshall[5.155] (From Bangash[3.3])

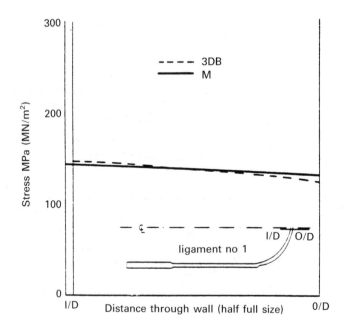

Fig. E1.14 Main hoop stress plotted through the wall at the section shown. Basic pressure load of 17.24 MPa (MN/m^2). 3DB: Bangash; M: Marshall[5.155] (From Bangash[3.3])

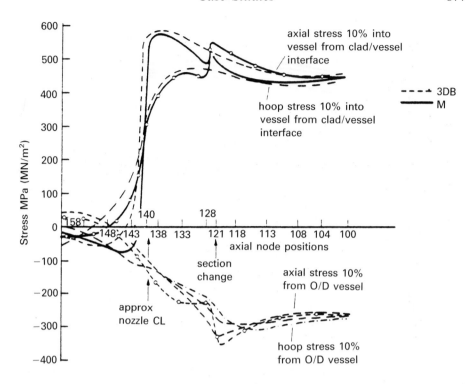

Fig. E1.15 Axial and hoop stress plotted along the nozzle course and belt-line region for a large LOCA at 600 s. 3DB: Bangash; M: Marshall.[5.155] (From Bangash.[3.3])

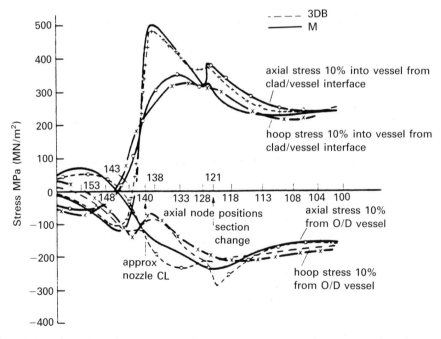

Fig. E1.16 Axial and hoop stress plotted along the nozzle course and belt-line region for a large LOCA at 2000 s. 3DB: Bangash; M: Marshall.[5.155] (From Bangash.[3.3])

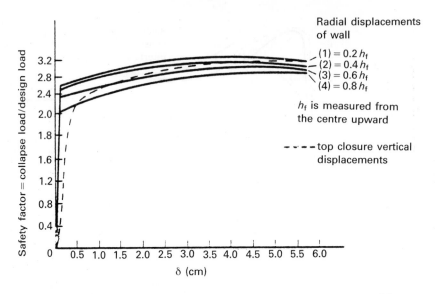

Fig. E1.17 Safety factors versus displacement (δ). (From Bangash[3.3])

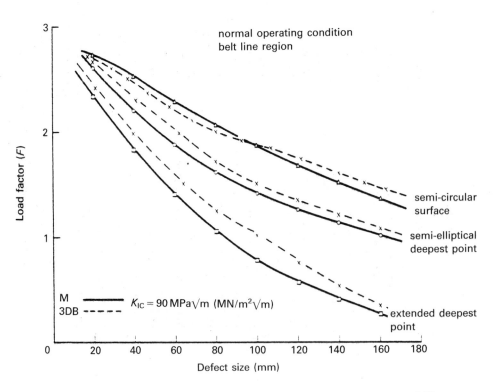

Fig. E1.18 Load factor versus defect size. 3DB: Bangash; M: Marshall[5.155] (From Bangash[3.3])

Table E1.1 Vessel material properties and parameters.

Proposed parameters for Sizewell B:

Vessel overall height	13 660 mm
Inside diameter	4394 mm
Wall thickness opposite core	215 mm
Wall thickness at the flange	500 mm
Normal clad thickness	6 mm
Thickness of the dome top	178 mm
Thickness of the dome bottom	127 mm
Inside diameter of inlet nozzle	700 mm
Inside diameter of outlet nozzle	737 mm
Number of closure studs	54 (each 1466 mm high)
	(nut 268 × 203)
	(washer 268 × 38)
Diameter of closure studs	173 mm
Dry weight of the pressure vessel	434.8×10^3 kg
Normal operating pressure	15.98 MPa
Design pressure	17.13 MPa
Initial hydraulic pressure	21.43 MPa
Normal operating inlet temperature	288°C
Normal operating outlet temperature	327°C
Design temperature	343°C
No load temperature	292°C
Design life	40 years at 80% load factor
$E_s = C \times E \times 10^5$ MPa	C varies with temperature
(1) Material SA533B	C = 0.218, Mn = 1.367, Ni = 0.547, Mo = 0.547, Si = 0.236, Cr = 0.074,
	P = 0.009, S = 0.014
(weight per cent)	C = 0.117
(2) Submerged arc-welding	C = 0.16, Mn = 2.20, Ni = 0, Mo = 0.6, Si = 0.05, Cr = 0, P = 0.025, S = 0.035
(electrode wire content)	C = 0.15

continued

Table E1.1 *Continued.*

(3) Mechanical properties	Yield stress (N/mm²)	Ultimate tensile stress (N/mm²)	Charpy test (minimum value)
(a) SA533B plates	at 20°C 345	at 20°C 555	34 J at 4.4°C
	at 400°C 280	at 400°C 520	flanges, shell or
(b) SA508 forging	at 400°C 280	at 20°C 555	rings and nozzles
		at 400°C 430	average values
(c) Weld metal	483	593	150, 138, 100, respectively
(d) Stresses after 288°C irradiation unirradiated	These are given before from 3(a) to (3c) Add 2% of the above values		
(e) Fracture toughness of plates	$K_{Ic} = 106\,\text{N/mm}^2$, $\sqrt{m} = 153$		
(f) Bonding material	Yield = 900–1050 N/mm²; ultimate tensile stress = 1050 N/mm²; impact energy = 60.9–81.2 J		

Table E1.2 R6 method of fracture assessment.

The failure assessment diagram for the R6 method (courtesy of CEGB, UK).

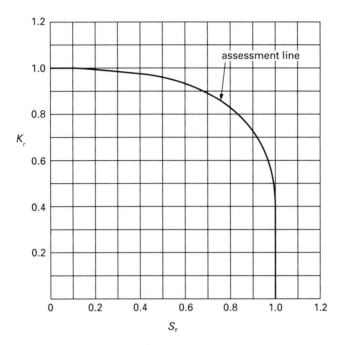

$$K_r = S_r \left\{ \frac{8}{\pi^2} \log_e \sec \left(\frac{\pi}{2} S_r \right) \right\}^{-1/2}$$

$$K_r = K_I^P / K_{IC}$$

where K_r is a measure of how close the vessel is to linear elastic failure $= 0.59$ and K_I^P is the stress intensity factor due to σ stresses and is given by

$$K_I^P = Y\sigma\sqrt{\pi a} = 39.42\,\mathrm{MPa}\sqrt{m}$$

where $Y =$ magnification factor when applying unflaw stresses obtained to postulated flawed vessel; a Y value of 1.25 has been taken

$\sigma =$ applied stress

$a =$ crack height

$K_{IC} =$ fracture toughness, with a lower limit of $170\,\mathrm{MPa}\sqrt{m}$ at 288°C

Defect length 2C parallel to component surface
Defect depth S distance between nearest edge of the defect and component surface (normal distance)
Defect height 2a distance between the nearest and furthest extremities of a defect normal to the surface (buried 2a, surface breaking a)

continued

Table E1.2 *Continued.*

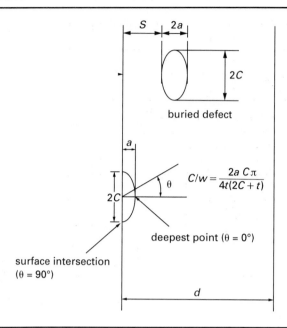

buried defect

$$C/w = \frac{2a\,C\pi}{4t(2C+t)}$$

deepest point ($\theta = 0°$)

surface intersection
($\theta = 90°$)

Table E1.3 Semi-elliptic surface breaking floor configuration: sample calculations.

For $0.1 < a/2c < 0.5$ and $a/t \le 0.8$. When $a/2c < 0.1$ then $c/w = a/t$ (extended crack). Therefore the S_r value is given by

$$S_r = \{1/\bar{\sigma}(1 - c/w)^2\}\{[\sigma_{bc}/4 + \sigma_{mc}'(c/w)]\} + [(\sigma_{bc}/4 + \sigma_{mc}(c/w))^2 + \sigma_{mc}^2(1 - c/w)^2]^{1/2}$$

$$a/c = \tfrac{1}{3} \text{ if } a = 70\,\text{mm}; \ c = 210\,\text{mm}$$

$$0.1 < 7/42 < 0.5 \text{ is satisfied}$$

$$c/w = 2ac\pi/4t(2c + t) = 2(70)(210)\pi/4(216)(420 + 216) = 0.168$$

$$\sigma_{bc} = 6M/t^2 = 7.56\,\text{MPa}$$

With $a/c = \tfrac{1}{3}$, $\sigma_{mc} = 171.5\,\text{MPa}$, if $a = 70\,\text{mm}$ and $c/w = 0.168 = $ area of flaw/area of rectangle.

$$S_r = [7.56E6/4 + 171.5E6(0.168)] + [(7.56E6/4 + 171.5E6(0.168)^2 + (171.5)^2(1 - 0.168)^2]^{1/2}/440E6(1 - 0.168)^2$$

or $S_r = 0.58$

and $K_I^P = Y\sigma_{mc}\sqrt{\pi a} = 1.25(171.5E6)\sqrt{\pi(0.07)} = 100.53\,\text{MPa}\,\sqrt{\text{m}}$

where $K_r = K_I^P/K_{IC} = 100.53E6/170E6 = 0.59$

$\sigma_{bc} = $ elastically calculated bending stress evaluated over the gross section containing the flaw.

$\sigma_{mc} = $ elastically calculated tensile stress evaluated over the gross section containing the flaw.

Table E1.4 Failure assessment of crack heights due to LOCA.

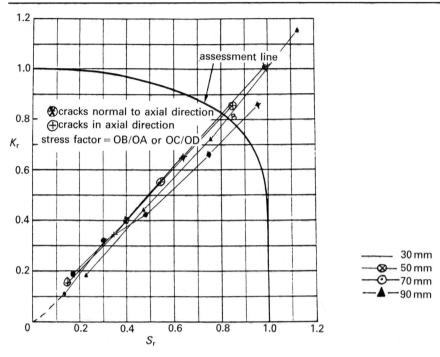

Computed data: normal operating conditions, belt-line region S_r and K_r values for flaws normal to the hoop stress.

Crack height a (mm)	Effective ratio c/w	Calculated values	
		S_r	K_r
70	0.168	0.58	0.59
90	0.234	0.70	0.67
110	0.301	0.86	0.74
130	0.371	1.10	0.81

Normal operating conditions, belt-line region S_r and K_r values for flaws normal to hoop stress.

Crack height a (mm)	Effective ratio c/w	Calculated values	
		S_r	K_r
70	0.168	0.25	0.23
90	0.234	0.30	0.26
110	0.301	0.38	0.29
130	0.371	0.48	0.32
150	0.440	0.64	0.34
170	0.510	0.88	0.36

E.2 Nuclear containment under hydrogen detonation

Hydrogen detonation has become an important issue after the Three-Mile-Island accident. The hydrogen burning occurred approximately 10 hours into the accident. The steam reacting with the Zircaloy cladding and the oxidation of the overheated steel vessel interiors created large quantities of hydrogen. This can also occur due

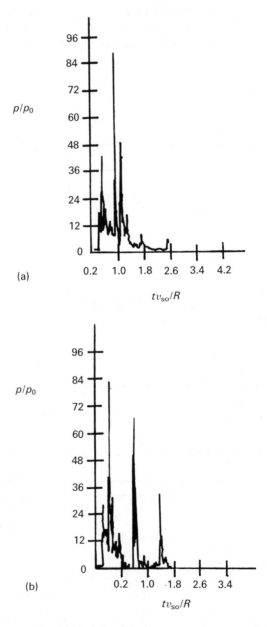

Fig. E2.1 (a) Pressure at the mid-height; (b) pressure at the apex of the dome.

to interaction of the molten core. In order to predict the wall pressures due to such detonations, non-linear gas dynamics equations for the entire volume of the containment vessel have to be solved. In the current analysis of the Sizewell B containment vessel, it is assumed that the wall pressure P_o is proportional to the containment pressure P. The vessel parameters are given in Table E1.1. It is assumed that the detonation starts approximately at the mid-height of the containment vessel. The spherical shock front generated obliquely converges at the dome, causing a strong reflection around the apex. The containment finite-element mesh scheme is unchanged. Bond slip and shear slip for the reinforced elements are considered. The following additional input data have been included:

Material
Liner: $\sigma_y = 390\,\text{MPa}$
Strain rate $= 0.1/\text{s}$
Pressures and shock front
Detonation strength $q/\bar{R}T_0 = 17$ (base) and 23 (mid-height)
where $q =$ energy release/unit volume
   $\bar{R} =$ gas constant $= 8.31\,\text{kJ/kg-mole K}$
   $T_0 =$ absolute temperature
Shock front
$v_{so} =$ shock wave velocity $= 2000\,\text{m/s}$
 $t = 0, 50, 100, 200, 250, 300\,\mu\text{s}$

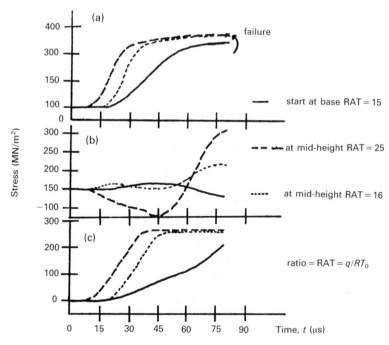

Fig. E2.2 Stress histories at 40% of the cylinder, with variation of steel properties included: (a) inside hoop bars; (b) inside vertical bars; (c) seismic diagonal bars.

Figure E2.1(a) and (b) shows the non-dimensional relationship for the pressure P/P_o and $(t/R)v_{so}$ for the Sizewell B vessel at the apex of the dome and at the mid-height. The parameter R is the radius of the containment vessel. Figures E2.2 to E2.4 illustrate the stress–time histories for the containment wall, dome springing and dome apex due to detonation.

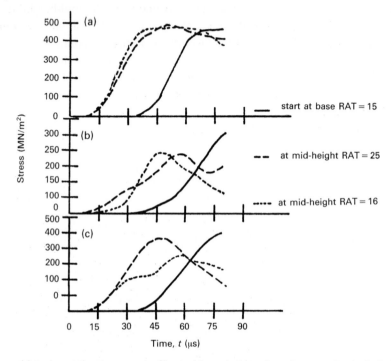

Fig. E2.3 Stress histories at the dome springline, with variation of steel properties included: (a) inside hoop bars; (b) inside vertical bars; (c) seismic diagonal bars.

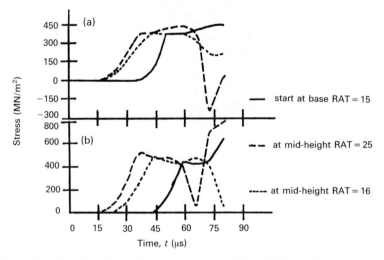

Fig. E2.4 Stress histories $10°$ from the dome apex, with variation of steel properties included: (a) inside hoop bars; (b) inside meridional bars.

E.3 Impact/explosion at a nuclear power station: turbine hall

The finite-element analysis of a concrete turbine hall is carried out for an impact/ explosion caused by a Tomahawk cruise missile. The finite mesh and damaged areas of the northeast wall are examined in Fig. E3.1 for the entire facility (shown in Fig. E3.2). The finite-element mesh of the soil beneath such a facility is shown in Fig. E3.2. Apart from the Tomahawk cruise missile data, given in Chapter 2, the following additional data have been considered:

Walls
33 m × 16 m × 2.5 m reinforced concrete walls
T25: 250 bars in both ways
Impact area 5 m × 3 m of centreline

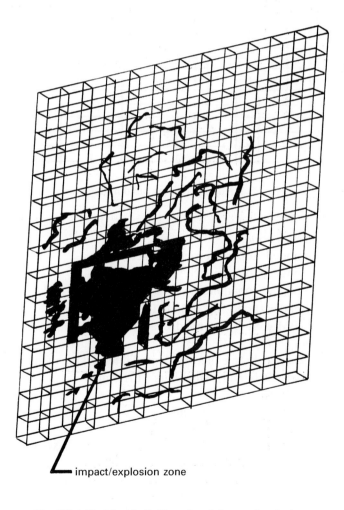

impact/explosion zone

Fig. E3.1 Turbine hall: Tomahawk impact/explosion.

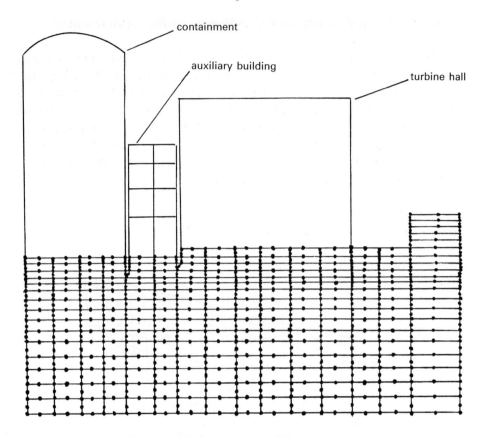

Fig. E3.2 Nuclear reactor building facilities.

Finite-element mesh scheme

360 elements, 20-noded isoparametric — concrete

2100 elements, 3-noded isoparametric — steel bars

1440 elements, 6-noded isoparametric — steel liners

Total number of nodes (bond slip included): 22140

Computer used: IBM 4381 front-ended Cray-2

Material properties

Concrete grade: 43 endochronic theory adopted interlocking of aggregates ignored

Yield strength: liner $250\,\text{N/mm}^2$; steel bars $410\,\text{N/mm}^2$

Impactor

Tomahawk cruise missile, $v_{so} = 1200\,\text{m/s}$

Damaged zones (two such zones were discovered)

(1) A complete perforation size of an oval shape of $9\,\text{m}^2$. Many bars burst and buckled in this area.

(2) Outside the perforated area, cracking and crushing extended to about 10 m diagonal. Crack depths ranged from 10 mm to 1.35 m inside to outside. Many bars yielded and bent.

E.4 Jet impingement forces on PWR steel vessel components

The safety of nuclear installations such as the pressure vessel and its piping systems requires strict measures. In the event of steel failure, the safety rules require the assessment of the jet impinging forces on the vessel nozzle areas. In the current analysis it was assumed that the transition from the cross-section of the discharging pipe to that of the outlet is abrupt. It is imperative to evaluate the structural behaviour under the jet impinging forces. The vessel chosen for this analysis was the Sizewell B PWR steel vessel. All parameters and finite-element mesh schemes are kept the same as in section E.1. The following data is used:

Nozzles and pipe
$P_w = 17.13\,\text{MPa}$; initial hydraulic pressure $= 21.42\,\text{MPa}$ design
Temperature: 327°C
Discharging pipe diameter: 133 mm
Outside nozzle diameter: 737 mm
Inside nozzle diameter: 730 mm
Hydraulic coefficient of resistance: 0.37 to 0.82
Nozzle−structure distance: 330 mm

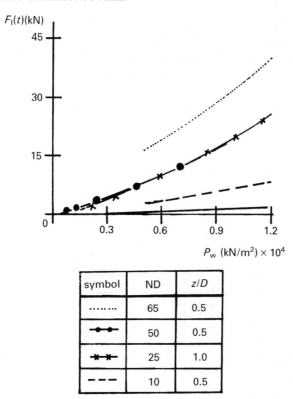

symbol	ND	z/D
....	65	0.5
●—●—	50	0.5
✳—✳—	25	1.0
− − −	10	0.5

Fig. E4.1 Measurements for jet impingement forces on a flat plate due to the discharge of saturated pressurized water from circular nozzles of different diameters.

At any height/diameter ratio ($z/D = z/737$), the pressure ratio $R'_p = p'T/P_w$ is computed, where p' is the saturation pressure at temperature T and P_w is the vessel pressure; $z/D = 0.85$, 0.95, 1, 2, 3, 4, 5 and 10; $p'T/P_w$ ranges between 0 and 1 and there is a total of 20 time steps (t), with time intervals of $0.24\,$s.

The jet impinging force against $p'T/P_w$ for failure conditions is given in Figs E4.1 and E4.2.

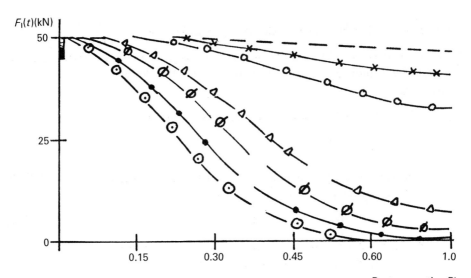

symbol	D (mm)	z/D
- - -	40	0.55
—×—×—	40	0.50
—o—o—	40	1.0
—▵—▵—	50	2.0
—φ—φ—	40	3.0
—•—•—	10	2.0
—⊙—⊙—	10	3.0

Fig. E4.2 R'_p versus $F_I(t)$.

F: CONCRETE NUCLEAR SHELTERS

F.1 Introduction

This section is devoted to the analysis and design of reinforced concrete nuclear shelters. Calculations are given for a particular study using both the British and American codes. Details are also given in this section regarding the Swedish Civil Defence Administration Code[6.79]

F.1.1 US code ultimate strength theory: general formulae

Figure F1.1 shows cracking, crushing and disengagement cases recommended in successive ACI building codes.

F.1.1.1 General equation: ultimate static moment capacity

Cross-section type I

(1) The ultimate unit resisting moment M_u of a rectangular section of width b, with tension reinforcement only, is given by

$$M_u = (A_s f_s / b)(d - a/2) \tag{F.1}$$

where A_s = area of tension reinforcement within the width b
$\quad f_s$ = static design stress for reinforcement
$\quad d$ = distance from the extreme compression fibre to the centroid of tension reinforcement
$\quad a$ = depth of equivalent rectangular block = $A_s f_s / 0.85 b f_c'$
$\quad b$ = width of compression face
$\quad f_c'$ = static ultimate compressive strength of concrete

The reinforcement ratio p is defined as

$$p = A_s / bd \tag{F.2}$$

(2) To ensure against sudden compression failures, p must not exceed 0.75 of the ratio p_b, which produces balanced conditions at ultimate strength and is given by

$$p_b = (0.85 K_1 f_c' / f_s)[87\,000/(87\,000 + f_s)] \tag{F.3}$$

where $K_1 = 0.85$ for f_c' up to 4000 psi and is reduced by 0.05 for each 1000 psi in excess of 4000 psi.

(3) For a rectangular section of width b with compression reinforcement, the ultimate unit resisting moment is

$$M_u = [(A_s - A_s')f_s / b](d - a/2) + [(A_s' f_s / b)(d - d')] \tag{F.4}$$

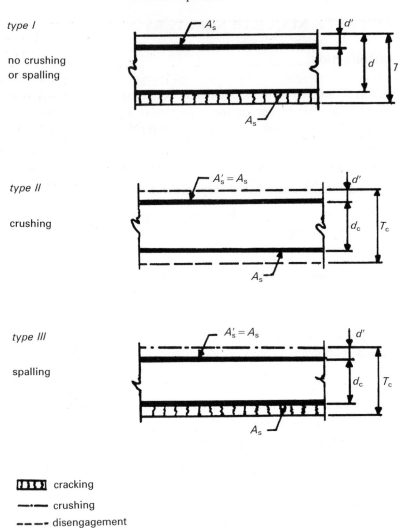

Fig. F1.1 Reinforced concrete cross-sections.

where A'_s = area of compression reinforcement within the width b

d' = distance from the extreme compression fibre to the centroid of compression reinforcement

a = depth of the equivalent rectangular stress block

$= (A_s - A'_s)f_s/0.85bf'_c$

The minimum area of flexural reinforcement is given in the following table.

Pressure design range	Reinforcement	Two-way elements	One-way elements
Intermediate and low	Main	$A_s = 0.0025bd$	$A_s = 0.0025bd$
	Other	$A_s = 0.0018bd$	$A_s + A'_s = 0.0020bT_c$
High	Main	$A_s = A'_s$ $= 0.0025bd_c$	$A_s = A'_s$ $= 0.0025bd_c$
	Other	$A_s = A'_s$ $= 0.0018bd^*_c$	$A_s = A'_s$ $= 0.0018bd^*_c$

* But not less than $A_s/4$ used in the main direction (see Fig. F1.2 for coefficients).

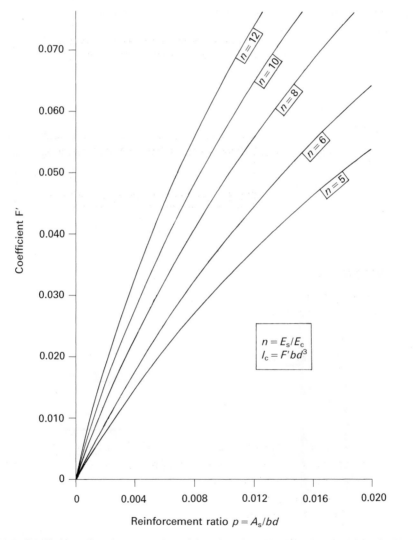

$$n = E_s/E_c$$
$$I_c = F'bd^3$$

Fig. F1.2 Coefficients for the moments of inertia of cracked sections with tension reinforcement only. (Courtesy of ACI.)

F.1.1.2 Ultimate static shear capacity

Diagonal tension

(1) The ultimate shear stress v_u, as a measure of diagonal tension, is computed for type I sections from

$$v_u = V_u/bd \tag{F.5}$$

and for type II and III sections from

$$v_u = V_u/bd_c \tag{F.6}$$

where V_u is the total shear on a width b at the section a distance d (type I) or d_c (type II and III) from the face of the support. The shear at sections between the face of the support and the section d or d_c therefrom need not be considered critical.

(2) The shear stress permitted on an unreinforced web is limited to

$$v_c = \phi[1.9\sqrt{f_c'} + 2500p] \le 2.28\phi\sqrt{f_c'} \tag{F.7}$$

where ϕ is the capacity reduction factor and is equal to 0.85 for all sections.

(3) When the ultimate shear capacity $v_u > v_c$, shear reinforcement must be provided. When stirrups are used, they should be provided for a distance d beyond the point theoretically required, and between the face of the support and the cross-section at a distance d. The required area for stirrups for type I cross-sections is calculated using

$$A_v = [(v_u - v_c)b_s s_s]/[\phi f_s(\sin\alpha + \cos\alpha)] \tag{F.8}$$

while for cross-sections conforming to types I, II and III, the required area of lacing reinforcement is (see Fig. F1.3):

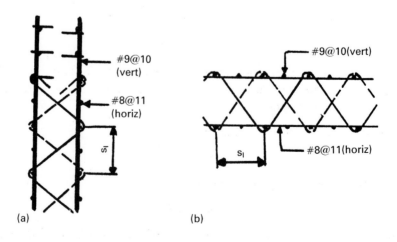

Fig. F1.3 Designs of lacings: (a) vertical and (b) horizontal.

$$A_v = [(v_u - v_c)b_\ell s_\ell]/[\phi f_s(\sin\alpha + \cos\alpha)] \tag{F.9}$$

where A_v = total area of stirrups or lacing reinforcement in tension within a width b_s, b_ℓ and distance s_s or s_ℓ

$(v_u - v_c)$ = excess shear stress

b_s = width of concrete strip in which the diagonal tension stresses are resisted by stirrups of area A_v

b_ℓ = width of concrete strip in which the diagonal tension stresses are resisted by lacing of area A_v

s_s = spacing of stirrups in the direction parallel to the longitudinal reinforcement

s_ℓ = spacing of lacing in the direction parallel to the longitudinal reinforcement

α = angle formed by the plane of the stirrups or lacing and the plane of the longitudinal reinforcement

The excess shear stress $v_u - v_c$ is as follows.

Limits	Excess shear stress $v_u - v_c$	
	Stirrups	Lacing
$v_u \leq v_c$	0	v_c
$v_c < v_u \leq 2v_c$	$v_u - v_c$	v_c
$v_u > 2v_c$	$v_u - v_c$	$v_u - v_c$

The ultimate shear stress v_u must not exceed $10\phi\sqrt{f_c'}$ in sections using stirrups. In sections using lacing there is no restriction on v_u because of the continuity provided by this type of shear reinforcement.

Wherever stirrups are required $(v_u > v_s)$, the area A_v should not be less than $0.0015bs_s$ and for type III rectangular sections of width b:

$$M_u = A_s f_s d_c/b \tag{F.10}$$

where A_s = area of tension or compression reinforcement within the width b

d_c = distance between the centroids of the compression and the tension reinforcement

The reinforcement ratios p and p' are given by

$$p = p' = A_s/bd_c \tag{F.11}$$

The reinforcement ratio p' is given by

$$p' = A_s'/bd \tag{F.12}$$

Equation (F.11) is valid only when the compression steel reaches the value f_s at ultimate stress, and this condition is satisfied when

$$p - p' \geq 0.85K_1\frac{f'_c d'}{f_s d}\left(\frac{87\,000}{87\,000 - f_s}\right) \tag{F.13}$$

Cross-section types II and III

(1) The ultimate unit resisting moment of type II.

F.1.1.3 Modulus of elasticity

Concrete

The modulus of elasticity of concrete, E_c, is given by

$$E_c = w^{1.5}\,33\sqrt{f'_c} \text{ psi} \tag{F.14}$$

The value of w, the unit weight of concrete, lies between 90 and 155 lb/ft^2.

Reinforcing steel

The modulus of elasticity of reinforcing steel, E_s, is

$$E_s = 30 \times 10^6 \text{ psi} \tag{F.15}$$

Modular ratio

The modular ratio, n, is given by

$$n = E_s/E_c \tag{F.16}$$

and may be taken as the nearest whole number.

F.1.1.4 Moment of inertia

The average moment of inertia, I_a, to be used in calculating the deflection is

$$I_a = (I_g + I_c)/2 \tag{F.17}$$

where I_g is the moment of inertia of the gross concrete cross-section of width b about its centroid (neglecting steel areas) and is equal to

$$I_g = bT_c^3/12 \tag{F.18}$$

and I_c is the moment of inertia of the cracked concrete section of width b considering the compression concrete area and steel areas transformed into equivalent concrete areas and computed about the centroid of the transformed section. I_c is calculated from

$$I_c = Fbd^3 \tag{F.19}$$

The coefficient F varies as the modular ratio n and the amount of reinforcement used. For sections with tension reinforcement only, F is given in Fig. F1.2.

F.2 Design of a concrete nuclear shelter against explosion and other loads based on the Home Office Manual

Figure F2.1 shows a typical layout of a domestic nuclear shelter for a family of six.

F.2.1 Basic data (Home Office code[6.80])

For a 1 megatonne ground burst at a distance of 1.6 km from ground zero:

Ductility ratio, μ: 5
Main reinforcement $\not< 0.25\%$ *bd*
Secondary reinforcement $\not< 0.15\%$ *bd*
Ultimate shear stress $\not> 0.04 f_{cu}$
Dynamic shear stress (mild steel) $\not> 172\,\text{N/mm}^2$
Protective factor: 4000
Concrete f_{cu} (static): 30 N/mm^2 (grade 30)
Concrete f_{cu} (dynamic): $1.5 f_{cu} = 37.5\,\text{N/mm}^2$
Reinforcement f_y (static): 420 N/mm^2
Reinforcement f_{yd} (dynamic): $1.10 f_y = 462\,\text{N/mm}^2$
Young's modulus, E_c: 20 GN/m^2
Young's modulus, E_s: 200 GN/m^2
Clear span: 3 m
Slab thickness: 300 mm (with minimum cover 50)
Blast load: 0.17 N/mm^2, $F_1(t) = P_{do}$

F.2.2 Additional data for designs based on US codes

Dynamic increase factors (DIF)
Concrete: compression   1.25
        diagonal tension   1.00
        direct shear   1.10
Reinforcement: bending   1.10
            shear   1.00
Dynamic stresses:
concrete f'_c (cylindrical strength) $= 0.87 f_{cu}$
                            $= 3000\,\text{lb/in}^2$ (psi)
concrete f_y (static) $= 60\,000\,\text{lb/in}^2$ (psi)

$$R_m = r_u = \left(\frac{1}{1 - \dfrac{1}{2\mu}}\right) F_1(t) = 1.1 F_1(t) = 0.187\,\text{N/mm}^2$$

Deadload of concrete plus soil $= 0.014\,\text{N/mm}^2$

$$r_u = 0.187 + 0.014 = 0.201\,\text{N/mm}^2$$

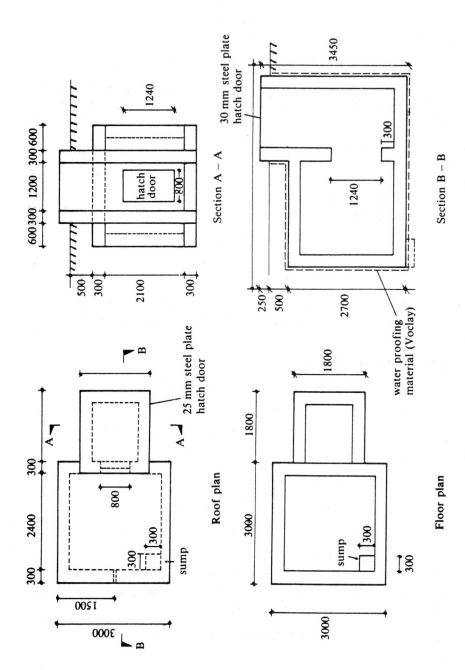

Fig. F2.1 Domestic nuclear shelter: general arrangement.

For a two-way slab

$$M_u = r_u L^2/16 = 0.201(3000)^2/16$$
$$= 113\,062.5\,\text{Nmm/mm}$$

300 mm thick slab
T16−200 bars; $A_s = 1005\,\text{mm}^2/\text{m}$; $d = 300 - 50 - 8 = 242$

$z = d - (0.84 f_{yd} A_s/f_{cu}(\text{dyn})$
$= 242 - (0.84(462)(1005)10^{-3}/37.5)$
$= 231.58\,\text{mm}$

Note: Later on, based on finite-element analysis, the T20−200 bars adopted were checked

Area of the roof $= 9\,\text{m}^2 = A_t$; $\sqrt{A_t} = 3\,\text{m}$
$H - x = 2.7 - 0.3 = 2.4$ or $3.4 - 0.3 = 3.1$
$\sqrt{A_t}/(H - x) = 1.25$ and 0.97
Weight of overhead material $= 1340\,\text{kg/m}^2$
$R = 0.025\%$ (roof contribution)

$$PF = 100/(R + G_T)$$
$$= 100/(0.025 + 0)$$
$$= 4000 \quad \text{(safe)}$$

where G_T is the percentage wall contribution, ignored in the worst case.
Figure F2.2 gives structural details of the reinforced concrete shelter.

Steel blast doors
Clear opening 800 mm × 1200 mm.

$$F_1(t) = p_{do} = 2.3 p_{so} = 2.3(0.17) = 0.39\,\text{N/mm}^2$$
$$r_u = 1.1 F_1(t) = 0.43\,\text{N/mm}^2$$
$$M_u \text{ (simply supported)} = 0.43(800)^2/8 = 34\,400\,\text{Nmm/mm}$$

20 mm thick steel door

$$z = bd^2/4 = 1(20^2)/4 = 100\,\text{mm}^3$$

Also
$$z_p = M_u/1.1(265) = 118\,\text{mm}^2$$

Calculated thickness of steel doors $= (118/100)20 = 23.6 \approx 25\,\text{mm}$

A 25 mm thick door was adopted.
The thickness of the glass door may have to be increased for protection against radiation fall-out. One possibility is a steel−concrete sandwich construction. One possible steel door design is given in Fig. F2.3.

$$\text{or } z = 0.95(242) = 229.9\,\text{mm} \approx 230\,\text{mm}$$

Walls: 300 mm thick
Blast load on walls $= p_{do} \times 0.5 = 0.085\,\text{N/mm}^2$
$r_u = 1.1 F_1(t) = 1.1(p_{do}) = 0.0935\,\text{N/mm}^2$
Total (including soil) $= 0.0935 + 0.08 = 0.1735\,\text{N/mm}^2$

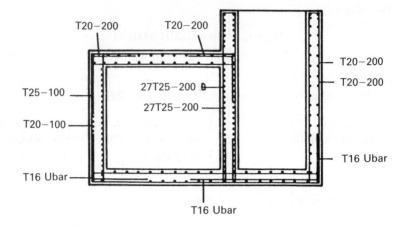

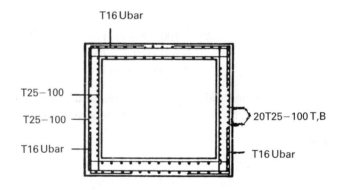

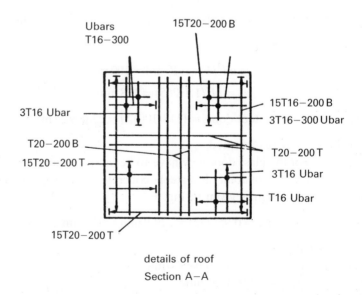

details of roof

Section A–A

Fig. F2.2 Domestic nuclear shelter (reinforced concrete): detail.

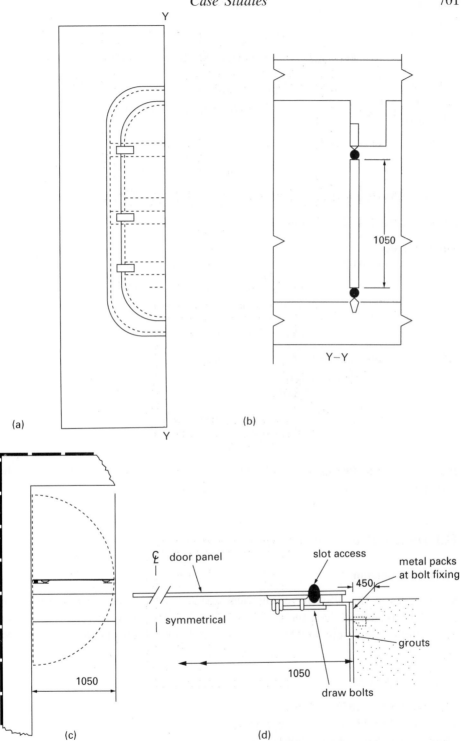

Fig. F2.3 Design of steel blast doors. (a) Elevation; (b) vertical section; (c) door location; (d) horizontal section: structural details.

Two-way slab

$$M_u = r_u L^2/16 = 0.1735(2700)^2/16 = 79\,050.941$$

Both walls
(2700 mm and
3400 mm) $\left.\begin{array}{l}\\\\\\\end{array}\right\}$ also $M_u = [(3400)^2/(2700)^2](79\,050.94)$
$= 125\,353.75\,\text{Nmm/mm}$ (adopted)

$$M_u = 125\,353.75 = A_s(230)(462)$$

$$A_s = 1.18\,\text{mm}^2/\text{mm} = 1180\,\text{mm}^2/\text{m}$$

adopted T20−200 (in some critical areas T20−100 and T25−100)

Shaft wall bars: T12−200 links T16−300 U-bars

Minimum steel:

Main→ $0.25\% \times 1 \times 242 = 0.605\,\text{mm}^2/\text{mm}$ (605 mm²/m)

$1005\,\text{mm}^2/\text{m} > 605$ (T16−200) adopted

Secondary→ $0.15\% \times 1 \times 242 = 0.363\,\text{mm}^2/\text{mm}$ (363 mm²/m)

(T16−200 or 300) adopted

Shear: allowable shear $= 0.04f_{cu} = 1.2\,\text{N/mm}^2$
shear $= r_u[(L/2 - d)/d]$
$= (2700/2 - 242)/2700$
$= 0.41 < 1.2\,\text{N/mm}^2$ (safe)
or $= (3400/2 - 242)/3400$
$= 0.43 < 1.2\,\text{N/mm}^2$

Protective factor (PF) in the middle of the shelter and at 0.25−0.30 m above the floor level.

F.3 Design of a nuclear shelter based on the US codes

F.3.1 Introduction

Many codes in the USA have empirical equations which are based on imperial units. The reader is given the conversions in SI units. However, the bulk of the calculations given here are based on imperial units (conversion factors shown below).

Conversion factors

1 ft = 0.3048 m; 1 lb/ft² = 47.88 N/m²;
1 lbf = 4.448 N; 1 lb/ft² = 16.02 kg/m²;
1 lbin = 0.113 Nm; 1 kg = 9.806 N;
1 lb/in² = 6895 N/m²; 1 in = 25.4 mm

Dynamic stresses

Concrete:
Comp $-$ 1.25(3000) = 3750 psi
Diagonal tension $-$ 1.00(3000) = 3000 psi
Direct shear $-$ 1.10(0.18)(3000) = 600 psi

Reinforcement:
Bending $-$ 1.10(60 000) = 66 000 psi
Shear $-$ 1.10(60 000) = 60 000 psi

since f'_c = 3000 psi and f_y(static) = 60 000 psi.

F.3.2 Wall design

Figure F2.1 shows a one-way slab fully restrained at the supports.
Wall thickness (T_c) = 300 mm (12 in) (see Fig. F3.1).
The US recommended covers are 0.75 in and 1.5 in (37 mm) rather than 50 mm (adopted by the Home Office).
For a negative moment, d = 12 $-$ 1.5 $-$ 0.3125 = 10.1875 in (assuming # 5 bars). For a positive moment, d = 12 $-$ 0.75 $-$ 0.3125 = 10.935 in.

$$A_s = 0.0025 \times 12 \times 10.935 = 0.328 \text{ in}^2/\text{ft}$$

5 bars at 11 in (275 mm), A_s = 0.34 > 0.328 in^2. The wall blast load = 0.085 N/mm^2 = 12.33 lb/in^2. The ultimate moment is given by

$$M_u = (A_s f_{yd}/b)(d - a/2)$$

where $a = A_s f_{yd}/0.85bf'_c(\text{dyn}) = 0.586$ in
$\qquad b = 12$ in

$\qquad M_u$ (positive) = M'_P = 19 900 in-lb/in
$\qquad M_u$ (negative) = M'_N = 18 500 in-lb/in

$\qquad E_c$ for concrete = $D^{1.5}$ 33$\sqrt{f'_c}$
$\qquad\qquad = (150 \text{ lb/in}^3)^{1.5} \times 33(3000)^2$
$\qquad\qquad = 3.32 \times 10^6$ psi

$\qquad \rho = D$ = density of concrete = 150 lb/in^3 (23.6 kN/m^3)

$\qquad E_s$ for steel = 30 $\times$ 10^6 psi (200 GN/m^2)

$\qquad\qquad n = E_s/E_c = 9.03$

Average moment of inertia for a 1 inch strip.
I_g (gross) = $bT_c^3/12 = 144$ in^4
T_c (thickness of the wall) = d
d (average) = 10.5625 in

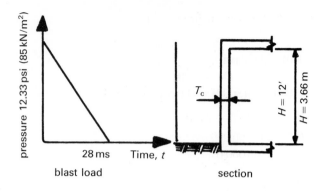

blast load   section

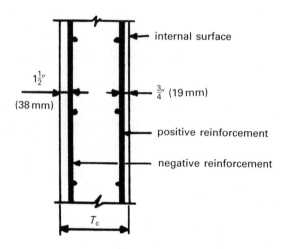

Fig. F3.1 Wall analysis and design.

p (average) $= A_s/bd = 0.00268 = \rho_s$

I (cracked section) using Fig. 2.1

$F' = 0.0175; I_{cracked} = I_c = F'bd^3 = 20.6\,in^4$

I_a = average moment of inertia $= (I_g + I_c)/2 = 82.3\,in^4$

Elastic (k_e) and elasto-plastic (k_{ep}) stiffness

$$k_e = (384E_cI_a)/bL^4 = 244\,lb/in^3; \qquad b = 1$$

$$k_{ep} = (384E_cI_a)/5bL^4 = 48.8\,lb/in^3$$

Elastic and elasto-plastic deflection

$$\delta_e = X_e = r_e/k_e = 10.71/244 = 0.0439\,in$$

$$\delta_{ep} = X_p - X_e = (r_u - r_e)/k_{ep} = 0.084\,in$$

$$X_p = 0.1279\,in$$

Equivalent elastic deflection and stiffness

$$X_E = X_e + X_p(1 - r_e/r_u)$$
$$= 0.0793 \, \text{in}$$

$$K_E = r_u/X_E = 186.8 \, \text{lb/in}^3$$

Load−mass factors and effective mass
Figure F3.2 gives:

K_{LM}	Range
0.77	elastic
0.78	elasto-plastic
0.66	plastic

K_{LM} (elastic and elasto-plastic) = 0.78 (average)
K_{LM} (elastic and plastic) = 0.72 (average)

$$M = \rho \, T_c/\mathbf{g} = 150 \times 1 \times 10^6/32.3(1728)$$
$$= 2700 \, \text{lb-ms}^2/\text{in}^3$$

$$M_{\text{effective}} = K_{LM} \times M = 1944 \, \text{lb-ms}^2/\text{in}^3$$

$$\text{natural period} = T_N = 2\pi \sqrt{(M_e/K_E)} = 20.3 \, \text{ms}$$

where $\mathbf{g} = 32.2 \, \text{ft/s}^2$; $K_E = 186.8$.

Response chart parameters
Reference is made to Fig. F3.2.

Peak pressure $B = 12.33 \, \text{psi}$
Peak resistance $r_u = 14.81 \, \text{psi}$
The chart $B/r_u = 0.8325 \rightarrow T/T_N = 28/20.3 = 1.38$

$X_m/X_E = 1.50$, as this is <3 the section is safe

The corresponding $t_m/T_N = 0.50 \rightarrow t_m/t_o = t_m/T$
$$= (t_m/T_N)/(T/T_N) = 0.50/1.38$$
$$= 0.3623$$

This lies within the range $3.0 > t_m/t_o > 0.1$, hence the response is satisfactory.

Diagonal tension at a distance d *from the support*

$$v_u = r_u[(L/2) - d_e]/d_e = 14.81(72 - 10.1875)/10.1875$$
$$= 89.9 \, \text{psi}$$

The allowable shear stress, v_c, is given by

$$v_c = \phi[1.9\sqrt{f_c'} + 2500p] \leq 2.28\phi\sqrt{f_c'}$$

where $\phi = 0.85$

$$= 94.4 \, \text{psi, as this is} > 89.9 \, \text{psi, OK with no stirrups}$$

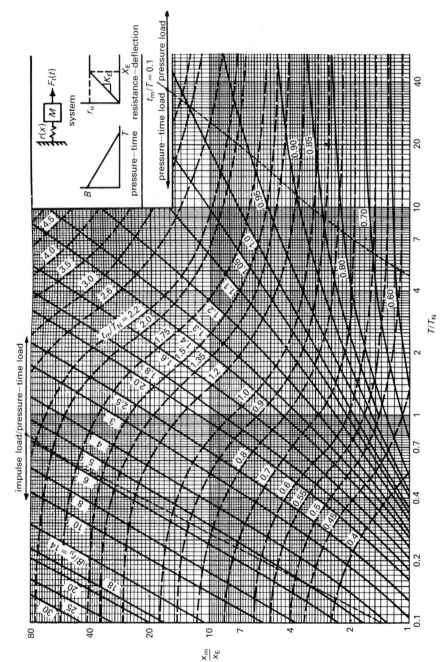

Fig. F3.2 Maximum response of simple spring-mass system.

Ultimate shear

$$V_s = r_u L/2.0 = 14.81 \times 144/2.0 = 1066 \, \text{lb/in}$$

Allowable shear

$$V_d = 0.18 f'_c (\text{dyn}) \, bd = 6050 \, \text{lb/in} > 1066 \, \text{lb/in}$$

Hence the 300 mm (12 in) wall designed against the same blast load in both codes (British and US) is safe. The roof slab can be checked in the same way as for the gas explosion, described earlier in the text.

F.4 Lacing bars

When a ring forced concrete element is subject to a blast load, the element deflects far beyond the stage of well defined cracking until:

(1) The strain energy of the element is developed sufficiently to balance the kinetic energy created by the applied load when it comes to rest.
(2) Fragmentation of the concrete element results in either its partial or total collapse.

For the development of the available energy of the concrete elements, it is necessary to make changes in the reinforcement layouts and details. Each element is reinforced symmetrically. They and the intervening concrete are laced together, as shown in Figs F4.1 and F4.2, with continuous bent diagonal bars. This system offers forces which will contribute to the integrity of the protective element. Where structural elements are located outside the immediate high blast intensity, they should be designed without lacing. All other types are given in Figs F4.3 to F4.7.

Design of lacing bars
Where lacing bars are needed, the following calculations will help in the design of nuclear shelters. Equations beginning from equation (F.7) are needed. The lacings can be in both the vertical and horizontal directions.

Vertical lacing bars
The wall thickness is kept the same. Data: $d_\ell = 10 \, \text{in}$; $S_\ell = 22 \, \text{in}$; no of bars $= 6$; $D_0 = 0.75 \, \text{in}$

$$d_\ell = 21 + 1.13 + 2 + 0.75 = 24.88 \, \text{in}$$

$$R_{\ell_{min}} = 3D_0$$

for $S_\ell / d_\ell = 0.884$

$$(2R_\ell + D_0)/d_\ell = 7D_0/d_\ell = 0.211$$

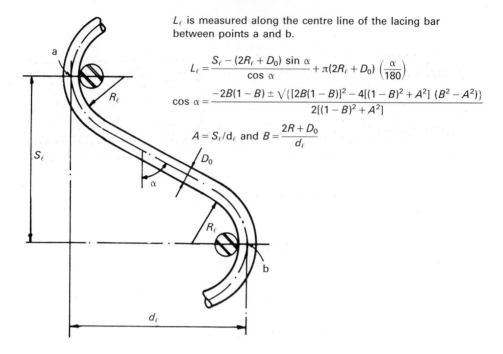

L_ℓ is measured along the centre line of the lacing bar between points a and b.

$$L_\ell = \frac{S_\ell - (2R_\ell + D_0)\sin\alpha}{\cos\alpha} + \pi(2R_\ell + D_0)\left(\frac{\alpha}{180}\right)$$

$$\cos\alpha = \frac{-2B(1-B) \pm \sqrt{\{[2B(1-B)]^2 - 4[(1-B)^2 + A^2]\,(B^2 - A^2)\}}}{2[(1-B)^2 + A^2]}$$

$$A = S_\ell/d_\ell \text{ and } B = \frac{2R + D_0}{d_\ell}$$

Fig. F4.1 Length of lacing bar.

$\alpha = 51.5^0$

$$A_v = (v_{uv} - v_c)b_\ell S_\ell/\phi f_s\,(\sin\alpha + \cos\alpha) = 0.378\,\text{in}^2$$

$A_{v_{min}} = 0.0015\,b_\ell S_\ell = 0.330\,\text{in}^2$; no of bars = 6; $A_s = 0.44\,\text{in}^2$,   OK

Horizontal lacing bars
No of bars = 6; $D_0 = 0.75\,\text{in}$

$$d_\ell = 21.0 + 1.13 + 0.75 = 22.8\,\text{in}$$

$$R_{\ell_{min}} = 3D_0$$

for $S_\ell d_\ell = 20/22.88 = 0.874$

$$(2R_\ell + D_0)/d_\ell = 7D_0/d_\ell = 0.229$$

$A_v = 0.339\,\text{in}^2$
$A_{v_{min}} = 0.0015\,b_\ell S_\ell = 0.0015 \times 11 \times 20 = 0.330\,\text{in}^2$; no of bars = 6; $A_s = 0.44\,\text{in}^2$,
still OK

Additional reinforcement details from the British code are given in Figs F4.8 and F4.9.

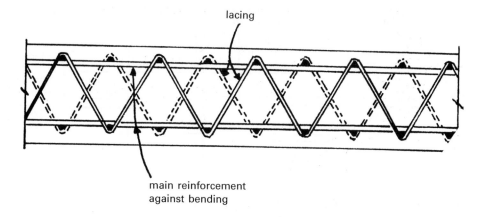

Fig. F4.2 Lacing reinforcement.

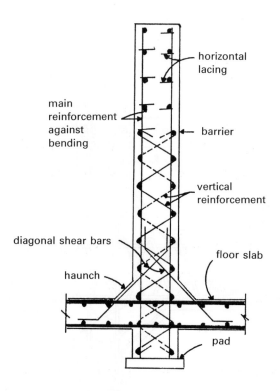

Fig. F4.3 Typical laced wall.

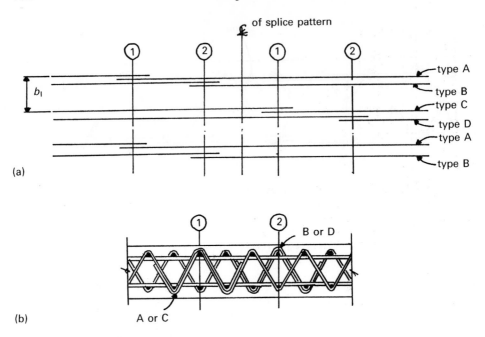

Fig. F4.4 Typical details for splicing of lacing bars: (a) splice pattern; (b) lacing splice.

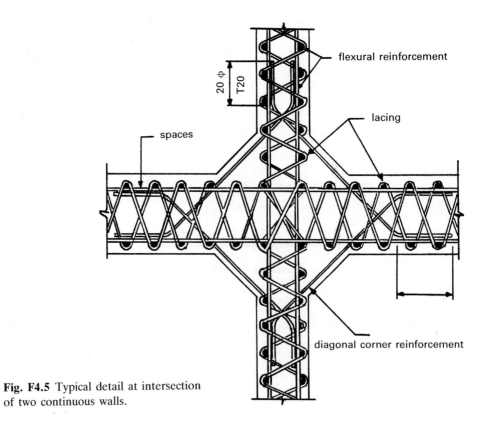

Fig. F4.5 Typical detail at intersection of two continuous walls.

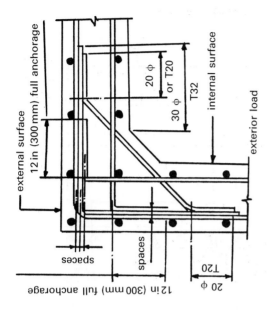

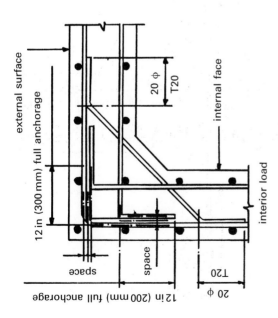

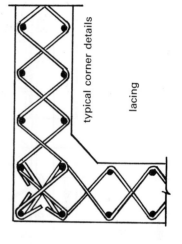

Fig. F4.6 Flextural reinforcement in lacings.

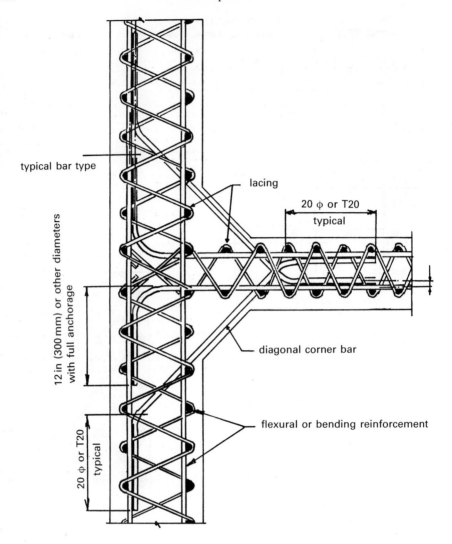

typical bar type

lacing

20 φ or T20
typical

12 in (300 mm) or other diameters
with full anchorage

diagonal corner bar

20 φ or T20
typical

flexural or bending reinforcement

Fig. F4.7 Typical detail at intersection of continuous and discontinuous walls.

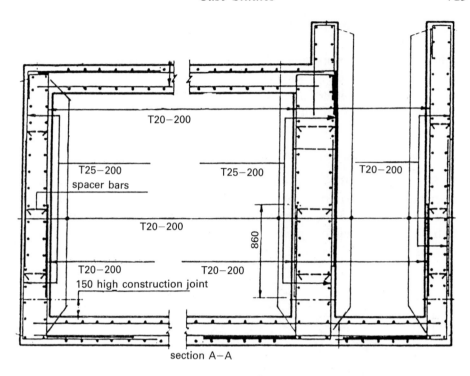

section A–A

Fig. F4.8 Reinforcement through section.

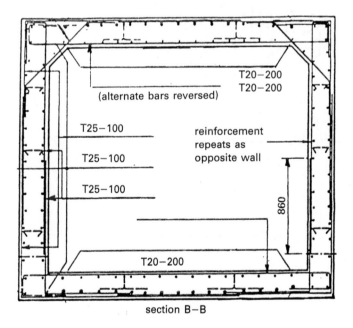

section B–B

Fig. F4.9 Section B–B through shelter.

F.5 Finite-element analysis

A three-dimensional isoparametric, finite-element analysis has been carried out by Bangash[1.149] Figure F5.1 shows the finite-element mesh scheme for a dynamic model for a nuclear shelter. Figure F5.2 gives the relationships between pressure and time. The results are given in Fig. F5.3.

F.5.1 The Swedish design and details

The Swedish code TB78E provides novel details of the nuclear shelter. They are presented here by courtesy of the Civil Defence Administration of Sweden. Figures F5.4 and F5.5 show structural details of a roof slab and sectional details illustrating various reinforcements.

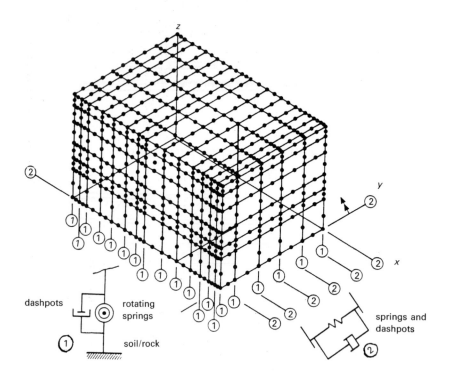

Fig. F5.1 Dynamic model for a nuclear shelter.

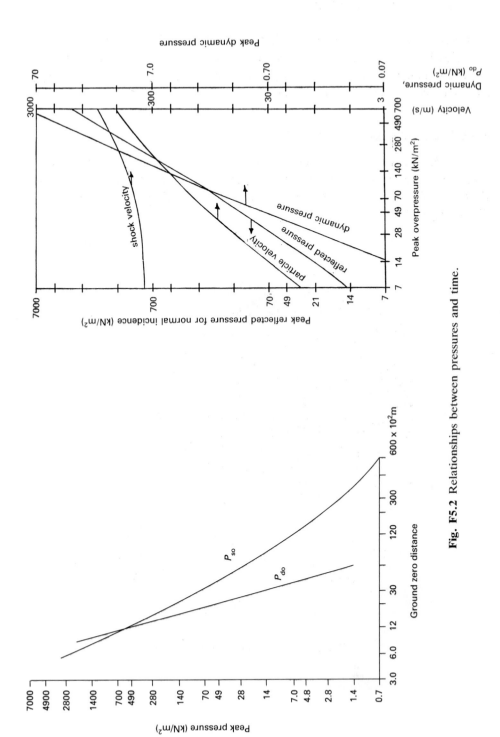

Fig. F5.2 Relationships between pressures and time.

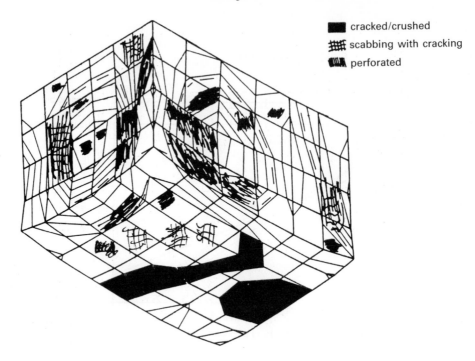

cracked/crushed
scabbing with cracking
perforated

stress trajectories/cracking

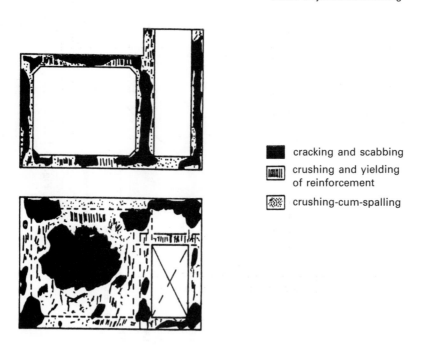

cracking and scabbing
crushing and yielding of reinforcement
crushing-cum-spalling

Fig. F5.3 Typical results from the finite-element analysis.

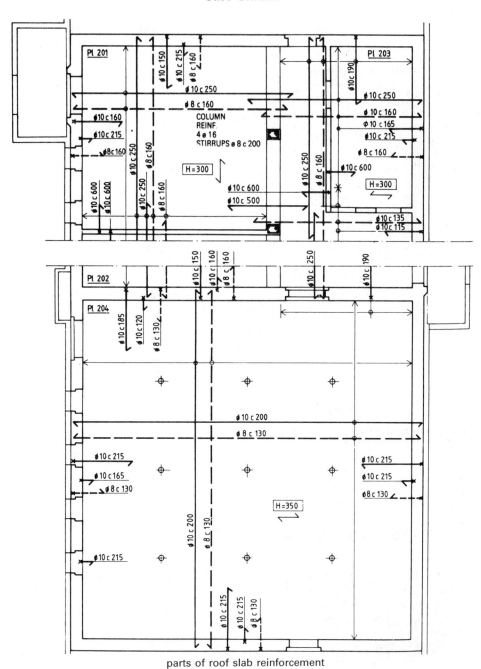

parts of roof slab reinforcement

Fig. F5.4 Structural details — I Swedish code TB78E.

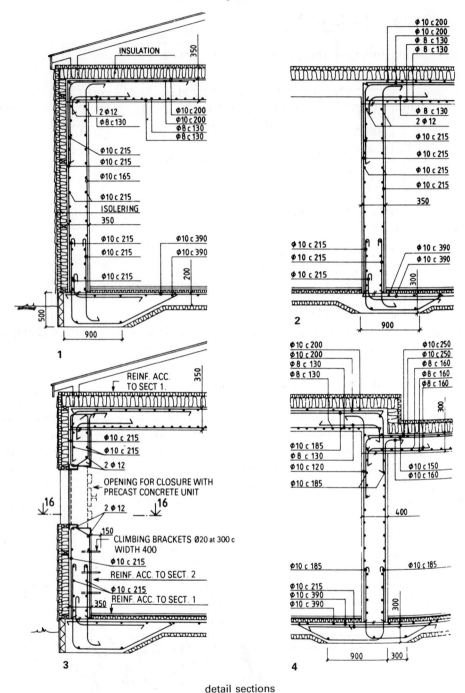

detail sections

Fig. F5.5 Structural details — II Swedish code TB78E.

G: SEA ENVIRONMENT

G.1 Multiple wave impact on a beach front

A multiple wave impact phenomenon is considered on a stone wall with a slope of 60° to the horizontal, as shown in Fig. G1.1. The following are the major input parameters included in the overall impact analysis:

Wall dimensions: 10 m high × 5 m wide × 300 m long
$E = 2.8 \times 10^7 \, \text{kN/m}^2$
$v = 0.2$
ρ (density of rock) $= 2.5 \, \text{Mg/m}^3$
Joints: $\phi = 45^0$
$\quad\quad c = \text{variable} \approx 23 \, \text{kN/m}^2$ and at collapse $21 \, \text{kN/m}^2$

$\quad\quad \sigma_t$ (tensile strength of joint) $= \frac{1}{10} c$

Intact rock: $\phi = 45^0$
$\quad\quad c = 5600 \, \text{kN/m}^2$
Wind associated with waves (100 waves with phase differences assumed): 120 mph
Total number of 20-noded elements: 700 (size $1 \, \text{m} \times 1 \, \text{m} \times 1 \, \text{m}$)
Total number of nodal points: 12 700
Total number of joint elements: 550 (3 mm thick)

The 20-noded, isoparametric finite-element mesh is given for the wall in Fig. G1.2. After 70 continuous wave impacts, the horizontal displacement of the stone wall was as shown in Fig. G1.3. At the hundredth wave impact the stones were dislocated and some rocks/stones slid on the plane. Tension cracks as depicted in Fig. G1.4 were developed. The total duration was 330 seconds and the time interval, t, was 30 seconds. The stress trajectory for this wall corresponding to

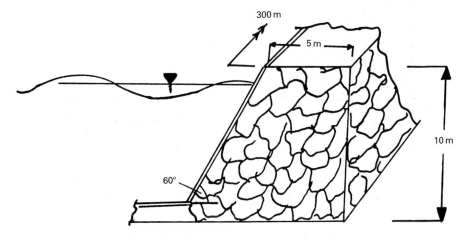

Fig. G1.1 Beach front sectional elevation.

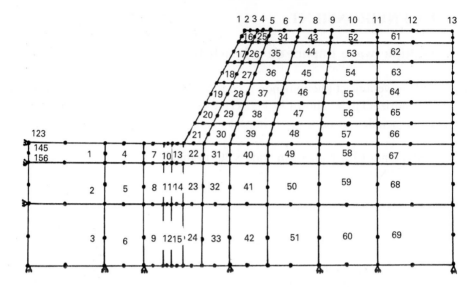

Fig. G1.2 Finite-element idealization of the beach wall.

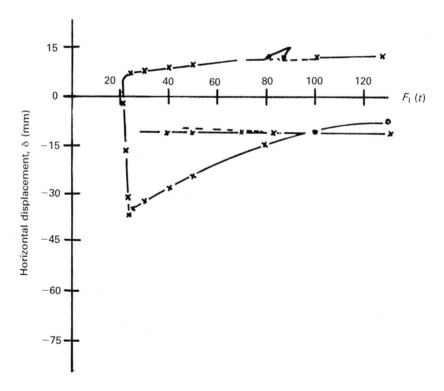

Fig. G1.3 Variation of the horizontal displacement with the decreasing value of the cohesion on the joints.

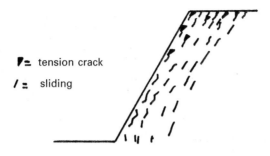

Fig. G1.4 Sliding and cracking.

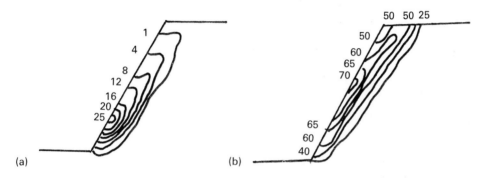

Fig. G1.5 Stress contours (MN/m²): (a) shear; (b) tensile.

these cracks is plotted in Fig. G1.5. The associated flow rule was adopted. A gap element was considered for the slipping joint. All the rock properties given in Chapter 2 were examined. The limestone wall considered in the analysis was examined for its tri-axial behaviour. Endochronic theory associated with a flow rule was the main objective behind this analysis.

Another beach front was investigated against multiple wave impacts. The beach defences were formed and the armour units of tetropods and doloses were located. A specific placement of units is given in Fig. G1.6 and is based on the progressive formation of a stack assembly. Since their positions are of a stochastic nature, the forces acting on the units will show a large variability. A typical formation of a breakwater is shown in Fig. G1.7. It is generally accepted that the most extensive breakage occurs in the vicinity of the still water level. Wave parameters, on the other hand, determine the new bearing conditions. The finite-element mesh scheme for a dolos is shown in Fig. G1.8. The following parameters are included in the finite-element analysis:

Layers   1   2   3   4   5   6

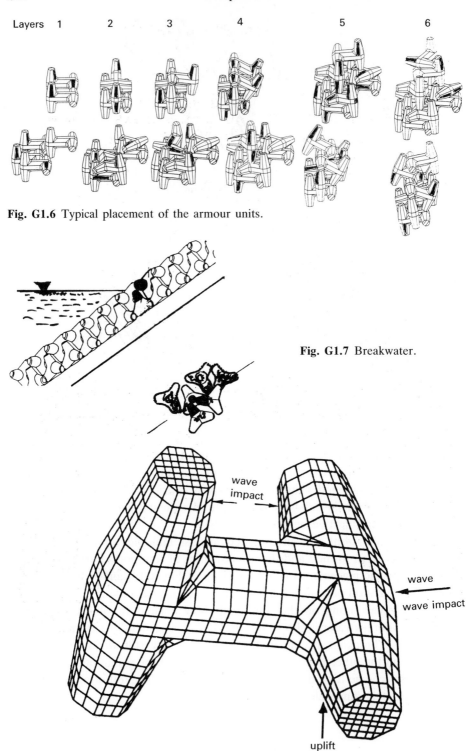

Fig. G1.6 Typical placement of the armour units.

Fig. G1.7 Breakwater.

wave
impact

wave

wave impact

uplift

Fig. G1.8 Finite-element mesh scheme of the armour unit.

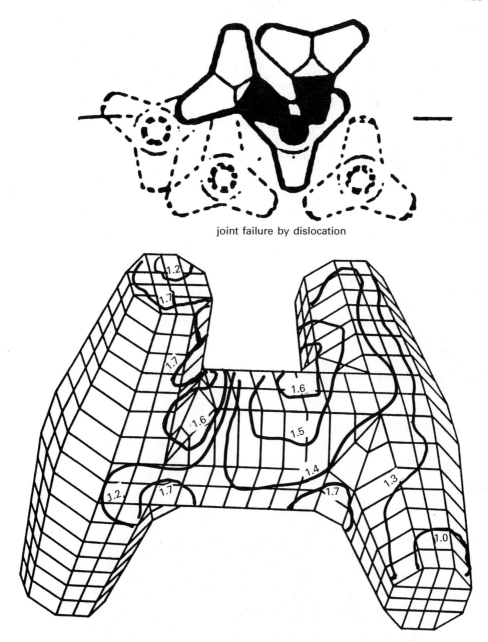

joint failure by dislocation

Fig. G1.9 Stress-trajectory principle tensile stress (N/mm²).

Mass: 1440 kg
Waist ratio: 0.329
Height: 1660 mm
Density, ρ: 3×10^{-6} kg/mm^3
Dynamic modulus, E_{dy}: 4.3×10^4 N/mm^2
Tensile strength, σ_{tu}: 3.58 N/mm^2
Number of solid isoparametric 8-node elements: 440/dolos
Number of prism elements: 25/dolos
Total number of elements for a 3-layer stack 3 m wide area/unit height: 1450/dolos
Initial water depth: 2.8 m
Wave height, H: 1.4 m
Wave period, T_p: 2.5 s
Breakwater slope: 1:2.5

The Jonswap normalized spectra are given by

$$h_{rms} = \sqrt{[(1/m) \sum_{i=1}^{m} h_i^2]}$$

where h_1 = surface elevation for a one-hour storm with a critical velocity of 2 m/s.

Analysis

One hundred and fifty multiple wave impacts were considered. The analysis indicated dislocation of the front units and cracking in the middle zones. The zone damaged by dislocation is shown in Fig. G1.9. The principle stress-trajectory is given on the 150th wave impact. Dashpots were used along with the gap element to evaluate the dislocation. The main units themselves were not damaged because the tensile strains developed were below the allowable values.

G.2 Explosions around dams

Underwater explosions and their effects on adjacent structures are of great concern to naval architects and hydraulic/dam engineers. Typical underground explosion charts, shown in Figs G2.1 and G2.2 give relationships for shock wave pressure versus slant range and peak pressure versus time t. A 50 kt burst in deep water was assumed at a distance of 12 000 ft (3658 m) from a concrete arch dam. The phenomenon is given in Fig. G2.1. The peak pressure $p(t)$ at 0.1 second at a slant range of 12 000 ft (3658 m) was found to be 470 psi. The time constant θ was 0.05 second. For the normalized case $t/\tau = 2$ and thus $p(t)/p$ is about 0.149. Therefore the $p(t)$ value will be $470 \times 0.149 \approx 70$ psi. A mixed finite-element analysis was carried out using 20-node isoparametric brick and prism elements, as shown in Fig. G2.3; $p(t)$ was applied as a dynamic load on the dam. The stress trajectories are shown in Fig. G2.4. The total number of brick and prism elements was 300. The number of gap elements was 98. The frequency and modal amplitude, taking into account the hydrostatic coupling and the total displacement of the dam, were

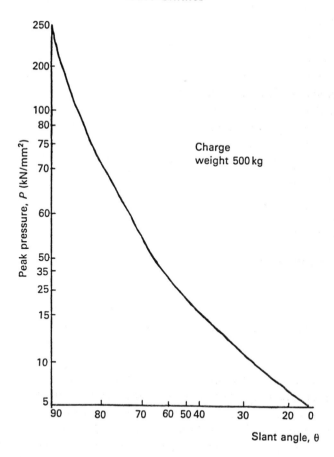

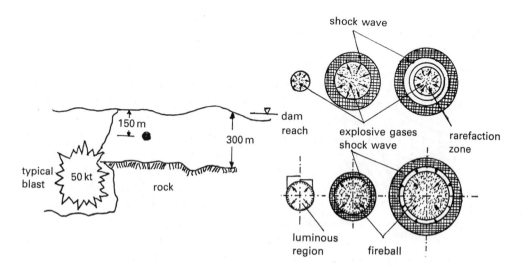

Fig. G2.1 Shock wave peak pressure versus slant range.

Chapter 7

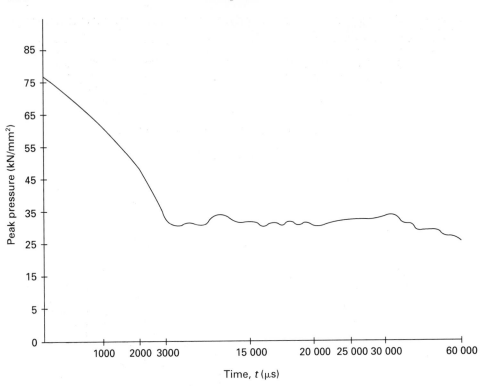

Fig. G2.2 Peak pressure versus time.

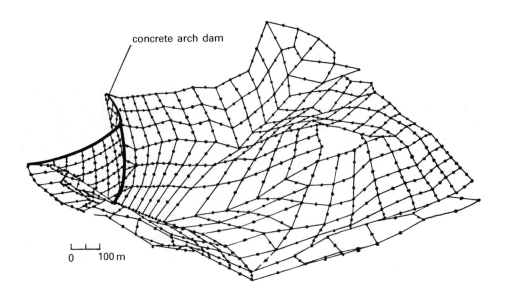

Fig. G2.3 Finite-element mesh scheme for an arch dam and the surrounding medium.

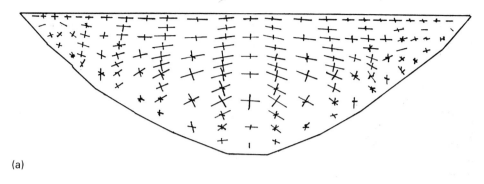

(a)

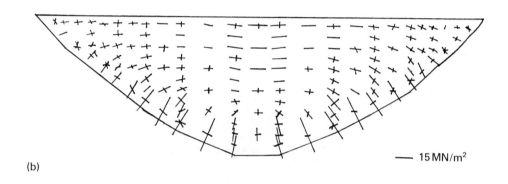

(b)

—— 15 MN/m²

Fig. G2.4 Principle stresses for the dam: (a) upstream face; (b) downstream.

calculated by recomposition. The final damaged zones are given in Fig. G2.5. The vertical construction joints were cut by the tensile stresses. It is assumed that the self-weight of the dam has not changed. A frontal solution procedure mixed with the Wilson-θ method was adopted for evaluation of the results. The convergence solution was limited to 0.01. Endochronic theory was adopted for the behaviour of concrete under an explosion. Figures G2.6 and G2.7 illustrate the displacement–time and acceleration–time relationships, respectively.

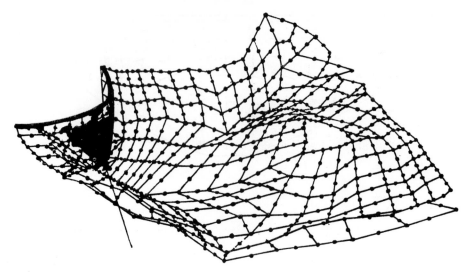

damaged zones and deformation of the dam and surroundings

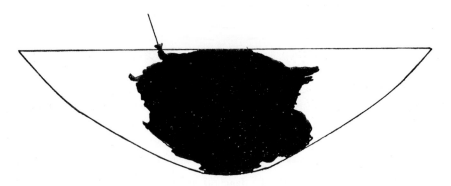

Fig. G2.5 Damaged zones.

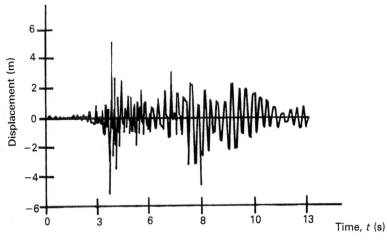

Fig. G2.6 Displacement–time relationship.

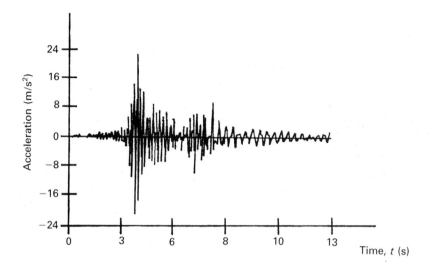

Fig. G2.7 Acceleration−time relationship.

G.3 Ship-to-ship and ship-to-platform: impact analysis

The effects of collisions involving ships and tankers, carrying military equipment and hazardous materials in particular, are of growing public interest. Tables G3.1 and G3.2 give useful formulae and data for ships/aircraft carriers impacting against each other or against oil and other installations at sea. They include ship impact mechanics in the sea environment. On the basis on these data, dynamic finite-element analysis has been carried out for a cargo ship impacting an aircraft carrier. The numbers of finite elements chosen were:

4000 8-noded isoparametric elements for the aircraft carrier
1500 8-noded isoparametric elements for the cargo ship
500 gap elements

As the impact force is quasi-static, the number of load increments was taken as 21. The Newmark β-method of the implicit type was chosen for the solution procedure. The computers used were an IBM 360/75 and an IBM 4381 with front-ended Cray-2. Another sophisticated output was achieved using Cyber 72. The total time taken to analyze this impact problem was 6 hours, 31 minutes and 31 seconds. Some initial work was done on VAX. Figure G3.1 shows the damaged zones of both the ship and the aircraft carrier. The extent of the damaged areas can be scaled out from this figure. The time increment was 0.04 second. The material behaviour was represented by associated Prandl−Reuss, von Mises and octahedral methods. The strain rate concept was included in the overall analysis. The area parts and structural details were provided by the US Navy.

Table G3.1 Ship impact mechanics.

Ship/platform impact mechanics are based on two criteria, namely the conservation of momentum and the conservation of energy. For a short impact duration:

$$E_s + E_p = \tfrac{1}{2}(m_s + \Delta m_s) \, v_t^2 \, \{(1 - v_t/v_s)^2/[1 + (m_s + \Delta m_s)/(m_t + \Delta m_t)]\}$$

where m_s = mass of the striking ship (m_s = added mass (40% of vessel displacement for sideways collision and 10% for bow or stern collision))

m_t = mass of the platform including the added mass or any other target

v_s = velocity of the striking ship immediately before collision

v_t = velocity of the semi-submersible platform immediately before collision

E_s = energy absorbed by the striking ship

E_p = energy absorbed by the ship/tanker or target struck

$F_I(t)$ = impact force

$$= E/e_t = \tfrac{1}{2}m_s \, v_s^2 \left[\frac{(m_s + m_t)}{m_t}/1 + \frac{(m_s + m_t)}{m_t} \right]/e_t$$

where m_s, v_s = mass and velocity of the striking ship, respectively

m_t = equivalent added mass of the target

e_t = penetration = x_p

E = energy of deformation = $\tfrac{1}{2}m_s \, v_s^2 \, \phi_1 \, \phi_2 \, \phi_4 + \tfrac{1}{2}m_t \, v_t^2 \, \phi_5 - v_s v_t m_t \phi_6$

ϕ stands for the non-dimensional functions representing the colliding ship/tanker, the impact location, the angle of encounter, the speed of the target and the product of both speeds.

(1) Duration for constant collision force:

$$T = \phi_1 \phi_2 (m_t/F_I)v_s$$

(2) Linear collision force, duration:

$$T = (\pi/2) \, \sqrt{(\delta(2 - \delta)/\gamma\eta^2 \, \tan\theta)} \, \phi_1 \phi_2 m_t$$

The acceleration at the time of collision is given by

$$\ddot{\delta}_{max(x)} = F_I(t)_x/m_t; \quad \ddot{\delta}_{(y)} = [F_I(t)_y/(m_t + m_s)] + [\delta x F_I(t)_x e/I']$$

where
δx = distance between the centre of gravity of the target and the point at which acceleration is computed

$F_I(t)_x$, $F_I(t)_y$ = maximum components of the collision force in the transverse and longitudinal directions

I' = mass moment of inertia of the target

Storm tide

A maximum tide of 2.15 m is assumed and finally combined with a storm surge of 0.6 m to produce a total storm tide rise of 2.75 m.

Operating tide

A maximum tide of 2.15 m is combined with a storm surge of 0.12 m to produce a total operating tide rise of 2.27 m.

Table G3.1 *Continued.*

Storm current

The maximum storm current velocity varies with water depth according to the following profile:

At the water surface: 1.38 m/s
Above the bottom: 0.54 m/s at 15 m
At the bottom: 0.24 m/s

Operating current

The maximum current velocity under the operating conditions varies with the water depth according to the following profile:

At the water surface: 1.17 m/s
Above the bottom: 0.45 m/s
At the bottom: 0.24 m/s

Temperatures

The minimum water temperature is assumed to be $+7.0°C$. The crude oil temperature in storage is taken as $40°C$.

Wind velocities

Storm wind: the maximum wind velocity of 125 mph, sustained for one minute, is used in conjunction with the storm wave.
Operating wind: a wind velocity of 60 mph is used in conjunction with the operating wave.
Instantaneous gust: a maximum instantaneous gust velocity of 160 mph is used throughout.

Ship masses

Aircraft carrier 60 000 tonnes; impacting ship 5000 tonnes.

Table G3.2 Additional data on ship impact mechanics. Angle of encounter versus obliquity and influence mass and eccentricity.

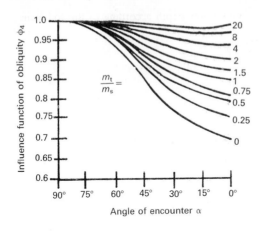

14 megajoules (MJ) for a side collision equal to 40% of added mass.

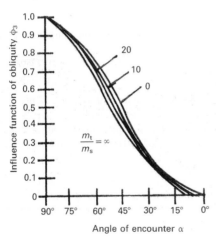

11 megajoules (MJ) for a bow or stern collision equal to 10% of added mass.

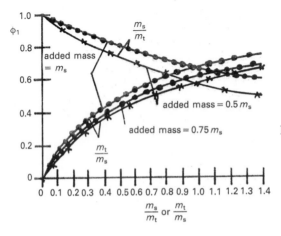

Influence of mass ratio ϕ_1.

Table G3.2 *Continued.*

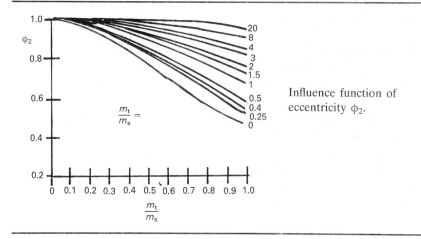

Influence function of eccentricity ϕ_2.

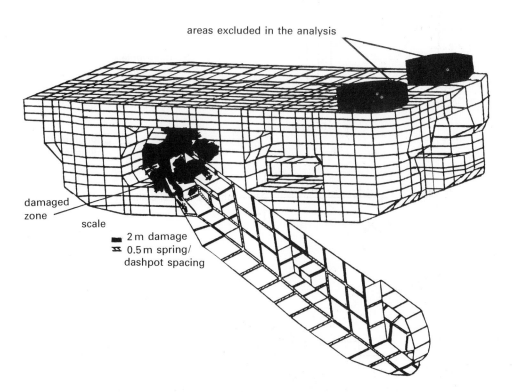

Fig. G3.1 Ship/carrier impact.

G.4 Jacket platform: impact and explosion

G.4.1 Ship impact at a jacket platform

Collisions between ships/tankers and platforms or other installations in offshore oil and gas fields constitute a type of accident which can have catastrophic effects. There is a growing concern about explosions caused by gas leaks and by terrorist bombs/missiles.

   Figure G4.1 gives a sketchy view of the type of impact which might occur at any level of the platform. Section G.3 (Tables G3.1 and G3.2) gives the relevant input, and if one of the objects becomes stationary the entire data can easily be modified to suit the relevant conditions. In this case the platform is assumed to be non-moveable and fixed firmly at the supports. The cargo ship of section G.3 is

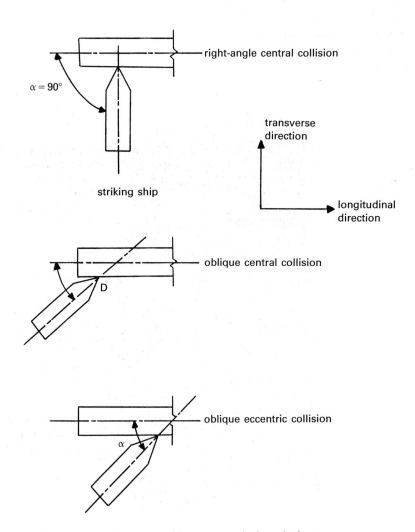

Fig. G4.1 Typical types of impact at a jacket platform.

assumed to be involved in the collision. The platform chosen for this analysis is the famous Heerema platform, shown in Fig. G4.2. In addition to the data in section G.3, the following are also required:

Operational impact
Maximum impact force, $F_1(t)$: 2.8 MN
Total kinetic energy, KE: 0.5 MJ
Velocity of the ship, v_s: 0.5 m/s
Average stiffness, K: 10.0 MN/m
Damping ratio: $\gamma = 0.065$
Accidental impact
Maximum impact force, $F_1(t)$: 10.3 MN
Total kinetic energy, KE: 7.8 MJ
Velocity, v_s: 2.0 m/s
Material and other parameters
Brace diameter, d_0: 2000 mm
Thickness, t: 80 mm

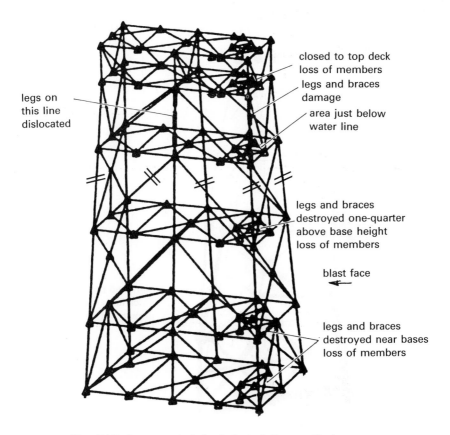

Fig. G4.2 A segment of the jacket platform with damage.

Brace span, L: 31.6 m
Uni-axial yield stress, f_{yi}: 360 N/mm^2
Indentation depth/radius, Δ/r: 0.50

T-Joint in out-of-plane bending is given by

$$K_r = 0.0016ER^3 \ (215 - 135\beta) \ ((1/\gamma_0) - 0.02) \ ^{(2.45-1.6\beta)}$$

where E = Young's modulus = 200 GN/m^2
       R = chord radius
       β = brace radius/chord radius ≤ 0.6
       γ_0 = chord radius/chord thickness
       K_r = rotational spring constant

The deformed region $> 3.4d$.
Finite elements: 100 000 3-noded line elements of the platform match with the 20-noded elements of the ship for this problem.
   A typical damaged finite-element mesh close to the ship impact is shown in Fig. G4.3. Figures G4.4 and G4.5 show the relationships for impact versus time

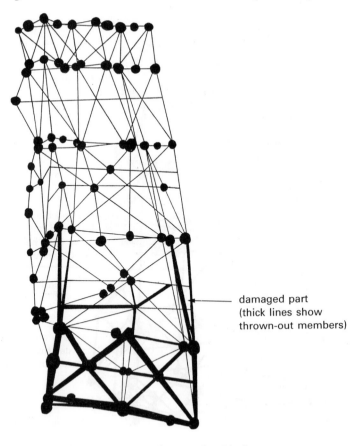

damaged part
(thick lines show
thrown-out members)

Fig. G4.3 Damaged part close to the ship impact zones.

and deflection versus time, respectively. Figures G4.6 and G4.7 give the impact load versus dent per diameter and impact versus deflection per length, respectively.

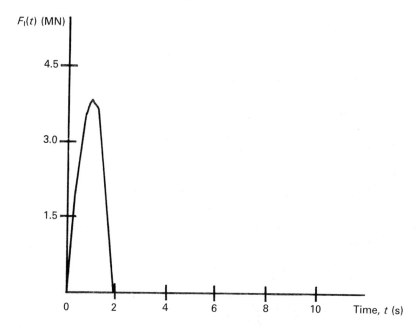

Fig. G4.4 Force−time function.

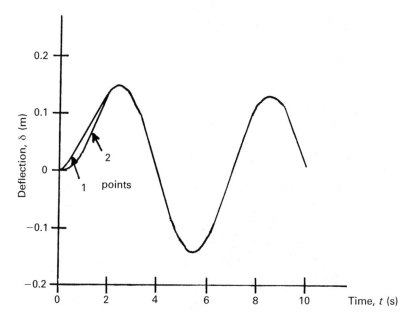

Fig. G4.5 Deflection−time relationship.

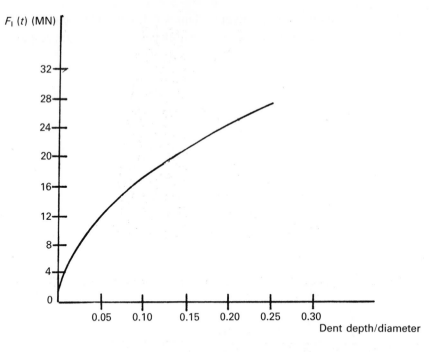

Fig. G4.6 Impact versus the ratio of dent depth/diameter.

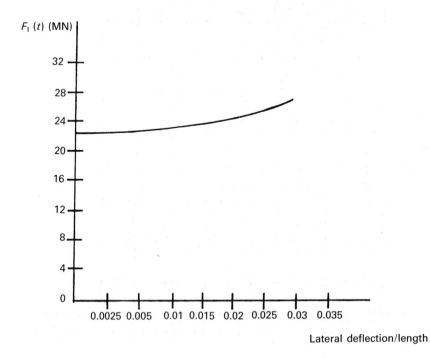

Fig. G4.7 Impact versus deflection per length.

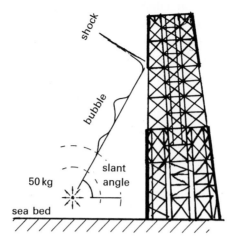

Fig. G4.8 Layout of the scheme for blast load against the jack-up platform.

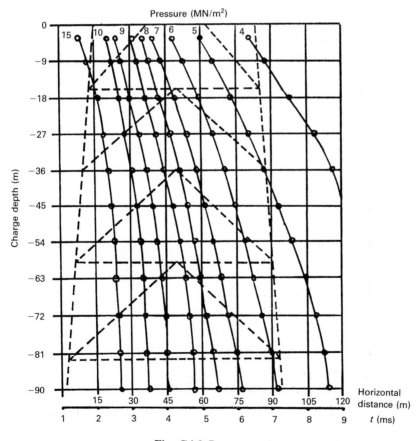

Fig. G4.9 Pressure pulse.

A 50 kg bomb was placed at a depth of 100 m at 30 m from the jacket platform. Figure G4.8 gives the layout of the scheme. Figures G4.9 to G4.11 give the pressure pulse, displacement−time and velocity−time relationships, respectively, for the platform. Figure G4.12 shows the finite-element scheme of a typical joint. The damage caused by the explosion is demonstrated in Fig. G4.13.

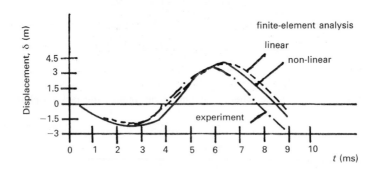

Fig. G4.10 Displacement−time relationship.

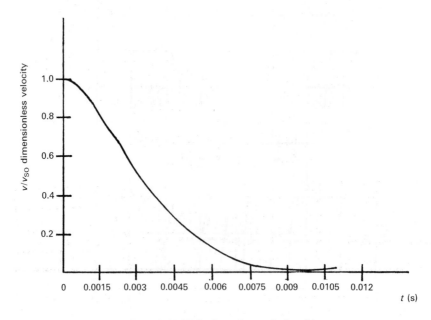

Fig. G4.11 Velocity−time relationship.

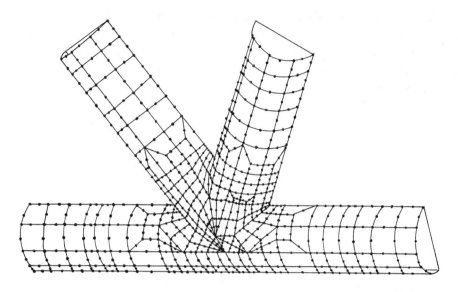

Fig. G4.12 The finite-element scheme for a joint subject to explosion pressure.

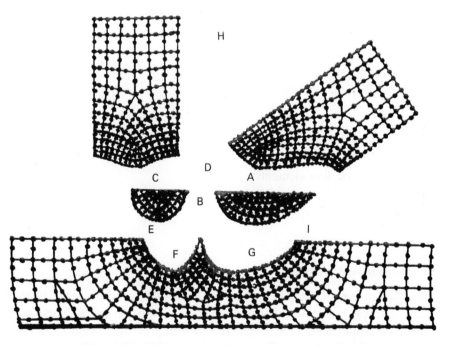

Fig. G4.13 Damaged joint and maximum ejected distances ($A = 5 \times 10^6$ mm; $B = 4 \times 10^6$ mm; $C = 9 \times 10^6$ mm; $D = 8 \times 10^6$ mm; $E = 6 \times 10^6$ mm; $F = 12 \times 10^6$ mm; $G = 20 \times 10^6$ mm; $H = 50 \times 10^6$ mm; $I = 6 \times 10^6$ mm).

G.5 Impact of dropped objects on platforms

The simultaneous drilling and production operation increases the risk of damage to completed well heads from accidentally dropped drilling equipment. In general, for the sake of protecting the platform against damage, heavy wooden mats encased in steel plate are laid out. The critical objects generally fall vertically. Heavy objects and equipment with large contact areas during impact represent severe cases for stringers and girders. For this reason, restrictions may be imposed on crane operators with regard to lifting heights and frequencies. It is essential, therefore, to evaluate initially the most critical impacted zones. Figure G5.1 shows a covered impacted area determined by the extent of the crane radius. Figures G5.2 and G5.3 give the structural layouts of the zones under investigation. The following data are additional input for the finite-element analysis.

Main platform dimensions: see Figs G5.2 and G5.3 for the deck.
Drop objects: see Fig. G5.4
Deck plate thickness: <50 mm.

Loading and areas
80 kN dropped 2 m on to a 1 m² area   { skid base lay down areas drill floor
88 kN dropped 1 m on to a 0.02 m² area { pipe rack and conductor hatches
50 kN dropped 3 m on to a 0.02 m² area { cantilever walkway

Material stresses
Minimum yield strength, f_y: 355 N/mm²
Minimum ultimate strength, f_u: 510−610 N/mm²
Young's modulus, E_s: 200 GN/m²

Finite elements
20-node isoparametric: 1578
Gap elements: 700
3-node isoparametric line elements:
stringers: 410
girders: 200

G.5.1 Finite-element analysis

Dynamic finite-element analysis was carried out on the top deck of the platform. The time interval was 0.04 s. The finite-element mesh scheme is based on the actual gridwork indicated in Fig. G5.5. The total time for the execution of the job was 3 hours, 15 minutes and 31 seconds on an IBM 4381 computer using the ISOPAR program.

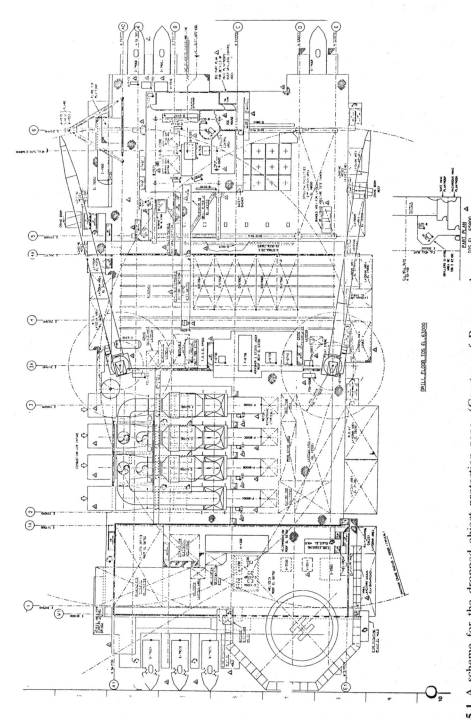

Fig. G5.1 A scheme for the dropped object protection zone. (Courtesy of Brown and Root, Collierswood, London, UK.)

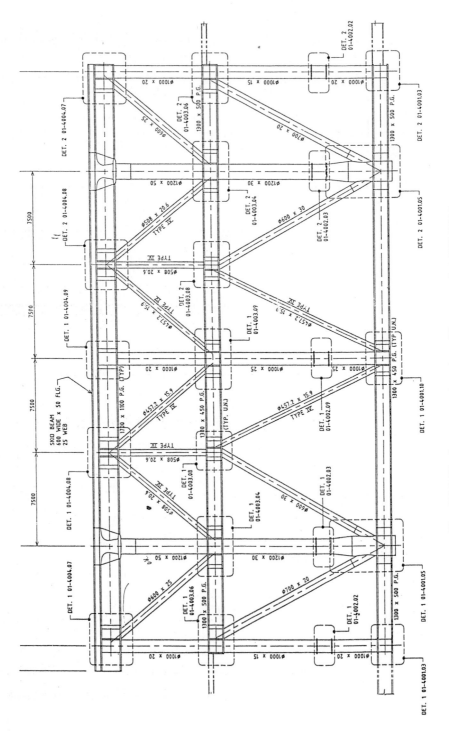

Fig. G5.2 Platform truss elevation. (Courtesy of AGIP UK.)

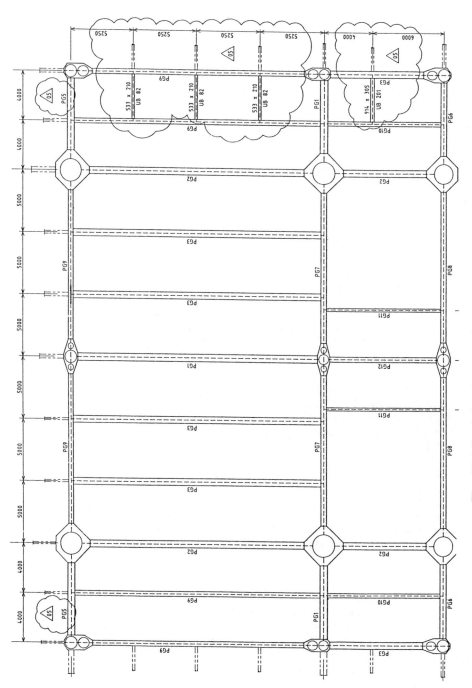

Fig. G5.3 Deck framing plan. (Courtesy of AGIP UK.)

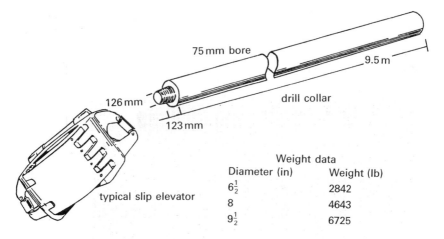

Fig. G5.4 A typical drill collar as a dropped object.

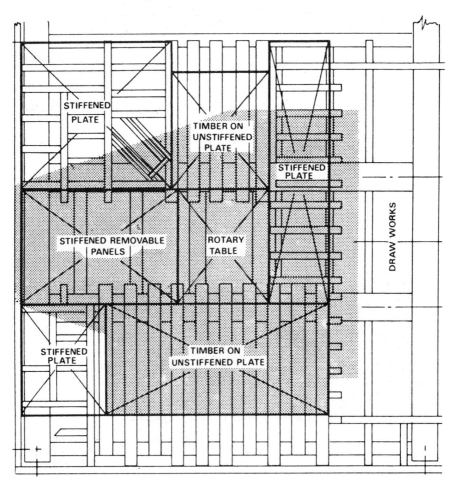

Fig. G5.5 A grid for the final preparation of the finite-element mesh scheme.

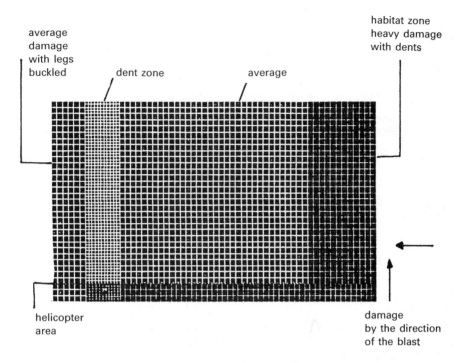

Fig. G5.6 Damage zones with dents (plotted on the finite-element mesh scheme).

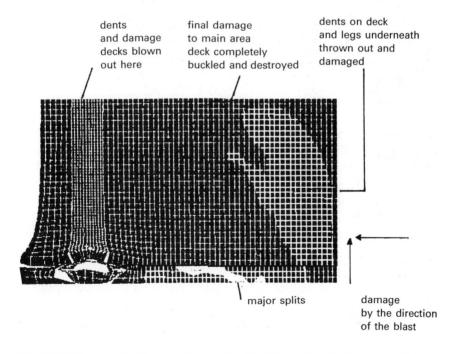

Fig. G5.7 Damage location on deck under direct impact − final post-mortem.

G.5.2 Results

The results are summarized as follows:

Size of ruptured area: $175 \times 150\,mm$
Maximum displacement: 310 mm edge; 455 mm edge; 726 mm central; 435 mm corner
Maximum indentations: drill floor: 75 mm; pipe bridge: 101 mm; lay-down area: 98 mm; pipe rack: 125 mm

Figure G5.6 illustrates the damaged zones with dents. Figure G5.7 gives the damage location.

H: SOIL/ROCK SURFACE AND BURIED STRUCTURES

H.1 General introduction

A great deal of information is given in this text on impact and explosion in the field of ground engineering. In this section a few examples are given on impact/ explosion occurring at ground level or below, with special emphasis on the prediction of the penetration of missiles/projectiles/bombs in soil/rock media. Table H1.1 lists properties and dimensions of some well known combinations of soils forming particular strata.

H.2 Soil strata subject to missile impact and penetration

The most important objective is to model in detail both the target soil and the penetrating missile. The modelling of the soil as a target material presents a problem which stems from the fact that *in situ* soil/rock is a non-linear anistropic material and is further complicated by variations in the existing layered strata. A 20-node, three-dimensional, isoparametric, dynamic finite-element method was adopted for the target material, as shown in Fig. H2.1. Each layer was modelled with respective material properties. Two types of expansion cavities, namely cylindrical and spherical, were modelled. Using the cylindrically-expanding cavity, the target material was assumed to move radially onwards as the missile penetrated. The radial pressure in the cylindrical cavity was then assumed to be acting normally on the penetrator surface. Where there was an angle of attack, apart from axial velocities, an angular velocity was introduced to the penetrator as and when the angular orientation of the penetrator occurs.

Where a spherical cavity is assumed, the target material was then assumed to move in a direction normal to the surface of the missile/bomb penetrator. The ISOPAR program uses both methods simultaneously in the analysis of the three-

Table H1.1 Selected strata as targets.

Target type	Soil description	Depth (m)	Soil constant
I	(a) Clayey silt, silty clay, hard and dry	0–8	5.18
	(b) Sandy silty, dense dry and well cemented	8–15	2.48
	(c) Clay, silty and soft	0–1.6	40.00
	(d) Sand, silty, clayey dense dry to damp	15–24	5.10
II	Clay, soft, wet, varied, medium to high plasticity	0–4	48.00
III	Ice glacier	0–20	4.10
IV	Silt, clayey permafrost	0–10	3.55
V	Sand loose to medium, very moist	0–22	6.70
VI	Stiff clay	0–5	9.75
VII	Sandstone	0–10	1.35
VIII	Fine grained rock	0–4	1.06
IX	Limestone	0–10	2.86

Typical density (ρ) calculation
$1\,N \approx 1\,kg \times 1\,m/s^2$; $1\,kN \approx 1\,Mg\,m/s^2$
Bulk density of $2\,Mg/m^3$; γ_w = unit weight = $\rho\,g$ = $19.62\,kN/m^3$

ρ (Mg/m^3):

loose sand	dense sand	glacial till	soft organic clay	peat
1.44–1.79	1.75–2.08	2.11–2.32	0.68–1.43	0.09–1.091

dimensional behaviour of the target material. The soil/rock behaviour was assumed to follow the bulk modulus method coupled with soil plasticity and critical state. Soil constants were used for top layers and the impact velocity is taken for those layers as discussed in Chapter 2. The velocity changes as the missile/bomb penetrates through various layers. The depth of penetration at any level is computed as the missile/bomb penetrates the layer corresponding to a particular soil constant. The target I(d) of Table H1.1 was chosen for the analysis.

H.2.1 Finite-element analysis

A dynamic finite-element analysis was carried out on target I(d) of Table H1.1. The missile/bomb was assumed to be initially elastic. It was also assumed that as the penetrator struck the earth, the volumetric strain was concentrated in the first column of the element. The time step δt was originally controlled by the smallest element in the missile/bomb. As the analysis progressed, the soil elements in the first row collapsed and the time step dropped by a factor ranging between 2.5 and 4. A subroutine was added to control this aspect and to maintain the same size

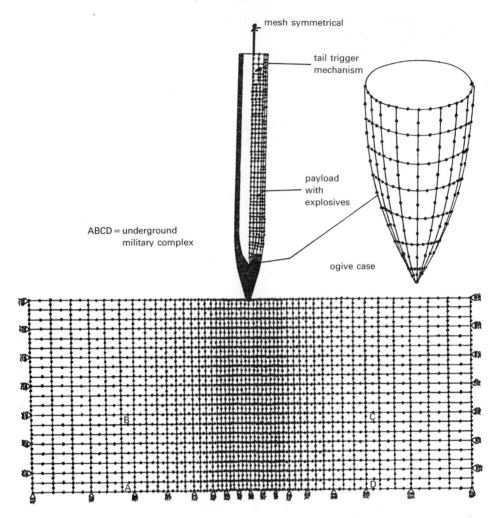

Fig. H2.1 Missile with warhead penetrating soil/rock strata (20 m deep).

step throughout. An implicit solution was introduced as well, to treat the smallest elements in particular. Manual rezoning was allowed each time a new soil element passed by the nose of the penetrator. All other large elements were advanced explicitly. The finite-element method gives a useful comparison with Petry's formula for assessing penetration depths. The following data were included in the finite-element analysis:

Number of finite elements
Target: 1750
Scud missile: 385
Master nodes: 3400
Slab nodes: 639

Length and width of target 100 m and 35 m
Depth of target: 20 m
The material properties are as described in Chapter 2.

Additional properties
Target:
    soil/rock density varies: see Table H2.1
    bulk modulus of elasticity: $7 \times 10^6 \, kN/m^2$
Missile:
    initial velocity: 250 m/s
    weight: 500 lb (227 kg)
    inclination to vertical: 15°
    nose CRH: 6
Scud type:
    case length and diameter: 8.3 m and 2.15 m

H.2.2 Results

Figures H2.2 and H2.3 show, for various missile velocities, the relationship between the impact force and the target sand/clay layers. As the missile penetrates through various layers at various time intervals, Fig. H2.4 gives the summary, in dimensionless form, for the penetration depth and velocity against time in sand, clay and rock layers. Figure H2.5 shows the missile cavity formation and the position of the gap elements in conjunction with the master and slave nodes of the elements. Figure H2.6 shows the final penetration depth for a velocity of 250 m/s. Figures H2.7 and H2.8 show the stress trajectories and the stress–time history,

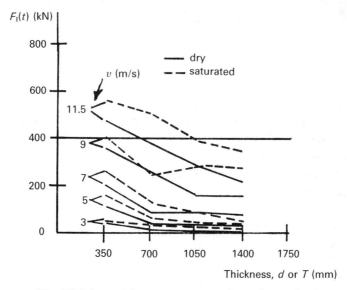

Fig. H2.2 Impact force versus target layer for sand.

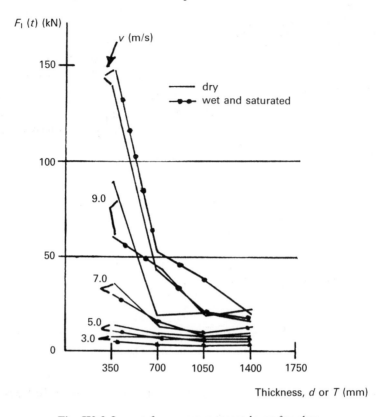

Fig. H2.3 Impact force versus target layer for clay.

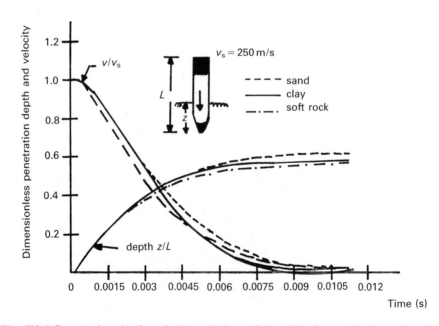

Fig. H2.4 Penetration depth, velocity and time relationships for sand, clay and rock.

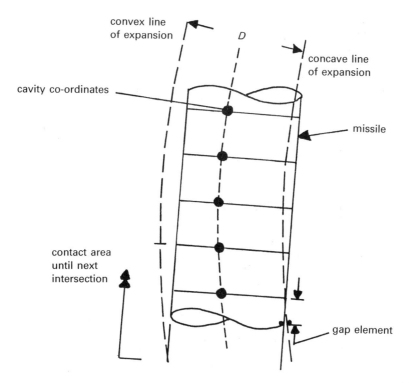

Fig. H2.5 Position of the missile cavity gap elements. (After Davie & Richgels[3.170])

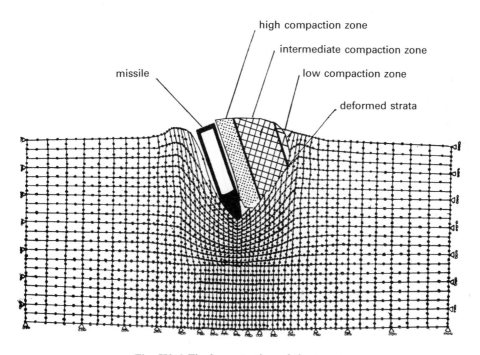

Fig. H2.6 Final penetration of the target.

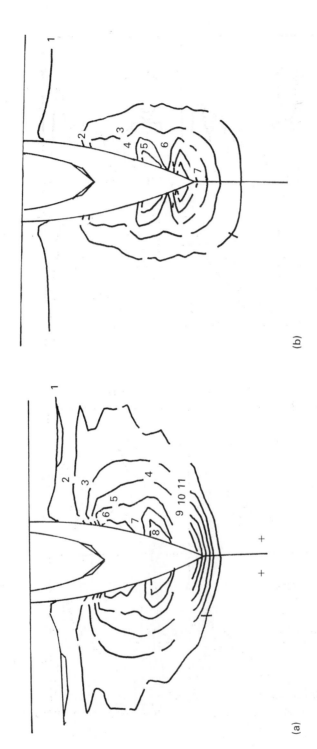

Fig. H2.7 Principal stresses in (a) the axial and (b) the radial directions. (a) (1) 12.75, (2) 14.93, (3) 16.71, (4) 18.35, (5) 20.75, (6) 25.91, (7) 26.81, (8) 28.37 MN/m², (9), (10) and (11) tip almost plastic; (b) (1) 12.75, (2) 14.59, (3) 17.13, (4) 23.75, (5) 21.81, (6) 16.35 MN/m², (7) plastic.

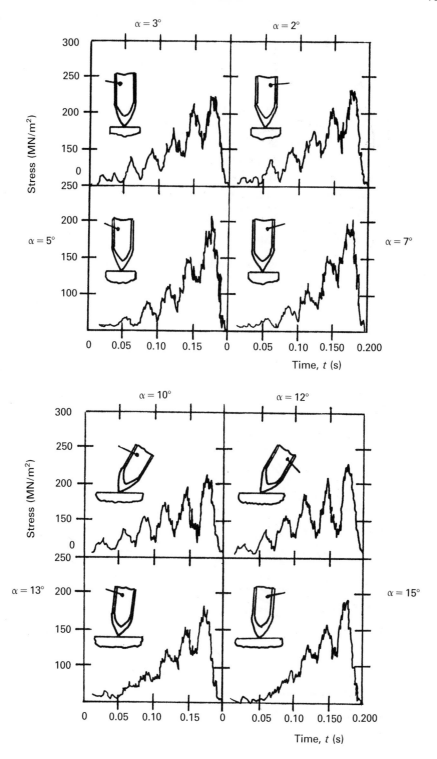

Fig. H2.8 Stress−time history of the missile at various angles.

respectively, for a scud missile impacting the ground at an angle varying between 0° and 15°.

H.2.3 Explosions in soil strata

The finite-element mesh given in Fig. H2.1 for soil/rock strata may also be used for other types of soils. The soils chosen are sand, clay and loam. The bomb yields considered are within the range of 45 kg to 900 kg. The altitude varies for different bombs. Figure H2.9 gives the final results for all three soil strata. The depth of penetration from the ground level to the centre of the bomb was evaluated for the three soils for different bombs dropped from different heights.

H.2.4 Craters resulting from explosions

The plowshare program was used to predict crater dimensions when an explosive of non-yield is buried at a given depth for a specific soil stratum. This work was

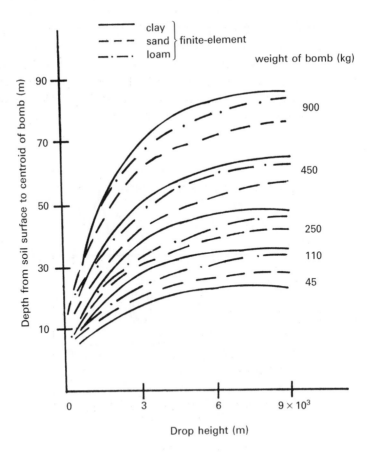

Fig. H2.9 Bomb explosions in soils.

carried out by Lawrence Radiation Laboratory at Livermore in California[3.171] A numerical method was developed for the calculation of crater development, including mound and cavity growth. Continuum mechanics (involving stress tensors) was the basis of the calculation. The medium was represented by the bulk modulus approach. Figures H2.10 to H2.13 summarize some of the achievements. Project Danny Boy was a nuclear cratering experiment in basalt. A 0.4 kt device was placed 33 m below the surface, yield 1.3 m basalt. The cavity and mound configurations at zero time are given in Fig. H2.10. The calculated cavity and mound configurations at various time intervals are shown in Fig. H2.11. For the scooter event, the tensor zoning is given in Fig. H2.12. The scooter mound is finally shown in Fig. H2.13.

H.2.5 Explosions in boreholes

A typical soil stratum, as given in Table H2.1, layout (c), was chosen for the finite-element analysis. The finite-element scheme given in Fig. H2.1 was adopted for economic reasons. The missile line was replaced by a borehole 15 m deep. The pulse duration was maintained from 0.5 to 2 seconds. The burden was assumed to be 2.10 m. The velocity at the initial stage was 0.7 m/s. Four different explosive pressure intensities were considered: 7, 15, 40 and 75 kg/m explosive weight per metre of the hole, thus giving explosive weights of 105, 225, 600 and 1125 kg. The radii R ranged between 10 and 50 m. Figure H2.14 gives the particle acceleration against R for various explosives. The total cavity extended by 350% laterally. The maximum frequency was 250 Hz. The maximum strain was 5000 μs.

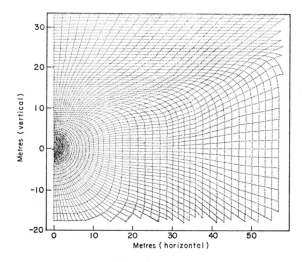

Fig. H2.10 Cavity and mound configurations for the Danny Boy event at zero time. (Courtesy of J.T. Cherry[3.171])

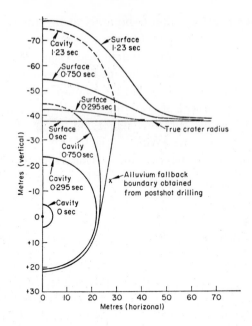

Fig. H2.11 Calculated cavity and mound configurations at 0, 0.295, 0.750 and 1.23 s. (Courtesy of J.T. Cherry[3.171])

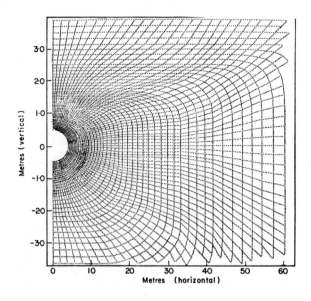

Fig. H2.12 Tensor zoning for the scooter event at the start of calculation. (Courtesy of J.T. Cherry[3.171])

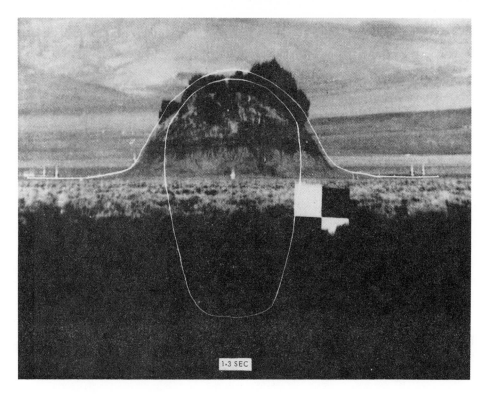

Fig. H2.13 Scooter mound (compared with tensor calculations) at 1.3 s.

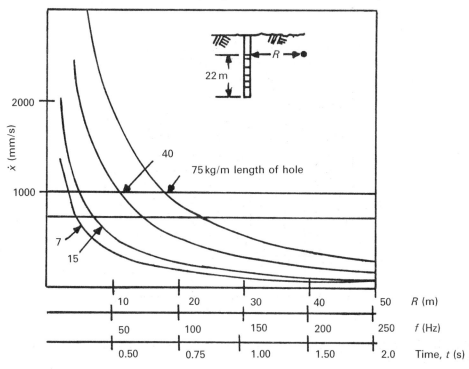

Fig. H2.14 Borehole particle acceleration for various radii R and explosives.

Table H2.1 Examples of soil profile.

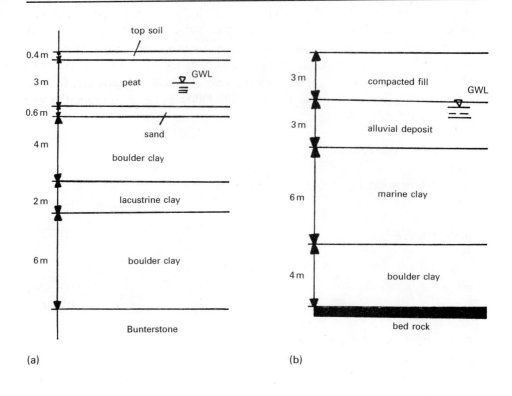

(a)

(b)

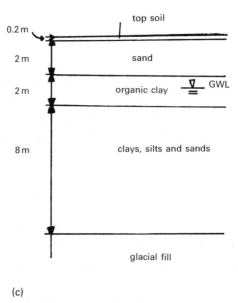

(c)

H.2.6 Explosions in an underground tunnel

Figure H2.15 shows a typical tunnel layout and Fig. H2.16 gives the tunnel cross-section. A typical finite-element mesh scheme is shown in Fig. H2.17. A 10 kt nuclear explosion causing a pulse–time relationship was assumed inside the tunnel, with a rock overburden of 6 m carrying set and bedding planes. A total number of 1180 20-node isoparametric finite elements were chosen, which gave rise to 46 230 nodes. Twenty increments were adopted. The following material properties were chosen for the finite-element input:

Maximum overburden 280 m
Intact rock
$E = 2.8 \times 10^7$ kPa; E_m rock mass = 7500 MPa
$K_o = \frac{1}{3}$
$\nu = 0.2$
$\rho = 0.27$ kN/m^3
Joints

$\left.\begin{array}{l} \text{Normal } \sigma_n \\ \text{Oblique} \\ \text{Shear } \tau_s \end{array}\right\}$ compliances $\begin{array}{l} = 1 \times 10^{-7}\,\text{m/kPa} \\ \\ = 2 \times 10^{-7}\,\text{m/kPa} \end{array}$ $\left.\begin{array}{l} \text{bedding} \\ \text{planes} \end{array}\right\}$ $\left.\begin{array}{l} \text{N100°E/60°W} \\ \text{joint set 1 N280°E/60°W} \\ \text{joint set 2 N170°E/90°W} \end{array}\right.$

Cohesion, $c = 50$ kPa or 0.05 MPa
Friction angle, $\phi = 45°$ and 0° for bedding planes
Spacing = 1 m
Reinforcement rock bolts
$E = 0.2 \times 10^9$ kPa
$G = 0.769 \times 10^8$ kPa
$\sigma_y = 0.25 \times 10^6$ kPa
$p = 0.003$

The non-associated flow rule reduced the safety factor; rock bolts were included as bonded elements. The stresses were obtained at Gauss points. The properties of the joints were incorporated with the blocks into a super element of an anisotropic material, reflecting the joint directions. The failure criteria of the rock joint act as bench marks.

$$F = \{\tau\} + (n\tan\phi/R_f) - (c/R_f) = 0$$

$$\begin{aligned} f_R &= \text{flexibility ratio} \\ &= E(1+\nu)/(1/R^3)[6E_\ell I_\ell/(1-\nu_l^2)] \end{aligned}$$

where the subscript ℓ is for the liner.

Figure H2.18 shows the adjacent rock tunnel failure. Figure H2.19 shows displacements for various iterations.

The soil properties around the tunnel were assumed to be non-linear, visco-plastic and anisotropic. The movement between the tunnel and the surrounding medium was allowed for using the 600 gap elements between the tunnel and the

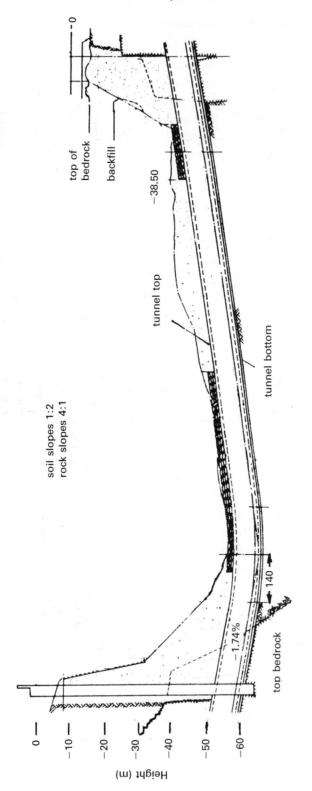

Fig. H2.15 Typical tunnel layout.

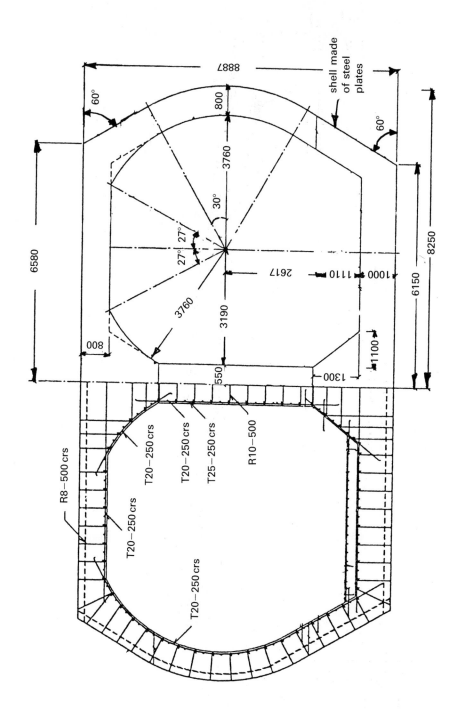

Fig. H2.16 Tunnel cross-section and typical reinforcement.

assumed tunnel under explosion

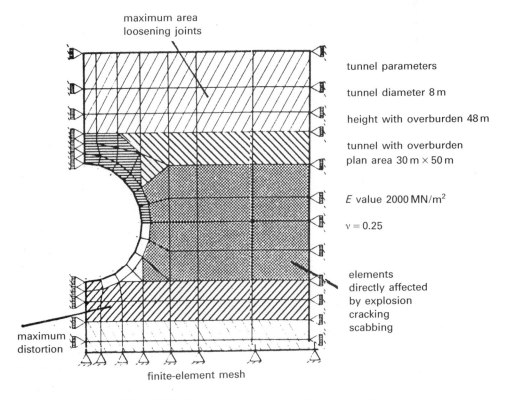

maximum area
loosening joints

tunnel parameters

tunnel diameter 8 m

height with overburden 48 m

tunnel with overburden
plan area 30 m × 50 m

E value 2000 MN/m²

$\nu = 0.25$

elements
directly affected
by explosion
cracking
scabbing

maximum
distortion

finite-element mesh

Fig. H2.17 Finite-element scheme for the tunnel.

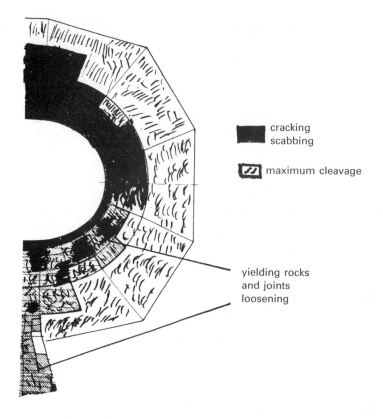

cracking
scabbing

maximum cleavage

yielding rocks
and joints
loosening

Fig. H2.18 Tunnel post-mortem.

surroundings. Figures H2.20 and H2.21 show the strain–time relationships and stress trajectories after the explosions.

H.2.7 Rock fractures caused by water jet impact

A granite rock mass was chosen for the study of fractures produced by the shear impact of a water jet. The extreme dimensions around the granite rock, which lies in the North-West Frontier Province of Pakistan, were found to be 10.4 m by 9.3 m by 8.8 m. Four boreholes were assumed, each of diameter 300 mm and at 2.7 m distances. The chosen pressure–time pulse is shown in Fig. H2.22. The injection rate given by various experts was $0.03\,m^3/s$. Figure H2.23 gives the finite-element mesh scheme for the rock. The increase of K_{IC} with fracture size was noted. The crack tip approached zero in a number of solutions when it reached the interface. However, the non-linear fracture mechanics approach did not support this elastic analysis. The material properties, including *in situ* toughness, were as follows:

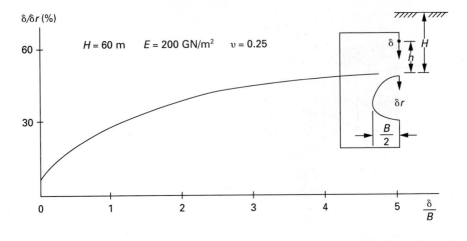

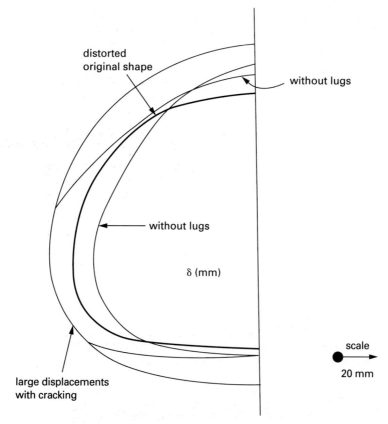

Fig. H2.19 Tunnel displacements.

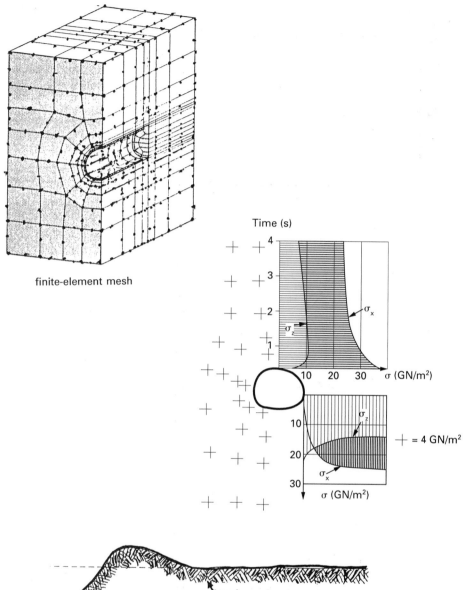

finite-element mesh

Time (s)

σ_x

σ_z

10   20   30 σ (GN/m^2)

σ_z

$+ = 4$ GN/m^2

σ_x

σ (GN/m^2)

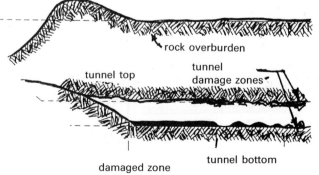

rock overburden

tunnel top

tunnel
damage zones

tunnel bottom

damaged zone

Fig. H2.20 Stress−time relationship after explosion.

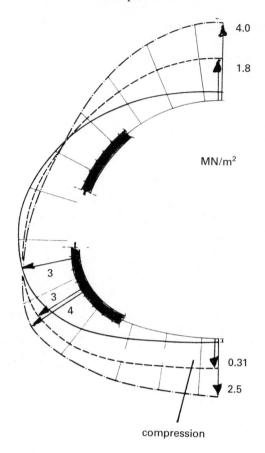

Fig. H2.21 Tunnel stress trajectory after explosion.

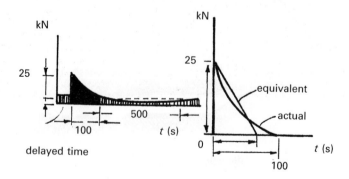

Fig. H2.22 Pressure–time pulse for fluid/rock impact.

Young's modulus, E_{rock}: 45 000 MPa
Poisson's ratio: 0.31
Fracture toughness: 4.5 MPa√m
Confining pressure: 0 to 20.9 MPa
K_{IC}: 1.11 to 3.3 MPa√m

Figure H2.24 gives a mesh plot showing the post-mortem after 10 minutes.

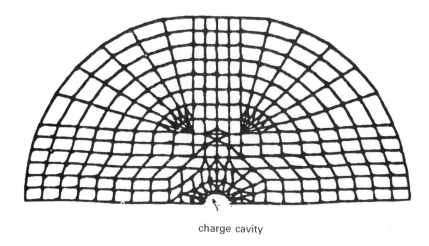

charge cavity

Fig. H2.23 Finite-element mesh scheme for the granite rock.

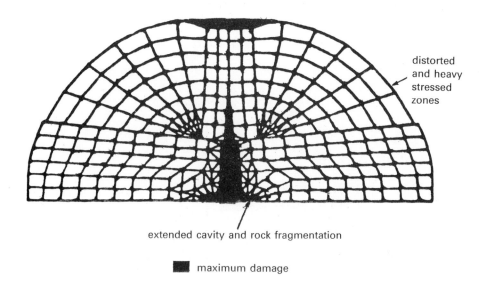

distorted
and heavy
stressed
zones

extended cavity and rock fragmentation

■ maximum damage

Fig. H2.24 Rock post-mortem.

BIBLIOGRAPHY

1 Material properties

1.1 Richart F.E., Brandzaeg A. & Brown R.L. (1928) A study of the failure of concrete under combined compressive stresses. *Bulletin No 185*, Vol. 26. Engineering Experiment Station, University of Illinois.

1.2 Mazars J. & Bazant Z.P. (1989) *Cracking and damage (Strain Localization and Size Effect)*. Elsevier Applied Science.

1.3 Hannant D.J. & Frederick C.D. (1968) Failure for concrete in compression. *Magazine of Concrete Research*, **20**, 137–44.

1.4 Link J. (1975) Numerical analysis oriented biaxial stress–strain relation and failure criterion of plain concrete. Third International Conference on Structural Mechanics in Reactor Technology (SMiRT), London, September 1975, paper H1/2.

1.5 Chinn J. & Zimmerman R.M. (1965) Behaviour of plain concrete under various high triaxial compression loading conditions. *Technical Report No WL TR 64–163 (AD468460)*. Air Force Weapon Laboratory, New Mexico.

1.6 Reimann H. (1965) *Kritische Spannungszustande der Betons bei mehrachsiger, ruhender Kurzzeitbelastung*. Deutscher Ausschuss für Stahlbeton, Berlin, Heft 175.

1.7 Willam K.J. (1974) Constitutive model for the trixial behaviour of concrete. IABSE Seminar, ISMES, Bergamo, Italy.

1.8 York G.P., Kennedy T.W. & Perry E.S. (1971) An experimental approach to the study of the creep behaviour of plain concrete subjected to triaxial stresses and elevated temperatures. *Union Carbide, Report No 28642*. The University of Texas at Austin.

1.9 Linse D., Aschl H. & Stockl S. (1975) Concrete for PCRV's strength of concrete under triaxial loading and creep at elevated temperatures. Third International Conference on Structural Mechanics in Reactor Technology (SMiRT), London, September 1975, paper H1/3.

1.10 Gardner J.J. (1969) Triaxial behaviour of concrete. *Proc. Am. Concr. Inst.*, no 2, 136–47.

1.11 Kasami H. (1975) Properties of concrete exposed to sustained elevated temperature. Third International Conference on Structural Mechanics in Reactor Technology (SMiRT), London, September 1975, Vol. 3, paper H1/5.

1.12 Browne R.D. (1958) Properties of concrete in reactor vessels. *ICE Conference on Prestressed Concrete Pressure Vessels*, pp. 131–51. Institution of Civil Engineers, London.

1.13 England G.L. & Jardaan I.J. (1975) Time-dependent and steady state stresses in concrete structures with steel reinforcement at normal and raised temperatures. *Magazine of Concrete Research*, **27**, 131–42.

1.14 Bangash Y. & England G.L. (1980) The influence of thermal creep on the operational behaviour of complex structures. International Conference on Fundamental Creep and Shrinkage, Lausanne, Switzerland.

1.15 Mohr O. (1914) *Abhandlungen aus dem Gebiet der Technischen Mechanik*, 2nd edn, p. 192. Wilhelm Ernst und Sohn, Berlin.

1.16 Kupfer H. (1973) Das Verhalten des Betons Unter Mehrachsigen Kurzzeitbelastung der Zweiachsigen Beanspruchung. Deutscher Ausschuss für Stahlbeton, Berlin, Heft 229.

1.17 Evans R.J. & Pister K.S. (1966) Constitutive equations for a class of on-linear elastic solids. *Int. J. Solids Struct.*, **2**, 427−5.

1.18 Chinn J. & Zimmerman R.M. (1965) Behaviour of plain concrete under various high triaxial compression loading conditions. *Technical Report No WL TR 64−163 (AD4684460)*. Air Force Laboratory, New Mexico.

1.19 Mills L.L. & Zimmerman R.M. (1970) Compressive strength of plain concrete under multiaxial loading conditions. *ACI J.*, **67**, 802−7.

1.20 Chen E.S. & Buyulozturk O. (1983) *Damage model for concrete in multiaxial cyclic stress*. Department of Civil Engineering, MIT, Cambridge, Massachusetts.

1.21 Bangash Y. (1985) PWR steel pressure vessel design and practice. *Progress in Nuclear Energy*, **16**, 1−40.

1.22 Hobbs D.W. (1977) Design stresses for concrete structures subjected to multiaxial stresses. *Struct. Eng.*, **55**, 157−64.

1.23 Kotsovos M. (1974) *Failure criteria for concrete under generalised stress states*, p. 284. PhD thesis, University of London.

1.24 Chen, A.C.T. & Chen W.F. (1975) Constitutive relations for concrete. *J. Eng Mech. Div., ASCE*, **101**, 465−81.

1.25 Chen W.F. *et al.* (1983) *Constitutive Equations for Engineering Materials*. Wiley, New York.

1.26 Bazant Z.P. (1978) Endochronic inelasticity and incremental plasticity. *Int. J. Solids Struct.*, **14**, 691−714.

1.27 Cedolin L., Crutzen Y.R.J. & Dei Foli S. (1977) Triaxial stress−strain relationships for concrete. *J. Eng Mech. Div., ASCE*, **103**, 423−39, no EM3, proc. paper 12969.

1.28 Bazant Z.P. & Bhat P.D. (1976) Endochronic theory of inelasticity and failure of concrete. *J. Eng Mech. Div., ASCE*, **102**, 701−21.

1.29 Bazant Z.P. & Shieh H. (1978) Endoelectronic model for non-linear triaxial behaviour of concrete. *Nucl. Eng Des.*, **47**, 305−15.

1.30 Pfiffer P.A. *et al.* (1983) Blunt crack propagation in finite element analysis for concrete structures. Seventh International Conference on Structural Mechanics in Reactor Technology (SMiRT), Amsterdam, North Holland, paper H/2.

1.31 Johnson W. (1972) *Impact Strength of Materials*. Edward Arnold, London.

1.32 Ford H. & Alexander J.M. (1977) *Advanced Mechanics of Materials*, 2nd edn. Ellis Norwood Chichester, UK.

1.33 Achenbach J.D. (1973) *Wave propagation in elastic solids*. North Holland/American Elsevier.

1.34 Freakley P.K. & Payne A.R. (1978) *Theory and Practice of Engineering with Rubber*. Applied Science Publishers.

1.35 Williams D.J. (1971) *Polymer Science and Engineering*. Prentice Hall.

1.36 Campbell J.D. (1970) *Dynamic Plasticity of Metals*. Springer Verlag.

1.37 Knott J.F. (1973) *Fundamentals of fracture mechanics*. Butterworth.

1.38 Lawn B.R. & Wilshire T.R. (1975) *Fracture of Brittle Solids*. Cambridge University Press.

1.39 François D. (Ed.) (1981) Advances in fracture research. In *Proceedings of Fifth International Conference on Fracture*, Vols 1 and 2. Pergamon.

1.40 Cottrell A.H. (1957) Deformation of solids at high rates of strain. In *The Properties of Materials at High Rates of Strain*. Institution of Mechanical Engineering.

1.41 Grundy J.D. *et al.* (1985) Assessment of crash-worthy car materials. *Chartered Mech. Eng.*, April, 31−5.

1.42 Oxley P.L.B. (1974) Rate effects in metal working processes. In *Mechanical Properties at High Rates of Strain*. Institute of Physics, London.

1.43 Radon J.C. & Fitzpatrick N.P. (1973) Deformation of PMMA at high rates of strain. In *Dynamic Crack Propagation* (Ed. by G.C. Sih). Noordhoff.

1.44 Vigness I. *et al.* (1957) Effect of loading history upon the yield strength of a plain carbon steel. In *The Properties of Materials at High Rates of Strain.* Institution of Mechanical Engineers, London.

1.45 Wigley D.A. (1971) *Mechanical Properties at Low Temperatures.* Plenum Press.

1.46 Al-Haidary J.T., Petch M.J. & De Los Rios E. (1983) The plastic deformation of polycrystalline aluminum. In *Yields, Flow and Fracture of Polycrystals,* (Ed. by T.N. Baker), p. 33. Applied Science Publishers, London.

1.47 Armstrong J.T. (1983) The yield and flow stress dependence on polycrystal grain size. In *Yield Flow and Fracture of Polycrystals,* (Ed. by T.N. Baker), p. 1. Applied Science Publishers, London.

1.48 Bloom T.A., Kocks U.F. & Nash P. (1985) Deformation behaviour of Ni−Mo alloys. *Acta Metall.,* **33,** 265.

1.49 Conrad H. (1964) Thermally activated deformation in metals. *J. Metals,* **16,** 582.

1.50 Duffy J. (1983) Strain-rate history effects and dislocation structure. In *Material Behavior under High Stress and Ultrahigh Loading Rates,* (Ed. by J. Mescall & V. Weiss), p. 21. Plenum Press, New York.

1.51 Hart E.W. (1970) A phenomenological theory for plastic deformation of polycrystalline metals. *Acta Metall.,* **18,** 599.

1.52 Klepaczko J.R. (1968) Thermally activated flow and strain rate history effects for polycrystalline aluminum and theory of intersections. *J. Mech. Phys. Solids,* **16,** 255.

1.53 Klepaczko J.R. (1975) Thermally activated flow and strain rate history effects for some polycrystalline F.C.C. metals. *Materials Sci. Eng,* **18,** 121.

1.54 Klepaczko J.R. & Duffy J. (1982) Strain rate history effects in body-centered-cubic metals. In *Proceedings of the Conference on Mechanical Testing for Deformation Model Development,* (Ed. by R.W. Rhode & J.C. Swearengen), p. 251. ASTM STP no 765, Philadelphia.

1.55 Klepaczko J.R. (1986) Constitutive modelling of strain rate and temperature effects in F.C.C. metals. In *Proc. Int. Symp., Beijing, China, Intense Dynamic Loading and its Effects,* p. 670. Pergamon Press, Oxford.

1.56 Klopp R.W. & Clifton R.J. (1984) Pressure-shear impact and the dynamic viscoelastic response of metals. In *Proceedings of the Workshop on Inelastic Deformation and Failure Modes.* Northwestern University, Evanston.

1.57 Rybin V.V. (1986) Large plastic deformations and fracturing of metals. *Metallurgy, Moscow* (in Russian). MIR, Georgia (former USSR).

1.58 Swearengen J.C. & Holbrook J.H. (1985) Internal variable models for rate-dependent plasticity: analysis of theory and experiment. *Res. Mechanica,* **13,** 93.

1.59 Shah S.P. (Ed.) (1985) Application of fracture mechanics to cementitious composites. In *Proceedings, NATO Advanced Research Workshop.* Martinus Nijhoff, The Hague.

1.60 Wittman F.H. (Ed.) (1986) *Fracture Toughness and Fracture Energy of Concrete.* Elsevier Science, The Netherlands.

1.61 Suaris W. & Shah S.P. (1982) Mechanical properties of materials under impact and impulsive loading. In *Introductory Report for the Interassociation (RILEM, CEB, IABSE, IASS) Symposium on Concrete Structures under Impact and Impulsive Loading.* Berlin (West), BAM 33−62.

1.62 Banthia N.P. (1987) *Impact resistance of concrete.* PhD thesis, University of British Columbia, Canada.

1.63 Mindess S. (1985) Rate of loading effects on the fracture of cementitious materials. In *Application of Fracture Mechanics to Cementitious Composites* (Ed. by S.P. Shah), pp. 617−36. Martinus, Nijhoff, The Hague.

1.64 Mayrhofer A.S. & Thor S. (1982) Dynamic response of fibre and steel reinforced concrete plates under simulated blast load, in concrete structures under impact and impulsive loading. In *Proceedings, RILEM-CEB-IABSE-IASS Interassociation Symposium.* BAM, Berlin (West), 279−88.

1.65 Gopalaratnam V.S. & Shan S.P. (1985) Properties of steel fiber reinforced concrete subjected to impact loading. *ACI J.,* **83**(8), 117−26.

1.66 Naaman A.E. & Gopalaratnam V.S. (1983) Impact properties of steel fiber reinforced concrete in bending. *International Journal of Cement Composites and Lightweight Concrete*, **5**(4), 225–33.

1.67 Tinic C. & Bruhwiler E. (1985) Effects of compressive loads on the tensile strength of concrete at high strain rates. *International Journal of Cement Composites and Lightweight Concrete*, **7**(2), 103–8.

1.68 Wecharatna M. & Roland E. (1987) Strain rate effects as properties of cement-based composites under impact loading. Engineering Mechanics Sixth Conference, ASCE, Buffalo, NY, USA.

1.69 Takeda J. & Tachikawa H. (1971) Deformations and fracture of concrete subjected to dynamic load. In *Proceedings, International Conference on Mechanical Behaviour of Materials, Vol. IV, Concrete and Cement Paste, Glass and Ceramics*, pp. 267–77, Kyoto, Japan.

1.70 Takeda J., Tachikawa H. & Fujimoto L. (1982) Mechanical properties of concrete and steel in reinforced concrete structures subjected to impact or impulsive loads. In *Proceedings of RILEM-CEB-IABSE-IASS Interassociation Symposium*. BAM, Berlin (West), 83–91.

1.71 Evans A.G. (1974) Slow crack growth in brittle materials under dynamic loading conditions. *International Journal of Fracture*, **10**(2), 251–9.

1.72 Charles R.J. (1958) Dynamic fatigue of glass. *J. Appl. Phys.*, **29**(12), 1657–62.

1.73 Mihashi H. & Wittmann F.H. (1980) Stochastic approach to study the influence of rate of loading on strength of concrete. The Netherlands, *Heron*, **25**(3).

1.74 Jenq Y.S. & Shah S.P. (1986) Crack propagation resistance of fiber reinforced concrete. *J. Struct. Eng., ASCE*, **112**(10), 19–34.

1.75 Jenq Y.S. (1987) *Fracture mechanics of cementitious composites*. PhD dissertation, Northwestern University, Evanston, Il, USA.

1.76 Goldsmith W., Polivka M. & Yang T. (1966) Dynamic behavior of concrete. *Experimental Mechanics*, February, 65–79.

1.77 Read H.E. & Maiden C.J. (1971) *The Dynamic Behavior of Concrete*. National Technical Information Service, Springfield, Va, no AD894240, August.

1.78 Bazant Z.P. & Oh B.H. (1982) Strain-rate in rapid triaxial loading of concrete. *Proceedings, ASCE*, **108**(EM5), 764–82.

1.79 Oh B.H. (1987) Behaviour of concrete under dynamic loads. *ACI Mat. J.*, **84**(1), 8–13.

1.80 Loland K.E. (1980) Continuous damage model for load response estimation of concrete. *Cement and Concrete Research*, **10**, 395–402.

1.81 Mould J.C. & Levine H. (1987) A three invariant visco-plastic concrete model. In *Constitutive Laws for Engineering Materials: Theory and Applications* (Ed. by C.S. Desai *et al.*), pp. 707–716. Elsevier Science, New York.

1.82 Brook N. (1977) The use of irregular specimens for rock strength tests. *Int. J. Rock Mech. Min. Sci.*, **14**, 193–202.

1.83 Clark G.B. (1981) Geotechnical centrifuges for model studies and physical property testing of rock structures. *Colo. School Mines Q.*, **76**(4).

1.84 Haller H.F., Pattison H.C. & Baldonado O.C. (1973) Interrelationship of *in-situ* rock properties, excavation method and muck characteristics. *Technical Report HN-8121.2*. Holmes and Narver.

1.85 Kreimer G.S. (1968) *Strength of Hard Alloys*. Consultant's Bureau, New York.

1.86 Protodiakonov M.M. (1962) Mechanical properties and drillability of rock. In *Proceedings of the Fifth Symposium on Rock Mechanics, University of Minnesota*, pp. 103–18.

1.87 Somerton W.H., Esfandari F. & Singhal A. (1957) Further studies of the relation of physical properties of rock to drillability. Fourth Conference on Drilling and Rock Mechanics, University of Texas, Austin.

1.88 Sandia Corp (1957) Energy-absorbing characteristics of several materials. *Technical Memorandum no SCRM 284–57(51)*. California, USA.

1.89 Seed H.B. & Idriss I.M. (1970) Soil moduli and damping factors for dynamic response analyses. *Report no EERC 70−10*. University of California Earthquake Eng. Research Centre.

1.90 Hardin B.O. & Drnevich V.P. (1972) Shear modulus and damping soils: measurement and parameter effects. *Journal of the Soil Mechanics and Foundations Division, ASCE*, **98**, no SM6, proc. paper 8977, 603−24.

1.91 Matsukawa E. & Hunter A.M. (1956) The variation of sound velocity with stress in sand. *Proceedings of the Physical Society (London) Series B*, **69**, no 8, 847−8.

1.92 Theirs G.R. & Seed H.B. (1968) Cyclic stress−strain characteristics of clay. *Journal of the Soil Mechanics and Foundations Division, ASCE*, **94**, no SM2, proc. paper 5871, 555−69.

1.93 Swain R.J. (1962) Recent techniques for determination of *in-situ* properties and measurement of motion amplification in layered media. *Geophysics*, **27**, no 2, 237−41.

1.94 Maxwell A.A. & Fry Z.B. (1967) A procedure for determining elastic moduli of *in-situ* soils by dynamic technique. In *Proceedings, International Symposium on Wave Propagation and Dynamic Properties of Earth Materials*, pp. 913−19. University of New Mexico Press.

1.95 Seed H.B. & Lee K.L. (1966) Liquefaction of saturated sands during cyclic loading. *Journal of the Soil Mechanics and Foundations Division, ASCE*, **92**, no SM6, proc. paper 4972, 105−34.

1.96 Kovacs W.D., Seed H.B. & Chan C.K. (1971) Dynamic moduli and damping ratios for a soft clay. *Journal of the Soil Mechanics and Foundations Division, ASCE*, **97**, no SM1, proc. paper 7813, 59−75.

1.97 Zeevaert L. (1967) Free vibrations torsion tests to determine the shear modulus of elasticity on soils. In *Proceedings, Third Pan-American Conference on Soil Mechanics and Foundation Engineering*, Vol. 1, pp. 111−29. Caracas.

1.98 Florin V.A. & Ivanov, P.L. (1961) Liquefaction of saturated sandy soils. In *Proceedings, Fifth International Conference on Soil Mechanics and Foundation Engineering*, Vol. 1, pp. 107−11. Paris, France.

1.99 Prakash S. & Mathur J.N. (1965) Liquefaction of fine sand under dynamic loads. In *Proceedings, Fifth Symposium of the Civil and Hydraulic Engineering Department*. Indian Institute of Science, Bangalore.

1.100 Rocker K. (1968) The liquefaction behavior of sands subjected to cyclic loading. *Progress Report no 3*. Repeated load and vibration tests upon sand. *Research Report R68-36, Soils Pub. no 221*. Department of Civil Engineering, Massachusetts Institute of Technology, Cambridge.

1.101 Seed H.B. & Peacock W.H. (1971) Test procedure for measuring soil liquefaction characteristics. *Journal of the Soil Mechanics and Foundations Division, ASCE*, **97**, no SM8, proc. paper 8330, 1099−119.

1.102 Schnable P.B., Lysmer J. & Seed H.B. (1972) SHAKE − a computer program for earthquake response analysis of horizontally layered sites. *Report no EERC 72−12*. Earthquake Engineering Research Center, University of California, Berkeley.

1.103 Wilson E.L. (1968) A computer program for the dynamic stress analysis of underground structures. *Report no 68−1*. Structural Engineering Laboratory, University of California, Berkeley.

1.104 Dibaj M. & Penzien, J. (1969) Nonlinear seismic response of earth structures. *Report no EERC 69−2*. Earthquake Engineering Research Center, University of California, Berkeley.

1.105 Konder R.L. (1963) Hyperbolic stress−strain response: cohesive soils. *Journal of the Soil Mechanics and Foundations Division, ASCE*, **89**, no SM1, proc. paper 3429, 115−53.

1.106 Konder R.L. & Zelasko J.S. (1963) A hyperbolic stress−strain formulation for sands. In *Proceedings, Second Pan-American Conference on Soil Mechanics and*

Foundation Engineering, Vol. 1, pp. 289–324. Brazil.

1.107 Singh A. & Mitchell J.K. (1968) General stress–strain time function for soils. *Journal of the Soil Mechanics and Foundations Division, ASCE*, **94**, no SM1, proc. paper 5728, 21–46.

1.108 Miller R.P., Troncoso J.H. & Brown F.R. jr (1975) *In-situ* impulse test for dynamic shear modulus of soils. In *Proceedings, Conference on In-Situ Measurement of Soil Properties, ASCE*. Vol. 1, pp. 319–35. North Carolina State University, Raleigh North Carolina.

1.109 DeAlba P., Chan C.K. & Seed H.B. (1975) Determination of soil liquefaction characteristics by large-scale laboratory tests. *Report no EERC 75–14*. Earthquake Engineering Research Center, University of California, Berkeley.

1.110 Kipp M.E., Grady D.E. & Chen E.P. (1980) Strain-rate dependent fracture initiation. *Int. J. Fract.*, **16**, 471–8.

1.111 Carroll M.M. & Holt A.C. (1972) Static and dynamic pore-collapse relations for ductile porous materials. *J. Appl. Phys*, **43**, 1626–36.

1.112 Davison L., Stevens A.L. & Kipp M.E. (1977) Theory of spall damage accumulation in ductile metals. *J. Mech. Phys Solids*, **25**, 11–28.

1.113 Zlatin N.A. & Ioffe B.S. (1973) Time dependence of the resistance to spalling. *Sov. Phys.–Tech. Phys.*, **17**, 1390–93. (English translation.)

1.114 Shockey D.A., Curran D.R., Seaman L., Rosenberg J.T. & Peterson C.F. (1974) Fragmentation of rock under dynamic loads. *Int. J. Rock Mech. Sci.*, **11**, 303–17.

1.115 Birkimer D.L. (1971) A possible fracture criterion for the dynamic tensile strength of rock. In *Dynamic Rock Mechanics*, (Ed. by G.B. Clark), pp. 573–90. Balkema, Holland.

1.116 Johnson J.N. (1983) Ductile fracture of rapidly expanding rings. *J. Appl. Mech.*, **50**, 593–600.

1.117 Regazzoni G., Johnson J.N. & Follansbee P.S. (1986) Theoretical study of the dynamic tensile test. *J. Appl. Mech.*, **53**, 519–28.

1.118 Johnson G.R. & Cook W.H. (1985) Fracture characteristics of three metals subjected to various strains, strain rates, temperature and pressures. *Eng Fract. Mech.*, **21**, 31–48.

1.119 Kinloch A.J. & Young R.J. (1983) *Fracture Behaviour of Polymers*. Applied Science Publishers, London and New York.

1.120 Samuelides E. & Frieze P.A. (1983) Rigid bow impacts on ship-hull models. In *Proceedings of the IABSE Colloquium on Ship Collision with Bridges and Offshore Structures, Preliminary Report*, pp. 263–70. Copenhagen.

1.121 Lorenz R.A. (1982) The vaporization of structural materials in severe accidents. In *Proceedings of the International Meeting on Thermal Nuclear Reactor Safety, NUREG/CP-0027*, Vol. 2, p. 1090. Chicago.

1.122 Atchley B.L. & Furr H.L. (1967) Strength and energy absorption capacity of plain concrete under dynamic and static loading. *Journal of the American Concrete Institute*, 745–56.

1.123 Birkimer D.L. (1971) A possible fracture criterion for the dynamic tensile strength of rock. In *Proceedings of the Twelfth Symposium on Rock Mechanics*, pp. 573–89.

1.124 Birkimer D.L. & Lindeman R. (1971) Dynamic tensile strength of concrete materials. *Journal of the American Concrete Institute*, **68**, 47–9.

1.125 Banthia N. & Ohama Y. (1989) Dynamic tensile fracture of carbon fibre reinforced cements. *Int. Conf. on Recent Developments in Fibre Reinforced Cements and Concretes, Cardiff September 1989*.

1.126 Alford N. McN. (1982) Dynamic considerations for fracture of mortars. *Journal of Materials Science*, **56**(3), 279–87.

1.127 John R. & Shah S.P. (1986) Fracture of concrete subjected to impact loading. *Cement Concrete and Aggregates*, **8**, no 1, 24–32.

1.128 Mindess S., Banthia N. & Cheng Y. (1987) The fracture toughness of concrete under impact loading. *Cement and Concrete Research*, **1**, no 17, 231−41.

1.129 ASCE (1964) Design of structures to resist nuclear weapons effects. *ASCE, Manual of Engineering Practice*, no 42. American Society of Civil Engineers, New York.

1.130 Crawford R.E., Higgins O.H. & Bultmann E.H. (1974) *The Air Force Manual for Design and Analysis of Hardened Structures*. AFWL TR 74−102, Air Force Weapons Laboratory, Kirtland AFB, New Mexico.

1.131 Meunier Y., Sangoy L. & Pont G. (1988) Dynamic behaviour of armoured steels. In *International Conference on Ballistics* (Ed. by East China Institute of Technology), pp. IV/74−IV/81. In *Proceedings of the International Conference on Ballistics*. Nanjing, China, 25−28 October 1988. ECIT, Nanjing.

1.132 Schulz J., Heimdahl E. & Finnegan S. (1987) Computer characterisation of debris clouds resulting from hypervelocity impact. *Int. J. Impact Eng.*, **5**, 577−84.

1.133 Meyers M.A. & Aimone C.T. (1983) Dynamic fracture (spalling) of metals. *Prog. Mat. Sci.*, **28**, 1−96.

1.134 Buchar J., Bilek Z. & Dusek F. (1986) *Mechanical Behaviour of Materials at Extremely High Strain Rates*. Trans Tech Publications, Switzerland.

1.135 Davison L., Stevens A.L. & Kipp M.E. (1977) Theory of spall damage accumulation in ductile metals. *J. Mech. Phys. Solids*, **25**, 11−28.

1.136 Johnson J.N. (1981) Dynamic fracture and spallation in ductile solids. *J. Appl. Phys.*, **52**, 2812−25.

1.137 Seaman L., Curran D.R. & Shockey D.A. (1976) Computational models for ductile and brittle fracture. *J. Appl. Phys*, **47**, 4814−26.

1.138 British Standards Institution (BSI) (1985) *BS 8110: The Structural Use of Concrete, Part 1: Code of Practice for Design and Construction*. BSI, London.

1.139 British Standards Institution (BSI) (1985) *BS 5950: Part 1 Structural Use of Steelwork in Buildings. Code of Practice for Design in Simple and Continuous Construction: Hot Rolled Sections*. BSI, London.

1.140 UCCS, European Convention for Constructional Steelwork (1981) *European Recommendation for Steel Construction*. The Construction Press, London.

1.141 British Construction Steelwork Association (BCSA) (1983) *International Structural Steelwork Handbook*. Chameleon Press, UK.

1.142 CONSTRADO *Steelwork Design Guides to BS 5950: Part 1* (1985), Vol. I, Constructional Steel Research and Development Organisation, London. (1986) Vol. II, The Steel Construction Institute, London.

1.143 Gran J.K., Florence A.L. & Colton J.D. (1987) Dynamic triaxial compression experiments on high-strength concrete. In *Proceedings of the ASCE Sixth Annual Structures Congress*. Orlando, Florida.

1.144 Malvern L.E., Tang D., Jenkins D.A. & Gong J.C. (1986) Dynamic compressive strength of cementitious materials. *Mat. Res. Soc. Symp.*, **64**, 119−38.

1.145 Frantziskonis G. & Desai C.S. (1987) Elastoplastic model with damage for strain softening geomaterials. *Acta Mechanica*, **68**, 151−70.

1.146 Pijaudier-Cabot G. & Bazant Z.P. (1987) Nonlocal damage theory. *Report No 86−8/428n*. Center for Concrete and Geomaterials, Northwestern Univ., Il.

1.147 Schreyer H.L. & Bean J.E. (1987) Plasticity models for soils, rock and concrete. Report to the New Mexico Engineering Research Institute, Albuquerque, NM.

1.148 Soroushian P. & Ghoi K.B. (1987) Steel mechanical properties at different strain rates. *J. Struct. Eng, ASCE*, **113**, no 4, 663−72.

1.149 Bangash M.Y.H. (1989) *Concrete and Concrete Structures − Numerical Modelling and Applications*. Elsevier Applied Science, Barking, UK.

1.150 Ottoson N.S. (1979) Constitutive model for short-time loading of concrete. *J. Eng. Mech. Div., ASCE*, **105**, EM1 proc. paper 14375.

1.151 Pietruszczak S. & Pande C.N. (1989) *Numerical Models in Geomechanics NUMOG III*. Elsevier Applied Science, Barking, UK.

2 Impact dynamics

2.1   Davie T.N. & Richgels A.M. (1983) An earth penetrator code. *Sandia Report (SAND-82-2358)*. Sandia National Laboratories, Livermore, California.

2.2   Kar A.K. (1978) Projectile penetration into buried structures. *J. Struct. Div., ASCE*, paper 13479STI.

2.3   Henrych J. (1979) *The Dynamics of Explosion and its Use*. Elsevier, Amsterdam.

2.4   Glasstone S. & Dolan P.J. (1977) *The Effects of Nuclear Weapons*. US Department of Defense, Washington.

2.5   Jones N. & Wierzbicki T. (1983) *Structural Crash Worthiness*. Butterworths, London.

2.6   Morrow C.T. & Smith M.R. (1961) Ballistic missile and aerospace technology. *Proc. 6th Symposium*. Academic Press, New York.

2.7   FIP (1987) Concrete for hazard protection. First Int. Conf. Concrete Society, Edinburgh.

2.8   Abaham L.M. (1962) *Structural Design of Missiles and Space-craft*. McGraw Hill, New York.

2.9   BAM (1982) *Concrete Structures under Impact and Impulsive Loading*. Bundesanstalt für Materialprüfung, Berlin.

2.10 Macmillan R.H. *Dynamics of Vehicle Collisions. Special Publications SP5*. Inderscience Enterprises, Jersey, UK.

2.11 Taylor D.W. & Whitman R.V. (1953) The behaviour of soils under dynamic loadings. 2. Interim report on wave propagation and strain-rate effect. *MIT Report AFSWP-117*.

2.12 Maurer W.C. (1959) *Impact crater formation in sandstone and granite*. Master's thesis submitted to Colorado School of Mines.

2.13 Rinehart J.S. (1980) Stresses associated with lunar landings. *Contribution no 11*. Mining Research Laboratory Colorado School of Mines.

2.14 Tolch N.A. & Bushkovitch A.V. (1947) Penetration and crater volume in various kinds of rocks as dependent on caliber, mass and striking velocity of projectile. *BRL Report no. 641*. UK.

2.15 NDRC (1946) Effects of impact and explosion. Summary Technical Report of Div. 2, National Design and Research Corporation (NDRC), Vol. 1. Washington DC.

2.16 Lang H.A. (1956) Lunar instrument carrier — landing factors. *Rand RM-1725, ASTIA Doc. no AD 112403*.

2.17 Palmore J.I. (1961) Lunar impact probe. ARS J., 1066–73.

2.18 Rinehart J.S. (1960) Portions of traite de balistique experimentale which deal with terminal ballistics. Chapters III, IV, V and XIV, Félix Hélie, Paris (1884), Naval Ordnance Test Station TM RRB-75.

2.19 Dennington R.J. (1958) Terminal ballistics studies. *Proj. Doan Brook Tech. Memo. no 20 Case Inst.*, ASTIA AD 146 913.

2.20 Kornhauser M. (1958) Prediction of cratering caused by meteoroid impacts. *J. Astron. Sci.*, **V**, no 3–4.

2.21 Rinehart J.S. & Pearson J. (1954) *Behaviour of Metals Under Impulsive Loads*. American Society for Metals, Cleveland.

2.22 Rinehart J.S. (1950) Some observations on high-speed impact. Naval Ordnance Test Station RRB-50. Félix Hélie, Paris, France.

2.23 Rinehart J.S. & White W.C. (1952) Shapes of craters formed in plaster-of-paris by ultra-speed pellets. *Am. J. Phys*, **20**, no 1.

2.24 Gazley C. jr (1957) Deceleration and heating of a body entering a planetary atmosphere from space. *RAND Corporation Report P-955*. USA.

2.25 Lees L., Hartwig F.W. & Cohen C.B. (1959) Use of aerodynamic lift during entry into the earth's atmosphere. *ARS J.*, 633–41.

2.26 Lowe R.E. & Gervais R.L. (1961) Manned entry missions to Mars and Venus. In *ARS Space Flight Report to the Nation*, paper 2158–61. New York.

2.27 Turnacliff R.D. & Hartnett J.P. (1957) Generalized trajectories for free-falling bodies of high drag. ARS paper 543−57, 12th annual meeting, New York, 2−5 December 1957.

2.28 Creighton D.C.(1982) Non-normal projectile penetration in soil and rock: user's guide for computer code PENCO2D. *Technical Report SL-82-7*. US Army Engineering Waterways Experiment Station, Vicksburg, Miss.

2.29 Young C.W. (1972) Empirical equations for predicting penetration performance in layered earth materials for complex penetrator configurations. *Development Report No SC-DR-72−0523*. Sandia National Laboratory, Albuquerque, NM.

2.30 Young C.W. (1985) Simplified analytical model of penetration with lateral loading. *SAND84−1635*. Sandia National Laboratory, Albuquerque, NM.

2.31 Matuska D., Durrett R.E. & Osborn J.J. (1982) Hull user's guide for three-dimensional linking with EPIC 3. *ARBRL-CR-00484*. Orlando Technology, Shalimar, Fl.

2.32 Ito Y.M., Nelson R.B. & Ross-Perry F.W. (1979) Three-dimensional numerical analysis of earth penetration dynamics. *Report DNA 5404F*. Defense Nuclear Agency, Washington DC, USA.

2.33 Johnson G.R. & Stryk R.A. (1986) Dynamic-Lagrangian computations for solids, with variable nodal connectivity for severe distortions. *Int. J. Num. Methods Eng*, **23**, 509−22.

2.34 Hallquist J.O. (1984) User's manual for DYNA2D − an explicit two-dimensional hydrodynamic finite element code with interactive rezoning. *Report UCID-18756, Revision 2*. University of California, Lawrence Livermore National Laboratory, USA.

2.35 Hallquist J.O. & Benson D.J. (1986) DYNA3D user's manual (nonlinear dynamic analysis of structures in three dimensions). *Report UCID-19592, Revision 2*. University of California, Lawrence Livermore National Laboratory.

2.36 Carr H.H. (1948) Analysis of head failure in aircraft torpedoes. *NOTS Navord Report 1019*, Patent Office Library, London.

2.37 Reissner E. (1946, 1947) Stresses and small displacements of shallow spherical shells. *J. Math. Phys*, **XXV**, 1 and 4.

2.38 Friedrichs K.O. (1941) On the minimum buckling load for spherical shells. In *Theodore von Kàrmàn Anniversary Volume*, pp. 258−72. CIT, Pasadena, USA.

2.39 Tsein H.S. (1942) A theory for the buckling of thin shells. *J. Aero. Sci.*, **9**, 373−84.

2.40 Payton R.G. (1962) Initial bending stresses in elastic shells impacting into compressible fluids. *Quart. J. Mech. Appl. Math.*, **15**, 77−90.

2.41 Office of the Scientific Research and Development (1946) Aircraft torpedo development and water entry ballistics, *OSRD Report 2550*. CIT, Pasadena, USA.

2.42 Kornhauser M. (1958) Impact protection for the human structure. Am. Astron. Soc., Western Regional Meeting, 18−19 August 1958, Palo Alto, California.

2.43 Kornhauser M. (1960) Structural limitations on the impulsiveness of astronautical maneuvers. ARS Conference on Structural Design of Space Vehicles, 6 April 1960, Santa Barbara, California.

2.44 ANC-5; Depts of Air Force, Navy and Commerce (1955) *Strength of Metal Aircraft Elements*. Republican Press, Hamilton, Ohio.

2.45 Coppa A. (1960) On the mechanics of buckling of a cylindrical shell under longitudinal impact. Tenth International Congress App. Mech., September 1960, Stressa, Italy.

2.46 Polentz L.M. (1959) Linear energy absorbers. *Machine Design*.

2.47 Martin E.D. (1961) A design study of the inflated sphere landing vehicle, including the landing performance and the effects of deviations from design conditions. *NASA TN D-692*.

2.48 Esgar J.B. & Morgan W.C. (1960) Analytical study of soft landings on gas-fill bags. *NASA TR R-75*.

2.49 Fisher L.J. jr (1961) Landing−impact−dissipation systems. *NASA TN D-975*.

2.50   Schrader C.D. (1962) Lunar and planetary soft landings by means of gas-filled balloons. ARS Lunar Missions Meeting, Cleveland, 17−19 July 1962, paper 2480−62.

2.51   Bethe H.A. (1941) Attempt of a theory of armor penetration. Frankford Arsenal Rep., Philadelphia.

2.52   Taylor G.I. (1941) Notes on H.A. Bethe's theory of armor penetration, II. Enlargement of a hole in a flat sheet at high speeds. Home Ministry Security Report, R.C. 280 UK.

2.53   Taylor G.I. (1948) The formation and enlargement of a circular hole in a thin plastic sheet. *Quart. J. Mech. Appl. Math.*, **1**.

2.54   Kineke J.H. & Vitali R. (1963) Transient observations of crater formation in semi-infinite targets. In *Proceedings of the Sixth Symposium on Hypervelocity Impact*, Vol. II, part 2. Springer Verlag, Berlin, Germany.

2.55   Eichelberger R.J. & Kineke J.H. (1967) Hypervelocity impact. In *High Speed Physics* (Ed. by K. Vollrath & G. Thomer). Springer-Verlag Wien, New York.

2.56   Turpin W.C. & Carson J.M. (1970) Hole growth in thin plates perforated by hypervelocity pellets. *Air Force Materials Lab. Rep. 70−83*. Wright-Patterson AFB, Ohio.

2.57   Holloway L.S. (1963) Observations of crater growth in wax. *BRL Rep., USABRL-M-1526*.

2.58   Freiberger W. (1952) A problem in dynamic plasticity: the enlargement of a circular hole in a flat sheet. *Proc. Cambridge Phil. Soc.*, **48**.

2.59   Fraiser J.T., Karpov B.G. & Holloway L.S. (1965) The behavior of wax targets subjected to hypervelocity impact. In *Proceedings of the Seventh Hypervelocity Impact Symposium*, Vol. V. Springer Verlag, Berlin, Germany.

2.60   Rolsten R.F., Wellnitz J.N. & Hunt H.H. (1964) An example of hole diameter in thin plates due to hypervelocity impact. *J. Appl. Phys*, **35**.

2.61   Allison F.E. (1965) Mechanics of hypervelocity impact. In *Proceedings of the Seventh Hypervelocity Impact Symposium*, Vol. I. Springer Verlag, Berlin, Germany.

2.62   Wilkins M. & Guinan M. (1973) Impact of cylinders on a rigid boundary. *J. Appl. Phys*, **44**.

2.63   Whiffin A.C. (1948) The use of flat-ended projectiles for determining dynamic yield stress, II, tests on various metallic materials. *Proc. R. Soc. London Ser. A*, **194**.

2.64   Goldsmith W. (1960) *Impact, the Theory and Physical Behaviour of Colliding Solids*. Edward Arnold, London.

2.65   Johnson W. (1972) *Impact Strength of Materials*. Edward Arnold, New York.

2.66   Walters W. (1979) Influence of material viscosity on the theory of shaped-charge jet formation. *BRL Rep., BRL-MR-02941*.

2.67   Chhabildas L.C. & Asay J.R. (1979) Rise-time measurements of shock transitions in aluminum, copper and steel. *J. Appl. Phys*, **50**.

2.68   Birkhoff G., MacDougall D., Pugh E. & Taylor G. (1948) Explosives with lined cavities. *J. Appl. Phys*, **19**.

2.69   Scott B. & Walters W. (1984) A model of the crater growth-rate under ballistic impact conditions. In *Proceedings of the Southeastern Conference on Theoretical and Applied Mechanics*, Georgia.

2.70   Rohani B. (1975) Analysis of projectile penetration into concrete and rock targets. *S-75−25*. US Army Engineer Waterways Experimental Station.

2.71   Housen K., Schmidt R. & Holsapple K. (1983) Crater ejecta scaling laws: fundamental forms based on dimensional analysis. *J. Geophys. Res.*, **88**.

2.72   Glass J. & Bruchey W. (1982) Internal deformation and energy absorption during penetration of semi-infinite targets. In *Proceedings of the Eighth International Symposium on Ballistics*, NASA, Langely, USA.

2.73   Wang W.L. (1971) Low velocity projectile penetration. *Journal of the Soil Mechanics and Foundations Division*, ASCE **97**, no SM12, proc. paper 8592.

2.74   Young C.W. (1969) Depth prediction for earth-penetrating projectiles. *Journal of the Soil Mechanics and Foundations Division, ASCE*, **95**, no SM3, proc. paper 6558.

2.75 Petry L. (1910) *Monographies de Systèmes d'Artillerie*. Brussels.

2.76 Hakala W.W. (1965) *Resistance of a granular medium to normal impact of a rigid projectile*. Thesis presented to the Virginia Polytechnic Institute, Blacksburg, Va, in partial fulfilment of the requirements for the degree of doctor of Philosophy.

2.77 Thigpen L. (1974) Projectile penetration of elastic-plastic earth-media. *Journal of the Geotechnical Engineering Division, ASCE*, **100**, no GT3, proc. paper 10414.

2.78 Butler D.K. (1975) An analytical study of projectile penetration into rock. *Technical Report S-75−7*. Soils and Pavements Laboratory, US Army Engineer Waterways Experiment Station, Vicksburg, Miss.

2.79 Butler D.K. (1975) Pretest penetration for DNA rock penetration experiments at a sandstone site near San Ysidro, New Mexico. Soils and Pavements Laboratory, US Army Engineer Waterways Experiment Station, Vicksburg, Miss.

2.80 Butler D.K. (1975) Development of high-velocity powder gun and analysis of fragment penetration tests into sand. *Misc. paper S-75−27*. Soils and Pavements Laboratory, US Army Engineer Waterways Experiment Station, Vicksburg, Miss.

2.81 Bernard R.S. & Hanagud S.V. (1975) Development of a projectile penetration theory, Report 1, Penetration theory for shallow to moderate depths. *Technical Report S-75−9*. Soils and Pavements Laboratory, US Army Engineer Waterways Experiment Station, Vicksburg, Miss.

2.82 Bernard R.S. (1976) Development of a projectile penetration theory, Report 2, Deep penetration theory of homogeneous and layered targets. *Technical Report S-75−9*. Soils and Pavements Laboratory, US Army Engineer Waterways Experiment Station, Vicksburg, Miss.

2.83 Hammel J. (1979) Impact loading on a spherical shell. Fifth International Conference on Structural Mechanics in Reactor Technology (SMiRT), Berlin.

2.84 Barber E.M. *et al.* (1972) A study of air distribution in survival shelters using a small-scale modelling technique. *Report 10689*, USA National Bureau of Standards.

2.85 Barneby H.L. *et al.* (1963) Toxic gas protection. Paper to ASHRAE Symposium on Survival Shelters, 1962.

2.86 Barthel R. (1965) Research on the climate in an underground shelter. *Ingenieur*, **77**, 143−7 (in Dutch).

2.87 Barthel R. (1965) Theoretical and experimental research regarding the indoor climate in an underground shelter. *Ingenieur*, **77**, 6143−7 (in Dutch).

2.88 Cooper J. (1980) After the bomb. *Journal of the Chartered Institution of Building Services*, **2**, 48−9.

2.89 Dasler A.R. & Minrad D. (1965) Environmental physiology of shelter habitation. *ASHRAE Trans*, **71**, 115−24.

2.90 Drucker E.E. & Cheng H.S. (1963) Analogue study of heating in survival shelters. ASHRAE Symposium on Survival Shelters, 1962. ASHRAE.

2.91 Bigg J.M. (1964) *Introduction to Structural Dynamics*. McGraw Hill, New York.

2.92 Gates A.S. (1963) Air revitalization in sealed shelters. ASHRAE Symposium on Survival Shelters, 1962. ASHRAE.

2.93 Gessner H. (1961) The ventilation of air-raid shelters. *Schweizerische Blätter für Heizung und Luftung*, **28**, 1−12 (in German).

2.94 Hanna G.M. (1962) Ventilation design for fallout and blast shelters. *Air Engineering*, **4**, 19−21.

2.95 Home Office (1981) *Domestic Nuclear Shelters (DNS) − Technical Guidance*. HMSO.

2.96 Home Office (1981) *Domestic Nuclear Shelters*. HMSO.

2.97 Eibl J. (1982) Behaviour of critical regions under soft missile impact and impulsive loading. *IBAM*, June.

2.98 Hughes G. & Speirs D. (1982) An investigation of the beam impact problem. *Technical Report 546*, C & CA, UK (now British Cement Association).

2.99 Bate S. (1961) The effect of impact loading on prestressed and ordinary reinforced concrete beams. *National Building Studies Research Paper 35*.

2.100 Billing I. (1960) *Structure Concrete*. Macmillan, London.

2.101 Watson A. & Ang T. (1982) Impact resistance of reinforced concrete structures. In *Designs for Dynamic Loading*. Construction Press, London, UK.

2.102 Watson A. & Ang T. (1984) Impact response and post-impact residual strength of reinforced concrete structures. International Conference and Exposition of Structural Impact and Crashworthiness, Imperial College, London, UK.

2.103 Perry S. & Brown I. (1982) Model prestressed slabs subjected to hard missile loading. In *Design for Dynamic Loading*. Construction Press, London, UK.

2.104 Perry S., Brown I. & Dinic G. (1984) Factors influencing the response of concrete slabs to impact. International Conference and Exposition of Structural Impact and Crashworthiness, Imperial College, London, UK.

2.105 Kufuor K. & Perry S. (1984) Hard impact of shallow reinforced concrete domes. International Conference and Exposition of Structural Impact and Crashworthiness, Imperial College, London, UK.

2.106 Burgess W. & Campbell-Allen D. (1974) Impact resistance of reinforced concrete as a problem of containment. *Research Report no R251*. School of Civil Engineering, University of Sydney.

2.107 Stephenson A. (1976) Tornado-generated missile full-scale testing. In *Proceedings of the Symposium on Tornadoes, Assessment of Knowledge and Implications for Man*. Texas University.

2.108 Jankov Z., Turnham J. & White M. (1976) Missile tests of quarter-scale reinforced concrete barriers. In *Proceedings of the Symposium on Tornadoes, Assessment of Knowledge and Implications for Man*. Texas University, June 1976.

2.109 Stephen A. & Silter G. (1977) Full-scale tornado-missile impact tests. Fourth International Conference on Structural Mechanics in Reactor Technology (SMiRT), Berlin, 1977, paper J10/1.

2.110 Jonas W. & Rudiger E. (1977) Experimental and analytical research on the behaviour of reinforced concrete slabs subjected to impact loads. Fourth International Conference on Structural Mechanics in Reactor Technology (SMiRT), Berlin, paper J7/6.

2.111 Beriaud C. *et al.* (1977) Local behaviour of reinforced concrete walls under hard missile impact. Fourth International Conference on Structural Mechanics in Reactor Technology (SMiRT), Berlin, 1977, paper J7/9.

2.112 Gupta Y. & Seaman L. (1977) Local response of reinforced concrete to missile impacts. Fourth International Conference on Structural Mechanics in Reactor Technology (SMiRT), Berlin, 1977, paper J10/4.

2.113 Barr P. *et al.* (1982) An experimental investigation of scaling of reinforced concrete structures under impact loading. In *Design for Dynamic Loading*. Construction Press, London, UK.

2.114 Barr P. *et al.* (1983) Experimental studies of the impact resistance of steel faced concrete composition. Seventh International Conference on Structural Mechanics in Reactor Technology (SMiRT), 1983, paper J8/4, Chicago, USA.

2.115 Det Norske Veritas (1981) *Rules for Classification of Mobile Offshore Units*.

2.116 Minorsky V.V. (1959) An analysis of ship collisions with reference to nuclear power plants. *Journal of Ship Research*, **3**, 1−4.

2.117 Soreide T.H. (1981) *Ultimate Load Analysis of Marine Structures*. Tapir, Trondheim.

2.118 JABSF (1983) Ship collision with an offshore structure. IABSE Colloquium, Copenhagen.

2.119 Woisin G. (1977) Conclusion from collision examinations for nuclear merchant ships in the FRG. In *Proceedings of the Symposium on Naval Submarines*. Hamburg.

2.120 Reckling K.A. (1977) On the collision protection of ships. PRADS Symposium, Tokyo.

2.121 Nagasawa H., Arita K., Tani M. & Oka S. (1977) A study on the collapse of ship structure in collision with bridge piers. *J. Soc. Nav. Arch. Japan*, **142**.

2.122 Macaulay M.A. & Macmillan R.H. (1968) Impact testing of vehicle structures. In *Proceedings of the 12th FISITA Congress*. Barcelona.

2.123 Emori R.I. (1968) Analytical approach to automobile collisions. *SAE Paper 680016.* Society of Automobile Engineers, USA.

2.124 Neilson I.D. *et al.* (1968) Controlled impact investigations. *TRRL Report LR 132.* Transport Road Research Laboratory, Crowthorne, UK.

2.125 Grime G. & Jones I.S. (1969) Car collisions — the movement of cars and their occupants. *Proc. I. Mech. Eng.*, **184**, no 5.

2.126 Emori R.I. & Tani M. (1970) Vehicle trajectories after intersection collision impact. *SAE Paper 700176.* Society of Automobile Engineers.

2.127 Wall J.G. *et al.* (1970) Comparative head-on impact tests. *TRRL Report LR 155.* Transport Road Research Laboratory, Crowthorne, UK.

2.128 Jones I.S. (1975) Mechanics of roll-over as the result of curb impact. *SAE Paper 750461.* Society of Automobile Engineers.

2.129 Wagner R. (1978) Compatibility in frontal collisions. In *Proceedings of the 17th FISITA Congress*, Budapest.

2.130 Rouse H. & Howe J.W. (1953) *Basic Mechanics of Fluids.* Wiley.

2.131 Kinslow R. (Ed.) (1970) *High Velocity Impact Phenomena.* Academic Press.

2.132 Nowacki W.K. (1978) *Stress Waves in Non-Elastic Solids.* Pergamon.

2.133 Hurty W.C. & Rubinstein M.F. (1964) *Dynamics of Structures.* Prentice-Hall.

2.134 Morrow C.T. (1963) *Shock and Vibration Engineering.* Wiley.

2.135 Biggs J.M. (1964) *Introduction to Structural Dynamics*, chapter 2. McGraw-Hill.

2.136 Snowdon J.C. (1968) *Vibration and Shock in Damped Mechanical Systems.* Wiley.

2.137 Craig R.R. (1981) *Structural Dynamics: An Introduction to Computer Methods.* Wiley.

2.138 Davies G.A. (1984) *Structural Impact and Crashworthiness*, Vol. 1, chapter 7. Elsevier.

2.139 Andrews K.P.F. *et al.* (1983) Classification of the axial collapse of cylindrical tubes under quasi-static loading. *International Journal of Mechanical Sciences*, **25**, 678–96.

2.140 Bodner S.R. & Symonds P.S. (1972) Experimental and theoretical investigation of the plastic deformation of cantilever beams subjected to impulsive loading. *Journal of Applied Mechanics*, December, 719–28.

2.141 Conway M.D. & Jakubowski M. (1969) Axial impact of short cylindrical bars. *Journal of Applied Mechanics*, **36**, 809.

2.142 Hagiwara K. *et al.* (1983) A proposed method of predicting ship collision damage. *International Journal of Impact Engineering*, **1**, 257–80.

2.143 Lee E.H. & Morrison J.A. (1956) A comparison of the propagation of longitudinal waves in rods of viscoelastic materials. *Journal of Polymer Science*, **19**, 93–110.

2.144 Samuelides E. & Frieze P.A. (1983) Strip model simulation for low energy impacts on flat-plated structures. *International Journal of Mechanical Sciences*, **25**, 669–86.

2.145 Backman M.E. & Goldsmith W. (1978) The mechanics of penetration of projectiles into targets. *Int. J. Eng. Sci.*, **16**, 1–99.

2.146 Bernard R.S. & Hanagud S.V. (1975) Development of a projectile penetration theory. *Technical Report S-75–9.* US Army Waterways Experiment Station.

2.147 Bernard R.S. (1976) Development of a projectile penetration theory report 2: deep penetration theory for homogeneous and layered targets. *Technical Report S-75–9.* US Army Waterways Experiment Station, Virginia, USA.

2.148 Byers R.K. & Chabai A.J. (1977) Penetration calculations and measurements for a layered soil target. *International Journal for Numerical and Analytical Methods in Geomechanics*, **1**, 107–38.

2.149 Bjork R.L. (1975) Calculations of earth penetrators impacting soils. *AD-AO46 236.* Pacifica Technology Hawaii, USA.

2.150 DNA/SANDIA (1975) Soil penetration experiment at DRES: results and analysis. *SAND.75.0001.* Sandia Laboratories, Albuquerque.

2.151 Hadala P.F. (1975) Evaluation of empirical and analytical procedures used for predicting the rigid body motion of an earth penetrator. *Technical Report S-75–15.* US Army Waterways Experiment Station, Virginia, USA.

2.152 Norwood F.R. & Sears M.P. (1982) A nonlinear model for the dynamics of penetration into geological targets. *Journal of Applied Mechanics*, **49**, 26–30.

2.153 Rohani B. (1972) High velocity fragment penetration of a soil target. In *Proceedings of the Conference on Rapid Penetration of Terrestrial Materials*. Texas A & M University.

2.154 Triandafilidis G.E. (1976) State of the art of earth penetration technology, 2 volumes. *Technical Report CE-42(76) DNA-297*. The University of New Mexico.

2.155 Wagner M.H., Kreyenhagen K.N. & Goerke W.S. (1975) Numerical analysis of projectile impact and deep penetration into earth media. *WES Contract Report S-75–4*. California Research and Technology Inc.

2.156 Yankelevsky D.Z. (1979) Normal penetration into geomaterials. *Research Report 30*. Faculty of Civil Engineering, Technion, Israel.

2.157 Yankelevsky D.Z. (1983) The optimal shape of an earth penetrating projectile. *International Journal of Solids and Structures*, **19**, 25–31.

2.158 Yankelevsky, D.Z. (1985) Cavitation phenomena in soil projectile interaction. *Int. J. Imp. Eng*, **3**, 167–78.

2.159 Young C.W. & Ozanne G.M. (1966) Low velocity penetration study. *SC-RR-66–118*. Sandia Laboratories, USA.

2.160 Young C.W. (1969) Depth prediction for earth penetrating projectiles. *J. Soil Mech. Found. Div., Proc. Am. Soc. Civ. Eng*, **95**, 803–17.

2.161 Young C.W. (1976) Status report on high velocity soil penetration. *SAND 76–0291*. Sandia Laboratories, USA.

2.162 Yarrington P. (1977) A one dimensional approximate technique for earth penetration calculations. *SAND 77–1126*. Sandia Laboratories, USA.

2.163 Alekseevskii V.P. (1966) Penetration of a rod into a target at high velocity. *Comb. Expl. Shock Waves*, **2**, 99–106 (English translation).

2.164 Backman M.E. & Goldsmith W. (1978) The mechanics of penetration of projectiles into targets. *Int. J. Eng Sci.*, **16**, 1–99.

2.165 Forrestal M.J., Rosenberg Z., Luk V.K. & Bless S.J. (1986) Perforation of aluminum plates with conical-nosed rods. *SAND 86–0292J*. Sandia National Laboratory, Albuquerque, USA.

2.166 Frank K. & Zook J. (1986) Energy-efficient penetration and perforation of targets in the hyper-velocity regime. Hypervelocity Impact Symposium, San Antonio, 1986.

2.167 Levy N. & Goldsmith W. (1984) Normal impact and perforation of thin plates by hemispherically-tipped projectiles – I. Analytical considerations. *Int. J. Impact Eng*, **2**, 209–29.

2.168 Pidsley P.H. (1984) A numerical study of long rod impact onto a large target. *J. Mech. Phys Sol.*, **32**, 315–34.

2.169 Tate A. (1986) Long rod penetration models – part I. A flow field model for high speed–long rod penetration. *Int. J. Mech. Sci.*, **28**, 525–48.

2.170 Zukas J.A., Jones G.H., Kinsey K.D. & Sherrick T.M. (1981) Three-dimensional impact simulations: resources and results. In *Computer Analysis of Large Scale Structures*, (Ed. by K.C. Park & R.F. Jones) AMD-49, pp. 35–68. New York.

2.171 Asay J.R. & Kerley G.I. (1987) The response of materials to dynamic loading. *Int. J. Impact Eng*, **5**, no 1–4, 69–99.

2.172 Holian K.S. & Burkett M.W. (1987) Sensitivity to hypervelocity impact simulation to equation of state. *Int. J. Impact Eng*, **5**, no 1–4, 333–41.

2.173 Maiden C.J. (1963) Experimental and theoretical results concerning the protective ability of a thin shield against hypervelocity projectiles. In *Proceedings of the Sixth Symposium on Hypervelocity Impact*, III, pp. 69–156. Springer Verlag, Berlin.

2.174 Ravid M., Bodner S.R. & Holeman I. (1987) Analysis of very high speed impact. *Int. J. Eng Sci.*, **25**, no 4, 473–82.

2.175 Tillotson J.H. (1962) *Metallic equations of state for hypervelocity impact*. General Atomic, Division of General Dynamics, GA-3216, San Diego.

2.176 Chen J.K. & Sun C.T. (1985) On the impact of initially stressed composite laminates. *J. Comp. Mat.*, **19**, 490−504.

2.177 Tsai S.W. & Hahn H.T. (1980) *Introduction to Composite Materials*. Technical Publishing Company, Pennsylvania.

2.178 Brook N. & Summers D.A. (1969) The penetration of rock by high-speed water jets. *International Journal of Rock Mechanics and Mining Science*, **6**, 249−58.

2.179 Haimson B. (1965) *High velocity, low velocity, and static bit penetration characteristics in Tennessee marble*. Master's thesis, University of Minnesota.

2.180 Haimson B.C. & Fairhurst C. (1970) Some bit penetration characteristics in pink Tennessee marble. In *Proceedings of the Twelfth Symposium of Rock Mechanics*, pp. 547−59. University of Missouri School of Mines, USA.

2.181 Simon R. (1963) Digital machine computations of the stress waves produced by striker impact in percussive drilling machines. In *Fifth Symposium on Rock Mechanics*. Pergamon, Oxford.

2.182 Vijay M.M. & Brierly W.H. (1980) Drilling of rocks with rotating high pressure water jets: an assessment of nozzles. In *Proceedings of the Fifth International Symposium on Jet Cutting Technology*. Hanover.

2.183 Watson R.W. (1973) Card-gap and projectile impact sensitivity measurements, a compilation. *US Bureau of Mines Information Circular No 8605*.

2.184 Wells E.S. (1949) Penetration speed of percussion drill bits. *Chem. Eng Mineral Rev.*, **41**(10), 362−4.

2.185 Winzer R.R. & Ritter A.P. (1980) Effect of delays in fragmentation in large limestone blocks. *Report No MML TR 80−25*. Martin Marietta Laboratories, USA.

2.186 O'Connell W.J. & Fortney R.A. (1974) Turbine missile impact analysis: a detailed treatment. *Transactions of the American Nuclear Society*, **19**, 31; EDS Nuclear Inc Report, 27 October 1974.

2.187 Moody F.J. (1969) Prediction of blowdown thrust and jet forces. *ASME Transaction 69-HT-31*. American Society of Mechanical Engineers.

2.188 Norris C.H. *et al.* (1959) *Structural Design for Dynamic Loads*. McGraw-Hill, New York.

2.189 Emori R.I. (1968) Analytical approach to automobile collisions. Automotive Engineering Congress, Detroit, Michigan, paper no 680016.

2.190 Ivey D.L., Ruth E. & Hirsch T.J. (1970) Feasibility of lightweight cellular concrete for vehicle crash cushions. Paper presented at the Annual Meeting of the Highway Research Board, Washington DC.

2.191 Chen E.P. & Sih G.C. (1977) Transient response of cracks to impact loads. In *Elasto-Dynamic Crack Problems*, Vol. 4. Noordhoff, Gröningen, Leyden.

2.192 Davison L. & Graham R.A. (1979) Shock compression of solids. *Phys Rep.*, **55**, 255−379.

2.193 Meyer M.A. & Aimone C.T. (1983) Dynamic fracture (spalling) of metals. *Prog. Mat. Sci.*, **28**, 1−96.

2.194 Grady D.E. (1981) Fragmentation of solids under impulsive stress loading. *J. Geophys. Res.*, **86**, 1047−54.

2.195 Grady D.E. (1981) Application of survival statistics to the impulsive fragmentation of ductile rings. In *Shock Waves and High-Strain-Rate Phenomena in Metals*, (Ed. by M.A. Meyers & L.E. Murr) pp. 181−92. Plenum, New York.

2.196 Grady D.E., Bergstresser T.K. & Taylor J.M. (1985) Impact fragmentation of lead and uranium plates. *SAND85−1545 (technical report)*. Sandia Laboratories.

2.197 Grady D.E. & Kipp M.E. (1979) The micromechanics of impact fracture of rock. *Int. J. Rock Mech. Min. Sci.*, **16**, 293−302.

2.198 Cour-Palais B.G. (1987) Hypervelocity impact in metals, glass and composites. *Int. J. Impact Eng*, **5**, 221−38.

2.199 Holsapple K.A. (1987) The scaling of impact phenomena. *Int. J. Impact Eng*, **5**, 343−56.

2.200 Johns Hopkins University (1961) The resistance of various metallic materials to perforation by steel fragments; empirical relationships for fragment residual velocity and residual weight. Ballistic Res. Lab. Proj. Thor TR no 47. Johns Hopkins University, Baltimore, Md.

2.201 Johns Hopkins University (1961) The resistance of various non-metallic materials to perforation by steel fragments; empirical relationships for fragment residual velocity and residual weight. Ballistic Res. Lab. Proj. Thor TR no 47. Johns Hopkins University, Baltimore, Md.

2.202 Yatteau J.D. (1982) High velocity penetration model. NSWU TR, 82–123. New South Wales University, Australia.

2.203 Piekutowski A.J. (1987) Debris clouds generated by hypervelocity impact of cylindrical projectiles with thin aluminum plates. *Int. J. Impact Eng*, **5**, 509–18.

2.204 Rinehart J.S. & Pearson J. (1965) *Behaviour of Metals under Impulsive Loading*. Dover, New York.

2.205 Al-Hassani S.T.S., Johnson W. & Nasim M. (1972) Fracture of triangular plates due to contact explosive pressure. *J. Mech. Eng Sci.*, **14**(3), 173–83.

2.206 Al-Hassani S.T.S. & Silva-Gomes J.F. (1979) Internal fracture paraboloids of revolution due to stress wave focussing. *Conf. Ser. – Inst. Phys.*, **47**, 187–96.

2.207 Al-Hassani S.T.S. & Silva-Gomes J.F. (1986) Internal fractures in solids of revolution due to stress wave focussing. In *Shock Waves and High-strain-rate Phenomena in Metals*, (Ed. by M.A. Meyers & L.E. Murr), chapter 10. Plenum, New York.

2.208 Al-Hassani S.T.S. (1986) Fracturing of explosively loaded solids. In *Metal Forming and Impact Mechanics*, (Ed. by S.R. Reid), chapter 18. Pergamon, Oxford.

2.209 Swain M.V. & Lawn B.R. (1976) Indentation fracture in brittle rocks and glasses. *Int. J. Rock Mech. Min. Sci. Geomech. Abstr.*, **13**, 311–19.

2.210 Jones N. & Liu J.H. (1988) Local impact loading of beams. In *Intense Dynamic Loading and its Effects*, pp. 444–9. Science Press, Beijing, and Pergamon Press.

2.211 Liu J.H. & Jones N. (1988) Dynamic response of a rigid plastic clamped beam struck by a mass at any point on the span. *Int. J. Solids Struct.*, **24**, 251–70.

2.212 Liu J.H. & Jones N. (1987) Experimental investigation of clamped beams struck transversely by a mass. *Int. J. Impact Eng*, **6**, 303–35.

2.213 Liu J.H. & Jones N. (1987) Plastic failure of a clamped beam struck transversely by a mass. Report ES/31/87. Department of Mechanical Engineering, University of Liverpool.

2.214 Jones N. (1979) Response of structures to dynamic loading. *Conf. Ser. – Inst. Phys.*, **47**, 254–76.

2.215 Recht R.F. & Ipson T.W. (1963) Ballistic perforation dynamics. *J. Appl. Mech.*, **30**, 384–90.

2.216 Lindholm U.S., Yeakley L.M. & Davidson D.L. (1974) Biaxial strength tests on beryllium and titanium alloys. *AFML-TR-74-172*. Air Force Systems Command, Wright-Patterson Air Force Base, Oh.

2.217 Jones N. (1970) The influence of large deflections on the behavior of rigid-plastic cylindrical shells loaded impulsively. *J. Appl. Mech.*, **37**, 416–25.

2.218 Lindberg, H.E., Anderson D.L., Firth R.D. & Parker L.V. (1965) Response of re-entry vehicle-type shells to blast loads. *SRI project FGD-5228*. Stanford Research Institute, Menlo Park, Ca.

2.219 Cronkhite J.D. & Berry V.L. (1983) Investigation of the crash impact characteristics of helicopter composite structure. *USA AVRADCOM-TR-82 D-14*.

2.220 Farley G.L. (1987) Energy absorption of composite materials and structures. Forty-third American Helicopter Society Annual Forum, New York.

2.221 Hull D. (1983) Axial crushing of fibre reinforced composite tubes. In *Structural Crashworthiness*, (Ed. by N. Jones & T. Wierzbider), pp. 118–35. Butterworth, London.

2.222 Schmeuser D.W. & Wickliffe L.E. (1987) Impact energy absorption of continuous fibre composite tubes. *Trans ASME*, **109**, 72–7.

2.223 Thornton P.H. & Edwards P.J. (1982) Energy absorption in composite tubes. *J. Comp. Mat.*, **16**, 521–45.

2.224 Hull D. (1982) Energy absorption of composite materials under crush conditions. In *Progress in Science and Engineering of Composites*, (Ed. by T. Hayashi, K. Kawata & S. Umekawa), Vol. 1, ICCM-IV, pp. 861–70. Tokyo University.

2.225 Fairfull A.H. (1986) *Scaling effects in the energy absorption of axially crushed composite tubes*. PhD thesis, University of Liverpool.

2.226 Price J.N. & Hull D. (1987) The crush performance of composite structures. In *Composite Structures*, (Ed. by I.H. Marshall), pp. 2.32–2.44. Elsevier, Amsterdam.

2.227 Price J.N. & Hull D. (1987) Axial crushing of glass fibre-polyster composite cones. *Comp. Sci. Technol.*, **28**, 211–30.

2.228 Berry J.P. (1984) *Energy absorption and failure mechanisms of axially crushed GRP tubes*. PhD thesis. University of Liverpool.

2.229 de Runtz J.A. & Hodge P.G. (1963) Crushing of tubes between rigid plates. *J. Appl. Mech.*, **30**, 381–95.

2.230 Abramowicz W. & Jones N. (1984) Dynamic axial crushing of square tubes. *Int. J. Impact Eng*, **2**(2), 179–208.

2.231 Soreide T.H. & Amdahl J. (1982) Deformation characteristics of tubular members with reference to impact loads from collisions and dropped objects. *Norw. Marit. Res.*, **10**, 3–12.

2.232 Soreide T.H. & Kavlie D. (1985) Collision damage and residual strength of tubular members in steel offshore structures. In *Shell Structures Stability and Strength* (Ed. by R. Narayanan), pp. 185–220. Elsevier Applied Science, London.

2.233 Taby J. & Moan T. (1985) Collapse and residual strength of damaged tubular members. BOSS '85, paper B8.

2.234 Durkin S. (1987) An analytical method for predicting the ultimate capacity of a dented tubular member. *Int. J. Mech. Sci.*, **29**, 449–67.

2.235 Ellinas C.O. & Valsgård S. (1985) Collisions and damage of offshore structure: a state-of-the art. In *Proc. Int. Symp. OMAE, 4th*. Organisation of Motor Association for Engineering, USA.

2.236 de Oliveira J.G. (1981) Design of steel offshore structures against impact loads due to dropped objects. *Report No 81–6*. Department of Ocean Engineering, Massachusetts Institute of Technology, Cambridge, Ma.

2.237 Adamchak J. (1984) An approximate method for estimating the collapse of a ship's hull in preliminary design. In *Proc. Ship Struct. Symp., SNAME*. Arlington, Va.

2.238 Furnes O. & Amdahl J. (1979) Computer simulation study of offshore collisions and analysis of ship platform impacts. Paper presented at The Second International Symposium on Offshore Structures, Brazil, October 1979 (also in *Applied Ocean Research*, **2**, no 3, 119–27, 1980).

2.239 Furness O. & Amdahl J. (1980) Ship collisions with offshore platforms. DnV paper 80-P080, presented at Intermaritec '80 Conference, Hamburg, September 1980.

2.240 Brown & Root Incorp. Design of steel offshore structures against impact loads due to dropped objects. Private Communication 1990.

2.241 Knapp A.E., Green D.J. & Etie R.S. (1984) Collision of a tanker with a platform. OTC paper 4734, presented at the Sixteenth Annual Offshore Technology Conference, Houston, Texas.

2.242 Kochler P.E. & Jorgensen L. (1985) Ship impact ice analysis. In *Proceedings of the Fourth Symposium on Offshore Mechanics and Arctic Engineering, ASME-Omae*, pp. 344–50. Dallas, Texas, February 1985.

2.243 Guttman S.I., Pushar F.J. & Bea R.G. (1984) Analysis of offshore structures subject to arctic ice impacts. In *Proceedings of the Third Symposium on Offshore Mechanics and Arctic In Engineering*, pp. 238–45. New Orleans, Louisiana.

2.244 Hamza H. & Muggeridge D.B. (1984) A theoretical fracture analysis of the impact of a large ice flow with a large offshore structure. In *Proceedings of the Third Symposium on Offshore Mechanics and Arctic Engineering*, pp. 291−7. New Orleans, Louisiana.

2.245 Roland B., Olsen T. & Skaare T. (1982) Ship impact on concrete shafts. Paper EUR 305, presented at the Third European Petroleum Conference, London.

2.246 Hetherington W.G. (1979) Floating jetties − construction, damage and repair. Paper no 3 from the Concrete Ships and Floating Structures Convention.

2.247 Larsen C.M. & Engseth A.G. (1978) Ship collision and fendering of offshore concrete structures. Paper EUR 17, presented at The First European Petroleum Conference, London.

2.248 Hjelde E., Nottveit A. & Amdahl J. (1978) Impacts and collisions offshore report no 2 − pilot tests with pendulum impacts on fendered/unfendered concrete cylinders. *DnV Report No 78−106.*

2.249 Carlin B. & Ronning B. (1979) Impacts and collisions, offshore report no 9: punching shear tests on concrete shells state-of-the-art. *DnV Report No 79−0690.*

2.250 Birdy J.N., Bhula D.N., Smith J.R. & Wicks S.J. (1985) Punching resistance of slabs and shells used for arctic concrete platforms. OTC paper 4855, presented at the Annual Offshore Technology Conference, Houston, Texas.

2.251 Jiang C.W. (1984) Dynamic interaction of ship−fender system. In *Proceedings of the Fourth OMAE Symposium*, pp. 192−7. New Orleans.

2.252 Hunter S.C. (1957) Energy absorbed by elastic waves during impact. *J. Mech. Phys. Sol.*, **5**, 162−71.

2.253 de Oliveira J.G. (1982) Beams under lateral projectile impact. In *Proc. ASCE J. Eng. Mech Div.*, **108**, no EM1, 51−71.

2.254 Jenssen T.K. (1978) Structural damage from ship collision. *Det Norske Veritas Report 78−063.*

2.255 Van Mater P.R. (1979) Critical evaluation of low energy ship collision damage theories and design methodologies. Vol. 1, evaluation and recommendations. *Ship Structure Committee Paper SSC-284.*

2.256 IABSE (1983) Ship collisions with bridges and offshore structures. In *Proceedings of the IABSE Colloquium Copenhagen*, Vol. 42.

2.257 Sorensen K.A. (1976) Behaviour of reinforced and presented concrete tubes under static and impact loading. First International BOSS Conference, Trondheim.

2.258 Jakobsen B., Olsen T.O., Roland B. & Skare E. (1983) Ship impact on the shaft of a concrete gravity platform. In *Proceedings of the IABSE Colloquium Copenhagen*, Vol. 42, pp. 235−43.

2.259 Blok J.J. & Dekker J.N. (1983) Hydrodynamic aspects of ships colliding with fixed structures. In *Proceedings of the IABSE Colloquium Copenhagen*, Vol. 42, pp. 175−85.

2.260 Haywood J.H. (1978) Ship collisions with fixed offshore structures. *Technical Memo TM 78282.* Admiralty Marine Technology Establishment (Dunfermline).

2.261 Corneliussen C. & Laheld P. (1982) Risk for collisions with oil and gas installations on the Norwegian Shelf − collection and preparation of data. *DnV Technical Note 200−002−92.* Norway.

2.262 Laheld P. (1982) Risk for collisions with oil and gas installations on the Norwegian Continental Shelf. Collection and preparation of data. Part report no 4a. Reports on incidents of drifting ships, barges, anchor buoys and other objects. *DnV Report No 20−003−92.* Norway.

2.263 Laheld P. (1982) Risk for collisions with oil and gas installations on the Norwegian Continental Shelf. Collection and preparation of data. Part report no 4b. Incidents of drifting vessels. *DnV Report FDIV/20−82−018.* Norway.

2.264 Laheld P. (1982) Risk for collisions with oil and gas installations on the Norwegian Continental shelf. Collection and preparation of data. Part report no 7. Reported ship impacts with fixed and mobile offshore platforms. *DnV Report FDIV/20−82−021.*

2.265 Brink-Kjoer O. *et al.* (1983) Modelling of ship collision against protected structures. In *Proceedings IABSE Colloquium Copenhagen*, Vol. 42, pp. 147–64.

2.266 Veritec A.S. (1984) Worldwide Offshore Accident Databank (WOAD). *Annual Statistical Report 83*. Government of Norway, Oslo, Norway.

2.267 Low H.Y. & Morley C.T. (1983) Indentation tests of simplified models of ships' structures. In *Proceedings of the IABSE Colloquium on Ship Collision with Bridges and Offshore Structures*, pp. 213–20. Preliminary report, Copenhagen.

2.268 Chang P.T. *et al.* (1980) A rational methodology for the prediction of structural response due to collisions of ships. SNAME Annual Meeting, New York, 13–15, paper no 6.

2.269 Wierzbicki T. *et al.* (1982) Crushing analysis of ship structures with particular reference to bow collisions. *Det Norske Veritas, Progress Report No 16 on Impacts and Collisions Offshore, Report No 82–079.*

2.270 Blok J.J. & Dekker J.N. (1979) On hydrodynamic aspects of ship collision with rigid or non-rigid structures. OTC paper 3664, presented at the Eleventh Annual Offshore Technology Conference, Houston, Texas.

2.271 Acres International Ltd (1983) Influence of small icebergs on semi-submersibles. Report prepared for the Petroleum Directorate, Government of Newfoundland.

2.272 Acres International Ltd (1987) Northumberland Strait Bridge: ice forces report, draft copy of Acres International Ltd. Report prepared for Public Works, Canada.

2.273 Afanasev V.P., Dolgopolov Y.V. & Shvaishstein Z.I. (1973) Ice pressure on individual marine structures. In *Ice Physics and Ice Engineering*, (Ed. by G.N. Yakovlev). Leningrad. (Translated by the Israel Program for Scientific Translation, Jerusalem.)

2.274 Alaska Oil and Gas Association project 309 (1986) Computer software to analyze ice interaction with semi-submersibles. Alaskan Government, USA.

2.275 Allyn M. (1986) Keynote address on global and local ice loads including dynamic effects. Ice/Structure Interaction State-of-the-art Research Needs Workshop, Calgary.

2.276 Arctic Petroleum Operators Association (1975) Preliminary modeling of the process of penetration of pressure ridges by conical structures. APOA Project no 86.

2.277 Arctic Petroleum Operators Association (1975) Computer program to evaluate the forces generated by a moving ice field encountering a conical structure. APOA Project no 87.

2.278 Ashby M.F., Palmer A.C., Thouless M., Goodman D.J., Howard M., Hallam S.D., Murrell S.A.F., Jones N., Sanderson T.J.O. & Ponter A.R.S. (1986) Nonsimultaneous failure and ice loads on arctic structures. In *Proceedings of the Eighteenth Annual Offshore Technology Conference*, paper no OTC 5127. Houston, Texas.

2.279 Bass D., Gaskill H. & Riggs N. (1985) Analysis of iceberg impacts with gravity based structures at Hibernia. In *Proceedings of the Fourth International Conference on Offshore Mechanics and Arctic Engineering (OMAE 1985)*. Dallas, Texas.

2.280 Bercha F.G. & Danys J.V. (1977) On forces generated by ice acting against bridge piers. In *Proceedings of the Third National Hydrotechnical Conference*. Canadian Society for Civil Engineering, L'Université Laval, Quebec City, 26–7.

2.281 Bercha F.G., Brown T.G. & Cheung M.S. (1985) Local pressure in ice–structure interactions. In *Proceedings ARCTIC '85.* pp. 1243–51. American Society of Civil Engineers, San Francisco.

2.282 Bruen F.J. & Vivitrat V. (1984) Ice force prediction based on strain-rate field. In *Third International Offshore Mechanics and Arctic Engineering Symposium OMAE '84*, Vol. 3, pp. 275–81.

2.283 Cammaert A.B. & Tsinker G.B. (1981) Impact of large ice floes and icebergs on marine structures. In *Proceedings of the Sixth International Conference on Port and Ocean Engineering under Arctic conditions*, Vol. II, pp. 653–67. Quebec City, Canada.

2.284 Cammaert A.B., Wong T.T. & Curtis D.D. (1983) Impact of icebergs on offshore gravity and floating platforms. In *Proceedings of the Seventh International Conference on Port and Ocean Engineering under Arctic Conditions (POAC '83)*. Helsinki, Finland.

2.285 Colbeck S.C. (Ed.) (1980) *Dynamics of Snow and Ice Masses*. Academic Press, New York.

2.286 Cox J., Lewis J., Abdelnour R. & Behnke D. (1983) Assessment of ice ride-up/pile-up on slopes and beaches. In *Proceedings of the Seventh International Conference on Port and Ocean Engineering under Arctic Conditions (POAC '83)*, Vol. II, pp. 971–81. Helsinki, Finland.

2.287 Croasdale K.R. (1975) Review of ice forces on offshore structures. Third International Ice Symposium, International Association of Hydraulic Research, Hanover, New Hampshire.

2.288 Croasdale K.R. (1980) Ice forces on fixed rigid structures. In *Working Group on Ice Forces on Structures: A State of the Art Report*, (Ed. by T. Carstens). USA CRREL Special Report 80–26, ADA-089674, Hanover, New Hampshire.

2.289 Croasdale K.R. (1985) Recent developments in ice mechanics and ice loads. In *Behaviour of Offshore Structures*, pp. 53–74. Elsevier, Amsterdam.

2.290 Croasdale K.R. (1986) Ice investigations at Beaufort Sea Caisson, 1985. K.R. Croasdale and Associates report prepared for the National Research Council of Canada and the US Department of the Interior.

2.291 Croasdale K.R. & Marcellus R.W. (1978) Ice and wave action on artificial islands in the Beaufort Sea. *Canadian Journal of Civil Engineering*, 5, no 1, 98–113.

2.292 Croteau P., Rojansky M. & Gerwick B.C. (1984) Summer ice floe impacts against Caisson-Type exploratory and production platforms. In *Third International Offshore Mechanics and Arctic Engineering Symposium OMAE '84*, Vol. 3, pp. 228–37.

2.293 Daoud N. & Lee F.C. (1986) Ice-induced dynamic loads on offshore structures. In *Proceedings of the Fifth Offshore Mechanics and Arctic Engineering Symposium, OMAE '86*, Vol. 4, pp. 212–18.

2.294 Danielewicz B.W., Metge M. & Dunwoody A.B. (1983) On estimating large scale ice forces from deceleration of ice floes. In *Proceedings of the Seventh International Conference on Port and Ocean Engineering, POAC '83*, Vol. 4, pp. 537–46, Helsinki, Finland.

2.295 Dickins D.F. (1981) Ice conditions along arctic tanker routes. D.F. Dickens and Associates Ltd, report prepared for Dome Petroleum Ltd.

2.296 Edwards R.Y., Wallace W.G. & Abdelnour R. (1974) Model experiments to determine the forces exerted on structures by moving ice fields (comparison with small prototype test results). Arctic Canada Limited APOA project no 77.

2.297 Edwards R.Y. & Croasdale K.R. (1976) Model experiments to determine ice forces on conical structures. Symposium of Applied Glaciology, International Glaciological Society, Cambridge, UK, 12–17 September 1976.

2.298 El-Tahan H., Swamidas A.S.J., Arockiasary M. & Reddy D.V. (1984) Strength of iceberg and artificial snow ice under high strain rates and impact loads. In *Proceedings of the Third Offshore Mechanics and Arctic Engineering Symposium*, Vol. 3, pp. 158–65.

2.299 El-Tahan H., Swamidas A.S.J. & Arockiasary M. (1985) Response of semi-submersible models to bergy-bit impact. In *Proceedings of the Eighth International Conference on Port and Ocean Engineering under Arctic Conditions, POAC '85*. Narssarssuaq, Kalaallitt Nunaat.

2.300 Eranti E., Hayes F., Maattanen M. & Soong T. (1981) Dynamic ice–structure interaction analysis for narrow vertical structures. *Proceedings of the Sixth International Conference on Port and Ocean Engineering under Arctic Conditions, POAC '81*. Quebec City.

2.301 Frederking R.M.W. (1979) Dynamic ice forces on an inclined structure. In *Physics and Mechanics of Ice*, (Ed. by P. Tryde), pp. 104–16. International Union of Theoretical and Applied Mechanics Symposium at the Technical University of Denmark, August 1979. Springer Verlag, New York.

2.302 Frederking R.M.W. & Timco G.W. (1983) On measuring flexural properties of ice using cantilever beams. *Annals of Glaciology*, **4**, 58–65.

2.303 Frederking R.M.W. & Nakawo M. (1984) Ice action on Nanisivik Wharf, Winter 1979–1980. *Canadian Journal of Civil Engineering*, **11**, no 4, 996–1003.

2.304 Guttman S.I., Puskar F.J. & Bea R.G. (1984) Analysis of offshore structures subject to Arctic ice impacts. In *Third International Offshore Mechanics and Arctic Engineering Symposium OMAE '84*, Vol. 3, pp. 228–45.

2.305 Hocking G., Mustoe G.G.W. & Williams J.R. (1985) Validation of CICE code for ice ride-up and ice ridge cone interaction. In *Proceedings, ARCTIC '85*, pp. 962–70. American Society of Civil Engineers, San Francisco.

2.306 Hocking G., Mustoe G.G.W. & Williams J.R. (1985) Dynamic global forces on an offshore structure from Multi-Year Floe Impacts. In *Proceedings, ARCTIC '85*, pp. 202–10. American Society of Civil Engineers, San Francisco.

2.307 Johansson P. (1981) Ice-induced vibration of fixed offshore structures. Part 1: review of dynamic response analysis. Marine structures — ice. *Norwegian Research Project, Report No 81–061*.

2.308 Jordaan I., Lanthos S. & Nessim M. (1985) A probabilistic approach to the estimation of environmental driving forces acting upon arctic offshore structures. Det Norske Veritas (Canada) Ltd.

2.309 Jordann I.J. (1986) Numerical and finite element techniques in calculation of ice–structure interaction. Third State-of-the-Art Report on Ice Forces, IAHR Ice Symposium, Iowa City.

2.310 Kashteljan V.I., Poznjak I.I. & Ryvlin A.Ja. (1968) *Ice resistance to motion of a ship.* (Translated from Russian by Marine Computer Applications Corp, USA.)

2.311 Kato K. & Sodhi D.S. (1983) Ice action on pairs of cylinders and conical structures. *CRREL Report 83–25*. Californian Regional Research Engineering Laboratory, California, USA.

2.312 Kitami E., Fujishima K., Taguchi Y., Nawata T., Kawasaki T. & Sakai F. (1984) Iceberg collision with semi-submersible drilling unit. In *Proceedings of the International Association of Hydraulic Research Symposium on Ice*. Hamburg, Germany.

2.313 Koehler P.E. & Jorgensen L. (1985) Ship ice impact analysis. In *Proceedings of the Fourth Offshore Mechanics and Arctic Engineering Symposium, OAME '85.*

2.314 Kry P.R. (1978) A statistical prediction of effective ice crushing stresses on wide structures. In *IAHR International Symposium on Ice Problems (IAHR 1978)*. Lulea, Sweden, part I, pp. 33–47.

2.315 Kry P.R. (1980) Implications of structure width for design ice forces. In *Physics and Mechanics of Ice*, (Ed. by P. Tryde), pp. 179–93. International Union of Theoretical and Applied Mechanics Symposium at the Technical University of Denmark, August 1979. Springer Verlag, New York.

2.316 Maattanen M. (1977) Stability of self-excited ice-induced structure vibrations. In *Proceeding of the Fourth International Conference on Port and Ocean Engineering under Arctic Conditions POAC '77.* Vol. 2, pp. 684–94. St John's, Newfoundland.

2.317 Maattanen M. (1979) Laboratory tests for dynamic ice–structure interaction. In *Proceedings of the Fifth International Conference on Port and Ocean Engineering under Arctic Conditions, POAC '79.* Trondheim, Norway.

2.318 Maattanen M. (1981) Ice structure — dynamic interaction. In *Proceedings of the International Association of Hydraulic Research Symposium on Ice.* Quebec City.

2.319 Matlock H. & Panak J. (1969) A model for the prediction of ice–structure interaction. Offshore Technology Conference, paper no 1066.

2.320 Matsuishi M. & Ettama R. (1986) Model tests on the dynamic behaviour of a floating, cable-moored platform impacted by floes of annual ice. Fifth Offshore Mechanics and Arctic Engineering Symposium, Houston, USA.

2.321 Neill C. (1976) Dynamic ice forces on piers and piles. An assessment of design guidelines in the light of recent research. *Canadian Journal of Civil Engineering*, 3, no 2, 305–41.

2.322 Nevel D.E. (1972) The ultimate failure of a floating ice sheet. In *Proceedings of the IAHR Symposium on Ice and Its Action on Hydraulic Structures*, pp. 17–23. The Netherlands.

2.323 Ralston T. (1977) Ice force design considerations for conical offshore structures. In *Proceedings of the Fourth International Conference on Port and Ocean Engineering under Arctic Conditions, POAC '77*. Memorial University of Newfoundland, St. John's, Newfoundland.

2.324 Ralston T. (1979) Ice force design considerations for conical offshore structures. In *Proceedings of the Fifth International Conference on Port and Ocean Engineering under Arctic Conditions, POAC '79*. Trondheim, Norway.

2.325 Ralston T. (1980) Plastic limit analysis of sheet ice loads on conical structures. In *Physics and Mechanics of Ice*, (Ed. by P. Tryde), pp. 289–308. International Union of Theoretical and Applied Mechanics Symposium at the Technical University of Denmark, August 1979. Springer Verlag, New York.

2.326 Reinicke K.M. (1980) Analytical approach for the determination of ice forces using plasticity theory. In *Physics and Mechanics of Ice*, (Ed. by P. Tryde), pp. 325–41. International Union of Theoretical and Applied Mechanics Symposium at the Technical University of Denmark, August 1979. Springer Verlag, New York.

2.327 Saeki B. *et al.* (1981) Mechanical properties of adhesion strength to pile structures. In *Proceedings of the International Association for Hydraulic Research Symposium*, Vol. 2, pp. 641–9. Quebec City.

2.328 Saeki H., Hamanika K. & Ozaki A. (1969) Experimental study on ice force on a pile. Offshore Technology Conference, Houston, Texas.

2.329 Sanderson T.J.O. & Child A.J. (1986) Ice loads on offshore structures: the transition from creep to fracture. *Cold Regions Science and Technology*, **12**, 157–61.

2.330 Semeniuk A. (1975) Computer program to evaluate the forces generated by a moving ice field encountering a conical structure. APOA project no 87. Association of Petroleum in Antarctica, San Diego, California.

2.331 Sinha N.K. (1982) Delayed elastic strain criterion for first cracks in ice. In *Proceedings of the International Union Theoretical and Applied Mechanics Symposium on Deformation and Failure of Granular Materials*, pp. 323–30. Delft, The Netherlands.

2.332 Sinha N.K. (1982) Acoustic emission and microcracking in ice. In *Proceedings of the 1982 SESA/Japan Society for Mechanical Engineers*, part II, pp. 767–72. Honolulu/Maui, Hawaii.

2.333 Sinha N.K. (1986) Young arctic frazil sea ice: field and laboratory strength tests. *Journal of Materials Science*, **21**, no 5, 1533–46.

2.334 Timco G.W. (1984) Ice forces on structures: physical modeling techniques. In *Proceedings of the IAHR International Ice Symposium*, Vol. IV, pp. 117–50. Hamburg, West Germany.

2.335 Timco G.W. (1986) Ice forces on multi-legged structures. Third state-of-the-art report on ice forces. In *Proceedings of the IAHR Ice Symposium*. Iowa City, Iowa.

2.336 Timco G.W. & Pratte B.D. (1985) The force of a moving ice cover on a row of vertical piles. Canadian Coastal Conference, Quebec.

2.337 Tsuchiya M., Kanie S., Ikejiri K., Yoshida A. & Saeki H. (1985) An experimental study of ice–structure interaction. Offshore Technology Conference, Houston, Texas.

2.338 Vaudrey K.D. (1977) Ice engineering: study of related properties of floating sea-ice sheets and a summary of elastic and viscoelastic analyses. *Technical Report R860*, p. 79. US Navy Civil Engineering Laboratory.

2.339 Vaudrey K.D. & Potter R.E. (1981) Ice defense for natural barrier islands during freeze up. In *Proceedings of the Sixth International Conference on Port and Ocean Engineering under Arctic conditions, POAC '81*. Quebec City, Quebec.

2.340 Wang Y. (1984) Analysis and model tests of pressure ridges failing against conical structures. In *Proceedings of the International Ice Symposium*. International Association of Hydrological Research, Hamburg.

2.341 Wierzbicki T., Chryssostomidis C. & Wiernicki C. (1984) Rupture analysis of ship plating due to hydrodynamic wave impact. Society of Naval Architects and Mechanical Engineers Ship Structure Symposium '84.

2.342 Zayas V.A., Dao B.V. & Hammett D.S. (1985) Experimental and analytical comparisons of semi-sumbersible offshore rig damage resulting from a ship collision. Offshore Technology Conference, Houston, Texas.

2.343 Banthia N. (1987) *Impact resistance of concrete*. PhD thesis, University of British Columbia, Vancouver.

2.344 Banthia N., Mindess S. & Bentur A. (1987) Impact behaviour of concrete beams. *Materials and Structures*, **20**, 293–302.

2.345 Ravid M. & Bodner S.R. (1983) Dynamic perforation of viscoplastic plates by rigid projectiles. *Int. J. Eng. Sci.*, **21**, 577–91.

2.346 Curran D.R., Seaman L. & Shockey D.A. (1987) Dynamic failure of solids. *Physics Report*, **147**, 254–387.

2.347 Anderson C.E. & Bodner S.R. (1988) Ballistic impact: the status of analytical and numerical modeling. *Int. J. Impact Eng*, **7**, 9–35.

2.348 Goldsmith W. & Finnegan S.A. (1986) Normal and oblique impact of cylindro-conical and cylindrical projectiles on metallics plates. *Int. J. Impact Eng*, **4**, 83–105.

2.349 Lambert J.P. (1978) *ARBRL-TR-02033*. Ballistic Research Laboratory, Alabama, USA.

2.350 Meunier Y., Sangoy L. & Pont G. (1989) Steels for ballistic protection. In *Proceedings of the Fourth Israel Materials Engineering Conference* (Ed. by J. Baram & D. Eliezer). Beer-Sheva, Israel, 7–8 December 1988.

2.351 Wilkins M.L. & Guinan M.W. (1973) Impact of cylinders on a rigid boundary. *J. Appl. Phys*, **44**, no 3, 1200–6.

2.352 Yatteau J. & Recht R. (1970) High speed penetration of spaced plates by computer fragments. In *Proceedings of the Ninth International Symposium on Ballistics 2*, pp. 365–74. Shrivenham.

2.353 Naz P. (1987) Spaced plated penetration by spherical high density fragments at high velocity. In *Proceedings of the Tenth International Symposium on Ballistics 2*. San Diego.

2.354 Richards A. & Bowden A. (1987) The development of a projectile velocity measuring system for the RARDE No. 2 Hypervelocity Launcher. *Int. J. Impact Eng*, **5**, 519–32.

2.355 Zhong Z.H. (1988) *On contact–impact problems*. Dissertation no 178, Linköping University, Sweden.

2.356 Zhong Z.H. & Nilsson L.A. (1988) A contact searching algorithm for general contact problems. *Computers & Structures*.

2.357 Zhong Z.H. (1988) New algorithms for numerical treatments of general contact–impact interfaces. In *Proceedings of the Fourth SAS–World Conference FEMCAD*, Vol. 1. China.

2.358 Chaudhary A.B. & Bathe K.J. (1986) A solution method for static and dynamic analysis of three-dimensional contact problems with friction. *Computers and Structures*, **24**.

2.359 Zhong Z.H. (1989) Evaluation of friction in general contact–impact interfaces. Communications in Applied Numerical Methods.

2.360 Zhong Z.H. (1989) Recent developments of the HITA and DENA contact–impact algorithms. *Publication LiTH.IKP.R.547*. Linköping University, Sweden.

2.361 Michie J.D. (1981) Collision risk assessment based on occupant flail-space model. *Transportation Research Record No 796*. Transportation Research Board, Washington DC.

2.362 Ray M.H. & Carney J.F. III (1989) *An Improved Method for Determining Vehicle and Occupant Kinematics in Full-scale Crash Tests*. Transportation Research Board, Washington DC.

2.363 Hargrave M.W., Hansen A.G. & Hinch J.A. (1989) A summary of recent side impact research conducted by the Federal Highway Administration. *SAE Technical Paper No 890377*. Society of Automotive Engineers, Warrendale, Pa.

2.364 Ray M.H., Mayer J.B. & Michie J.D. (1987) Replacing the 4500-lb passenger sedan in report 230 tests, evaluation of design analysis procedures and acceptance criteria for roadside hardware. Final Report, Federal Highway Administration, Washington DC.

2.365 Shadbolt P.J. (1981) *Impact loading of plates*. DPhil thesis, Oxford University.

2.366 Salvatorelli F. & Rollins M.A. (1988) *Some Applications of ABAQUS*, ABAQUS User's Conference, pp. 377–96. HKS, Inc. Newport, Rhode Island.

2.367 Xia Y.R. & Ruiz C. (1989) Response of layered plates to projectile impact. In *Proceedings of the Fourth International Conference on Mechanical Properties of Materials at High Rates of Strain*. Institute of Physics, Oxford.

2.368 Xia, Y.R. & Ruiz C. (1989) Shear effects between the layers of laminated beams under impact loading. *Proceedings of the Institute of Mechanical Engineers*, London, 87–92.

2.369 Winter R. & Pifko A.B. (1988) Finite element crash analysis of automobiles and helicopters. In *Structural Impact and Crash Worthiness*, (Ed. by J. Morton), Vol. II, pp. 278–309. Conference papers. Elsevier.

2.370 Chan A.S.L. & Li F-N (1989) Impact of thin-walled shells using a simplified finite element model. Private communication.

2.371 Chan A.S.L. (1989) Reinforcements in the impact analysis of thin-walled shells using a simplified finite element model. Private communication.

2.372 Varpason P. & Kentalla J. (1981) The analysis of the containment building for global effects of an aircraft crash. Sixth International Conference, SMiRT, paper J9/10, 1981. Paris.

2.373 Ohnuma H. (1985) Dynamic response and local rupture of reinforced concrete beam and slab under impact loading. Eighth International SMiRT Conference, Brussels, 1985, paper J5/3.

2.374 Crutzen Y. (1979) *Non-linear transient dynamic analysis of thin shells using the SEMILOOF finite element*. PhD thesis, University of Brussels.

2.375 NDRC (1946) *Effects of Impact and Explosion, Summary Technical Report of Division 2*, Vol. 1. National Defense Research Committee, Washington DC.

2.376 Jenkov Z.D. *et al.* (1976) Missile tests of quarter-scale reinforced concrete barriers. Prepared for Stone & Webster Inc by Texas Technical University.

2.377 BPC (1974) Design of structures for missile impact. *Topical Report BC-TOP-9-A*. Bechtel Power Corporation.

2.378 Haldav A. Hatami M. & Miller F. (1983) Penetration and spalling depth estimation for concrete structures. Seventh International Conference on Structural Mechanics in Reactor Technology (SMiRT), paper J7/2. (Also Turbine missile – a critical review. *Nucl. Eng Des.*, **55**, 1979.)

2.379 Takeda J. *et al.* (1982) *Proc. Inst. Assoc. Symp. Concrete Structures Under Impact and Impulsive Loading*, pp. 289–95. BAM, Berlin.

2.380 Kar A.K. (1979) Impactive effects of tornado missiles and aircraft. *Trans. ASCE*, **105**, ST2, 2243–60. (Also Loading time – history for tornado generated missiles. *Nucl. Eng Des.*, **51**, 487–93, 1979.)

2.381 Rotz J. (1976) Evaluation of tornado missile impact effects on structures. In *Proceedings of the Symposium on Tornadoes, Assessment of Knowledge and Implications for Man*. USA.

2.382 PLA (1989) *Design Manual for Concrete Pavements*. Port of London Authority.

2.383 Kar A. (1979) Projectile penetration into steel. *J. Struct. Div., ASCE*.

2.384 ACE (1946) Fundamentals of protective design. *Report AT 1207821*. Army Corps of Engineers, USA.

2.385 Rinaldi L.J. (1972) *Containerization: the New Method of Intermodal Transport*. Stirling Publishing.

2.386 D'Arcangelo A.M. (Ed.) (1975) *Ship Design and Construction*. The Society of Naval Architects and Marine Engineers, New York.

2.387 Petry S. & Brown I. (1982) *Model Prestressed Concrete Slab Subjected to Hard Missile Loading. Design for Dynamic Loading*. Construction Press.

2.388 International Conference: Structural Impact and Crash Worthiness, July 1984, Imperial College, London. Elsevier, Vols 1, 2, 3.

2.389 Stevenson J. (Ed.) (1976) *Design for Nuclear Power Plant Facilities*. ASCE, New York.

2.390 Behaviour of reinforced concrete barriers subject to the impact of turbine missiles. Fifth Int. Conf. Struct. Mech. in Reactor Technology (SMiRT), Berlin, 1979, paper J7/5.

2.391 SRI (1960) *Report No 361*. Stanford Research Institute, California, USA.

2.392 Neilson A.J. (1985) Winfrith perforation energy model. *UK Atomic Energy Report No 31*. Winfrith, Dorset, UK.

2.393 Neilson A.J. (1980) Missile impact on metal structures. *Nucl. Energy*, **19**, no 3, 191−8.

2.394 Chellis R.D. (1961) *Pile Foundations*. McGraw Hill, New York.

2.395 Chandrasekaran V. (1974) *Analysis of pile foundations under static and dynamic loads*. PhD thesis, University of Roorkee, India.

2.396 Van der Meer J.W. & Pilarczyk K.W. (1984) Stability of rubble mound slopes under random wave attack. In *Proceedings of the 19th ICCE*, chapter 176. Houston, USA.

2.397 Chiesa M.L. & Callabressi M.L. (1981) Non-linear analysis of a mitigating steel nose cone. In *Computational Methods in Non-Linear Structural and Solid Mechanics* (Ed. by A.K. Noor & H.G. McComb, Jr), pp. 295−300. Pergamon, New York.

2.398 Shular W. (1970) Propagation of study shockwaves in polymethyl methocrylate. *J. Mech. Phys. Solids*, **18**, 277−93.

2.399 Jernström C. (1983) Computer simulation of a motor vehicle crash dummy and use of simulation in the design/analysis process. *Int. J. Vehicle Design*, **4**, no 2, 136−49.

2.400 Körmeling H.A. (1984) Impact tensile strength of steel fibre concrete. *Report 1.5−84−8*. Department of Civil Engineering, The Delft University of Technology.

2.401 Suaris W. & Shah S.P. (1982) Properties of concrete subjected to impact. *ASCE St. Div.*, **109**(7), 1727−41.

2.402 Biggs J.M. (1964) *Introduction to Structural Dynamics*. McGraw Hill.

3 Finite-element and other numerical methods

3.1 Hussain M. & Saugy S. (1970) Elevation of the influence of some concrete characteristics on non-linear behaviour of a prestressed concrete reactor vessel. Concrete for Nuclear Reactors, American Concrete Institute, Detroit, Michigan, paper SP-34-8.

3.2 Zienkiewicz O. & Watson M. (1966) Some creep effects in stress analysis with particular reference to concrete pressure vessels. *Nucl. Eng Des.*, no 4.

3.3 Bangash Y. (1982) Reactor pressure vessel design and practice. *Progress in Nuclear Energy*, **10**, 69−124.

3.4 Saugy B. (1973) Three-dimensional rupture analysis of a prestressed concrete pressure vessel including creep effects. Second International Conference on Structural Mechanics in Reactor Technology (SMiRT), Berlin.

3.5 Saouma V. (1981) *Automated nonlinear finite element analysis of reinforced concrete: a fracture mechanics approach*. Thesis presented to Cornell University, Ithaca, NY, in partial fulfilment of the requirements for the degree of Doctor of Philosophy.

3.6 Phillips D.V. & Zienkiewicz O.C. (1976) Finite element non-linear analysis of concrete structures. *Proc. Inst. Civ. Eng Res. Theory*, **61**.

3.7 Cook R.D. (1981) *Concepts and Applications of Finite Element Analysis*, 2nd edn. Wiley, New York.

3.8   Zienkiewicz O.C. (1977) *The Finite Element in Engineering Science.* McGraw Hill, London.
3.9   Martin H.C. & Carrey G.F. (1973) *Introduction to Finite Element Analysis.* McGraw Hill, New York.
3.10 Bathe K.J. & Wilson E.L. (1972) *Numerical Methods in Finite Element Analysis.* Prentice-Hall, Englewood Cliffs, New Jersey.
3.11 Desai C.S. & Abel J.F. (1972) *Introduction to the Finite Element Method.* Van Nostrand Reinhold, New York.
3.12 Ahmed S. (1969) *Curved finite element in the analysis of solid shell and plate structures.* PhD thesis, University of Wales, Swansea.
3.13 Marcal P.V. (1972) Finite element analysis with material non-linearities — theory and practice. In *Conference on Finite Element Methods in Civil Engineering,* pp. 71–113. Montreal, 1972.
3.14 Rashid Y.R. (1966) Analysis of concrete composite structures by the finite element method. *Nucl. Eng Des.,* **3**.
3.15 Rashid Y.R. & Rockenhauser W. (1968) Pressure vessel analysis by finite element techniques. In *Conference on Prestressed Concrete Pressure Vessels.* Institute of Civil Engineers, London.
3.16 Akyuz F.A. & Merwin J.E. (1968) Solution of non-linear problems of elastoplasticity by finite element method. *AIAA J.,* **6**.
3.17 Argyris J.H. (1965) Three-dimensional anisotropic and inhomogenous elastic media, matrix analysis for small and large displacements. *Ingenieur Archiv,* **34**.
3.18 Dupuis G.A. (1971) Incremental finite element analysis of large elastic deformation problems. Brown University, USA.
3.19 Bangash Y. (1981) The structural integrity of concrete containment vessels under external impacts. In *Proceedings of the Sixth International Conference on Structural Mechanics in Reactor Technology (SMiRT),* paper J7/6. Paris.
3.20 Bangash Y. (1981) The automated three-dimensional cracking analysis of prestressed concrete vessels. In *Proceedings of the Sixth International Conference on Structural Mechanics in Reactor Technology (SMiRT),* paper H3/2. Paris.
3.21 Bathe K.J. & Wilson E.L. (1971) Stability and accuracy analysis of direct integration method. *International Journal of Earthquake Engineering and Structural Dynamics,* **1**, 283–91.
3.22 Ahmed M. (1983) *Bond strength history in prestressed concrete reactor vessels.* PhD Thesis (CNAA), R-84, Thames Polytechnic.
3.23 Bicanic N. & Zienkeiwicz O.C. (1983) Constitutive models for concrete under dynamic loading. *Earthquake Engineering and Structural Dynamics,* **11**, 689–720.
3.24 Gunasekera J.S. *et al.* (1972) Matrix analysis of the large deformation of an elastic-plastic axially symmetric continuum. In *Proceedings of the Symposium on Foundations of Plasticity.* Warsaw.
3.25 Martin H.C. (1965) Derivation of stiffness matrices for the analysis of large deflection and stability problems. In *Proceedings of the First Conference on Matrix Methods in Structural Mechanics.* Ohio.
3.26 Washizu K. (1968) *Variational Methods in Elasticity and Plasticity.* Pergamon Press, Oxford.
3.27 Murray D.W. *et al.* (1969) An approximate non-linear analysis of thin-plates. In *Proceedings of the Second Conference on Matrix Methods in Structural Mechanics.* Berlin, Germany.
3.28 Bechtel Corporation (1972) *Containment Building Liner Plate Design Report.* San Francisco.
3.29 Mutzl J. *et al.* (1975) Buckling analysis of the hot liner of the Austrian PCRV Concept. In *Third International Conference on Structural Mechanics in Reactor Technology,* paper H3/9. London.
3.30 Koiter W.T. (1963) The effect of axisymmetric imperfections on the buckling of

cylindrical shells under axial compression. *Proc. Koninkl. Nederl. Akademie van Wenschappen*, (Series B) **66**.

3.31 Yaghmai S. (1969) *Incremental analysis of large deformations in mechanics of solids with applications to axisymmetric shells of revolution.* PhD thesis, University of California, Berkeley.

3.32 Bangash Y. (1972) Prestressed concrete reactor vessel. Time-saving ultimate load analysis. *J. Inst. Nucl. Eng*, **13**.

3.33 Bangash Y. (1972) A basis for the design of bonded reinforcement in the prestressed concrete reactor vessels. *Inst. Civ. Eng*, paper 7478S, supplement (8).

3.34 Smee D.J. (1967) The effect of aggregate size and concrete strength on the failure of concrete under multiaxial compression. *Civil Engineering Transaction, Inst. of Engineers, Australia*, **CE9**, no 2, 339–44.

3.35 Bangash Y. (1987) The simulation of endochronic model in the cracking analysis of PCPV. In *Proceedings of the Ninth International Conference on Structural Mechanics in Reactor Technology (SMiRT)*, Vol. 4, pp. 333–40. Lausanne.

3.36 Al-Noury S.I. & Bangash Y. (1987) Prestressed concrete containment structures — circumferential hoop tendon calculation. In *Proceedings of the Ninth International Conference on Structural Mechanics in Reactor Technology (SMiRT)*, Vol. 4, pp. 395–401. Lausanne.

3.37 Belytschko T., Yen H.J. & Mullen R. (1979) Mixed methods for time integration. *Computer Methods in Applied Mechanics and Engineering*, **17**, 259–75.

3.38 Belytschko T. (1980) Nonlinear finite element analysis in structural mechanics. In *Partitioned and Adaptive Algorithms for Explicit Time Integration*, (Ed. by W. Wunderlich *et al.*), pp. 572–84.

3.39 Belytschko T. & Lin J.I. (1987) A three-dimensional impact-penetration algorithm with erosion. *Computers and Structures*, **25**(1), 95–104.

3.40 Hallquist J.O., Gondreau G.L. & Benson D.J. (1985) Sliding interfaces with contact–impact in large-scale Lagrangian computations. *Computer Methods in Applied Mechanics and Engineering*, **51**, 107–37.

3.41 Oden J.T. & Martin J.A.C. (1985) Models and computational methods for dynamic friction phenomena. *Computer Methods in Applied Mechanics and Engineering*, **52**, 527–634.

3.42 Hallquist J.O. (1984) NIKE3D: an implicit, finite deformation, finite element code for analyzing the static and dynamic response of three-dimensional solids. *UCID-18822, Rev. 1.*, University of California.

3.43 Hallquist J.O. (1986) NIKE2D: A vectorized, implicit, finite deformation, finite element code for analyzing the static and dynamic response of 2-D solids. *UCID-19677, Rev. 1.* University of California.

3.44 Hughes T.J.R., Pister K.S. & Taylor R.L. (1979) Implicit-explicit finite elements in nonlinear transient analysis. *Computer methods in Applied Mechanics and Engineering*, **17/18**, 159–82.

3.45 Hughes T.J.R., Levit I. & Winget J.M. (1983) An element-by-element solution algorithm for problems of structural and solid mechanics. *Methods in Applied Mechanics and Engineering*, **36**, 241–54.

3.46 Hughes T.J.R., Raefsky A., Muller A., Winget J.M. & Levit I. (1984) A progress report on EBE solution procedures in solid mechanics. In *Numerical Methods for Nonlinear Problems*, (Ed. by C. Taylor *et al.*), Vol. 2, pp. 18–26. Pineridge Press, Swansea, UK.

3.47 Belytschko T.B. & Millen R. (1988) Two-dimensional fluid-structure impact computations with regularization. *Computer Methods in Applied Mechanics and Engineering*, **27**, 139–54.

3.48 Key S.W. (1985) A comparison of recent results from HONDO III with JSME nuclear shipping cask benchmark calculations. *Nucl. Eng Des.*, **85**, 15–23.

3.49 Kulak R.F. (1984) A finite element quasi-eulerian method for three-dimensional fluid-structure interactions. *Computers and Structures*, **18**, 319−32.

3.50 Kulak R.F. (1985) Adaptive contact elements for three-dimensional fluid-structure interfaces. In *Fluid−Structure Dynamics. PVP-98-7*, (Ed. by D.C. Ma & F.J. Moody) pp. 159−66, American Society of Mechanical Engineers, New York.

3.51 Kulak R.F. (1985) Three-dimensional fluid−structure coupling in transient analysis. *Computers and Structures*, **21**(3), 529−542.

3.52 Nichols B.D. & Hirt C.W. (1978) Numerical simulations of hydro-dynamic impact loads on cylinders. *EPRI NP-824 Interim Report*. Electric Power Research Institute, Palo Alto, California.

3.53 Hughes T.J.R. (1978) A simple scheme for developing upwind finite elements. *International Journal for Numerical Methods in Engineering*, **12**, 1359−65.

3.54 Hughes T.J.R. & Mallat M. (1986) A new finite element formulation for computational fluid dynamics: III. The generalized streamline operator for multidimensional advective diffuse systems. *Computer Methods in Applied Mechanics and Engineering*, **58**, 305−28.

3.55 Amdsden A.A., Ruppel H.M. & Hirt C.W. (1980) *SALE: A simplified ALE computer program for fluid flow at all speeds*. Los Almos Scientific Laboratory.

3.56 Donea J. (1983) Arbitrary Lagrangian-Eulerian finite element methods. In *Computational Methods for Transient Analysis*, (Ed. by T. Belytschko & T.J.R. Hughes), pp. 473−516. Elsevier.

3.57 Hughes T.J.R., Liu W.K. & Zimmerman T.K. (1981) Lagrangian-Eulerian finite element formulation for incompressible viscous flows. *Computer Methods in Applied Mechanics and Engineering*, **29**, 329−49.

3.58 Bathe K.J. & Wilson E.L. (1976) *Numerical Methods in Finite Element Analysis*. Prentice-Hall, New Jersey.

3.59 Belytschki Y. & Engelmann B.E. (1987) On flexurally superconvergent four-node quadrilaterals. *Computers and Structures*, **25**, 909−18.

3.60 Belytschko T., Ong J.S.-J., Liu W.K. & Kennedy J.M. (1984) Hourglass control in linear and non-linear problems. *Computer Methods in Applied Mechanics and Engineering*, **43**, 251−76.

3.61 Sun C.T. & Huang S.N. (1975) Transverse impact problems by higher order beam finite element. *Computers and Structures*, **5**, 297−303.

3.62 Tan T.M. & Sun C.T. (1985) Use of statical indentation laws in the impact analysis of laminated composite plates. *J. Appl. Mech.*, **107**, 6−12.

3.63 Zienkiewicz O.C., Irons B. & Nath B. (1965) Natural frequencies of complex free or submerged structures by the finite element method. Symposium on Vibrations in Civil Engineering, London.

3.64 Bathe K.J., Wilson E.L. & Peterson F.E. (1973) SAPIV − a structural analysis program for static and dynamic response of linear systems. *EERC Report No 73−11*. University of California, Berkeley.

3.65 NASA (1970) The NASTRAN theoretical manual. *Report SP-221*. National Aeronautics and Space Administration.

3.66 Swanson J.A. (1971) *ANSYS-Engineering Analysis System User's Manual*. Swanson Analysis Systems, Inc., Pittsburgh, Pennsylvania, USA.

3.67 Johnston P.R. & Weaver W. (1984) *Finite Elements for Structural Analysis*. Prentice-Hall, Englewood Cliffs, NJ.

3.68 Johnston P.R. & Weaver W. jr (1987) *Structural Dynamics by Finite Elements*. Prentice-Hall, Englewood Cliffs, NJ.

3.69 Derbalian G., Fowler G. & Thomas J. (1984) Three-dimensional finite element analysis of a scale model nuclear containment vessel. *Am. Soc. Mech. Eng*, paper 84-PVP-55.

3.70 Margolin L.G. & Adams T.F. (1982) Numerical simulation of fracture. In *Issues in Rock Mechanics*, (Ed. by R.E. Goodman & F.E. Heuze), pp. 637−44.

3.71 Kipp M.E. & Stevens A.L. (1976) Numerical integration of a spall-damage viscoplastic constitutive model in a one-dimensional wave propagation code. *SAND76-0061*. Sandia Lab. (technical report).

3.72   Fredriksson B. & Mackerle J. (1983) *Structural Mechanics Finite Element Computer Programs*, 4th edn. Advanced Engineering Corporation, Linköping, Sweden.

3.73   Noor A.K. (1981) Survey of computer programs for solution of nonlinear structural and solid mechanics problems. *Computers and Structures*, **13**, 425−65.

3.74   Yagawa G., Ohtsubo H., Takeda H., Toi Y., Aizawa T. & Ikushima T. (1984) A round robin on numerical analyses for impact problems. *Nucl. Eng Des.*, **78**, 377−87.

3.75   de Roevray A., Haug E. & Dubois J. (1983) Failure mechanisms and strength reduction in composite laminates with cut-outs − a 3-D finite element numerical autopsy. In *Composite Structures*, (Ed. by I.H. Marshall). Applied Science, London and New York.

3.76   Bhat S. (1985) *Analysis of large plastic shape distortion of shells*. PhD thesis, Massachusetts Institute of Technology, Cambridge.

3.77   ADINA Engineering (1984) *Automatic Incremental Nonlinear Analysis: Users Manual*. ADINA Engineering Inc, MIT, Massachusetts, USA.

3.78   Dyhrkopp F. (1984) Inspection of floating offshore platforms. *Proc. Int. Symp. Role Des., Inspect. Redundancy Mar. Struct. Reliabil.* National Academic Press, Washington DC.

3.79   Engseth A. (1984) *Finite element collapse analysis of tubular steel offshore structures*. Doctoral thesis, Report UR-85-46, Div. Marine Structures Norw. Inst. Technol., Trondheim.

3.80   Marshall P.W., Gates W.E. & Anagnostopoulos S. (1977) Inelastic dynamic analysis of tubular offshore structures. *Proc. of Offshore Technology Conference*, paper no 2908.

3.81   Bergen P.G. & Arnesen A. (1983) FENRIS − a general purpose finite element program. In *Proceedings of the Fourth International Conference on Finite Element Systems*. Southampton.

3.82   Plesha M.E. (1987) Eigenvalue estimation for dynamic contact problems. *J. Eng. Mech., ASCE*, **113**, 457−62.

3.83   Isaacson E. & Keller H.B. (1966) *Analysis of Numerical Methods*. Wiley, New York.

3.84   Kulak R.F. & Fiala C. (1983) NEPTUNE: a system of finite element programs for three-dimensional non-linear analysis. *Nucl. Eng Des.*, **106**, 47−68.

3.85   Kulak R.F. (1989) Adaptive contact elements for three-dimensional explicit transient analysis. *Computer Methods in Applied Mechanics and Engineering*, **72**, 125−51.

3.86   Madsen N. (1984) Numerically efficient procedures for dynamic contact problems. *Int. J. Num. Meth. Eng*, **20**, 1−14.

3.87   Kanto Y. & Yagawa G. (1980) Finite element analysis of dynamic buckling with contact. *Trans Japan Soc. Mech. Eng., Ser. A*, **51**, 2747−56.

3.88   Pascoe S.K. & Mottershead J.E. (1988) Linear elastic problems using curved elements and including dynamic friction. *Int. J. Num. Methods Eng*, **26**, 1631−42.

3.89   Bronlund O.E. (1969) Eigenvalues of large matrices. Symposium on Finite Element Techniques, University of Stuttgart, 10−12 June 1969.

3.90   Clint M. & Jennings A. (1970) The evaluation of Eigenvalues and Eigenvectors of real symmetric matrices by simultaneous iteration. *Comput. J.*, **13**(1).

3.91   Bathe K.J. (1971) Solution methods for large generalized Eigenvalue problems in structural engineering. *SESM Rep. 71−20*. Civil Engineering Department, University of California, Berkeley.

3.92   Truesdell C. & Noll W. (1965) The non-linear field theories of mechanics. In *Encyclopedia of Physics*, (Ed. by S. Flugge), Vol. III, Art. 3. Springer-Verlag, Berlin.

3.93   Murakawa H. & Atluri S.N. (1978) Finite elasticity solutions using hybrid finite elements based on a complementary energy principle. *J. Appl. Mech., Trans ASME*, **45**, 539–48.

3.94 Atluri S.N. (1979) On rate principles for finite strain analysis of elastic and inelastic non-linear solids. In *Recent Research on Mechanic Behavior of Solids* (Miyamoto Anniversary Volume), pp. 79–107. University of Tokyo Press, Tokyo.

3.95 Drucker D.C. (1951) A more fundamental approach to plastic stress strain relations. *Proc. 1st US National Congress on Applied Mechanics*, pp. 487–91. Pasadena, USA.

3.96 Atluri S.N. (1984) On constitutive relations at finite strain: hypoelasticity and elasto-plasticity with isotropic or kinematic hardening. *Comput. Meth. Appl. Mech. Eng.*, **43**, 137–71.

3.97 Watanabe O. & Atluri S.N. (1985) A new endochronic approach to computational elastoplasticity: example of a cyclically loaded cracked plate. *J. Appl. Mech.*, **52**, 857–64.

3.98 Reed K.W. & Atluri S.N. (1983) Analysis of large quasistatic deformations of inelastic bodies by a new hybrid-stress finite element algorithm. *Comp. Meth. Appl. Mech. Eng*, **39**, 245–95; II. Applications. **40**, 171–98.

3.99 Surana K.S. (1983) *FINESSE (Finite Element System for Nonlinear Analysis) Theoretical Manual*. McDonnell Douglas Automation Company, St Louis.

3.100 Moan T. (1974) A note on the convergence of finite element approximations for problems formulated in curvilinear coordinate systems. *Comp. Meth. Appl. Mech.*, **3**, 209–35.

3.101 Fried I. (1971) Basic computational problems in the finite element analysis of shells. *Int. J. Solids Struct.*, **7**, 1705–15.

3.102 Noor A.K. & Peters J.M. (1981) Mixed models and reduced/selective integration displacement models for non-linear analysis of curved beams. *Int. J. Num. Meth. Eng*, **17**, 615–31.

3.103 Cantin G. (1970) Rigid body motions in curved finite elements. *AIAA J.*, **8**, 1252–5.

3.104 Elias Z.M. (1972) Mixed finite element method for axisymmetric shells. *Int. J. Num. Meth. Eng*, **4**, 261–77.

3.105 Ahmad S., Irons B.M. & Zienkiewicz O.C. (1970) Analysis of thick and thin shell structures by curved finite elements. *Int. J. Num. Meth. Eng*, **2**, 419–51.

3.106 Pugh E.D.L., Hinton E. & Zienkiewicz O.C. (1978) A study of quadrilateral plate bending elements with reduced integration. *Int. J. Num. Meth. Eng*, **12**, 1059–79.

3.107 Argyris J.H. & Scharpf D. (1968) The SHEBA family of shell elements for the matrix displacement method. *Aeronaut. J.*, **72**, 873–83.

3.108 Sander G. & Idelsohn S. (1982) A family of conforming finite elements for deep shell analysis. *Int. J. Num. Meth. Eng*, **18**, 363–80.

3.109 Irons B.M. (1976) The semiloof shell element. In *Finite Elements for Thin Shells and Curved Members*, (Ed. by D.G Ashwell & R.H. Gallagher). Wiley, London.

3.110 Belytschko T., Liu W.K. & Ong J.S.J. (1984) Nine node Lagrange shell elements with spurious mode control. Twenty-fifth Structures, Structural Dynamics and Materials Conference, AIAA/ASME/AHS, Palm Springs, California.

3.111 Szabo B.A. & Mehta A.K. (1978) P-convergent finite element approximations in fracture mechanics. *Int. J. Num. Meth. Eng*, **12**, 551–60.

3.112 Babuska I. & Szabo B.A. (1982) On the rates of convergence of the finite element method. *Int. Num. Meth. Eng*, **18**, 323–41.

3.113 Babuska I. & Miller A. (1984) The post-processing approach in the finite element method: part I calculation of displacements, stresses and other higher derivatives of the displacements. *Int. Num. Meth. Eng*, **20**, 1085–109.

3.114 Babuska I. & Miller A. (1984) The post-processing approach in the finite element method: part 2 the calculation of stress intensity factors. *Int. J. Num. Meth. Eng*, **20**, 1111–29.

3.115 Babuska I. & Miller A. (1984) The post-processing approach in the finite element method: part 3 error estimates and adaptive mesh selection. *Int. J. Num. Meth. Eng*, **20**, 2311–24.

3.116 Parks D.M. (1975) *Some problems in elastic-plastic finite element analysis of cracks.* PhD thesis, Brown University, Providence, RI.

3.117 Barsoum R.S. (1976) On the use of isoparametric finite elements in linear fracture mechanics. *Int. J. Num. Meth. Eng*, **10**, 25–37.

3.118 *MARC User's Manuals, Version J.2.* MARC Analysis Research Corp, Palo Alto, California. (Published annually.)

3.119 Schulz J.C. (1978) Finite element analysis of a kinetic energy warhead penetrating concrete. In *Proceedings of the Fourth International Symposium on Ballistics.* Monterey, California.

3.120 Key S.W., Beising Z.E. & Krieg R.D. (1978) *HONDO II, a finite element computer program for the large deformation dynamic of axisymmetric solids.* Sandia Laboratories, Albuquerque.

3.121 Levy A. (1979) Development of the PLANS computer program for elastic-plastic creep analysis of nuclear reactor structural components. *Grumman Research Department Report RE-567, ORNL-Sub4485-2.* Grumman Aerospace Corporation, Bethpage, New York.

3.122 Felippa C.A. & Park K.C. (1978) Direct time integration methods in non-linear structural dynamics. PENOMECH 1978 Conference, University of Stuttgart.

3.123 Underwood P.G. & Park K.C. (1980) A variable step central difference method for structural dynamic analysis: implementation and performance evaluation. *Comput. Meth. Appl. Mech. Eng*, **23**, 259–79.

3.124 Geradin M., Hogge M. & Idelsohn S. (1983) In *Implicit Element Methods. Computational Methods for Transient Analysis*, (Ed. by T. Belytschko & T. Hughes), pp. 417–71. North-Holland, Amsterdam.

3.125 Mathies H. & Strang G. (1979) The solution of non-linear finite element equations. *Int. J. Num. Meth. Eng*, **14**, 1613–26.

3.126 Batoz J.L. & Dhatt G. (1971) Une évaluation des méthodes du type Newton-Raphson implicit l'accroissement d'un déplacement. Rapport Université Technologique de Compiègne.

3.127 Belytschko T. & Rughes T.J.R. (Eds) (1983) *Computational Methods for Transient Analysis.* Elsevier, New York.

3.128 Beckers P. & Sander G. (1979) Improvement of the frontal solution technique. *LTAS Report SA-72.* University of Liège, Belgium.

3.129 Ericsson T. & Ruhe A. (1980) The spectral transformation Lanczos method for the numerical solution of large sparse generalized symmetric Eigenvalue problems. *Math. Comput.*, **35**, 152.

3.130 Atkinson K.E. (1978) *An Introduction to Numerical Analysis*, 1st edn. Wiley, New York.

3.131 Wolf M.A. (1978) *Numerical Methods for Unconstrained Optimization — an Introduction.* Van Nostrand Rheinhold, New York.

3.132 Pian T.H.H. & Tong P. (1970) Variational formulations of finite displacement analysis. *IUTAM Symposium, High Speed Computing of Elastic Structures*, pp. 43–63. Liège, Belgium.

3.133 Riks E. (1979) An incremental approach to the solution of snapping and buckling problems. *Int. J. Solids Struct.*, **15**, 529–51.

3.134 Geradin M. & Fleury C. (1982) Unconstrained and linearly constrained minimization. In *Foundation of Structural Optimization: A Unified Approach*, (Ed. by A.J. Morris), chapter 8. Wiley, New York.

3.135 Gerardin M. & Carnoy E. (1979) On the practical use of Eigenvalue bracketing in finite element applications to vibration and stability problems. *Euromech 112*, pp. 151–71. Hungarian Academy of Science.

3.136 Wolfshtein M. (1967) *Conviction processes in turbulent impinging jet.* PhD thesis, University of London.

3.137 Baker A.J. (1983) *Finite Element Computational Fluid Mechanics.* Hemisphere/McGraw-Hill, New York.

3.138 Chung T.J. (1978) *Finite Element Analysis of Fluid Dynamics*. McGraw-Hill, New York.

3.139 Norrie D.H. & de Vries G. (1980) Admissibility requirements and the least squares finite element solution for potential flow. In *Proceedings of the Seventh Australasian Hydraulics and Fluid Mechanics Conference*, pp. 115–18. Nat. Conf. Pub. 80/4, Institution of Engineers (Australia).

3.140 Argyris J.H., Mareczek G. & Scharpf D.W. (1969) Two and three dimensional flow using finite elements. *Aero. J. R. Aero. Soc.*, **73**, 961–4.

3.141 Zienkiewicz O.C. & Heinrich J.C. (1978) The finite element method and convection potential fluid mechanics. In *Finite Elements in Fluids*, (Ed. by R.H. Gallagher *et al.*) Vol. 3, pp. 1–23. Wiley, New York.

3.142 Seto H. (1982) New hybrid element approach to wave hydrodynamic loadings on off-shore structures. In *Finite Element Flow Analysis*, (Ed. by T. Kawai), pp. 435–42. University of Tokyo Press, Tokyo.

3.143 Sakai F. (1981) Vibration analysis of fluid–solids systems. In *Interdisciplinary Finite Element Analysis*, (Ed. by J.F. Abel, T. Kawai & S.F. Shen), pp. 453–77. Cornell University Press, Ithaca, NY.

3.144 Bettess P. & Zienkiewicz O.C. (1977) Diffraction and refraction of surface waves using finite and infinite elements. *Int. J. Num. Meth. Eng*, **11**, 1271–90.

3.145 Ganeba M.B., Wellford L.C. jr & Lee J.J. (1982) Dissipative finite element model for harbour resonance problems. In *Finite Element Flow Analysis*, (Ed. by T. Kawai) pp. 451–9. University of Tokyo Press, Tokyo.

3.146 Kawahara M. (1980) On finite element methods in shallow water long wave flow analysis. In *Computational Methods in Non-Linear Mechanics*, (Ed. by J.T. Oden), pp. 261–87. North Holland, Amsterdam.

3.147 Kawahara M. (1978) Finite element method in two layer and multi-leveled flow analysis. In *Finite Elements in Water Resources*, (Ed. by S.Y. Wang *et al.*), pp. 5.3–5.19. University of Mississippi Press.

3.148 Kawahara M. & Hasegawa K. (1978) Periodic galerkin finite element method of tidalflow. *Int. Num. Meth. Eng*, **12**, 115–27.

3.149 Kawahara M., Hasegawa K. & Kawanago Y. (1977) Periodic tidal flow analysis by finite element perturbation method. *Comput. Fluid.*, **5**, 175–89.

3.150 Agnntaru V. & Spraggs L. (1982) A time integration technique for modelling of small amplitude tidal waves. In *Finite Elements in Water Resources*, (Ed. by K.P. Holz *et al.*), pp. 5.17–5.25. Springer-Verlag, Hanover.

3.151 Partridge P.W. & Brebbia C.A. (1976) Quadratic finite elements in shallow water problems. *Proc. ASCE 102 (HY9)*, 1299–1313.

3.152 Wong I.P. & Norton W.R. (1978) Recent application of RMA's finite element models for two dimensional hydrodynamics and water quality. In *Proceedings of the Second Conference on Finite Element Water Resources*, pp. 2.81–2.99.

3.153 Tanaka T., Ono Y., Ishise T. & Nakata K. (1982) Simulation analysis for diffusion of discharged warm water in the bay by finite elements. In *Finite Elements in Water Resources*, (Ed. by K.P. Holz *et al.*), pp. 15.31–15.41. Springer-Verlag.

3.154 Zienkiewicz O.C., Bettess P. & Kelly D.W. (1978) The finite element method for determining fluid loading on rigid structures — two- and three-dimensional formulations. In *Numerical Methods in Offshore Engineering*, (Ed. by O.C. Zienkiewicz *et al.*), pp. 141–83. Wiley, New York.

3.155 Austin D.I. & Bettess P. (1982) Longshore boundary conditions for numerical wave models. *Int. J. Num. Meth. Fluids*, **2**, 263–76.

3.156 Taylor C., Patil B.S. & Zienkiewicz O.C. (1969) Harbour oscillation — a numerical treatment for undamped natural modes. *Proc. Inst. Civ. Eng*, **43**, 141–56.

3.157 Connor J.J. & Brebbia C.A. (1976) *Finite Element Techniques for Fluid Flow*. Newnes, Butterworth, London.

3.158 Zienkiewicz O.C. & Heinrich J.C. (1979) A unified treatment of steady-state shallow water and two-dimensional Navier-Stokes equations — finite element penalty function approach. *Comput. Mech. Appl. Mech. Eng*, **17/18**, 673−98.

3.159 Byrne P.M. & Janzen W. (1981) SOILSTRESS: a computer program for non-linear analysis of stress and deformation in soil. *Soil Mech. Ser.*, no 52, University of British Columbia.

3.160 Duncan J.M. & Chang C.Y. (1970) Non-linear analysis of stress and strain in soils. *J. Soil. Mech. Found. Div., ASCE*, **96**(SM5), 1629−51.

3.161 Duncan J.M., Byrne P.M., Wong K.S. & Mabry P. (1980) Strength stress-strain and bulk modulus parameters for finite element analyses of stresses and movements in soil masses. *Report No UCB/GT/78−02*. Department of Civil Engineering, University of California, Berkeley.

3.162 Duncan J.M., D'Orazio T.B., Chang C.S., Wong K.S. & Namiq L.I. (1981) CON2D: a finite element computer program for analysis of consolidation. *Report No UCB/GT/81−01*. US Army Engineers Waterways Experiment Station, Vicksburg, Miss., USA.

3.163 Lee M.K.W. & Liam Finn W.D. (1978) DESRA-2: dynamic effective stress response analysis soil deposits with energy transmitting boundary including assessment of liquefaction parameters. *Report No 38, Soil Mech. Ser.* Department of Civil Engineering, University of British Colombia, Vancouver.

3.164 Martin P.P. (1975) *Non-linear methods for dynamic analysis of ground response*. PhD thesis, Department of Civil Engineering, University of California, Berkeley.

3.165 Princeton University Computer Centre (1981) DYNAFLOW — a nonlinear transient finite element analysis program. Princeton University, Princeton, NJ.

3.166 Ungless R.F. (1973) *An infinite element*. MSc thesis, Department of Civil Engineering, University of British Columbia, Vancouver.

3.167 Stevenson J. (1980) Current-summary of international extreme load design requirements for nuclear power plant facilities. *Nucl. Eng Des.*, **60**.

3.168 Riera J. (1980) A critical reappraisal of nuclear power plant safety against accidental aircraft impact. *Nucl. Eng Des.*, **57**.

3.169 Ribora B., Zimmermann Th. & Wolf J.P. (1976) Dynamic Rupture Analysis of R.C. Shells. *Nucl. Eng Des.*, **37**, 269−77.

3.170 Davie N.T. & Richgels M.A. (1983) GNOME: an earth penetrator code. *Sandia Report 82−2358*. Sandia Laboratories, USA.

3.171 Cherry J.T. (1967) Explosion in soils. *Int. J. Rock Mech.*, **4**, 1−22.

4 Blasts and Explosion Dynamics

4.1 Whitney C.S., Anderson B.G. & Cohen E. (1955) Design of blast construction for atomic explosions. *J. Am. Concrete Inst.*, **26**, no 7, 589−695.

4.2 Glasstone S. (Ed.) (1962) The effects of nuclear weapons. United States Atomic Energy Commissions, revised edition.

4.3 Azo K. (1966) *Phenomenon involved in presplitting by blasting*. Doctoral dissertation, Stanford University, Palo Alto, USA.

4.4 Banks D.C. (1968) Selected methods for analyzing the stability of crater slopes. *Miscellaneous Paper S-68-8*. USA Waterways Experiment Station.

4.5 Barker D.B., Fourney W.L. & Holloway D.C. (1979) Photoelastic investigation of flaw initiated cracks and their contribution to the mechanisms of fragmentation. 21st Symposium on Rock Mechanics, Austin, Texas, June 1979.

4.6 Bickel J.O. & Kuesel T.R. (Eds) (1982) *Tunnel Engineering Handbook*. Van Nostrand Reinhold, New York.

4.7 Blindheim O.T. (1976) Preinvestigations resistance to blasting and drillability predictions in hard rock tunnelling, mechanical boring or drill and blast tunnelling. First US — Swedish Underground Workshop, Stockholm, pp. 81–97.

4.8 Bowden F.P. & Yoffe A.D. (1962) *Initiation and Growth of Explosions in Liquid and Solids*. Cambridge University Press, Cambridge, UK.

4.9 Brown F.W. (1941) Theoretical calculations for explosives. *US Bureau of Mines Technical Publication No 632*.

4.10 Bullock R.L. (1975) Technological review of all-hydraulic rock drills. *Transactions of the American Institute of Mining, Metallurgical, and Petroleum Engineers — Society of Mining Engineers*, preprint no 75-Au-42.

4.11 Bullock R.L. (1976) An update of hydraulic drilling performance. *American Institute of Mining, Metallurgical, and Petroleum Engineers, Rapid Excavation and Tunnelling Conference*, pp. 627–48. Las Vegas.

4.12 Clark G.B. (1959) Mathematics of explosives calculations. Fourth Symposium on Mining Research. *University of Missouri School of Mines and Metallurgy Bulletin Technical Series No 97.32–80*.

4.13 Clark G.B. (1979) Principles of rock drilling. *Colo. School Mines Q.*, **74**(2).

4.14 Clark G.B., Bruzewski R.F., Yancik J.J., Lyons J.E. & Hopler R. (1961) Particle characteristics of Ammonium Nitrate and blasting agent performance. *Colo. School Mines Q.*, **56**, 183–98.

4.15 Clark G.B. & Maleki H. (1978) Basic operational parameters of an automated plug and feather rock splitter. Sponsored by National Science Foundation, NSF Ap73-07486-A02, Colorado School of Mines, USA.

4.16 Cook M.A. (1958) *The Science of High Explosives*. Reinhold, New York.

4.17 Crow S.C. & Hurlburt G.H. (1974) The mechanics of hydraulic rock cutting. Second International Symposium on Jet Cutting Technology, Cambridge, UK, paper E1.

4.18 Cunningham C. (1983) The Kuz-Ram model for prediction of fragmentation from blasting. *First International Symposium on Rock Fragmentation by Blasting*, Vol. 2, pp. 439–52. Lulea, Sweden.

4.19 Dick R.A. & Gletcher L.R. (1973) A study of fragmentation from bench blasting in limestone at reduced scale. *US Bureau of Mines Report of Investigations No 7704*.

4.20 Dinsdale J.R. (1940) Ground failure around excavations. *Trans. Inst. Min. Metall. L.*

4.21 Ditson J.D. (1948) Determining blow energy of rock drills. *Compressed Air Magazine*, **53**(1), 15–16.

4.22 Duvall W.I. (1953) Strain wave shapes in rock near explosions. *Geophysics*, **18**, 310–23.

4.23 Duvall W.I. & Atchison T.C. (1957) Rock breakage by explosives. *US Bureau of Mines Report of Investigations No 5356*.

4.24 Fairhurst C. (1961) Wave mechanics in precursive drilling. *Mining Quarry Eng*, **27**.

4.25 Fairhurst C. & Lacabanne W.D. (1957) Hard rock drilling techniques. *Mining Quarry Eng*, **23**, 157–74.

4.26 Field J.E. & Ladegaarde-Pedersen A. (1971) The importance of the reflected wave in blasting. *Int. J. Rock Mech. Mining Sci.*, **8**, 213–26.

4.27 Fogelson D.E., Duvall W.I. & Atchison T.C. (1959) Strain energy in explosive-generated strain pulses. *US Bureau of Mines Report of Investigations No 5514*.

4.28 Hino K. (1959) *Theory and Practice of Blasting*. Nippon Kayaku Co, Japan.

4.29 Hornsey E.E. & Clark G.B. (1968) Comparison of spherical elastic viogt and observed wave forms for large underground explosions. Tenth Symposium on Rock Mechanics, Austin, Texas.

4.30 Jost W. (1946) *Explosion and Combustion Processes in Gases*. McGraw-Hill, New York.

4.31 Kutter H.K. & Fairhurst C. (1971) On the fracture process in blasting. *Int. J. Rock Mech. Min. Sci.*, **8**, 181–202.

4.32 Lean D.J. & Paine G.G. (1981) Preliminary blasting for a bucket wheel excavator operation at COCA Coonyella Mine, Central Queensland. Australian Mineral Foundation Workshop Course, No 152/81.

4.33   Lee H.B. & Akre R.L. (1981) Blasting process (patent). *US Patent 2,703,528.* US Patent-Office, Washington.

4.34   Lounds C.M. (1986) Computer modelling of fragmentation from an array of shotholes. *First International Symposium on Rock Fragmentation by Blasting*, Vol. 2, pp. 455–68. Lulea, Sweden.

4.35   Lundquist R.G. & Anderson C.F. (1969) Energetics of percussive drills – longitudinal strain energy. *US Bureau of Mines Report of Investigation No 7329.*

4.36   Mason C.M. & Aiken E.G. (1972) Methods of evaluating explosives and hazardous materials. *US Bureau of Mines Information Circular No 8541.*

4.37   Morrell J.R. & Larsen D.A. (1974) Disk cutter experiments in metamorphic and igneous rocks – tunnel boring technology. *US Bureau of Mines Report of Investigations No 7691.*

4.38   National Fire Protection Association (1962) Blasting agents. In *Code for the Manufacture, Transportation, Storage and Use of Explosive and Blasting Agents*, no 495, pp. 25–9.

4.39   Nikonov G.P. & Goldin Y.A. (1972) Coal and rock penetration by fine, continuous high pressure water jets. First International Symposium on Jet Cutting Technology, BHRA, Coventry, UK, paper E2.

4.40   Obert L. & Duvall W.I. (1950) Generation and propagation of strain waves in rock. Part 1. *US Bureau of Mines Report of Investigations No 4683.*

4.41   Peele R. & Church J.H. (1941) *Mining Engineers Handbook.* Wiley, New York.

4.42   Pons L. *et al.* (1962) Sur la fragilisation superficielle au cours de frottement de carbures de tungstens frittes. *Academie des Sciences, Paris,* **225**, 2100.

4.43   Rad P.F. & Olson R.C. (1974) Tunneling machine research: size distribution of rock fragments produced by rolling disc cutters. *US Bureau of Mines Report of Investigations No 7882.*

4.44   Roxborough F.F. & Rispin A. (1972) Rock excavation by disc cutter. Report to Transport and Road Research Laboratory of the Department of Environment, University of Newcastle-upon-Tyne, UK.

4.45   Roxborough F.F. & Phillips H.R. (1975) Rock excavation by disc cutter. *Int. J. Rock Mech. Min. Sci.,* **12**, 361–6.

4.46   Ryd E. & Holdo J. (1953) Percussive rock drills. In *Manual on Rock Blasting.* Atlas Deisel and Saudvikens, Stockholm.

4.47   Sharpe J.A. (1942) The production of elastic waves by explosion pressure. I. Theory and empirical field observation. *Geophysics,* **17**(3), 144–55.

4.48   Tarkoy P.J.J. (1973) Predicting TBM penetration rates in selected rock types. In *Proceedings, Ninth Canadian Rock Mechanics Symposium*, pp. 263–74. Montreal.

4.49   Van Dolah R.W. & Malesky J.S. (1962) Fire and explosion in a blasting agent mix building, Norton Virginia. *US Bureau of Mines Report of Investigations No 6015.*

4.50   Van Dolah R.W., Gibson F.C. & Murphy J.N. (1966) Sympathetic detonation of ammonium nitrate and ammonium nitrate – fuel oil. *US Bureau of Mines Report of Investigations No 6746.*

4.51   Van Dolah R.W., Mason C.M., Perzak F.J.P. & Forshey D.R. (1966) Explosion hazards of ammonium nitrate under fire hazard. *US Bureau of Mines Report of Investigations No 6773.*

4.52   (1965) Structures to Resist the Effects of Accidental Explosions. *TM 5–1300.* Department of the Army, Washington DC.

4.53   Okamoto S., Tamura C., Kato K. & Hamada M. (1973) Behaviors of submerged tunnels during earthquakes. *Proceedings, Fifth World Conference on Earthquake Engineering*, Vol. 1, pp. 544–53. Rome, Italy.

4.54   Wyllie L.A., McClure F.E. & Degenkolb H.J. (1973) Performance of underground structures at the Joseph Jensen Filtration Plant. *Proceedings, Fifth World Conference on Earthquake Engineering.* Rome, Italy.

4.55   Lewis B. & Von Elbe G. (1961) *Combustion, Flames and Explosions of Gases.* Academic Press, USA.

4.56 Bradley J.N. (1969) *Flame and Combustion Phenomena.* Chapman and Hall.

4.57 Andrews G.E. & Bradley D. (1972) Determination of burning velocities, a critical review. *Combustion and Flame,* **18**, 133. See also The burning velocity of methane– air mixtures. *Combustion and Flame,* **19**, 275.

4.58 Perry R.H. & Chilton C.H. (1973) *Chemical Engineers' Handbook,* 5th edn. McGraw-Hill.

4.59 Zabetakis M.G. (1965) Flammability characteristics of combustible gases and vapours. *US Bureau of Mines Bulletin 627.*

4.60 Gibbs G.J. & Calcote H.F. (1959) Effect of molecular structure on burning velocity. *Journal of Chemical Engineering Data,* **5**, 226.

4.61 Egerton A. & Lefebvre A.H. (1954) The effect of pressure variation on burning velocities. *Proc. R. Soc. A222,* 206.

4.62 Dugger G.L. (1962) Effect of initial mixture temperature on flame speed of methane air, propane air and ethylene air mixtures. *NASA Report 1061.* National Aeronautics and Space Aministration, USA.

4.63 Cubbage P.A. (1959) Flame traps for use with town gas air mixtures. *Inst. Gas Engineers, Communication GC 63.*

4.64 Cubbage P.A. (1963) The protection by flame traps of pipes conveying combustible mixtures. *Second Symposium on Chemical Process Hazards, I. Chem. E.*

4.65 Rasbash D.J. & Rogowski Z.W. (1960) Gaseous explosions in vented ducts. *Combustion and Flame,* **4**, 301.

4.66 Rasbash D.J. & Rogowski Z.W. (1961) Relief of explosions in duct systems. *First Symposium on Chemical Process Hazards, I. Chem. E.,* 58.

4.67 Rasbash D.J. & Rogoswki Z.W. (1963) Relief of explosions in propane air mixtures moving in a straight unobstructed duct. *Second Symposium on Chemical Process Hazards, I. Chem. E.*

4.68 Leach S.J. & Bloomfield D.P. (1973) Ventilation in relation to toxic and flammable gases in buildings. *Building Science,* **8**, 289.

4.69 *British Standard 5925, Code of Practice for Design of Buildings. Ventilation Principles and Designing for Natural Ventilation.* (Formerly CP3 Chapter 1(c).) British Standards Institute (1980).

4.70 British Gas Engineering Standard PS/SHAI, Code of Practice for hazardous Area Classification, part 1 — Natural Gas. (Draft, 1981).

4.71 Jost W. (1946) *Explosion and Combustion Processes in Gases.* McGraw Hill.

4.72 Bradley D. & Metcheson A. (1976) Mathematical solutions for explosions in spherical vessels. *Combustion and Flame,* **26**, 201.

4.73 Nagy J., Conn W.J. & Verakis H.C. (1969) Explosion development in a spherical vessel. *US Bureau of Mines Report of Investigations 727.*

4.74 Donat C. (1977) Pressure relief as used in explosion protection. *Loss Prevention II,* p. 86. American Institute of Chemical Engineers.

4.75 Heinrich H.J. & Kowall R. (1971) Results of recent pressure relief experiments in connection with dust explosions. *Staub Reinhalting der Luft 31,* no 4.

4.76 Bartknecht W. (1981) *Explosions, Course Prevention, Protection.* Springer Verlag.

4.77 Maisey H.R. (1965) Gaseous and dust explosion venting — a critical review of the literature. *Chem. Process. Eng.,* **46**, 527.

4.78 Lee J.H.S. & Guirao C.M. (1982) Pressure development in closed and vented vessels. *Plant/Operations Progress,* **1**, 2, 75.

4.79 Harris G.F.P. & Briscoe P.G. (1967) The venting of pentane vapour air explosion, in a large vessel. *Combustion and Flame,* **11**, 329.

4.80 Zalosh R.G. (1979) Gas explosion tests in roomlike vented enclosures. *Thirteenth Loss Prevention Symposium, A. I. Chem. E.*

4.81 Dragosavic M. (1972) Structural measures to prevent explosions of natural gas in multi-storey domestic structures. *Institute TNO (Delft) Report No. B1–72–604302520.*

4.82 Zeeuwen J.P. (1982) Review of current research at TNO into gas and dust explosions.

Proceedings of the International Conference on Fuel-Air Explosions. McGill University, Montreal, Canada. University of Waterloo Press.

4.83 Palmer H.N. (1956) Progress review no 38. A review of information on selected aspects of gas and vapour explosions. *J. Inst. Fuel*, **29**, 293.

4.84 Butlin R.N. (1975) A review of information on experiments concerning the venting of gas explosions in buildings. *Fire Research Note No 1026.*

4.85 Bradley D. & Mitcheson A. (1978) The venting of gaseous explosions in spherical vessels I — theory. *Combustion and Flame*, **32**, 221.

4.86 Bradley D. & Mitcheson A. (1978) The venting of gaseous explosions in spherical vessels II — theory and experiment. *Combustion and Flame*, **32**, 237.

4.87 Cubbage P.A. & Simmonds W.A. (1957) An investigation of explosion reliefs for industrial drying ovens — II Back reliefs in box ovens, relief in conveyor ovens. *Trans Inst. Gas Eng.*, **107**.

4.88 Rasbash D.J. (1969) The relief of gas and vapour explosions in domestic structures. *Fire Research Note No 759.*

4.89 Rasbash D.J., Drysdale D.D. & Kemp N. (1976) Design of an explosion relief system for a building handling liquefied fuel gases. *I. Chem. E. Symposium Series No 47.*

4.90 Marshall M.R. (1977) Calculation of gas explosion relief requirements. The use of empirical equations. *I. Chem. E. Symposium Series No 49.*

4.91 Cubbage P.A. & Marshall M.R. (1972) Pressure generated in combustion chambers by the ignition of gas−air mixtures. *I. Chem. E. Symposium Series No 33.*

4.92 Cubbage P.A. & Marshall M.R. (1974) Explosion relief protection for industrial plant of intermediate strength. *I. Chem. E. Symposium Series No 39a.*

4.93 National Fire Protection Association (1974, revised 1978) *Booklet No 68 — Guide for Explosive Venting.*

4.94 Department of Employment (1972) *Health and Safety at Work Booklet No 22, Dust Explosion in Factories.* HMSO, London.

4.95 Runes E. (1972) Explosive venting. *A. I. Chem. E., Sixth Loss Prevention Symposium*, p. 63.

4.96 Yao C. (1973) Explosion venting of low strength equipment and structures. *A. I. Chem. E., Eighth Loss Prevention Symposium*, p. 1.

4.97 Rust A.E. (1979) Explosion venting for low pressure equipment. *Chemical Engineering*, p. 102.

4.98 Fairweather M. & Vasey M.W. (1982) A mathematical model for the prediction of overpressures generated in totally confined and vented explosions. *Nineteenth Symposium (International) on Combustion*, p. 645. The Combustion Institute.

4.99 Kinney G.F. (1962) *Explosive Shocks in Air.* Macmillan.

4.100 Glasstone S. & Dolan P.J. (1977) *The Effects of Nuclear Weapons*, 3rd edn. United States Department of Defense and United States Department of Energy.

4.101 Pritchard D.K. (1981) Breakage of glass windows by explosions. *J. Occ. Accidents*, **3**, 69.

4.102 Astbury N.F., West H.W.H., Hodgkinson H.R., Cubbage P.A. & Clare R. (1970) Explosions in load bearing brick structures. *British Ceramic Research Association Special Publication No 68.*

4.103 Astbury N.F., West H.W.H. & Hodgkinson H.R. (1972) Experimental gas explosions report of further tests at Potters Marston. *British Ceramic Research Association Special Publication No 74.*

4.104 Harris R.J., Marshall M.R. & Moppett D.J. (1977) The response of glass windows to explosion pressures. *I. Chem. E. Symposium Series No 49.*

4.105 Thomas M., Wachob H.F. & Osteraas J.D. (1984) Investigation of bellows failure in coal gasification plant. *Failure Anal. Assoc. Rep. FaAA-84-3-7.*

4.106 Grady D.E. & Kipp M.E. (1987) Dynamic rock fragmentation. In *Fracture Mechanics of Rock*, (Ed. by B.K. Atkinson), pp. 429−75. Academic Press, London.

4.107 Chou P.C. & Carleone J. (1977) The stability of shaped-charge jets. *J. Appl. Phys.*, **48**, 4187–95.

4.108 Grady D.F. (1982) Local inertia effects in dynamic fragmentation. *J. Appl. Phys.*, **53**, 322–5.

4.109 Kiang T. (1966) Random fragmentation in two and three dimensions. *Z. Astrophys.*, **64**, 433–9.

4.110 Silva-Gomes J.F., Al-Hassani S.T.S. & Johnson W. (1976) A note on times to fracture in solid perspex spheres due to point explosive loading. *Int. J. Mech. Sci.*, **18**, 543.

4.111 Lovall E., Al-Hassani S.T.S. & Johnson W. (1974) Fracture of spheres and circular discs due to explosive pressure. *Int. J. Mech. Sci.*, **16**, 193.

4.112 Al-Hassani S.T.S. & Johnson W. (1969) The dynamics of the fragmentation process for spherical shells containing explosives. *Int. J. Mech. Sci.*, **11**, 811.

4.113 Al-Hassani S.T.S. (1986) Explosive requirements and structural safety aspects in offshore decommissioning applications. Offshore Decommissioning Conference ODC86, Heathrow Penta.

4.114 Proctor J.F. (1970) Containment of explosions in water-filled right-circular cylinders. *Exp. Mech.*, **10**, 458.

4.115 McNaught L.W. (1984) *Nuclear Weapons and their Effects*. Brasseys Defence Publishers, London.

4.116 Baker W.E., Cox P.A., Westine P.S., Kulesz J.J. & Strehlow A. (1983) *Explosion Hazards and Evaluation*. Fundamental Study in Engineering. Elsevier, Amsterdam.

4.117 Theofanous T.G., Nourkbakshs H.P. & Lee C.H. (1984) Natural circulation phenomena and primary system failure in station blackout accidents. Sixth Information Exchange Meeting on Debris Coolability, UCLA.

4.118 Butland A.T.D., Turland B.D. & Young R.L.D. (1985) The UKAEA PWR Severe Accident Containment Study. Thirteenth Water Reactor Safety Research Information Exchange Meeting, Washington DC.

4.119 Wooton R.O. & Avci H.I. (1980) MARCH (Meltdown Accident Response Characteristics) code description and users manual. *NUREG/CR-1711*. Nuclear Regulatory Agency, Washington DC, USA.

4.120 Allison C.M. *et al.* (1981) Severe core damage analysis package (SCDAP). *Code Conceptual Design Report EEG-CDAP-5397*. USA.

4.121 Wright R.V., Silberberg M. & Marino G.P. (1984) Status of the joint program of severe fuel damage research of the USNRC and foreign partners. Fifth International Meeting on Thermal Reactor Safety, Karlsruhe, Germany.

4.122 Butland A.T.D., Haller J.P., Johns N.A., Roberts G.J. & Williams D.A. (1984) Theoretical studies of primary system retention in PWR severe accidents. In *Proceedings of the ANS Topical Meeting on Fission Product Behaviour and Source Term Research*. Snowbird, Utah.

4.123 Wooton R.O., Cybulskis P. & Quayle S.F. (1984) MARCH2 (Meltdown Accident Response Characteristics) Code Description and Users Manual. *NUREG/CR-3988*. Nuclear Regulatory Agency, Washington DC, USA.

4.124 Butland A.T.D., Turland B.D. & Young R.L.D. (1985) The UKAEA PWR Severe Accident Containment Study. Thirteenth Water Reactor Safety Research Information Exchange Meeting, Washington, USA.

4.125 Yue D.D. & Cole T.E. (1982) BWR 4/MARK I Accident Sequence Assessment. *NUREG/CR-2825*. Nuclear Regulatory Agency, Washington DC, USA.

4.126 Cook D.H. *et al.* (1981) Station Blackout at Brown's Ferry Unit One — Accident Sequence Analysis. *NUREG/CR-2182*. Nuclear Regulatory Agency, Washington DC, USA.

4.127 Cook D.H. *et al.* (1983) Loss of DHR Sequence at Brown's Ferry Unit One — Accident Sequence Analysis. *NUREG/CR-2973*. Nuclear Regulatory Agency, Washington DC, USA.

4.128 Silberberg M. (1984) Completion of BWR MARK I, MARK II Standard Problems. NCR Memorandum. Nuclear Commission and Regulation, Washington DC, USA.

4.129 Speis T.P. *et al.* (1985) Estimates of early containment loads from core melt accidents. *NUREG/1079* (draft). Nuclear Regulatory Agency, Washington DC, USA.

4.130 Muir J.F., Cole R.K., Corradini M.L. & Ellis M.A. (1981) An improved model for molten core−concrete interactions. *SAND 80−2415.* Sandia Laboratory, California, USA.

4.131 Harrington R.M. & Ott L.J. (1983) MARCH 1.1 Code Improvements for BWR Degraded Core Studies. *NUREG/CR-3179*, Appendix B. Nuclear Regulatory Agency, Washington DC, USA.

4.132 Ghorbani A. (1990) *Three dimensional finite element analysis of loss-of-coolant accident in PWR.* MPhil thesis, CNAA, UK.

4.133 Murfin W.B. (1977) A preliminary model for core/concrete interactions. *SAND-77/0370.* Sandia Laboratory, California, USA.

4.134 Industry Degraded Core Rulemaking, Nuclear Power Plant Response to Severe Accidents. Final Technical Summary, 1984. Sandia Laboratory, California, USA.

4.135 Berthion Y., Lhiaubet G. & Gauvin J. (1985) Aerosol behavior in the reactor containment building during a severe accident. Fission Product Behavior and source Term research (Proc. ANS Topical Meeting Snowbird, Utah, 1984), Electric Power Research Institute Report NP-4113-SR.

4.136 Bergeron K.D. *et al.* (1985) User's manual for CONTAIN 1.0. *Sandia Laboratories Reports, NUREG/CR-4085, SAND84−1204.* Nuclear Regulatory Agency, Washington DC, USA.

4.137 Fermandjian J., Bunz H., Dunbar I.H., Gauvin J. & Ricchena R. (1986) Comparison of computer codes relative to the aerosol behavior in the reactor containment building during severe core damage accidents in a PWR. Presented at the International Conference of the American Nuclear Society/European Nuclear Society ANS/ENS Topical Meeting on Thermal Reactor Safety, San Diego.

4.138 Braun W., Hassmann K., Hennies H.H. & Hosemann J.P. (1984) The reactor containment of Federal German PWRs of standard design. International Conference on Containment Design, Toronto.

4.139 Alysmeyer H., Reimann M. & Hosemann J.P. (1984) Preliminary results of the KFK molten core concrete experimental BETA facility. Transactions of the 12th Water Reactor Safety Research Information Meeting, United States Nuclear Regulatory Commission Report NUREG/CP-0057. Nuclear Regulatory Agency, Washington DC, USA.

4.140 Henrt R.E. & Fauske H.K. (1981) Required initial conditions for energetic steam explosions. *Proceedings of the ASME Winter Meeting.* Washington DC.

4.141 Corrandini M.L. & Moses G.A. (1983) A dynamic model for fuel−coolant mixing. *Proceedings of the International Meeting on LWR Severe Accident Evaluation.* Cambridge Ma, 6.3−1.

4.142 Berman M. (1984) Molten core−coolant interactions program. *Proceedings of the 12th Water Reactor Safety Research Information Meeting.*

4.143 Corradini M.L. *et al.* (1984) Ex vessel steam explosions in the Mark II containment. University of Wisconsin Research Staff Report 7321.

4.144 Haskin F.E., Behr V.L. & Smith L.N. (1984) Combustion-induced loads in large-dry PWR containments. *Proceedings of the Second Containment Integrity Workshop.* Crystal City, Va.

4.145 Pong L., Corradine M.L. & Moses G.A. (1985) HMC: A containment code for severe accident sequence analysis. University of Wisconsin. Private communication.

4.146 Evans N.A. (1983) Status of core-melt programs, July−August 1983, (M. Berman and R.K. Cole, memorandum to T.J. Walker and S.B. Burson USNRC). US Nuclear Regulatory Commission).

4.147 Green G. (1984) Corium/concrete interaction in the Mark I containment drywell and local liner failure. *NUREG-1079*, appendix II. Nuclear Regulatory Agency, Washington DC, USA.

4.148 Gasser R.D. *et al.* (1984) MARCON Results for Mark I and II Containment Standard Problems, Letter to T. Pratt, Brookhaven National Laboratory, Sandia National Laboratories, USA.

4.149 Clauss D.B. (1985) Comparison of analytical predictions and experimental results for a 1:8-scale containment model pressurized to failure. NUREG/CR-4209, SAND85−0679 Nuclear Regulatory Agency, Washington, DC, USA and Sandia Laboratory, California, USA.

4.150 The Reactor Safety Study (1975) WASH 1400 (NUREG 75/014). Nuclear Regulatory Agency, Washington DC, USA.

4.151 Hillary J.J. *et al.* (1966) Iodine removal by a scale model of the S.G.H.W. Reactor Vented Steam Suppression System. TRG Report 1256. UK Atomic Energy Commission, Winfrith, Dorset, UK.

4.152 Hobbs B., Watson A.J. & Wright S.J. (1989) Explosive tests on model concrete bridge elements. Fourth International Symposium on the Interaction of Conventional Munitions with Protective Structures, Florida.

4.153 Baker W.E., Cox P.A., Westine P.S., Kulesz J.J. & Strehlow R.A. (1983) *Explosion Hazards and Evaluation*, chapter 2. Elsevier, Amsterdam, Oxford, New York.

4.154 Fleischer C.C. (1983) A study of explosive demolition techniques for heavy reinforced and prestressed concrete structures. *CEC Report No EUR 9862 EN.*

4.155 Sheridan A.J. (1987) Response of concrete to high explosive detonation. In *Proceedings of the International Symposium on the Interaction of Non-Nuclear Munitions with Structures*. Mannheim.

4.156 Weerheijm J., van Zantyoort P.J.H. & Opschoor G. (1988) The applicability of the FE-technique to dynamic failure analysis of concrete structures. Twenty-third DoD Explosives Safety Seminar, Atlanta, Georgia, USA.

4.157 Cullis I. & Nash M. (1986) The use of fracture models in the theoretical study of explosive, metal interactions. In *Ninth International Symposium on Ballistics 2*, pp. 285−92. Shrivenham, UK.

4.158 Baylot J.T. *et al.* (1985) Fundamentals of protective design for conventional weapons. Structures Laboratory, US Army Engineer Waterways Experiment Station, Vicksburg, Miss.

4.159 Crawford R.E. *et al.* (1971) Protection from nonnuclear weapons. *Technical Report No AFWL-TR-70−127*. Air Force Weapons Laboratory, Kirtland AFB, NM, USA.

4.160 Whitney M.G. *et al.* (1986) Structures to resist the effects of accidental explosions Vol. II, blast, fragment and shock loads. *Special Publication ARLCD-SP-84001.* US Armament Research, Development and Engineering Center, Dover, NJ.

4.161 Baker W.E. *et al.* (1982) Manual for the prediction of blast and fragment loadings on structures. *DOE/TIC-11268.* US Department of Energy, Pantex Plant, Amarillo, Tx, USA.

4.162 Marchand K.A. *et al.* (1986) Impulsive loading of special doors: flyer plate impact of heavily reinforced concrete blast doors test program results. Southwest Research Institute.

4.163 Coltharp D.R. *et al.* (1985) Blast response tests of reinforced concrete box structures − methods for reducing spall. *Proceedings of the Second Symposium on the Interaction of Nonnuclear Munitions with Structures*. Washington DC, USA.

4.164 Baylot J.T., Kiger S.A., Marchand K.A. & Painter K.T. (1985) Response of buried structures to earth-penetrating conventional weapons. *ESL-TR-85−09.* Engineering & Services Laboratory, Tyndall AFB, Fl.

4.165 Marchand K.A. *et al.* (1988) Development of an alternate munition storage barrier system, phase II report tests of the barrier concepts. *Contract Report No DACA88−86D-0017.* Delivery Order Nos 002 002, 004 and 010. US Army Construction Engineering Research Laboratory. Vicksburg, Miss., USA.

4.166 Marchand K.A. & Garza L.R. (1988) Countering explosive threats — analysis report, SWRI project 06−1473−090, Tecolote Research Inc, Santa Barbara, CA (subcontract TR187−107, NCEL prime contract no 00123−86−D-0299).

4.167 Cockcroit F. (1982) The circumstances of sea collisions. *Journal of Navigation*, **35**, 100.

4.168 Marine Board (1983) Shop collisions with bridges: the nature of the accidents, their prevention and mitigation. Commission on Engineering and Technical Systems, National Research Council, Washington DC.

4.169 Frandsen P. (1982) Accidents involving bridges. IABSE Colloquium on Ship Collision with Bridges and Offshore Structures — Copenhagen. *Proceedings IABSE*, **41**, 11.

4.170 United States Coast Guard (1980−83) Statistics of casualties. In *Proceedings of the Marine Safety Council*.

4.171 Standing & Brendling (1985) *Collisions of Attendant Vessels with Offshore Installations*, parts 1 and 2. National Maritime Institute Ltd, London.

4.172 Sibul K. (1954) Laboratory studies of the motion of freely floating bodies in non-uniform and uniform long crested waves. In *Proceedings of the First Conference on Ships and Waves*. Hoboken, New Jersey.

4.173 Bradford G. (1971) A preliminary report of the observation of sea ice pressure and its effect on merchant vessels under icebreaker escort. In *Proceedings of International Conference on Sea Ice*. Reykjavik, Iceland.

4.174 Yankelevsky D.Z. (1985) Elasto-plastic blast response of rectangular plates. *International Journal of Impact Engineering*, **3**, no 2, 107−19.

4.175 Gürke G., Bücking P. *et al.* (1987) Elasto-plastic response of steel plates to blast loads. In *Proceeding of the Tenth International Symposium on Military Application of Blast Simulation*, pp. 382−94. Bad Reichenhall, Federal Republic of Germany.

4.176 Forsén R. (1987) Increase of in-plane compressive forces due to inertia in wall panels subjected to air-blast loading. In *Proceedings of the Tenth International Symposium on Military Applications of Blast Simulation*, pp. 369−80. Bad Reichenhall, Federal Republic of Germany.

4.177 Bishop V.J. & Rowe R.D. (1967) The interaction of a long Friedlander shaped blast wave with an infinitely long circular cylinder. *AWRE Report 0−38/67*. Atomic Weapons Research Establishment, Aldermaston, Berkshire.

4.178 Salvatorelli-D'Angelo F. (1988) *Structural stability under dynamic loading of LNG tanks*. PhD thesis, Oxford University.

4.179 Ross G.A., Strickland W.S. & Sierakowski R.L. (1977) Response and failure of simple structural elements subjected to blast loadings. *The Shock and Vibration Digest*, **9**(12).

4.180 Hall S.F., Martin D. & Mackenzie J. (1984) Gas cloud explosions and their effects on nuclear power plants, FLARE User Manual. EUR, first edition, EUR 8955EN, 1984.

4.181 Thompson V.K. (1985) *Structural integrity of liquid natural gas storage tanks*. PhD thesis, Oxford University.

4.182 Dobratz B.M. (1981) LLNL Explosives Handbook. Properties of Chemical Explosives and Explosive Simulants. *LLNL Report UCRL-52997*. University of California Research Laboratory, Berkeley, California, USA.

4.183 Bulson P.S. (1966) Stability of buried tubes under static and dynamic overpressure, part 1: circular tubes in compacted sand. *Research Report RES 47.5/7*. Military Engineering Experimental Establishment, Aldermaston, UK.

4.184 Marino R.L. & Riley W.F. (1964) Response of buried structures to static and dynamic overpressures. In *Proceedings of the Symposium on Soil−Structure Interaction*. University of Arizona, Tucson.

4.185 Albritton G.E., Kirkland J.L., Kennedy T.E. & Dorris A.F. (1966) The elastic response of buried cylinders. *Technical Report 1−750*. US Army Waterways Experiment Station, USA.

4.186 Dunns C.S. & Butterfield R. (1971) Flexible buried cylinders — part II: dynamic response. *Int. J. Rock Mech. Min. Sci.*, **8**, 601–12.

4.187 Gumbel J.E., O'Reilly M.P., Lake L.M. & Carder D.R. (1982) The development of a new design method for buried pipes. In *Proceedings of the Europipe Conference*, pp. 87–98. Basle.

4.188 Drake J.L., Frank R.A. & Rochefort M.A. (1987) A simplified method for the prediction of the ground shock loads on buried structures. In *Proceedings of the International Symposium on the Interaction of Conventional Munitions with Protective Structures*. Mannheim, West Germany.

4.189 Hinman E. (1989) Interaction of deformation and shock response for buried structures subject to explosions. In *Proceedings of the International Symposium on the Interaction of Non-Nuclear Munitions With Structures*. Panama City, Florida.

4.190 Hinman E. (1989) Shock response of buried structures subject to explosions. In *Proceedings of the ASCE Speciality Conference on Structures for Enhanced Safety and Physical Security*. Arlington, Va.

4.191 Hinman E. (1988) Shock response of underground structures to explosions. Presented at the Eighth International Colloquium on Vibration and Shock, 14–16 June 1988, Ecole Centrale de Lyon, Ecully, France.

4.192 Weidlinger P. & Hinman, E. (1988) Analysis of underground protective structures. *Journal of Structural Engineering, ASCE*, **114**, no 7, 1658–73.

4.193 Wong F.S. & Weidlinger P. (1979) Design of underground protective structures. *Journal of Structural Engineering, ASCE*, **109**, no 8, 1972–1979.

4.194 Mazzá G., Minasola R., Molinaro P. & Papa A. (1987) Dynamic structural response of steel slab doors to blast loadings. *Proceedings of the International Conference on Computational Plasticity*. Barcelona, Spain.

4.195 Adushkin V.V. (1961) On the formation of a shock wave and the outburst of explosion products. ИИМТФ, no 5, AH CCCP, Mockba.

4.196 Alekseenko V.D. (1963) Experimental investigation of a dynamic stress field in soft soil produced by an explosion. ИИМТФ no 5, AH CCCP, Mockba.

4.197 Alekseev N.A., Rakhmatulin KhA. & Sagomonyan A.Ya. (1963) On the fundamental equations of soil dynamics. ИИМТФ, no 2, AH CCCP, Mockba.

4.198 Andreev K.K. & Belyaev A.F. (1960) Theory of explosives. No 3, AH CCCP, Mockba.

4.199 Assonov V.A (1953) Blasting. No 1, AH CCCP, Mockba.

4.200 Barkan D.D. (1948) Dynamics of soils and foundations. Mockba.

4.201 Brode H.L. (1959) Blast wave from a spherical charge. *The Physics of Fluids*, no 2.

4.202 Enhamre R. (1954) The effects of underwater explosion on elastic structures in water. *Transactions of the Royal Institute of Technology*, bulletin no 42. The Institution Hydraulics, Stockholm.

4.203 Hauptman L. (1954) *Handbook for Blasters*. SNTL, Prague, Hungary.

4.204 Heer J.E. (1965) *Response of inelastic systems to ground shock*. PhD thesis, University of Illinois.

4.205 Henrych J. (1966) Plastická deformace oblouku pri dynamickém zatizeni (plastic deformation of arches subjected to a dynamic load). *Stavebnicky casopis SAV*, **XIV**, no 6, Bratislava.

4.206 Henrych J. (1966) Plastická deformace kopuli pri dynamickém zatizeni (plastic deformation of domes subjected to a dynamic load). *Stavebnicky casopis SAV*, **XIV**, Bratislava.

4.207 Henrych J. (1966) Intenzivni dynamické zatizeni kruhovych desek (intensive dynamic loading of circular plates). *Strojnicky casopis*, **XVII**, no 2, Bratislava.

4.208 Ivanov P.L. (1967) Consolidation of cohensionless soils by explosion. no 3, AH CCCP, Mockba.

4.209 Yakolev Yu S. (1961) Hydrodynamics of explosion. No 6, AH CCCP, Mockba.

4.210 Pokrovskil G.I. & Fedorov I.S. (1957) Effect on shock and explosion in deformable media. Mockba.

4.211 Salamakhin T.M. (1969) Effect of explosion on structure elements. Mockba.

4.212 Seismology and Explosion Craters of Underground Explosions. Collection of Essays, Mockba 1968.

4.213 Chou P., Karpp R.R. & Huang S.L. (1967) Numerical calculation of blast waves by the method of characteristics. *AIAA J.*, **5**, no 4, 618−23.

4.214 Cowan G. *et al.* (1970) The present status of scientific application of nuclear explosives. *Proceedings of a Symposium on Engineering with Nuclear Explosives*, Las Vegas, Nevada p. 216.

4.215 Vlasov O.E. (1965) Principles of explosion dynamics. Mockba.

4.216 Butkovich T.R. (1965) Calculation of the shock wave from an underground nuclear explosion granite. *J. Geophys. Res.*, **70**, 885−92.

4.217 Johnson G.W. & Higgins G.H. (1965) Engineering applications of nuclear explosives: project plowshare. *Rev. Geophys.*, **13**, no 3, 365−85.

4.218 Johnson G.W. (1964) Excavating with nuclear explosives. *Discovery*, **25**, no 11, 16.

4.219 Kestenbolm KH.S., Roslyakov G.S. & Chudov L.A. (1974) Point explosion. Hayka, Mockba.

4.220 Kirkwood J.G. & Bethe H.A. (1967) In *John Gamble Kirkwood Collected Works − Shock and Detonation Wayes* (Ed. by W.W. Wood *et al.*), p. 1. John Wiley & Sons New York.

4.221 Knox J. (1969) Nuclear excavation, theory and applications. *Nuclear Applications and Technology*, **7**, no 3, 189−231.

4.222 Kot C.A. (1970) *Point source underwater explosions*. PhD thesis, Illinois Institute of Technology.

4.223 Kyrpin A.V., Solovev V.Ya., Sheftel N.I. & Kobelev A.G. (1975) Deformation of metals by explosion. M CCCP 6, Mockba.

4.224 Le Mehaute B. (1971) Theory of explosions-generated water waves. *Advan. Hydrosci.*, **7**, 1.

4.225 Lorenz H. (1955) *Dynamik im Grundbau*. Grundbau Taschenbuch, B.I. Berlin.

4.226 Mueller R.A. (1969) Prediction of seismic motion from contained and excavation nuclear detonations. In *Proc. Symp. Eng. Nuclear Explosives, US Atomic Energy Commission.*

4.227 Mueller R.A. (1969) Seismic energy efficiency of underground nuclear detonations. *Bull. Soc. Am.*, **59**, 2311−23.

4.228 Parkin B.R., Gilmore F.R. & Brode H.L. (1961) Shock waves in bubbly water. Memorandum RM-2795-PR (abridged). US Atomic Energy, Washington DC, USA.

4.229 Sternberg H.M. & Walker W.A. (1970) Artificial viscosity method calculation of an underwater detonation. Fifth Symp (Intl) on Detonation, ONR Report DR-163.

4.230 Shurvalov L.V. (1971) Calculation of large underwater explosions. M CCCP No 5, Mockba.

4.231 Newmark N.M. (1956) An engineering approach to blast-resistant design. *Trans ASCE*, no 2786.

4.232 Taylor J. (1952) *Detonation in Condensed Explosives*. Clarendon Press, Oxford.

4.233 Cook M.-A. (1958) *The Science of High Explosives*. Reinhold, New York.

4.234 Duvall W.I. & Atchison T.C. (1957) Rock breakage by explosives. *Bureau of Mines Report No 5356.*

4.235 Atchison T.C. & Tournay W.E. (1959) Comparative studies of explosives in granite. *Bureau of Mines Report No 5509.*

4.236 Atchison T.C., Duvall W.I. & Pugliese J.P. (1964) Effect of decoupling on explosion-generated strain pulses in rocks. *Bureau of Mines Report No 6333.*

4.237 Bur T.R., Colburn L.W. & Slykhouse T.E. (1967) Comparison of two methods for studying relative performance of explosives in rock. *Bureau of Mines Report No 6888.* USA.

4.238 Livingston C.W. (1956) Fundamental concepts of rock failure. *Colorado School of Mines Q*, **51**, no 3. USA.

4.239 Bauer A. (1984) Crater method formulas — a criticism. *E and M. J.*, **166**, no 7.

4.240 Grant C.N. (1964) Simplified explanation of crater method. *E and M. J.*, **165**, no 11, 86−9.

4.241 Slykhouse T.E. (1965) Empirical methods of correlating explosive cratering results. *Seventh Symposium on Rock Mechanics*, Vol. 1, pp. 22−47. The Pennsylvania State University, University Park, Pa, 14−16 June 1968.

4.242 Hino K. (1956) Fragmentation of rock through blasting and shock wave theory of blasting. *Quarterly Colorado School of Mines*, **51**, no 3.

4.243 Pearse G.E. (1955) Rock blasting — some aspects on theory and practice. *Mines and Quarry Engineering*, **21**, 25−30.

4.244 Noren C.H. (1956) Blasting experiments in granite rock. *Quarterly Colorado School of Mines*, **51**, no 3, 210−25.

4.245 Clay R.B., Cook M.A., Cook V.O., Keyes R.T. & Udy L.L. (1965) Behaviour of rock during blasting. *Seventh Symposium on Rock Mechanics*, Vol. 2 pp. 438−61. The Pennsylvania State University, University Park, Pa.

4.246 Langefors U. & Kihlström K. (1963) *The Modern Technique of Rock Blasting*. John Wiley, New York.

4.247 Langefors U. (1965) Fragmentation in rock blasting. *Seventh Symposium on Rock Mechanics*, Vol. 1, pp. 1−21.

4.248 Johnson J.B. (1962) Small-scale blasting in mortar. *Bureau of Mines, Report No 6012*.

4.249 Bergmann O.R., Riggle J.W. & Wu F.C. (1973) Model rock blasting — effectiveness of explosive properties and other variables on blasting results. *Int. J. Rock Mech. Min. Sci. Geomech. Abstr.*, **10**, 585−612.

4.250 Hilliar H.W. (1949) *Report R.E. 142/119*. British Scientific Research, UK.

4.251 Hudson G.E. (1943) TMB Report 509, 1943, United Kingdom.

4.252 Jones H. Report RC-166 1941, United Kingdom.

4.253 Jones H. Report RC-212 1941, United Kingdom.

4.254 Jones H. Report RC-383 1943, United Kingdom.

4.255 Kirkwood J.G. *et al.* The pressure wave produced by an underwater explosion:
   (a) Basic Propagation Theory. *OSRD*, 588, 1942.
   (b) Properties of Pure Water at a Shock Front. *OSRD*, 676, 1942.
   (c) Properties of Salt Water at a Shock Front. *OSRD*, 813, 1942.
   (d) Calculations of initial conditions, superseded by V and *OSRD*, 3949, 1942.
   (e) Calculations for thirty explosives. *OSRD*, 2022, 1943.
   (f) Calculations for cylindrical symmetry. *OSRD* 2023, 1943.
   (g) Final report on propagation theory for cylindrical symmetry. *OSRD*, 3950, 1944.

4.256 Kirkwood J.G. & Brinkley S.R. jr (1945) Theory of propagation of shock waves from explosive sources in air and water. *OSRD*, 4814.

4.257 Kirkwood J.G. & Richardson J.M. (1944) The plastic deformation of circular diaphragms under dynamic loading by an underwater explosion wave. *OSRD*, 4200 1944. Plastic deformation of marine structures by an underwater explosion. *OSRD*, 793, 1115.

4.258 Penney W.G. & Dasgupta H.K. (1942) *British Report RC-333*.

4.259 The NFPA Fire Code 68, National Fire Protection Association, USA, 1981.

4.260 The NFPA Fire Code 70, National Fire Protection Association, USA, 1983.

4.261 *Report TM55−1300*. Nuclear Regulatory Agency, USA, 1983.

4.262 Coles J.S. (1946) *Report OSRD 6240*.

4.263 Coles J.S. (1946) *Report OSRD 6241*.

4.264 Taylor G.I. (1946) The air wave surrounding an expanding sphere. *Proc. Roy. Soc. A.*, **186**, 273.

4.265 Cole R.H. & Coles J.S. (1947) Propagation of spherical shock waves in water. *Phys. Rev.*, **71**, 128.

4.266 Cole R.H. (1948) *Underwater Explosions*. Princeton University Press, USA.

5 Related studies

5.1   Forrestal M.J. *et al.* (1980) An explosive loading technique for the uniform of 304
        stainless steel cylinders at high strain rates. *Journal of Applied Mechanics*, **47**, 17.
5.2   Kuznetsov V.M. (1973) The mean diameter of the fragments formed by blasting
        rock. *Soviet Mining Science*, **9**, 144–8.
5.3   Lyakhav G.M. (1964) Principles of explosion dynamics in soils and in liquid media.
        Nera, Mockba.
5.4   *Underwater Explosion Research*, Vol. 1, p. 273. Office of Naval Research, Department
        of the Navy, Washington 1950.
5.5   Department of Energy (1985) *Offshore Installations Guidance on Design and Con-
        struction*, 4th edn. HMSO, London.
5.6   Onoufriou A., Harding J.E. & Dowling P.J. (1987) Impact damage on ring-stiffened
        cylinders. In *Proceedings of the International Colloquium on Stability of Plate and
        Shell Structures*. Ghent, Belgium.
5.7   Onoufriou A. & Harding J.E. (1985) Residual strength of damaged ring-stiffened
        cylinders. In *Proceedings of the Fourth OMAE Symposium*. ASME, Dallas.
5.8   Ronalds B.F. & Dowling P.J. (1987) Residual compressive strength of damaged
        orthogonally stiffened cylinders. In *Proceedings of the International Colloquium on
        Stability of Plate and Shell Structures*, p. 503. Ghent, Belgium.
5.9   Onoufriou A., Elnashai A.S., Harding J.E. & Dowling P.J. (1987) Numerical
        modelling of damage to ring stiffened cylinders. In *Proceedings of the Sixth OMAE
        Symposium*. ASME, Houston.
5.10   Department of Energy (1987) *Study on Offshore Installation Protection against
        Impact. Offshore Technology Report*. HMSO, London.
5.11   Det Norske Veritas (1984) *Rules for Classification of Mobile Offshore Units*.
5.12   Det Norske Veritas (1981) Impact loads from boats. *Technical Note TNA 202*.
5.13   Williams K.A.J. & Ellinas C.P. (1987) Design guidelines and developments in
        collision and damage of offshore structures. In *Proceedings of the International
        Conference on Steel and Aluminium Structures*. University College, Cardiff.
5.14   NMI Ltd, Collision of attendant vessels with offshore installations. Part 1: General
        Description and Principal Results. *Offshore Technology Report, OTH 84 208*. HMSO
        London.
5.15   NMI Ltd (1985) Collisions of attendant vessels with offshore installations. Part 2:
        detailed calculations. *Offshore Technology Report OTH 84 209*. HMSO, London.
5.16   UNO (1990) Code for the construction and equipment of mobile offshore drilling
        units (MODU CODE). Inter-Government Maritime Consultative Organisation
        (IMCO).
5.17   Atwood (1796) Disquisition on the stability of ships. *Philosophical Transactions of
        the Royal Society of London*.
5.18   Moseley C. (1850) On the dynamical stability and oscillation of floating bodies.
        Philosophical Transactions of the Royal Society of London.
5.19   Harrori Y., Ishihama T., Matsumoto K., Sakata N. & Ando A. (1982) Full scale
        measurement of natural frequency and damping ratio of jack-up rigs and some
        theoretical considerations. OTC Conference, paper no 4287.
5.20   Trickey J.C. (1981) A simplified approach to the dynamic analysis of self-elevating
        drilling units. *Noble Denton Report L8292/NDA/JCT*. USA.
5.21   Wordsworth A.C. & Smedley G.P. (1978) Stress concentrations at unstiffened
        tubular joints. Presented at the European Offshore Steels Research Seminar,
        Cambridge, UK.
5.22   Kinnaman S.C. (1986) The workover task impact on subsea well servicing and
        workover vessels. Norwegian Petroleum Society (NPF) Conference, Kristiansand.
5.23   Standing R.G. (1987) The sensitivity of structure loads and responses to environmental

modelling. Conference on Modelling the Offshore Environment, Society for Underwater Technology, London.

5.24   Takagi M., Arai S.I., Takezawa S., Tanaka K. & Takarada N. (1985) A comparison of methods for calculating the motion of a semi-submersible. *Ocean Eng*, **12**, no 1, 45–97.

5.25   Eatok Taylor T. & Jefferys E.R. (1986) Variability of hydrodynamic load predictions for a tension leg platform. *Ocean Eng*, **13**, no 5, 449–90.

5.26   Standing R.G., Rowe S.J. & Brendling W.J. (1986) Jacket transportation analysis in multi-directional waves. Offshore Technology Conference, Houston, 1986, paper no OTC 5283.

5.27   Rowe S.J., Brendling W.J. & Davies M.E. (1984) Dynamic wind loading of semi-submersible platforms. International Symposium on Developments in Floating Production Systems, RINA, London.

5.28   Ikegami K., Watenabe Y. & Matsuura M. (1986) Study on dynamic response of semisubmersible platform under fluctuating wind. Third International Conference on Stability of Ships and Ocean Vehicles (STAB '86), Gdansk.

5.29   Standing R.G. & Dacunha N.M.C. (1982) Slowly-varying and mean second-order wave forces on ships and offshore structures. In *Proceedings of the Fourteenth ONR Symposium on Naval Hydrodynamics*. Ann Arbor.

5.30   Pinkster J.A. (1980) Low frequency second order wave exciting forces on floating structures. NSMB, Wageningen, report no 650. Germany.

5.31   Huse E. (1977) Wave induced mean force on platforms in direction opposite to wave propagation. *Norwegian Maritime Res.*, **5**, no 1, 2–5.

5.32   Pinkster J.A. & Wichers J.E.W. (1987) The statistical properties of low-frequency motions of nonlinearly moored tankers. Offshore Technology Conference, paper no OTC 5457, Houston.

5.33   McClelland B., Young A.G. & Remmes B.D. (1981) Avoiding jack-up rig foundation failures. In *Proceedings of the Symposium on Geotechnical Aspects of Offshore and Nearshore Structures*. Bangkok.

5.34   Young A.G., Remmes B.D. & Meyer B.J. (1984) Foundation performance of offshore jack-up drilling rigs. *Journal of Geotechnical Engineering*, **110**, no 7, 841–59.

5.35   Davie J.R. & Sutherland H.B. (1977) Uplift resistance of cohesive soils. *ASCE Journal of the Geotechnical Engineering Division*, 935–53.

5.36   Rapoport V. & Young A.G. (1985) Uplift capacity of shallow offshore foundations. In *Proceedings of a Session on Uplift Behavior of Anchor Foundations in Soil, ASCE Convention*, pp. 73–85. Detroit, Michigan.

5.37   Smith C.S., Davidson P.C., Chapman J.C. & Dowling P.J. (1988) Strength and stiffness of ships plating under in-plane compression and tension. *Trans. R. Inst. Naval Arch.*, **130**.

5.38   Akita Y. (1982) Lessons learned from failure and damage of ships. Jt. Sess. 1, Eight ISSC, Gdansk.

5.39   Amdahl J. (1980) Impact capacity of steel platforms and tests on large deformations of tubes under transverse loading. *Det Norske Veritas, Prog. Rep. No 10, Rep. No 80–0036*.

5.40   Taby J. & Moan T. (1987) Ultimate behaviour of circular tubular members with large initial imperfections. In *Proceedings of the Annual SSRC Conference*.

5.41   Marshall P.W. (1984) Connections for welded tubular structures. In *Welding of Tubular Structures*. International Institute of Welding, Pergamon, Oxford.

5.42   Hoadley P.W. *et al.* (1985) Ultimate strength of tubular joints subjected to combined loads. *Proc. of Offshore Technology Conference*, paper no 4854.

5.43   Mahinom Y., Kurobane Y., Takizawa S. & Yamamoto N. (1986) Behaviour of tubular T- and K-joints under combined loads. *Proc. of Offshore Technology Conference*, paper no 5133.

5.44 *Risk Assessment of Buoyancy Loss, Ship Model Collision Frequency*. Technica, London, 1987.

5.45 Norwegian Petroleum Directorate (NPD) (1981) *Guidelines for Safety Evaluation of Platform Conceptual Design*. NPD, Stavanger.

5.46 American Petroleum Institute (API) (1985) *Planning, Designing and Constructing Fixed Offshore Platforms*, 15th edn. API RP2A, Washington DC.

5.47 Det Norske Veritas (1983) *Classification Note 31.5, Strength Analysis of Main Structures of Self Elevating Units*. Hovik, Norway.

5.48 Barratt M.J. (1981) Collision risk estimates in the English Channel and western approaches. *Offshore Technology Report OT-R-8144 to Department of Energy*. HMSO.

5.49 Hathaway R.S. & Rowe S.J. (1981) Collision velocities between offshore supply vessels and fixed platforms. *Offshore Technology Report OT-R-8201 to Department of Energy*. HMSO, UK.

5.50 NMI (1985) Collision of attendant vessels with offshore installations. Part 2: detailed calculations. *Offshore Technology Report OTH 84 209*.

5.51 Borse E. (1979) Design basis accidents and accident analysis with particular reference to offshore platforms. *Journal of Occupational Accidents*, **2**, 227–43.

5.52 Amdahl J. & Heldor E. (1978) Impacts and collisions offshore. Progress report no 1. Simulation applied to analysis of the probability of collision between ships and offshore structures. *DnV Report 78–037*. Norway.

5.53 Amdahl J. & Anderson R. (1979) Impacts and collisions offshore. Progress report no 5. Computer simulation analysis of the collision probability offshore. *DnV Report 78–624*.

5.54 Amdahl J. & Anderson E. (1979) Impacts and collisions offshore. Progress report no 6. Computer simulation study of collision probability between tanker/platform. *DnV Report 79–0192*. Norway.

5.55 Heldor E. & Amdahl J. (1979) Impacts and collisions offshore. Progress report no 7. Collision accidents occurred to offshore structures during the period 1970–1978. *DnV Report 79–0538*. Norway.

5.56 Tviet O.J. & Evandt O. (1981) Experiences with failures and accidents of offshore structures. Third International Conference on Structural Safety and Reliability (ICOSSAR '81), Trondheim, Norway, June 1981 (also published as *DnV Paper 81-P038*). Norway.

5.57 Macduff T. (1974) Probability of collision — a note on encounters between ships, terra firma and offshore structures. *Fairplay International Shipping Weekly*, 21 March 1974, 28–32.

5.58 Kjeoy H. (1982) Impacts and collisions offshore, phase II. Progress report no 12. Overall deformation of braced offshore structures subjected to ship impact loads. Part 1. *DnV Report 81–1213*.

5.59 Frieze P.A. & Samuelides E. (1982) Summary of research on collisions and other impacts on ships and offshore platforms. University of Glasgow, Department of Naval Architecture of Ocean Engineering *Report No NAOE-82–15*.

5.60 Jones N. (1971) A theoretical study of the dynamic plastic behaviour of beams and plates with finite deflections. *Int. J. Solids Struct.*, **7**, 1007–29.

5.61 Haskell D.F. (1970) Large impact deformation response of spherical shells. *AIAA J.*, **8**, 2128–9.

5.62 Updike D.P. (1972) On the large deformation of a rigid plastic spherical shell compressed by a rigid plate. *Journal of Engineering for Industry*, 949–55.

5.63 Minorsky V.U. (1983) An analysis of ship collisions with reference to protection of nuclear power plants. *J. Ship Res.*, 1–4.

5.64 Taby J., Moan T. & Rashed S.M. (1981) Theoretical and experimental study of the behaviour of damaged tubular members in offshore structures. *Journal of Norwegian Maritime Research*. no 2, 26–33.

5.65   Ronalds B.F. & Dowling P.J. (1986) Finite deformations of stringer stiffened plates and shells under knife edge loading. Fifth OMAE Symposium, Tokyo.

5.66   Walker A.C. & Kwok M. (1986) Process of damage in thin-walled cylindrical shells. Fifth OMAE Symposium, Tokyo.

5.67   Walker A.C. & McCall S. (1985) Combined loading of damaged cylinders. Appendices 3–8. University of Surrey.

5.68   Cho S.R. & Frieze P.A. (1985) Lateral impact tests on unstiffened cylinders. Final Report, University of Glasgow.

5.69   Rhodes P.S. (1974) The structural assessment of buildings subjected to bomb damage. *Structural Engineer*, **52**, 329–39.

5.70   Hopkirk R.J. & Rybach L. (1989) Modelling pressure behaviour in artificially stimulated rock masses — matching field pressure records. *Int. J. Rock Mech. Min. Sci. Geomech. Abstr.*, **26**, no 314, 341–9.

5.71   Kirkwood J.G. & Montroll E. (1942) Properties of pure water at a shock. *Report OSRD*, 676.

5.72   Powel K.A. (1986) *The hydraulic behaviour of shingle beaches under regular waves of normal incidence.* Doctoral thesis, University of Southampton.

5.73   PIANC (1976) Final report of the International Commission for the Study of Waves, second part. *Bulletin No 25*, Vol. III (annex).

5.74   PIANC (1980) Final report of the International Commission for the Study of Waves, third part. *Bulletin No 36*, Vol. II (annex).

5.75   Burcharth H.F. (1981) A design method for impact-loaded slender armour units *Bulletin H2 18*. Aalborg University Center, Denmark.

5.76   Coastal Engineering Research Center (1984) *Shore Protection Manual.* US Army Corps of Engineers, USA.

5.77   Dai Y.B. & Kamel A.M. (1969) Scale effect tests for rubble-mound breakwaters. *Research Report H-69–2.* WES, USA.

5.78   Delft Hydraulics-ENDEC (1986) *ENDEC, Wave Propagation Model. User's Manual.*

5.79   Delft Hydraulics-H24 (1987) Golfoverslag Afsluitdijk. Verslag modelonderzoek (Wave overtopping at the Afsluitdijk. Report on model investigation) (in Dutch).

5.80   Delft Hydraulics-M1216 (1974) Grindstranden. Evenwichtsprofiel bij loodrechte regelmatige golfaanval. Verslag modelonderzoek, deel I. (Gravel beaches. Equilibrium profile under monochromatic and perpendicular wave attack. Report on model investigation, Part I) (in Dutch).

5.81   Delft Hydraulics-M1216 (1975) Grindstranden. Evenwichtsprofiel bij loodrechte regelmatige golfaanval. Verslag modelonderzoek, deel II. (Gravel beaches. Equilibrium profile under monochromatic and perpendicular wave attack. Report on model investigation, Part II) (in Dutch).

5.82   Delft Hydraulics-M1216 (1979) Grindstranden. Evenwichtsprofiel bij loodrechte onregelmatige golfaanval. Verslag modelonderzoek, deel III. (Gravel beaches. Equilibrium profile under irregular and perpendicular wave attack. Report on model investigation, Part III) (in Dutch).

5.83   Delft Hydraulics-M1216 (1981) Grindstranden. Evenwichtsprofiel en langstransport bij scheve onregelmatige golfaanval. Verslag modelonderzoek, deel IV. (Gravel beaches. Equilibrium profile under irregular and oblique wave attack. Report on model investigation, Part IV) (in Dutch).

5.84   Delft Hydraulics-M1968 (1983) Golfbrekers. Sterkte betonnen afdekelmenten — Belastingen. (Breakwaters. Strength of concrete armour units — loads) (confidential report in Dutch).

5.85   Delft Hydraulics-M1968 (1983) Golfbrekers. Sterkte betonnen afdekelementen — Bloksterkte. (Breakwaters. Strength of concrete armour units — strength of block) (confidential report in Dutch).

5.86   Delft Hydraulics-M1809 (1984) Taluds van losgestorte materialen. Hydraulische aspecten van stortsteen, grind en zandtaluds onder golfaanval. Verslag literatuurstudie. (Slopes of loose materials. Hydraulic aspects of rock, gravel and sand slopes under wave attack. Report on literature) (in Dutch).

5.87 Delft Hydraulics-M2006 (1986) Taluds van losgestorte materialen. Stabiliteit van stortsteen-bermen en teenkonstrukties. Verslag literatuurstudie en modelonderzoek. (Slopes of loose materials. Stability of rubble mound berm and toe structures. Report on literature and model investigation) (in Dutch).

5.88 Delft Hydraulics-M1983 (1988) Taluds van losgestorte materialen. Statische stabiliteit van stortsteen taluds onder golfaanval. Onwerp formules. Verslag modelonderzoek, deel I (Slopes of loose materials. Static stability of rubble mound slopes under wave attack. Design formulae. Report on model investigation, part I) (in Dutch).

5.89 Delft Hydraulics-M1983 (1988) Taluds van losgestorte materialen. Dynamische stabiliteit van grind-en stortsteen taluds onder golfaanval. Model voor profielvorming. Verslag modelonderzoek, deel II. (Slopes of loose materials. Dynamic stability of gravel beaches and rubble mound slopes under wave attack. Model for profile formation. Report on model investigation, part II), (in Dutch).

5.90 Gravesen H., Jensen O.J. & Sörensen T. (1979) Stability of rubble mound breakwaters II. Conference on Coastal Structures 79, Virginia, USA.

5.91 IAHR/PIANC (1986) List of sea state parameters. *Bulletin No 52* (suppl).

5.92 Kobayashi N. & Jacobs B.K. (1985) Riprap stability under wave action. *Proc. ASCE, J. WPC OE*, **111**, no 3.

5.93 Kobayashi N. & Jacobs B.K. (1985) Stability of armor units on composite slopes. *Proc. ASCE, J. WPC OE*, **111**, no 5.

5.94 Kobayashi N., Roy I. & Otta A.K. (1986) Numerical simulation of wave runup and armor stability. In *Proceedings of the Eighteenth Annual OTC*, pp. 51–60. Houston, USA.

5.95 Kobayashi N. & Otta A.K. (1987) Hydraulic stability analysis of armor units. *Proc. ASCE, J. WPC OE*, **113**, no 2.

5.96 Kobayashi N., Otta A.K. & Roy I. (1987) Wave reflection and run-up on rough slopes. *Proc. ASCE, J. WPC OE*, **113**, no 3.

5.97 Losada M.A. & Giménez-Curto L.A. (1981) Flow characteristics on rough, permeable slopes under wave action. *Coastal Eng*, **4**, 187–206.

5.98 Losada M.A. & Giménez-Curto L.A. (1982) Mound breakwaters under oblique wave attack; a working hypothesis. *Coastal Eng*, **6**, 83–92.

5.99 Pilarczyk K.W. & Den Boer K. (1983) Stability and profile development of coarse materials and their application in coastal engineering. In *Proceedings of the International Conference on Coastal and Port Engineering in Developing Countries*. Colombo, Sri Lanka.

5.100 Popov I.J. (1960) Experimental research in formation by waves of stable profiles of upstream faces of earth dams and reservoir shores. In *Proceedings of the Seventh ICCE*, chapter 16. The Hague, The Netherlands.

5.101 Ryu C.R. & Sawaragi T. (1986) Wave control functions and design principles of composite slope rubble mound structures. *Coastal Engineering in Japan*, **29**, 227–40.

5.102 Sawaragi T., Ryu C.R. & Kusumi M. (1985) Destruction mechanism and design of rubble mound structures by irregular waves. *Coastal Engineering in Japan*, **28**, 173–89.

5.103 Siggurdson G. (1962) Wave forces on breakwater capstones. *Proc. ASCE, J. WHC CED*, **88**, no WW3.

5.104 Tautenhain E., Kohlhase S. & Partenscky H.W. (1982) Wave run-up at sea dikes under oblique wave approach. In *Proceedings of the Eighteenth ICCE*, chapter 50. Cape Town, South Africa.

5.105 Thompson D.M. & Shuttler R.M. (1975) Riprap design for wind wave attack. A laboratory study in random waves. *Report EX 707*. Hydraulic Research Station, Wallingford, UK.

5.106 Van der Meer J.W. (1985) Stability of rubble mound revetments and breakwaters under random wave attack. In *Development in Breakwaters, ICE, Proceedings of the Breakwaters '85 Conference*. London, UK.

5.107 Van der Meer J.W. & Pilarczyk K.W. (1986) Dynamic stability of rock slopes and gravel beaches. In *Proceedings of the Twentieth ICCE*, chapter 125. Taipei, Taiwan.

5.108 Walton I.L. & Weggel J.R. (1981) Stability of rubble-mound breakwaters. *Proc. ASCE, J. WPC OE*, **107**, no WW3, 195–201.

5.109 Whillock A.F. & Price W.A. (1976) Armour blocks as slope protection. In *Proceedings of the Fifteenth ICCE*, chapter 147. Honolulu, Hawaii.

5.110 Thompson D.M. & Shuttler R.M. (1976) Design of riprap slope protection against wind waves. *CIRIA Report 61*. Hydraulic Research Station, Wallingford.

5.111 The Gas Council (1960) *Industrial Gas Handbook*.

5.112 Chappell W.G. (1973) Pressure time diagrams for explosion vented spaces. *A.I. Chem. E., Eighth Loss Prevention Symposium*, p. 76.

5.113 Harvey J.F. (1985) *Theory and Design of Pressure Vessels*. Van Nostrand-Reinhold, New York.

5.114 Haug E., Dowlatyari P. & de Rouvray A. (1986) Numerical calculation of damage tolerance and admissible stress in composite materials using the PAM-FISS Bi-PHASE material model. *ESA SP-238*. European Space Agency (special publication). Belgium.

5.115 de Rouvray A., Vogel F. & Haug E. (1986) Investigation of micromechanics for composites, Phase 1 Rep. Vol. 2. *ESI Rep. ED/84−477/RD/MS* (under ESA/ESTEC contract). Eng. Sys. Int., Rungis-Cedex, France.

5.116 de Rouvray A., Dowlatyari P. & Haug E. (1987) Investigation of micromechanics for composites, Phase 2a Rep. *WP2, ESI Rep. ED/85−521/RD/MS* (under SEA/ASTEC contract). Eng. Syst. Int., Rungis-Cedex, France.

5.117 Magee C.L. & Thornton P.H. (1978) Design considerations in energy absorption by structural collapse. *SAE Trans*, **87**, 2041−55.

5.118 Thornton P.H., Mahmood H.F. & Magee C.L. (1983) Energy absorption by structural collapse. In *Structural Crashworthiness*, (Ed. by M. Jones & T. Wierzbicki). pp. 96–117. Butterworth, London.

5.119 Dye L.C. & Lankford B.W. (1966) A simplified method of designing ship structure for air blast. *Nav. Eng J.*, **693**.

5.120 JCSS (1981) General principles on reliability for structural design. *Jt. Comm. Struct. Saf.*, **35**, part II.

5.121 Bach-Gansmo O. *et al.* (1985) Design against accidental loads in mobile platforms. *Project Summary Report, Rep. No 15, Veritec Rep. 85−3094*. Veritec/Otter, Oslo, Norway.

5.122 Amdahl J. *et al.* (1987) Progressive collapse analysis of mobile platforms. *Proc. Int. Symp. PRADS, 3rd*.

5.123 Fjeld S. (1983) Design for explosion pressure/structural response. *Seminar, Gas Explosions − Consequences and Measures*. Norw. Soc. Char. Eng., Geilo.

5.124 Ellis G.R. & Perret K.R. (1980) The design of an impact resistant roof for platform wellhead modules. In *Proceedings of the OTC, paper no 3907*. Houston, USA.

5.125 Bach-Gansmo O. (1982) Selection of relevant accidental loads based on reported rig accidents. Project report no 3. Design against accidental loads on mobile platforms. *DnV Report 82−0764*. Norway.

5.126 Kjeoy H. & Amdahl J. (1979) Impacts and collisions offshore. Progress report no 8. Ship impact forces in collision with platform legs. *DnV Report 79−0691*.

5.127 de Oliveira J.G. & Mavrikios Y. (1983) Design against collision for offshore structures. *Report MITSG 83−7*. Massachusetts Institute of Technology, USA.

5.128 Wenger A., Edvardsen G., Olafssen S. & Alvestad T. (1983) Design for impact of dropped objects. Fifteenth Annual Offshore Technology Conference, Houston, Texas, May 1983, OTC paper 4471.

5.129 Blok J.J., Brozius L.H. & Dekker J.N. (1983) The impact loads of ships colliding with fixed structures. Fifteenth Annual Offshore Technology Conference, Houston, Texas, May 1983, OTC paper 4469.

5.130 Foss G. & Evardsen G. (1982) Energy absorption during ship-impact on offshore steel structures. Fourteenth Annual Offshore Technology Conference, Houston, Texas, May 1982, OTC paper 4217.

5.131 Chryssanthopoulos M. & Skjerven E. (1983) Progressive collapse analysis of a jack-up platform in intact and damaged condition. Project report 9 in design against accidental loads on mobile platforms. *DnV Technical Report 83–1026*. Norway.

5.132 Valsgard S. (1982) Design against accidental loads on mobile platforms. Presented at Closure Conference of Safety Offshore Project, Stavanger, Norway, November 1982. (Also issued as *DnV Paper 82–P080*.)

5.133 Hagiwara K., Tahanabe H. & Kawano H. (1983) A proposed method of predicting ship collision damage. *International Journal of Impact Engineering*, 1, 257–79.

5.134 Richards D.M. & Andronicou A. (1985) Residual strength of dented tubulars impact energy correlation. In *Proceedings of the Fourth International Symposium of Offshore Mechanics and Arctic Engineering*, pp. 1–10. OMAE-ASME, Dallas, Texas.

5.135 Jones N., Jouri W.S. & Birch R.S. (1984) On the scaling of ship collision damage. In *Proceedings of the Third International Congress*. Athens, Greece.

5.136 Ronalds B.F. & Dowling P.J. (1985) Damage of orthogonally stiffened shells. In *Proceedings of the BOSS 85 Conference*. Delft.

5.137 Griffiths D.R. & Wickens H.G. (1984) The effects of damage on circular tubular compression members. In *Proceedings of the Third International Space Structures Conference*. University of Surrey.

5.138 Ellingwood B. (1981) Treatment of accidental loads and progressive failures in design standards. In *Proceedings of the Third International Conference on Structural Safety and Reliability (ICOSSAR 3)*, pp. 649–65. Trondheim, Norway.

5.139 Davies I.L. (1980) A method for the determination of the reaction forces and structural damage arising in ship collisions. In *Proceedings of the European Offshore Petroleum Conference and Exhibition*, paper EUR237, pp. 245–54. London, 1980. (Also in *J. Petroleum Technology*, 2006–14, 1981.)

5.140 Valsgard S. & Pettersen E. (1983) Simplified non-linear analysis of ship/ship collisions. *Norwegian Maritime Research*, 10, 2–17.

5.141 Woisin G. (1979) Design against collision. In *International Symposium on Advances in Marine Technology* Vol. 2, pp. 309–36. Trondheim.

5.142 Faulkner D. (1979) Design against collapse for marine structures. In *International Symposium on Advances in Marine Technology*, pp. 275–308. Trondheim.

5.143 Braun P. (1983) Mooring and fendering rational principles in design. *Eighth International Harbour Congress*, paper 2/21. Antwerp. Belgium.

5.144 Donegan E. (1982) New platform designs minimise ship collision damage. *Petroleum Engineer International*, February, 79–88.

5.145 Soreide T. (1985) Ultimate load analysis of marine structures. Tapir, University of Denmark.

5.146 Pettersen E. (1981) Assessment of impact damage by means of a simplified non-linear approach. *Second International Symposium on Integrity of Offshore Structures*, p. 317. Glasgow.

5.147 Donegan E.M. (1982) Appraisal of accidental impact loadings on steel piled North Sea structures. Fourteenth Annual Offshore Technology Conference, OTC paper 4193. Houston, Texas.

5.148 *Suppressive Shields, Structural Design and Analysis Handbook*. (1977) US Army Corps of Engineers, Huntsville Division, HNDM-1110–1–2 (AD A049 017).

5.149 Dobbs N. & Caltigirone J.P. (1987) Structures to resist the effects of accidental explosions. Vol. 1 Introduction. *Special Publication ARLCD-SP-84001*. Government Printing Press, Washington DC, USA.

5.150 Dobbs N. *et al.* (1986) Structures to resist the effects of accidental explosions. Vol II Blast, fragment and shock loads. *Special Publication ARLCD-SP-84001*. Government Printing Press, Washington DC, USA.

5.151 Dobbs N. *et al.* (1984) Structures to resist the effects of accidental explosions. Volume III Principles of dynamic analysis. *Special Publication ARLCD-SP-84001*. Government Printing Press, Washington DC, USA.

5.152 Dobbs N. *et al.* (1987) Structures to resist the effects of accidental explosions. Volume IV Reinforced concrete design. *Special Publication ARLCD-SP-84001*. Government Printing Press, Washington DC, USA.

5.153 Dobbs N. *et al.* (1987) Structures to resist the effects of accidental explosions. Volume V Structural steel design. *Special Publication ARLCD-SP-84001*. Government Printing Press, Washington DC, USA.

5.154 Dobbs N. *et al.* (1985) Structures to resist the effects of accidental explosions. Vol. VI Special considerations in explosive facility design. *Special Publication ARLCD-SP-84001*. Government Printing Press, Washington DC, USA.

5.155 Marshall W. (1982) An assessment of the integrity of PWR pressure vessels. Report of the Study Group of UKAEA, Risley, Cheshire.

6 Design for impact and explosion

6.1   Getzler F., Komomek A. & Mazwicks A. (1968) Model study on arching above buried structures. *J. Soil Mech. Found., ASCE.*

6.2   Simiu E. (1975) Probabilistic models of extreme wind speeds: uncertainties and limitation. In *Proceedings of the Fourth International Conference on Wind Effect on Structures and Buildings*, pp. 53–62. London, UK.

6.3   Tryggvason B.V., Surry D. & Davenport A.G. (1976) Predicting wind-induced response in hurricane zones. *J. Struct. Div., ASCE*, **102**, no ST12, 2333–51.

6.4   Sklarin J. (1977) *Probabilities of hurricanes*. Research report, Department of Civil Engineering, MIT, Cambridge, Massachusetts.

6.5   Fujita T.T. (1970) Estimates of a real probability of tornadoes from inflationary reporting of their frequencies. *SMRP Research Paper No 89*, Satellite and Meso-meteorology Research Project, University of Chicago.

6.6   Thom H.C.S. (1963) Tornado probabilities. *Monthly Weather Review*, October–December, pp. 73–736.

6.7   Pautz M.E. (Ed.) (1969) Severe local storm occurrences 1955–1967. *ESSA Tech. Memo. WBTM FCST 12*, US Department of Commerce, Environmental Science Services Administration, Washington DC, USA.

6.8   Markee E.H. jr, Beckerley J.G. & Sanders K.E. (1964) *Technical basis for interim regional tornado criteria*. US Atomic Energy Commission, Office of Regulation, WASH 1300.

6.9   Fujita T.T. (1972) F-scale classification of 1971 tornadoes. *SMRP Research Paper No 100*. University of Chicago.

6.10 Singh M.P., Morcos A. & Chu S.L. (1973) Probabilistic treatment of problems in nuclear power plant design. In *Proceedings of the Speciality Conference on Structural Design of Nuclear Plant Facilities*, Vol. 1, pp. 263–89. Chicago, Illinois.

6.11 Garson R.C., Catalan J.M. & Cornell C.A. (1975) Tornado design winds based on risk. *J. Struct. Div., ASCE*, **101**, no ST9, 1883–97.

6.12 Hoecker W.H. jr (1960) Wind speed and air flow patterns in the Dallas tornado of April 2, 1957. *Monthly Weather Review*, **88**, no 5, 167–80.

6.13 Hoecker W.H. jr (1961) Three dimensional pressure patterns of the Dallas tornado and some resultant implications. *Monthly Weather Review*, **89**, no 12, 533–42.

6.14 Wen Y.K. (1975) Dynamic wind loads on tall buildings. *J. Struct. Div., ASCE*, **101**, no ST1, 169–85.

6.15 McDonald J.R., Minor J.E. & Mehta K.C. (1973) Tornado generated missiles. In *Proceedings of the Speciality Conference on Structural Design of Nuclear Power Plant Facilities*, Vol. II, pp. 543–56. Chicago, Illinois.

6.16 Abbey R.F. & Fujita T.T. (1975) Use of tornado path length and gradations of damage to assess tornado intensity probability. Ninth Conference on Severe Local Storms, Norman, Oklahoma.

6.17 VanDorn W.G. (1965) Tsunamis. In *Advances in Hydroscience*, Vol 2. Academic Press, New York.

6.18 Johnson B. (1976) Tornado missile risk analysis. Study conducted for Boston Edison Power Company by Science Applications Inc. Private communication.

6.19 Bush S.H. (1973) Probability of damage to nuclear components due to turbine failure. *Nuclear Safety*, **14**, no 3, 187–201.

6.20 Rankin A.W. & Seguin B.R. (1973) Report of the investigation of the turbine-wheel fracture at Tanners Creek, *Nuclear Safety*, **14**, no 3, 47.

6.21 Kalderton D. (1972) Steam turbine failure at Hinkley Point 'A'. *Proc. Inst. Mech. Eng.*, **186**, 341.

6.22 Gray J.L. (1972) Investigation into the consequences of the failure of a turbine-generator at Hinkley Point 'A' Power Station. *Proc. Inst. Mech. Eng.*, **186**, 379.

6.23 Downs J.E. (1973) Hypothetical turbine missiles — probability of occurrence. Memo report, General Eletric Co.

6.24 Hagg A.C. & Sankey G.O. (1974) The containment of disk burst fragments by cylindrical shells. *Transactions of the ASME, Journal of Engineering for Power*, **96**, 114–23.

6.25 Semanderes S.N. (1972) Method of determining missile impact probability. *Trans. Am. Nucl. Soc.*, **15**, no 1, 401.

6.26 Bhattacharyya A.K. & Chaudhuri S.K. (1976) The probability of a turbine missile hitting a particular region of a nuclear power plant. *Nuclear Technology*, **28**, 194–8.

6.27 Johnson B. *et al.* (1976) Analysis of the turbine missile hazard to the nuclear thermal power plant at Pebble Springs, Oregon. Science Applications Inc. Report to the Portland General Electric Company, PGE-2012.

6.28 USNRC (1975) *Standard Review Plan, Section 3.5.1.6., Aircraft Hazards*. US Nuclear Regulatory Commission, Office of Nuclear Reactor Regulation.

6.29 Chalapathi C.V., Kennedy R.P. & Wall I.B. (1972) Probabilistic assessment of aircraft hazard for nuclear power plants. *Nucl. Eng. Des.*, **19**, 333–64.

6.30 Chalapathi C.V. & Wall I.B. (1970) Probabilistic assessment of aircraft hazard for nuclear power plants — II. *Trans. Am. Nucl. Soc.*, **13**, 218.

6.31 Wall I.B. (1974) Probabilistic assessment of aircraft risk for nuclear power plants. *Nuclear Safety*, **15**, no 3.

6.32 Hornyik K. (1973) Airplane crash probability near a flight target. *Trans. Am. Nucl. Soc.*, **16**, 209.

6.33 Hornyik K. & Grund J.E. (1974) Evaluation of air traffic hazards at nuclear power plants. *Nuclear Technology*, **23**, 28–37.

6.34 USNRC (1975) *NRC Regulatory Guide 1.70, Standard Format and Content of Safety Analysis Reports for Nuclear Power Plants*, LWR edn, revn 3, NUREG-75/094. US Nuclear Regulatory Commisssion, Office of Standards Development.

6.35 Anon (1976) Aircraft crash probabilities. *Nuclear Safety*, **17**, 312–14.

6.36 Stevenson J.D. (1973) Containment structures for pressurized water reactor systems, past, present and future — state of the art. Second International Conference on Structural Mechanics in Reactor Technology, Berlin, September 1973, paper 72/1.

6.37 ACI–ASME Joint Technical Committee (1975) Code for concrete vessels and containments, ASME boiler and pressure vessel code section III — division 2 and ACI standard 359–74.

6.38 US Atomic Energy Commission (1973) Regulatory guide 1.57. Design limits and loading combinations for metal primary reactor components. US Atomic Energy Commission, Directorate of Regulatory Standards.

6.39 US Atomic Energy Commission (1976) Standard review plan, section 3.8.2. Steel containment. US Atomic Energy Commission, Directorate of Licensing.

6.40 American Concrete Institute (ACI) (1976) ACI standard 349—76. Code requirements for nuclear safety related concrete and structures.

6.41 US Atomic Energy Commission (1976) Standard review plan, section 3.8.3. Concrete and steel internal structures of steel or concrete containments. US Atomic Energy Commission, Directorate of Licensing.

6.42 American Concrete Institute (1975) Proposed code requirements for nuclear safety-related concrete structures, draft of section C.3.3.3., rotational limits. American Concrete Institute, Committee 349.

6.43 ASCE (1961) Design of structures to resist nuclear weapons effects. *American Society of Civil Engineers Manual No 42*.

6.44 American Concrete Institute (1975) Proposed code requirements for nuclear safety-related concrete structures. Report by the American Concrete Institute, Committee 349.

6.45 Amirikian A. (1950) Design of protective structures. *Report NT-3726*. Bureau of Yards and Docks, Department of the Navy Washington DC, USA.

6.46 Gwaltney R.C. (1968) Missile generation and protection in light-water-cooled power reactor plants. *ORNL NSIC-22*. Oak Ridge National Laboratory, Oak Ridge, Tennessee, for the US Atomic Energy Commission.

6.47 Kennedy R.P. (1966) *Effects of an Aircraft Crash into a Concrete Building*. Holmes & Narver Inc, Anaheim, Ca.

6.48 Vassalo F.A. (1975) Missile impact testing of reinforced concrete panels. *HC-5609-D-1*. Calspan Corporation, Buffalo, New York, prepared for Bechtel Power Corporation.

6.49 United Engineers and Constructors Inc (1975) Seabrook Station aircraft impact analysis. *Docket Nos 50—433/444*. Prepared for Public Service Company of New Hampshire, Seabrook, New Hampshire.

6.50 United Engineers and Constructors Inc (1975) Supplemental information to Seabrook Station aircraft impact analysis. *Docket Nos 5—443/444*. Prepared for Public Service Company of New Hamshire, Seabrook, New Hampshire.

6.51 Lorenz H. (1970) Aircraft impact design. *Power Engineering*, November, 44—6.

6.52 Gilbert Associates Inc (1968) Three-Mile Island preliminary safety analysis report. *Docket No 50—289*, supplement W5. Report prepared for Metropolitan Edison Company.

6.53 Stevenson A.E. (1975) Tornado vulnerability of nuclear production facilities. Sandia Laboratories Los Angeles, California, USA.

6.54 Chalapathi C.V. (1970) Probability of perforation of a reactor building due to an aircraft crash. *HN-212*. Holmes & Narver, Anaheim, Ca.

6.55 White R.W. & Botsford N.B. (1963) Containment of fragments for a runaway reactor. *Report SRIA-113*. Stanford Research Institute, USA.

6.56 American Society of Mechanical Engineers (1974) Rules for construction of nuclear power plant components, ASME boiler and pressure vessel code section III division 1.

6.57 US Atomic Energy Commission (1974) Reactor safety study, an assessment of accident risks in the US commercial nuclear power plants. *WASH-1400*. Washington, USA.

6.58 1976 Quality assurance criteria for nuclear power plants and fuel reprocessing plants. Federal Regulation 10 CFR 50 appendix B.

6.59 American National Standards Institute (1978) Status report of projects under the Nuclear Standards Management Board. *NSMB SR-15*.

6.60 Anthony E.J. (1977/8) The use of venting formulae in the design and protection of buildings and industrial plant from damage by gas or vapour explosions. *Journal of Hazardous Materials*, **2**, 23.

6.61 West H.W.H., Hodgkinson H.R. & Webb W.F. (1973) The resistance of brick walls to lateral loading. *Proc. Brit. Ceramic Soc.*, **21**, 141.

6.62 Hendry A.W., Sinha B.P. & Maurenbracher A.H.P. (1971) Full scale tests on the lateral strength of brick cavity walls with precompression. *Proc. Brit. Ceramic Soc.*, 33.

6.63 Astbury N.F. & Vaughan G.N. (1972) Motion of a brickwork structure under certain assumed conditions. *British Ceramic Research Association Technical Note No 191.*

6.64 Mainstone R.J. (1971) The breakage of glass windows by gas explosions. *Building Research Station Current Paper 26/71.*

6.65 West H.W.H. (1973) A note on the resistance of glass windows to pressures generated by gaseous explosions. *Proc. Brit. Ceramic Soc.*, **21**, 213.

6.66 British Gas Corporation (1977) Code of practice for the use of gas in atmosphere gas generators and associated plant, parts 1, 2 and 3. *Publication IM/9.*

6.67 British Gas Corporation (1982) Code of practice for the use of gas in low temperature plant (includes aspects of dual fuel burners previously covered by *IM/7*). *Publication IM/18.*

6.68 Marshall M.R. (1980) Gaseous and dust explosion venting: determination of explosion relief requirements. Third International Symposium on Loss Prevention and Safety Promotion in the Process Industries, Basle, Switzerland.

6.69 Howard W.B. & Karabinis A.H. (1980) Tests of explosion venting of buildings. Third International Symposium on Loss Prevention and Safety Promotion in the Process Industries, Basle, Switzerland.

6.70 Hattwig M. (1977) Selected aspects of explosion venting. Second International Symposium on Loss Prevention and Safety Promotion in the Process Industries, Heidelberg.

6.71 Croft W.M. (1981) Fires involving explosion — a literature review. *Fire Safety Journal*, **3**, 21.

6.72 Butlin R.N. & Tonkin P.S. (1974) Pressures produced by gas explosions in a vented compartment. *Fire Research Note No 1019.*

6.73 Timoshenko S. & Goodier G.N. (1951) *Theory of Elasticity*, 2nd edn. McGraw-Hill.

6.74 Barkan D.D. (1962) *Dynamics of Bases and Foundations*, McGraw Hill.

6.75 Bycroft G.N. (1956) Forced vibration of a rigid circular plate on a semi-infinite elastic space and on an elastic stratum. *Phil. Trans R. Soc. London, Ser. A*, **248**, 327−68.

6.76 Reissner E. (1936) Stationare Axial Symmetricke durch eine schuttiond massive Erregt schwingiuingen eines Homogenen Blastischen Halbraumes. *Ingenieur−Archiev*, **7**, no 6, Berlin, 381−96.

6.77 Nowaki W. (1961) *Dynamika Budowli*. Arkady, Warsaw.

6.78 Richart F.E. *et al.* (1970) *Vibrations of Soils and Foundations*. Prentice Hall, New Jersey, USA.

6.79 Technical Regulations for Shelters TB78I. Publication no 6.06. Swedish Government Press, Stockholm, Sweden.

6.80 Home Office (1975) *Home Office Manual: Domestic Nuclear Shelters*. HMSO, UK.

APPENDIX

Program structural layout

Andrew Watson main programmer (supervised by M.Y.H. Bangash)

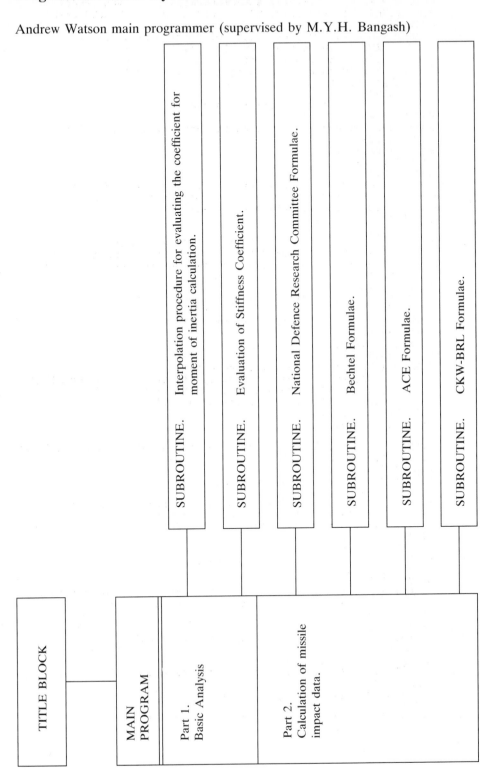

```
       REAL I,B,T,F,D,K,V,P,N,PN,PT,AY,AX,EC,ES,X,Y,Q,Q1,S,ME,MPUA,C
       REAL NTP,KL,NSF,FC,W,DIA,VEL,FCI,KCPFI,WI,DIAI,VELI,PENEI
       REAL PENET,RAT,PERF,SCAB,SSCAB,HSCAB,ACEP,ACPER,CKPEN,CKPER
       WRITE(6,1)
1      FORMAT (1H/,'Put in your values of B(cm),T(cm),D(cm)'/
      +         1H/,'B (Unit width of slab ,cm)'/
      +         1H/,'T (Overall depth of slab ,cm)'/
      +         1H/,'D (Depth to reinforcing steel ,cm)')
       READ (5,*) B,T,D
       CALL INTN(F)
       I=0.5000*(((B*(T*T*T))/12)+F*B*(D*D*D))
       WRITE (6,2)
2      FORMAT(1H/,'The average moment of inertia,Ia(cm4/cm),is')
       WRITE (6,3) I
3      FORMAT(1H/,F30.2)
       CALL INT (Q)
       WRITE (6,17)
17     FORMAT (1H/,'Key in the value of Q that corresponds to'/
      +         1H/,'your calculated value of X/Y.This is the '/
      +         1H/,'required Stiffness Coefficient')
       READ (5,*) Q
       WRITE (6,12)
12     FORMAT (1H/,'Put in your values of V,EC(MPa),Y(m)'/
      +         1H/,'V (Poissons Ratio for concrete ,usually 0.17'/
      +         1H/,'EC (Elastic modulus of concrete ,MPa)       '/
      +         1H/,'Y (Length of slab ,m)                       ')
       READ (5,*) V,EC,Y
       K=((12*EC*I)/(((Q*Y*Y)*(1-(V*V)))*1000000000))
       WRITE (6,5)
5      FORMAT(1H/,'The value of K (MN/mm) is')
       WRITE (6,6) K
6      FORMAT(1H/,F30.2)
       WRITE (6,7)
7      FORMAT (1H/,'Input the following data :             '/
      +         1H/,'WD (The Weight Density of the concrete ,Kg/m3)'/
      +         1H/,'T (The overall depth of the slab ,m)'/
      +         1H/,'X (The width of the slab ,m)')
       READ (5,*) WD,T,X
       MPUA=((WD*T)/9.81)
       WRITE (6,8)
8      FORMAT (1H/,'The mass per unit area of the slab is,(Kg.sec2/m3')
       WRITE (6,9) MPUA
9      FORMAT (1H/,F5.1)
       ME=((MPUA*3.142*X*X*9.81)/(6*4*1000))
       WRITE (6,10)
10     FORMAT (1H/,'The effective mass is one sixth of the mass'/
      +         1H/,'within the circular yield pattern.          '/
      +         1H/,'Effective mass is ,N.sec2/mm')
       WRITE (6,11) ME
11     FORMAT (1H/,F4.2)
450    FORMAT(1H/,'Key in the corresponding F value')
       READ (5,*) Y1
       WRITE (6,500)
500    FORMAT (1H/,'Key in the value of PN from the table that is'/
      +         1H/,'just higher than your calculated value of PN')
       READ (5,*) X3
       WRITE (6,550)
550    FORMAT(1H/,'Key in the corresponding F value')
       READ (5,*) Y2
       WRITE (6,600)
600    FORMAT (1H/,'Key in your calculated value of PN')
       READ (5,*) X2
       F=((((X2-X1)/(X3-X1))*(Y2-Y1))+Y1)
       WRITE (6,700)
700    FORMAT(1H/,'The value of F you require is')
       WRITE (6,800) F
```

```
 800   FORMAT(1H/,F8.5)
       RETURN
       END
       SUBROUTINE INT (Q)
       WRITE (6,1000)
1000   FORMAT (1H/,'Key in the length,X (m),& width,Y (m),of the slab')
       READ (5,*) X,Y
       Q1=X/Y
       WRITE (6,1100)
1100   FORMAT (1H/,'You require a Q value that corresponds with this'/
      +        1H/,'calculated value of X/Y')
       WRITE (6,1200) Q1
1200   FORMAT (1H/,F4.2)
       WRITE (6,1300)
1300   FORMAT (1H/,'The table you use is dependant on support'/
      +        1H/,'conditions at the sides')
       WRITE (6,1400)
1400   FORMAT(1H/,'SIMPLY SUPPORTED ON          FULLY FIXED ON ALL   '/
      +        1H/,'ALL FOUR SIDES               FOUR SIDES'/
      +        1H/,'X/Y VALUES |Q VALUES        X/Y VALUES |Q VALUES'/
      +        1H/,'========== |========        ========== |========'/
      +        1H/,'    1.0    | 0.1391            1.0     | 0.0671 '/
      +        1H/,'    1.1    | 0.1518            1.2     | 0.0776 '/
      +        1H/,'    1.2    | 0.1624            1.4     | 0.0830 '/
      +        1H/,'    1.4    | 0.1781            1.6     | 0.0854 '/
      +        1H/,'    1.6    | 0.1884            1.8     | 0.0864 '/
      +        1H/,'    1.8    | 0.1944            2.0     | 0.0866 '/
      +        1H/,'    2.0    | 0.1981         INFINATE   | 0.0871 '/
      +        1H/,'    3.0    | 0.2029                             '/
      +        1H/,' INFINATE  | 0.2031                             ')
       RETURN
       END
       SUBROUTINE NDRC (PENET,PERF,SCAB)
       REAL NSF,KCPFI,FC,W,DIA,VEL,FCI,WI,DIAI,VELI,PENEI,PENET,RAT,T
       WRITE (6,1998)
1998   FORMAT (1H/,'Input,T,the overall depth ,mm')
       READ (5,*) T
       WRITE (6,2000)
2000   FORMAT (1H/,'Input the relevant missile shape factor ,NSF'/
      +        1H/,'For flat nosed missiles, NSF=0.72'/
      +        1H/,'For blunt nosed missiles, NSF=0.84'/
      +        1H/,'For sperical nosed missiles, NSF=1.00'/
      +        1H/,'For very sharp nosed missiles, NSF=1.14')
       READ (5,*) NSF
       WRITE (6,2100)
2100   FORMAT (1H/,'Input the following :'/
      +        1H/,'FC The ultimate concrete compressive strength,N/mm2'/
      +        1H/,'W (The weight of the missile, N)'/
      +        1H/,'DIA (The circular section diameter, mm)'/
      +        1H/,'VEL (The impact velocity, m/sec)')
       READ (5,*) FC,W,DIA,VEL
       FCI=(FC/0.007)
       KCPFI=(180/SQRT(FCI))
       WI=((2*W)/9)
       DIAI=(DIA/25.4)
       VELI=(VEL/0.3048)
       PENEI=(SQRT((4*KCPFI*NSF*WI*DIAI)*((VELI/(1000*DIAI))**1.8)))
       PENET=(PENEI*25.4)
       RAT=(PENET/DIA)
       IF (RAT.LT.2.0) THEN
       GO TO 2200
       ELSE IF (RAT.GT.2.0) THEN
       GO TO 2400
       END IF
2200   WRITE (6,2300)
```

```
2300   FORMAT (1H/,'The missile penetration using the NDRC (National'/
      +          1H/,'Defence Research Committee) Formula for'/
      +          1H/,'"x/d" less than or equal to 2.0 is ,(mm)')
       WRITE (6,2350) PENET
2350   FORMAT (1H/,F6.1)
       GO TO 2700
2400   PENEI=(((KCPFI*NSF*WI)*((VELI/(1000*DIAI))**1.8))+DIAI)
       PENET=(PENEI*25.4)
       WRITE (6,2450)
2450   FORMAT (1H/,'The missile penetration using the NDRC (National'/
      +          1H/,'Defence Research Committee) Formula for'/
      +          1H/,'"x/d" greater than 2.0 is ,(mm)')
       WRITE (6,2470) PENET
2470   FORMAT (1H/,F6.1)
2700   WRITE (6,2710)
2710   FORMAT (1H/,'The "x/d"ratio is')
       WRITE (6,2720) RAT
2720   FORMAT (1H/,F5.3)
       IF (RAT.LT.1.35) THEN
       GO TO 2800
       ELSE IF (RAT.GT.1.35) THEN
       GO TO 2900
       END IF
2800   PERF=(DIA*((3.19*RAT)-(0.718*RAT*RAT)))
       GO TO 3000
2900   PERF=(DIA*(1.32+(1.24*RAT)))
3000   WRITE (6,3100)
3100   FORMAT (1H/,'The Perforation,calculated using the NDRC,is ,mm')
       WRITE (6,3200) PERF
3200   FORMAT (1H/,F5.1)
       IF (PERF.LT.T) THEN
       GO TO 3300
       ELSE IF (PERF.GT.T) THEN
       GO TO 3400
       END IF
3300   WRITE (6,3350)
3350   FORMAT (1H/,'This value is less than the overall depth.The'/
      +          1H/,'slab adequately resists collapse due to perforation')
       GO TO 3500
3400   WRITE (6,3450)
3450   FORMAT (1H/,'This value is greater than the overall depth.The'/
      +          1H/,'slab will collapse due to perforation')
3500   IF (RAT.LT.0.65) THEN
       GO TO 3600
       ELSE IF (RAT.GT.0.65) THEN
       GO TO 3700
       END IF
3600   SCAB=(DIA*((7.91*RAT)-(5.06*RAT*RAT)))
       GO TO 3800
3700   SCAB=(DIA*(2.12+(1.36*RAT)))
3800   WRITE (6,3850)
3850   FORMAT (1H/,'The Scabbing thickness,calculated using NDRC,is ,mm')
       WRITE (6,3860) SCAB
3860   FORMAT (1H/,F5.1)
       IF (SCAB.LT.T) THEN
       GO TO 3900
       ELSE IF (SCAB.GT.T) THEN
       GO TO 4000
       END IF
3900   WRITE (6,3950)
3950   FORMAT (1H/,'This value is less than the overall depth.The'/
      +          1H/,'slab will not collapse due to scabbing')
       GO TO 4100
4000   WRITE(6,4050)
4050   FORMAT (1H/,'This value is greater than the overall depth.The'/
      +          1H/,'slab will collapse due to scabbing')
```

```
4100  RETURN
      END
      SUBROUTINE BTEL (SSCAB,HSCAB)
      REAL W,VEL,DIA,FC,FCI,WI,DIAI,VELI,SSCI,HSCI,SSCAB,HSCAB
      WRITE (6,5000).
5000  FORMAT (1H/,'Input the following:'/
     +          1H/,'W (The weight of the missile,N)'/
     +          1H/,'VEL (The impact velocity of the missile,m/sec)'/
     +          1H/,'DIA (The diameter of the missile,mm)'/
     +          1H/,'FC (The concrete compressive strength,N/mm2)')
      READ (5,*) W,VEL,DIA,FC
      FCI=(FC/0.007)
      WI=((2*W)/9)
      DIAI=(DIA/25.4)
      VELI=(VEL/0.3048)
      SSCI=((15.5*(WI**0.4)*(VELI**0.5))/(SQRT(FCI)*(DIAI**0.2)))
      HSCI=((5.42*(WI**0.4)*(VELI**0.65))/(SQRT(FCI)*(DIAI**0.2)))
      SSCAB=(SSCI*25.4)
      HSCAB=(HSCI*25.4)
      WRITE (6,5100)
5100  FORMAT (1H/,'Using the BECHTEL formula the slab thickness'/
     +          1H/,'to prevent scabbing from a solid missile is,mm')
      WRITE (6,5200) SSCAB
5200  FORMAT (1H/,F6.1)
      WRITE (6,5300)
5300  FORMAT (1H/,'Using the BECHTEL formula the slab thickness to'/
     +          1H/,'prevent scabbing from a hollow missile is,mm')
      WRITE (6,5400) HSCAB
5400  FORMAT (1H/,F5.1)
      RETURN
      END
      SUBROUTINE ACE (ACEP,ACPER)
      REAL W,DIA,VEL,FC,FCI,DIAI,WI,VELI,ACEPI,APFI,ACEP,ACPER,EPI
      WRITE (6,6000)
6000  FORMAT (1H/,'Input the following:'/
     +          1H/,'W (The weight of the missile,N)'/
     +          1H/,'DIA (The diameter of the missile,mm)'/
     +          1H/,'VEL (The impact velocity of the missile,m/sec)'/
     +          1H/,'FC (The concrete compressive strength,N/mm2)')
      READ (5,*) W,DIA,VEL,FC
      FCI=(FC/0.007)
      WI=((2*W)/9)
      DIAI=(DIA/304.8)
      VELI=(VEL/0.3048)
      ACPI=((282*WI*(DIAI**0.215)*((VELI/1000)**1.5))/(FCI*DIAI**2))
      EPI=(0.5*DIAI)
      ACEPI=(ACPI+EPI)
      APFI=((1.23*DIAI)+(1.07*ACEPI))
      ACEP=(ACEPI*25.4)
      ACPER=(APFI*25.4)
      WRITE (6,6100)
6100  FORMAT (1H/,'The penetration depth using the ACE formula is,mm')
      WRITE (6,6200) ACEP
6200  FORMAT (1H/,F5.1)
      WRITE (6,6300)
6300  FORMAT (1H/,'Thickness to prevent perforation using ACE is,mm')
      WRITE (6,6400) ACPER
6400  FORMAT (1H/,F5.1)
      RETURN
      END
      SUBROUTINE CKW (CKPEN,CKPER)
      REAL W,DIA,VEL,WI,DIAI,VELI,CKPEI,CKPRI,CKPEN,CKPER
      WRITE (6,7000)
7000  FORMAT (1H/,'Input the following:'/
     +          1H/,'W (The weight of the missile,N)'/
     +          1H/,'DIA (The diameter of the missile,mm)'/
```

```
      +          1H/,'VEL (The impact velocity of the missile,m/sec)')
      READ (5,*) W,DIA,VEL
      WI=((2*W)/9)
      DIAI=(DIA/25.4)
      VELI=(VEL/0.3048)
      CKPEI=((6*WI*(DIAI**0.2)*((VELI/1000)**1.333333))/(DIAI**2))
      CKPRI=(1.3*CKPEI)
      CKPEN=(CKPEI*25.4)
      CKPER=(CKPRI*25.4)
      WRITE (6,7100)
7100  FORMAT (1H/,'The penetration using the CKW-BRL formula is,mm')
      WRITE (6,7200) CKPEN
7200  FORMAT (1H/,F5.1)
      WRITE (6,7300)
7300  FORMAT (1H/,'Thickness to prevent perforation using CKW-BRL,mm')
      WRITE (6,7400) CKPER
7400  FORMAT (1H/,F5.1)
      RETURN
      END
```

Blast loading program

```
>LIST10,410
   10 REM "INITIALIZE PRINTER"
   20 VDU2,1,27,1,64,3
   30 REM "DISABLE PAPER END DETECTOR"
   40 VDU2,1,27,1,56,3
   50 REM "SELECT PRINT STYLE  24"
   60 VDU2,1,27,1,33,1,56,3
   70 REM "SET LEFT MARGIN - 4 SPACES"
   80 VDU2,1,27,1,108,1,4,3
   90 REM "SET LINE SPACING - 35/216INCHES"
  100 VDU2,1,27,1,51,1,38
  110 PRINT:VDU3:INPUT "DO YOU REQUIRE PRINT-OUT ON PAPER? ENTER Y FOR Y
ES AND N FOR NO ";A1$:IF A1$="Y" THEN VDU2
  120 PRINT "BLAST LOADING PROGRAM:"
  130 VDU2,1,27,1,33,1,53,3:IF A1$="Y" THEN VDU2
  140 PRINT "BY N.M. ALAM (1987)"
  150 VDU2,1,27,1,106,1,10,3:IF A1$="Y" THEN VDU2
  160 PRINT "_____":PRINT:GOTO 180
  170 PRINT:VDU3:INPUT "DO YOU REQUIRE PRINT-OUT ON PAPER? ENTER Y FOR Y
ES AND N FOR NO ";A1$
  180 VDU2,1,27,1,33,1,0,3:VDU2,1,27,1,108,1,8,3:IF A1$="Y" THEN VDU2
  190 PRINT "_____
_____":VDU2:PRINT:VDU2,1,27,1,106,1,18,3
  200 INPUT "OPERATOR'S NAME";N$;"RUN NUMBER";N1:INPUT "DATE";N1$
  210 IF A1$="Y" THEN VDU2:VDU21:PRINT "OPERATOR'S NAME: ";N$;TAB(36);"R
UN NUMBER: ";N1;TAB(52);"DATE: ";N1$:VDU6
  220 VDU2,1,27,1,106,1,15,3:IF A1$="Y" THEN VDU2
  230 PRINT "_____
_____":PRINT:IF A1$="Y" THEN VDU2
  240 PRINT "DESIGN OF A WALL, IN A HIGH EXPLOSIVE ENVIRONMENT. THE DESI
GN AIM IS"
  250 PRINT "TO LIMIT THE DAMAGE RESULTING FROM BLAST LOADS IN CONNECTIO
N WITH AN"
  260 PRINT "ACCIDENTAL EXPLOSION."
  270 PRINT:VDU2:PRINT:VDU3:IF A1$="Y" THEN VDU2
  280 REM DETERMINE THE WORST CASE LOADING ON THE WALL. THE WALL WILL
  290 REM BE LOADED BY BLAST WAVES AND BY THE BUILD-UP OF QUASI-STATIC
  300 REM PRESSURE WITHIN THE ENCLOSED VOLUME.
  310 PRINT "B L A S T   W A V E  LOADING:"
```

```
  320 VDU2,1,27,1,106,1,20,3:IF A1$="Y" THEN VDU2
  330 PRINT "_____"
  340 PRINT
  350 REM LOADING FROM THE BLAST WAVE IS INFLUENCED BY THE CHARGE
  360 REM LOCATION. A CHARGE LOCATED ADJACENT TO A SIDE WALL WILL
  370 REM GIVE A REFLECTION OFF THE SIDE WALL AS WELL AS THE FLOOR
  380 REM AND PRODUCE HIGHER LOADS.
  390 PRINT "FOR   W O R S T   CASE LOADING A CHARGE REFLECTION FACTOR O
F 2,"
  400 PRINT "FROM BOTH FLOOR AND WALL SHOULD BE USED."
  410 PRINT
>LIST420,860
  420 VDU2,1,27,1,106,1,15,3
  430 INPUT "CHARGE REFLECTION FACTOR (WALL) ";C1
  440 INPUT "CHARGE REFLECTION FACTOR (FLOOR)";C2
  450 INPUT "CHARGE WEIGHT (kg of TNT)";W
  460 IF A1$="Y" THEN VDU2
  470 VDU21:PRINT "CHARGE REFLECTION FACTOR (WALL) ? ";C1
  480 PRINT "CHARGE REFLECTION FACTOR (FLOOR)? ";C2
  490 PRINT "CHARGE WEIGHT (kg of TNT)? ";W:VDU6
  500 LET W1=C1*C2*W
  510 PRINT
  520 VDU2,1,27,1,106,1,15,3:IF A1$="Y" THEN VDU2
  530 PRINT "EFFECTIVE CHARGE WEIGHT = ";W1;"kg of TNT"
  540 PRINT:PRINT
  550 PRINT "*** CALCULATION OF CHARGE STAND OFF ***"
  560 PRINT
  570 PRINT "CHARGE STANDOFF IS THE DISTANCE FROM THE WALL BEING DESIGNE
D TO THE"
  580 PRINT "EDGE OF THE HIGH EXPLOSIVE AREA, PLUS THE CHARGE RADIUS."
  590 PRINT
  600 VDU2,1,27,1,106,1,15,3
  610 INPUT "DISTANCE FROM WALL TO EDGE OF HIGH EXPLOSIVE AREA (m)";D
  620 INPUT "SPHERICAL CHARGE RADIUS (m)";D1
  630 IF A1$="Y" THEN VDU2
  640 VDU21:PRINT "DISTANCE FROM WALL TO EDGE OF HIGH EXPLOSIVE AREA (m)
? ";D
  650 PRINT "SPHERICAL CHARGE RADIUS (m)? ";D1:VDU6
  660 LET R=D+D1
  670 PRINT
  680 VDU2,1,27,1,106,1,15,3:IF A1$="Y" THEN VDU2
  690 PRINT "STANDOFF DISTANCE = ";R;"m"
  700 PRINT:PRINT
  710 PRINT "*** SCALED STANDOFF DISTANCE ***"
  720 PRINT
  730 LET R1=R/(W1^(1/3))
  740 LET R7=INT(R1*1000+0.5)/1000
  750 PRINT "SCALED STANDOFF DISTANCE = ";R7;"m/kg";
  760 VDU2,1,27,1,83,1,0,3:IF A1$="Y" THEN VDU2
  770 PRINT "1/3":VDU2,1,27,1,84,3:IF A1$="Y" THEN VDU2
  780 PRINT
  790 VDU2,1,27,1,106,1,15,3:IF A1$="Y" THEN VDU2
  800 PRINT "FOR THIS SCALED STANDOFF DISTANCE-REFER TO FIGURE 4 FOR THE
 REFLECTED"
  810 PRINT "PRESSURE AND REFLECTED IMPULSE VALUES."
  820 PRINT
  830 VDU2,1,27,1,106,1,15,3
  840 INPUT "REFLECTED PRESSURE Pr (kPa) ";P
  850 INPUT "VALUE  FOR  ir/W^1/3 (kPa.sec/kg^1/3)   ";I
  860 IF A1$="Y" THEN VDU2
```

```
>LIST870,1340
  870 VDU21:PRINT "REFLECTED PRESSURE Pr (kPa)? ";P
  880 PRINT "VALUE FOR ir/W";
  890 VDU6:VDU2,1,27,1,83,1,0,3:IF A1$="Y" THEN VDU2
  900 VDU21:PRINT "1/3";
  910 VDU6:VDU2,1,27,1,84,3:IF A1$="Y" THEN VDU2
  920 VDU21:PRINT " (kPa.sec/kg";
  930 VDU6:VDU2,1,27,1,83,1,0,3:IF A1$="Y" THEN VDU2
  940 VDU21:PRINT "1/3";
  950 VDU6:VDU2,1,27,1,84,3:IF A1$="Y" THEN VDU2
  960 VDU21:PRINT ")? ";I:VDU6
  970 LET I1=W1^(1/3)*I
  980 LET I2=INT(I1*1000+0.5)/1000
  990 PRINT
 1000 VDU2,1,27,1,106,1,15,3:IF A1$="Y" THEN VDU2
 1010 PRINT "REFLECTED IMPULSE ON WALL ir = ";I2;"kPa.sec"
 1020 PRINT
 1030 PRINT "BLAST WAVE LOADING IS IDEALIZED AS A TRIANGULAR PULSE WITH
ZERO RISE"
 1040 PRINT "TIME."
 1050 PRINT
 1060 PRINT "*** LOAD DURATION ***"
 1070 PRINT
 1080 LET T=2*I1/P
 1090 LET T5=INT(T*1E5+0.5)/1E5
 1100 PRINT "LOAD DURATION = ";T5;"sec"
 1110 VDU2,1,27,1,51,1,210,3:IF A1$="Y" THEN VDU2
 1120 PRINT:PRINT
 1130 VDU2,1,27,1,51,1,42,3:IF A1$="Y" THEN VDU2
 1140 PRINT "QUASI-STATIC LOADING:"
 1150 VDU2,1,27,1,106,1,27,3:IF A1$="Y" THEN VDU2
 1160 PRINT "_____ "
 1170 PRINT
 1180 PRINT "*** VOLUME OF ROOM ***"
 1190 PRINT:VDU3
 1200 INPUT "LENGTH OF ROOM (m)";L
 1210 INPUT "WIDTH OF ROOM  (m)";L1
 1220 INPUT "FLOOR TO CEILING HEIGHT (m)";L2
 1230 IF A1$="Y" THEN VDU2
 1240 VDU21:PRINT "LENGTH OF ROOM (m)? ";L
 1250 PRINT "WIDTH OF ROOM  (m)? ";L1
 1260 PRINT "FLOOR TO CEILING HEIGHT (m)? ";L2:VDU6
 1270 LET V=L*L1*L2
 1280 LET V1=INT(V*1000+0.5)/1000
 1290 PRINT
 1300 VDU2,1,27,1,106,1,15,3:IF A1$="Y" THEN VDU2
 1310 PRINT "VOLUME OF ROOM = ";V1;"m";
 1320 VDU2,1,27,1,83,1,0,3:IF A1$="Y" THEN VDU2
 1330 PRINT "3":VDU2,1,27,1,84,3:IF A1$="Y" THEN VDU2
 1340 LET D3=W/V
>LIST3600,3990
 3600 LET T6=INT(T3*1E5+0.5)/1E5
 3610 PRINT "PERIOD OF THE SYSTEM = ";T6;"sec"
 3620 PRINT
 3630 PROCdisplay
 3640 VDU2,1,27,1,51,1,210,3:IF A1$="Y" THEN VDU2
 3650 PRINT:PRINT:VDU2,1,27,1,106,1,20,3:IF A1$="Y" THEN VDU2
 3660 VDU2,1,27,1,51,1,31,3:IF A1$="Y" THEN VDU2
 3670 PRINT "*** NUMERICAL INTEGRATION ***"
 3680 PRINT "*** FOR ONE DEGREE OF FREEDOM SPRING-MASS SYSTEM ***"
```

```
3700 PRINT
3720 PRINT "TO INTEGRATE THE EQUATION OF MOTION, A TIME STEP LESS THAN
OR EQUAL"
3730 PRINT "TO ONE-TENTH OF THE FUNDAMENTAL PERIOD IS ADEQUATE IN MOST
INSTANCES."
3740 PRINT:PRINT
3750 VDU2,1,27,1,106,1,25,3
3760 INPUT "CHOSEN VALUE FOR TIME STEP (sec)";T4
3770 IF A1$="Y" THEN VDU2
3780 VDU21:PRINT "CHOSEN VALUE FOR TIME STEP (sec)? ";T4:VDU6
3790 PRINT
3800 VDU2,1,27,1,106,1,10,3
3810 INPUT "SPECIFY TIME AT WHICH CALCULATIONS SHOULD TERMINATE (sec) "
;N
3820 IF A1$="Y" THEN VDU2
3830 VDU21:PRINT "SPECIFY TIME AT WHICH CALCULATIONS SHOULD TERMINATE (
sec)? ";N:VDU6
3840 VDU2,1,27,1,33,1,15:VDU2,1,27,1,108,1,10,3:IF A1$="Y" THEN VDU2:PR
INT:PRINT:VDU2,1,27,1,106,1,20,3:IF A1$="Y" THEN VDU2
3850 PRINT TAB(1);"COL 1";TAB(14);"COL 2";TAB(27);"COL 3";TAB(41);"COL
4";TAB(55);"COL 5";TAB(66);"COL 6";TAB(76);"COL 7"
3860 PRINT
3870 VDU2,1,27,1,106,1,15,3:IF A1$="Y" THEN VDU2
3880 PRINT TAB(1);"TIME:";TAB(55);"ACC.:";TAB(66);"VEL.:";TAB(76);"DISP
.:"
3890 VDU2,1,27,1,106,1,20,3:IF A1$="Y" THEN VDU2:PRINT "_____
_____"
3900 PRINT:VDU2,1,27,1,106,1,15,3:IF A1$="Y" THEN VDU2
3910 LET E1=0
3920 LET E2=0
3930 FOR S=0 TO N STEP T4
3940 LET E=F4*1000-((F4*1000/T)*S)
3950 IF E<0 THEN LET E=0
3960 IF S=0 THEN LET Q=0:Q1=0:Q2=0
3970 LET U=K1*1000*(Q+T4*Q1+((T4^2/4)*Q2))
3980 LET U1=E-U
3990 LET U2=U1/(0.66*M7+0.25*K1*1000*T4^2)
>LIST4000,4460
4000 LET U3=Q1+0.5*(U2+Q2)*T4
4010 LET U4=Q+0.5*(U3+Q1)*T4
4020 IF S=0 THEN LET U3=0:U4=0
4030 LET O1=INT(S*1E5+0.5)/1E5
4040 LET O2=INT(E*100+0.5)/100
4050 LET O3=INT(U*100+0.5)/100
4060 LET O4=INT(U1*100+0.5)/100
4070 LET O5=INT(U2*1000+0.5)/1000
4080 LET O6=INT(U3*1E4+0.5)/1E4
4090 LET O7=INT(U4*1E5+0.5)/1E5
4100 LET G=O1:PROCtable:PRINT TAB(2-G1);G;
4110 LET G=O2:PROCtable:PRINT TAB(19-G1);G;
4120 LET G=O3:PROCtable:PRINT TAB(32-G1);G;
4130 LET G=O4:PROCtable:PRINT TAB(46-G1);G;
4140 LET G=O5:PROCtable:PRINT TAB(57-G1);G;
4150 LET G=O6:PROCtable:PRINT TAB(66-G1);G;
4160 LET G=O7:PROCtable:PRINT TAB(76-G1);G
4170 LET Q=U4
4180 LET Q1=U3
4190 LET Q2=U2
4200 LET E1=U4
4210 IF E2<E1 THEN LET E2=E1
4220 NEXT S
```

```
 4230 VDU2,1,27,1,51,1,210,3:IF A1$="Y" THEN VDU2
 4240 PRINT:PRINT
 4250 VDU2,1,27,1,51,1,40,3:IF A1$="Y" THEN VDU2
 4260 PRINT
 4270 VDU2,1,27,1,33,1,0,3:IF A1$="Y" THEN VDU2
 4280 VDU2,1,27,1,108,1,8,3:IF A1$="Y" THEN VDU2
 4290 PRINT "*** MAXIMUM DISPLACEMENT OBTAINED FROM INTEGRATION PROCEDUR
E ***"
 4300 LET E3=E2*1000
 4310 LET E4=INT(E3*100+0.5)/100
 4320 PRINT
 4330 VDU2,1,27,1,106,1,15,3:IF A1$="Y" THEN VDU2
 4340 PRINT "MAXIMUM  CENTRE  DISPLACEMENT OF WALL = ";E4;"mm"
 4350 PRINT:PRINT
 4360 PRINT "*** DUCTILITY RATIO ***"
 4370 LET H=E2/(R5/K)
 4380 LET H4=INT(H*100+0.5)/100
 4390 PRINT
 4400 VDU2,1,27,1,106,1,15,3:IF A1$="Y" THEN VDU2
 4410 PRINT "DUCTILITY RATIO = ";H4
 4420 PRINT:PRINT
 4430 PRINT "*** MAXIMUM HINGE ROTATION AT THE SUPPORT ***"
 4440 PRINT
 4450 VDU2,1,27,1,106,1,15,3:IF A1$="Y" THEN VDU2
 4460 LET H1=E2/(L2/2)
>LIST4470,4900
 4470 LET H2=ATN(H1)
 4480 LET H3=H2*360/(2*3.141592654)
 4490 LET H5=INT(H3*100+0.5)/100
 4500 PRINT "MAXIMUM HINGE ROTATION AT THE SUPPORT = ";H5;" DEGREES"
 4510 PRINT:PRINT:PRINT
 4520 VDU3
 4530 INPUT "DO YOU WANT TO RUN THE INTEGRATION PROCEDURE AGAIN? ENTER Y
 FOR YES AND N FOR NO ";Z1$
 4540 IF A1$="Y" THEN VDU2
 4550 VDU21:PRINT "DO YOU WANT TO RUN THE INTEGRATION PROCEDURE AGAIN? "
;Z1$:VDU6
 4560 IF Z1$="Y" THEN GOTO 3650
 4570 PRINT:PRINT
 4580 VDU3
 4590 INPUT "DO YOU WANT TO RUN THE PROGRAMME AGAIN FROM THE START? ENTE
R Y FOR YES AND N FOR NO ";Z$
 4600 IF A1$="Y" THEN VDU2
 4610 VDU21:PRINT "DO YOU WANT TO RUN THE PROGRAMME AGAIN FROM THE START
? ";Z$:VDU6
 4620 IF Z$="Y" THEN GOTO 170
 4630 PRINT:PRINT
 4640 PRINT "              * * *   E   N   D   * * * "
 4650 VDU3
 4660 END
 4670 DEF PROCtable
 4680 LET G$=STR$(G)
 4690 LET J=0
 4700 LET J=J+1
 4710 LET G1$=RIGHT$(G$,J)
 4720 LET G2$=LEFT$(G1$,1)
 4730 LET G7$=LEFT$(G$,1)
 4740 LET G3$=MID$(G$,2,1)
 4750 IF G7$="-" THEN LET G3$=MID$(G$,3,1)
 4760 IF G2$="." THEN GOTO 4790
 4770 IF LEN(G1$)=LEN(G$) THEN GOTO 4810
```

Appendix

Program ISOPAR

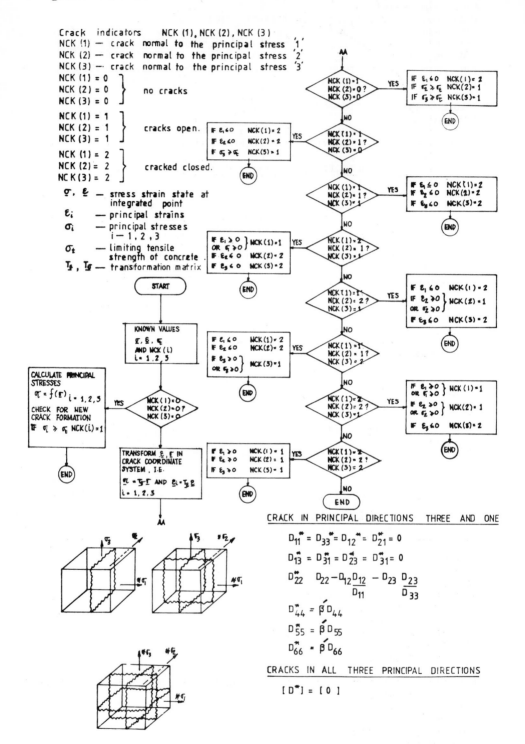

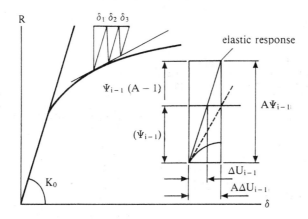

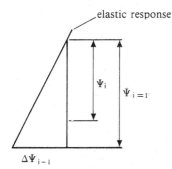

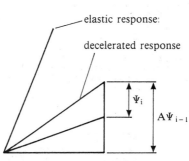

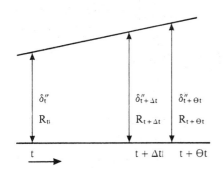

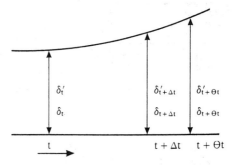

```
4780 GOTO 4700
4790 LET G1=LEN(G$)-LEN(G1$)
4800 IF G2$="." THEN GOTO 4820
4810 LET G1=LEN(G$)
4820 IF G3$="E" THEN LET G1=G1-3
4830 ENDPROC
4840 DEF PROCdisplay
4850 VDU3
4860 PRINT "PRESS ANY KEY TO CONTINUE(EXCEPT CONTROL KEYS SUCH AS >BREA
K<, >ESCAPE<, ETC."
4870 key$=GET$
4880 CLS
4890 IF A1$="Y" THEN VDU2
4900 ENDPROC
```

Ottosen model

```
      IMPLICIT REAL*8(A – H,O – Z)
      COMMON /MTMD3D/ DEP(6,6),STRESS(6),STRAIN(6),IPT,NEL
      DIMENSION PAR(3,5),FS(6,6),FSTPOS(6,6),PROP(1),SIG(1),
     @          DVI1DS(6),DVJ2DS(6),DVJ3DS(6),DVTHDS(6)
      OPEN (UNIT = 5,FILE = 'PARAMETERS',STATUS = 'OLD')
      READ (5,*,END = 3700)((PAR(IF,JF),JF = 1,5),IF = 1,3)
3700  CLOSE (5)
      PK = PROP(3)/PROP(4)
      IP = 0
      JP = 0
      IF (PK .LE. 0.08) IP = 1
      IF (PK .EQ. 0.10) IP = 2
      IF (PK .GE. 0.12) IP = 3
      IF (PK .LT. 0.10) JP = 1
      IF (PK .GT. 0.10) JP = 2
      IF (IP .EQ. 0) GOTO 3800
      A = PAR(IP,2)
      B = PAR(IP,3)
      PK1 = PAR(IP,4)
      PK2 = PAR(IP,5)
      GOTO 3909
3800  SUB1 = PK – PAR(JP,1)
      SUB2 = PAR(JP + 1,1) – PAR(JP,1)
      A = SUB1*(PAR(JP*1,2) – PAR(JP,2))/SUB2 + PAR(JP,2)
      B = SUB1*(PAR(JP*1,3) – PAR(JP,3))/SUB2 + PAR(JP,3)
      PK1 = SUB1*(PAR(JP*1,4) – PAR(JP,4))/SUB2 + PAR(JP,4)
      PK2 = SUB1*(PAR(JP*1,5) – PAR(JP,5))/SUB2 + PAR(JP,5)
3900  VARI1 = SIG(1) + SIG(2) + SIG(3)
      VARJ2 = 1.0/6.0*((SIG(1) – SIG(2))**2 + (SIG(2) – SIG(3))**2 +
     @        (SIG(3) – SIG(1))**2) + SIG(4)**2 + SIG(5)**2 + SIG(6)**2
      VARI13 = VARI1/3.0
      VI131 = SIG(1) – VARI13
      VI132 = SIG(2) – VARI13
      VI133 = SIG(3) – VARI13
      VARJ3 = VI131*(VI132*VI133 – SIG(5)**2) – SIG(4)*(SIG(4)*VI133
     @        – SIG(5)*SIG(5)) + SIG(6)*(SIG(4)*SIG(5) – SIG(6)*VI132)
      VAR3TH = 1.5*3.0**(0.5)*VARJ3/VARJ2**1.5
      IF (VAR3TH .GE. 0.0) GOTO 4000
      ALAM = 22.0/21.0 – 1.0/3.0*ACOS( – PK2*VAR3TH)
      TOTLAM = PK1*COS(ALAM)
      DFD3TH = PK1*PK2*VARJ2**0.5*SIN(ALAM)/(3.0*PROP(4)*
     @         SIN(ACOS( – PK2*VAR3TH)))
      GOTO 4100
4000  ALAM = 1.0/3.0*ACOS(PK2*VAR3TH)
      TOTLAM = PK1*COS(ALAM)
      DFD3TH = PK1*PK2*VARJ2**0.5*SIN(ALAM)/(3.0*PROP(4)*
     @         SIN(ACOS(PK2*VAR3TH)))
4100  DFDI1 = B/PROP(4)
      DFDJ2 = A/PROP(4)**2 + TOTLAM/(PROP(4)*VARJ2**0.5)
      DVI1DS(1) = 1.0
      DVI1DS(2) = 1.0
      DVI1DS(3) = 1.0
      DVI1DS(4) = 0.0
      DVI1DS(5) = 0.0
      DVI1DS(6) = 0.0
      DVJ2DS(1) = 1.0/3.0*(2.0*SIG(1) – SIG(2) – SIG(3))
      DVJ2DS(2) = 1.0/3.0*(2.0*SIG(2) – SIG(1) – SIG(3))
      DVJ2DS(3) = 1.0/3.0*(2.0*SIG(3) – SIG(1) – SIG(2))
      DVJ2DS(4) = 2.0*SIG(4)
      DVJ2DS(5) = 2.0*SIG(5)
      DVJ2DS(6) = 2.0*SIG(6)
      DVJ3DS(1) = 1.0/3.0*(VI131*( – VI132 – VI133)) + 2.0*VI132*VI131 –
     @            2.0*SIG(5)**2*SIG(4)**2*SIG(6)**2
      DVJ3DS(2) = 1.0/3.0*(VI132*( – VI131 – VI133)) + 2.0*VI131*VI133 –
     @            2.0*SIG(6)**2*SIG(4)**2*SIG(5)**2
      DVJ3DS(3) = 1.0/3.0*(VI133*( – VI131 – VI132)) + 2.0*VI131*VI132 –
     @            2.0*SIG(4)**2*SIG(5)**2*SIG(6)**2
      DVJ3DS(4) = – 2.0*VI133*SIG(4) + 2.0*SIG(5)*SIG(6)
      DVJ3DS(5) = – 2.0*VI131*SIG(5) + 2.0*SIG(4)*SIG(6)
```

```
        DVJ3DS(6) = -2.0*VI132*SIG(6) + 2.0*SIG(4)*SIG(5)
        CONVJ2 = 3.0*3.0**0.5/(2.0*VARJ*1.2)
        VJ3J2 = VARJ3/VARJ2**0.5
        DVTHDS(1) = CONVJ2*(-0.5*VJ3J2*(2.0*SIG(1)-SIG(2)-SIG(3))+
     @             DVJ3DS(1))
        DVTHDS(2) = CONVJ2*(-0.5*VJ3J2*(2.0*SIG(2)-SIG(1)-SIG(3))+
     @             DVJ3DS(2))
        DVTHDS(3) = CONVJ2*(-0.5*VJ3J2*(2.0*SIG(3)-SIG(1)-SIG(2))+
     @             DVJ3DS(3))
        DVTHDS(4) = CONVJ2*(-3.0*VJ3J2*SIG(4)+DVJ3DS(4))
        DVTHDS(5) = CONVJ2*(-3.0*VJ3J2*SIG(5)+DVJ3DS(5))
        DVTHDS(6) = CONVJ2*(-3.0*VJ3J2*SIG(6)+DVJ3DS(6))
        DO 4200 IS = 1,6
        FS(IS,1) = DFDI1*DVI1DS(IS)+DFDJ2*DVJ2DS(IS)*
                   DFD3TH*DVTHDS(IS)
 4200   FSTPDS(1,IS) = FS(IS,1)
        RETURN
        END
```

Main program for non-linear analysis

```
C
        SUBROUTINE NONSTR(IEL,IGAUS,TEM)
C-------------            THIS SUBR. CALC. THE INCREMENTAL
C-------------            AND UPDATED STRESS, LOADING AND UNLOADING
C-------------            AND CRACK FORMATION
C
£INSERT COMMON.FF
C
C-----    IF POINT IS ALREADY CRUSHED DO NOT DO ANY CALCULATION
        IF(NCRK(IGAUS,IEL).EQ.999)GO TO 555
C
C
C               CRACK INDICATOR
        NCR=NCRK(IGAUS,IEL)
C               LOADING - UNLOADING INDICATOR
        IUNL=IUNLOD(IGAUS,IEL)
C
C--------        CRACK WIDTH(IN TERMS OF STRAINS)
C       DO 5 J=1,3
C       CRW(J)=CWI(J,IGAUS)
C5      CONTINUE
C
C----------        COPY ANGL INTO DC
        DO 7 J=1,9
        DC(J)=ANGL(J,IGAUS,IEL)
 7      CONTINUE
C
        CALL RZERO(CET,6)
C
C---------------        STRESSES CURRENT AND TOTAL
        DO 10 J=1,6
        STG(J)=SIGT(J,IGAUS,IEL)
        STB(J)=STG(J)+SIG(J)
 10     CONTINUE
C
        IF(NCR.EQ.0)GO TO 30
C
C----------        TRANSFORMATION MATRIX IN CRACK DIRECTIONS
        DO 20 J=1,3
        DC1(J)=ANGL(J,IGAUS,IEL)
        DC2(J)=ANGL(J+3,IGAUS,IEL)
        DC3(J)=ANGL(J+6,IGAUS,IEL)
 20     CONTINUE
C
        CALL TRANSF(3)
C----------------        TRANSFORM STB STRESS IN CRACK DIRECTION
C----------------        ALSO INCREMENTAL STRESS SIG
        CALL MVECT(QM,STB,CET,6,6)
C
        CALL MVECT(QM,SIG,STC,6,6)
C
C----------------        TRANSF. TOTAL STRAIN IN CRACK DIR.
C
        CALL TRANSF(1)
        DO 21 J=1,6
        AJ(J)=ECT(J,IGAUS)
 21     CONTINUE
C
        CALL MVECT(QM,AJ,ECA,6,6)
```

```
C
C
      GO TO 50
C
30    CONTINUE
C-----------                         CALC. PRINCIPAL STRESSES DUE TO STB
      CALL PRINCL(2,IGAUS,STB)
      CET(1)=PS1(IGAUS)
      CET(2)=PS2(IGAUS)
      CET(3)=PS3(IGAUS)
      KK00=1
      IF(KK00.EQ.1)GO TO 40
C-----------             PRINCIPAL STRAINS
      DO 38 J=1,6
      ECB(J)=ECT(J,IGAUS)
38    CONTINUE
C
      CALL PRINCL(1,IGAUS,ECB)
      ECA(1)=EP1(IGAUS)
      ECA(2)=EP2(IGAUS)
      ECA(3)=EP3(IGAUS)
40    CONTINUE
C
C
C------         CALCULATE EQUIVALENT STRAIN
C
      EGSTN=SIGEFF(ECB)
C--------------             CHECK FOR CONCRETE CRUSHING
C     EC1=ECA(1)+ECU
C     EC2=ECA(2)+ECU
C     EC3=ECA(3)+ECU
C     IF(EC1.LT.0.0 .OR. EC2.LT.0.0 .OR. EC3.LT.0.0)GO TO 888
C
      CRUSH=EGSTN - ECU
      IF(CRUSH .GT. 0.0) GO TO 888
C
   50 CONTINUE
C
C-----        CALC. AND UPDATE CRACK INDICATOR
C
C
      CALL CRACK(CET,ECA,NCR,CRW)
C
C-----------              STORE UPDATED VALUES IN ARRAYS
      NCRK(IGAUS,IEL)=NCR
      DO 41 J=1,9
      ANGL(J,IGAUS,IEL)=DC(J)
   41 CONTINUE
C
C     DO 42 J=1,3
C     CWI(J,IGAUS)=CRW(J)
C 42     CONTINUE
C
      IF(NCR.EQ.0)GO TO 110
      DO 105 J=1,3
      DC1(J)=ANGL(J,IGAUS,IEL)
      DC2(J)=ANGL(J+3,IGAUS,IEL)
      DC3(J)=ANGL(J+6,IGAUS,IEL)
  105 CONTINUE
      CALL TRANSF(1)
      DO 106 J=1,6
      ECB(J)=EC(J)
  106 CONTINUE
C
      CALL MVECT(QM,ECB,EC,6,6)
C
C
  110 CONTINUE
C--------      GO TO APPROPRIATE CONCRETE COMPRESSION CRITERION
      IF(ICOMP.EQ.1)GO TO 98
C---------          CALC. UNIAXIAL STRAINS
      DO 308 J=1,3
      IF(ENU(J).LT.1.E-15)GO TO 306
      EIU(J)=SIG(J)/ENU(J)
      GO TO 308
  306 CONTINUE
      EIU(J)=0.0
  308 CONTINUE
C--------------            EQUIV. STRESS AT PREVIOUS UNLOADED POINT
      SEQ=SIGY(IGAUS,IEL)
C--------------                CALC. EFFECTIVE STRESS DUE TO CURRENT
C--------------                AND TOTAL STRESS
      SIGEF2=SIGEFF(ECT(1,IGAUS))
      DO 109 J=1,6
      ECB(J)=ECT(J,IGAUS)-ECRT(J)
  109 CONTINUE
      SIGEF1=SIGEFF(ECB)
      FLOA=SIGEF2-SIGEF1
      IF(IUNL .EQ. 1) GO TO 60
C        IUNL=0  --- ON THE EQUIV. CURVE
```

```
C
        IF(SIGEF2 .GE. SEQ)GO TO 43
C-------------            UNLOADING AT THIS POINT
        RFACT=(SEG-SIGEF1)/FLOA
C-------------                  NON-LINEAR STRAIN
        DO 35 J=1,6
        ECB(J)=RFACT*EC(J)
   35   CONTINUE
C-----            MEAN NON-LINEAR STRAIN
        DO 201 J=1,3
        ETU(J,IGAUS)=ETU(J,IGAUS)+0.5*RFACT*EIU(J)
  201      CONTINUE
        CALL DMATL(STB,IEL,IGAUS,TEM)
C-------------            INCREMENTAL STRESS ASSOCIATED WITH -ECB
        CALL MVECT(DDS,ECB,AJ,6,6)
C-------------                  ELASTIC STRAIN
        DO 36 J=1,6
        ECB(J)=(1.0-RFACT)*EC(J)
   36   CONTINUE
        IUNLOD(IGAUS,IEL)=1
C-------------            ELASTIC STRESS INCR.
        CALL DMATL(CET,IEL,IGAUS,TEM)
        CALL MVECT(DDS,ECB,STA,6,6)
C---------- TOTAL STRESS INCREMENT
        DO 37 J=1,6
        SIG(J)=STA(J)+AJ(J)
   37   CONTINUE
C

        DO 202 J=1,3
        ETU(J,IGAUS)=ETU(J,IGAUS)+EIU(J)-0.5*RFACT*EIU(J)
  202      CONTINUE
        GO TO 99
   43   CONTINUE
C-------------            LOADING AT THIS POINT
C-----            MEAN UNIAXIAL STRAIN
        DO 203 J=1,3
        ETU(J,IGAUS)=ETU(J,IGAUS)+EIU(J)*0.5
  203      CONTINUE
        CALL DMATL(STB,IEL,IGAUS,TEM)
C-------------                  STRESS INCREMENT
C

        CALL MVECT(DDS,EC,SIG,6,6)
C
C-----            ACCUMULATE TOTAL UNIAXIAL STRAIN
        DO 205 J=1,3
        ETU(J,IGAUS)=ETU(J,IGAUS)+EIU(J)*0.5
  205      CONTINUE
        GO TO 99
C
   60   CONTINUE
C-------------            NOT ON THE EQUIV. CURVE
        IF(SIGEF2 .GT. SEG)GO TO 70
C-------------            ELASTIC UNLOADING
        CALL DMATL(CET,IEL,IGAUS,TEM)
C-------------            ELASTIC STRESS
        CALL MVECT(DDS,EC,SIG,6,6)
C-----            ACCUMULATE UNIAXIAL STRAIN
        DO 206 J=1,3
        ETU(J,IGAUS)=ETU(J,IGAUS)+EIU(J)
  206      CONTINUE
        GO TO 99
   70   CONTINUE
C-------------            LOADING PARTLY ELASTIC PARTLY NON-LINEAR
        FRAC=(SEG-SIGEF1)/FLOA
C-------------            ELASTIC STRAIN
        DO 71 J=1,6
        ECB(J)=FRAC*EC(J)
   71   CONTINUE
        CALL DMATL(CET,IEL,IGAUS,TEM)
        CALL MVECT(DDS,ECB,STA,6,6)
C-------------                  STRESS AT THE CURVE
        DO 72 J=1,6
        AJ(J)=STG(J)+STA(J)
   72   CONTINUE
C-----            MEAN UNIAXIAL STRAIN
        DO 207 J=1,3
        ETU(J,IGAUS)=ETU(J,IGAUS)+0.5*EIU(J)*(1.+FRAC)
  207      CONTINUE
C

        IUNLOD(IGAUS,IEL)=0
C-------------            STRAIN ASSOCIATED WITH NON-LINEAR CURVE
        DO 73 J=1,6
        ECB(J)=(1.0-FRAC)*EC(J)
   73   CONTINUE
        CALL DMATL(AJ,IEL,IGAUS,TEM)
C-------            STRESS INCR.
        CALL MVECT(DDS,ECB,STB,6,6)
C-------------------            TOTAL INCREMENTAL STRESS
        DO 74 J=1,6
```

```
         SIG(J)=STA(J)+STB(J)
   74    CONTINUE
C
         DO 208 J=1,3
         ETU(J,IGAUS)=ETU(J,IGAUS)+0.5*EIU(J)*(1.-FRAC)
  208    CONTINUE
         GO TO 99
   98    CONTINUE
C-------          STRESS INCREMENT ON THE BASIS OF ENDOCHRONIC THEORY
C
         DO 1223 J=1,6
         SIGG(J)=SIG(J)
         ECC(J)=EC(J)
 1223 CONTINUE
         CALL ENDOST(IEL,IGAUS,SIGG,ECC)
C
   99    CONTINUE
         IF(NCR.EQ.0)GO TO 50
         DO 91 J=1,6
         AJ(J)=SIGT(J,IGAUS,IEL)
   91    CONTINUE
         CALL TRANSF(3)
         CALL MVECT(QM,AJ,STG,6,6)
C----------          TRANSFORM LOCAL STRESSES IN GLOBAL DIRN
         CALL TRANSF(2)
         DO 92 J=1,6
         STG(J)=STG(J)+SIG(J)
92    CONTINUE
C----------          RELEASE STRESSES ACROSS THE OPEN CRACKS
         CALL GETNCK(NCR,NCK)
         IF(NCK(1).EQ.1)STG(1)=0.0
         IF(NCK(2).EQ.1)STG(2)=0.0
         IF(NCK(3).EQ.1)STG(3)=0.0
         CALL MVECT(QM,STG,STA,6,6)
C
         DO 94 J=1,6
         SIGT(J,IGAUS,IEL)=STA(J)
   94    CONTINUE
C
         GO TO 999
90    CONTINUE
         DO 100 J=1,6
         SIGT(J,IGAUS,IEL)=SIGT(J,IGAUS,IEL)+SIG(J)
100    CONTINUE
         GO TO 999
888    CONTINUE
C-----------          CRUSHING OF CONCRETE
         NCRK(IGAUS,IEL)=999
C----------          RELEASE STRESSES
         DO 101 J=1,6
         SIGT(J,IGAUS,IEL)=0.0
101    CONTINUE
C
         GO TO 555
999    CONTINUE
C-----          CHECK THAT THE CURRENT STATE OF STRESS IS
C-----          INSIDE THE FAILURE SURFACE
         CALL PRINCL(2,IGAUS,SIGT(1,IGAUS,IEL))
         CET(1)=PS1(IGAUS)
         CET(2)=PS2(IGAUS)
         CET(3)=PS3(IGAUS)
         IF(CET(1).GT.0.0 .OR.CET(2).GT.0.0 .OR.CET(3).GT.0.0)
     1                                            GO TO 555
         CALL CONCR1(CET)
C        IF(ICOMP.EQ.2)          CALL CONCR3(CET)
         BRING=1.0
         IF(FF.GT.1.0001)BRING=BRING/FF
         DO 553 J=1,6
         SIGT(J,IGAUS,IEL)=BRING*SIGT(J,IGAUS,IEL)
553    CONTINUE
C
555    CONTINUE
         RETURN
         END
C
         SUBROUTINE ASSLOD(IEL,NER,ELOD)
£INSERT COMMON.FF
C
C-----------          TO ASSEMBLE LOAD VECTOR
C
         DO 95 J=1,NER
         M1=(MCODE(J,IEL)-1)*NDF
C------          ELASTO-PLASTIC STRAIN INCR.
         DO 88 J=1,3
         ECM(J)=FPROP*ECM(J)
   88    CONTINUE
         CALL SBINST(ECM,NSUB,SGMT(1,IGAUS,I1))
C----- ADD ELASTIC STRESS INCR
         DO 94 J=1,3
         SGMT(J,IGAUS,I1)=SGMT(J,IGAUS,I1)+SGM(J)
```

```
   94 CONTINUE
      NYM(IGAUS,I1)=2
99    CONTINUE
      RETURN
      END
C
      SUBROUTINE MEMDAT(I1,IGAUS,STD)
£INSERT COMMON.FF
C
C------------                 TO CALC. ELASTO - PLASTIC MATERIAL MATRIX
C------------                 AT STRESS LEVEL,STD FOR MEMBRANE ELEMENTS
C
C
C------------                 CALC. ELASTIC MATERIAL MATRIX
C
      CALL DMEMB
C------------                 CHECK WHETHER CURRENT POINT IS PLASTIC
.     IF(NYM(IGAUS,I1).NE.1)GO TO 50
C
C
      EFF=ZMISE(STD)
      SX=(2.*STD(1)-STD(2))/3.
      SY=(2.*STD(2)-STD(1))/3.
      FAC=EFF/1.5
C
C------------                 CALC. (DF/D(STD) = AJ
      AJ(1)=SX/FAC
      AJ(2)=SY/FAC
      AJ(3)=2.*STD(3)/FAC
C
C------------                 CALC. DENOMINATOR OF PLASTIC MATRIX
C                        '  DENOM=AJ(T)*DJ*AJ + HARDG
C
      CALL MVECT(DJ,AJ,STC,3,3)
C
      DENOM=0.0
      DO 10 J=1,3
      DENOM=DENOM+AJ(J)*STC(J)
10    CONTINUE
C
      DENOM=DENOM+HARDG
C
C------------                 CALC. ELASTO-PLASTIC MATERIAL MATRIX AND
C------------                 STORE IT INTO DJ
      DO 30 J=1,3
      DO 20 K=1,3
      DJ(J,K)=DJ(J,K)-STC(J)*STC(K)/DENOM
20    CONTINUE
30    CONTINUE
C
50    CONTINUE
C
      RETURN
      END
C
C
      SUBROUTINE SBINST(ECL,NSUB,STA)
C-----      STRESS INCR IS CALCULATED USING SUB-INCREMENTAL
C-----      METHOD,ALSO STRESS STA IS UPDATED
£INSERT COMMON.FF
      DIMENSION ECL(1)
      RSUB=NSUB
      DO 3 J=1,3
      ECB(J)=ECL(J)/RSUB
   3  CONTINUE
      SIGY1=ZMISE(STA)
C-----      LOOP OVER SUB-INCREMENTS
      DO 70 ISUB=1,NSUB
C------------                 CALC. ELASTO-PLASTIC MATERIAL MATRIX -DJ
C
      SX=(2.*STA(1)-STA(2))/3.
      SY=(2.*STA(2)-STA(1))/3.
      FAC=SIGY1/1.5
C
C------------                 CALC. (DF/D(STA) = AJ
      AJ(1)=SX/FAC
      AJ(2)=SY/FAC
      AJ(3)=2.*STA(3)/FAC
C
C------------                 CALC. DENOMINATOR OF PLASTIC MATRIX
C                        DENOM=AJ(T)*DJ*AJ + HARDG
C
      CALL MVECT(DJ,AJ,STC,3,3)
C
      DENOM=0.0
      DO 10 J=1,3
      DENOM=DENOM+AJ(J)*STC(J)
10    CONTINUE
C
      DENOM=DENOM+HARDG
```

```
C-----------                     CALC. ELASTO-PLASTIC MATERIAL MATRIX AND
C-----------                     STORE IT INTO DJ
         DO 30 J=1,3
         DO 20 K=1,3
         DJ(J,K)=DJ(J,K)-STC(J)*STC(K)/DENOM
 20      CONTINUE
 30      CONTINUE
C-----        CALC. DLAMB AND EQUIV. PLASTIC STRAIN INCREMENT
         DLAMB=0.0
         DO 64 J=1,3
         DLAMB=DLAMB+STC(J)*ECB(J)
 64      CONTINUE
         DLAMB=DLAMB/DENOM
C-----        UNLOADING IS PLASTIC INSIDE A SUBINCREMENTS
C        IF(DLAMB.LT.0.0)DLAMB=0.0
         BB=0.0
         DO 65 J=1,3
         BB=BB+AJ(J)*STA(J)
 65      CONTINUE
         EGSTN=DLAMB*BB/SIGY1
         CALL MVECT(DJ,ECB,STB,3,3)
         DO 60 J=1,3
         STA(J)=STA(J)+STB(J)
 60      CONTINUE
C-----        CALC. UPDATED YIELD SURFACE
         SIGY2=ZMISE(STA)
         SIGY1=SIGY1+HARDG*EQSTN
         FACT=1.0
         IF(SIGY2.GT.SIGY1)FACT=SIGY1/SIGY2
C-----        UPDATE STRESS VECTOR
         DO 62 J=1,3
         STA(J)=STA(J)*FACT
 62      CONTINUE
C
 70      CONTINUE
         RETURN
         END
C
         SUBROUTINE STELST(I)
£INSERT COMMON.FF
C
C---------                     CALC. ELASTO-PLASTIC STRESS INCR AND UDATE
C---------                     CURRENT STRESS AND PLASTIC INDICATOR
         N=LRF(I)
C
         I1=I-(NTE1+NTE2)
C
         T1=SIGGT(I1)
         T2=T1+STRV
         T1=RABS(T1)
         T2=RABS(T2)
C
         FACL=T2-T1
         IF(FACL.EQ.0.0)GO TO 99
C-----------                     CHECK FOR LOADING OR UNLOADING AT THIS POINT
C
         IF(ISPL(I1).EG.1)GO TO 40
C-----        POINT ELASTIC BEFORE
         IF(T2.LT.YIELST(I1))GO TO 50
C-----        TRANSITION ZONE  - LOADING
C---------        FRACTION OF ELASTIC STRAIN INCR.
         FRAC=(YIELST(I1)-T1)/FACL
C
         ISPL(I1)=1
C*--------        ELASTIC STRESS INCREMENT
         STRV=STRV*FRAC
C*--------        STRESS AT YIELD SURFACE
         SIGGT(I1)=SIGGT(I1)+STRV
C*--------        PLASTIC STRAIN INCR.
         STRNV=(1.0-FRAC)*STRNV
         GO TO 45
C
 40      CONTINUE
C-----        POINT PLASTIC BEFORE
         IF(T2.GT.YIELST(I1))GO TO 45
         GO TO 50
 45      CONTINUE
         EPSMOD=ZESBAR(I)
C---------        CALC. PLASTIC STRESS INCR. AND UPDATE STRESS
         STRVPL=STRNV*EPSMOD
         STRV=STRV+STRVPL
         SIGGT(I1)=SIGGT(I1)+STRVPL
C
         GO TO 99
C
 50      CONTINUE
C---------                     UNLOADING AT THIS POINT
C
C-------        CHECK IF POINT WAS PLASTIC IN PREVIOUS ITERATION
         IF(ISPL(I1).EQ.1)GO TO 55
```

```
      SIGGT(I1)=SIGGT(I1)+STRV
      GO TO 99
   55 CONTINUE
C-----      CHECK THAT THE UNLOADING IS REAL
      STRV=ZESBAR(I)*STRNV
      T2=SIGGT(I1)+STRV
      T2=RABS(T2)
      IF(T2.GT.YIELST(I1))GO TO 45
      FACL=T2-T1
      IF(FACL.EQ.0.0)GO TO 99
      FRAC=(YIELST(I1)-T1)/FACL
C--------------            PLASTIC STRESS INCR.
      STRPL=FRAC*STRV
      ISPL(I1)=2
C
C--------------            ELASTIC STRESS INCR.
      STREL=ZESBAR(I)*(1.0-FRAC)*STRNV
      STRV=STRPL+STREL
C
      SIGGT(I1)=SIGGT(I1)+STRV
   99 CONTINUE
      RETURN
      END
C
      SUBROUTINE INCLNE(I,NER)
£INSERT COMMON.FF
C-------------------------------------------------------------------
C
      IF(JRADL.NE.0)GO TO 10
      IF(NBC.GT.0)GO TO 188
   10 CONTINUE
      JP=NER*NDF
      I1=I-(NTE1+NTE2)
      LET=IDENT(I)
      DO 13 J=1,NER
      IJ=MCODE(J,I)
      IF(LET.GT.2)IJ=LCODE(J,I1)
   13 LC(J)=IJ
      DO 3 J=1,3
      DO 3 K=1,3
    3 C(J,K)=0.0
      IF(JRADL.NE.0)GO TO 31
      C(1,1)=RCOS(SHY)
      C(2,2)=RCOS(SHY)
      C(3,3)=1.0
      C(1,2)=-RSIN(SHY)
      C(2,1)=RSIN(SHY)
   31 DO 132 J=1,NER
      M1=LC(J)
      IF(JRADL.NE.0)GO TO 17
      M1=(M1-1)*NDF
      JS=0.
      DO 26 K=1,6
      Q(K,J)=QM(J,K)
   26 CONTINUE
C
      CALL MPRODT(Q,D,DDS,6,6,6)
   25 CONTINUE
      RETURN
      END
C
      SUBROUTINE PRINCL(NEP,M,CET)
£INSERT COMMON.FF
C
C--------------------------------------------------------------------
C              THIS SUBROUTINE CALCULATES THE PRINCIPAL STRAINS
C              PRINCIPAL STRESSES AND DIRECTION COSINES
C              NEP=2 - PRINCIPAL STRESS AND D.C.
C              NEP=1 - PRINCIPAL STRAINS ONLY
      LL=0
      DO 10 J=1,6
      IF(RABS(CET(J)).GT.1.0E-15)LL=1
   10 CONTINUE
C
      IF(NEP.EQ.1)GO TO 20
      PS1(M)=0.0
      PS2(M)=0.0
      PS3(M)=0.0
      GO TO 30
   20 CONTINUE
      EP1(M)=0.0
      EP2(M)=0.0
      EP3(M)=0.0
   30 CONTINUE
C
      IF(LL.EQ.0)GO TO 999
C
      G1= CET(1)
      G2= CET(2)
      G3= CET(3)
```

```
      G4= CET(4)
      G5= CET(5)
      G6= CET(6)
      ZNV1 = G1 + G2 + G3
      ZNV2 = G1*G2 + G2*G3 + G3*G1 - G4*G4 - G5*G5 - G6*G6
      ZNV3 = G1*G2*G3+2.0 *G4*G5*G6  -G1*G5*G5 - G2*G6* G6
     1          - G3*G4*G4
      BB = - ZNV1
      CW =   ZNV2
      CD = - ZNV3
C     FIND ALL ROOTS OF CUBIC EQUATION AA*X*3 + BB*X*2 + CC*X + DD
C     FIRST ROOT (X5) IS FOUND BY NEWTON'S METHOD USING 0 AS
C     FIRST APPROX. THEN SOLVE QUADRATIC BY STANDARD FORMULA
C     ERR IS THE ACCURACY REQUIRED FOR ROOT X5
      ERR = 1E-6
      X1 = 0.0
      CORT=2.0*ERR
      MGG=0
1000  B1 = BB + X1
      MGG=MGG+1
      IF(MGG.GT.35)GO TO 2000
      B2 = CW + X1*B1
      IF(RABS(CORT).LT. ERR)GO TO 2000
      B3 = CD+ X1 * B2
      C3 = ( X1 + B1) * X1 + B2
      IF( RABS(C3) .LT.   1E-30) C3 = 1.0
      CORT=B3/C3
      X1=X1-CORT
      GO TO 1000
2000  X5 = X1
C
C     SECOND PART - FIND ROOTS OF QUADRATIC
C                   X**2 + B1*X + B2 = 0.0
C
      DIP = B1*B1 - 4.0*B2
      IF(DIP  .LT.  0.0) GO TO 3000
      SD = RSQRT(DIP)
      X6 = (SD - B1) * 0.5
      X7 = - (SD + B1) * 0.5
      GO TO 335
3000  X6 = - 0.5 * B1
      X7 = 0.5 * RSQRT(-DIP)
      WRITE(JOUT,800) I,M
800   FORMAT(/,15X, 9HCONJUGATE, 2I5)
335   CONTINUE
C                   PRINCIPAL STRESSES AND DIRECTION COSINES
C         DC1,DC2,DC3 ARE THE DIRECTION COSINES OF
C         PRINCIPAL STRESSES  PS1, PS2, PS3
      IF (X5 .GE. X6  .AND. X6 .GE. X7) GO TO   430
      IF(X5 .GE. X7  .AND.  X7 .GE. X6) GO TO 431
      IF(X6 .GE. X6  .AND. X5 .GE. X7) GO TO 432
      IF(X6 .GE. X7 .AND. X7 .GE. X5) GO TO 433
      IF(X7 .GE. X5 .AND. X5 .GE. X6) GO TO 434
      IF(X7 .GE. X6 .AND. X6 .GE. X5) GO TO 435
430   X1 = X5
      X2 = X6
      X3 = X7
      GO TO 438
431   X1 = X5
      X2 = X7
      X3 = X6
      GO TO 438
432   X1 = X6
      X2 = X5
      X3 = X7
      GO TO 438
433   X1 = X6
      X2 = X7
      X3 = X5
      GO TO 438
434   X1 = X7
      X2 = X5
      X3 = X6
      GO TO 438
435   X1 = X7
      X2 = X6
      X3 = X5
438   CONTINUE
      IF(NEP.EQ.1)GO TO 99
C---------        PRINCIPAL STRESSES
      PS1(M)=X1
      PS2(M)=X2
      PS3(M)=X3
      DO 440 IS = 1,3
      GO TO (443 , 445 ,447) ,IS
443   AS1 =  G1 - X1
      AS2 =  G2 - X1
      AS3 =  G3 - X1
      GO TO 444
```

```
445     AS1 =   G1  - X2
        AS2 = , G2  - X2
        AS3 =   G3  - X2
        GO TO 444
447     AS1 =   G1  - X3
        AS2 =   G2  - X3
        AS3 =   G3  - X3
444     CONTINUE
        AK=G4
        BK= G5
        CK= G6
        YAP1=AS2*CK-BK*AK
        YAP2=AK*AK-AS1*AS2
        IF(YAP1 .EQ. 0.0 ) YAP1=1.0
        IF(YAP2 .EQ. 0.0 ) YAP2=1.0
        BJM1=     (BK*BK-AS2*AS3)/YAP1
        BJM2= (AS1*BK-AK *CK)  /YAP2
        BJ1 =  BJM1*BJM1
        BJ2 =  BJM2*BJM2
        ZIP = RSQRT( BJ1 + BJ2 + 1.0)
        IF ( ZIP .LT. 0.0 ) ZIP=1.0
        DC3(IS)= 1.0 / ZIP
        DC1(IS)= BJM1 * DC3(IS)
        DC2(IS)= BJM2 * DC3(IS)
440     CONTINUE
        GO TO 999
 99 CONTINUE
C-----------          PRINCIPAL STRAINS
        EP1(M)=X1
        EP2(M)=X2
        EP3(M)=X3
999 CONTINUE
        RETURN
```

INDEX

absorbers
  shock, 82
  steel encased, 82
acceleration, 235, 239, 402, 408
accidents, 55, 57, 58
added mass, 407, 408, 409, 410
adfreeze, 425
air blast, 219
aircraft
  bird strike, 32
  bomb, 35
  civilian, 125−40
  collision, 30, 31, 32
  crashes, 17−21, 24
  electrical, 38
  ground level, 26, 27, 34
  hail, 38
  ice/snow, 39
  military, 141−65
  over-running runway, 40−44
  stealth, 59
  water, 34, 36, 37
ambient pressure, 433, 451
amplitude
  acceleration, 239
  decaying motion, 269
  displacement, 238
  force, 273
  velocity, 239
angular displacement, 241
angular velocity, 240
armour, 105
attenuation coefficient, 470
AWACS, 59

beach
  rock slope, 413
  unstable profile, 719−21
  wave impact, 411, 719−24
blades, 83
blast and blast loads
  buildings and structures, 641−4

construction and demolition, 479
  dynamics, 427
  load−time, 435
  loads, 219, 220
  pressure, 220−25
  resistance and design, 616−25, 626−37
  rock, 473
  velocity, 437
  wave, 432, 433
  wind, 434
blows, 353
Blow-Pipe missile, 122−4
bomb explosion, 655−60
bombs, 45, 56, 103, 105, 106, 154, 196, 471
  Matra, 121
boulders, vii
Bowbelle, 3
brickwork, finite-element analysis, 651−5
bulk modulus, 529
bulldozer, vii

camouflet, 465
cargo vessel, 4
cartridges, 103, 105
charge
  cavity, 428, 465−8
  chemical, 487
  cylindrical, 463−5, 492
  slab, 427, 481
  spherical, 465
  volume, 443
  wedge shaped, 483
  weight, 465, 466, 470
circular frequency, 268, 427
coefficient
  attenuation, 470
  contact, 418
  curve fitting, 414
  drag, 404
  reflection, 418
  restitution, 335, 488

shape, 418
  water entry, 407
  wave force, 412
coefficient for springs/soils, 263
collision, 7, 8–14, 45, 336, 406, 407
components, 331–6
compressive strength, 419
concrete nuclear shelter, 691–718
conservation of momentum, 467
containers, 78, 79, 81
containment, 338
core
  impermeable, 413, 415
  permeable, 413, 415
coupling factor for weapon, 468–73
cranes, 76–8, 80
crashes, 17, 21, 646–51
crater
  dimensions, 210–12, 368
  predictions, 465
  profile, 472, 473
cruiser, 3
cycle, 269

damage
  local, 367
  ships/vessels, 408, 409
damping
  coulomb, 270, 271
  critical, 260, 270, 275
  factor, 355
  force, 265, 269
  friction coulomb, 272
  material, 273
  matrix, 409
  over, 267, 270, 275, 277
  ratio, 266, 428
  small, 269
  stress, 428
  under, 268, 270, 275, 277
  viscous, 265, 269, 270, 275, 304
dashpots, 353
deceleration, 395, 404
decrement, logarithmic, 268, 269
deflection
  shock, 439
  static, 235
demolition, 473
density, 261
  air shock front, 433
  explosives, 473

fluids, 408
  loading, 474
  water, 408
depth, concentrated charge, 481
detonation, nuclear, 60
diameter, 261
directivity throw, 479
disasters, 1
discharge coefficient, 451
displacement/vessel, 236, 240, 265, 345, 408
dissipation factor, 490
distance-scaled, 429
drag coefficients, 404, 434, 435
drag energy, 436
driver, Bodine resonant, 355
dropped weights, 76, 313, 353, 363
dust explosion methods, 229, 230
dynamic
  basic, 235
  direct, 321
  displacement, 345
  impact, 331
  stability, 413
  stable profile, 411
dynamite, 197

earth penetration, 39
empirical formulae, 391–4, 609, 610
energy, 236, 237, 304, 339
explosion (general), 3, 7–14, 45, 201–23
explosion analysis
  cylindrical, 428
  dynamic, 426
  in air, 432
  in gas, 436
  spherical, 428
explosion around dams, 724–8
explosion – dust
  Heinrich, 451
  K_{st} factor, 448, 449
  Maisey, 451
  methods, 448–51
  Schwal/Othmer, 450
  vent ratio or coefficient, 448
explosion – gas, 442–8
  empirical formulae, 442–8
  fully vented, 442, 444
  multiple, 443
  open-air, unconfined, 445, 446
  partially vented, 445–8
explosion – list, major occurrences, 45–56,
    57, 60, 191, 196, 197, 216–30, 231

explosion − soil, 458
  contact, 462
  contained, 462
  parameters, 461−3
  strata, 757−60
  underground, 761−5
explosion − water
  conditions in vicinity, 485
  initial parameters, 486
  pressure distribution, 485
  shock front, 486
  shock wave, 487
explosives − air, 223
  crater, 198−205
  gas, 212−14
  pressure−time, 231−4
  surface, 223
  underground, 220
  underwater, 231
explosives − containment, 476−9, 487
explosive types, 45, 56, 462, 463
Euler relation, 241

factor
  ballistic density, 404
  charge size, 93
  dissipation, 490
  permeability, 413
  spring, 261, 262
failure envelope (ice), 426
finite-element, 496
  blunt-crack propagation, 518−22
  brick, 516
  buckling/slip phenomenon, 512−13
  concrete, 500−12
  concrete modelling, 517, 518
  contact/gap element, 527−9
  dynamic linear analysis, 507−509
  dynamic non-linear/plastic, 510−13
  impact/explosion, 527
  interface element, 425, 496
  isoparametric, 498
  material compliance matrices, 500−502, 508, 515
  material properties, 503−507
  noded elements, 530, 531
  solution procedures
    frequency-domain, 534
    load−time functions, 535−40
    mesh schemes, 541
    modal analysis, 535
    time-domain, 533

spectrum analysis, 531−3
  strain rate, 514, 515, 523−7
  transformation matrices, 513
  rock models, 515, 516
flight, 331, 333
footings/foundations, 236, 264
forced vibration, amplitude, 276, 296
formulae
  deformable missiles, 380−89
  non-deformable missiles, 370−89
fragments
  crushing, 482
  primary, 476
  size, 475
  velocity, 476−8
frequency, 249−61, 274, 277, 305, 306
friction, 270, 272, 425

gas−air
  constant, 451
  mixture, 445
gas burning
  pressure functions, 212−14
  velocity functions, 212−14
gas explosion, 191
gauge hole pressure, 475
grenades, 45, 60
Gulf war, 59
guns, 157

Hamburg ferry, 3
heat capacity, 436
heights
  crest, 414
  step, 414
  transition, 414
helicopters, 3, 164−9
hurricanes, 2, 4−6
hydrodynamic reactions, 408
hydrogen detonation, 684−6

ice, 72−5 (*see also* snow/ice)
impact (general)
  aircraft, 7−24, 344
  buildings, 83, 103
  cars, 62−8
  direct, 16, 337
  lorries, 83
  missile, 14
  rock blasting, 68
  sea-going vessels, 8−14
  surface, 163

trucks, 83
impact (specific), on
  concrete, 367
  shot peening surface, 391
  steel, 367
impact oblique, 414
impact due to dropped weights, 742−8
impact − explosion, 645, 646, 687−8
impact − ice/snow
  concrete platform, 661−6
  finite-element analysis, 661
  ice, due to, 416−24, 523
impact on structures, 337, 340−61
impact − soil/rock, 68, 391, 748−56
impact on water surface
  jet, 765−9
  jet fluid, 68
  ocean surface, 406
  water surface, 390, 399
impactor, 331, 349
impedance, 292, 404
impedance isolation, 295
impulse, 331, 349
incipient motion, 411, 412

Javelin missile, 122−3
jet fluids, 60, 396−8
jet impingement forces, 689, 690
JONSWAP spectrum, 399−401
JONSWAP surface, 406, 407

kinetic energy, 236

length
  crest, 414
  crushed, 347
  run-up, 414
  step, 414
  uncrushed, 347
load−time functions and models, 346, 347, 367,
    370
lorries, 83
lumped mass, 347, 353

Mach numbers, 433, 436−438, 442
magnification factor, 276, 283, 288, 289
Marchioness, 7
mass, 235, 428
mechanical impedance, 292
missile
  building, 1
  concrete, 370

Cruise, 59
Gulf, 51
high trajectory, 61
hurricanes, 1
military, 103, 108−18, 121, 122
Patriot, 59, 108
plant generated, 61, 69
steel, 62
tornado, 1
wind generated, 1
wood, 62
missile types
  conical, 383
  deformable, 380
  non-deformable, 370
  spherical, 383
modified, Petry formula, 371
moles, 489
momentum, 331−43
motion, 235, 237
multi-element bottle, 363

Newton's law, 235
normal waveforce, 412
nuclear containment, 684
nuclear detonation, 60
nuclear explosion, 219−25
nuclear flask, 82
nuclear fractures, 681−3
nuclear loss-of-coolant accident, 667−83
nuclear peak pressure, 220−24
nuclear reactors, 667

oblique shock, 437
oblique wave, 414
offshore
  floating/mobile, 191
  semi-submersible, 191, 192
  platform, 192, 195
orthogonality principle, 320, 324
oscillating support, 302
overdamping, 266, 267
overpressure
  chemical, 429
  nuclear, 429
  shock, 433, 434

Palmer's equation, 454, 455
parameter
  similarity, 413
  wave impact, 416, 462−4
particle velocity, 404

peak displacement, 344
penetration, 367
perforation, 367
permeability, 413
phase, 237, 239, 286
pile, 353, 356−60
pipe targets, 387
platform, 406
Poisson's ratio, 429
position (equilibrium), 237
pressure
    ambient, 433, 451
    burning rate, 229
    coefficients, 407
    dynamic, 220, 221
    leakage, 443
    overpressure, 227−9
    peak, 469
    pulse, 442
    reflected, 433
    stagnating, 403
    vent, 227−9
profile, 406
projectile, 331, 403
propagation of velocities, 427
pulse load, 278
punching shear, 367, 368

radiation, 55
railways, 83, 87−103
reactions, hydro elements, 408, 409
reflection
    coefficients, 442
    stress, 472
reinforced concrete slabs/walls, 610−15
relations
    displacement−time, 287
    force−time, 287
resistance factor, 408
resonant, 288, 305
response, 278−9
Reynolds number, 433
roadways, 637−41
rock
    blasting, 473
    fall, 362
    fracture, 785
rockets, 45, 115−20
rotating
    unbalance, 293, 306, 307
    vectors, 240, 247
runways, impact and explosion, 637−41

Rust method, 454−8

scabbing, 367, 375
scaling, 430
sea-environment, 719
seismic velocity, 470
shape factor, 457
shear punching, 367
shells, 60, 103, 107, 196
ship-to-ship impact/explosion, 729−37
ship-to-platform impact/explosion, 737−41
shock, 437, 467, 472, 489−91
shock reflection, 440, 441, 470
shrapnell, 60, 103
simulation − impact, 331
slope − wave, 414
snow/ice
    empirical impact formulae, 417−23
    properties (static and dynamic), 418−23, 426,
        427
    strain and strain rates, 418, 419, 603
soils − data, 208, 209, 458, 459, 460
soil/rock, 70, 71
spalling, 368
spring and spring constants, 242, 243, 246, 262
stagnation pressure, 403, 436
steel targets, 382, 383
    car collisions and explosion, 578−84
    human body and skull damage, 578−84
    train crash, 586−90
stiffness coefficient, 482
storm duration, 414
stress, 472
structures
    composite, 552−62
    concrete
        fibre-reinforced beams, 600−606
        pre-stressed concrete, 596−9
        slabs/walls, 606−616
        steel-reinforced beams, 590−96
    steel
        beams, 542−4
        hollow spherical cavities and domes, 570−73
        pipes, 562−70
        plates, 545−51
systems
    multi-degree-of-freedom, 321−30
    single-degree-of-freedom, 235, 280, 281,
        308−309
    two-degrees-of-freedom, 314−20

tanker, 7

tanks, 170−74
target
  concrete, 370
  deformable, 351
  pipe, 387
  response, 369
  rigid, 346, 351
  slow speed indentations, 383
  steel, 382
terminal, 7
thermal radiation, 55
time, 402
trains, 83, 87−103, 104
  impact, 365
  time-history, 366, 367
transportation, 82
tornado, 2−6
torque, 241
torsional vibrations, 242
transmissibility, 296−304
transmission, 295
transmission factor, 430, 431
trucks, 83−6
typhoon, 2−6

unclamped free vibrations, 235, 242−4
uncrushed length, 347
underdamped system, 273
underwater
  contact explosion, 493
  pressure, 494
  shock wave reflection, 493
  temperature, 494
  velocity, 494
unshocked air, 440

vectors, 324
  rotating, 240
  translating, 240
vehicles
  collision, 336
  impact, 336
velocity, 238, 239, 265, 337, 395, 427, 434
  detonation, 473
  downstream, 438
  seismic, 470
  sonic, 474
  upstream, 438
  vectors, 438
vent area, 444, 445
vented explosion, 444
vessel
  marine, 175−90

SAMO, 3
  striking, 406−408
  struck, 406−408
vibration
  free damped, 265, 270
  natural, 305
  torsional, 241
  transverse, 246
  unclamped, 235
  undamped, 242
vibration factor, 261, 262
viscous damping, 265
volume
  actual, 430
  standard, 430

wall, containment, 338
weights, dropped, 60, 76, 353
water
  displacement, 405
  force, 405
  jet, 765−9
  pressure, 405
  time, 405
  velocity, 405
wave impact
  beach, 411
  rock slope, 411
  slope angle, 411−13
  unstable profile, 413
wave measures
  damage, 415
  density, 405
  height, 413
  number, 415
  steepness, 413
waves
  celerity, 402
  composite, 427
  deformation, 426
  forces, 408
  fronts, 426
  hydrodynamic, 427
  Love or Q, 426
  plunging, 413, 415
  pressure, 426
  Rayleigh or R, 426
  shapes, 404
  stress, 404, 428
  surging, 413, 415
  transverse, 426

Wilson-θ, 327–9
winds
  blast, 434
  natural, 434

Yield, weapon, 430, 462, 463
Young's modulus, 418

Zone
  crushing, 458
  elastic, 458
  function, 490
  highly deformed, 458
  rupture, 458